EDUCATION AND CONTINUING DEVELOPMENT FOR THE CIVIL ENGINEER

Setting the Agenda for the 90's and Beyond

Proceedings of the National Forum

Sponsored by the American Society of Civil Engineers in cooperation with:

Accreditation Board for Engineering & Technology
American Association of State and Highway Transportation Officials
American Association of Engineering Societies
American Concrete Institute
American Institute of Architects
American Institute of Steel Construction
American Plywood Association
American Public Works Association
American Society of Photogrammetry and Remote Sensing
American Society for Testing and Materials
American Society for Engineering Education
American Society of Mechanical Engineers
American Society of Agricultural Engineers
American Society of Certified Engineering Technicians
American Water Works Association
Association Schools of Construction
Association of Engineering Geologists
Association of Environmental Engineering Professors
Association of Soil and Foundation Engineers
Canadian Society for Civil Engineering
Construction Specifications Institute
Institute of Transportation Engineers
International Right of Way Association
National Action Council for Minorities in Engineering
National Concrete Masonry Association
National Council of Engineering Examiners
National Safety Council
National Science Foundation
National Society of Professional Engineers
Portland Cement Association
Prestressed Concrete Institute
Society of American Military Engineers
Society of Women Engineers
Water Pollution Federation

Las Vegas, Nevada
April 17-20, 1990

ASCE 1852

Published by the
American Society of Civil Engineers
345 East 47th Street
New York, New York 10017-2398

ABSTRACT

The 1990 Forum Proceedings is a collection of papers that were written to provide a basis for deliberation and conclusions at the ASCE 1990 National Forum on Education and Continuing Development for the Civil Engineer, held in Las Vegas, Nevada, April 17-20, 1990. They address major areas of interest and concern to the future of civil engineering education. The overall theme of the forum—"Setting the Agenda"—is reflected in the writings so as to generate recommendations for how educators, practitioners, and the American Society of Civil Engineers can influence education for the next decade and into the twenty-first century. Curricula and accreditation; continuing education; design creativity and safety; ethics, communications and legal aspects; student quality and quantity; career guidance; computers; and management are several of the topics discussed. In addition, educational aspects of technical specialty areas, such as construction, structures & mechanics, and environmental engineering are covered.

Library of Congress Cataloging-in-Publication Data

Education and continuing development for the civil engineer: setting the agenda for the 90's and beyond: proceedings of the national forum.

p. cm.

Sponsored by the American Society of Civil Engineers in cooperation with Accreditation Board for Engineering & Technology . . . and others.

ISBN 0-87262-748-9

1. Civil engineering—Study and teaching—Congresses. I. American Society of Civil Engineers.

T62.E29 1990

624'.071'1—dc20 90-207

CIP

Library of Congress Catalog Card No: 90-207

ISBN 0-87262-748-9

Manufactured in the United States of America

FOREWORD

The papers in this volume summarize the authors' opinions and experiences concerning a wide range of topics and issues to be addressed at the fifth quinquennial ASCE-sponsored conference focusing on the education of the civil engineer. While the printed Proceedings will serve as a valuable resource for the reader interested in any aspect of education or continuing professional development for the civil engineer, attendees are requested to read the papers associated with the areas or tracks of their particular interest prior to the Forum. Individual papers will not be presented formally at the conference, but rather are intended to provide data and ideas to stimulate discussions. As the American Society of Civil Engineers launches its new Education Activities Division, it is especially appropriate that we begin the decade of the 1990s by bringing together both educators and practitioners to focus on the future of education within our profession.

Planning and coordinating a conference of this magnitude requires the efforts of a very large number of individuals and groups. The Forum Steering Committee wishes to express its sincere gratitude to all involved including authors of papers; track chairs and vice chairs; invited speakers; the ASCE Committee on Research in Civil Engineering Education; and Charles Day and Rosalie Rothenstein of the ASCE Education Activities Staff. The Steering Committee extends a special thank you to those firms listed elsewhere in the Proceedings who demonstrated their commitment to the future of our profession by providing substantial funding for conference activities.

STEERING COMMITTEE FOR THE 1990 FORUM:

Richard J. Scranton, Forum Chairman—Northeastern University
George K. Wadlin, Chairman, Subcommittee on Papers—ASCE Education Consultant
Donn E. Hancher—Texas A & M University
Carl Selinger—Port Authority of New York and New Jersey
Thomas S. Tarpy—S.D. Lindsey & Associates
Fred F. Videon—Montana State University
William A. Welsh, Jr.—Pennsylvania State University at Harrisburg
Richard V. Wyman—University of Nevada—Las Vegas
Carl L. Monismith—University of California—Berkeley
Charles Day—Staff Contact

Funding for the Forum has been made available by the following:

PATRONS

Bechtel Civil, Inc.
Brown & Root, Inc.
Camp Dresser & McKee, Inc.
CH2M Hill
Holmes & Narver, Inc.
Howard Needles Tammen & Bergendoff
Reynolds Electrical & Engineering Co., Inc.
Woodward-Clyde Group, Inc.

SPONSORS

Alvord Burdick & Howson
Bissell & Karn, Inc.
Blauvelt Engineers, P.C.
Brown and Caldwell
Delon Hampton & Associates
Exxon Company, U.S.A.
Goldberg-Zoino & Associates, Inc.
Graef, Anhalt, Schloemer & Associates
Greiner Engineering, Inc.
HKM Associates
James M. Montgomery, Inc.
Kaiser Engineers, Inc.
Kimball Chase Company, Inc.
Kimley-Horn and Associates, Inc.
Marnell Corrao Associates, Inc.
Parsons Brinckerhoff Quade & Douglas, Inc.
Strand Associates, Inc.
The ERM Group
The Sear-Brown Group
The Terracon Companies, Inc.
Vulcan Materials Company

SURVEYING AND MAPPING

ENGINEERING MANAGEMENT

MATERIALS

TRACK GROUP II

COMPUTERS

STUDENT QUALITY/QUANTITY

PROFESSIONAL DEVELOPMENT

ACCREDITATION AND CURRICULA

DESIGN/CREATIVITY: QA/QC

COLD REGIONS

URBAN PLANNING

AEROSPACE

Survey of Civil Engineering Education - 1989
by
Colby V. Ardis,[1] F.ASCE

ABSTRACT

Changes are happening in civil engineering education! This survey, taken in the Spring and Summer of 1989, is the third point on our linear analysis since the first survey in 1978. The 1984 survey provided us with the first picture of changes (Wadlin, 1985). Tables 1 and 2 summarize significant changes. In each table, the first set of parentheses refer to percentages in 1989, 1984, and 1978 respectively. The second set of parentheses refer to the specific question asked of civil engineering department heads. You may find many of the trends surprising.

Table 1
Up Trends In Civil Engineering Education
(Significant changes since 1984 and 1978)

WHAT'S UP!

Enrollment in civil engineering (see question 50)
Selective admissions (81,74,73) (6)
Master's degree evening programs (See questions 18 & 19)
Co-op programs (74, 66, 62) (20)
Credit TV courses (23, 15, 13) (24)
% of those who are full professor (48, 46, 46) (25)
Graduate Teaching Assistants per department (26)
Graduate Research Assistants per department (26)
Use of part-time faculty (64,60,56) (27)
Research only faculty (13, 8, 8) (35)
Faculty who belong to ASCE (87, 81, 71) (38)
Faculty engaged in research (84, 72, 63) (39)
Part-time students: undergraduate & graduate (47-49)
External industrial advisory committees (48,36,26) (53)
Required course in Ethics & Professionalism (80,38,30) (68)

---

[1] Dean of Engineering, Southern Illinois University at Edwardsville, Edwardsville, IL 62026-1804

Required course in Technical Report Writing (82,48,31) (68)
Required course in Probability & Statistics (93,49,41) (68)
Required course in Hydrology (86, 49, 44) (68)
Computers in courses (99.9, 92, 91) (68)
Newsletters to alumni (75, 66, 17) (75)
Surveys of alumni (72, 67, 61) (77)

Table 2
Down Trends in Civil Engineering Education
(Significant changes since 1984 and 1978)

WHAT'S DOWN!

Irregular allocation for equipment (29% from 37%) (11)
Evening degree courses (56, 59, 63) (17)
Bachelor's degree evening programs (See questions 18 & 19)
Off-campus extension courses (24, 27, 43) (23)
Faculty who are PE's (62, 69, 66) (29)
Faculty who belong to NSPE (16, 20, 19) (38)
Faculty who belong to ASEE (30, 34, 34) (38)
Faculty with external research support (46, 59, 75) (40)
Faculty without an earned doctorate (7, 12) (45)
New faculty without a doctorate (56)
Required soil mechanics course (57, 98, 93) (68)
Required structural analysis course (42, 98, 95) (68)
Required water/wastewater system course (48,85,80) (68)
Required thermodynamics course (75, 81, 80) (68)

PROCEDURE

The survey was first mailed to department heads/chairmen on March 22, 1989 with a requested response date of April 14 . A second request was necessary. It was mailed April 25, 1989. A total of 161 responses were received from 214 programs -- a decrease of 30 since 1984. ASCE Districts 1, 11, and 14 had 6, 7, and 6 fewer responses than in 1984. District 4's participation improved, 11 compared to 9 in 1984. Results are presented in the Appendix for each of the 81 questions asked.

SUMMARY

The 1984 concern for decreasing CE enrollments has turned around. Enrollments are up. Selective admission procedures continue. Thanks, presumably to our accreditation procedures, we're doing better with regular equipment budgets.

More universities are involved in co-operative education, evening programs for a master's degree in CE, credit TV courses, and research only faculty. More faculty are involved in research; but external support is down drastically. As more faculty are pressured into research, fewer are registered Professional Engineers requiring departments to hire part-time faculty who are registered to teach design courses. Part-time faculty are also used in lower division courses and to teach graduate courses.

The use of industrial advisory committees for the department has doubled from one in four in 1978 to one in two in 1989. If there isn't an advisory committee in the department, over half will have one at the college level. In all, 88% have an advisory committee in the department or in the college. (Question 54)

The changes in required coursework are noteworthy (See question 29). Instruction in Geology, Hydrology, Probability and Statistics, Technical Report Writing, and Ethics and Professionalism are up the most. However, students are now being given more opportunities to design their own programs. Courses that used to be required by almost all programs, e.g. Soil Mechanics, Structural Analysis, Fluid Mechanics, Water/Waste Water Systems, Surveying, Electric Circuits, and Thermodynamics are now only required by about 50 to 75 percent of the programs. I believe our programs are evolving to meet the needs of society due to the increasing interaction between educators and practitioners, industrial advisory committees, ASCE education activities, and the accreditation process. Congratulations.

ACKNOWLEDGMENTS

This survey was designed by ASCE's Committee on Research in Civil Engineering Education (CORCEE) in 1978 under the leadership of George Wadlin, then Administrator of Education Services for ASCE. A few additional questions were added in 1984 and 1989. Two questions (16 and 17) were dropped from the 1984 survey; they related to student faculty FTE ratios that were difficult to interpret, answer and analyze. The 1989 survey was summarized by the author, currently the Secretary of CORCEE. The assistance of CORCEE Chairman Garrett Evans and Vice Chairman Robert Taylor in reviewing the original questionnaire and editing the draft summary of this survey is gratefully appreciated. ASCE mailed the survey to department chairmen. The assistance of Charles Day, Manager of Education Services is also appreciated.

APPENDIX
AMERICAN SOCIETY OF CIVIL ENGINEERS
SURVEY OF ABET ACCREDITED CIVIL ENGINEERING PROGRAMS

UNIVERSITY DATA

1. NAME OF COLLEGE OR UNIVERSITY: 161 RESPONDENTS

2. LOCATION BY ASCE DISTRICT AND NUMBER OF RESPONDENTS:

DIST.#--No.	DIST.#--No.	DIST.#--No.	DIST.#--No.
1-----10	2-----18	4-----11	5------ 6
6----- 9	7-----15	8-----11	9------ 9
10----- 5	11-----22	12----- 8	14------ 7
15-----14	16-----16		

3. OFFICIAL NAME OF THE DEPARTMENT:

	1989	1984	1978
Civil Eng. Dept......	100(62.1%)	142(74.3%)	134(69.1%)
Civil & Env.Eng. Dept.	15(09.3%)	15(7.9%)	12(6.2%)
Other..................	46(28.6%)	34(17.8%)	48(24.7%)

4. MAJOR SOURCE OF FINANCIAL SUPPORT:

	1989	1984	1978
State(Public)......	99(61.6%)	121(63.4%)	121(62.4%)
Private............	62(38.4%)	61(31.9%)	63(32.5%)
Both of above......	not asked	7(3.7%)	8(4.2%)
Federal............	not asked	2(1.0%)	2(1.0%)

5. TYPES OF DEGREES AWARDED:

	1989	1984	1978
B.S.C.E.......	123 (77.3%)	147 (77.0%)	138 (71.1%)
B.S.E.........	20 (12.6%)	33 (17.3%)	36 (18.6%)
B.S...........	36 (22.6%)	38 (19.9%)	49 (25.3%)
M.S.C.E.......	89 (55.9%)	99 (51.8%)	89 (45.9%)
M.Eng.........	40 (25.1%)	57 (29.8%)	53 (27.3%)
M.S...........	62 (38.9%)	80 (41.9%)	92 (47.4%)
C.E...........	12 (7.5%)	13 (6.8%)	17 (8.8%)
Ph.D..........	92 (57.9%)	110 (57.6%)	108 (55.7%)
Sc.D..........	3 (1.9%)	6 (3.1%)	4 (2.1%)
D.Eng.........	12 (7.5%)	18 (9.4%)	13 (6.7%)

6. WHAT KIND OF ADMISSION POLICY DOES YOUR SCHOOL HAVE FOR ENTRANCE INTO THE BACHELOR ENGINEERING PROGRAMS?

	1989	1984	1978
Open admission for graduates of accredited high schools in your state.	20(12.5%)	39(20.4%)	47(24.2%)
Selective admission standards with no restriction on the size of the freshman class.	64(39.7%)	59(30.9%)	74(38.1%)
Selective admission standards with restriction on the size of the freshman class.	66(40.9%)	83(43.5%)	67(34.5%)
Other	11(6.9%)	10(5.2%)	6(3.1%)
Total percent selective	(80.6%)	(74.4%)	(72.6%)

DEPARTMENTAL ADMINISTRATION

7. TITLE OF DEPARTMENT ADMINISTRATION:

	1989	1984	1978
Head..................	41(27.3%)	53(27.7%)	52(27.1%)
Chairman............	105(65.2%)	125(65.4%)	126(65.6%)
Director............	4(2.5%)	4(2.1%)	not asked
Other................	8(5.0%)	9(4.7%)	14(7.3%)

8. HOW IS THE DEPARTMENT HEAD SELECTED ?

	1989	1984	1978
Appointed by the dean without advice from departmental faculty.	11(6.8%)	13(6.8%)	15(7.9%)
Appointed by the Dean with advice from departmental faculty.	121(75.1%)	135(71.1%)	157(83.1%)
Elected by the departmental faculty.	12(7.4%)	11(5.8%)	17(9.0%)
Other	17(10.7%)	31(16.3%)	not asked

9. WHAT IS THE NORMAL TERM OF OFFICE FOR THE DEPT. HEAD ?

	1989	1984	1978
Permanent appointment subject to periodic higher administrative review.	79(49.1%)	94(49.5%)	74(38.5%)
One term-fixed no. of yrs.	23(14.3%)	23(12.1%)	26(13.5%)
Two terms of a fixed no. of years each.	12(7.4%)	15(7.9%)	8(4.2%)
More than two terms of a fixed no. of years each.	30(18.6%)	30(15.8%)	not asked
Other	17(10.6%)	28(14.7%)	36(18.8%)
Indeterminate term	not asked	not asked	48(25.0%)

10. DO YOU HAVE AN INDEPENDENT DEPARTMENTAL NON-SALARY BUDGET FOR OPERATIONS ALLOCATED EACH FISCAL YEAR ?

	1989	1984	1978
Yes..........	146(90.7%)	174(91.6%)	164(85.0%)
No...........	15(9.3%)	16(8.4%)	29(15.0%)

11. CHOOSE THE STATEMENT WHICH BEST DESCRIBES YOUR DEPARTMENTAL EQUIPMENT BUDGET:

	1989	1984	1978
Regular allocation each year.	56(34.7%)	49(25.7%)	60(31.4%)
Department bargains for its fair share of the engineering college's equipment budget each year.	47(29.1%)	59(30.9%)	69(31.4%)
Irregular allocation each year which sometimes is zero.	46(28.6%)	70(36.6%)	62(32.5%)
Other	12(7.6%)	13(6.8%)	not asked

12. DO YOU HAVE AN ASST./ASSOC. DEPARTMENT HEAD ?

	1989	1984	1978
Yes........	33(20.5%)	35(18.3%)	not asked
No.........	128(79.5%)	156(81.7%)	not asked

13. HOW MANY BUDGETED FULL-TIME-EQUIVALENT (FTE) FACULTY DO YOU HAVE?

1989

* 155 schools responded.
* Total FTE reported was 2438.
* The average was 15.7 per school.
* Highest FTE reported was 70 & lowest 2.5.

1984

* 188 schools responded. Total FTE reported was 2,826.

1978

Question was not asked.

14. HOW MANY BUDGETED FULL-TIME-EQUIVALENT (FTE) CLERICAL STAFF DO YOU HAVE ?

1989

* 155 schools responded.
* Total FTE reported was 520.6 .
* 3 schools reported zero.
* 5 schools reported < one.
* The average was 3.3 per school.
* Highest value reported was 19 .

1984

189 schools responded. Total FTE reported was 532.2
4 schools reported zero.
11 schools reported < one.
The average was 2.8 per school.
Highest value reported was 21.5 .

1978

Question was not asked.

15. HOW MANY BUDGETED FULL-TIME-EQUIVALENT (FTE) TECHNICAL STAFF DO YOU HAVE ?

1989

* 156 Schools responded.
* Total FTE reported was 302.93 .
* 16 schools reported zero.
* 13 schools reported < one.
* The average was 1.9 per school.
* Highest value reported was 19.39 .

1984
187 schools responded . Total FTE reported was 303.5
26 schools reported zero
26 schools reported < one
The average was 1.6 per school
Highest value reported was 21.5

1978
Question was not asked.

NOTE:

THERE WERE TWO MORE QUESTIONS AT THIS POINT IN THE 1984 SURVEY, WHICH WERE NOT REPEATED IN 1989.

INSTRUCTIONAL PROGRAMS

16. WHAT IS YOUR SCHOOL YEAR BASIS ?

	1989	1984	1978
Semester..........	122(75.8%)	141(73.8%)	134(69.1%)
Quarter or term...	37(22.9%)	46(24.1%)	54(27.8%)
Other.............	2(1.3%)	4(2.1%)	6(3.1%)

17. DO YOU OFFER DEGREE COURSES IN THE EVENING ?

	1989	1984	1978
Yes...........	90(55.9%)	112(58.6%)	123(63.4%)
No............	71(44.1%)	79(41.4%)	71(36.6%)

18. DO YOU OFFER DEGREE PROGRAMS IN THE EVENING ?

	1989	1984	1978
Yes..........	53(32.9%)	62(32.5%)	75(38.7%)
No...........	108(67.1%)	129(67.5%)	119(61.3%)

19. IF YES IN #18, AT WHAT LEVELS ?

	1989	1984	1978
Bachelors			
Yes.........	14(26.4%)	22(35.5%)	37(49.3%)
No..........	37(73.6%)	40(64.5%)	38(50.7%)

	1989	1984	1978
Masters			
Yes..............	52(98.1%)	58(93.5%)	69(92.0%)
No...............	1(1.9%)	4(6.5%)	6(8.0%)
Professional			
Yes..............	9(16.9%)	10(16.1%)	12(16.0%)
No...............	43(83.1%)	52(83.9%)	63(84.0%)

Doctoral			
Yes............	13(24.5%)	21(33.9%)	27(36.0%)
No.............	38(75.5%)	41(66.1%)	48(64.0%)

20. DO YOU HAVE A CO-OP PROGRAM WITH INDUSTRY ?

	1989	1984	1978
Yes..........	119(73.9%)	126(66.0%)	121(62.4%)
No...........	42(26.1%)	65(34.0%)	73(37.6%)

21. IF YES ,IS IT:

	1989	1984	1978
Mandatory.......	7(5.9%)	7(5.6%)	18(14.9%)
Optional.......	112(94.1)	119(94.4%)	103(85.1%)

22. IF OPTIONAL, WHAT PERCENTAGE OF YOUR FULL-TIME STUDENTS ARE INVOLVED IN THE PROGRAM ?

1989
112 Responses showed an average answer of 10.8%
Answers ranged from < 1% to a high of 100% .

1984
117 Responses showed an average answer of 9.6%
Answers ranged from < 1% to a high of 80% .

1978
Question was not asked.

23. DO YOU OFFER (FOR DEGREE CREDIT) EXTENSION COURSES AT OFF-CAMPUS LOCATIONS ?

	1989	1984	1978
Yes...........	38(23.6%)	52(27.4%)	83(42.8%)
No............	123(76.4%)	138(72.6%)	111(57.2%)

24. DO YOU OFFER COURSES (FOR DEGREE CREDIT) BY TELEVISION TO REMOTE AREAS ?

	1989	1984	1978
Yes............	37(23.0%)	29(15.2%)	26(13.4%)
No.............	124(77.0%)	162(84.8%)	168(86.6%)

25. PLEASE STATE NUMBER OF AUTHORIZED 1988-89 FACULTY APPOINTMENTS.

FULL-TIME

	1989	1984	1978
Professor.........	1184(48.1%)	46.5%	45.6%
Associate professor...	654(26.6%)	27.0%	29.0%
Assistant professor...	531(21.3%)	21.4%	10.9%

Instructor............	52(2.1%)	2.4%	2.9%
Lecturer..............	34(1.4%)	2.3%	1.6%
Other.................	42(1.7%)	0.5%	0.8%
Total.................	2494(100.0%)		
PART-TIME			
Professor...........	114(24.1%)	18.1%	11.3%
Associate professor.	45(9.4%)	11.6%	9.6%
Assistant professor.	52(10.9%)	8.6%	9.2%
Instructor..........	41(8.6%)	12.1%	14.5%
Lecturer............	133(28.1%)	36.3%	40.8%
Other...............	91(18.9%)	12.3%	14.7%
Total...............	474(100.0%)		

26. HOW MANY GRADUATE STUDENTS (AVERAGE PER SEMESTER OR TERM) ARE EMPLOYED AS:

	1989		1984		1978	
	No.	School	No.	School	No.	School
Grad. Teaching Assts.	1298	128	1345	149	1034	135
Grad. Research Assts	2928	125	2207	140	1865	124

27. DO YOU REGULARLY UTILIZE PART-TIME TEACHING APPOINTMENTS WITH ENGINEERS FROM PRACTICE ?

	1989	1984	1978
Yes............	104(64.6%)	115(60.2%)	108(56.0%)
No.............	57(35.4%)	76(39.8%)	85(44.0%)

28. HOW MANY FULL-TIME-EQUIVALENT FACULTY ARE REPRESENTED BY YOUR ANSWER TO #27 ABOVE ?

1989: Total of 169 reported at 104 schools responding.
1984: Total of 175 reported at 107 schools responding.
1978: Total of 216 reported at 108 schools responding.

29. HOW MANY <u>FULL-TIME</u> FACULTY ARE REGISTERED PROFESSIONAL ENGINEERS ?
The 1989 responses showed a percentage of 62.4
The 1984 responses showed a percentage of 68.8
The 1978 responses showed a percentage of 66.4

30. HOW MANY <u>FULL-TIME</u> FACULTY ARE REGISTERED LAND SURVEYORS?
The 1989 responses showed a percentage of 4.6
The 1984 responses showed a percentage of 8.8
The 1978 responses showed a percentage of 9.0

31. HOW MANY FULL-TIME FACULTY WHO ARE NOT REGISTERED P.E.'S HOLD AN EIT CERTIFICATE ?

The 1989 responses showed a percentage of 9.9
The 1984 responses showed a percentage of 10.3
The 1978 responses showed a percentage of 8.7

32. HOW MANY FULL-TIME FACULTY ARE NOT ELIGIBLE FOR REGISTRATION AS AN ENGINEER OR SURVEYOR ?

The 1989 responses showed a percentage of 5.1
The 1984 responses showed a percentage of 4.2
The question was not asked in 1978.

33. WHAT PERCENT OF YOUR FACULTY IS TENURED ?

* The 1989 responses came from 156 schools and they showed an average of 66.2% were tenured. The median value was 71.5%. 8 schools reported their faculty were 100% tenured and 3 schools reported a percentage of zero.

* The 1984 responses came from 187 schools and they showed an average of 66.1% were tenured. The median value was 70.0%. 12 schools reported their faculty were 100% tenured and 3 schools reported a percentage of zero.

* The question was not asked in 1978.

34. AT THE START OF THE 1988-89 ACADEMIC YEAR, HOW MANY FACULTY VACANCIES DID YOU HAVE IN YOUR AUTHORIZED POSITIONS?

The 1989 responses showed 143 vacancies at 81 schools.
The 1984 responses showed 173 vacancies at 102 schools.
The question was not asked in 1978.

35. DO YOU HAVE FACULTY WHO ARE APPOINTED SOLELY TO CONDUCT RESEARCH ?

	1989	1984	1978
Yes..........	21 (13.1%)	16 (8.4%)	16 (8.2%)
No...........	140 (86.9%)	175 (91.6%)	178 (91.8%)

36. DOES YOUR UNIVERSITY HAVE A SABBATICAL LEAVE POLICY ?

	1989	1984	1978
Yes.........	149 (92.5%)	168 (88.0%)	166 (96.5%)
No..........	12 (7.5%)	23 (12.0%)	6 (3.5%)

37. WHAT IS THE HEAD COUNT OF THE NUMBER OF FACULTY WHO HAVE BEEN GRANTED A SABBATICAL LEAVE DURING THE LAST FIVE YEARS?

1989 Responses: 124 schools reported a total of 515 leaves.

1984 Responses: 139 schools reported a total of 534 leaves.

Question was not asked in 1978.

38. HOW MANY FULL-TIME FACULTY BELONG TO:

	1989	1984	1978
A.S.C.E.....	1711(86.7%)	1953(81.3%)	1956(70.6%)
N.S.P.E.....	326(16.5%)	461(20.5%)	537(19.4%)
A.S.E.E.....	595(30.1%)	749(33.6%)	941(33.9%)

These percentages were based on the reported sum of full-time faculty less non-engineering faculty. Excludes FTE totals for schools that did not answer

39. HOW MANY FULL-TIME FACULTY ARE ENGAGED IN RESEARCH AS PART OF THEIR REGULAR DUTIES ?

1989 responses: for 161 schools the average was 83.95%
14 schools reported zero

1984 responses: for 189 schools the average was 72.3%
27 schools reported zero

1978 responses: for 158 schools the average was 62.8%

40. WHAT PERCENT OF THOSE INCLUDED IN THE ANSWER TO #39 PERFORM RESEARCH WHICH IS EXTERNALLY SPONSORED?

1989 responses: for 147 schools the average was 46.2%
28 schools reported 100%
10 schools reported zero

1984 responses: for 163 schools the average was 59.0%
49 schools reported 100%
8 schools reported zero.

1978 responses: for 157 schools the average was 75.4%
70 schools reported 100%

41. DOES YOUR SCHOOL HAVE A POLICY ON CONSULTING ACTIVITIES OF THE FACULTY ?

	1989	1984	1978
Yes..........	153(95.0%)	187(97.9%)	186(95.9%)
No...........	8(5.0%)	4(2.1%)	8(4.1%)

42. IF YES, IS IT A WRITTEN POLICY ?

	1989	1984	1978
Yes.......	122(75.8%)	146(78.5%)	144(77.4%)
No........	39(24.2%)	40(21.5%)	42(22.6%)

43. IF A CONSULTING POLICY EXISTS, IS IT:

	1989	1984	1978
One day per week.......	137(85.0%)	160(85.1%)	153(83.2%)
Less than one day/week..	5(3.1%)	7(3.7%)	11(6.0%)
More than one day/week..	0(0.0%)	2(1.1%)	2(1.1%)
Up to the dept. head to supervise amt. of cons..	9(5.6%)	9(4.8%)	18(9.8%)
Other....................	10(6.3%)	10(5.3%)	N/A

44. IRRESPECTIVE OF THE POLICY, WRITTEN OR UNWRITTEN, WHAT DO YOU FEEL IS THE ATTITUDE OF THE DEAN TOWARD FACULTY CONSULTING?

	1989	1984	1978
Encouraged.......	104(64.6%)	125(66.1%)	116(59.8%)
Tolerated........	56(34.8%)	61(32.3%)	76(39.2%)
Discouraged......	1(0.6%)	3(1.6%)	2(1.0%)
Forbidden........	0(0.0%)	0(0.0%)	0(0.0%)

45. AT THE RANK OF ASSISTANT PROFESSOR AND HIGHER, HOW MANY FACULTY DO NOT HOLD AN EARNED DOCTORAL DEGREE ?

1989 responses: 68 schools reported a total of 161.5 faculty or about 6.8% .

1984 responses: 94 schools reported a total of 333 faculty or about 12% .

The question was not asked in 1978.

DEPARTMENTAL STATISTICS

46. HOW MANY CIVIL ENGINEERING DEGREES DID YOU AWARD BETWEEN JULY 1, 1987 AND JUNE 30, 1988 ?

	Total	Schools	Avg.	High
Bachelors degrees..	6102	159	37.9	201
Masters degrees....	2762	136	20.3	138

Engineers degrees...	50	10	5.0	24
Doctors degrees.....	532	82	6.5	42

47. WHAT WERE YOUR FALL, 1988 ENROLLMENT FIGURES ?

FULL-TIME

	Total		High		Avg.	
	1989	1984	1989	1984	1989	1984
Freshman.......	4935	6239	632	205	95	51
Sophomore......	5350	6999	383	173	38	45
Junior.........	6896	9530	621	215	46	54
Senior.........	7726	13218	227	395	52	74
5th Year.......	884	994	249	155	32	34
Master.........	3250	6461	142	204	22	43
Engineer.......	375	143	43	68	31	10
Doctor.........	2581	2107	111	128	24	19
Totals.........	31997	45691				
Schools reporting	161	191				

PART-TIME

	Total	
	1989	1984
Freshman...........................	189	174
Sophomore..........................	300	250
Junior.............................	595	466
Senior.............................	601	547
5th Year...........................	296	153
Master.............................	2611	2751
Engineer...........................	95	62
Doctor.............................	395	308
Total..............................	5082	4711
Schools reporting	161	191

48. WHAT WERE YOUR COLLEGE OF ENGINEERING (OR EQUIVALENT) ENROLLMENT FIGURES FOR FALL, 1988 ?

Undergrad.-Full-Time...235,167 total at 161 schools.
Undergrad.-part-Time... 17,623 total at 86 schools.
Graduate-Full-Time..... 33,695 total at 152 schools.
Graduate-Part-Time..... 15,631 total at 93 schools.

49. WHAT WERE YOUR UNIVERSITY ENROLLMENT FIGURES FOR FALL, 1988 ?

Full-Time.........2,131,020 total at 156 schools.
Part-time......... 297,361 total at 132 schools.

50. WHICH STATEMENT BEST DESCRIBES YOUR PRESENT DEPARTMENTAL UNDERGRADUATE ENROLLMENT SITUATION ?

	1989	1984	1978
Increasing rapidly....... (6% / yr. & up)	29(18.0%)	15(7.9%)	93(48.2%)
Incresing slightly (<6%).	47(29.2%)	26(13.7%)	60(31.1%)
Approximately constant...	46(28.6%)	56(29.5%)	31(16.1%)
Decreasing slightly(<6%).	27(16.8%)	62(32.6%)	7(3.6%)
Decreasing rapidly....... (6% / yr. & up)	12(7.4%)	31(16.3%)	2(1.0%)

51. IN THE 1987-88 ACADEMIC YEAR (JUNE TO JUNE), HOW MANY STUDENT CREDIT-HOURS DID YOUR DEPARTMENT GENERATE ?

Undergrad lower div.161,683 at 114 schools (Avg.=1418)
Undergrad upper div.245,900 at 111 schools (Avg.=2215)
Masters & Eng........72,334 at 90 schools (Avg.= 803)
Doctoral level.......29,193 at 44 schools (Avg.= 663)

INDUSTRIAL RELATIONS

52. PLEASE LIST THE NUMBER OF FULL-TIME FACULTY WHO HAVE EXPERIENCE IN ENGINEERING PRACTICE ACCORDING TO THE FOLLOWING CATEGORIES :

	1989 # %	1984 # %	1978 # %
Full-time annual employment........	1152(59%)	1578(57%)	1562(56%)
Part-time annual employment.........	233(12%)	513(19%)	553(20%)
Summers only......	567(29%)	611(22%)	579(21%)
Consulting while teaching..........	1283(66%)	1634(59%)	1678(61%)

53. DOES YOUR DEPARTMENT HAVE AN EXTERNAL INDUSTRIAL ADVISORY COMMITTEE OR SIMILAR GROUP?

	1989	1984	1978
Yes................	76(47.5%)	69(36.3%)	51(26.3%)
No.................	84(52.5%)	121(63.7%)	143(73.7%)

54. IF YOUR ANSWER IS "NO" TO #53, IS THERE AN INDUSTRIAL ADVISORY COMMITTEE WHICH SERVES THE ENGINEERING COLLEGE ON YOUR CAMPUS?

	1989	1984	1978
Yes..............	54(33.8%)	80(42.1%)	64(33.0%)
No...............	30(18.8%)	40(21.1%)	77(39.7%)

NOTE: The percentages in question #54 are based on the total of all school responses, so that when the "yes" answers are added from questions #53 & #54 the results show that there are advisory committees at 87.5% of the schools in 1989 compare to 78.4% in 1984 and 59.3% in 1978.

PROMOTION AND TENURE

55.(a) WHO INITIATES ACTION?

	1989	1984	1978
Committee..........	25(15.5%)	35(18.3%)	37(19.1%)
Individual.........	74(46.0%)	75(39.3%)	51(26.3%)
Dept. Head.........	62(38.5%)	81(42.4%)	106(54.6%)

55.(b) DOES YOUR DEPT. HAVE A COMMITTEE TO CONSIDER RECOMMENDATIONS?

	1989	1984	1978
Yes............	128(79.5%)	157(82.2%)	147(75.8%)
No............	33(20.5%)	34(17.8%)	47(24.2%)

55.(c) DOES THE DEPARTMENT HEAD HAVE THE POWER TO REJECT RECOMMENDATIONS AND THEREBY STOP THE PROCESS?

	1989	1984	1978
Yes.............	19(11.8%)	37(19.6%)	60(31.2%)
No..............	49(30.4%)	43(22.8%)	132(68.8%)
Yes, can reject but progress does not stop...........	93(57.8%)	109(57.7%)	not asked

55.(d) FROM WHAT LEVEL DOES FINAL APPROVAL OR DENIAL COME?

	1989	1984	1978
College (Dean)...........	4(2.5%)	6(3.2%)	6(3.1%)
Administration(Vice Pres).	31(19.2%)	30(15.8%)	41(21.1%)
Administration(President)	43(26.7%)	62(32.6%)	65(33.5%)
Board of Trustees........	81(51.6%)	92(48.4%)	82(42.3%)

56. ARE YOU ALLOWED TO HIRE FULL-TIME TENURE FACULTY WHO DO NOT HOLD A DOCTORAL DEGREE (AT THE RANK OF ASSISTANT PROFESSOR OR HIGHER)?

	1989	1984	1978
Yes.............	75(46.6%)	102(54.0%)	106(55.2%)
No..............	86(53.4%)	87(46.0%)	86(44.8%)

a. IF YES, HAVE YOU DONE SO ?

Yes.............	30(19.0%)	50(49.5%)	not asked
No..............	45(81.0%)	51(50.5%)	not asked

UNIONISM

57. DOES YOUR FACULTY BELONG TO A COLLECTIVE BARGAINING UNIT?

	1989	1984	1978
Yes..............	31(20.0%)	36(18.8%)	25(12.9%)
No..............	130(80.0%)	155(81.2%)	169(87.1%)

58. IF YES IN #57, ARE DEPT. HEADS INCLUDED IN THE BARGAINING GROUP?

	1989	1984	1978
Yes..........	23(74.2%)	19(55.9%)	16(64.0%)
No...........	8(25.8%)	15(44.1%)	9(36.0%)

CURRICULUM

59. HOW MANY SEMESTER CREDITS DO YOU REQUIRE FOR GRADUATION WITH A BACHELOR'S DEGREE?

	1989	1984	1978
Average............	133.6	132.2	131.8

60. DOES THE ANSWER IN #59 INCLUDE PHYS. ED. CREDITS?

1989: 37(23.0%) of 161 schools answered " Yes ".
1984: 55(30.1%) of 183 schools answered " Yes ".

61. IF YES, HOW MANY CREDITS?

The answer to this question were subtracted from the school's answers to question # 59 to get the net requirement for graduation.

147 cr. @ 1 sch.	134 cr. @ 18 sch. (median)
144 cr. @ 1 sch.	133 cr. @ 16 sch.
143 cr. @ 1 sch.	132 cr. @ 11 sch.

140 cr. @ 7 sch.	131 cr. @ 8 sch.
139 cr. @ 6 sch.	130 cr. @ 7 sch.
138 cr. @ 4 sch.	129 cr. @ 6 sch.
137 cr. @ 7 sch.	128 cr. @ 10 sch.
136 cr. @ 20 sch.	127 cr. @ 2 sch.
135 cr. @ 14 sch.	126 cr. @ 4 sch.
	.
	.
	112 cr. @ 1 sch.

62. IF YOU DO NOT USE THE CREDIT SYSTEM, HOW MANY COURSES DO YOU REQUIRE FOR GRADUATION?

	1989	1984	1978
No. of schools answering.....	11	8	3
Avg. no of courses req'd.....	39.4	37.5	39.3

63. HOW MANY YEARS OF STUDY DOES YOUR CATALOG STATE ARE NORMALLY REQUIRED FOR A BACHELOR'S DEGREE?

	1989	1984	1978
4 years.............	154 (95.6%)	180	185
4.3 years...........	-	2	-
4.5 years...........	2	-	-
4.7 years...........	-	1	-
5.0 years...........	5 (3.2%)	5	7

64. BASED ON THE ABET DEFINATION, WHAT PERCENT (TO NEAREST WHOLE NUMBER) OF YOUR BACHELOR'S CURRICULUM IS DEVOTED TO THE FOLLOWING AREAS:

a. MATHEMATICS			
High.............	31%	18%	20%
Low..............	7%	10%	10%
Median...........	15%	13%	13%
b. BASIC SCIENCES			
High............	32%	25%	26%
Low.............	7	9	6
Median..........	14	13	13
c. ENGINEERING SCIENCES			
High..........	62%	57%	52%
Low...........	9	11	8
Median.........	33	27	25
d. HUMANITIES/SOCIAL SCIENCES			
High............	42%	35%	35%
Low..............	11	8	4
Median...........	15	14	15

e. ENGINEERING DESIGN

High.............	50%	38%	39%
Low..............	5	5	2
Median...........	16	14	16

f. UNSPECIFIED

High.............	34%	54%	35%
Low..............	0	0	0
Median...........	4	16	14

65. HOW MANY SEMESTER CREDITS OF TECHNICAL ELECTIVE COURSES CAN BE USED TO MEET YOUR BACHELOR'S DEGREE REQUIREMENT?

	1989	1984	1978
High.........	54	40	50
Low..........	0	0	0
Median.......	15	12	14

66. HOW MANY CREDITS CAN BE LISTED AS REQUIRED BY YOUR UNIVERSITY IN THE FOLLOWING CATEGORIES:

	High	Low	Median	# Answering
Gen'l. Educ. Req'ts.	138	0	18	107
ROTC or Military Sci..	18	0	0	5
Religion or Theology..	24	0	0	17
Other.................	78	0	9	9

67. DO YOU OFFER UNDERGRADUATE COURSES IN THE CIVIL ENG. DEPARTMENT IN ANY OF THE FOLLOWING AREAS? LIST NUMBER OF CREDITS BOTH REQUIRED AND ELECTIVES.

	1989/1984			
	AVG. CREDITS REQ'D	% OF YES	AVG. CREDITS ELECT	% OF YES
Structural Eng...	7.6/9.3	97.5	12.2/11.9	96.1
Soils Eng........	3.8/4.2	97.3	6.8/ 6.5	70.0
Environ. Eng.....	4.5/4.5	92.1	7.8/ 9.5	75.3
Transp. Eng......	4.8/3.5	97.5	4.4/ 6.8	90.0
Hydraul. & Hydro.	3.0/4.7	31.0	3.7/ 7.7	83.5
Surv. & Photog...	3.7/3.7	60.0	6.3/ 6.2	49.4
Eng. Mechanics...	3.5/8.3	3.7	3.7/ 5.4	17.4
Materials Eng....	2.7/3.2	16.7	3.0/ 4.3	23.6
Systems Eng......	8.3/3.1	72.0	6.1/ 4.9	18.2
Eng. Management.	3.7/3.1	90.1	6.7/ 3.9	55.9
Constr. Eng.....	3.0/3.4	1.3	5.6/ 6.2	56.6
Urban Planning..	3.3/2.4	62.3	2.4/ 3.9	22.9

Ocean Eng.......	3.Ø/3.Ø	26.7	2.8/ 4.9	18.4
Others..........	5.7/5.6	29.2	7.2/ 6.Ø	27.3

68. DOES YOUR BACHELOR'S DEGREE CURRICULUM HAVE ANY OF THE FOLLOWING COURSES REQUIRED FOR THE DEGREE?

	1989	1984	1978
Soil mechanics........	56.6%	98.4%	93.2%
Structural analysis...	42.5	98.4	94.8
Computer programming..	78.9	98.4	95.3
Fluid mechanics.......	77.Ø	95.2	92.7
Engineering graphics..	98.7	85.2	81.3
Water/wastewater sys.	47.8	84.7	8Ø.2
Surveying.............	52.2	83.6	82.3
Concrete design.......	78.3	82.Ø	79.2
Elect eng. fund'ls....	77.9	81.4	82.8
Thermodynamics........	75.2	81.Ø	79.7
Steel design..........	93.2	76.7	78.6
Hydraulics............	76.4	72.5	69.8
Highway eng...........	65.2	68.4	66.7
Engineering economics.	75.1	67.7	65.6
Conc/bitum. materials.	74.5	63.Ø	63.5
Economics..............	7Ø.8	59.3	56.3
Geology...............	75.1	51.3	49.5
Hydrology.............	86.3	49.2	44.3
Prob.& statistics.....	93.2	49.2	4Ø.6
Tech. report writing..	81.9	48.1	31.3
Ethics & profession'sm	79.5	38.1	3Ø.2
System analysis.......	12.4	33.9	31.3
Speech................	19.9	32.8	28.6
Eng. management.......	28.9	27.Ø	18.2
Contracts & specs.....	27.4	24.3	25.Ø
Timber design.........	13.9	13.2	14.1
Labor law & relations.	11.2	1Ø.1	1Ø.4
Photogrammetry........	6.7	5.3	7.3
Public works mang.....	2.6	.5	3.1

69. DO YOU HAVE DEPARTMENTAL LABORATORY COURSES IN THE FOLLOWING AREAS?

	%YES	
	1989	1984
Structural eng. lab..............	69.7%	59.2%
Materials testing lab............	92.4	89.Ø
Soil mechanics lab...............	99.8	99.5
Hydraulics lab...................	91.8	83.8
Env. eng (Sanitary) lab..........	87.3	74.3
Computer lab.....................	77.9	72.3
Drafting lab.....................	66.8	58.1
Other labs.......................	44.1	27.5

70. HOW ARE YOUR LAB EXERCISES CONDUCTED?

	1989 %YES	1984 % YES
"Hands on"......	100%	100%
Demonstration..	74%	75%

71. WITH REGARD TO LAB COURSE SCHEDULING, DO YOU:

	1989	1984
Have open labs?....	23.6%	33.0%
Have regularly sched. labs?.......	100.0%	100.0%
Do you utilize video tape instruction?.......	30.0%	25.7%

72. IS THE USE OF COMPUTERS AN INTEGRAL PART OF YOUR UNDERGRADUATE UPPER LEVEL CIVIL ENG. COURSES (IN ADDITION TO REGULAR COURSES IN COMPUTER PROGRAMMING)?

	1989	1984	1978
Yes...........	160(99.9%)	173(91.5%)	174(90.6%)
No............	1(0.1%)	16(8.5%)	18(9.4%)

CHECK OFF THE COMPUTER METHODS YOU USE.

	%CHECKED
Some homework assigned to be done by computer...	98.7%
Students write own programs(at upper level).....	89.4%
Commercial software only is utilized............	44.1%
" Hands on " computer work emphasized...........	86.9%

73. IS YOUR C.E DEPT RESPONSIBLE FOR TEACHING ENG. MECHS. COURSES SUCH AS STATICS & STRENGTH OF MATERIALS?

	1989	1984	1978
Yes...........	125(77.6%)	148(77.9%)	138(71.5%)
No...........	36(22.4%)	42(22.1%)	55(28.5%)

74. CHOOSE THE STATEMENT WHICH BEST DESCRIBES THE WAY YOU COVER THE TOPICS OF ETHICS & PROFESSIONALISM.

	PERCENT CHECKED		
	1989	1984	1978
A specific course is offered..	34.2%	3.2%	20.7%
No specific course is offered but topics are covered as part of other courses.............	62.1%	59.0%	70.5%
No coverage offered..........	4.9%	7.9%	8.8%

75. DO YOU PUBLISH AN ALUMNI NEWS LETTER AT LEAST ONCE A YEAR WHICH IS MAILED TO ALL DEPARTMENTAL ALUMNI?

	1989	1984	1978
Yes	121(75.2%)	126(66.3%)	32(16.6%)
No	20(24.8%)	64(33.7%)	161(83.4%)

76. IF YES, AT WHAT LEVELS?

	#	% YES
Departmental	59	48.8% (36.6% of all schools)
College/school	78	64.5% (48.4% 0f all schools)
University	62	51.2% (38.5% of all schools)

77. CHOOSE THE STATEMENT WHICH BEST DESCRIBES HOW YOU CONDUCT SURVEYS OF YOUR ALUMNI.

	1989	1984	1978
Conduct alumni surveys yearly	15(9.4%)	6.5%	1.6%
Conduct on an irregular basis	101(62.7%)	61.3%	59.9%
Do not conduct surveys	45(27.9%)	32.3%	38.5%

78. EXCLUSIVE OF INDUSTRIALLY SPONSORED RESEARCH, WHAT INDUSTRIAL GIFTS OR GRANTS HAVE BEEN RECEIVED BY YOUR DEPARTMENT IN THE LAST TWO YEARS?

GIFTS: 1989: YES...108 NO...53
Total dollars reported.....8,782,925
1984: YES...131 NO...44
Total dollars reported.....4,721,896

GRANTS: 1989: YES...66 NO...95
Total dollars reported....23,413,400
1984: YES...63 NO..104
Total dollars reported.....4,920,050

EQUIPMENT: 1989: YES...73 NO...88
Total dollars reported.....4,401,645
1984: YES...81 NO...88
Total dollars reported.....7,622,595

ENDOWED CHAIRS: 1989: YES...20 NO...141
Total $ amt. of all endowments................31,157,000
(Not asked in 1984)

OTHER:
Total dollars reported.......608,879
(Not asked in 1984)

Ardis

79. DID YOU RECEIVE FUNDS FROM ALUMNI GIVING THAT WERE SPECIFICALLY DEDICATED TO CIVIL ENG.(LAST 5 YEARS)?

	1989	1984
Alumni............... YES....	45	
Other Indiv..........YES.....	9	Total YES: 132
Both................. YES....	78	
No........................	29	56
	161	

Dollars equivalent reported........11,526,327
(Not asked in 1984)

STUDENT CHAPTER

80. HOW IS THE FACULTY STUDENT CHAPTER ADVISOR CHOSEN?

	1989	1984	1978
Duties assigned by dept.head...	59(36.6%)	48.4%	41.6%
Faculty mbr. volunteers	78(48.4%)	36.2%	44.2%
Chosen by the students.........	15(9.3%)	10.6%	-
Other.........................	9(5.7%)	4.8%	14.2%

81. HOW IS THE FACULTY MEMBER RECOGNIZED WHILE PERFORMING AS THE FACULTY ADVISOR?

	1989	1984	1978
Advisory duties considered part of the teaching load with an appropriate reduction of course teaching assignments..............	46(28.6%)	25.5%	20.0%
Considered as an overload to normal teaching assignments..	115(71.4%)	74.5%	80.0%

Reference

Wadlin, George K., "Survey of Civil Engineering Education-1984," in Challenges to Civil Engineering Educators and Practitioners-Where Should We Be Going?, ASCE, 1985, p. 1-18.

SURVEY OF PRACTITIONERS 1989-90
Education and Continuing Development for Civil Engineers

by Garrett H. Evans, M. ASCE*

Education survey data was a central part of past Education Conferences and continues to be important for the present 1990 Forum. The Civil Engineering Department Survey has been conducted for the 1979 and 1984 Conferences and the Survey of Civil Engineering Technology Programs was stared for the 1984 Conference. The Forum Steering Committee requested the Committee on Research in Civil Engineering Education to expand the education surveys to include practitioners. The objectives were to (1) assess practitioner attitudes toward university degree programs and (2) study the perceived adequacy of career long education systems.

The survey questionnaire was distributed to consulting firms employing a significant number of civil engineers. These firms regularly hire entry level engineers from the pool of graduates at the bachelor and master level. CORCEE recognized that if larger firms would be surveyed, the scope of the results would be somewhat limited. However practical considerations required this distribution of questionnaires. Notwithstanding some shortcomings, the committee believed a cross section of practicing civil engineers and ASCE membership would be represented. As a consequence of surveying consulting firms, the response was remarkable; out of 210 firms polled, 121 replied, a return percentage of about 60 percent. The lengthy questionnaires were completed carefully and thoughtfully considered comments were included with several returned forms.

The questionnaire included 15 questions on 5 pages. Questions were grouped in six categories: (1) information on responding firm, (2) employment patterns and practices, (3) satisfaction with entry level hires based on their demonstrated command of fundamentals, (4) the firm's policies for in Professional/Managerial Development, (5) continuing education programs available to the employee and continuing education program needs, and (6) computer facilities which the entry level engineer will encounter as

*Vice President, Soil and Materials Engineers, Livonia, Michigan. Chairman, CORCEE 1988-90.

a part of the transition from educational training to practice.

The results of the survey are tabulated in this paper. Data is presented for the number responding to an individual question and a proportional breakdown by percentage.

The full interpretation of the data is left to the individual judgment of each reader. However, major correlations are suggested by the survey data. The demand for civil engineers is high. Many firms indicated hiring levels are greater than usual the supply of civil engineers for entry level positions is short of demand. The demand is strong for both bachelor level and master level graduates; one master level graduate is hired for every two bachelor level graduates.

Employers are most satisfied with the proficiency of entry level hires in: computer skills, statics-dynamics, structures and mathematics; but least satisfied with their proficiency in management, communication, surveying and construction. Comparing the proficiency of bachelor level to the master level new hire, substantial improvement is noted in the subject areas of environmental, geotechnical and communication; whereas, little or no improvement is noted for: construction, ethics, surveying and drafting/visualization.

After the graduate is employed, the responsibility for professional/managerial development is recognized by employers. A majority of firms have policies to fund: participation in professional societies, preparation for the Professional Engineering exam and continuing education. The greatest challenges and needs in development of civil engineering employees are: the supply of entry-level engineers, improvement in communication aptitude, and improved business skills. Opportunity for additional training by short courses was indicated to be a moderate need. In-house training and seminars or short courses are the most popular delivery systems for professional development training. Other options for training are non-university sponsored short courses, which out ranks continuing education programs sponsored by universities. Remote learning systems, such as video and correspondence courses, are not popular. In terms of meeting the needs of employers, the selection of technical programs is adequate, but more management and inter-personal skills courses are desired. At the work place in practice, the new hire will encounter a wide range of computer facilities consisting of personal computer stations to mainframe hardware, and word processing to CADD software programs.

Firms were invited to record remarks with the questionnaire. Concerning the qualification of new hires, the responding firms remarked that the employee should be well rounded and capable of expressing complicated technical ideas in writing. Also the new employee should be more knowledgeable about management or basic business essentials.

The comprehensive survey was made possible by the Forum Steering Committee and the staff of the ASCE Education Division. The author wishes to acknowledge the assistance of Glen Martin, Luther Graef and David Hummel, and especially Bruce Jones. Last but not least, appreciation is expressed to the 121 firms who replied to the survey.

PART I INFORMATION ON RESPONDING PARTY

1. IS YOUR FIRM: (121 Respondents)
   a. Consulting engineering firm.......107
   b. Other: Architect-Engineer......... 14

2. WHAT IS THE APPROXIMATE NUMBER OF CIVIL ENGINEERS EMPLOYED BY YOUR FIRM

   Total in U.S. (average of 116)..........127
   Total outside U.S. (average of 23)...... 8.6

3. WHAT IS YOUR TITLE IN THE FIRM? (121)

   President or V.P. 74
   Personnel director............32
   Other.........................15

PART II EMPLOYMENT OF ENTRY-LEVEL CIVIL ENGINEERS

4. ABOUT HOW MANY ENTRY-LEVEL CIVIL ENGINEERS, TECHNOLOGISTS (4-YEAR DEGREE), AND TECHNICIANS (ASSOCIATED DEGREE) DID YOUR FIRM HIRE DURING THE PAST TWELVE MONTHS? (120)

   15.1 (average)

5. IS THIS NUMBER

   more than usual?...................32%
   less than usual?...................13%
   about average for one year?........55%

6. BASED ON YOUR EXPERIENCE, IS THE CURRENT LEVEL OF CIVIL ENGINEERING ENTRY-LEVEL JOB CANDIDATES (120)

abundant...................... 4%
adequate......................51%
in short supply...............45%

7. WHAT PERCENTAGE OF YOUR ENTRY-LEVEL HIRES INTO CIVIL ENGINEERING-RELATED POSITIONS COMES FROM EACH OF THE FOLLOWING: (121)

Masters in engineering...........................24.9%
Bachelor's level engineering degrees.............52.7%
B.S. in civil engineering technology.............12.1%
Associated degree in civil engineering technology. 7.3%
Other: (Generally Ph.D. indicated)............... 1.0%

PART III SATISFACTION WITH ENTRY-LEVEL HIRES

8A. PLEASE INDICATE YOUR IMPRESSIONS OF THE BASIC EDUCATION IN FUNDAMENTALS AT THE BACHELOR DEGREE LEVEL RECEIVED BY YOUR ENTRY-LEVEL CIVIL ENGINEERING HIRES IN THE FOLLOWING AREAS. PLEASE RATE FROM 5 (EXCELLENT) TO 1 (POOR). USE N/A FOR NOT APPLICABLE.

	excel-lent 5	4	3	2	poor 1
a. Computer Skills (115)	6.1	55.7	33.0	5.2	0.0
b. Computer Programming (96)	1.0	25.0	44.8	26.0	3.1
c. Statics-Dynamics (104)	2.9	36.5	55.8	4.8	0.0
d. Material Prop. (103)	1.0	25.2	60.2	12.6	1.0
e. Structures (108)	5.6	39.8	42.6	11.1	0.9
f. Construction (111)	0.9	7.2	35.1	45.0	11.9
g. Fluid Mechanics/ Hydraulics (103)	1.0	26.2	62.1	9.7	1.0
h. Environmental (106)	3.8	35.8	41.5	17.9	0.9
i. Hydrology (104)	0.0	23.1	55.8	18.3	2.9
j. Geotechnical (100)	1.0	18.0	51.0	28.0	2.0
k. Transportation (92)	1.1	21.7	54.3	21.7	1.1
l. Mathematics (108)	5.6	47.2	43.5	2.8	0.9
m. Oral Communications (117)	0.0	10.3	36.8	47.0	6.0
n. Ethics (114)	4.4	28.9	38.6	21.1	7.0
o. Surveying (98)	1.0	12.2	34.7	31.6	20.4
p. Drafting/ visualization (110)	0.0	10.9	30.9	45.5	12.7
q. Management/legal (101)	0.0	0.0	12.9	49.5	37.6
r. Writing skills (116)	0.0	0.0	24.1	48.3	27.6

8B. PLEASE INDICATE YOUR IMPRESSIONS OF THE BASIC EDUCATION IN FUNDAMENTALS AT THE MASTERS DEGREE LEVEL RECEIVED BY YOUR ENTRY-LEVEL CIVIL ENGINEERING HIRES IN THE FOLLOWING AREAS. PLEASE RATE FROM 5 (EXCELLENT) TO 1 (POOR). USE N/A FOR NOT APPLICABLE.

	excel-lent 5	4	3	2	poor 1
a. Computer Skills (83)	21.7	68.7	7.2	2.4	0.0
b. Computer Programming (73)	4.1	46.6	41.1	6.8	1.4
c. Statics-Dynamics (74)	10.8	50.0	37.8	1.4	0.0
d. Material Prop. (76)	1.3	42.1	50.0	5.3	1.3
e. Structures (79)	17.7	44.3	32.9	5.1	0.0
f. Construction (81)	0.0	14.8	34.6	35.8	14.8
g. Fluid Mechanics/ Hydraulics (71)	5.6	42.3	47.9	4.2	0.0
h. Environmental (77)	16.9	51.9	24.7	6.5	0.0
i. Hydrology (74)	6.8	47.3	37.8	6.8	1.4
j. Geotechnical (74)	14.9	39.2	35.1	10.8	0.0
k. Transportation (64)	1.6	40.6	43.8	14.1	0.0
l. Mathematics (81)	12.3	51.9	34.6	1.2	0.0
m. Oral Communications (85)	3.5	18.8	55.3	21.2	1.2
n. Ethics (81)	6.2	24.7	40.7	22.2	6.2
o. Surveying (68)	0.0	5.9	38.2	32.4	23.5
p. Drafting/ Visualization (79)	1.3	6.3	36.7	43.0	12.7
q. Management/Legal (76)	0.0	1.3	30.3	53.9	14.5
r. Writing Skills (84)	0.0	17.9	40.5	33.3	8.3

PART IV PROFESSIONAL/MANAGERIAL DEVELOPMENT

9. WHAT IS YOUR FIRM'S POLICY REGARDING: (120)

a. career assessment, planning and goal-setting for professional employees?

Fund and encourage................46.7%
Encourage.........................45.0%
Neutral........................... 7.5%
Discourage........................ 0.0%
No policy......................... 0.8%

b. professional education in preparation for the P.E. exam?

Fund and encourage................68.3%
Encourage.........................30.0%
Neutral........................... 1.7%
Discourage........................ 0.0%
No policy......................... 0.0%

c. active participation in professional societies and associations?

Fund and encourage	80.8%
Encourage	18.3%
Neutral	0.8%
Discourage	0.0%
No policy	0.0%

d. professional development after the P.E. exam? (e.g., seminars, courses, continuing education)

Fund and encourage	87.6%
Encourage	10.7%
Neutral	1.7%
Discourage	0.0%
No policy	0.0%

10. PLEASE RATE EACH OF THE FOLLOWING CHALLENGES FACING MANY FIRMS IN TERMS OF YOUR FIRM'S PERCEPTION OF ITS SERIOUSNESS. (PLEASE RATE ON SCALE OF 5-1, WITH 5=VERY SERIOUS.) (120)

	Very Serious 5	4	3	2	Not Serious 1
entry-level engineers (next 5 years)	29.8	39.7	24.8	1.7	4.1
b. High turnover of mid-career personnel	18.3	27.5	40.0	11.7	2.5
c. Mid-career lethargy	3.4	27.7	46.2	21.0	1.7
d. Need for better computer capability among mid-career personnel	5.8	37.5	38.3	17.5	0.8
e. Need for younger professionals to upgrade their qualifications by getting advanced degrees	6.7	24.2	35.8	27.5	5.8
f. Need for better communications skills on the part of professionals	42.5	48.3	8.3	0.8	0.0
g. Need for improved business skills: marketing, financial management, etc.	33.3	47.5	15.8	2.5	0.8
h. Need for additional training (e.g., short courses and seminars in technical subjects)	3.3	35.0	48.3	10.0	3.3

PART V DELIVERY SYSTEMS

11. TO WHAT EXTENT DO YOU LOOK TO THE FOLLOWING FOR CAREER-LONG TRAINING AND PROFESSIONAL DEVELOPMENT: (PLEASE RATE ON SCALE OF 1-5, WITH 5=VERY GREAT EXTENT) (121)

	Very Great Extent 5	4	3	2	Not At All 1
a. University graduate study program	12.4	26.4	40.5	17.4	3.3
b. University non-credit continuing education programs	2.5	32.2	43.0	19.8	2.5
c. Conferences and conventions	14.9	37.2	36.4	11.6	0.0
d. Seminars, short courses, and workshops sponsored by professional societies and other non-university sources	24.8	58.7	13.2	3.3	0.0
e. Periodicals and journals	15.7	40.5	33.9	9.9	0.0
f. Remote learning via tele-courses, video and audio cassettes, correspondence courses, etc.	0.8	8.3	35.5	41.3	14.0
g. In-house training, by either in-house training staff or outside consultants	27.3	38.0	21.5	12.4	0.8
h. Management programs of more than one-week duration	2.5	12.4	19.8	32.2	33.1

12. DOES YOUR FIRM KEEP FORMAL RECORDS OF CREDITS EARNED FROM THE ABOVE KINDS OF PROGRAMS? (121)

Degrees earned..............................74.4%
Continuing Education Units
 (CEU's) awarded by your firm................14.9%
Continuing Education Units
 (CEU's) awarded by
 outside entities............................26.4%
Programs attended............................50.4%
No formal record-keeping.....................31.4%

13. IN TERMS OF MEETING YOUR NEEDS, ARE YOU FINDING CAREER-LONG EDUCATION PROGRAMS IN THE FOLLOWING AREAS MORE THAN ENOUGH, ADEQUATE, OR INADEQUATE: (120)

	More Than Enough		Ade-quate		Inad-equate
	5	4	3	2	1
a. Technical	4.1	33.1	49.6	10.7	2.5
b. Management	5.0	20.0	30.0	36.7	8.3
c. Inter-personal skills	2.5	11.6	30.6	42.1	13.2
d. Computer skills	8.4	26.1	45.4	13.4	6.7

PART VI COMPUTER FACILITIES

14. WHAT TYPE OF COMPUTER FACILITIES DO YOU HAVE IN-HOUSE? (121)

None.. 0.8%
PC's..96.7%
Mini..66.1%
Mainframe...................................47.1%
Local area networks.........................52.9%
Connection to outside systems...............38.0%
Other.......................................12.4%

15. WHAT TYPE OF SOFTWARE DO YOU EXPECT A SIGNIFICANT NUMBER OF CIVIL ENGINEERING STAFF TO USE? (121)

Technical programs..........................97.5%
CADD..71.1%
Graphics....................................47.1%
Word Processing.............................48.8%
Desktop publishing..........................18.2%
Spreadsheet-data base management............86.8%
Other....................................... 7.4%

SURVEY OF CIVIL ENGINEERING TECHNOLOGY PROGRAMS
1989

William A. Welsh, Jr., *M.ASCE

Abstract

Results of a survey of civil engineering technology programs conducted by the Committee on Technology Curricula and Accreditation of ASCE and CORCEE in 1989 are reported and discussed. Where possible the survey results are compared with the results of a survey conducted by Richard Phelps [(1)] in 1983-84 which was summarized in the Proceedings of the 1985 Civil Engineering Education Conference.

The survey requested information on the following: university data, departmental administration, instructional programs, faculty data, departmental statistics, industrial relations, promotion and tenure, curriculum, alumni activities, student clubs, and opinion on ABET program criteria.

Procedures

The survey questionnaire was mailed to all institutions listed in the 1987 ABET Annual Report [(2)] who offered TAC accredited civil/ construction engineering technology programs. One institution indicated that they had phased out their baccalaureate CET program since the publication of the 1987 ABET Annual Report and three institutions with both baccalaureate and associate degree programs indicated that they no longer sought accreditation of the associate degree programs. Three institutions with accredited associate degree programs reported offering non-accredited baccalaureate degree programs. Twenty-nine of forty-five (64%) of accredited baccalaureate degree programs completed the questionnaire and forty-two of sixty-six (64%) of accredited associate degree programs completed the questionnaire after two mailings.

*Head, Division of Science, Engineering and Technology
Penn State Harrisburg, Middletown, PA 17057

Comparison with 1983 Survey

Data reported in 1983 was for twenty associate and twenty-nine baccalaureate programs. The complete survey results were not reported or published. The average faculty in 1983 consisted of four full-time members, fifty percent of which were tenured, as compared to four and a half faculty, 63% tenured in 1989. Seventy percent of the faculty were registered professional engineers in 1989 compared to only fifty percent in 1983. The percent of faculty registered as land surveyors has not changed in the last six years while the percent holding and EIT certificate has decreased from ten to four percent.

The number of faculty engaged in research increased from sixteen to nineteen percent while the percent engaged in funded research remained approximately constant. The number of faculty belonging to ASCE increased from twenty-seven percent in 1983 to fifty percent in 1989.

The length of both the associate degree and baccalaureate degree curricula increased by approximately four semester hours. Hands on laboratory experiences which are regularly scheduled continue to be predominant. Most programs continue to cover ethics and professionalism as part of the technical courses with very few specific courses being offered.

There were 576 bachelors degrees and 523 associate degrees reported in 1983. The average number of associate degrees per program declined from twenty six to just over thirteen. Total program enrollment increased from 4000 to 5326 but the enrollment per program stayed relatively constant at approximately 83.

Only forty-three percent of the baccalaureate programs reported having an ASCE student club in 1989 as compared to sixty-two percent in 1983.

Summary

It is hoped that this survey instrument or a similar instrument can be used to collect data on civil engineering technology programs prior to the next ASCE conference or civil engineering education so that additional important trends can be identified and the few trends occurring from 1983 to 1989 monitored.

There has been a slight growth in the number of faculty per program and significant growth in the percent who are registered professional engineers during the last six years. Membership of faculty in ASCE also increased significantly while the number of institutions reporting ASCE student clubs declined.

References

1. Phelps, Richard R., "Survey of Civil Engineering Technology Programs," Challenges to Civil Engineering Educators and Practitioners - Where Should We Be Going?, ASCE, 1985, pp. 19-21.

2. 1987 Annual Report, Accreditation Board for Engineering and Technology, Inc.

UNIVERSITY DATA

1. NAME OF COLLEGE OR UNIVERSITY 63 Institutions Responded
   - Associate Degree Only - 31 Respondents
   - Baccalaureate Degree Only - 20 Respondents
   - Associate and Baccalaureate Degrees Both - 12 Respondents

2. LOCATION BY ASCE DISTRICT.

District #	District #	District #	District #
1. 9A,2B	6. 7A,2B	10. 7A,2B	15. 1B
2. 5A,4B	7. 7A,2B	11. 1B	16. 2A,4B
4. 1A,3B	8. 1A	12. 2A,2B	
5. 1A,1B	9. 6A,5B	14. 1A,2B	

3. OFFICIAL NAME OF THE DEPARTMENT

	Assoc.	Bacc.
Engineering Technology	4	7
Civil Engr. Tech.	7	7
Engr. & Engr. Tech.	0	4
Other	21	13

4. OFFICIAL NAME OF THE PROGRAM

	Assoc.	Bacc.
Civil Engr. Tech.	27	19
Const. Engr. Tech.	4	5
Other	6	8

5. IS THE PROGRAM ACCREDITED BY TAC/ABET?

	YES	NO
Associate Degree	41	2
Baccalaureate Degree	29	3

6. MAJOR SOURCE OF FINANCIAL SUPPORT

	Assoc	Bacc
State	41	27
Private	1	4
Both of the Above	1	1
No Response	0	0

7. TYPE OF PROGRAM 2 YR 43 2+2 7 4YR 23 +2 2

TYPES OF DEGREES AWARDED

AAS	29
ASET	11
BS	12
BSET	13
BET	5
MSET	
OTHER	2
NO RESPONSE	3

8. WHAT KIND OF ADMISSIONS POLICY DOES YOUR SCHOOL HAVE FOR ENTRANCE INTO THE CIVIL ENGINEERING TECHNOLOGY PROGRAM.

	Assoc	Bacc
- Open Admissions for graduates of accredited high schools in your state	23 (66%)	13 (45%)
- Selective admission standards with no restrictions the size of the freshman class.	7 (20%)	13 (45%)
- Selective admissions standards with restrictions on the size of the other freshman class.	5 (14%)	3 (10%)

DEPARTMENTAL/PROGRAM ADMINISTRATION

9. TITLE OF DEPARTMENT/PROGRAM ADMINISTRATOR

	Assoc	Bacc.
Head	8 (24%)	8 (26%)
Chairman	17 (50%)	13 (42%)
Director	3 (9%)	1 (3%)
Other	6 (17%)	9 (29%)

10. HOW IS THE DEPARTMENT/PROGRAM HEAD SELECTED?

	Assoc.	Bacc.
Appointed by administration without advice from department faculty	0	4 (13%)
Appointed by administration with advice from faculty	19 (76%)	25 (78%)
Elected by the department/program faculty	6 (24%)	2 (6%)
Other	0	1 (3%)

11. WHAT IS THE NORMAL TERM OF OFFICE FOR THE DEPARTMENT/PROGRAM HEAD?

	Assoc.	Bacc.
Permanent appointment subjected to periodic review by higher administration	19 (54%)	16 (52%)
One term for a fixed number of years	2 (6%)	2 (6%)
Two terms for a fixed number of years each	0	2 (6%)
More than two terms of a fixed number of years each	5 (14%)	3 (10%)
Indeterminate term	7 (20%)	5 (16%)
Other	2 (6%)	3 (10%)

12. DO YOU HAVE AN INDEPENDENT DEPARTMENTAL NON-SALARY BUDGET FOR OPERATIONS ALLOCATED FOR EACH FISCAL YEAR?

	Yes	No
Assoc.	30 (86%)	5 (14%)
Bacc.	29 (91%)	3 (9%)

13. CHOOSE THE STATEMENT WHICH BEST DESCRIBES YOUR DEPARTMENTAL EQUIPMENT BUDGET

	Assoc.	Bacc.
Regular allocation each year.	7 (19%)	11 (35%)
Department/program bargains for its fair share of the college's equipment budget each year.	15 (42%)	16 (52%)
Irregular allocation each year which sometimes is zero	13 (36%)	4 (13%)
Other	1 (3%)	0

14. HOW MANY BUDGETED FULL-TIME-EQUIVALENT (FTE) FACULTY DO YOU HAVE?

Assoc.	33 Responses, Range 1 to 10.5,	Total 122
Bacc.	32 Responses, Range 2 to 18,	Total 171

15. HOW MANY BUDGETED FULL-TIME-EQUIVALENT (FTE) CLERICAL STAFF DO YOU HAVE?

Assoc.	33 Responses, Range 0 to 2,	Total 18.8
Bacc.	32 Responses, Range 0 to 3,	Total 27.6

16. HOW MANY BUDGETED FULL-TIME EQUIVALENT (FTE) TECHNICAL STAFF TO YOU HAVE?

Assoc.	34 Responses, Range 0 to 4,	Total 14.7
Bacc.	32 Responses, Range 0 to 2,	Total 12.8

17. WHAT IS YOUR SCHOOL YEAR BASIS?

	Assoc.	Bacc.
Semester	24 (69%)	24 (75%)
Quarter or term	11 (31%)	8 (25%)
Other	0	0

18. DO YOU OFFER DEGREE COURSES IN THE EVENING

	Assoc.	Bacc.
Yes	26 (74%)	20 (63%)
No	9 (26%)	12 (37%)

19. DO YOU OFFER DEGREE PROGRAMS IN THE EVENING?

	Assoc.	Bacc.
Yes	21 (60%)	12 (37%)
No	14 (40%)	20 (63%)

20. IF YES IN #19, AT WHAT LEVELS?

Associate Degree	Yes 20	(65%)
Bachelors Degree	Yes 11	(35%)

21. DO YOU HAVE A CO-OP PROGRAM WITH INDUSTRY?

	Assoc.	Bacc.
Yes	20 (57%)	26 (81%)
No	15 (43%)	6 (19%)

22. IF YES, IS IT:

	Assoc.	Bacc.
Mandatory	3 (15%)	1 (4%)
Optional	17 (85%)	24 (96%)

23. IF OPTIONAL, WHAT PERCENTAGE OF YOUR FULL-TIME STUDENTS ARE INVOLVED IN THE PROGRAM?

Assoc.	Range 0 to 100%,	Avg. 15%
Bacc.	Range 0 to 60%,	Avg. 15%

24. DO YOU OFFER (FOR DEGREE CREDIT) EXTENSION COURSES AT OFF CAMPUS LOCATIONS?

	Assoc.	Bacc.
Yes	4 (11%)	5 (16%)
No	31 (89%)	26 (84%)

25. DO YOU OFFER COURSES (FOR DEGREE CREDIT) BY TELEVISION TO REMOTE AREAS?

	Assoc.	Bacc.
Yes	2 (6%)	1 (3%)
No	31 (94%)	31 (97%)

26. PLEASE STATE THE NUMBER OF AUTHORIZED 1988-89 FACULTY APPOINTMENTS?

FULL-TIME	Assoc.	Bacc.
- Professor	31 (25%)	42 (26%)
- Associate Professor	36 (28%)	59 (37%)
- Assistant Professor	26 (21%)	47 (29%)
- Instructor	28 (22%)	11 (7%)
- Lecturer	1 (1%)	1 (1%)
- Other	4 (3%)	0
Total	**126(100%)**	**160(100%)**

PART-TIME	Assoc.	Bacc.
- Professor	1 (2%)	5 (7%)
- Associate Professor	6 (11%)	7 (10%)
- Assistant Professor	6 (11%)	6 (9%)
- Instructor	23 (41%)	25 (36%)
- Lecturer	17 (30%)	25 (36%)
- Other	3 (5%)	1 (1%)
TOTAL	**56 (100%)**	**69 (100%)**

27. DO YOU EMPLOY GRADUATE STUDENTS AS TEACHING ASSISTANTS?

	Assoc.	Bacc.
Yes	1 (3%)	8 (25%)
No	33 (97%)	24 (75%)

28. DO YOU REGULARLY UTILIZE PART-TIME TEACHING APPOINTMENTS WITH ENGINEERS FROM PRACTICE?

	Assoc.	Bacc
Yes	20 (59%)	26 (81%)
No	14 (41%)	6 (19%)

29. HOW MANY FULL-TIME-EQUIVALENT FACULTY PER YEAR ARE REPRESENTED BY YOUR ANSWER TO #28 ABOVE?
Assoc. 35.9 **TOTAL** Bacc. 29.8 **TOTAL**

30. HOW MANY FULL-TIME FACULTY ARE REGISTERED PROFESSIOAL ENGINEERS?
Assoc. 86(69%) Bacc. 114(72%)

31. HOW MANY FULL-TIME FACULTY ARE REGISTERED LAND SURVEYORS?
Assoc. 22(17%) Bacc. 29 (18%)

32. HOW MANY FULL-TIME FACULTY WHO ARE NOT REGISGERED P.E.'S HOLD AN EIT CERFTIFICATE?
Assoc. 7 Total Bacc. 5 Total

33. HOW MANY FULL-TIME FACULTY ARE NOT ELIGIBLE FOR REGISTRATION AS AN ENGINEER OR SURVEYOR?
Assoc 26 Total Bacc. 9 total

34. WHAT PERCENTAGE OF YOUR FACULTY IS TENURED?
Assoc. Range 0 to 100% Avg. 60%
Bacc. Range 0 to 100% Avg. 65%

35. AT THE START OF THE 1988-89 ACADEMIC YEAR, HOW MANY FACULTY VACANCIES DID YOU HAVE IN YOUR AUTHORIZED POSITIONS?
Assoc. 5 Total Bacc. 9 Total

36. DOES YOUR UNIVERSITY/COLLEGE HAVE A SABBATICAL LEAVE POLICY?

	Assoc.	Bacc.
Yes	29 (83%)	28 (88%)
No	6 (17%)	4 (12%)

37. HOW MANY FACULTY HAVE BEEN GRANTED A SABBATICAL LEAVE DURING THE LAST FIVE YEARS?
Assoc. 21 from 26 institutions
Bacc. 20 from 30 institutions

38. HOW MANY FULL-TIME FACULTY BELONG TO:

	Assoc.	Bacc.
ASCE	48 (38%)	87 (54%)
NSPE	10 (8%)	42 (26%)
ASEE	30 (24%)	71 (44%)

39. HOW MANY FULL-TIME TEACHING FACULTY ARE ENGAGED IN RESEARCH AS PART OF THEIR REGULAR DUTIES?

Assoc. 9(7.1%) Bacc. 42(26%)

40. WHAT PERCENT OF THOSE INCLUDED IN THE ANSWER TO #39 PERFORM RESEARCH WHICH IS EXTERNALLY FUNDED?

Assoc. 13% Avg. Bacc. 40% Avg.

41. DOES YOUR SHOOL HAVE A POLICY ON CONSULTING ACTIVITIES OF THE FACULTY?

	Assoc.	Bacc.
Yes	28 (80%)	29 (94%)
No	7 (20%)	2 (6%)

42. IF YES, IS IT A WRITTEN POLICY?

	Assoc.	Bacc.
Yes	21 (70%)	23 (72%)
No	9 (30%)	9 (28%)

43. IF A CONSULTING POLICY EXISTS, IS IT?

	Assoc.	Bacc.
- One day per week	8 (47%)	21 (75%)
- Less than one day/week	2 (12%)	2 (7%)
- More than one day/week	0	0
- Up to the department head to supervise amount of consulting	1 (6%)	5 (18%)
- Other	6 (35%)	

44. IRRESPECTIVE OF POLICY, WRITTEN OR UNWRITTEN, WHAT DO YOU FEEL IS THE ATTITUDE OF THE DEAN TOWARD FACULTY CONSULTING?

	Assoc.	Bacc.
- Encouraged	16 (47%)	24 (75%)
- Tolerated	17 (50%)	8 (25%)
- Discouraged	1 (3%)	0
- Forbidden	0	0

45. AT THE RANK OF ASSISTANT PROFESSOR AND HIGHER, HOW MANY FACULTY DO NOT HOLD AN EARNED DOCTORAL DEGREE?

Assoc 85 Total(91%) Bacc. 99 Total (67%)

46. AT THE RANK OF ASSISTANT PROFESSOR AND HIGHER, HOW MANY FACULTY DO NOT HOLD AN EARNED MASTERS DEGREE?

Assoc. 24 Total (26%) Bacc. 14 Total (9%)

DEPARTMENTAL STATISTICS

47. HOW MANY CIVIL ENGINEERING TECHNOLOGY DEGREES DID YOU AWARD BETWEEN JULY 1, 1987 AND JUNE 30, 1988?

	Total	Schools	Avg.	High.
Bachelors Degrees	616	32	19.3	60
Associate Degrees	552	41	13.5	60

48. WHAT WERE YOUR FALL 1988 ENROLLMENT FIGURES?

Full-Time	Total	High	Avg.
Freshman	1577	140	27.2
Sophomore	1058	100	18.2
Junior	667	80	22.2
Senior	700	75	23.3

Part-Time	Total	High	Avg.
Freshman	556	188	9.8
Sophomore	344	113	6.0
Junior	203	65	6.8
Senior	221	65	7.4

49. WHAT WAS THE TOTAL ENGINEERING TECHNOLOGY ENROLLMENT AT YOUR INSTITUTION FOR FALL 1988?

31,514 Total at 60 institutions

50. WHAT WAS THE TOTAL UNDERGRADUATE ENROLLMENT AT YOUR INSTITUTION FOR FALL 1988?

421,870 Total at 59 institutions

51. WHICH STATEMENT BEST DESCRIBES YOUR PRESENT DEPARTMENTAL UNDERGRADUATE ENROLLMENT SITUATION?

	Assoc.	Bacc.
Increasing rapidly (6%/yr and up)	3 (9%)	2 (6%)
Increasing slightly (<6%/Yr)	12 (35%)	9 (28%)
Approximately constant	14 (41%)	13 (41%)
Decreasing slightly (<6%/yr)	5 (15%)	5 (16%)
Decreasing rapidly (6%/yr and up)	0	3 (9%)

52. IN THE 1987-88 ACADEMIC YEAR (JUNE TO JUNE), HOW MANY STUDENT-CREDIT-HOURS DID YOUR DEPARTMENT GENERATE?

Associate degree 35,787 at 20 institutions (1789 avg.)
Baccalaureate degree 40,709 at 19 institutions (2143 avg.)

INDUSTRIAL RELATIONS

53. PLEASE LIST THE NUMBER OF FULL-TIME FACULTY WHO HAVE HAD EXPERIENCE IN ENGINEERING PRACTICE ACCORDING TO THE FOLLOWING CATEGORIES:

	Assoc	Bacc.
Full-time annual employment	105 (83%)	129 (81%)
Part-time annual employment	29 (23%)	26 (16%)
Summers only	20 (16%)	16 (10%)
Consulting while teaching	51 (41%)	58 (36%)

54. DOES YOUR PROGRAM HAVE AN INDUSTRIAL ADVISORY COMMITTEE?

	Assoc.	Bacc.
Yes	35 (100%)	30 (94%)
No	0	2 (6%)

55. IF YOUR ANSWER TO #54 IS NO, IS THERE AN INDUSTRIAL ADVISORY COMMITTEE WHICH SERVES ALL ENGINEERING TECHNOLOGY PROGRAMS ON YOUR CAMPUS?

	Assoc.	Bacc.
Yes	0	2 (100%)
No	0	0

PROMOTION & TENURE

56. PROMOTION AND TENURE

a. WHO INITIATES ACTION:

	Assoc.	Bacc.
Committee	5 (14%)	4 (11%)
Individual	17 (49%)	26 (70%)
Department Head	13 (37%)	7 (19%)

b. DOES YOUR DEPARTMENT HAVE A FACULTY COMMITTEE TO CONSIDER RECOMMENDATION?

	Assoc.	Bacc.
Yes	18 (58%)	27 (87%)
No	13 (42%)	4 (13%)

c. DOES THE DEPARTMENT HEAD HAVE THE POWER TO REJECT RECOMMENDATIONS AND THEREBY STOP THE PROCESS?

	Assoc.	Bacc.
Yes	4 (13%)	4 (13%)
No	12 (39%)	10 (31%)
Yes, can reject but progress does not stop	15 (48%)	18 (56%)

d. FROM WHAT LEVEL DOES FINAL APPROVAL OR DENIAL COME?

	Assoc.	Bacc.
College (Dean)	1 (4%)	4 (13%)
Administration (Vice President)	2 (7%)	3 (9%)
Administration (President)	14 (48%)	10 (31%)
Board of Trustees	12 (41%)	15 (47%)

e. IS RESEARCH REQUIRED FOR PROMOTION AND TENURE?

	Assoc.	Bacc.
Yes	3 (9%)	12 (37%)
No	30 (91%)	20 (63%)

f. IS PUBLICATION REQUIRED FOR PROMOTION AND TENURE?

	Assoc.	Bacc.
Yes	6 (18%)	18 (56%)
No	27 (82%)	14 (44%)

57. DOES YOUR INSTITUTION RECOGNIZE THE MASTERS DEGREE AS THE APPROPRIATE TERMINAL DEGREE FOR TEACHING ENGINEERING TECHNOLOGY?

	Assoc.	Bacc.
Yes	26 (76%)	26 (81%)
No	8 (24%)	6 (19%)

58. DOES YOUR FACULTY BELONG TO A COLLECTIVE BARGAINING UNIT?

	Assoc.	Bacc
Yes	17 (49%)	8 (26%)
No	18 (51%)	23 (74%)

59. IF YES, ARE DEPT. HEADS INCLUDED IN THE BARGAINING GROUP?

	Assoc.	Bacc.
Yes	11 (65%)	4 (50%)
No	6 (35%)	4 (50%)

60. HOW MANY SEMESTER CREDITS DO YOU REQUIRE FOR GRADUATION?

Associate Degree	Range 62 to 108	Avg. 72
Bachelors Degree	Range 127 to 140	Avg. 133

61. HOW MANY PHYS ED CREDITS ARE INCLUDED IN THE ANSWER TO #60?
Range - 0 TO 8 CREDITS

62. IF YOU DO NOT USE THE CREDIT SYSTEM, HOW MANY COURSES DO YOU REQUIRE FOR GRADUATION?

All programs responding used the credit system.

63. HOW MANY YEARS OF STUDY DOES YOUR CATALOG STATE ARE NORMALLY REQUIRED FOR:

An Associate Degree	2.0	Avg.
A Bachelors Degree	4.0	Avg.

64. BASED ON THE TAC/ABET DEFINITIONS, WHAT PERCENT (TO THE NEAREST WHOLE NUMBER) OF YOUR CURRICULUM IS DEVOTED TO THE FOLLOWING AREAS?

	Associate			Baccalaureate		
	High	Low	Avg.	High	Low	Avg.
a) Mathematics	17%	10%	13%	13%	5%	10%
b) Basic Sciences		2%	11%	17%	5%	10%
c) Technical sciences		2%	12%	28%	2%	14%
d) Technical specialities	64%	6%	40%	59%	16%	32%
e) Technical electives	28%	0%	2%	19%	0%	8%
f) Written and Oral Communication	20%	6%	11%	15%	5%	8%
g) Social Science/ g) Humanities	15%	0%	7%	42%	8%	15%

65. DO YOU OFFER UNDERGRADUATE COURSES IN THE CIVIL ENGINEERING TECHNOLOGY DEPARTMENT IN ANY OF THE FOLLOWING AREAS? LIST NUMBER OF CREDITS BOTH REQUIRED AND ELECTIVE.

	Associate (32 resp)		Baccalaureate (23 resp)	
	Average Credits	Number Responses	Average Credits	Number Responses
Structural	5.7	25	6.8	21
Soils	3.8	23	3.8	18
Environmental	3.5	11	3.5	12
Transportation	3.6	12	3.0	13
Surveying	7.6	30	6.5	18
Engineering Mechanics	5.6	30	7.6	19
Materials	3.9	25	3.6	18
Const. Management	4.2	16	7.2	18

66. DOES YOUR CURRICULUM HAVE ANY OF THE FOLLOWING COURSES REQUIRED FOR THE DEGREE?

	Associate		Baccalaureate	
	Yes	No	Yes	No
Soil Mechanics	32(74%)	11(26%)	27(87%)	4(13%)
Structural analysis	26(60%)	17(40%)	24(75%)	8(25%)
Computer programming	42(95%)	2(4%)	30(100%)	0
CAD	33(77%)	10(23%)	16(53%)	14(47%)
Fluid mechanics	19(45%)	23(55%)	21(66%)	11(34%)
Engineering graphics	38(86%)	6(14%)	29(94%)	2(6%)
Water/waste water systems	18(43%)	24(57%)	16(52%)	15(48%)
Surveying	44(100%)	0	29(94%)	2(6%)
Concrete design	28(65%)	15(35%)	23(77%)	7(23%)
Steel design	31(74%)	11(26%)	24(80%)	6(20%)
Electrical Engr. Fund.	4(10%)	38(90%)	20(65%)	11(35%)
Thermodynamics	1(2%)	40(98%)	6(19%)	26(81%)
Hydraulics	29(69%)	13(31%)	19(63%)	11(37%)
Highway Engr.	16(39%)	25(61%)	15(48%)	16(52%)
Engineering Economics	11(26%)	32(74%)	27(84%)	5(16%)
Concrete/bitum. materials	34(79%)	9(21%)	26(87%)	4(13%)
Geology	5(12%)	35(88%)	11(35%)	20(65%)
Hydrology	13(30%)	30(70%)	10(32%)	21(68%)
Probability & statistics	3(7%)	40(93%)	19(61%)	12(39%)
Technical report writing	28(65%)	15(35%)	30(97%)	1(3%)
Ethics & professionalism	4(9%)	9(91%)	14(45%)	17(55%)
Speech	29(67%)	14(33%)	26(81%)	6(19%)
Contracts & specs.	20(47%)	23(53%)	18(58%)	13(42%)
Estimating	22(52%)	20(48%)	25(81%)	6(19%)
Scheduling	13(30%)	30(70%)	22(71%)	9(29%)
Labor law & relations	4(9%)	39(91%)	13(42%)	18(58%)
Dynamics	0	42(100%)	12(37%)	20(63%)

67. DO YOU HAVE DEPARTMENTAL LABORATORY COURSES IN THE FOLLOWING AREAS?

	Yes		No	
	Assoc.	Bacc.	Assoc.	Bacc.
Strength of Materials	28(82%)	23(74%)	6(18%)	8(26%)
Soil Mechanics	31(89%)	28(90%)	4(11%)	3(10%)
Concrete & bitum. materials	28(82%)	26(87%)	6(18%)	4(13%)
Fluids/Hydraulics	17(52%)	21(68%)	16(48%)	10(32%)
Surveying	35(100%)	29(94%)	0	2(6%)
Environmental Sanitary	8(25%)	12(39%)	26(75%)	19(61%)
Computer	34(100%)	28(93%)	0	2(7%)
Drafting	35(100%)	29(97%)	0	1(3%)
Other				

68. HOW ARE YOUR LAB EXERCISES CONDUCTED?

	% OF THE TIME			
	Assoc.		Bacc.	
Hands on		84%		89%
Demonstration		16%		11%

69. WITH REGARD TO LAB COURSE SCHEDULING, DO YOU:

	Yes		No		% of time	
	Assoc.	Bacc.	Assoc.	Bacc.	Assoc.	Bacc.
Have open labs	8	6	25	24	6%	8%
Have regularly scheduled labs	33	27	0	1	94%	92%

70. WHICH OF THE FOLLOWING METHODS DO YOU USE TO INTEGRATE COMPUTERS INTO THE TECHNICAL COURSES?

	Assoc.	Bacc.
Some homework assigned to be done on computers	29 (83%)	30 (94%)
Students write own programs	19 (54%)	25 (78%)
Commercial or faculty software only is used	23 (66%)	19 (59%)
Lab reports prepared with computer (word processing, spread sheet, etc.)	29 (83%)	27 (84%)

71. IS YOUR CIVIL ENGINEERING TECHNOLOGY DEPARTMENT RESPONSIBLE FOR TEACHING ENG. MECHS. COURSES, SUCH AS STATICS & STRENGTH OF MATERIALS?

	Assoc..	Bacc.
Yes	30 (86%)	24 (77%)
No	5 (14%)	7 (23%)

72. CHOOSE THE STATEMENT WHICH BEST DESCRIBES THE WAY YOU COVER THE TOPICS OF ETHICS & PROFESSIONALISM.

	Assoc.	Bacc.
A specific course is offered	2 (5%)	5 (15%)
Topics are covered as part of other courses	24 (67%)	24 (73%)
No coverage offered	10 (28%)	4 (12%)

ALUMNI ACTIVITIES

73. DO YOU PUBLISH AN ALUMNI NEWS LETTER AT LEAST ONCE A YEAR WHICH IS MAILED TO ALL DEPARTMENTAL/PROGRAM ALUMNI?

	Assoc.	Bacc.
Yes	8 (23%)	13 (43%)
No	27 (77%)	17 (57%)

74. IF YES, AT WHAT LEVELS?

	Assoc.	Bacc.
Program	1 (10%)	2 (9%)
Department	2 (18%)	7 (32%)
College/school	5 (45%)	8 (36%)
University	3 (27%)	5 (23%)

75. CHOOSE THE STATEMENT WHICH BEST DESCRIBES HOW YOU CONDUCT SURVEYS OF YOUR ALUMNI.

	Assoc.	Bacc.
Conduct alumni surveys yearly	8 (23%)	9 (29%)
Conduct on a irregular basis	23 (66%)	21 (68%)
Do not conduct surveys	4 (11%)	1 (3%)

76. WHAT EXTERNAL SUPPORT HAS YOUR DEPARTMENT/PROGRAM RECIEVED IN THE LAST TWO YEARS IN THE FOLLOWING AREAS:

<u>GIFTS</u>
Industrial, Total amount:
- 21 institutions reported total of $233,750

<u>GRANTS</u>
Industrial Total amount:
- 6 institutions reported total of $39,500

Government, Total amount:
- 12 institutions reported total of $1,321,160

<u>EQUIPMENT</u>
Industrial, Total value:
- 21 institutions reported total of $368,355

Government Total value:
- 10 institutions reported total of $390,500

77. DID YOU RECEIVE FUNDS FROM ALUMNI GIVING THAT WERE SPECIFICALLY DESIGNATED TO CIVIL ENGINEERING TECH. (LAST 5 YEARS)?

	Assoc.	BACC.
Yes	9 (26%)	13 (42%)
No	26 (74%)	18 (58%)

STUDENT CLUB

78. DO YOU HAVE AN ASCE STUDENT CLUB?

	Assoc.	Bacc.
Yes	4 (11%)	13 (43%)
No	31 (89%)	17 (57%)

79. HOW IS THE FACULTY ADVISOR FOR THE STUDENT CLUB CHOSEN?

	Assoc.	Bacc.
Duties assigned by dept. Head:	3	9
Faculty member volunteers:	11	9
Chosen by students	2	3
Other		1

80. HOW IS THE FACULTY MEMBER RECOGNIZED WHILE PERFORMING AS THE FACULTY ADVISOR?

	Assoc.	Bacc.
Advisory duties considered part of teaching load with an appropriate reduction of course teaching assignments	7 (44%)	8 (33%)
Considered as an overload to normal teaching assignments	9 (56%)	16 (67%)

OPINIONS

81. TAC/ABET PROGRAM CRITERIA FOR ASSOCIATE DEGREE PROGRAMS IN CIVIL ENGINEERING TECNOLOGY SHOULD REQUIRE SOME COVERAGE OF THE FOLLOWING TOPICS:

	Assoc.		Bacc.	
	Yes	No	Yes	No
Statics	32(97%)	1(3%)	23(100%)	0
Strength of Materials	33(97%)	1(3%)	22(100%)	0
Engineering Graphics	34(97%)	1(3%)	22(96%)	1(4%)
Surveying	35(100%)	0	23(100%)	0
Civil Engineering Materials	27(96%)	1(4%)	23(100%)	0
Other (please list)				

82. TAC/ABET PROGRAM CRITERIA FOR BACCALAUREATE DEGREE PROGRAMS IN CIVIL ENGINEERING TECHNOLOGY SHOULD REQUIRE SOME COVERAGE OF THE FOLLOWING TOPICS:

	Assoc.		Bacc.	
	Yes	No	Yes	No
Statics	16(100%)	0	29(100%)	0
Dynamics	11(65%)	6(35%)	17(59%)	12(41%)
Stren. of Mat.	16(100%)	0	29(100%)	0
Hydraulics	15(94%)	1(6%)	26(90%)	3(10%)
Engr. Graphics	16(100%)	0	28(97%)	1(3%)
Surveying	16(100%)	0	29(100%)	0
C.E. Materials	16(100%)	0	29(100%)	0
Engr. Economics	15(94%)	1(6%)	28(97%)	1(3%)
Soils & Found.	15(94%)	1(6%)	27(94%)	2(6%)

83. TAC/ABET PROGRAM CRITERIA FOR BACCALAUREATE DEGREE PROGRAMS IN CONSTRUCTION ENGINEERING TECHNOLOGY SHOULD REQUIRE SOME COVERAGE OF THE FOLLOWING TOPICS:

	Assoc		Bacc.	
	Yes	No	Yes	No
Statics	17(100%)	0	28(100%)	0
Dynamics	9(53%)	8(47%)	10(38%)	16(62%)
Stren. OF Mat.	16(94%)	1(6%)	27(100%)	0
Hydraulics	13(81%)	3(19%)	19(68%)	9(32%)
Engr. Graphics	16(94%)	1(6%)	27(96%)	1(4%)
Surveying	16(94%)	1(6%)	28(100%)	0
CE Materials				
Engr. Economics	14(88%)	2(12%)	28(97%)	1(3%)
Soils & Found.	15(94%)	1(6%)	27(94%)	2(6%)

84. CIVIL ENGINEERING TECHNOLOGY AND CONSTRUCTION ENGINEERING TECHNOLOGY SHOULD HAVE SEPARATE AND DISTINCT TAC/ABET PROGRAM CRITERIA?

	Assoc.	Bacc.
Yes	21 (68%)	19 (63%)
No	10 (32%)	11 (37%)

A Course in Construction Field Testing and Monitoring

Walter Konon[1]
Member ASCE

Abstract

The quality control of materials in the field and the monitoring of environmental and worker safety at construction sites is now controlled through the use of field testing equipment. Students should be familiarized with the use and limitations of this testing and monitoring equipment and the associated OSHA, EPA and ASTM standards. A new laboratory course devoted to this topic is presented.

Introduction

During the last ten years, there has been a significant increase of affordable portable field testing equipment available to contractors. As a result, the quality control of materials in the field and the monitoring of environmental and worker safety at construction sites is now controlled through the use of field testing equipment.

Testing procedures presented in Strength of Materials courses are usually geared to demonstrating basic material properties using conventional laboratory testing equipment.

In 1986, an existing undergraduate course at New Jersey Institute of Technology (NJIT), "Construction Testing," was modified and expanded as shown in Appendix A. The original course was a follow-up to Strength of Materials and was intended to familiarize BSET Construction and Contracting majors with portable field testing procedures and equipment that they were likely to encounter on construction sites.

At about this time, the newly emerging field of hazardous and toxic waste collection, extraction and confinement work was becoming important. This work involved many activities that required extensive field testing and monitoring by construction personnel. The establishment at NJIT for the NSF/University Cooperative Center for Research in Hazardous and Toxic Substances made the Institute a focal point for hazardous clean-up study. New Jersey

[1]Professor, Head of Construction Division, Department of Civil Engineering, New Jersey Institute of Technology, Newark, N.J. 07102

contractors engaged in site remediation work naturally looked to and expected NJIT's construction graduates to have some familiarity with the special environmental field monitoring and testing requirements of this type of work.

Clearly, training students in the proper use of field testing and monitoring equipment would help to insure a safer environment and a healthier work force. With the increased emphasis on toxic and hazardous clean-ups in New Jersey, and Nationally, it was felt that a strict materials field testing approach was too limited and that environmental field monitoring should be included in a revised course. The course material was expanded to include correct usage and limitations of the equipment, EPA, OSHA and ASTM standards and procedures associated with the equipment, and familiarization with the different types of portable field testing and monitoring equipment available. A list of new equipment purchased for the revised course through a grant in 1986, and associated reference standards, is shown in Appendix B.

The above described course was part of the required course work in a BSET Construction Engineering Technology program. While this type of hands-on course may not be appropriate for civil engineering students, a state-of-the-art course in construction field testing and monitoring, makes for a more qualified and sought after BSET graduate and facilitates his entrance into the job market. In the first few years after graduation, this student will most likely be involved in field supervision and construction and, as such, may be called upon to perform or supervise field testing and monitoring. After a few years, when the graduates have reached a managerial level, they would be familiar with the types of equipment available, the correct use and limitations of the equipment, and the standards and procedures associated with equipment.

APPENDIX A

PROPOSED CHANGES TO COURSE OUTLINE
CET 431 - Construction Testing

	Present Topic	Proposed Topic
1	Introduction to Material Properties	No Change
2	Soils, Aggregates, and Proctor Compaction Test	No Change
3	Field Testing of Soils	Field Testing of Soils (including Hazardous & Toxic Waste Detection and Nuclear Density Measurements)
4	Portland Cement Concrete Mix Design and Testing	No Change
5	"	No Change
6	"	No Change
7	Sound and Noise Measurement, Ground Vibration Measurement	Sound and Noise Measurement, Ground Vibration Measurement (including Seismic/ Acoustical triggered Seismo.
8	Mid-Term Exam	No Change
9	Brick Specifications and Testing	No Change
10	Wood Grading and Testing	No Change
11	Construction Safety Testing	Testing of the Construction Environment and other safety tests (including Hazardous Atmospheric Testing and use of Personal Dosimeters)
12	Welding and Weld Testing	Subsurface Detection of Hazardous and Toxic Wastes using Conductivity, Resistivity, and pH
13	Iron and Steel Testing	Welding and Weld Testing; Iron and Steel Testing
14	Asphalt and Asphalt Concrete Testing	No Change
15	Final Exam	No Change

APPENDIX B

NEW EQUIPMENT LIST AND REFERENCE STANDARDS

	Equipment Description	Applicable Standard(s)*
1.	Organic Puff Sampler	OSHA 1926.55; EPA Tested Apr. '84 under EPA Contract 6802-3745 (WA-9)
2.	Dust Sampler	OSHA 1910.1001; OSHA 1926.55
3.	Dust Impactor	OSHA 1910.1001; OSHA 1926.55
4.	Air Monitoring Kit	OSHA 1926.55
5.	Mini pH Meter	No standard established. Presently used in soil and groundwater field testing
6.	Turbiditymeter	ASTM D1889
7.	Conductivitymeter	No standard established. Presently used in soil and groundwater field testing
8.	Resistivitymeter	ASTM G57
9.	Organic Vapor Dosimeter	OSHA 1926.55; MSA Data 08-00-35
10.	Explosimeter	OSHA 1926.800 (c) (i); MSA Data 08-00-03; UL approved
11.	Oxygen Indicator	OSHA 1926.800 (c) (ii); MSHA Approval - No. 2G3336
12.	High Volume Air Sampler	EPA Approved - Federal Reference Method for TSP; Fed. Reg Dec. 6, 1982, Vol. 47, #234, pp. 54896-54921, 40 CFR Part 50; EPA Proposed New Method, March 20, 1984 for Ambient PM10
13.	Gas, Oxygen and Combustible Meter	OSHA 1926.800 (c)(i); MSA Data 00-00-38
14.	Personal Noise Dosimeter	OSHA 1926.52; ANSI S1.4 -1971
15.	Sound Level Meter	OSHA 1926.52; MSHA Approved 2G2933; ANSI S1.4 - 1971
16.	Seismograph	U.S. Bur. of Mines RI 8506, RI8507; N.J. Dept. of Labor
17.	Nuclear Density/ Moisture Meter	ASTM D2922; ASTM D2950; ASTM D3017
18.	Concrete Test Hammer	ASTM C805
19.	Concrete Air Volume Meter	ASTM C231; AASHTO T152

APPENDIX B (cont'd)

*Abbreviations:

OSHA	Occuptational Safety and Health Administration
ASTM	American Society for Testing and Materials
ANSI	American National Standards Institute
AASHTO	American Association of State Highway and Transportation Officials
MSHA	Mine Safety and Health Administration

Construction Engineering and Management Challenges and Opportunities

Fred Moavenzadeh, MASCE[1]

Abstract

Changes in demand for and supply of construction and engineering services and products have necessitated a new look at our educational programs in Construction Engineering and Management. Traditional emphasis on technology of management of products (construction technology) and the arrangement of resources necessary for that assemblage (project management) seem no longer adequate to cope with recent challenges facing the industry. More emphasis is needed on organizational issues facing construction firms as well as industrial issues emerging in the industry as a whole. Topics such as finance, strategic management, information technology, niche strategy and market segmentation, as well as productivity, competitiveness and technological innovation seems to be equally important to be included in construction engineering and management programs.

The Scope of Construction Engineering and Management

Construction engineering and management is concerned with assembling and transforming manpower, materials, equipment and economic resources into the buildings, industrialplants, and infrastructure facilities required to serve a variety of societal needs. It involves accomplishing planning, design, construction, operations, and maintenance activities through a complex network and set of interrelationships among a diversity of private companies and governmental organizations. In addition to the construction contractors and subcontractors which do the actual construction work, other participants in the process include: the client or owner who defines the need for a facility; the investor or lender who provides the capital; the engineers and architects who design the facility; the labor organizations which supply manpower resources and skills, the manufacturer and supplier who provide construction materials products and equipment; regulatory agencies which prescribe and enforce codes, standards, and regulations; and other organizations and individuals

[1]George Macomber Professor of Construction Management and Director of the Center for Construction Research and Education, M.I.T., 77 Massachusetts Avenue, Cambridge, MA 02139

involved in research, development, education and information exchange.

The combination of these three types of diversities - in markets, activities, and participants - is unique to the construction industry. Therefore a specialized education is needed for those who aspire to positions of technical and managerial leadership in the industry.

The Changing Nature of Construction

Over the past several decades, while construction engineering and management was becoming an established discipline in civil engineering, both the market for construction and the nature of the industry and its basic resources were undergoing significant change and restructuing. During the 1980's increasing public concern for the environment resulted in regulations and review processes which added to the time and greatly increased the complexity of planning, design, and construction activities. In addition, it opened the construction process to an even wider range of participants and forced the consideration of social and environmental factors together with traditional economic and technical factors.

The energy crisis of the 1970's gave rise to construction of superprojects of unprecedented technical complexity, scale, and costs as the world sought out new sources of oil, or began to develop alternative energy resources such as synfuels. Increased revenues in the oil-producing countries created and sustained construction booms which greatly increased the size, geographic scope, and complexities of the international construction market.

The declining condition of infrastructure facilities which came to light in the early years of the 1980's, particularly in the U.S., has become the major change agent of this decade and for the remainder of this century. The multi-trillion dollar cost of corrective reconstruction, rehabilitation, and maintenance of infrastructure is providing major incentive for the development of new materials and equipment technology, construction methods, analytic techniques and models, information systems and management methods which are specifically tailored to the many unique characteristics of maintenance, rehabilitation and reconstruction activities.

Changes on the supply side of construction are as significant as those on the demand side. Construction industry firms are becoming involved in a wider range of construction process activities. For example, through construction management, design-build, and turnkey contracting approaches, firms are integrating and managing planning, design, construction, and start-up operations activities. In addition, participation of construction firms in financial packaging and/or equity financing of commerical and industrial facilities is occuring with increasing frequency. There are signs that this practice could extend to public infrastructure type facilities in the future, through Build-Operate-Transfer (BOT) or outright privatization of public infrastructures.

The nature and technology of resources are also undergoing continued change and development. The ability to deal with migrant labor resources has recently become a major factor in the increasingly competitive international market place. In the U.S., contractors employing non-union labor are increasing their share of construction relative to those relying on union labor. New construction materials and equipment are continually being made available to the industry by foreign as well as domestic suppliers. However, despite growing pressures to improve construction productivity, the acceptance and use of these technological advances are often confronted by a host of institutional barriers. Increased availability of low cost computational capabilities, personal computers, workstations, and knowledge-based software programs combined with modern communication technology offer an unprecedented opportunity for enhancing the efficiency and effectiveness of a wide range of engineering and management planning and control tasks.

A noticeable characteristic of E&C firms is the division of functions such as design, engineering, cost estimation, construction management, subcontracting, procurement, etc. Although many of these functions can be embodied together in a single firm, many of these functions and various physical works (such as foundation and structure, mechanical, roofing, etc.) have become independent and commercialized in separate firms. Each has thus developed a human capital resource that is unqiue.

In this emerging picture of the E&C industry, the relative importance of knowledge becomes greater and the accumulation of knowledge in the firm becomes a major factor in the client's selection process. Knowledge and information can be embodied in E&C firms in three ways: in its employees, in its organization, or in its equipment. It is this propriety of knowledge that gives the firms their comparative advantage. In the past, most knowledge was simply embodied in individuals and were mostly gained through apprenticeship and on-the-job training. Today, however, a significant amount of knowledge is being generated and embodied in the organization and in intelligent data processing and computer systems which are proprietary and subject to licensing and patent rights. This accumulated knowledge in firms needs upkeep, development, and advancement. Morover, the human capital component of knowledge needs both a rigorous formal education as well as a frequent updating and upkeeping of that knowledge.

More importantly, the structure of the construction industry, and extensive knowledge content in many engineering and construction funtions, lend themselves to proprietary investments in human capital and in knowledge-intensive physical capital which can gain immediate recogniton among clients and be converted to increased contracts. The goal of these investments seems to be to shift the embodiment of routine knowledge from individuals to firms with proprietary machines and organizations which can be controlled. These new capabilities with their inherently higher productivities can then provide the comparative advantages that firms need for expansion in sales and growth in profit. This would free the human capital resources of a firm to devote more time to innovative and creative portions of engineering and management.

Implications for Graduate Education

The inherently complex and changing nature of the construction industry and construction markets has significant implications for construction engineering and management education. Technical and managerial leadership in the industry requires the ability to perceive in totality the interrelationships that exist within and between the industry, its market and the construction process. It requires recognizing the economic, political, and social forces which strongly affect the industry at the project, firms, and industry levels. It also requires analytic skills and specialized technical knowledge of the construction process and the basic construction resources of manpower, materials, and equipment. Thus the graduate education program must provide students the opportunity to broaden their conceptual perspectives, managerial skills, and technical knowledge along many dimensions. It must also impart to students a lasting motivation to attempt to discern future contemporary challenges and opportunities facing the industry, we well as the analytic skills and conceptual frameworks required to shape effective and reasoned responses to them.

The educational program in construction engineering and management has four major thrusts: (1) technology of assembly; (2) management of assembly on site; (3) management of organizations specializing in that assembly; namely, E&C firms; and (4) interaction of assembled systems with economics, social, environmental, cultural, and legal systems of the society. The primary emphasis of most current programs has been on the first two thrusts: construction technology and project management. Lack of adequate emphasis in our educational systems on the other two items; namely, organization and management of construction firms, and the broad and pervasive societal interactions has resulted in a rather benign image on the part of the public of our built environment, stiff international competition, loss of technological leadership and decay of our infrastructure which is the backbone of our economic prosperity.

Over the past few years a few academic institutions have recognized the importance of these issues and have developed academic programs that emphasize more construction management as distinct from construction technology and project management elements of the program. The current graduate program in construction engineering and management at M.I.T. is built on these features and maintains a distinctive educational environment.

The M.I.T. Master's Degree Program in Construction Engineering and Management should continue to build on these features and maintain its distinctive educational environment. In particular, high priority and emphasis is given to developing research opportunities which can attract academics from a broad range of disciplines and which can provide support for students in the form of thesis supervision and funding. This is essential in order to offer encouragement and incentives to students to spend three full semesters so as to take more than the minimum number of subject units required and to become more deeply involved in thesis research endeavors. In addition, enrichment of this environment is encouraged by renewed attention and effort to involve industry professionals and firms as instructors of complete subject offerings, and as partners in various research activities.

Structures and Topics of the Academic Program at M.I.T.

The formal academic program addresses the construction process and the construction industry in its broadest context. Accordingly, it retains its focus on the industry as a whole and emphasizes the diversity of industry participants and the roles they play in the realization of constructed facilities. The fundamental educational objective of the program is to impart to its students the conceptual framework, analytic capabilities, technical knowledge, and effective knowledge of the operation in the industry, and prepare them for continued learning, growth, and contribution to the advancement of the industry.

In order to implement this philosophy, the program consists of our major areas - construction organization, management of construction, construction technology, and decision sciences/quantitative techniques.

Construction Organization: Topics in this area are meant to give students an understanding of the total context of construction and the environment in which it operates, and a meaningful framework which highlights the many unique characteristics which sets the industry apart from other industrial and service activities. Topics include: the structure and nature of the construction market and the larger social and economic environments which create demand for construction; the nature and structure of the industry which serves the construction market; the basic operating methods, labor, materials, equipment, and financial resources used by the industry to fulfill construction demand; the economic and technological performance of the industry as a whole; and the diversity of firms and organizations which participate in the construction process.

Management of Construction: Two categories of management topics are included. First are those which are fundamental to business enterprises in general. These topics include organization design, financial management, management accounting and control, management information systems, and human resource management. Second are the topics which deal with unique or particularly important aspects of the construction industry and construction process such as: project planning and control, construction contracting and bidding, construction law, performance and productivity measurement, and construction firm strategy and policy formulation.

Construction Technology: This area consists of topics which develop an understanding and appreciation of the needs and opportunities for technological innovations and advancements within the industry. Thus topics include: automation and robotics, the nature and applicability of software technology with respect to engineering analysis and design, quality control methods and techniques, and the selection and control of factors of productions.

Decision Sciences/Quantitative Techniques: Topics in this area are meant to give students a solid grounding in economics, mathematics, optimization, probability, statistics, computer modeling and risk assessment and an understanding of how these quantitative techniques can be applied to both technical and managerial aspects of construction. These topics reflect and address the high degree of risk and uncertainty, and the time and economic pressure which characterize decision making in the industry.

The topics identified in the above sections can be treated in the academic program either in general applied methodology subjects or in industry specific application methods subjects. However, in order to maintain the specialized focus of the construction engineering and management program for its students, and to achieve an appropriate degree of external visibility and credibility, a set of specific application and case study subjects are offered by experts from industry or by a team of experts from E&C firms. Currently over ten subjects of this nature are offered on an annual basis. Such industry-wide subjects provide an identifiable focus, continuity for committed students, and strong external visibility for the program.

Finally, the program places an important emphasis on thesis and independent research. All candidates for the M.S. degree are required to spend an equivalent of three months on a research topic of interest to them. The thesis requirement encourages students to concentrate on developing an in-depth understanding of the material presented in the various subjects - because of its potential value to their thesis work - rather than race through the program in an effort to maximize the accumulation of course units in the minimum time possible. It gives them an opportunity to become familiar with the research and do an independent investigation of a particular aspect of construction.

TECHNOLOGY'S ROLE IN CONSTRUCTION EDUCATION

By Boyd C. Paulson, Jr.[1], M.ASCE and
Leonhard E. Bernold[2], M.ASCE

"So long as we represent technology as an instrument, we remain held fast in the will to master it. We press on past the essence of technology."

Martin Heidegger, *The Question Concerning Technology,* 1955

Abstract

Whether by design or default, technology influences the economic well-being of the construction industry. This paper examines the role that advanced technologies — in particular, computers and automation — should play in educating students for careers in construction. For the mainstream of students, emphasis should remain on the fundamentals of science and engineering, the liberal arts, and on practice-oriented courses. Exposure to advanced technologies should come mainly in the context of courses central to their applications in construction. However, we must also identify a few students who can be educated to become strong change-agents to advance construction's technology. Students suited for this role should be free to explore in depth advanced courses throughout their universities and receive more personal mentorship for independent study. Even then there remains the problem of how best to employ such graduates to society's long-term advantage. This paper asks pertinent questions and outlines relevant issues, but offers only general guidance toward designing programs to achieve the needed educational objectives.

Introduction

Advancing technology influences the construction industry's ability to compete, even if the industry does little to exploit the advances, and even if the technologies at first do not seem directly applicable to the industry's needs. When some sectors of the economy move ahead more rapidly, others are left behind. Similarly, if in a given market sector one nation's firms choose to pursue technology more aggressively than another's, that nation's firms eventually will have a larger share of the market. Even

[1]Ohbayashi Professor, Dept. of Civil Engineering, Stanford University, Stanford, CA 94305-4020.
[2]Assistant Professor, Dept. of Civil Engineering, University of Maryland, College Park, MD 20742.

legal barriers and political systems designed to restrain artificially the forces of free trade will not prevail in the long run. Whether in a stagnant or a growing economy, resources are finite and re-allocations will take place to favor more effective producers. The underlying forces are immutable.

It is not our purpose here to repeat the tired statistics attesting to the so-called "decline and fall" of the once-proud American construction industry. That refrain has become so trite and is so often self-serving that it fails to motivate many people any more. Yes, the industry has a lesser share of the U.S. economy than it used to, and yes, it faces stronger competition from abroad than it did in the first decades following World War II. In large measure this should have been expected as the broken nations recovered from their wars of the first half-century, and in fact it is the course of most nations — not just the United States — to shift resources away from construction once they have established their basic infrastructure for housing, industry and public works. As educators, safe within the walls of academe, we too have been guilty of heaping more ashes upon the industry's head, usually while trying to sell graduate education and research as the miracle cure for all of the industry's ills. We have not been humbled by the fact that the decline — in statistics, at least — has continued unabated while more and more of our graduates have become the industry's leaders. Again, the underlying forces remain immutable.

On the other hand, technology can indeed be used to advantage, whether we develop it ourselves or adapt the innovations of others. We live in an era where technology in some sectors is advancing at bewildering speed — for computers, communications, biotechnology and aerospace. Construction moves more slowly — improving the strengths of materials, developing tools and methods to work in more difficult conditions — but it is moving ahead nonetheless, declining statistics notwithstanding. While construction may not be leading the charge, it can advance through its own efforts and it can gain from the advances of others. This is not a zero-sum game. To do well at the game, however, the industry's working professionals and its future leaders need to have the fundamental education upon which to build new knowledge so that they can adapt to what will probably be even more rapidly changing conditions in the future.

Construction Education and the Needs of Industry

Most of the civil engineering schools that have programs specializing in construction call these programs *Construction Engineering and Management.* This choice of words contains a significant balance. *Construction,* of course, refers to the industry toward which the programs are focused. Some courses thus expound upon the characteristics of the industry — its language, culture, practices, beliefs and even its history.

Engineering conveys the underlying scientific principles and technical methods that give the field its uniquely physical nature. Engineering in the undergraduate construction programs ranges from a smattering of simplified "real engineering" courses from civil engineering and mechanical and electrical technology to some which build from first principles and try to address the construction process as a scientific discipline. Graduate programs, for the most part, just say that they expect that their students became engineers as undergraduates, and assume that the residual knowledge will be applied in courses on equipment and methods, estimating, etc. Some programs may require an additional course or two in related technical fields, and perhaps offer one or more electives to provide an accelerated introduction to or survey of computer applications, expert systems or robotics. However, there is seldom enough depth in these courses to have a major technical impact of the programs' overall content.

And so we come to *Management*. Many if not most of the students want to become projects managers and vice presidents and presidents and chairmen and own their own companies and . . . , but they often only have a vague idea of what all this means and about how to prepare themselves for whatever it turns out to be. Their educators often seem similarly confused and proceed to offer a number of construction-flavored business courses such as accounting, planning and scheduling, organizational behavior, labor relations, contract law, etc. Mid-career managers and senior executives — typically when waxing nostalgic — most often express the wish that they had taken more courses in the *liberal arts* — whatever that means to them personally — so that they could do a better job of understanding and managing *people,* and perhaps so that they could discover what else there is to life after they have made it to President!

There is a persuasive school of thought that the very best university education one can obtain for a career in *engineering,* regardless of discipline, is a basic grounding in mathematics, science and the theoretical principles behind engineering. Technologies and methods evolve so rapidly that an education shallowly based on mastering current and exciting tools rapidly becomes obsolete. Today's efforts to teach microcomputer CAD systems, spreadsheets and applied robotics may be like yesterday's courses on drafting, the slide-rule and steam-engines, whereas an understanding of the geometry, mathematics and physics behind them will last a lifetime. The most respected senior engineers and scientists do indeed seem to be the grand conceptualizers and the analysts, not the masters of tools. A similarly persuasive school argues that the best education for *management* is, in fact, the liberal arts, more specifically those courses that broaden one's understanding of ideas and thought processes, give a historical and cultural perspective on the various peoples one encounters in life, and teach one to read well, write effectively and speak clearly. The aforementioned typical senior-executive refrain attests to this.

There are many pressures facing construction education today, not the least of which is technological change. It is too tempting to be caught up in short-term trends. Construction needs bright people and new ideas to advance, and these people will get a head-start with a proper education. But what is a proper education? How much of it can be accomplished in a four-year undergraduate program or a master's degree program? Where should such programs best focus their efforts? A quality liberal arts education takes four years in its own right, and many would argue that one can't begin to appreciate it until one has lived a bit longer and gained more experiences in the world than most undergraduates have by age 20 or so. Four years is clearly insufficient to catch up with what has taken place in biology, chemistry and physics, let alone understand the mathematics that is behind much of their progress. Does one need a Ph.D. in science to gain the "fundamentals that will last a lifetime"? And what about the practical employment realities that demand job skills like estimating, scheduling, computer programming and some knowledge of the vocabulary, tools and methods of a business? Beyond college, what is the role of continuing education? How can educators equip and motivate their students to take responsibility for their own self-education, both liberal and technical?

Frankly, we have mostly questions, some comprehension of the issues, but few if any answers. Any meaningful inquiry, however, begins with questions, so perhaps in that sense this paper is justified. Our main focus is on how to accommodate — if at all — some of the more rapidly advancing technologies in construction education, particularly those that might support construction automation. But, as the reader has seen already, we may wander from this objective where philosophy seems more important.

Advanced Technology's Role in Construction Education

Numerous engineering educators today — the authors among them — seem convinced that they can breathe new life into basic, traditional industries by embarking on research, offering courses and eventually infusing the new ideas and methods into their target firms. Construction has not been immune to these efforts, and some of its major companies have actually set up their own in-house research and development departments. Once content with their monumental structures, powerful machines and their near-immortal ability to "leave something concrete in place where nothing existed yesterday," civil engineers and construction professionals are now exploring such abstract fields as artificial intelligence, expert systems and automation for their potential to revolutionize the industry. Some of them have developed respectable levels of expertise in these areas, and a few seem determined that all their brethren should similarly be indoctrinated. For those who worried while their discipline fell behind others in trying to recruit students to courses little changed in several decades, and failed to attract grants for "research" to achieve unimaginatively small advances in the status quo, it is an exciting time. There now is actually considerable promise for achieving some of their newfound dreams. However, for educators the problem remains much as Edward R. Murrow said it in October 1964:

> *"The newest computer can merely compound, at speed, the oldest problem in the relations between human beings, and in the end the communicator will be confronted with the old problem, of what to say and how to say it."*

What to say? Should we cut back on the core courses of civil engineering and construction to offer new ones in computer applications, robotics and perhaps space technologies? Even within traditional fields there is an ever-growing body of knowledge that is also demanding more attention to give would-be professionals a "minimum level of competence." There have been numerous efforts to add a survey of robotics here, an introduction to expert systems there, and computer applications courses are proliferating everywhere; although some students are uncertain and take the courses "just in case" they might turn out to be useful, the courses usually are quite popular. However, there has been little effort to obtain a consensus among leading construction educators and practitioners about how their curriculums ought to evolve to accommodate advanced technologies; mainly there have just been incremental additions and displacements.

How to say it? For students whose primary ambition is to build buildings or bridges or tunnels, and to manage these works rather than engineer them, how does one convey enough information about the potential for applying the advanced technologies in their profession so that they can make intelligent decisions to support development and acquisition of useful new technologies at the appropriate times? In coming to civil engineering or construction, they have chosen not to major in computer science or electrical engineering or physics. Will they become confused or disillusioned if we coerce them toward broader views of technology? Even if they went along, how many such technically educated change-agents should come from within an application discipline like construction? Somebody must attend to the day-to-day business of just getting things built with technology that exists.

Would it be easier to teach advanced technologists about construction and motivate them to focus their talents on the needs of our industry? Perhaps not. In a peculiar symmetry of parallel career worlds, these folks have set their sights on high-technology in its own right and by implication have chosen not to go into fields like construction. The contrast is evident enough between those who build the structures of Silicon Valley, and those who work within them. Somewhere there is a middle ground, and it is usually those entrepreneurs who reach out from within the application areas who find this ground and go on to find their counterparts in the disciplines that can provide the needed technology. We do need a few of these change-agents to come from within construction, and the industry as a whole needs a better idea about what these people might do for it.

Civil engineering and construction programs can do something to help educate these people, and certainly they can support the research ambitions of faculty members and students who want to explore new technologies to advance their field. But in so doing they must use caution not to damage the curriculums that serve their primary mission, and not to compromise the needs of the mainstream students going into practice. In effect, there must be multiple paths.

In the first — let's call it the *traditional* path — there should be a gradual infusion of new technology, but educators should exercise caution to avoid side-tracking students along trendy technological dead-ends at the expense of knowledge central to their field. A safe bet, particularly in undergraduate programs, is to emphasize fundamental principles and lasting concepts of science and technology, and to be sure that there is also a solid core of liberal arts. These students will want and should have practice-oriented courses, but somehow in the process should be taught about the obsolescence of such knowledge and become aware of the means for regeneration. Advanced technologies — microcomputers, software, instruments of various types — can assist in the *process* of education, but should not be a primary focus in their own right. While they can enhance classroom presentations, facilitate homework and enable laboratories to cover more ground — and in the process give the students a feel for what they are about — they can become addictive, substituting ephemeral tool knowledge for lasting principles, if they are too greatly extolled as ends in themselves.

Yet, the traditional path should maintain flexibility to incorporate new theories and principles which evolve and are expected to stay. If one accepts the belief that automation in construction will play an increasingly important role, a basic understanding of the underlying theories and technologies will be necessary for construction engineers. Teaching construction automation, however, risks becoming mere training in tool application. On the other hand, properly taught it provides an excellent opportunity for learning based on Jean Piaget's constructivist theory, which involves three processes: logico-mathematical experience, physical experience, and socially transmitted information. Using this theory, students would learn best through a mix of formal teaching and physical hands-on exposure. Today's educational hardware in robotics provides the opportunity to establish such a nurturing learning environment. Intelligent small-scaled laboratory equipment permits physical testing of theoretical designs, thus allowing the ultimate gratification of a learning experience which integrates logical comprehension and physical experimenting. Teaching construction automation could also create an environment for intra- and interdisciplinary work to include other engineering specialties, as well as architecture. However, the present compartmentalization of these disciplines may make this wishful thinking, at least for the near future.

The second path — is *innovative* the word? or *creative?* — should be designed to identify those individuals — perhaps the individualists — who are clearly the self-starters, the ones who motivate themselves and work well independently, the ones who generate ideas and who teach themselves. They must also be tenacious, persistent and willing to suffer setbacks in pursuing their objectives. They too should be grounded in the fundamental principles of science and engineering, and they will most probably steer their own way into the avenues of liberal education — these are the idea people, after all, the future philosopher-statesmen of our profession. However, less emphasis will be placed here on courses dealing with current practice. In advanced research universities they will find courses at the cutting edge of technology in whatever specialties interest them; it doesn't greatly matter which — it is the *process* of conquering new fields that is most important. They are also ripe candidates for directed reading courses if they have proper mentorship. It does not matter as much if these people occasionally become side-tracked with the tools of modern technology — it is their job to learn to explore and evaluate, to sort that which works from that which won't, to take nothing at face value. They are also less likely to resent time spent chasing down blind alleys, as long as the chase itself was fun or shed new insights.

This second path is a special one; it cannot be found or established in every program. It is not one that should be institutionalized into a formal curriculum with many course requirements, at least not beyond the fundamentals. Traditional students should largely be shielded from its pitfalls. It is along this path, however, that we are most likely to find the change-agents who can infuse new technologies into the construction industry, and it is in nurturing such people that we can deliberately advance, not just accept what comes. The main question is, however, what will industry do with these people when they get there?

Industry's Role in Technological Advancement

It is not easy to be a change-agent, yet most institutions consciously or unwittingly make the job even tougher than it needs to be. Nicholas Machiavelli, whose very name seems to be be a synonym for manipulative change, still said it well half a millennium ago:

> *" . . . There is nothing more difficult to plan, more doubtful of success, nor more dangerous to manage than the creation of a new system. For the initiator has the enmity of all who would profit by the preservation of the old institutions and merely lukewarm defenders in those who would gain by the new ones. The hesitation of the latter arises in part from the fear of their adversaries, who have the laws on their side, and in part from the general skepticism of mankind which does not really believe an innovation until experience proves its value. So it happens that whenever his enemies have occasion to attack the innovation they do so with the passion of partisans while the others defend him sluggishly, so that the innovator and his party are alike vulnerable."*
>
> *The Prince,* 1513

In brief, industries as a whole, and traditional ones in particular, have a reputation for resisting change. Ironically, the people with the ideas and the intellect to stimulate progress are likely to be viewed as threats or be dismissed as eccentrics. Most of the really motivated ones thus go to other industries, or else start their own businesses where their philosophy might last until the next generation of management takes over. They do exist in construction, sometimes at the very top levels and probably more than in other traditional industries in view of the risks of this business. Some of our most prestigious firms trace their roots back to such individuals — Henry Kaiser, Allen Christensen, Warren Bechtel, Harry Morrison and others. But even their firms evolved to corporate conservatism and some have already disappeared.

How to integrate technological change-agents into the culture of modern construction is not a problem that we can adequately describe here, let alone solve. But those who find the answer will be well on their way to moving civil engineering and construction back into the mainstream of technological progress. With intelligent business strategies to complement these efforts, they will be on their way to competitive domination as well. Nobody else is even coming close.

Conclusion

What can one conclude about such a topic? Is this really the time to conclude or to commence? The authors not only recognize the importance of developing and applying new technologies to improve the well-being of construction, they have also been groping among those at the forefront in trying to do so. Perhaps it is this experience that has caused them to turn cautious, to say that perhaps we ought to pause and reflect on where we are headed. We are concerned that the incremental and almost serendipitous approach to introducing advanced technologies to construction education programs may not be as well thought out as it should be, may not be the best way to serve many of our students, nor the most effective means to move the needed technologies into industry.

There is little doubt that the new technologies are needed and must be accommodated, and to some extent this goal can be met with new courses in formal curriculums. But what should go into these courses? How should they be taught? By whom and for whom? What is their relationship to the established courses in basic science and engineering, the liberal arts and to professional courses for today's practice? It is appropriate that we conclude with a series of questions. There is need for a much broader discussion — involving colleagues from industry as well as academe — to come to a better understanding of the issues and to seek enduring answers.

References

The authors must confess that, while they vaguely remember reading Heidegger and Machiavelli during the liberal part of the undergraduate studies they too little appreciated at the time, and did listen to some of Edward R. Murrow's newscasts, they cannot take credit for finding the three quotations featured in this paper. This credit belongs to the editors of *Computing Reviews,* published by the Association for Computing Machinery, who feature one thoughtful quote each month on their front cover. The pertinent issues here are December 1986 for Heidegger, June 1988 for Murrow, and December 1988 for Machiavelli. Jean Piaget's theory was published in *The Construction of Reality in the Child,* Basic Books, New York, 1937.

CIVIL ENGINEERS MUST HAVE A CONSTRUCTION EDUCATION

D. T. (Tom) Iseley,[1] M. ASCE

Abstract

This paper summarizes the status of the U.S. construction industry, identifies major problem areas, describes recent research endeavors, and explains why civil engineering education needs to be modified to provide a stronger background in construction methods.

The cost of all goods and services produced is a function of the cost of physical facilities necessary to conduct research and development, manufacturing, storage and distribution. Thus, for the United States to be able to compete domestically and on a global basis, the construction industry must be able to produce facilities efficiently and cost effectively.

Introduction

Civil engineers are the key ingredient to the life cycle of the construction process. The construction process begins long before the actual construction phase begins. The process is initiated when conceptual planning starts. Eighty to 85 percent of the total budget may be spent during the construction phase. Therefore, if decisions are made that impact the cost of the construction phase by individuals not familiar with construction methods, then the net result will have an adverse impact on the construction phase. This impact will not only affect the direct project cost, but will manifest itself in many indirect ways (i.e., delays, adversary relationships, etc.).

Figure 1 illustrates the ability to influence cost which each phase has over the life of the construction process.

----

[1]Assistant Professor of Civil Engineering, Louisiana Tech University, Ruston, LA 71272.

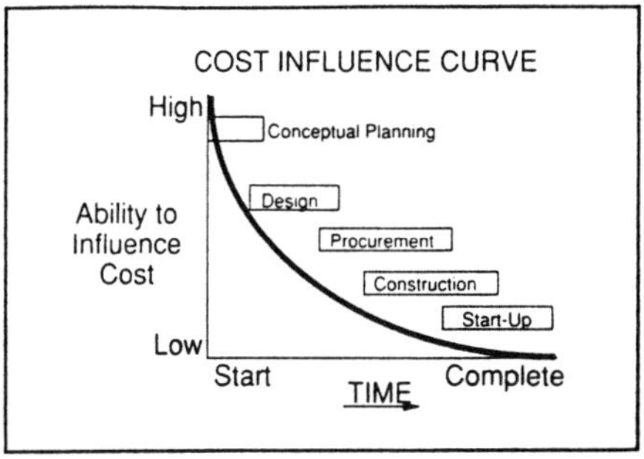

Figure 1 - Cost Influence Curve (CII, 1988)

The Problem

The problems associated with the construction industry are not new. The Business Roundtable, composed of the chief executive officers of approximately 200 major U.S. corporations who meet to focus and act on a wide range of public issues, in the early 80's determined that "by common consensus and every available measure, the United States no longer gets its money's worth in construction, the nation's largest industry." (BR, 1983) The Business Roundtable's CICE Task Force conducted an in-depth look at the industry from the owner's point of view. This project generated 23 separate reports by the study team.

As a result, the CICE Task Force Summary Report emphasized the impact by stating "the creeping erosion of construction efficiency and productivity is bad news for the entire U.S. economy. Construction is a particularly seminal industry. The price of every factory, office building, hotel, or power plant that is built affects the price that must be charged for the goods or services produced in it or by it. And that effect generally persists for decades. At a time when the nation has been losing its erstwhile share of world markets for export goods, plunging productivity in construction, by driving up the cost of doing business and making products in the U.S., makes it increasingly difficult for the rest of the nation's business and industry to retain-or regain-foreign markets and to compete against imports. These markets have become increasingly important to the U.S. economy." (BR, 1983)

Background

The CICE Task Force Project gave birth to the concept of a need for an organization to promote construction research. The Construction Industry Institute (CII) was formed in 1984 with a mission "to improve the cost effectiveness of the construction industry, thereby strengthening the competitive position of American industry in the international marketplace." (CII, 1988) CII is founded on the premise that needed improvement will be accomplished through research. This research is promoted in six thrust areas (i.e., technology, organization, people, information, controls, and sigma).

At the fifth anniversary, CII was composed of 72 member companies. These companies represent 35 owners and 37 contractors. While the annual budget exceeds $2 million, the estimated volume of research conducted is in excess of $12 million. This phenomenal success is indicative of the commitment of the construction industry to address the issues. CII incorporates representatives from academic institutions on all task forces. Currently 28 academic institutions are involved with the CII programs.

In summary, the CICE Task Force identified the problems associated with the construction industry. CII subsequently conducted extensive research on problems directly related to the construction industry. Many of the results of the CII research are now being implemented with outstanding success.

The Challenge

While CII accomplished great things during its first five years of existence, much more remains to be investigated. Smaller organizations need to be encouraged to participate in this type of organization. The owner/contractor/academic forum is healthy because all entities involved in the industry freely share the results of CII's research.

CII has established an excellent example for members of the civil engineering profession to follow because the civil engineer is involved in every phase of the construction process. Thus he/she is a key decision maker impacting construction cost significantly. The civil engineer should follow the example established by the construction industry to determine if he/she is part of the problem or part of the solution.

Even with what has been accomplished to date, the construction industry continues to receive much negative publicity which addresses such sensitive areas as: (1) low and decreasing levels of productivity; (2) poor work quality; (3) employee morale and turnover problems; (4) decreasing competitiveness; (5) high and increasing levels of rework; (6) increase in project delays, claims, and disputes; (7) high volume of business failures. There is a tendency for the design profession to consider these problems as strictly the construction industry's problems rather than conducting a self examination to determine if these problems may be results of the designer's decisions which are simply manifested in the construction phase.

It is suggested that the total construction process needs to be evaluated in depth. The majority of previous and current research has been focused on how to control the symptoms rather than how to deal with the root problem.

The Basis For The Challenge

Recently, the author completed a two year research project which was funded by the Indiana Department of Highways (IDOH) to develop construction specifications for highway projects requiring horizontal earth boring and/or pipe jacking techniques. (Hancher, 1989) One of the primary objectives of this project was to obtain maximum input from all segments of the industry (i.e., contractor, engineer, equipment manufacturer, material manufac-turer, highway agencies, etc.). This endeavor re-quired extensive interviews with many representatives from each segment of the industry. One major conclusion was that the major problem had to do with communications involving all parties.

During the two year duration of this research, numerous field problems were investigated. Litigation and arbitration proceedings were reviewed for many previous and current problems. The purpose for such an extensive review of problems was that it was felt that for one to develop construction specifications one must have a clear understanding of previous problems in order to prevent such reoccurrence. All problems reviewed had their root cause associated with the specifications.

The plans and specifications are the primary communication tools which serve to link the design phase to the construction phase. The importance of the this link was recognized by the Construction (CO) Division of ASCE in the ASCE News when it stated that "because incomplete, inaccurate and faulty construction specifications often result in a large number of field problems, a CO Division group is trying to do something about the situation. One reason may be a lack of emphasis on specifications at the undergraduate level." (ASCE, 1988) Most current civil engineering curriculums concentrate on design with little emphasis on how to communicate that design to the contractor charged with the responsibility to construct the facility. As a result, "the single largest source of professional liability claims is specifications. This may be because designers often regard specification writing as a nuisance, preferring the more creative technical work involving calculations and drawings. However, effec-tive specification writing requires special attention. Superficial treatment of specification writing can easily lead to conflicts among the various sections of the specifications, or omission of an item critical to the project." (DPIC, 1988)

Another factor associated with the large amount of incomplete, inaccurate and faulty specifications currently written was recently summarized very effectively in a recent issue of the ENR magazine stating that "today we have engineers designing bridges who do not fully understand how to build them, and contractors bidding bridgework without the engineering expertise needed to appreciate the design requirements that affect construction. The solution to the problems ... is to have engineers come out from behind closed doors and become engaged in construction. ... Engineers should have at least one year of construction engineering experience in order to qualify for advanced licenses, and engineers should be tested on elementary contract law. With better education, we would come to understand that contracting is (or should be) an equitable quid pro quo, not a war game." (Goodyear, 1989)

Since the specifications are the primary communication tools of the designer to the contractor, problems will occur if they are defective. In many instances the specifications are defective simply because they have not been updated to permit more advanced technology. In several cases investigated by the author for the IDOH research project, contractors had bid projects as per plans and specifications. After being awarded the projects, they approached the engineer concerning the use of new technology. The new technology offered major savings to the owners in terms of social costs, but the methods would not be considered unless the contractors were willing to reduce the base bid by ten percent. Had the engineer been current on the state-of-the-art and this new technology addressed in the original specifications then this method would have been selected and the owner would have received a far superior end product.

When field problems occur as a result of defective specifications then symptoms previously described will be manifested. A contractor can implement all of the recommendations which have resulted from the CII research, but if he/she is given defective specifications then delays and confusions will characterize the construction phase. These delays and confusions will be demotivating factors which will directly impact worker morale, absenteeism, turnovers, work quality, productivity, etc. Therefore, the problems typically associated with the construction operation may be a function of the quality of specifications developed by the designer. The same is true with promoting the team (i.e., owner, designer, contractor) concept. The best team will have major problems when the primary communication tool is defective.

The most convincing evidence relating to the importance of the specification to the construction phase occurred during the IDOH research project. (Hancher, 1989) As previously stated, one of the primary objectives was to obtain maximum industry input. To obtain contractor input, the author contacted the Trenchless Excavation Committee (formerly the Horizontal Earth Boring and Pipe Jacking Committee) of the National Utility Contractors Association (NUCA). After this committee was made aware of the objectives of the IDOH project, they became very excited concerning the opportunity to provide input into the development of specifications. As a result over 40 firms contributed $500 to support the activities of a Special Task Force that was formed to assist the author with contractor input. In addition, NUCA contributed $5000 to the Task Force. This level of support is indicative of the degree of importance associated with specifications.

Conclusions

This paper has recognized that significant problems do exist in the construction industry. It reviewed the results from the CICE Task Force which identified the nature of these problems. In the process of putting the house in order, the root causes of the problems plaguing the industry must be analyzed. These problems are not just industry-wide, but they impact all goods and services produced. This is a national problem.

This paper suggests that communications are the root problem, and the specifications are the primary communication tool for the construction process. Should these be defective (i.e., incomplete, inaccurate, faulty, etc.) then all efforts being made by the contractor to implement cost effectiveness will fail. The primary reasons for the large number of defective specifications being introduced into the construction process were identified as the lack of formal training; emphasis on the development and writing of specifications; and the lack of understanding that engineers have concerning the construction process.

Recommendations

Academic institutions charged with the responsibility of training civil engineers must realize that this training must be broader in scope than simply concentrating on pure design. Future civil engineers must be trained to be effective leaders in each phase of the construction process. That mandates a construction education as well as a design education. A construction education requires courses in project management (i.e., planning and scheduling), contracts and specifications, construction methods and equipment.

References

ASCE News, Construction Technical Activities News, American Society of Civil Engineers, Volume 13, No. 2, New York, NY, February 1988.

The Business Roundtable, More Construction for the Money, Summary Report of the Construction Industry Cost Effectiveness Project, January 1983.

The Construction Industry Institute, 1988 Annual Report, (Fifth Anniversary), 3208 Red River, Suite 300, Austin, Texas, 1988.

Design Professionals Insurance Company, Lessons in Professional Liability, A Notebook For Design Professionals, 2959 Monterey-Salinas Highway, Monterey, CA 93940, 1988.

Goodyear, David, Get Engineers Into Field, Letter to the Editor, ENR Magazine, NY, July 27, 1989.

Hancher, D.E., White, T.D., Iseley, D.T., Construction Specifications for Highway Projects Requiring Horizontal Earth Boring and/or Pipe Jacking Techniques, Draft Final Report, Joint Highway Research Project, Purdue University and the Indiana Department of Highways, July 1989.

A Comparison of the Educational and Training Needs of Civil Engineers & Construction Managers

John H. Erion, Jr., P.E., Member ASCE[1]
Saeid Y. Eidgahy[2]

Abstract:

This paper reviews the status of Civil Engineering and Construction Management today and their development over recent times. This analysis includes accreditation of both ABET and ACCE and the effects of increasing specialization within the field. A regional survey of educational programs is presented, so that common threads of these programs may be identified. And furthermore, continuing education and professional development needs are also identified through this survey. Finally, future developments and trends are discussed.

Introduction:

Civil Engineering is the oldest recognized division of engineering studies, tracing its origin in the United States to Norwich University in 1819. The criteria for Civil Engineering accreditation have weathered the years and evolved through the old Engineer's Council for Professional Development (ECPD) into the Accreditation Board for Engineering and Technology (ABET), serving as a time-tested model for colleges and universities offering that program. The study of Construction Management traces its roots through two very different paths, one from the Civil Engineer's approach as either an arm of Civil or Construction Engineering, or as an Engineering Technology still under the purview of ABET (1986). The other, much

----

[1]Assistant Professor, College of Technology, Bowling Green State University, Bowling Green, Ohio 43403.

[2]Doctoral Fellow, College of Technology, Bowling Green State University, Bowling Green, Ohio 43403.

younger, development in Construction Management has founded its curricula without the classical bounds of engineering. The American Council for Construction Education (ACCE), organized in 1974, serves for many institutions as the establishing and reviewing authority for accreditation in the construction field (ACCE, 1988). This paper seeks neither to argue nor settle the differences between these two schools of thought. Rather, the authors will demonstrate a growing need for professional development and continuing education for graduates of both educational foundations who have selected construction as their career. This paper will also attempt to demonstrate that a technology-based master's degree is a viable alternative to engineering and business administration graduate programs not only for geographical exigencies, but also to fulfill a demonstrated need for specific educational pursuits.

Accreditation and Specialization Influences:

Engineering schools a quarter century ago made a major movement from five-year to four-year curricula as a result of many factors, competition for the shrinking pool of engineering freshmen not being the least. The reasons for that change are not significant herein, but the resulting loss of twenty five percent of the studies of subsequent graduates is. Allowing that the ECPD criteria were not frivolously approached in the five-year programs (and one of the authors recalls no such frivolity in his studies), significant reduction in the education of later engineers could only be recovered through three channels; graduate studies, continuing education (either resident or non-resident), or the "school of hard knocks". While continuing education was not suddenly invented then, there did emerge a real need for those affected engineers to be able to develop professionally.

The American Society for Engineering Education (ASEE) in 1972 identified a constrictive trend in the breadth of all engineering education. While identifying an increasing need for baccalaureate programs in engineering technologies to fill the shortage of engineering graduates in the middle 70's, the report cited a narrowing of engineering education which those scholars felt could only be corrected through graduate studies for the engineers, and further education for technologists. About this same time, curricula in Construction Management, and similar titles, began to mature and seek recognition for their graduates. Since many lacked the engineering rigor, at least in the eyes of many engineers, the void was soon filled by ACCE with support from several prestigious groups including Associated General Contractors (AGC), National Association of Women In Construction (NAWIC) American Institute of Constructors (AIC) , and Associated Builders and Contractors (ABC), to mention only a few. A recent and significant addition to their corporate membership was the Construction Management Association of America (CMAA).

The trend to more limited education has not abated since and, thus, an even stronger case for this paper evolves. Dr. Mac Van Valkenburg (1989) wrote of the jamming of curricula by basing an editorial on a chastisement he had received. Valkenburg stated "(You engineers have) never learned that you can't teach everything" (p. 456). Citing three options of: 1) returning to five-year curricula, 2) accepting jamming in the four-year curricula, or 3) reducing the content of curricula in a four year time frame, he made a very strong case for the latter, demonstrating that engineering education should provide more flexibility in student options and that accreditation should be less concerned with bean counting courses and hours and strive for excellence in engineering programs without necessarily credit hour guidelines. While it is noteworthy to this paper that Dr. Valkenburg included statistics on the number of practicing engineers, including 25% membership in the National Academy of Engineers, that do not hold undergraduate degrees in engineering, the issue at all points is that our field's young graduates are emerging not necessarily less educated but definitely with less education in certain subjects. This shortcoming can be satisfied through continuing education as well as graduate studies.

Regional Survey of Programs:

To further assess and quantify this need and to identify strategies to assist educators in developing professional development programs, the authors conducted a survey of colleges and universities across the United States that offer Civil Engineering, Construction Engineering, and Construction Management (or titles that are identifiably associated with one of those) undergraduate programs. Twenty colleges and programs in each category were surveyed. The first significant result, one that was anticipated but allowed to manifest itself, was that those schools categorizing their curricula as Construction Engineering were, when analyzed, totally identifiable with either Civil Engineering or Construction Management. Thus, the subsequent discussion of the survey will be limited to those remaining two categories.

The authors defined Construction Manager (CM) for the purpose of this study as

> "that member of the construction team (along with the owner and the designer) which protects the owner's interests in all phases from conception through design to construction. A professional manager, the CM is concerned with all contracting activities with primary responsibility on project cost, time, and quality."

This definition orients on the person rather than the field because our objective is the professional development of that person. The wording is actually a composite of definitions of the American Institute of Architects (1976 & 1984), The Associated

General Contractors of America (1979), the Construction Management Association of America (1984), and Facility Planning News (1984). The term Construction Manager, we felt, required defining so that the respondents to the survey and, ultimately, the readers of this paper would have a common frame of reference.

The most striking attitude found in this survey, yet one which most any disinterested party might postulate, is that what each program group views as one of their greater shortfalls is the strength of the other. That is, engineers cited management subjects as one of their continuing education needs while construction managers pointed to design subjects. A total of 28 surveys were returned with an even distribution of Civil Engineering and Construction Management programs responding. The ability of faculty from either curricula to cross into courses for students of the other was evidenced by the number of engineering degrees (96% for C-E's and 55% for C-M's) and the number of professional licenses (80% and 53%, respectively). While a few schools in the C-M group reported very few or no faculty with the above qualifications, none were without either. This indicates that the Construction Management schools have the talent to conduct Continuing Education courses for Civil Engineers. And Civil Engineering faculty likewise are qualified to offer structural analysis and other design topics to graduates of Construction Management programs.

Investigation into the curricular breakdown showed some interesting facts. One is that the great majority of respondents are on the semester system. Especially in the Civil Engineering curricula, the distribution of courses in all the curricula varies little from the respective accreditation guidelines. This makes a very strong supportive statement of Dr. Van Valkenburg's (1989) remark about accreditation being more a system of model duplication (bean counting, as he phrased it) than one of defensible concern for excellence. One very noteworthy fact to take to the continuing education classroom is that there continues to be only one computer course in each curricula except two, one of whom has none and the other several. That subject area alone is ripe for alumni of both backgrounds as evolution has incorporated computer usage within the courses rather than to see specific computer courses. The challenge to interested faculty is to pluck the computer oriented portions from their classes and offer a computer applications course on a continuing education basis.

Professional Development Needs:

Both Civil Engineering and Construction Management programs identified the following group of subjects as critical for professional development, with the percentage of that response in the brackets following the content area:

- Financial and economics issues (60.7%)
- Ethics (25.0%)
- Civil works construction (46.4%)
- Marketing (25.0%)
- Contract law, labor law, and labor relations (39.3%)

Subjects listed by several Civil Engineering educators for professional development were:

- Social issues inc. pollution, safety, waste disposal (64.3%)
- General construction (57.1%)
- Infrastructure (35.7%)
- CADD & MIS (21.4%)
- Project Management (64.3%)

Subjects listed by several Construction Management educators were:

- Statics and Dynamics (21.4%)
- Quantitative analysis (85.7%)
- Mechanical and electrical system design (57.1%)
- Real estate (21.4%)
- Technical communications (28.6%)

A complete listing of the survey results is available upon request from the authors or can be found in the proceedings of the 1989 Great Lakes Regional Meeitng of the Associated Schools of Construction (Erion & Eidgahy, 1989). The fifteen subjects above were mentioned because of their frequency of mention in the responses. A short, thoughtful review would, in several instances, cause us to react that there just isn't enough time in four years. But that is just the point - simply because the graduates walk through commencement after their four years doesn't mean that educational opportunity has ended. We do have the wherewithal to help our alumni develop professionally. And the list of potential students extends far beyond our alumni office records.

The existence in both groups' faculties of engineering graduates and registered professionals points to skills that can serve many of the listed shortcomings. Several others can be addressed by the faculty of the same institutions' Colleges or Schools of Business. None of the subjects were exotic nor do they call for unusually specialized faculty. The greatest challenge is all to often the time demands for the faculty to extract portions of their regular offerings or to gear up for new ground when structuring a continuing education course.

Future Outlook:

Baccaulaureate programs are not going to return to five year curricula, in part because of the rapidly increasing cost of education and because neither the Construction Industry nor society in general have accused us of educating to an inferior level. The educators' reasoning for postulating five year degrees is also satisfied by graduate and continuing education studies.

There is also an emergence of Construction Management masters degrees among a majority of the existing undergraduate programs with one institution (University of Florida) just starting a doctorate in Construction. These programs, along with the well-established graduate programs in Civil Engineering, need a dual front strategy. The one strategy is the natural association of bachelor degreed engineers to study in graduate engineering and the bachelor degreed construction managers to study in their graduate programs. But the geographical availability is not universal to either.

If the common needs are viewed along with the very associative abilities of the faculties of both groups, the second strategy involves the crossing of students into the other groups' programs either on a graduate basis or on a continuing education basis. Thus, with both groups of students able to avail themselves of area institutions' programs, the educators of both groups can extend professional development opportunities to a much wider base of students and can likewise improve the classroom experience through that broader base.

References

Valkenburg, M.V. (1989). Unjamming our curricula. Engineering Education. 79(3), p. 456.

Accreditation Board for Engineering and Technology. (1986). Criteria for accrediting programs in Engineering Technology (AB-8). New York: Author.

Erion, J.H. & Eidgahy, S.Y. (1989). Professional Development Opportunities for Schools of Construction. Proceedings of Associated Schools of Construction: Meeting of the Great Lakes Region. Cincinnati, Ohio.

Accreditation Board for Engineering and Technology. (1988). Criteria for accrediting programs in engineering. (AB-7). New York: Author.

American Society for Engineering Education. (1972). Engineering technology education study: Final report. Washington, DC: Author.

American Council on Construction Education. (1988). Annual report. Washington, DC: Author.

Construction Management Association of America. (1984). CMAA 1984 directory. Reston, VA: Author.

Haviland, D.S. (Ed.). (1976). Project delivery approaches: An AIA guide. Washington, DC: American Institute of Architects.

Associated General Contractors of America. (1979). Construction Management Guidelines. Washington, DC: Author.

American Institute of Architects. (1984). Design-Build-Bid task force report. Washington, DC: Author.

Facility Planning News. (1984). Words to know: A glossary of key facility planning terms. Orinda, CA: Author.

Advanced Computer Techniques for Construction Education

Trefor P. Williams[1]

Abstract

Knowledge-based expert systems have the potential to transfer expert knowledge of construction field operations to students and novice engineers. Educational expert systems can be developed to provide information on construction techniques and expected durations and productivity of construction operations. Potentially, a large "library" of expert systems could be developed which could transfer expert advice to students.

Introduction

Much attention has recently been focused on improving construction productivity. Recent estimates indicate that productivity in the construction industry has been decreasing (Business Roundtable 1982). One method of improving construction productivity is to use artificial intelligence techniques to transfer expertise from construction experts to students and inexperienced engineers. Particular attention should be given to techniques which can be implemented on microcomputers and are available to the widest range of users. Recently, knowledge-based expert systems have emerged from artificial intelligence research as useful areas of microcomputer software development.

Expert Systems for Construction Education

Expert systems can be defined as computer programs, based on artificial intelligence techniques, designed to reach the level of performance of a human expert in some specialized problem solving domain (Rehak and Fenves 1985). Expert systems are particularly well suited to construction problems which have no algorithmic solution, but are characterized by having experts with considerable field experience.

---

[1]Assistant Professor, Department of Civil and Environmental Engineering, Rutgers University, P.O. Box 909, Piscataway, NJ 08855.

Education using expert systems should focus in two specific areas. The first area is to capture the knowledge of field experts and make it available to students and less experienced engineers. To control complex construction operations the correct management techniques are learned over years of trial and error. For many areas of construction, little documentation is available in textbooks. The expert system can serve as a vehicle for documenting the expert knowledge of construction methods, and foster discussion on how construction techniques could be improved. If expert systems addressing specific construction problems could be made available to students, it could speed their learning process and increase productivity by reducing the number of management errors that novice engineers make. These expert systems could be used in college classes, and more practically by individual construction firms and organizations wishing to more quickly make their expertise available to new employees.

The second area of focus should be to provide construction students with the capability to produce expert systems so they can apply the technique in the construction industry after graduation. The proliferation of powerful expert system shell programs capable of running on microcomputers will make the use of expert systems possible for even small construction firms. If engineers entering the work force are capable of producing expert systems, numerous expert systems will be produced which will aid in controlling and managing construction processes. Potentially, the increased availability of scarce expert knowledge of construction processes and techniques could increase productivity at the construction site by allowing managers to make better informed decisions.

An important topic of educating students in the development of expert systems is the knowledge acquisition process. Knowledge acquisition is the most difficult and time consuming step of the experrt system development process. One of the major difficulties in developing construction expert systems is the considerable amount of time which must be expended by busy experts during knowledge acquisition. Engineers who understand the knowledge acquisition process will be able to make better judgments concerning when expert system development is appropriate, and what manpower resources will be required to develop a useful expert system.

Potential Expert System Applications in Construction Education

There are two specific problem areas where knowledge-based expert systems can be used to provide information to students and inexperienced engineers. They are construction techniques, and planning and control of construction operations.

In the area of construction techniques, inexperienced engieners require information concerning the correct techniques for performing operations, how to recognize problems in the construction operation, and remedial actions to correct deficiencies in quality.

For planning and control of construction operations, expert information which can be transferred to students is the expected productivity, required crew and equipment, and effect of cosntruction problems on productivity.

Example Expert System for Construction Techniques

An example application in the area of construction techniques is for the asphalt paving construction operation. Monitoring the construction of an asphalt concrete pavement requires considerable expertise to recognize and correct paving problems. Decisions made by the inspector during construction can have significant effects on the performance and durability of the pavement. A microcomputer-based expert system was developed to assist highway engineers in inspecting the construction of asphalt pavements. The goal of the expert system is to transfer scarce expert knowledge to inexperienced personnel. The information included in the expert system is designed to inform inexperienced engineers of correct paving techniques, and how to quickly identify deficiencies in the paving operation. The information which is included in the asphalt paving expert system includes a number of different subject areas. The major areas of concern to inexperienced personnel were identified, through interviews with experts, as conformance with standard specifications, preparation of existing pavements for overlay, quality control of the arriving asphalt material, placement techniques and compaction.

The construction engineer must insure that paving is performed in accordance with standard specifications, such as those used by a state department of transportation which may be unfamiliar to inexperienced engineers. Standard specifications describe information such as environmental conditions when paving is not allowable, and required methods of placement and compaction. When preparing existing pavements for overlay, the inspector must know the required rules for patching, sealing and cosntruction of trueing and leveling courses. An inexperienced engineer may not have enough experience to make correct judgments concerning the appropriate actions to prepare an existing pavement for repaving. The expert system can provide students with the detailed knowledge of the expert.

An additional area of application of the asphalt expert system is to provide information on how to interpret variations in the quality of the paving, and corrective actions which can be taken. An example of the information contained in the expert system to aid inexperienced engineers in inspecting the operation, is for interpreting variations in the surface of the paving mat during construction. Someone with little experience in controlling the asphalt paving process will need information on how to identify the problems that area occurring and what can be done to correct them. Two important factors which control the quality of paving are the level of material around the auger of the paving machine and the amount of time the auger operates. Figure 1 shows an example table of expert advice of varying conditions which could be observed during construction. The system

Nationwide data were gathered during 1988 and were analyzed in late 1988 and 1989. Distribution of survey matrices was done through the Construction Industry Institute and by individual members of the Task Force, which is made up of representatives of contractors, building owners and professors. Analysis was done at the University of Cincinnati. All returns are on file there for further review if desired.

The Study

Ten skill/traits were recognized as being required, in varying degrees, to perform in eleven identified position/functions:

Skill/Traits	Position/Functions
Oral Communication	Senior Executive
Written Communication	Mid-level Executive
Graphic Communication	Project Manager
Financial Management	Project Engineer
Numerical (math)	Estimator
Personnel (People Mgmt.)	Site Superintendent
Manual	Field Engineer
Planning and Control	Trade Superintendent
Leadership	Foreman
Ethical Decision Making	Journeyman
	Subjourneyman

A survey matrix instrument was designed to combine these variables on one page to provide a visual inter-relation of all skill/traits to all position/functions. Further, respondents were asked to give their perceptions of the portion of each skill/trait to be acquired through three sources:

Formal education
Special training
On-the-job experience

274 responses were received, of which 266 were usable.

The survey matrix form contained a cover page on which occupational identification was asked plus two basic questions regarding construction education and training. In response to a question on the adequacy of current education and training, 74% of the total responded in the negative. 152 written comments, related to that question, cited deficiences throughout the industry, with the largest single group of responses, 64, pointing to critical craft skill training needs. 40 responses cited deficiencies in higher education. They ranged from inadequate rigor to insufficient coverage of the business of construction.

In response to the other question, 48% indicated that management education was the highest priority in their companies, 24% indicated technical training and 18% craft training. A few gave equal weights to more than one category.

Overall the largest batches of responses were from mid-level management and engineering segments of the construction industry, with representative samples from both owners and contractors. There were a few responses from foremen and journeymen but none from subjourneymen. Thus the results are most valid for middle upper level position/functions.

The data from the 266 matrices were processed first into statistical means by position/function to give a cross referenced view of each position/function by occupants of that level and then by occupants of each other level. Next, the data were refined into weighted averages for each position/function and priorities were established for each for 1) formal education; 2) special training and; 3) on-the-job experience. Salient findings include:

1. The broad perception that current education and training at all levels are not adequate to meet the needs of the construction industry.

2. Education and training are a continuum, with upper level position/functions requiring roughly equal portions of formal education, special training and on-the-job experience; and lower level position/functions requiring proportionately greater special training and on-the-job experience.

3. All formal education at all levels should be strong on the fundamental skills of communications and mathematics. (Civil engineering education is strong on the latter but not necessarily on the former).

4. The subject matter of continuing education and on-the-job experience are closely correlated, thus implying that there should be more coordination of the two through formal ongoing training programs.

Rankings

The following listings give the consensus rankings of the respondents, determined by weighted means analysis, for each position function. The relative value column relates to the overall ranking of each skill/trait, and the other three columns give the respondents' perceived priorities in each of three acquisition modes.

ing practice, it exists in both.

The courts and arbitrators are not impressed with the contract language used to exclude design professionals from construction liability and third party claims continue to be successful against them. Without certain protection in their contracts many design practitioners are taking a pro-active approach to the liability problem by getting better at what they intended to avoid.

The Gap in Education

Influenced by the isolation attitude of the design profession, but certainly more so by economics and faculty inclinations, the separation of design and construction in civil engineering education is wider than in practice.

The financial base of technical universities is vested in research, not teaching. A professed dedication to undergraduate programs notwithstanding, research is the principal goal of university administrators, the criteria for measuring success at every level and the prime consideration for faculty tenure, promotion and reward. In spite of promotional efforts to put undergraduate teaching in a position of high regard the real world of university funding forces reversion to research priorities.

The dedicated search for research dollars adversely effects construction education. The construction area of knowledge is not oriented toward research. As a consequence, career minded faculty divert their interests to other areas where funded research is readily available. As long as research is the major criteria for success in academia, the construction area will be spurned by civil engineering faculty.

Based on current hiring standards, and due to the acute shortage of PhD construction faculty, all newcomers to the rank are in demand and employable immediately upon receiving their doctorate. Many PhD construction candidates interview for positions six months to a year before completing their degree requirements. The acute need and the acute shortage precludes industrial experience. As a result many who end up teaching construction in civil engineering programs have little or no practical experience to draw upon. This situation satisfies university standards but falls far short of satisfying the real needs of civil engineering education and the construction area.

Construction education is much more dependent on practice and design than on theory, analysis and research. This does not imply in any way that the latter pursuits do not contribute to the education of the construction engineer. They most certainly do. However, exposure to them is found in other civil engineering coursework. Construction courses apply them.

Properly specified construction courses are essentially capstone courses, comprehensive courses which apply the undergraduate's accumulative knowledge and broaden the perspective of the construction industry where he/she will vest a career. The construction engineer in practice, more than any other civil engineer in practice, will be required to constantly draw on his/her aggregate civil engineering education.

In support of the trend in practice to close the gap between design and construction, it seems prudent for civil engineering programs to do the same. All undergraduate civil engineers would profit from an exposure to construction coursework and benefit from the advice and counsel provided by experienced construction faculty.

Available Construction Research

In the constant search for research, funding must come from wherever it can be found. Construction's share of research funding being almost negligible, construction researchers must be aggressive and creative in its pursuit. As is appropriate, the larger construction users and practitioners, those with a major financial stake in the construction industry and money to invest in research, provide the funding. Unfortunately their research needs are unique to their high participation level in the construction industry and not easily translated to the average industry participant. It is not practical to scale down and accept their problems and solutions as being universal to the average construction industry user or participant.

In essence it's extremely difficult to find funded research which actually contributes to undergraduate construction course work in a meaningful way. Consequently, construction faculty hard pressed to find funded research, reach outside of the general definition of the civil engineering area to find opportunities. A review of construction research topics and publications during recent years amply documents this.

There are reasons for the lack of funded research which are easy to comprehend but not so easy for academia to accept. First of all with the exception of environmental engineering (which is rapidly finding its own identity apart from the civil area) most scholarly civil engineering research has been accomplished.

Secondly, civil engineering is essentially a practice oriented profession rather than a scholarly research oriented scientific field of study. What we do as civil engineers is rigidly governed by the very forces of nature which we, by definition, utilize. We apply our technology to the materials and technology researched and developed by other disciplines. We react to man-made codes and employ proven design and construction technology which has withstood the test of time.

Although there is still research to be done, it is mostly applied research relating to the broad construction industry we exclusively serve. Unfortunately, that industry is not dedicated to research as fully as the materials and manufacturing industries. Nor for that matter are divisions of government. Construction research dollars are made obvious by their meagerness in every tabulation of private and government research funding.

Faculty Hiring Criteria

At universities dominated by a school of engineering, faculty hiring criteria aligns with the research oriented engineering disciplines. Little if any consideration is given to the unique nature of civil engineering or its important subset, construction. Incoming faculty must have a proven or promising ability to obtain and conduct scholarly research. A PhD degree is essential. Teaching ability is assumed, and practical experience is desirable.

Based on the needs of civil engineering education current faculty hiring criteria fall short. As previously explained, civil engineering is a practice oriented profession, unique among engineering disciplines. Most civil engineers are connected directly or remotely with the construction industry.

Civil engineers in design practice are more often entrepreneurs who function in direct contact with clients and the public. Civil engineers in construction practice are essentially technical managers employed by either owners, contractors, construction managers or construction consultants. Engineers in other disciplines are less visable as a group and are usually employed deep in the work force with little or no public visibility. From this perspective, civil engineering education should be more practice oriented than other engineering disciplines.

Practice should be taught by faculty with credible experience in the appropriate industry. In civil engineering this means the construction industry. However, current university hiring criteria is not amicable to this type of faculty member. Credible experience is seldom accepted as a substitute for the PhD requirement and new PhDs generally have less than credible experience in industry.

The rigid blanket position on the PhD requirement for incoming civil engineering faculty is enigmatic to say the least. The rationale behind the mandatory PhD requirement is difficult to understand. There is no doubt the PhD recipient has demonstrated both intelligence and fortitude by completing rigorous requirements. However, the PhD is an academic degree granted without a test of satisfactory performance in engineering practice or a demonstration of teaching ability in the classroom. (Walker 1989, Corotis 1989, Laursen 1989)

It is incomprehensible that the selection of quality faculty is solely dependent on such a single, narrow requirement. One must question why it is permitted to exist without flexibility. The writer has yet to hear a convincing argument in support of the inflexible PhD requirement with regard to civil engineering education.

Summary and Conclusions

Civil engineering programs provide the educational link to the nation's construction resource. The credibility and strength of that resource is vital to the nation's economic growth and national security. The construction industry, the custodian of the resource, is currently undergoing constructive change after experiencing an extended period of criticism by its users.

Construction education within civil engineering programs provide an important link between design and its implementation. The construction industry, under the pressures of failures and criticism have joined with its users in reestablishing the temporarily lost continuity between design and construction. The message from industry to reestablish the connection between design and construction has not been decisively received by civil engineering educators. After three decades of teaching separation, civil engineering programs have established a pattern which is difficult to change.

There are increasing tensions in many civil engineering programs across the country. Universities realize civil engineering programs have difficulty meeting the expanded research criteria they have set. Civil engineering programs reacting to increased research pressure, are concluding that construction education within their programs dilute their research status with the university.

The writer firmly believes, that unless universities reevaluate their responsibility to undergraduate education in terms of national interest rather than research economics, the possibility exists that universities will drop traditional civil engineering programs. Before that point is reached, construction education within civil programs will have totally vanished.

The author suggests the following.

1- Universities, Colleges of Engineering and Civil Engineering faculty should bring CE education into proper cooperative focus with the profession's obligation to public safety and welfare and the nation's well-being.

2- Civil engineering faculty must accept the fact (without loss of any professional pride) that civil engineering was a high tech science and practice early in this century but is a low tech science (but still a high tech practice) on the current technology scale.

3- All parties should conclusively recognize civil engineering education and practice (with partial exception of the environmental area) as an inextricable part of the construction industry.

4- Faculty should move closer to professional practice and more sincerely discover, consider and help to solve the problems of the construction industry in the classroom and through applied research.

5- Practitioners should be more aggressive in their effort to influence change in civil engineering education.

6- Serious consideration should be given to an accrediting rule requiring a practitioner on each civil engineering faculty.

Appendix. References

Committee on Construction Management. (1987) "Qualifications and Selection of Construction Managers with Suggested Guidelines for Selection Process" J. of Const. Engrg. and Mgmt. ASCE 113(1) 51-89.

Corotis, Ross B., Scanlon, Robert H. (1989). "Future of Civil Engineering Profession and Role of Education" J. of Issues in Engrg. ASCE, 115(2) 117-124.

Laursen, Emmet M. (1989). "First, Second, and Third Thoughts on Civil Engineering Education". J. of Pof. Issues in Engrg. ASCE, 115(2) 129-147.

Walker, Eric A. (1989) "Engineering Schools Share the Blame for Declining Productivity". The Bent of Tau Beta Pi. Spring. 12-14.

STRUCTURAL ENGINEERING EDUCATION PRACTICE AND LIABILITY

PAUL M. LAHOUD[1], M. ASCE

Abstract

Undergraduate engineering education has demonstrated a gradual trend away from applied engineering over the last 30 years. The addition of computer topics to course work and faculty commitment to basic research have reduced the exposure of students to the practice elements of structural engineering. Significant changes in the legal climate have further modified the practice of structural design. Graduates are ill-prepared to adapt to this environment. Transition training is needed.

Introduction

The practice of structural engineering under todays conditions requires knowledge and training that are not addressed in most undergraduate engineering programs. Some of the shortcomings result from the long-term changes in undergraduate education objectives. Others can be traced to the escalation in liability resulting from changing legal interpretation of the responsibilities of the engineer in the design and construction process. This paper discusses one point of view of these issues and some possible solutions.

Traditional Undergraduate Engineering Education

In the context of this discussion, the term "traditional" is used to describe engineering curriculum and faculty that existed from the end of World War II until the early 1960s. The required curriculum in an undergraduate degree with a structural engineering major usually included:

---

[1] Chief, Structural Engineering Section, Huntsville Division, U.S. Army Corps of Engineers P.O. Box 1600 Huntsville Alabama 35807-4301

Statics	Dynamics
Engineering drawing	Strength of Material
Determinate structures	Indeterminate Structures
Steel Design	Concrete Design

Faculty members who taught during this period were often active in the practice of structural engineering. Discussion of procedures in the design office, dealing with fabricators and on-site inspection of projects, provided a glimpse of the practical aspects of the design profession. Design courses in steel and concrete involved use of the AISC Steel Handbook and ACI 318 in some fairly detailed manner. Moderate amounts of "hands-on" connection design and rebar scheduling were invaluable experiences to the student since practice, in both the design office and the fabrication shop, was not far removed in either form or substance. In any case the importance of detailing and connection design was strongly demonstrated in the educational process.

Traditional Practice of Structural Engineering

During the era of this "traditional" education, the practice of structural engineering was more straight-forward than today. Involvement was more complete and allowed more latitude in guiding the design from preparation of drawings and specifications through review of fabrication drawings and erection. The responsible design professional exercised considerably more influence in the implementation of the design by both the fabricator and the erector. The presence of the engineer on the job site was accepted practice. Public perception of the "Professional Engineer" was considerably more positive than today. Relationships between the owner, the engineer and the contractor were much less confrontational.

The Traditional Graduate in Practice

The new graduate employed by a design firm usually started work with basic responsibilities. He performed simple procedural tasks including drafting, detailing and simple repetitive load and member sizing calculations. The engineering tasks he performed were, however, very similar to those he was exposed to during the latter stages of his undergraduate education. The observed design/construct process was logical in the light of his training. The parallel of practice with education during this period was reasonably consistent.

Evolution of Undergraduate Education

In the early 1960s undergraduate engineering education began to experience a significant change. Two principal factors are suggested by most observers as the cause of this change. The first was the arrival of the digital computer as a powerful analysis tool. The second was the emergence of the engineering faculty career path dedicated to sustained basic research activities. Each of these factors influenced the undergraduate engineering curriculum. While there have been benefits from these changes, some troubling problems have also appeared.

Impact of Computer technology on Engineering Curriculums

It was clear that the power of the computer would improve the practice of structural engineering. Complex analysis tasks which previously required crude approximations and judgement could now be resolved with much greater accuracy and speed. Iterative analysis to optimize structure performance was now feasible. There was no question that undergraduate engineering must incorporate this new technology. The issue was how could it be accommodated within the limits of existing curriculum. Considerable discussion centered around requiring extension of the four-year degree to five years. This was not found to be a viable alternative. The only option, therefore, was to begin removing some elements from the existing curriculum and replacing them with computer-related material. Topics that began to be reduced or eliminated were usually those viewed by educators as skills which could be acquired on the job after graduation. Engineering drawing and surveying courses were scaled back or even eliminated. Time allocated to instruction in the design and detailing of steel and concrete was reduced. In the place of this Material, the student was exposed to computer languages, matrix mathematics and the fundamentals of finite element analysis procedures.

Engineering Faculty Influences on Engineering Education

In conjunction with the demands of emerging computer technology came a fundamental shift in the career path for engineering faculty. Significant investments in basic research fostered by the federal government led to a demand for long-term commitment by faculty in these areas. This career path fostered research in highly specialized activities. Engineering faculties began to actively seek research funds, clearly with a priority over teaching. This pursuit required a great deal of effort and limited the exposure to (and interest in) day-to-day practical engineering issues. Research skills and experiences began to dominate the knowledge base of the staff responsible for undergraduate education. As the years have passed, the faculty experience conveyed to the student no longer represents the parallel with practice that existed previously. Consistent with human nature, course work is presented in the light of the instructor's educational background, experience and personal interests. As one would expect, graduating students reflect the background of their respective faculties. They are computer literate and comfortable in analytical procedures. They understand programming, automated analysis methods and the formulation of finite elements. Many have had training in the operation of microcomputers and computer aided design (CAD) equipment. On the other hand, their knowledge of correct engineering drawing procedures is very limited. Many are barely familiar with the applications of AISC and ACI codes to design problems. Knowledge of connection design is very limited. They can model a structure on any of the popular finite element codes and generate output. But they may not understand the limitations of the analysis and the effects of boundary condition and connection assumptions on the results.

Practice in the Current legal Environment

Even as the training of the graduate engineer has evolved, the practice of structural engineering has also changed. The United States has been called the most litigious nation in the history of the world. This fact is evident in the field of structural engineering. The essence of structural design is founded in life safety. Not withstanding the moral and ethical issues associated with a structural failure, the constant threat now exists of financial ruin or even criminal proceedings. Not only are corporate assets threatened but also professional careers as well. These risks have dramatically altered the practice of structural engineering. Designers must now strive to limit their exposure. Information in design drawings is minimized and frequently worded to place responsibilities on the contractor and the fabricator. Fabricators and contractors strive to limit their risk by assuring responsibility remains with the designer. Legal guidance and insurors discourage or carefully limit participation in on-site inspections as some recent court decisions have assigned wide liability to this practice. The overriding design objective is often to minimize risk of legal controversy. Structural designers struggle to adjust their practice to this environment. Innovative designs must be considered with caution or avoided to minimize risks. Blanket maximum member uniform loads are often specified to fabricators for connection design. This results in greatly overdesigned connections and higher cost, but provides a measure of protection from possible errors in analysis. Requests by fabricators to use alternate details are handled defensively or denied to avoid subsequent risk. Relationships among owner, designer and contractor are often aloof and controlled by legal staffs attempting to place or avoid liability rather than resolve problems.

The Modern Graduate in Practice

The modern graduate is now on his first job without some of the fundamental engineering skills he needs to perform effectively. Of equal importance, he has no perception of the legal and regulatory environment in which he must practice. He does not have the benefit of the correlation of practice and education that existed a few decades ago. He does not realize that if he write a computer code to solve a problem efficiently, he is creating a liability risk for his company and in some cases for himself. He does not understand the legal relationship that exists between the designer and the fabricator. One of the most common on-the-job training assignments for new graduates is shop drawing checking. He does not realize the importance of this work legally, nor does his academic training prepare him for the task. His lack of knowledge of engineering drawing practice and proper detailing is a weakness which contributes directly to one of the most common problem areas in the execution of a structural design. He has little understanding of the the relationship of drawings and specifications with regard to contractor performance. The greatest strength of recent graduates is in the area of computer analysis. However, this may represents only 20 to 30 percent of the design effort. The remainder of the effort is in preparation of drawings, details, specifications, and performing quality control. In these areas the recent graduate has no parallel in his education to assist him. He must struggle to adapt and become productive while at the same time trying to sort out all these "practical" issues. He is not aware that under some conditions his employer may even choose to litigate against him personally.

Possible solutions

There are two areas which must be addressed to improve the transition of new graduates to the practice of structural engineering. The first of these is to achieve a minimum level of training in certain basic engineering skills including, as a minimum, drawing procedures, steel and concrete detailing, and the relationship between drawings and specifications. The second is education in the legal responsibilities and limitations of the designer. These two elements should be provided in an integrated manner. Such training could be provided several ways. It could be added to existing undergraduate curriculum. Graduates could also acquire the training on their own. Finally, the employer could take an active role in assisting the employee in obtaining the necessary training.

Recommended Approach

Inclusion of additional training in existing curriculum is not considered viable for two reasons. First, there is simply no place to accomplish it within current four year programs. Second, The typical academically oriented engineering faculty simply does not have the interest or background to treat the subject properly. It is possible that the graduate might choose to accomplish the training personally. However, he would have to be convinced it was beneficial, and adequate guidance would be needed to select the training. The American Society of Civil Engineeers could be a factor in providing such guidance in the form of recommended training for new graduates as well as developing suitable short courses. This approach would also be attractive for smaller firms where full or partial compensation would be provided to the new employee. However, the author believes that the most effective solution would be for employers to develop their own training. These programs could be tailored to address the issues described as applied to their firm. The company's legal position on various phases of the design process could be easily and effectively incorporated in such training. The training could be spread out over several weeks or months in parallel with entry level job assignments. This would provide a certain amount of integration with work and promote a more rapid learning curve and early productivity for the employee. Such training could be considered as part of the compan'y overall quality control/assurance program and become a beneficial selling point for clients as well as insurors.

Conclusion

High quality structural design requires more practical skills and training in legal awareness than current engineering graduates possess. Coordinated training can provide the necessary background, resulting in increased quality, productivity and reduced liability. This training can best be achieved through company sponsored continuing education.

ENGINEERING - TAUGHT DIFFERENTLY

Kurt H. Gerstle [1],F. ASCE

Abstract

In the proposed approach to the teaching of structural engineering, the student is introduced to sufficiently simple, yet non-trivial, design problems, and rational mechanics is covered as needed for design purposes. In this way, students are exposed to structural systems and design-oriented decision-making at an early stage, and understand the purpose and use of theory, thus increasing their motivation. An example of a design-analysis project taught in this manner is presented.

Introduction

The primary purpose of professional engineering education is to turn out creative, productive professional engineers. Over the last hundred years, we have attempted to achieve this goal by sticking to one tried and true scheme, in spite of ever-changing conditions which have to do with different clientele with different background and expectations, increasing competition for the best minds, and an ever-increasing treasury of techniques and tools.

In particular, comparing structural engineering curricula over the years, there has been remarkably little change: an added math or computer course to replace a drawing course, and a general contraction of required credits - that's about all. Maybe it is time to do some rethinking along larger lines which address the needs of this day. In the following, we propose a different approach to training in structural engineering which we have tried with some success on a small scale. To focus the thinking, we will, in the next section, outline in a nutshell a typical current structures curriculum with its pros and cons. Following this, we will suggest an alternate sequence toward the same goal which appears to have some promise. Finally, we will itemize this approach in terms of one specific example.

Our purpose here is not to present a full-blown recipe, but rather a concept whose future and success depends on the reader's

[1]Professor of Civil Engineering, University of Colorado,
Boulder, Colo. 80309

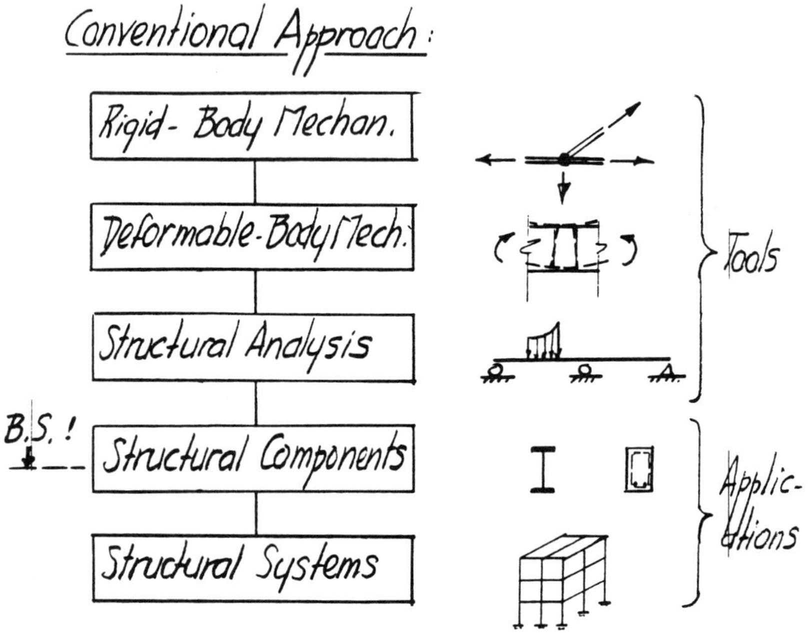

ADVANTAGES	DISADVANTAGES
LOGICAL	LITTLE MOTIVATION
STEP-BY-STEP	INSUFFICIENT TIME
	NO DECISION-MAKING
	ANALYSIS-DESIGN RELATION OBSCURE
	NO REAL STRUCTURES

Fig.1 The Classical Structural Engineering Training

willingness to consider, and experiment with, different, and possibly better, ways of achieving our goals.

The Conventional Approach

Fig.1 presents an overview of the classical sequence of topics in structural mechanics and engineering. It is a logical, step-by-step approach. Unfortunately, the long and arduous road through rigid- and deformable-body mechanics and structural analysis provides few outlooks toward the high peaks of structural creativity and only scanty resting places for meaningful applications.

Further, with the reduction of required undergraduate credits, this road stops prematurely: with only one "design" course - which should probably be more appropriately called "mechanics of structural members" - the entire field of structural systems, which includes the relation of analysis and design, the opportunity for decision-making which is the essence of real design, to develop judgement, and to experience real design situations falls by the wayside.

In the absence of these aspects, the analysis overpowers the design, and the curriculum becomes one of "structural mechanics", not "structural engineering". The analysis, instead of constituting a tool for good design, becomes an end in itself.

An Alternate Approach

The sequence suggested in Fig.2 is based on the following concepts: Integration of theory and application;

Motivation through purpose;

Judgement through decision-making.

We achieve these goals by reversing the classical approach: instead of laboriously constructing a scientific foundation prior to setting a real structure on it, we begin with a real, though simple, structure, try to decide what our problem is, primarily, how it can carry given loads safely to ground, and then, rationally and step by step, acquire the necessary tools to achieve this end. The structural need motivates the analytical tool.

Example

The building of Fig.3a, taken from Milo S. Ketchum's "Standard Structural Details for Buildings" (Prentice-Hall,1957) will serve as example to explain how this approach has been successfully

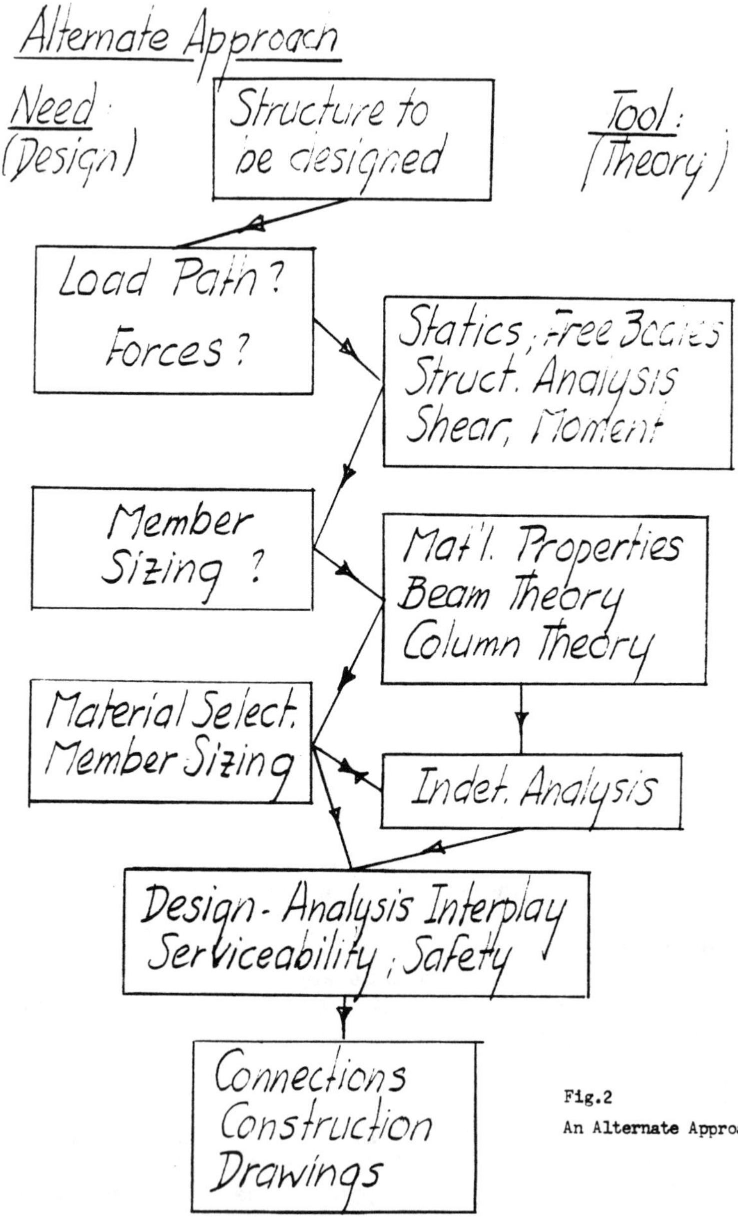

Fig.2
An Alternate Approach

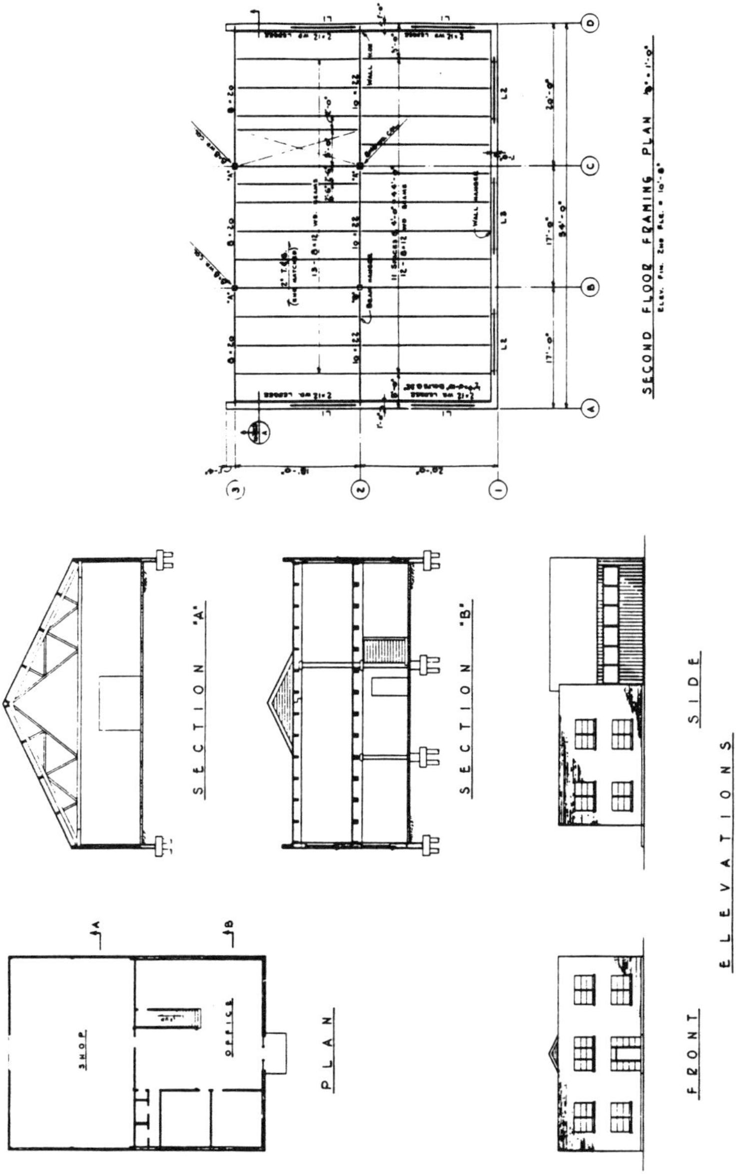

b, Mezzanine Framing Plan

a, Building Plans

Fig.3

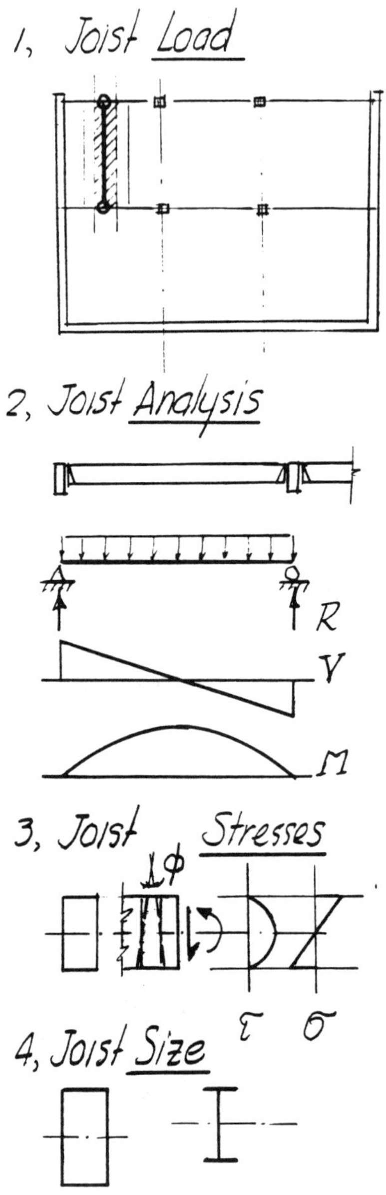

Concepts

Dead and Live Loads

Load Factors

Tributary Area

Structural Modeling

Statically Determinate Analysis:

Free Body

Equilibrium Equations

Reactions

Shear

Moment

Beam Theory:

Curvature

Bending Stresses

Shear Stresses

Section Modulus

Material Strength

Allowable Stresses

Beam Selection

Fig.4 Floor Framing Design

5, Girder Design

Concepts

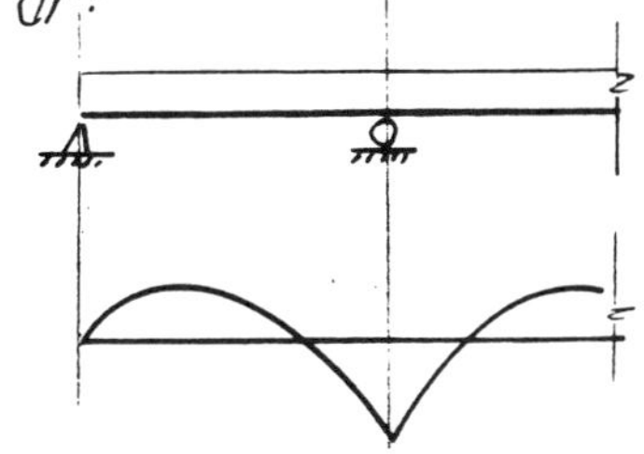

Load Transfer

Load Idealization

Material Selection

Statically Determinate

vs.

Statically Indeterminate

System and Analysis

Interrelation between

Analysis and Design

6, Column Design

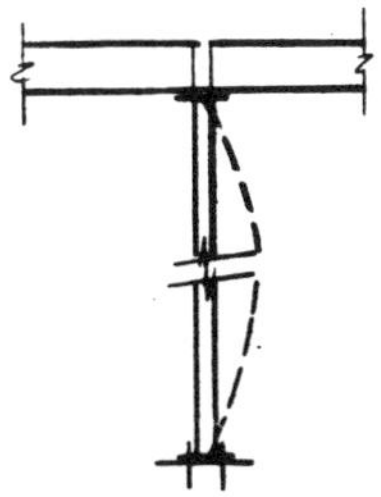

Axial Compression

Elastic and Inelastic

Column Theory

Column Design

Fig.4 (Cont'd.) Girder and Column Design

implemented. We start with drawings such as might come from an architect's office. The need to be able to read these drawings will emphasize drawing literacy, a skill often lacking among students.

From here on, we will consider only the structural design of the second floor shown in Fig.3b, a mezzanine extending over the office portion of the building. Our aim will be to size three members: a typical floor joist, a girder, and one interior post. The sizes specified in Fig.3b are to ignored; material choice is left to the designer. Only the design live load and allowable stresses are specified.

From a traditional viewpoint, it might seem an absurd task for a person who has not even been exposed to statics. However, Fig.4 indicates, step by step, how the design questions drive the need for a rational analytical answer.

It is clear that such an approach precludes the all-encompassing, systematic coverage of topics of traditional mechanics courses. We hope that our suggested approach will have the student sufficiently convinced of the need for sound analysis so that with added maturity he will be rushing toward more systematic analytical courses which will fill the gaps left by this approach. With this motivation, we may be able to turn out engineers who are better equipped to cope with the realities of structural practice.

Teaching Aids for Timber Engineering Design

Lawrence A. Soltis [1], Member, ASCE

Abstract

Structural engineering curricula and courses are often based on the premise that steel and concrete are the primary construction materials and timber and masonry are secondary. The objective of this paper is to encourage teaching of timber engineering and to describe the teaching materials that are available. Data are presented on where and how each material is used in the construction industry. Trends in timber engineering indicate that the wood industry has targeted nonresidential building and bridge construction as potential expanding markets. One factor opposing this trend is the lack of engineers trained in timber design. Information is presented on available textbooks and on the Clark Heritage Memorial Series of educational modules to help engineering faculty prepare and teach timber engineering.

Introduction

The objective of this paper is to encourage teaching of timber engineering and to describe the teaching materials that are available. In addition, this information may help educators to plan the engineering curriculum and coursework.

---

[1] Supervisory General Engineer, U.S. Department of Agriculture, Forest Service, Forest Products Laboratory, One Gifford Pinchot Drive, Madison, WI 53705-2398. The Forest Products Laboratory is maintained in cooperation with the University of Wisconsin.

Structural engineering education is currently based on the premise that steel and concrete are the primary construction materials and timber and masonry are the secondary materials. A closer look at how and where construction materials are used indicates that a specific material may be either primary or secondary, depending on the type of construction project and the structural system within a project.

Wood has traditionally been used for residential construction. In the past several years, the wood industry has increased its effort to use wood in nonresidential building and bridge construction. If properly designed, wood is equally durable and usually more economical than competing materials. One problem with the increased use of wood in engineered structures is the lack of undergraduate training in timber design. A 1984 survey (Wadlin 1984) found that over 75 percent of the curricula leading to a bachelor's degree in civil engineering required a course in either steel or concrete design, but only 13 percent required a course in timber design.

Materials and Uses

Data on how and where different materials are used should be considered for curriculum and course content. Census data (U.S. Department of Commerce 1989) for new construction in 1987 are given in Table 1. Residential construction accounted for about half (US$171 billion) the total 1987 construction (US$349 billion). Wood is the primary material used in residential construction. Nonresidential building construction (public and private) accounted for US$99 billion. Information on nonresidential building size (Table 2) for 1982 (Spelter and Anderson 1985) indicates 78 percent of the floor area was in one- or two-story buildings. Wood is often very cost competitive in low-rise nonresidential construction. This explains the timber industry's effort to gain a larger share of the 78 percent of this US$99 billion industry.

Data from a 1982 survey (Spelter and Anderson 1985) on primary and secondary materials used for various components of nonresidential buildings are given in Table 3. Wood was used in over one-third of the roof and interior wall systems. The use of wood in roof systems was split between truss and glued-laminated beam systems. Wood was also used for a significant share of the upper floor systems. Concrete was the primary material for the ground floor systems (no data were given for foundation systems), metal (steel) the primary material for roof systems, and a composite of steel and concrete the primary material for upper floor systems. All materials were used for exterior wall systems.

Data on how and where different materials are used in bridge construction are not available, although several wood systems for short-span bridge construction are used.

In summary, Tables 1, 2, and 3 can be used as tools for structural engineering curriculum and course planning. Timber structures or components are often either the primary material in construction or have the potential to be used for an increased share of structural systems.

Trends in Timber Engineering

Wood is an anisotropic material with highly variable and time-dependent material properties. It shrinks and swells with changing moisture content, and it is susceptible to decay and fire. Many of these properties are unique to wood construction and, therefore, they are not well understood by students schooled in steel and concrete design.

The wood industry produces composite products (such as I-beams and laminated-veneer lumber) that reduce variability; preservatives exist to prevent decay, and fire-rated construction is possible. The wood industry has targeted a larger share of the nonresidential building construction and the short-span bridge market as potential expanding markets. They are currently developing a Load and Resistance Factor Design (LRFD) manual that will parallel the steel and concrete LRFD manuals. Engineers trained in timber engineering can design reliable timber structures.

Timber Engineering Education

Wadlin (1984) found that only 13 percent of the civil engineering curricula required a timber design course for a bachelor's degree, whereas more than three-fourths of the curricula required either a steel or concrete design course. Moody and Freas (1987) discussed the present timber engineering education for undergraduates. They suggested increasing student exposure to timber design either by combining timber design with a steel design course or by developing a timber design project for senior students.

Several reasons exist for the shortage of timber design course requirements. First, timber is thought of as a secondary material (discussed previously). Second, only a limited number of courses can be taught within a curriculum. Third, many engineering faculty are not well-schooled in timber engineering. The remainder of this paper focuses on the third issue.

One problem perceived by engineering faculty is the lack of comprehensive technical information. Several excellent recently published textbooks are available (Breyer 1988; Faherty and Williamson 1989; Hoyle and Woeste 1979; Stalnacker and Harris 1989) as well as various design specifications (American Institute of Timber Construction 1985; National Forest Products Association (NFPA) 1986) and product literature. The NFPA specification will be updated shortly and a Load and Resistance Factor Design Manual is in preparation.

To complement these books, four volumes of educational materials have been developed based on a series of workshops sponsored by the USDA Forest Service, Forest Products Laboratory (FPL), in cooperation with the University of Wisconsin (Clark Heritage Memorial Series 1981, 1982, 1983, 1986). Each volume is divided into modules suitable for developing course material; each module contains an objective statement, technical material including sample problems, and self-test questions and answers. Each module has also been published separately in the *Journal of Materials Education* and its predecessor, the *Journal of Educational Modules for Materials Science and Engineering*.

Volume I of the Clark Heritage Memorial Series, *Wood, Its Structure and Properties*, concentrates on the physical characteristics of wood. Volume II, *Wood as a Structural Material*, is directed at the use of wood in structures. Volume III, *Adhesive Bonding of Wood and Other Structural Materials*, looks at the bonding necessary for many wood products. Volume IV, *Wood Engineering Design Concepts*, is directly applicable to structural engineering design. This volume covers the design of beams (solid, laminated, and composite), columns, connections, trusses, diaphragms, and curved members including arches and domes.

In addition to developing these four volumes, the FPL, in cooperation with Marquette University, sponsored a workshop in 1988 to educate university faculty in timber engineering. Thirty participants were selected from about 120 applications. One product resulting from the workshop was a "class notes" style teaching manual (Faherty 1988) to complement Volume IV of the (Wood Engineering Design Concepts) Clark Heritage Memorial Series.

Concluding Remarks

Economics favor the use of wood in many nonresidential applications, but improved education is necessary to bring about expanded wood use. Engineering education in timber design is currently not proportional to the actual or potential use of wood in nonresidential construction. The Clark Heritage Memorial Series of teaching modules are designed to help the educator integrate timber engineering into the university curriculum and to overcome the perceived lack of comprehensive technical data on wood.

Table 1--New construction in 1987[a]

Type of construction	Cost (billion US$)
Private	
Residential	171
Nonresidential	78
Farm	2
Public utilities	30
Miscellaneous	2
Subtotal	283
Public	
Buildings	21
Highways	20
Military facilities	4
Conservation and development	5
Sewer and water	12
Miscellaneous	4
Subtotal	66
Total	349

[a]U.S. Department of Commerce 1989.

Table 2--Distribution of non-residential building sizes by relative floor area[a]

Nonresidential building size	Percent
One story	55
Two story	23
Three or more stories	22
Total	100

[a]Spelter and Anderson 1985.

Table 3--Materials used in nonresidential buildings[a]

Building component	Material (percent)				
	Wood	Metal	Concrete	Concrete and metal	Masonry
Roof	36[b]	60	3	--	--
Exterior wall	18	35	19	--	27
Interior wall	35	52	2	--	11
Ground floor	4	--	92	4	--
Upper floor	18	3	15	63	--

[a]Office, store, industrial, school, hospital, religious, recreational, and public buildings (Spelter and Anderson 1985).

[b]Wood: 20 percent trusses, 16 percent glue-laminated beams.

Appendix

American Institute of Timber Construction. (1985). Timber Construction Manual, 3rd edition, John Wiley and Sons, New York.

Blomquist, R.F., A.W. Christiansen, R.H. Gillespie, and G.E. Meyers, eds. (1983). Clark Heritage Memorial Series: Adhesive Bonding of Wood and Other Structural Material, Vol. III, Educational Modules for Materials Science and Engineering, Pennsylvania State University, University Park, Pennsylvania.

Breyer, D.E. (1988). Design of Wood Structures, 2nd edition, McGraw Hill, Inc.

Dietz, A., E.L. Schaffer, and D.S. Gromala, eds. (1982). Clark Heritage Memorial Series: Wood as a Structural Material, Vol. II, Educational Modules for Materials Science and Engineering, Pennsylvania State University, University Park, Pennsylvania.

Faherty K. (1988). "Summer institute on wood engineering design principles for college faculty notes. Department of Civil Engineering, Marquette University, Milwaukee, Wisconsin.

Faherty, K. and T. Williamson. (1989). Wood Engineering and Construction Handbook, McGraw Hill, Inc.

Freas, A., R.C. Moody, and L.A. Soltis, eds. (1986). Clark Heritage Memorial Series: Wood Engineering Design Concepts, Vol. IV, Educational Modules for Materials Science and Engineering, Pennsylvania State University, University Park, Pennsylvania.

Hoyle, R.J. and F. Woeste. (1979). Wood Technology in the Design of Structures, 5th edition, Iowa State University Press, Ames, Iowa.

Moody, R.C. and A.D. Freas. (1987). "Education of engineers on structural use of wood. In White, J.L., ed., Wood Markets: Alternatives to Residential Construction: Proceedings 47347; 1985 November 18-20; Portland, Oregon. Forest Products Research Society, Madison, Wisconsin, p. 25-32.

National Forest Products Association. (1986). National Design Specification for Wood Construction, NFPA, Washington, D.C.

Spelter, H. and R.G. Anderson. (1985). "A profile of wood use in nonresidential building construction." FPL-RB-15. U.S. Department of Agriculture, Forest Service, Forest Products Laboratory, Madison, Wisconsin.

Stalnacker, J.J. and E.C. Harris. (1989). Structural Design in Wood, Van Nostrand Reinhold.

U.S. Department of Commerce. (1989). "Value of new construction put in place." C-30-89-06, April. U.S. Department of Commerce, Bureau of Census, Washington, D.C.

Wadlin, G.K. (1984). "Survey of civil engineering education." In Proceedings of Challenge to Civil Engineering Education Practioners--Where Should We Be Going? 1984 April 11-13, Ohio State University. ASCE, New York. p. 1-16.

Wangaard, F., ed. (1981). Clark Heritage Memorial Series: Wood, Its Structure and Properties, Vol. I, Educational Modules for Materials Science and Engineering, Pennsylvania State University, University Park, Pennsylvania.

OHIO NORTHERN UNDERGRADUATE COMPUTER APPLICATIONS

Authors: Marlin D. Minich[1], M.ASCE
Jonathan S. Smalley[2], M.ASCE
Robert L. Swart[3]

ABSTRACT

This paper presents implementation for teaching the finite element method of analysis at the undergraduate level at Ohio Northern University using a large scale industrial computer program. Descriptions of the content and format of the finite element analysis courses and the use of the finite element method in other courses are given. Three student projects are presented as examples of typical project topics. The present curricular coverage provides the student with a good practical background in the finite element method as preparation for entering professional practice.

INTRODUCTION

The topic of finite element analysis has undergone rapid changes in engineering education. During the last twenty years, it has evolved from nearly a pure theoretical course of study at the doctoral level to a universally employed technique, requiring its presentation to engineering students at all levels. This paper will outlines how the undergraduate engineering faculty at Ohio Northern University has implemented the teaching of finite element analysis in general and the application of the method through the ANSYS® computer program[4] developed and supported by Swanson Systems Analysis, Inc. in particular.

---

[1] Professor, and Chairman Civil Engineering Department, Ohio Northern University, Ada, Ohio, 45810

[2] Associate Professor, Civil Engineering Department, Ohio Northern University, Ada, Ohio, 45810

[3] Professional Engineering Consultant, Ada, Ohio, 45810

BACKGROUND

The first offering of the finite element method at Ohio Northern University was in 1981-82, as part of the third course in Structural Analysis, an elective three credit hour, senior level course in the Civil Engineering Department. Since it carried previous structural analysis courses as a prerequisite, the course was taken only by civil engineering majors with a strong interest in structures. As a result, class size was relatively small with an average enrollment of about fifteen students. The next year the name of the course was changed to Finite Element Analysis (FEA) and the prerequisite was changed to permit others to take the course with instructor permission.

In the mid 80's the FEA course became a requirement course for civil engineering majors. With all civil engineering majors and a large proportion of the mechanical engineering students taking the course as an elective, average class size increased considerably to about thirty-five. Then the Mechanical Engineering Department initiated a required finite element course at the junior level for their students. Seniors from both departments often use ANSYS as part of their senior design project.

FINITE ELEMENT COURSE STRUCTURE

Senior Civil Engineering Course

Finite Element Analysis (202 555) is presently a three credit hour senior level course required of all civil engineering majors. The course is taught in the spring quarter of the students' final year. The course encompasses the development of the finite element method as applied to various topics from continuum mechanics. Topics covered include one-, two-, and three-dimensional problems in structures, seepage, and fluid mechanics which require the use of several different element types. Solution algorithms for linear and nonlinear static and dynamic analyses are developed. The finite element method is implemented by the use of a large scale production computer program, ANSYS, from the Swanson Analysis Systems, Inc. on a Data General ECLIPSE MV/10000.®

The course is team taught by two civil engineering faculty members. One's expertise is primarily structural applications of the ANSYS program while the other's is the theoretical development of the finite element method for non-structural problems. The major course objective is to develop the finite element method as a solution technique. To achieve this end, each student is required to present an oral and written report on a term project using the finite element method. The project must include as a minimum the following:

- a statement of the generalized problem under consideration,
- a statement of the specific case under consideration,
- a listing of the computer input and output,
- an appropriate presentation of results,
- a comparison of results to either expected theoretical results or actual experimental results,
- numerical experimentation with the finite element method to determine the effect on the results of such parameters as mesh fineness, high order elements and aspect ratios, and
- discussion of the results.

Topics for student projects range from bending of reinforced concrete beams and automobile suspension vibrations to leachate seepage through sanitary landfills.

Junior Mechanical Engineering Course

Finite Element Analysis (204 445) is a three credit, junior level course required of all mechanical engineering majors and is taught during the winter quarter. The instructor is a professional engineer, who has expertise in mechanical analysis and considerable industrial experience with the ANSYS program. The finite element techniques are used to solve solid mechanics stress-deformation problems under various modeling strategies through the use of the ANSYS computer program.

Topics include an introduction to finite element theory, the solution of some simple spring and beam problems using classical analytical methods, and long hand solutions of finite element problems using the direct stiffness method involving small matrices. This is followed by a study of various element types, such as spars or struts that make up a simple truss, beam segments that can be built into a rigid frame, plate elements to construct shells and thin-walled pressure vessels, and solid brick elements to build more general machine parts.

A number of in-class and homework assignments to build simple models are given. This is primarily a geometrical problem of converting drawing dimensions into node locations, which in turn define corners or ends of elements. In addition, material properties, loads, and support constraints are also defined. Some of the typical homework problems include indeterminate beams, two-dimensional frames and trusses, thick-walled pressure vessels, spur gear teeth, and flat plates.

Topics for student projects are selected from areas where theoretical or experimental results can be used to validate the solutions. Some recent examples are curved beams, threaded connections, thin multi-curved pressure vessels, stress concentrations, and simple tools or household items, such as a spanner wrench, coat hanger, or snow shovel.

PROJECTS TOPICS:

Civil Engineering Seepage Analysis

The finite element method can be used for analyzing the seepage of fluids through porous media. Under a set of reasonable assumptions, the governing partial differential equation is the well-known Laplace equation where the unknown is the hydraulic potential subject to either essential boundary conditions of known hydraulic potential or natural boundary conditions of known flows. Such problems can be solved by ANSYS using a two- or three-dimensional isoparametric element type with proper interpretation of the element parameters. Nodal values of the hydraulic potential are directly obtained and other quantities of interest, such as hydraulic gradients, velocities, and seepage flowrates, can then be computed during postprocessing.

A popular project in the senior CE course is the seepage under a row of impervious sheetpiling. Finite element solutions for flowrate, exit gradient, and pressures along the sheetpile are compared to theoretical values as given by Harr's Method of Fragments. Numerical experimentation is performed to ascertain the effect on the solution of mesh fineness, aspect ratio, or changed boundary conditions.

Structural Engineering Bridge Design

The American Institute of Steel Construction has initiated a steel bridge competition to give our students hands-on analysis, design, and construction experience. The objective of the competition was to design and erect a model bridge with a 20 foot span which would best satisfy six areas of judging in the AISC competition. A senior civil engineering student chose to enter the competition as his senior design project and to use the ANSYS finite element program to analyze trial designs. Seven trusses were initially rated to determine optimum design, conservation of steel, speed of erection, and aesthetic value. A deck Warren truss, a camber deck truss and the A arch were selected for further study. The ANSYS program was initially used for a two dimensional analysis using spar elements and then expanded for the three dimensional analysis using frame elements to predict the stresses for design and the displacements. The displacements on the actual structure were later proved to be within acceptable limits.The bridge was erected in 18 minutes and won the "Most Unique Bridge" prize.

Mechanical Engineering: Static Stress Analyses

One group selected the static stress analysis of the blade of a snow shovel for its project. The students started by analyzing the blade as a simple cantilever beam to get some preliminary estimates. Next they widened the beam so that it had the proportions of the snow shovel blade and fixed it all along one edge rather than only at the mid point where the handle would attach to the blade. This analysis was followed by a finite element analysis of a flat plate supported in a similar manner. Thus they could compare the elementary theory and the first numerical approximation using finite elements.

This gave them confidence that they were on the right track as far as element definition and overall geometry. They then selected specific material properties, realistic dimensions for the blade, and a coarse grid for the mesh defining the blade geometry. They continued to add complexity and more accurate detail to the configuration of the snow shovel blade.

The next step in the analysis was to add the curvature of the blade so it deviated from a flat plate to represent the configuration of the actual snow shovel. They proceeded to add the reinforcing ridges along the back of the blade. This final analysis using finite elements including the detail of the curvature and the reinforcing ridges turned out to be a very interesting phase of their project. They were pleased with the results of the final analysis using the finite element method to evaluate an everyday component which turned out to be somewhat complex structure.

Other groups who worked on problems, such as curved beams, stress concentrations, and threaded connections, were able to investigate other variables. They broadened their understanding of the finite element approach by experimenting with size, number, and shape of elements. Some groups used different element types, such as plates, bricks, or beams to model the same structure while others looked at a variety of load conditions. These projects proved to be especially valuable when they were shared with the rest of the class as report presentations and for further discussion.

OTHER IMPLEMENTATIONS

The application of the finite element method is employed in courses other than the two previously discussed. The students are introduced to computer applications as freshmen in an introductory course using WordPerfect® as the wordprocessing package, LOTUS 1-2-3® as the spreadsheet, dBASE III Plus® as the database, and FORTRAN for engineering computations. FORTRAN is continued in the sophomore year. During the junior year computer graphics is taught on a DEC 11/44. The finite element method, using ANSYS, is introduced as a "black box" tool for indeterminate structural analysis. The students draw moment diagrams using the tenth-point moment output. At the senior level the finite element method is used in the second soils course for seepage analysis. It is also used in the senior and capstone design sequence for both mechanical and civil applications through UNIGRAPHICS or directly on the Data General.

CONCLUSIONS

After a decade of teaching the finite element method to undergraduate students using the ANSYS program, several observations can be made. First, the choice of project topic is critical to ensure the proper combination of realism and scope consistent with the students' lack of FEA experience and the time constraint of the quarter system. It would be unfair to expect students to be experts while learning the method. The best FEA projects are those where the student investigates the results of various modeling strategies and compareS them to results of other methods of analysis. Yet, even in senior capstone design projects, few, if indeed any, students have utilized the full capabilities of the finite element method as implemented through the ANSYS program. Therefore, a micro-based version would be adequate and perhaps desirable.

The Engineering College's computer analysis coverage provides good preparation at the bachelor of science level for entry positions in engineering, as well as for advanced study. The use of a large-scale production finite element program such as ANSYS complements this coverage nicely and provides students an opportunity to use a real-life FEA program on practical problems.

REFERENCE

[4] ANSYS Engineering Analysis System User's Manual, Vol 1,2, Swanson Analysis Systems, Inc., Rev. 4.4 Houston, PA 15342, 1989

REGISTERED TRADEMARKS

AN EXPERT DESIGN SYSTEM IN TEACHING

Yin Mao[1] and Bruce G. Rogers[2]

Abstract

Structural design is difficult not only because it requires broad knowledge and years of experience in structural engineering but also because it is a laborious process. Obtaining a higher level of automation in design and finding new ways to train new engineers are worthy challenges. This paper discusses the development of a new structural design tool called "expert design system", by applying the expert system technique together with conventional programming. Structural masonry design has been chosen for the model of developing such a system and part of the system has been employed in the teaching of masonry design.

Introduction

Structural design involves broad knowledge and years of experience in structural engineering. It is also a laborious process. For years, engineers have been trying to reduce the work in design by developing various design aids such as tables, charts and, more recently, computer methods. Meanwhile, they have longed for a more efficient way to reduce the time and effort in training new engineers in the classroom or on the job.

Conventional programs (e.g., computer programs for computational and drafting purposes) are widely used by

---

1 Graduate student, Civil Engineering Department, Lamar University, P.O. Box 10024, Beaumont, Texas 77710
2 Professor of Civil Engineering, Civil Engineering Department, Lamar University, P.O. Box 10024, Beaumont, Texas 77710

structural engineers and large amounts of work can now be done automatically; however, there is still time-consuming work such as checking for requirements within the complex building code standards. This work is not easily accomplished by conventional programs which are deficient in dealing with logic within text information (e.g., decision making).

The new concept of computer programming, "expert systems" was investigated (Mao, Luo & Rogers, 1988) and found to be a powerful tool for obtaining a higher level of automation in design and for educating new engineers. Expert systems can deal with problems involving knowledge, heuristics, and decision making that conventional programs can not. Unlike conventional programs, where the reasoning is not known to users (Figure 1 (a)), expert systems can show their line of reasoning; hence the entire design process represented by such programs is "transparent" to its users (Figure 1 (b)).

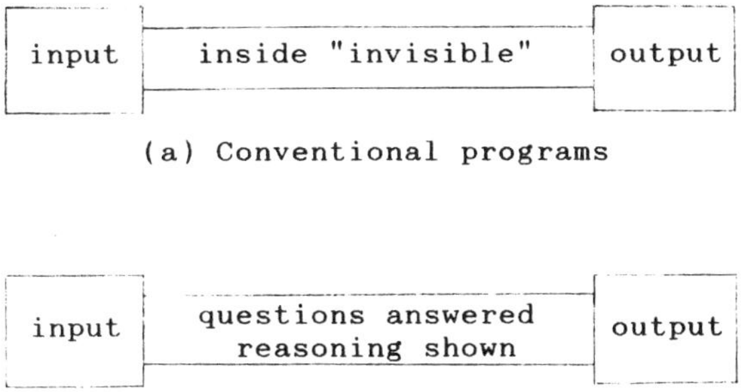

(b) Expert systems

Figure 1. "Visibility" to user

A differently structured computer program for structural design is being developed. Since it has the advantages of both expert systems and conventional computation techniques, and it is neither a simple expert system nor a conventional design program, it is called "Expert Design System" by the authors. This system not only performs engineering computations and makes decisions like a human expert but also teaches the user by guiding him or her to the appropriate solution and explaining the line of reasoning.

Structural masonry design has been chosen for the model of developing such a system. Part of the system has been tested in the teaching of a masonry design course and the result will be discussed later.

2. Expert Systems

Expert systems are defined as computer programs that use logical relationships to incorporate knowledge and expertise about a specific problem area to perform specialized tasks that typically require human judgment (Palmer, 1987). The most attractive characteristics of expert systems that conventional programs do not have can be summarized as the ability to explain and justify answers, the ability to grow gradually by adding new pieces of knowledge, the ability to deal with uncertain or incomplete information, and the use of verbal or symbolic encoding for knowledge (Parsaye and Chignell, 1988). Expert system applications are especially suitable for those types of problems that have complex fact structures, with large volumes of specific items of information. The knowledge in an expert system is stored in one or more knowledge bases.

3. The Expert Design System

Building designs are governed by standards through building codes which contain a large amount of knowledge. Great effort has been made to aid the usage of codes. This includes commentaries, handbooks, textbooks and design courses offered by universities. Yet, finding and interpreting the necessary information in a code is often difficult and time consuming, especially for students and new engineers. This difficulty can be solved by building an expert system for code checking.

In the teaching of engineering courses, after the basic concepts are taught, it is important for students to obtain a good sense of design and become familiar with the design codes. Students often have difficulty in putting pieces of knowledge together in design. This requires a lot of practice. On the other hand, the design practice is often limited by the time needed for the large amount of calculation and the effort of tutoring. There are many existing structural design programs and it is good for students to know how to use one or more of these programs. However, like using a black box, although speed is high, these programs never show the user what is going on inside these programs but only accept input and give output. It is now possible to build an expert system to help students to practice efficiently.

The intention of developing an "expert design system" is to replace the manual work in code checking and to reduce the time and effort in learning and teaching structural design. The knowledge in the design codes and the design procedures is represented by production rules, which is one of the expert system techniques, and stored in the knowledge bases.

The expert design system for masonry design consists of three parts: one knowledge base for code checking (called code checking KB); one knowledge base for design procedures (called design KB); and one computing tools bank. They are stored in a housing called "expert system shell". Each part of them will be described below.

The Code Checking KB

The code checking KB contains the knowledge in a building code such as standard loads including vertical, wind, and seismic loads, allowable stresses including axial, flexural, and shearing stresses, and other limitations such as height to thickness ratio, etc.

The code can be generally expressed with a "tree" structure. Figure 2 shows one of the branches in such a "tree". It is the branch for finding the allowable stresses. Each path in this tree represents a particular situation and each number in the circles represents a particular formula for the allowable stress. The system can well lead the user to travel along the appropriate path from the root to an end, i.e., to the appropriate formula to be used in this particular case.

This knowledge base can be used independently to 1) perform a pre-design code checking --provide the necessary data such as standard live loads, allowable stresses, steel requirements; 2) carry out a post-design code checking --check for the consistency of a finished design and the code specifications/standards; 3) serve as a subroutine to be called by the design KB or other design program for data.

The Design KB

The design KB contains design procedures for masonry members as well as for some typical masonry buildings. Data may be provided by the user or obtained from other parts of the program, for example, from the code checking KB.

The purpose is to let the user see the flow of design and participate in the whole design process while still making use of the high speed computation. In

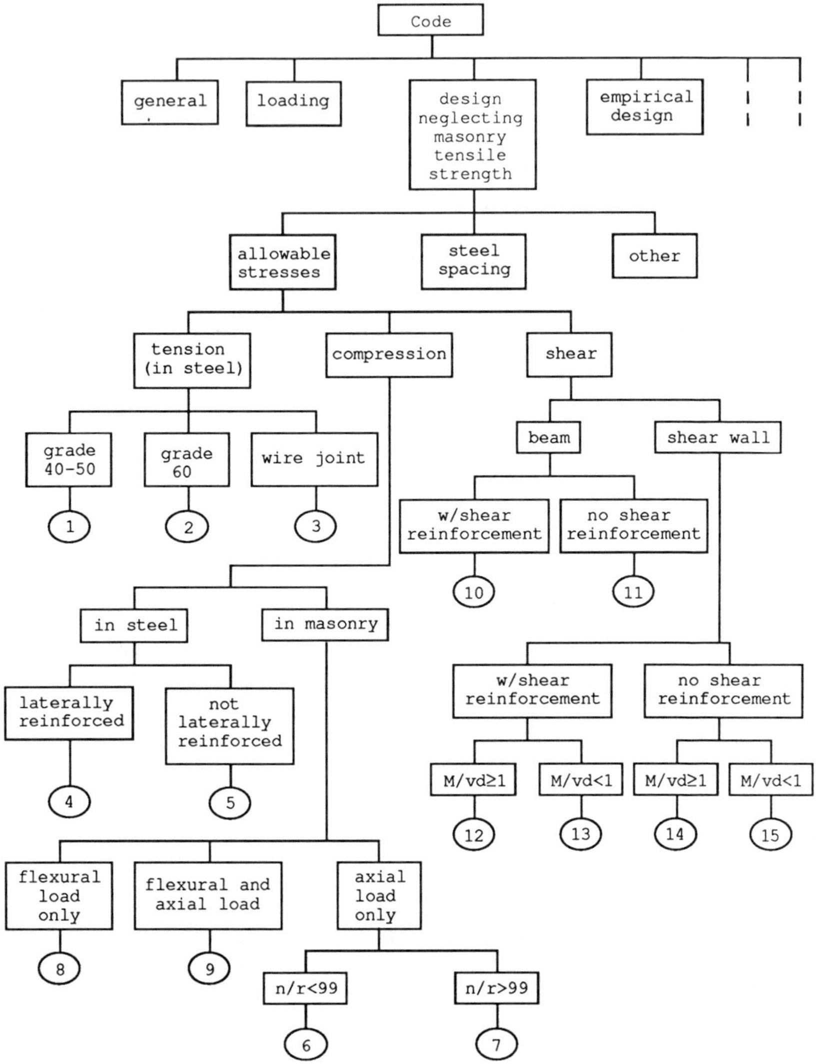

Figure 2. A branch of the "code checking tree"

order to simulate the structural design procedures and guide the user, the flow of design computation is split into sections and levels. At each point, the user can choose to determine a piece of information on his own knowledge or by assistance of the system. When a question is asked by the system, the user can specify the piece of information as "known" and then input the value (or fact), or "unknown"; then the system will direct him to find the information, either through the code checking KB, the computing bank, or by assigning a commonly used value.

Since the expert system can back track and can guide the user, and the computation tools bank can take care of the time consuming calculations, the time needed and effort of tutoring on design practice can be greatly reduced and students can become more familiar with the design process and the design codes.

The Computing Tools Bank

Although expert systems have great advantages, conventional programming methods are more efficient in taking care of the heavy computation. The computing tools bank contains many computational subroutines. When the interest is not in how a particular calculation is carried out but only in results, this calculation can be left to the conventional computation method. For example, in designing a masonry wall, the students are normally not interested in how to calculate the member forces (they are often assumed to have sufficient knowledge about this calculation before taking a design course); they may let the system get the member forces. The system will call the subroutine for member forces in the computing tools bank and these requested values will then be returned to the design KB.

4. Use of The System in Teaching

Part of the expert design system for masonry design, the code checking KB, was employed in teaching a masonry design course at Lamar University. It was found that the time and effort in both learning and teaching were reduced.

The Uniform Building Code (ICBO, 1988) and the ASCE-ACI Standards (ACI-ASCE, 1988) were used together with the text book (Schneider & Dickey, 1987) in teaching the structural masonry design course. Students usually have to turn pages back and forth trying to follow the details provided in design codes through tables, formulas and specifications. This is a time consuming and sometimes confusing process to many of them. Use of the system made it much easier to follow and to understand the design codes.

5. Conclusions

Expert systems are excellent in reasoning and are "transparent" to users but weak in computation; while conventional programs are efficient in computation. Consequently, the combination of these two techniques can be very powerful. This combination in the authors' recent work brought in a higher level of automation in design through replacing the manual code checking work with a computer method, and improvements in teaching through reducing the time and effort in both learning and teaching.

References

ACI-ASCE 530 (1988). *Building Code Requirements for Masonry Structures*, ASCE

International Conference of Building Officials (1988). *Uniform Building Code*, International Conference of Building Officials

Mao, Y., Luo, D. and Rogers, B. G. (1988). "Expert Systems in Structural Engineering", *Proceedings of the 4th Chinese Conference on Computer Applications in Architectural Engineering*, Guangzhou, China, pp.31-34

Palmer, R.N. (1987). Editorial, *J. of Computing in Civil Engineering*, Vol.1, No.4, Oct., 1987, pp.223-224

Parsaye, K. and Chignell, M. (1988). *Expert Systems for Experts*, John Wiley & Sons, Inc., NY

Porter, M.L. and Klingner, R.E. (Compiled) (1988). "Needed Masonry Research", *The Masonry Society Journal*, Vol.7, No.2, December, 1988

Composite Design Education in
Undergraduate Civil Engineering Curricula

C. Dale Buckner[1] M. ASCE, Lewis B. Burgett[2] M.ASCE,
and W. Samuel Easterling[3] A.M. ASCE

Abstract

A survey of educators was conducted during the Spring of 1989 to determine the course content, specifications used and the amount of coverage given to composite design in undergraduate steel design courses. Some highlights are summarized from the 122 responses received from civil engineering programs.

Introduction

Composite steel-concrete structural members are used in many buildings, bridges and other structures. Composite action is achieved by attaching shear connectors (e.g., welded headed studs) to the steel section and casting concrete around the connectors. After the concrete hardens, the shear connectors cause the steel section and a portion of the concrete to act together , i.e., compositely, to resist applied loads. The composite action produces a significant increase in strength and stiffness when compared to the noncomposite section.

The advantages of composite action have been recognized for many years, but advancement in this field has been particularly rapid in recent years. Extensive research, much of it conducted at Lehigh University during the past 20 years, has produced better understanding of behavior and resulted in more rational design criteria (Lu, et.al. 1987). With improvements in design and construction procedures, composite beams have become the best choice for virtually all steel framed buildings and bridges having concrete slabs.

Despite its practical importance, composite design receives little attention in undergraduate civil engineering curricula. Most civil engineering students are required to complete design courses in structural steel and reinforced concrete, but most graduate without a basic knowledge of the structural and economical advantages of composite design.

In 1986, the American Institute of Steel Construction (AISC)

[1]Prof., Dept. of Civ.Engr., Virginia Military Institute, Lexington, VA 24450.
[2]Assoc. Dir. of Education, Am. Inst. of Steel Const.,Inc., Lilburn, GA 30247.
[3]Asst.Prof., Via Dept. of Civ.Engr., VPISU, Blacksburg, VA 24061.

published a new design specification entitled "Load and Resistance Factor Design Specification for Steel Buildings". The load and resistance design (LRFD) approach is generally superior to the traditional approach, allowable stress design (ASD), in that it provides more consistent treatment of safety and more accurately predicts member behavior. The LRFD usually results in more economical designs, especially for composite beams where savings of up to 15% in steel weight have been demonstrated (Zahn 1987). Additionally, the LRFD approach significantly simplifies composite beam and column design.

The ASCE Committee on Composite Construction recognized several years ago that the LRFD specification would likely lead to revisions in many undergraduate steel design courses. This seemed an opportune time to examine course content and to encourage the introduction of composite design principles into undergraduate courses. A subcommittee on composite design education was appointed for these purposes.

The initial task of the subcommittee was to ascertain the status of composite design education in undergraduate programs. For this purpose, a questionnaire was prepared and mailed to 243 professors who teach undergraduate steel design courses in the United States. Sixty percent of the questionnaires were completed and returned, and the data from these responses are summarized and discussed in the following sections.

Survey Questionnaires

Information was solicited on the type and number of credit hours of undergraduate structural design and the number of credit hours of steel design electives available to undergraduates. Of particular interest were the content of undergraduate steel design courses, whether the steel courses were currently taught using LFRD or ASD, whether schools currently teaching LRFD planned to increase the LRFD coverage in the near future, and the amount of coverage given to composite design.

Questionnaires were mailed in late January, 1989 using a list compiled and furnished by AISC. A questionnaire was sent to only one professor from each department offering an undergraduate structural steel design course. Of 243 questionnaires mailed, 143 (60%) were completed and returned.

Responses were received from schools in 48 states with the geographical distribution roughly proportional to the distribution mailed. There was no apparent bias in the number of responses from schools that emphasize graduate study and research as compared to those that emphasize undergraduate education. Of the 143 responses, 21 (15%) were from architecture, construction or other programs and 122 (85%) were from civil engineering programs. A copy of the questionnaire and a list of respondents are included in the appendices of the subcommittee's final report (Buckner, et.al. 1989). In what follows, only the responses related to civil engineering programs will be discussed.

Survey Responses

Responses from CE programs were separated into two broad categories: 1) programs that require courses in both structural steel and reinforced concrete design and 2) those that do not require a course in structural steel design. For the 122 CE responses, 99 (81%) require both steel and concrete design, and the remaining 23 (19%) either did not require structural steel (13%); required either steel or concrete (3%); or required a combined course in steel and concrete (3%). The programs requiring both steel and concrete were further divided into those providing undergraduate electives beyond the first required course and those that did not. The responses were separated into these categories on the assumption that course content and choice of design approach (LRFD or ASD) might be affected by whether or not a course was required, and whether or not an elective was available. However, analysis of responses in these various categories indicated no significant differences in course content or design approach.

The percentage of programs emphasizing the new LRFD specification during the Spring of 1989 was somewhat surprising. Only 10% of the respondents indicated that LRFD was used exclusively in their introductory design course, and 18% indicated that both ASD and LRFD were used in the first course. The majority, 72%, were emphasizing the ASD specification. However, in at least 60% of those programs that had not changed to LRFD exclusively, the respondents indicated that they planned to revise their courses in the near future to incorporate more LRFD. Several respondents indicated that they were postponing the change until LRFD textbooks became available.

The average topic coverage for the required courses in structural steel is tabulated in Table 1. Similar tables for other categories of responses are included in the subcommittee's final report. Certain topics receive some coverage in virtually all first courses in steel design. For example, Columns 2 and 6 of Table 1 indicate that the following topics are included in more than 90% of the courses:

- o Introductory Material (mean coverage = 5%);
- o Material Properties (4%);
- o Tension Members (10%);
- o Compression Members (13%);
- o Laterally Braced Beams (10%);
- o Laterally Unbraced Beams (11%);
- o Combined Axial and Bending (12%);
- o Simple Connections (12%).

These 8 topics account for 77% of coverage in the first course.

The remaining 12 topics listed in Table 1 are less frequently covered. To assess their relative importance, a factor was calculated that weights the percentage (decimal) of courses covering the topic

Table 1. Topic Coverage in Required Structural Steel Design Courses in Undergraduate Civil Engineering Programs.

	For Programs Offering at Least One Elective Steel Design Course Beyond the Required First Course (56/93 or 60% of Programs)				For Programs that do not offer an Elective Design Course Beyond the Required First Course (37/93 or 40% of Programs)			
	% Providing Some Coverage	Extent of Coverage			% Providing Some Coverage	Extent of Coverage		
(1)	(2)	Range % (3)	Mean % (4)	σ_{n-1} % (5)	(6)	Range % (7)	Mean % (8)	σ_{n-1} % (9)
1. Introductory Material	95%	2-to-15	5	±2.8	100%	1-to-12	5	±2.6
2. Material Properties	98%	1-to-7	4	±1.7	97%	2-to-10	4	±1.6
3. Tension Members	100%	3-to-25	11	±4.3	100%	4-to-21	9	±3.5
4. Compression Members	100%	5-to-30	14	±5.2	100%	4-to-25	12	±4.6
5. Laterally Supported Beams	98%	2-to-20	10	±3.6	100%	2-to-20	9	±3.8
6. Laterally Unbraced Beams	100%	4-to-20	11	±4.3	97%	3-to-20	10	±3.6
7. Torsion of Open Sections	23%	1-to-8	4	±2.1	49%	1-to-10	4	±2.6
8. Combined Axial and Bending	93%	3-to-25	12	±4.3	97%	5-to-20	11	±3.5
9. Multistory Buildings	41%	1-to-11	6	±3.2	38%	2-to-11	6	±4.1
10. Mixed Construction	7%	2-to-6	4	±2.1	8%	2-to-13	8	±5.0
11. Continuous Beams and Frames (Elastic)	70%	2-to-20	7	±3.6	73%	2-to-30	6	±5.2
12. Continuous Beams and Frames (Plastic)	32%	1-to-13	6	±3.0	43%	2-to-10	6	±2.8
13. Simple Connections	95%	5-to-21	11	±5.1	100%	5-to-29	14	±6.6
14. Rigid Connections	61%	2-to-30	7	±5.1	76%	5-to-21	8	±4.6
15. Lightgage Cold-formed Members	2%	1	1	--	3%	2	2	--
16. Plate Girders (Buildings)	20%	1-to-15	6	±4.2	35%	2-to-14	9	±3.7
17. Steel Bridges	16%	1-to-11	5	±2.7	19%	1-to-10	7	±3.6
18. Composite Steel-Concrete Beams	25%	1-to-11	5	±2.9	27%	3-to-10	7	±2.0
19. Composite Steel-Concrete Columns	9%	1-to-5	2	±1.7	11%	2-to-8	5	±3.2
20. Others (e.g., Fatigue; Fracture; Base Plates; Open-Web Joists; Field Trips; Design Projects; etc.	38%	3-to-25	11	±6.2	38%	4-to-35	13	±8.7

and the mean percentage of time devoted to the topic. For example, the relative importance, RI, of rigid connections was calculated to be

$$RI = (.60 \times .61 \times .07) + (.40 \times .76 \times .08) = 0.050, \qquad (1)$$

in which 61% of 60% of CE respondents devoted an average 7% of the course to the topic and 76% of the remaining 40% of respondents devoted 8% to the topic. On the basis of this factor, the remaining 12 topics rank:

- o Rigid Connections ------------------------- RI = 0.050;
- o Cont. Beams and Frames (Elastic Design) --- RI = 0.047;
- o Others (Fatigue, Fracture, etc.)----------- RI = 0.045;
- o Multistory Buildings---------------------- RI = 0.024;
- o Cont. Beams and Frames (Plastic Design)---- RI = 0.022;
- o Plate Girders (Buildings) ----------------- RI = 0.020;
- o <u>Composite Beams</u> ------------------------ RI = 0.015;
- o Torsion of Open Sections ----------------- RI = 0.013;
- o Plate Girders (Bridges) ------------------ RI = 0.010;
- o <u>Mixed Construction</u> --------------------- RI = 0.004;
- o <u>Composite Columns</u> ---------------------- RI = 0.003;
- o Lightgage, Cold-formed Members ----------- RI = 0.000.

It is evident that topics related to composite design receive little attention in an introductory steel design course, accounting -- on average -- for less than 2% of the topic coverage in the CE programs surveyed.

When asked their opinion as to the best place to introduce design fundamentals for composite steel-concrete beams, 26% favored a first course in steel design; 9% indicated a first course in reinforced concrete design, and 3% indicated mechanics of materials courses. The majority of respondents, 62%, felt that the topic should be introduced in a subsequent course following courses in both steel and concrete design.

In 60% of the CE programs responding, at least one undergraduate steel elective is available beyond the introductory course. An analysis of the topic coverage reported for elective courses indicates that composite design is covered in about two-thirds of these courses. When topics in the elective courses are weighted as indicated by Eqn 1, composite beams and columns account for approximately 10% of the course coverage in the courses.

<u>Summary</u>

Despite its practical importance, composite design receives little attention in introductory structural steel design courses. Thus, many civil engineering students graduate without an awareness of the types and benefits of composite steel-concrete members.

The survey summarized in this paper indicates that only about 10% of the structural steel design courses taught in civil engineering were emphasizing the new LRFD specification during the Spring of 1989. However, in those programs that had not changed to the LRFD specification by the Spring of 1989, the majority (more than 60%), indicate that they plan to change in the near future to incorporate more LRFD into their courses.

Composite design by LRFD is simpler, more rational and results in material cost savings when compared to designs made using earlier specifications. Simplification in design criteria makes it easier to provide some coverage of composite design in an introductory steel design course. It is the hope of the writers that professors who plan to update their structural steel design courses will consider the importance of composite design, and include an introduction to this topic in their introductory course.

Acknowledgements

The survey described in this paper was conducted by a subcommittee of the ASCE Committee on Composite Construction, which is chaired by Ivan M. Viest. The mailing list for the survey was furnished by the American Institute of Steel Construction (AISC).

References

American Institute of Steel Construction, Inc., Manual of Steel Construction -- Load and Resistance Factor Design, 1st Edt., Chicago,IL, 1986.

Buckner, C.D., Burgett, L.B., and Pugh, T., "Educator Survey on Composite Steel-concrete Construction," Final Report, Committee on Composite Construction, American Society of Civil Engineers, June, 1989, 12pp.

Lu, L.W., Slutter, R.G., and Yen, B.T., "Research on Composite Structures for Buildings and Bridge Applications," Proceedings of the U.S.-Japan Seminar on Composite and Mixed Structural Systems, Tokyo, Japan, 1978, pp 3-18.

Zahn, M.C., "The Economies of LRFD in Composite Floor Beams," Engineering Journal, American Institute of Steel Construction, Inc.,Vol. 2, No. 2, 2nd Quarter, 1987, pp. 87-92.

Modern Methods and Technology for Teaching Structures

Dr. Hany J. Farran Ph.D., P.E.[1]

Abstract

The purpose of this paper is to highlight the significance of modern teaching methods and the impact of advanced technology in Civil Engineering education. The discussion is limited to the field of structural analysis and design. In view of the fact that structural engineering is a very competitive and dynamic area, the author foresees the need for upgraded programs of study across all American Universities. During a recent conference that was held during last May in Japan, the author had a chance to visit many research facilities, as well as the sites of ongoing projects across the country. It is the belief of the author that the Japanese construction enterprises will be capable of attracting large scale projects across the world in the near future. It is also clear to see that the Japanese have the intention of becoming the experimental test facility of the pacific rim nations in the future. Such great achievements are direct indications of a dynamic academic environment and a supportive industry. It is imperative to note that if the United States is to maintain the leading edge in structural civil engineering projects on an international level, a change is to occur in our teaching methods and programs. The author also foresees the importance of developing a national policy toward quality education rather than generating excessive full-time equivalents across our Universities and Colleges.

Introduction

In the past 30 years, the engineering profession as a whole has witnessed a dramatic development in analytical methods and numerical simulation techniques. The introduction of digital computers has enhanced the engineering profession. In the field of structural analysis and design, complex problems can now be solved with ease. In spite of all these noteworthy developments, a need exists to introduce the fundamental concepts of modern design into all our engineering programs. Our engineers need to become more aware of modern structural materials and

---

[1]Professor & Consulting Engineer, Civil Engineering Department, California State Polytechnic University, Pomona, CA 91768–4064.

techniques. We need to emphasize value engineering and train our engineers to design for quality and economy. These issues are of paramount importance for us to dominate future markets in the construction industry. A long-term strategy and vision are in need across all engineering programs of study. We also need to be more selective and demanding as far as students are concerned.

Modern Technology

In addition to emphasizing the fundamental concepts of conventional materials in structure design, like structural steel and reinforced concrete, we need to teach our students prestressed concrete and segmental erection techniques. The academic community is in agreement that a Master's degree is needed to practice structural engineering. The author feels that we need to teach these advanced topics at the undergraduate level as required courses for students declaring a major in structures. A closer look at the typical flow-sheet of a Civil Engineering program reveals that there are so many courses that it is very difficult to add any more units and get University approval of such a modification. What needs to be done is one of two things: Either make the program a 5-year plan of study, or require the Master's degree for acquiring the status of a registered professional engineer. The point to be made here is that courses like Finite Elements, Matrix Methods, Prestressed Concrete, and Segmental Bridges should be taught as required courses at the undergraduate level. Why leave the burden of education on industry when we have the time to teach the student these advanced topics. In brief, we advocate an upgraded program at the cost of additional education time. This should be done without sacrificing essential general education course requirements. We need to advocate the concepts of a well rounded engineer but also enhance his technical knowledge base.

A need exists to familiarize our students with modern experimental structural mechanics as well as modern testing equipment. Most testing facilities today have data acquisition systems and computers to manage and display the large data arrays. A serious shortage of such modern equipment exists at most undergraduate departments. On occasions we find that such sophisticated equipment is reserved for graduate research and development work. The timely training of students to use such equipment is a key element to success and productivity. It is very important that engineering students learn prior to going to industry. European and Japanese Universities implement this doctrine by advocating an extensive teaching program at the undergraduate level.

A Return to Classical Fundamentals

It is the opinion of the author that a need exists across our Universities to advocate a fundamental understanding of classical theories.

The introduction of such basic scientific principles in our lectures and presentations is of utmost value. The idea of graduating a large number of engineers to meet current market demands is a questionable concept and is short-sighted. The engineering academic community in cooperation with the practicing professionals in industry needs to draft a proposal to upgrade engineering education along lines similar to what the American Medical Association has. The purpose of such an effort is to upgrade the social status of the engineer and the engineering educator. Practically, it is very important in today's standards. It is the author's opinion also that we need to look at University teaching not only as a business. We need to have the will and determination to invest in innovative teaching methods and to offer advanced courses even though the student enrollment in such classes is small.

The universities in this country will be charged with the task of producing more doctoral candidates in the future. We ought to encourage an international exchange of scholars in the various engineering programs.

Professional Activities

Faculty members in various universities in the United States compliment their salaries by participating in consulting engineering work. Such professional involvement is recommended highly by school administrators but there is a time limit on how much consulting work a faculty member should undertake. Consulting at no time should interfere with the teaching duties of a professor. If it does then a faculty member will not perform very well in the classroom. A similar situation arises at major universities where professors are assigned a research related project. Adequate release time should be allocated to allow for proper teaching. Teaching should not be permitted for graduate assistants, such a procedure will bring about eventual decrease of quality and respect for a program. Consulting engineering is good for a faculty member because in performing such a task, exposure to current practices and methods is only anticipated. A faculty member can make a lecture more interesting by sharing some of his recent experiences with his students.

A faculty member should belong to various professional societies serve to promote the knowledge and keep the faculty member abreast of recent developments. One of the very important criterion in promoting faculty is the fact that they themselves are registered professional engineers. A faculty member is required by most major universities to be a licensed professional engineer.

Institutional Related Activities

Involvement with college committees is very important for a faculty member. Having a chance to serve on various school committees gives the engineer a chance of seeing politics at work. We, as engineers, have been trained and taught not to meddle with politics. We really should to a certain extent because refusal to participate in politics can be evidence of not caring. Traditionally speaking, engineers have not participated in the political machine. Engineers should participate in efforts intended to promote their welfare. There are many school committees that carry a certain amount of power that can change institutional policies. We, as educators, should participate in such committees if we are going to change to the better. The drafting of admission requirements in the task assigned to one school committee. If we are interested in promoting quality education, service on such a committee can give us the chance to change certain questionable admission policies. The important thing to note is that we really should participate in such committees for reforming admission requirements. and making our students a select group.

Being a sponsor of the student chapter of professional societies is another very important aspect of faculty evaluation. For example the American Society of Civil Engineers Student Chapter contributes significantly in promoting knowledge and team spirit to students. Being a sponsor of such an activity is time consuming and demanding. Scheduling events alone is a full-time job.

Faculty Sabbatical

A faculty member should demonstrate willingness to go on a sabbatical leave to earn a higher degree or participate in some graduate program of relevance. I feel such a sabbatical is justified once every 5 years. Technology is advancing at a very fast pace, if we do not keep track we shall become obsolete. We have to grow out of this dilemma of teaching excellence being the only criterion for success as an educator. We have to accept the renewal of the mind at any cost. Financially speaking, sabbaticals might present the faculty member with problems in budgeting. Here is where the institution should help in every way possible. For faculty members who possess the doctorate, a post-doctoral fellowship or a visiting professor position at another university might prove very beneficial and rewarding. We should be different from regular teachers. Faculty members should continuously try to better their educational background. Having tenure should not be portrayed as retirement. A lot of faculty members think that the only thing they have to do is teach two lectures a day. Such an attitude can spell professional death, stagnation and ineffectiveness regardless of how outstanding they may be as teachers. Tenure requirements need to be revised and the philosophy of tenure I think is questionable. We should look at this issue in a manner similar to private industry. A productive engineer stays because he promotes company objectives. The public might have more respect to teachers if we would revise our tenure criterion. Tenure, on many occasions, is viewed as semi-retirement. A progressive productive faculty member should have no concern for tenure because there is always a place for a good university professor. Participation in short courses and seminars is recommended very highly for faculty members. We should portray an interest in being active members not only in being a part of the audience.

Note on Research

A university professor has traditionally been considered a teacher as well as a researcher. We really cannot separate research from teaching. Teaching for a certain period of time can make the faculty member an expert in a particular field. The expansion of that field of knowledge is our responsibility as educators. Time will show us more effective ways for teaching not to mention the expansion of the technical boundary of our field of expertise. A faculty member should not be pushed into conducting research and publishing papers. Such a practice will only flood the journals with low quality material. Good research is a timely effort based on studying and searching for the genuine motive of advancing the state of the art. I think that major universities should also have the same philosophy on research. But in the case of institutions not participating in funded research, the publication of papers should display taste and originality and a documented effort for promoting the state-of-the-art of the engineering profession. The publication of a book by a faculty member is a very important achievement. Such a publication leaves students and engineers a record of knowledge gained through years of experience and teaching engineering. Such a publication is definitely an asset to the university or institution where the professor is a faculty member. Universities should provide faculty with time and the resources to bring about such an original creation.

Teaching and research are not separable. Research and eventual publication should be the outcome of years spent in teaching. People can improve the state-of-the-art of a science only if they understand the essentials of that particular science. The difference in quality between original research and duplicated second order research is only time. Similar analogies can be drawn to various industrial products and advances in technology. In the case of institutions offering a professionally oriented engineering program, faculty should demonstrate a certain level of competence substantiated by a certain number of years of experience. Professional experience should be comparable to research experience at colleges not participating in funded research. Such institutions might offer programs leading to a Master of Engineering Degree. Such a program should have some affiliation with industry, similar to an internship. Engineering schools offering a practice oriented program should have adequate laboratories and experimental support equipment. Faculty members at non-research oriented colleges can work on some creative projects if experimental equipment is available to them. Faculty can participate in creative work to get a relief from the monotony of undergraduate teaching.

Conclusion

Technology has advanced dramatically in the past 30 years. Structural engineering is a very competitive field. It is our duty as engineering educators to keep in mind the importance of staying abreast of all developments in our areas of expertise. It is of utmost importance that we observe developments around us and study their impact on our economy in the future. Having the privilege of living in a country with abundant natural resources and human talent, we are optimistic that the years to come will serve as a platform for us to establish technological superiority. We believe that Universities will play a leading role in this task in the future.

Undergraduate Environmental and Civil Engineering Degrees within One Department

Neil J. Hutzler[1]

Abstract

Few civil engineering departments administer both civil engineering with environmental engineering emphasis and environmental engineering degrees. One such program has been successful at Michigan Technological University. Not only has overall enrollment increased, but the two programs have been shown to complement one another.

Introduction

Most environmental engineers have graduated from traditional civil engineering programs. As environmental contamination problems have become more widespread and more complex, however, the typical civil engineering student is not well-prepared to deal with the chemical and biological aspects of pollution control because of the heavy emphasis on physical sciences, engineering mechanics, and structural engineering in traditional curricula. Recognizing the need for a broader scientific and engineering base, the student at most universities has two choices, either continue his or her education in pursuit of an advanced degree or find a school that offers a baccalaureate degree in environmental engineering.

As of 1988, there are only nine ABET-accredited programs strictly in environmental engineering, two in civil and environmental engineering, and one in environmental resources engineering (ABET, 1988). Over thirty universities, including Michigan Technological University, have civil engineering that allow some specialization in environmental engineering, and some universities teach environmental engineering as a part of some other curriculum such as chemical engineering. Over one hundred universities offer M.S. and Ph.D. degrees related to environmental science and engineering (AEEP, 1989).

There has been a reluctance of some civil engineering departments to develop special programs in environmental engineering for

----

[1] Associate Professor of Civil and Environmental Engineering, Michigan Technological University, Houghton, Michigan 49931

perhaps the fear that they will dilute the existing programs or because they fear it will decrease the number of students entering graduate programs. This paper describes a program where a B.S. in environmental engineering and B.S. civil engineering are administered in the same department.

Environmental Engineering

According to the "Guide for Environmental Engineering Visitors on Accreditation Teams" published by the Accreditation Board for Engineering and Technology (ABET), Environmental Engineering is defined as: "...that branch of engineering which is concerned with 1) the protection of human populations from the effects of adverse environmental factors, 2) the protection of the environment, both local and global, from the potentially deleterious effects of human activities, and 3) the improvement of environmental quality for man's health and well-being." Major subject areas within environmental engineering are: water quality engineering, environmental health engineering, air pollution control engineering, and solid wastes engineering. All of these areas are covered in both the existing degree programs at Michigan Tech. However, the primary emphasis is on water quality engineering and treatment of hazardous wastes.

Some universities have created environmental engineering programs within established departments to conserve resources and to develop data on student interest and demand for graduates. Michigan Tech has followed this course with the Bachelor of Science in Environmental Engineering (BSEnE) program within the Department of Civil and Environmental Engineering. There has been an increasing demand for the degree program and graduates have had little trouble finding employment of their choice in spite of the fact that little recruiting has been done to attract students.

Evolution of the B.S. in Environmental Engineering

In 1972, the College of Engineering developed a Bachelor of Science in Engineering (BSE) degree program for the purpose of enabling students with special interests not emphasized in any of the established curricula at Michigan Tech to develop a program that matched those individual interests. This flexibility has been achieved without sacrificing the rigor of the engineering degree. This Program was accredited in 1974 and reaccredited in 1980. At each of these accreditations, most of the graduates from the BSE program selected the environmental engineering curriculum.

In the BSE program, an individualized program is developed when a student requests it and when, in the judgment of the program coordinator (the associate dean of engineering or the assistant to the dean of engineering), existing courses and faculty make it possible to do so. When it appears that such a request is meritorious and feasible, a professional interest group of at least two faculty are appointed. The group meets with the student and develops a suitable course of study within the framework of the BSE program. The program must be approved by the College of Engineering.

One of the first curricula to evolve out of the BSE program was environmental engineering. It arose out of a need and desire on the part of some students to specialize in environmental sciences and engineering by taking coursework that did not readily fit into the traditional civil, chemical, or mechanical engineering curricula. The BSE (environmental engineering) program attracted high quality, highly motivated students with diverse backgrounds. Over 60 BSE students graduated from the environmental engineering program. Based on graduates that have been tracked, about one third work for consulting firms, another third work for some level of government, one fifth work in industry, and the remainder have other occupations such as insurance underwriter, research engineer, and plant engineer. In addition, at least one fifth have gone on for advanced degrees. Only one student has ever expressed any regrets about getting the BSE instead of a BSCE.

Recognizing the quality of the graduates of the BSE program and their ability to find the work of their choice, the University had the degree program formally accredited in 1986. Since most of the faculty involved with the environmental engineering program held appointments in the Civil Engineering Department and because of a need to control costs, the BSEnE degree is administered through the Civil Engineering Department in much the same way as the previous BSE degree was. The Department formally changed its name to the Department of Civil and Environmental Engineering to acknowledge the two parallel degree programs.

Related Programs

The Michigan Tech program is not unique. The ABET Roster of Accredited Programs (1988) lists accredited BSEnE degree programs at the following institutions:

California Polytechnic State Univ, San Luis Obispo
Florida Institute of Technology
Michigan Technological University
Montana College of Mineral Science & Technology
Northwestern University
Pennsylvania State University
Renselear Polytechnic Institute
University of Central Florida
University of Florida

Of these, four (MTU, Northwestern, Penn State, and Central Florida) administer the BSEnE with their civil engineering department.

BSEnE Curriculum and Comparison to the BSCE

Table 1 compares the BSEnE and BSCE/environmental engineering curricula as offered at Michigan Tech. Examination of these programs shows a strong foundation in mathematics and physical science which is embellished by engineering sciences and capped by design and analysis courses in civil and environmental engineering. A substantial content of general education as well as written and oral communications courses is also reflected in the curriculum. Both programs are built upon a basic civil engineering program. The BSEnE program drops all or a portion of the courses dealing with structures, transportation, construction, and soil mechanics and adds courses in the chemical and biological sciences. There is much more flexibility in selecting additional coursework in the earth sciences, water resources, chemical engineering, and chemistry. The BSEnE program is similar to one proposed by Gloyna and Sorber (1986).

Impact on Civil Engineering Department and the University

The impact on the University was positive in several ways. First, there was a net increase in students as shown in Table 2. The program attracted students who were not interested in the traditional civil engineering curriculum but were interested in environmental protection and water quality. The program grew from the approximately 6 graduates per year in the BSE program to over 20 graduates per year in the BSEnE program. Of these, it is estimated that less than half were drawn from those who might have been BSCE.

Special efforts have been made to avoid confusion on the part of students between the BSEnE Degree and the environmental specialty within the BSCE curriculum. It is carefully explained to students that one program is not better than the other but that they are different. The BSEnE graduate is more competent in environmental engineering but will not have exposure to all areas of civil engineering. The choice depends on the students' interests.

Administration of Program

The environmental engineering faculty bear primary responsibility for the BSEnE degree within established ABET guidelines in the same way that the Department Curriculum Committee is responsible for the BSCE degree. At least one faculty member not associated with environmental engineering is a member of the Environmental Engineering Curriculum Committee. The Chairman of the Department of Civil and Environmental Engineering reviews both curricula and has to sign the degree schedules of all graduating students.

Table 1. Comparison of BSEnE and BSCE Curricula.

Course Classification	BSEnE Credits[1]	BSCE/EnE Credits
<u>Math and Computer Science</u>		
Calculus & Anal. Geometry	20	20
Diff. Equations	3	3
Probability & Statistics	3	3
Computer Science	3	3
<u>Basic Science</u>		
General Chemistry	10-14	10-14
Organic Chemistry	3-6	0-3
Biological Sciences	10-14	0-8
General Physics	10	10-15
Earth Sciences	0-7	3-6
Physical Chemistry	0-6	0
<u>Engineering Science and Design</u>		
Des. Geometry, CAD	2-5	5
Surveying	4	7
Engineering Mechanics	12-14	14
Thermodynamics	3	3
Environmental Engineering	25-28	17
Water Resources	8-14	8
Structural Engineering	0-4	12
Soil Mechanics	0-4	7
Transportation	0	4
Electrical Engineering	4	4
Chemical Engineering	0-6	0
Engineering Economy	0	3
Construction	0	4
<u>Communication Skills</u>		
Freshman English	6	6
Oral Communication	3	3
Technical Writing	3	3
<u>Humanities and Social Sciences</u>		
Government	3	3
Economics	0	3
Humanities	9-12	9-12
Social Sciences	9-12	6-9
<u>Other</u>		
Thematic Courses	6	6
Physical Education	4	4
Free Electives	10	3-4

[1] Quarter Credits

Table 2. Civil and Environmental Engineering Enrollment Statistics.

Academic Year	BSCE Enrollment	BSEnE Enrollment
1985-86	407	--
1986-87	350	17
1987-88	364	32
1988-89	395	54
Fall 89	408	83

References

Accreditation Board for Engineering, 1988. "Annual Report", New York.

Association of Environmental Engineering Professors, 1989. "Register of Environmental Engineering Graduate Programs".

Gloyna, E.F., and C.A. Sorber, 1986. "A Look to the Future: Environmental Education," in "Proceedings, Fifth Conference on Environmental Engineering Education", C.R. Baillod, ed., American Academy of Environmental Engineers, Association of Environmental Engineering Professors, Houghton, MI.

Environmental Engineering and Science

Graduate Student Enrollment

Bill T. Ray[1], M. ASCE

Abstract

This report summarizes the graduate student enrollment in the area of environmental engineering and science for the academic year 1988-89. It was compiled for the Association of Environmental Engineering Professors (AEEP) as part of the 1988-1989 curriculum survey. Contained within is information on the number of students, their areas of research, and their funding status. The survey was sent only to universities which have environmental engineering graduate programs.

Introduction

The intent of this report is to indicate the number of students presently seeking advanced degrees in the field of environmental engineering, to present a summary of the research areas of graduate students enrolled in environmental engineering curricula, to estimate the fraction of students receiving financial support, and to estimate the number of individuals entering the work force in the environmental field.

Data Collection

The survey was mailed to approximately 150 universities in the U.S. and Canada during the month of April, 1989. Universities were selected on the basis of membership in AEEP. Names and addresses were obtained from the AEEP Register. There were 75 responses, or 50 per cent, by the time this report was written.

The survey was sent only to universities which have environmental engineering graduate programs. However, programs such as environmental management, public health, and radiological health were included by the respondents.

---

[1]Assistant Professor, Department of Civil Engineering and Mechanics, Southern Illinois University at Carbondale, Carbondale, IL 62901

When such areas were included, they were tallied in the "Other" category. Since the survey was directed at engineering programs with AEEP membership, environmental programs associated only with science or agriculture academic units were excluded.

Results

This year's survey identified 2570 graduate students: 1840 MS and 730 PhD students. Of the 2570 students, 1648 were funded. There were 739 MS graduates and 216 PhD graduates. The data are summarized in Table 1, which provides a listing of graduate enrollment by degree sought and research interest. Table 1 also includes data for PhD and MS degrees granted by research area.

Table I. Summary of Research Interests.

Area	Students			Graduates		
	MS	PhD	Total	MS	PhD	Total
Water/Wastewater	844	330	1174	339	81	420
Air Pollution	101	37	138	28	12	40
Solid & Hazardous	216	64	280	89	28	117
Water Research	322	149	471	112	37	149
Envr Sci & Mgmt	160	86	246	63	19	82
Other	197	64	261	108	39	147
Total	1840	730	2570	739	216	955

Student Research Areas

A breakdown of the research or specialty areas for all graduate students is shown in Figure 1. This chart includes both masters and doctoral students. The breakdown between areas shows that approximately 46 per cent of the graduate students are enrolled in the area of water and wastewater. The second largest area of enrollment is water resources with about 18 per cent, followed by solid and hazardous waste with 11 per cent. The environmental science and management area represents about 10 per cent of the graduate students, and air pollution studies represents about 5 per cent. The "Other" classification represents the remaining 10 per cent of the students. There is little change in this year's breakdown when compared to that for previous years.

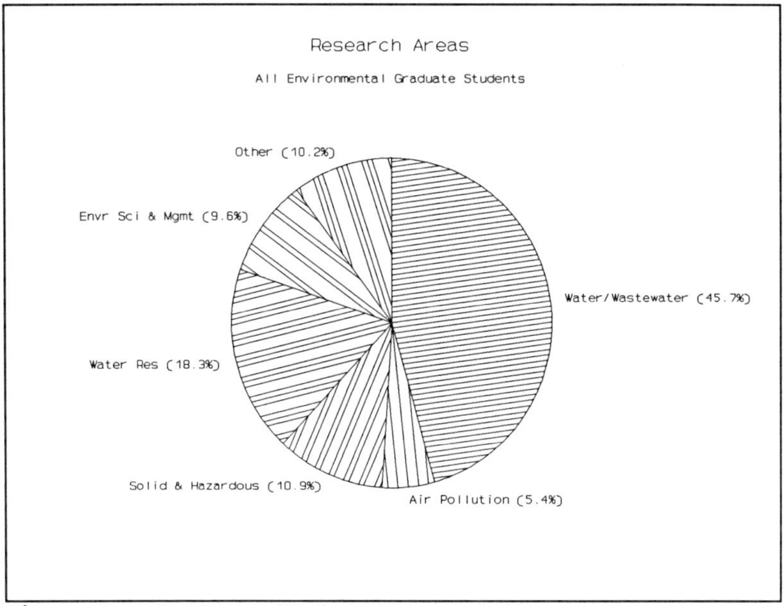

Figure 1. Graduate Student Research Areas.

Graduate Research Areas

There were 739 MS degrees granted and 216 PhD degrees granted, compared to 508 MS and 117 PhD degrees granted for last year's survey. The research areas are shown in Figure 2. As with currently enrolled students, the water and wastewater category represents the largest segment, followed by water resources. As would be expected, the trends in research areas for graduates closely parallels that of currently enrolled students.

Funding

Approximately 64 per cent of the 2500 graduate students were funded during the 1988-89 academic year. 58 per cent of the MS students were funded and 80 per cent of the PhD students were funded.

Yearly Trends

The change in enrollment over the past several years is shown in Table 2. The data for all previous years is for

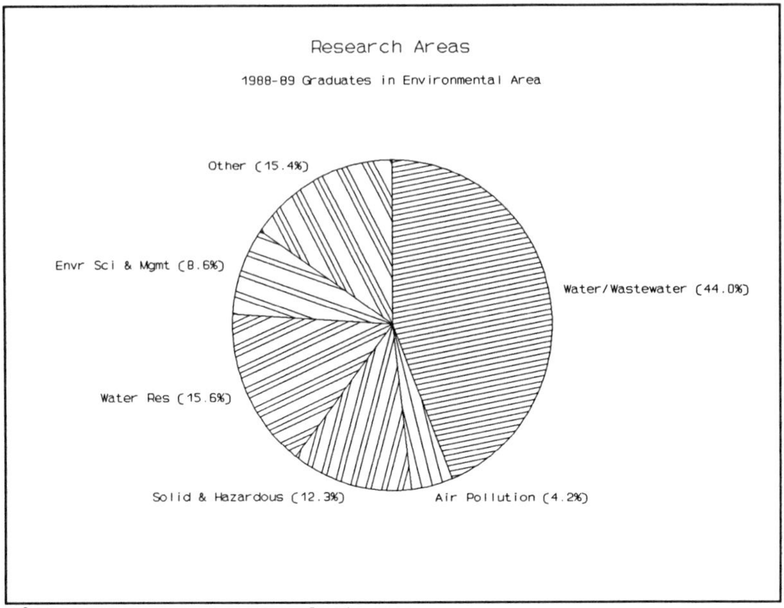

Figure 2. 1988-89 Graduates.

full-time graduate students. However, the data for 1988 is for funded and unfunded graduate students as opposed to full-time and part-time graduate students. Considering that last year's survey had 84 responses and this year's survey had 75, there does not appear to be a significant increase or decrease in the total enrollment. The total numbers presented here are probably more a function of the number of responses than of any yearly trend in total enrollment. It is, however, possible to detect trends within the different areas. For instance, there is an unmistakable and consistent increase in the percentage of students studying solid and hazardous waste disposal. This area has grown from only 2.7 per cent of graduate enrollment in 1982 to 15.2 per cent for 1988. About half of the graduate students accounted for in the survey have consistently remained in the water and wastewater treatment specialization. Air pollution also appears to be holding steady with about 5 to 7 per cent of the students. The enrollment distribution is shown for the past several years in Figure 3.

Table 2. SUMMARY OF GRADUATE STUDENT ENROLLMENT

YEAR	1982	1983	1984	1985	1986	1987	1988
NUMBER OF RESPONSES	88	67	59	85	62	84	75
Water and Wastewater	935	733	631	770	743	931	726
Air Pollution	81	71	100	127	75	106	89
Solid & Hazardous Waste	46	42	76	126	112	185	204
Water Resources	436	225	219	418	261	426	302
Envr. Science & Management	594	226	344	543	284	395	152
Other	58	206	162	181	61	102	175
Total Full Time Enrollment	2150	1503	1532	2165	1536	2145	1648[1]
Part Time Enrollment	540	398	391	397	405	542	922[2]
TOTAL ENROLLMENT	2690	1901	1923	2562	1441	2687	2570

[1]Funded graduate students.
[2]Unfunded graduate students.

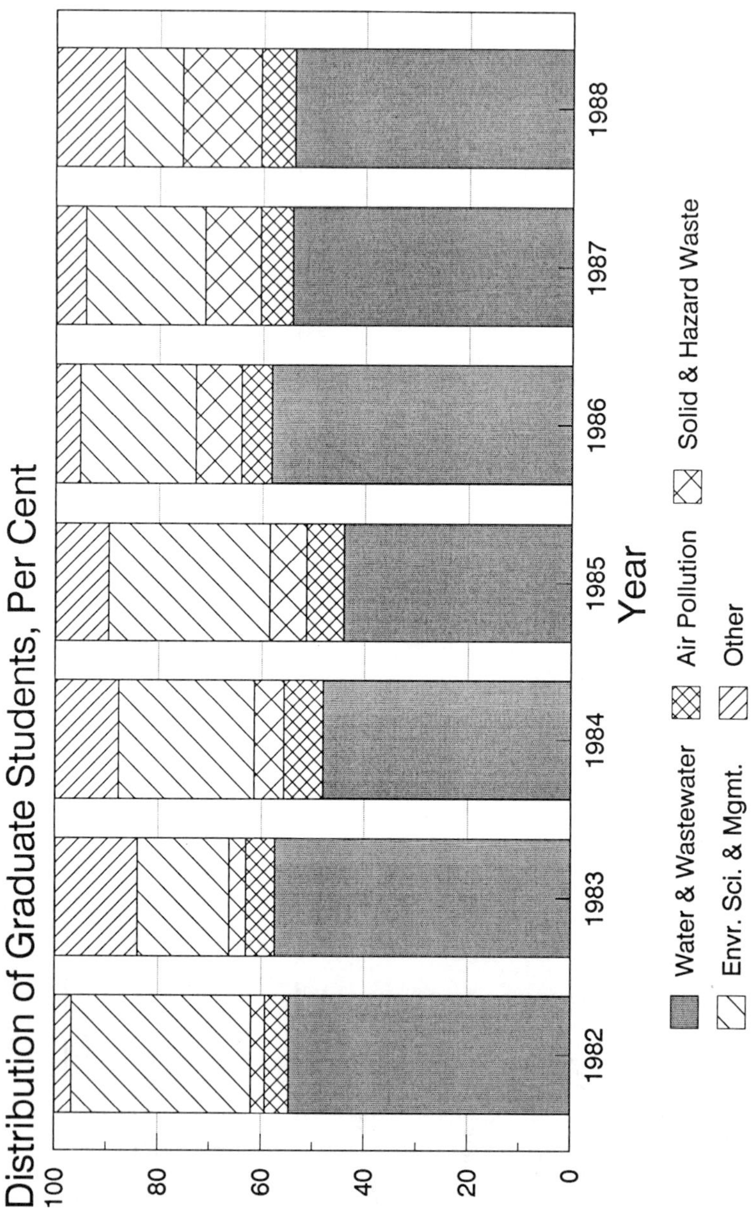
Distribution of Graduate Students, Per Cent
100
80
60
40
20
0
1982
1983
1984
1985
1986
1987
1988
Year
Water & Wastewater
Air Pollution
Solid & Hazard Waste
Envr. Sci. & Mgmt.
Other

Summary

Graduate enrollment in the environmental field appears to be holding relatively steady although it is difficult to determinetotal enrollment numbers due to the variability in the number of responses over the years. Trends in areas of study are apparent though. The most significant trend is that the area of solid and hazardous waste has been steadily climbing over the past seven years. Also, the area of water and wastewater treatment is the most common field of study.

An interesting question, and one which this survey does not address, is whether current environmental engineering education is supplying engineers with expertise in the needed areas. As an example, the solid and hazardous waste area only comprises 15 per cent of graduate enrollment, but it may represent a much higher market area for graduates. This could only be addressed by surveying not only graduate programs, but prospective employers.

AEEP/AAEE ENVIRONMENTAL ENGINEERING EDUCATION CONFERENCE SERIES

C. Robert Baillod[1]

Over the past three decades, the Association of Environmental Engineering Professors (AEEP) and the American Academy of Environmental Engineers (AAEE) have sponsored a series of conferences on environmental engineering education. The purpose of this paper is to present a brief overview of the conferences and to summarize the positions reached at the most recent such conference.

The First National Conference (1), held in 1960 at Harvard University was a landmark event which initiated the drive toward the teaching of unified concepts in environmental engineering and science.

The Second National Conference (2) was held at Northwestern University in 1967 and emphasized the multi-disciplinary characteristics of environmental engineering education.

The Third National Conference (3) convened at Drexel University in 1973. This meeting dealt with issues ranging from technician training to graduate education to continuing education programs.

The Fourth Conference (4), held in 1980 at the University of Toronto dealt with the issues of excellence, curricular balance and the relationship between undergraduate and graduate environmental engineering education.

The Fifth Conference (9), held in 1986 at Michigan Technological University, is the most recent of this series and dealt with environmental engineering education at the advanced undergraduate and graduate levels focusing on: Integrated, Air, Water, Land Approaches in Environmental Engineering, the Role of Computers in Environmental Engineering Education and Research, and the Role of Design and Operation in Environmental Engineering Education.

----

[1] Professor of Civil and Environmental Engineering, Michigan Technological University, Houghton, Michigan

Patterson (5), and Baillod (6) (7) described the relationship of these conferences to the environmental engineering profession and educational system. Table I outlines the issues addressed at each of the conferences.

TABLE I. SUMMARY OF ISSUES ADDRESSED AT CONFERENCES ON ENVIRONMENTAL ENGINEERING EDUCATION

First Conference - Harvard, 1960

- Graduate Courses, Curricula and Terminology
- Period of Study and M.S. Thesis Requirement
- Accreditation of the M.S. Degree

Second Conference - Northwestern, 1967

- Role of the Engineer and Other Professionals
- Chemistry, Biology and Social Sciences in Curricula
- Graduate Curricula for Research and Professional Practice
- Non-engineering Students in Environmental Engineering Programs
- Accreditation of Professional Curricula

Third Conference - Drexel, 1973

- Environmental Quality Goals and Challenges
- Manpower Needs in Environmental Engineering
- Educational Needs for Graduate, Undergraduate, Technician and Continuing Education Programs

Fourth Conference - Toronto, 1980

- Excellence in Environmental Engineering Education
- Comprehensive Versus Specialists Programs
  - Curricular Balance
- Relationship of Baccalaureate to Graduate Education in Environmental Engineering

Fifth Conference - Michigan Tech, 1986

- Role of Computers in Environmental Engineering Education
- Integrated, Air, Water, Land Approaches in Environmental Engineering
- Role of Design and Operation in Environmental Engineering

FORMAT AND PROGRAM OF THE FIFTH CONFERENCE

The Fifth Conference followed the format of previous conferences. Committees of educators and practitioners developed and distributed position papers on the focal issues prior to the Conference. The program itself centered on the focal issues. Each position paper was summarized and discussed on the first day of the conference. The second day of the Conference began with concurrent forums on each focal issue. These were followed by Task Committee discussion and writing sessions designed to generate resolutions for presentation at the final plenary session of the Conference.

The Conference concluded on the morning of the third day with a final plenary session at which the reports and proposed resolutions of the Task Committees were received and presented for approval, modification or rejection by the Conference.

SUMMARY OF THE FIFTH CONFERENCE

One hundred thirty five environmental engineering educators and practitioners spent two and one-half days discussing educational needs, problems and solutions related to the conference issues. In this exercise, the conferees accomplished two things: First, a general consensus was reached regarding some things that ought to be done to improve environmental engineering education; these items are reflected in the resolutions passed at the final plenary session. Second, both the educators and practitioners became more aware of things they could do to improve instruction in the areas of integrated approaches, use of computers, and design and operation. Some items of discussion and consensus were:

General Concerns

There was general concern over how an increased emphasis on the fundamental topics necessary for an integrated multimedia approach to problems solving could be achieved in view of the limited amount of time and credits available in a typical master's degree program. There was also concern over how a baccalaureate environmental engineering program could be designed to include the broad basic science base necessary for environmental engineering as well as the amount of engineering science and design required for accreditation by ABET (8). This latter concern led to adoption of the overall conference resolution that up to one-fifth year of advanced quantitative physical or natural science courses which have environmental engineering applications can be counted as engineering science in satisfying ABET requirements for basic level accreditation of environmental engineering programs. It was also resolved that the American Academy of Environmental Engineers should take a leadership role in developing educational materials.

Integrated, Air-Water-Land Approaches:

The position paper defined an "integrated approach" as holistic, incorporating parts into a whole and based on an understanding of the multimedia behavior of substances in the environment. Traditionally, educational and regulatory approaches have not been integrated but have been segregated into air, water or land media.

The paper addressed how curricula and teaching style might be modified to better address the multimedia nature of environmental problems. Ways to incorporate an integrated, multimedia approach into existing environmental engineering introductory, design, modeling and industrial pollution control courses were discussed along with the need for new courses and the problems with students deficient in organic chemistry, geology and meteorology. The lack of good teaching materials and shortage professorial time to change direction were recognized as constraints to implementing a more integrated approach.

Conference resolutions declared that environmental engineering education should include elements focusing on multimedia problems with integrated approaches to their solutions. Graduates of environmental engineering programs should have strong backgrounds in science and engineering fundamentals, technical competence in a specialty, an appreciation for limitations on data for characterizing multimedia problems, and sufficient familiarity with other engineering and science specialties to be able to communicate and work together on multimedia problems.

Additional resolutions encouraged practitioners from the industrial, governmental and consulting sectors to participate in the development of multimedia educational programs. Faculty were encouraged to avail themselves of opportunities to broaden their backgrounds to more effectively deal with multimedia solutions.

Role of Computers:

Items addressed in the position paper included the availability of computers, micro versus mainframe computers, appropriate programming assignments for students, the role of numerical methods and special purpose "black box" programs, and pitfalls in using computers in education. Conference consensus was reflected in the plenary resolutions declaring that:

* It is desirable that students entering graduate environmental engineering programs (and presumably those graduating from preparatory baccalaureate programs) be able to prepare simple programs in a high level language, use work processor and spreadsheet programs, and understand the limitations, strengths and weaknesses of major numerical methods.

* Graduate curricula in environmental engineering should emphasize basic understanding of processes and mechanisms and the mathematical expression of these principles through models. Graduating M. S. students should be able to implement these models on a computer. In order to spend more time on the formulation and interpretation of the problem, students should also be exposed to commercially available programs that reduce the computational and solution effort.

* Easy access to both personal and mainframe computers is important. Accordingly, students should be encouraged to buy their own personal computers and Universities should establish programs to facilitate their acquisition. In order to facilitate the implementation of computing into coursework, Universities should provide personal computers to all engineering faculty and should provide easy access to computer labs for student use.

* AEEP should establish a standing committee on Computer Applications.

* University administrators should provide support for the development and integration of computers into the curriculum, namely, teaching assistantships, software developers, release time, summer support, hardware and purchase of commercial software. Developing software for classroom instruction should be recognized as a creative and demanding activity, for which due credit should be given in career evaluation.

* To avoid legal and ethical problems that may arise with the educational use of proprietary software, faculty should make sure students are aware of the ethical considerations relating to copy restrictions and the consequences of violating them. Classroom assignments should incorporate available freeware and shareware, where feasible; otherwise Universities should seek site licensing agreements that allow students to acquire proprietary software at an affordable cost. Software vendors should be encouraged to make software available to students at a reasonable price.

* In order to provide a variety of educational opportunities, graduate curricula in environmental engineering programs should include exercises in advanced computing applications such as simulation, process control, data acquisition, and artificial intelligence.

Role of Design and Operation:

The position paper discussed the ABET definition of engineering design (a process of devising a system, component, or process to meed desired needs... a decision making process.. in which the basic sciences, mathematics and engineering sciences are applied to convert resources optimally to meet a stated objective). In environmental engineering practice, this consists of conceptual design (flow diagrams, material balances, siting, preliminary costs), detailed design (plans, specifications, construction documents), and construction design (shop and fabrication drawings). Environmental engineering operation, on the other hand, was defined as the process of making systems and facilities function according to the intent of the design. Requisites for competence in design and and operation were viewed to be generally similar. However, certain traditional civil engineering subjects such as structures were judged to be less important for operation, whereas other subjects such as process dynamics, instrumentation, and control were judged to be more important for the operations engineer.

The present situation for instruction in design and operation at the undergraduate and graduate levels was reviewed. Based on this review, it was observed that, although much weight is placed on design in undergraduate program accreditation, there is little consensus on what constitutes a design course. It was also observed that little emphasis is placed on instruction in environmental engineering operation. At the graduate level, the problem of teaching engineering design to students without an engineering baccalaureate degree was considered. The traditional graduate "research" project was viewed as an opportunity for a student to define and solve practical problems related to facility design and operation.

Recommendations for improving education in design and operation centered on getting distinguished environmental engineering practitioners from the consulting and industrial sectors more involved in undergraduate and graduate educational programs. This might be effected through adjunct appointments and preparation of instructional materials.

Consensus resolutions reflected a concern that the situation with liability insurance and litigation could prevent faculty members from getting design experience. It was also resolved that design courses place additional emphasis be placed on the use of computers, hazardous waste management, operability, maintainability, and process dynamics. The conferees encouraged that functional design courses include the integration of forensics, regulations, finance, ethics, risk assessment and management, safety, and private practice of engineering.
A slim majority of the conferees also resolved that "design project courses" be taught by registered professional engineers.

THE SITUATION THREE YEARS LATER

Three years later, it appears that the issues addressed during the Fifth Conference were timely indeed. The past three years have witnessed an unprecedented evolution of computing practices in both the environmental engineering industry and environmental engineering education. In general, this evolution has been in accord with the positions taken at the Conference. The drive toward holistic, integrated solutions to environmental pollution problems has intensified. The current emphasis on hazardous waste management, waste reduction, waste minimization and volatile emission controls are examples of this. The environmental engineering industry is facing and dealing with these problems today. On the other hand, traditional civil/environmental engineering education has been slow to broaden its water/wastewater focus to a more balanced multi-media perspective.

Finally, the need for environmental engineers trained in the fundamentals of engineering design remains as strong as ever. Traditional civil/environmental engineering educational programs are adequately preparing students in these fundamentals. However, the complexity of environmental pollution control systems is demanding an engineering approach to their operation. It is not clear that operation of pollution control systems receives adequate emphasis in traditional civil/environmental engineering education.

LITERATURE CITED

1. "Report of the Study Conference on the Graduate Education of Sanitary Engineers," American Sanitary Engineering Intersociety Board, Cambridge, Mass., 1960.

2. "Report of the Second National Conference on Environmental and Sanitary Engineering Graduate Education," Environmental Engineering Intersociety Board, American Association of Professors in Sanitary Engineering, Evanston, Ill., 1967.

3. "Proceedings of the Third National Environmental Engineering Education Conference," American Academy of Environmental Engineers, Association of Environmental Engineering Professors, Philadelphia, PA., 1973.

4. "Proceedings, Fourth Conference on Environmental Engineering Education," edited by J.W. Patterson and R.A. Minear, American Academy of Environmental Engineers, Association of Environmental Engineering Professors, Toronto, Ont., 1980.

5. Patterson, J.W., "Environmental Engineering Education: Academia and an Evolving Profession," Environmental Science and Technology, 14:524–532, 1980.

6. Baillod, C.R., "Education for Environmental Engineering," Environmental Science and Technology, 20:432–437, 1986.

7. Baillod, C.R., "Fifth Conference on Environmental Engineering Education," The Diplomate, Vol. 21, No. 1, 1986.

8. Annual Report; Accreditation Board for Engineering and Technology: New York, 1984.

9. Proceedings, Fifth Conference on Environmental Engineering Education: AAEE, AEEP, edited by C. Robert Baillod, Michigan Technological University, 1986.

The USDHHS Study of Hazardous Waste Education

Ralph H. Kummler,* Catherine A. Witt**,
Robert W. Powitz*** and Barry Stern****

ABSTRACT

The United States Department of Health of Health and Human Services through Wayne State University has completed a study evaluating hazardous waste graduate education and training in the United States. The purpose, methodology and results of the study are summarized herein.

INTRODUCTION

The need for educated and trained personnel in hazardous waste and hazardous materials management in the United States is urgent and urgent and growing. Busch (1,2) cites the need for 22,500 professionals from 1990 to 1995 just to meet the hazardous waste clean up needs in the United States; the U.S. Department of Health and Human Services (3) estimates that 100,000 professionals overall will be needed by 1992.

Universities in the U.S. have responded slowly and cautionusly to this need. In 1986, Davis (4) published the results of a brief, informal survey of 69 selected universities in the U.S. Of the 42 respondents only 30 offered one or more courses in hazardous waste management at that time and only four had sufficient commitment to offer laboratory training. Davis concluded that the size of the university programs were minimal compared to the size of the problem. Busch (1) estimated that formal university programs are probably providing only 5% of the country's needs.

---

* Professor and Chairman of Chemical and Metallurigical Engineering, Wayne State University, Detroit, MI

** Masters Student, Hazardous WAste Management, Wayne State Univeristy, Detroit, MI

*** Director of Environmental Health and Safety, Wayne State University, Detroit, MI

****Project Officer, United States Department of Health and Human Services, Bureau of Health Professions, Rockville, MD

In this context, the U.S. Department of Health and Human Services awarded a contract to obtain current information on hazardous waste education to Wayne State University (WSU). WSU also volunteered to conduct a comprehensive survey of all universities in the U.S. to obtain a complete picture of hazardous waste formal education and continuing educations in 1989.

The purpose of the contract was to provide a limited evaluation of the adequacy and appropriateness of Hazardous Waste Management courses in the Unites States, both for: 1) Majors or track in graduate degree, and in 2) Non-degree professional development.

The method involved developing a set of criteria using an Advisory Panel and comparing a limited number of existing academic graduate level programs and non-degree programs against those criteria. The Advisory Panel made recommendations and a large expert review panel reacted to the recommendations.

The Advisory Panel consisted of Academicians, Employers, Consultants, Federal Agency Representatives, and Local/State Agency Representatives.

THE SPECIFIC OBJECTIVES OF THE CONTRACT WERE TO:

- Define and Identify a cross section of graduate degree programs (including schools of publich health) which grant degrees with a major or track in Hazardous Waste Management.
- Define and identify non-degree training programs and training courses (professional development, training institutes, and short courses) that are available to professionals on Hazardous Waste Management subjects.
- Develop a consensus among Advisory Panel educators, employers, and Federal agencies as to the criteria for Hazardous Waste Management Graduate Level Degree Programs and Non-degree short (training) courses.
- Evaluate existing graduate degree educational programs and non-degree Hazardous Waste Management courses comparing them with criteria developed by this contract.
- Develop specific recommendations for correcting inadequacies identified as a result of the evaluation.

THESE OBJECTIVES WERE ADDRESSED IN THE FOLLOWING WAYS:

- Degree programs and non-degree courses were identified via a survey and compilation of secondary data collected from academic Institutions, Professional Associations, and Federal Agencies. This material was provided to the Advisory Panel.
- The Advisory Panel developed draft criteria for evaluating Hazardous Waste Management programs/courses and suggested curriculum evaluation issues, programs, and courses for the contractor to explore.
- Recommendations were made for evaluating specific Hazardous Waste Management specialities in representative academic programs and non-degree courses.
- Using the finalized criteria, the contractor evaluated the prescribed programs and courses and provided the Advisory Panel with an "Evaluation" Report summarizing its findings.

THE PRODUCTS OF THE CONTRACT INCLUDED:

- Listings of academic graduate degree granting programs with a major or track in Hazardous Waste Management
- Listing of organizations with Hazardous Waste Management Professional Development courses
- List of recommended criteria for Hazardous Waste Management curricula
- Electronic Bulletin Board
- Final Report

<u>SURVEY RESULTS</u>

<u>Hazardous Waste Education Provided By Academic Institutions</u>

Two mailings were conducted, one to a select prototype group and the second to all four-year educational institutions. Of the 141 universities contacted in the selective mailing, 63 replied (47%). Of the 1469 institutions contacted through the comprehensive mailing, 684 replied (47%). Eliminating duplicates between the two mailing sessions, a total of 732 separate institutions replied to the survey out of 1469 separate institutions contacted (50%). Of those 732, 602 (82%) have no hazardous waste education of any kind.

One hundred and thirteen of those responding offer at least one course as part of traditional degree programs. Of the 113, slightly less than half (54) have hazardous waste courses that are required for the respective degree. The majority of degrees containing hazardous waste education are in engineering, particularly Civil, Environmental and Chemical Engineering. Public and Environmental Health are second. Figure 1 presents a breakdown of the number of departments which have been identified as offering at least one course in hazardous waste as part of a non-hazardous waste degree.

Figure 1. Departments Offering Credit Courses In Hazardous Waste

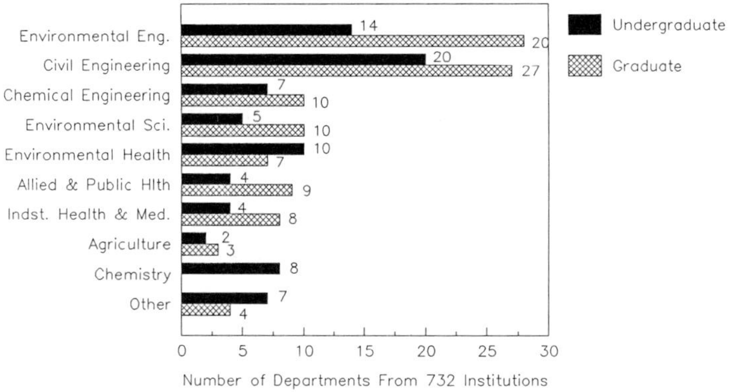

Figure 2 provides a profile of the types of hazardous waste instruction offered at both undergraduate and graduate levels.

Three universities (New Jersey Institute of Technology in Newark, Tufts in Boston and Wayne State in Detroit) have been identified as offering a Master's Degree entitled Hazardous Waste or Materials Management. A comparative analysis of the three degree programs is presented in reference 6.

Thirteen universities offer an option, minor, or emphasis in hazardous waste, at the graduate level; nine through Civil and Environmental Engineering, three through Environmental/Industrial/Public Health and one through both departments. Five universities offer certificate programs. Three offer non-credit continuing education certificates and two offer certificates for graduate credit.

Figure 2. Profile of Educational Institution Hazardous Waste Instruction Offerings

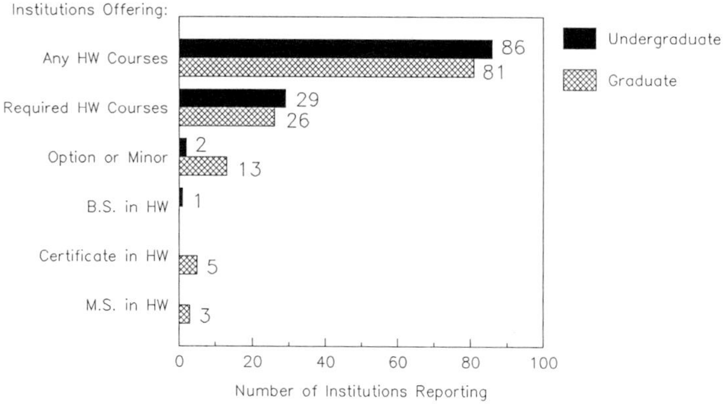

Fifty-two colleges and universities were identified as offering non-credit continuing education short courses and workshops.

None of the hazardous waste programs were accredited. However, most of the traditional undergraduate degree programs (such as engineering graduate programs in public health) which contained hazardous waste were accredited by accepted accrediting bodies.

Continuing Education

Universities and the government offer the greatest variety of different courses and the greatest number of courses overall (including annual replications). However, there are several private companies which provide a signficant number of courses. Table 2 provides a comparison between the number of different courses and the number of total courses offered by 10 of the largest course suppliers. The U.S. EPA Office of Solid Waste and Emergency Response has been identified as the largest single provider of hazardous waste related short courses, offering approximately 36 distinct short courses per quarter and 154 total courses per quarter including replications.

CONCLUSIONS

The demand for professionals with education and training in hazardous waste is large and growing. Universities have responded to the demand but the number of formal programs available remains quite small in terms of the projected need.

The survey of 732 responding colleges and universities described herein identified 113 programs offering some formal credit courses in hazardous waste management but only three with a Master's degree level of effort, and only 13 with an identified graduate option, track or emphasis. Programs incorporating hazardous waste tend to be concentrated in engineering; primarily in Civil, Environmental, or Chemical Engineering and in Schools of Environmental or Public Health.

The major delivery of training appears to be in the continuing education, non-credit, short course format. Fifty-two colleges and universities participate in that format and 499 separate and distinct short courses were identified. Over two thousand such courses are offered per year by the major government, professional, academic and private groups. The U.S. EPA, Georgia Tech, Texas A&M University and NIOSH offer the greatest number of these courses. The primary interest lies in courses on law and administration, asbestos handling, and emergency response. These continuing education courses seem to be the short term solution to the lack of hazardous waste professionals. After analyzing the data, the expert review panel has suggested that each course should provide an optional comprehensive exam. These exams would give the courses credibility beyond obtaining Continuing Education Units (CEU's). The exams may help justify the expense of the courses in the eyes of employers.

REFERENCES

1. Busch, P.L. And W.C. Anderson, "Education of Hazardous Waste Engineering Professionals," 116th Annual Meeting of the American Public Health Association, Boston, Mass., November, 1987.

2. Busch, P.L., "A Hazardous Waste Crisis: Too Few People," Waste Age, September, 1988.

3. Levine Associates, "Evaluating the Environmental Health Workforce," U.S. DHHS Report, 240-286-00076, January, 1988.

4. Davis, M., "A Survey of Graduate Education in Hazardous Waste Management," J. APCA 36, 1036 (1986).

5. Kummler, R.H., Witt, C.A., Powitz, R.W, and Stern, B., "A Comprehensive Survey of Graduate Education and Training in Hazardous Waste Management, J. APCA, January 1990.

LIST OF THE TOP 10 SHORT COURSE OFFERING ORGANIZATIONS

Organization Name	Number of Different Courses Offered Per Year	Total Number of Courses (including multiples) Offered Per Year
U.S. EPA, Office of Solid Waste and Emergency Response (OSWER)	36	600*
Georgia Institute of Technology, Environmental Health and Safety Division	38	100
Texas A&M University System, Texas Engineering Extension Service	23	257
National Institute for Occupational Safety and Health (NIOSH), Educational Resource Centers	33	100
University of Florida, Continuing Education Center for Training, Research and Education	30	129
University of California System, Department of Continuing Education	30	38
Eastern Michigan University Corporate Services	22	113
University of Wisconsin System, Department of Engineering Professional Development	22	31
Executive Enterprises, Inc.	21	38
Government Institutes, Inc.	17	21
TOTAL:	**272**	**1,427**

Based on data collected before 10-20-89
*estimation

Environmental Engineering Education for Groundwater Contamination and Hazardous Waste

Richard G. Luthy[1], M ASCE and Mark M. Benjamin[2], M ASCE

Abstract

This paper discusses the current status and trends in environmental engineering graduate education that prepares students for tackling groundwater pollution and related hazardous waste problems. This discussion is abbreviated from an article prepared for the Water Pollution Control Federation Groundwater Committee on the subject of graduate education for groundwater pollution and hazardous waste (Luthy and Benjamin, 1989). Hazardous waste is included in the discussion of groundwater contamination because much of the concern over groundwater pollution has resulted from the RCRA and Superfund programs, which focuses on the investigation and remediation of subsurface contamination.

Changing Curricula

Education and research on management of groundwater contamination requires interdisciplinary training that emphasizes fundamental principles. Environmental engineers are significantly involved in this effort owing to their background in environmental sciences, their attention to intermedia processes, and their presence in shaping environmental policy for the safeguarding of human health and the protection of the environment. Groundwater pollution education in environmental engineering curricula is undergoing change, largely as a result of the sequence of events arising from new environmental regulations which impact groundwater quality. Groundwater contamination is an interdisciplinary subject, and there is need to incorporate features from both the earth sciences and the engineering curricula in order to adequately address the subject.

Groundwater pollution education is necessarily a graduate-level activity because undergraduate instruction can provide only the introductory science and engineering fundamentals needed to understand groundwater problems. Presently there exist a variety of graduate courses at numerous institutions in which groundwater pollution is addressed. Some programs have elected to incorporate discussion of groundwater contamination processes as components of fundamental courses in subsurface hydrology and physico-chemical processes, while other programs have chosen to present groundwater contamination concepts as a separately-identified course. The orientation of such courses may fall anywhere on a spectrum from current practice and/or design techniques, to much more abstract or hypothetical analyses based on fundamental engineering-science principles. Frequently, this orientation reflects the relative emphasis between classroom teaching and research in the graduate program.

The combination of the need for trained professionals in hazardous waste and subsurface

[1]Professor and Head, Department of Civil Engineering, Carnegie Mellon University, Pittsburgh, PA 15213

[2]Professor of Civil Engineering, University of Washington, Seattle, WA 98195

contamination, coupled with the many unsolved problems in understanding groundwater pollutant transport, transformation and fate, has been the motivating force behind the development of graduate-level training and research programs in environmental engineering on these topics. However, research funds and grants to support and train graduate students are inadequate to satisfy demand (Luthy, 1989). Graduate programs in groundwater and hazardous waste may typically have a great deal of interest on the part of students and potential employers, but lack the resources to meet demand.

Elements of Groundwater Pollution Education

Groundwater pollution education has been approached from various disciplinary perspectives, as listed in Table I. While this list is not all-inclusive, it does recognize the general agreement that groundwater contamination issues span several disciplines, and may be appropriately addressed in part from each of these perspectives. Ideally, a "complete" program in groundwater contamination would synthesize subject matter and perspectives from hydrology, water chemistry, hydrogeology, soil science, geology, microbiology, and water and waste treatment. However, few, if any, graduate programs are sufficiently broad or well-integrated that each of these disciplines is adequately represented (Neuman, 1989). The curriculum which encompasses many of the disciplines listed in Table I is civil engineering, which includes hydrology, water resources, environmental engineering and geotechnical engineering. Hence, historically groundwater was typically in the realm of the civil engineer (Bouwer, 1989). Today, groundwater geology courses also play an important role in groundwater pollution education and research with emphasis on hydrogeology (Lehr, 1989), and related topics of mathematical modeling, microbiology and chemistry that are also addressed in environmental engineering.

Undergraduate studies in civil engineering or chemical engineering provide appropriate curricula leading to graduate education in environmental engineering and groundwater pollution. Undergraduate civil engineers are weak in chemistry, reactor engineering, and mass balance calculations, and these deficiencies are addressed during graduate education. For some aspects of groundwater pollution, such as porous media transport with reactions, a chemical engineering preparation is likely to be better than preparation in civil engineering. Some of the weaknesses of a chemical engineering background are the lack of exposure to soils, geology and hydrology, and inexperience in working with the complexities of the environment, and working with poorly-defined systems and with the public sector. Again, these topics can be included in graduate education, and chemical engineers are sought by many graduate environmental engineering programs.

Usually environmental engineering, geology, or hydrology serves as the nucleus of groundwater pollution teaching and research. Environmental engineering has an important role, as by necessity environmental engineering has been an integrator of interdisciplinary sciences and engineering for environmental protection and health. Hydrogeologists, currently in great demand, are at least M.S.-level geologists with some course work in groundwater. Hydrogeologists possess the geological understanding and the three dimensional geologic visualization that engineers usually lack, even engineers with specialty in subsurface transport. Environmental engineers, also M.S.-level graduates, possess understanding of the physical, chemical and microbial processes affecting water quality, the principles of mathematical modeling, and some of the quantitative aspects of groundwater flow. Groundwater problems are addressed successfully when both hydrogeologists and environmental engineers work together. For example, many of the models currently being used for groundwater problems were written by individuals with skills in both disciplines.

Table I: Disciplines Relevant to Groundwater Pollution Education

Discipline	Historical Focus with Regards to Groundwater Issues
Hydrology/Water Resources	Subsurface water flow and transport; water yield; water resource management; numerical and analytical solutions to flow and transport.
Hydrogeology/Hydrogeochemistry	Groundwater evolution; source determination; recharge and mixing; age dating; geologic settings; ore genesis.
Environmental Engineering	Microbial, chemical and physical processes affecting water quality; treatment systems; inter-media processes; environmental protection and health.
Soil Science	Soil fertility; soil physics and chemistry; vadose zone; pesticide chemistry; soil microbiology.
Agricultural Engineering	Irrigation and drainage; tillage, erosion and conservation.
Geotechnical Engineering	Seepage; dewatering; water containment.
Chemical/Petroleum Engineering	Flow and transport with reaction; multiphase flow; oil reservoir management.
Geological Sciences	Geophysics; geochemistry.
Public Policy	Water regulations; water rights; environmental risk; environmental regulation.

Environmental Engineering Curricula in Groundwater Pollution

Essentially all graduate environmental engineering programs now list some courses devoted to problems in groundwater contamination or related topics on management and treatment of hazardous substances, according to the new Sixth Edition of the Register of Environmental Engineering Graduate Programs (AEEP, 1989). As recently as 1981, when the Fourth Edition of the registry was published, comparatively few programs explicitly identified significant courses and/or research activities in these areas. Today there is more emphasis given to specific departmental-based activities in these areas, as well as cross-listing of faculty and courses in complementary programs in the earth sciences, soil science or agronomy.

Tables presented in Luthy and Benjamin (1989) provide a sampling of the current types of graduate courses which focus on the science or engineering aspects of groundwater pollution. The largest category of courses currently available on groundwater pollution deals with the broad subject of the fate of contaminants in the subsurface. These include, to varying extent, a discussion of physical and chemical processes with some introduction to microbial degradation. These courses usually have a water resources, hydrology or hydrogeology course as a prerequisite, as well as some requirement for familiarity with applied aquatic or soil chemistry. Most of these courses are not explicitly identified with a field-training component.

There are few specialized courses on subsurface microbiology or microbial ecology with an emphasis on transformation and degradation of organic contaminants. These specialized microbiology courses are often developed from an adaptation of subject matter from environmental engineering microbiology or soil microbiology. The comparatively few specialized courses devoted to subsurface microbial processes and/or microbial degradation of contaminants is an indication of the resource constraint faced by most academic departments. Another important category of courses on groundwater pollution are those that address some of the more fundamental aspects of contaminant reactions in aquatic/porous media. The subject matter in these courses is dependent principally on the particular research interests and orientation of individual faculty and programs. The range of topics covered in these courses includes heat and mass transfer, surface chemistry, environmental organic chemistry, modeling, and multiple-phase systems.

Environmental Engineering Curricula in Hazardous Wastes

Graduate-level training in groundwater pollution is linked to education and research on hazardous wastes through regulations and technical issues regarding waste disposal and containment, and waste site investigation and remediation. A recent survey conducted by the Association of Environmental Engineering Professors queried its membership about course offerings at their institutions in the general area of hazardous wastes. Owing to the timeliness of the subject matter, most programs are eager to state that they are involved with hazardous wastes, although the focus varies. About 10% of the 47 respondents indicated that their program offers a focused curriculum dealing with hazardous waste, and about half offer at least one course in which hazardous waste is the primary topic addressed. The remainder all indicated that hazardous waste treatment technology, and/or transport and transformations of hazardous wastes in the environment, was covered in many of their courses dealing with the fundamentals of environmental engineering. This acknowledges that environmental engineers have always dealt with hazardous wastes, although today's applications, such as with uncontrolled waste disposal, are different than addressed historically by environmental engineers, such as with industrial waste treatment. It is clear from the responses that there is no agreement as to what constitutes a "focused curriculum" in this field.

At institutions where a single or introductory course on hazardous waste was identified, the topics covered typically include an historical perspective on legislative and engineering approaches to the issue; extensive coverage of RCRA and CERCLA, their predecessors and amendments; definitions, classifications, and generation rates of hazardous wastes; landfill design and operation, with special emphasis on provisions for leachate and gas control/collection; an overview of pump-and-treat, stabilization, and incineration technologies; and several case studies. The dominant intent of these courses is clearly to familiarize undergraduate and graduate students with the field of hazardous waste engineering as it is currently practiced, and not to develop competence in the engineering/scientific principles governing contaminant transport and transformation.

At the other end of the spectrum are programs where hazardous compounds are discussed as appropriate within the context of general biological, physical, or chemical treatment/transport processes. Since virtually any process can be discussed using a hazardous compound as an example, the extent to which these courses describe the behavior of hazardous wastes may vary widely, depending on the instructor's interests and predilection. A reasonable way to locate a course on this spectrum might be to ascertain the extent to which it focuses on "hazardous wastes" as composite materials requiring overall engineering management, versus a focus on groups of "hazardous compounds" whose behavior can best be determined by understanding their chemical properties, often at the molecular level. There is an urgent need for better textbooks to use in these courses.

The Need for Interdisciplinary and Fundamental Approaches

Effective management of groundwater contamination, requires expertise from many disciplines. Adequate training requires integration of formal basic sciences, health sciences, engineering, mathematics, ecology, public policy and risk assessment. It is not possible for any individual to become expert in all of those topics. Hence graduate training must provide depth in a specific area of expertise, but also must emphasize the importance of interdisciplinary communication. Courses and research that emphasize fundamental principles allow one to deal effectively with the challenges of environmental problems, while providing skills that make an individual versatile and self-educable. Table II outlines a multidisciplinary environmental engineering and science curriculum for training in the management of groundwater contamination. An undergraduate engineering degree program including courses in chemistry and fluid mechanics is assumed. The core comprises four courses in scientific and engineering fundamentals, a contaminant transport course intended to synthesize the other engineering-science courses, and a course or seminar that addresses current practice. Of the courses dealing with fundamentals, the one on groundwater hydrology focuses on the behavior of the groundwater itself, and the other three deal with factors which affect groundwater quality.

The electives listed cover a wide range, including advanced training in molecular-scale processes, engineering processes, modeling, and public policy. Most of these courses are not traditionally offered in environmental engineering programs, but the majority would be available through other departments at a university. A curriculum which includes the core plus several of the electives would provide an excellent starting point for a student entering the workforce. Graduate curricula in environmental engineering rarely include practical field techniques for sampling, monitoring, and evaluating groundwater systems and surface-groundwater interactions. Such coursework is available at some universities in which there is a large program in hydrology or hydrogeology. While extensive on-site training in sampling is properly the province of the employer, an understanding of sampling techniques and protocols, the strengths and weaknesses of various approaches, and an appreciation of the factors which affect the reliability of such samples can be addressed at the university.

We believe that it is essential that today's environmental engineer develop the ability to evaluate information from different sectors in order to address groundwater pollution problems. This should include not only the skills to integrate technical data, but also the ability to assess realistic goals through understanding the intent of legislation with recognition of the true complexities of groundwater problems. It is becoming increasingly apparent that there are serious deficiencies in legislation and regulation of groundwater contamination and corrective measures (Freeze and Cherry,

1989). During his or her career, the environmental engineer/groundwater specialist may be called upon for advice about how to improve legislation or how to choose among projects competing for our limited technical and financial resources. Our educational process must serve to develop perspective and the ability to use all available information to discriminate fairly and wisely among these competing demands. Combining engineering judgment with an understanding of public policy issues so that technically achievable goals are established, and the projects which are most valuable to society receive top priority, is the ultimate objective of the interdisciplinary approach to groundwater problems. To the extent possible, practice in utilizing this approach should be provided as part of graduate training.

Table II: Multidisciplinary Environmental Engineering and Science Curriculum for Groundwater Professionals

Core	Groundwater Hydrology; Aquatic Chemistry; Fundamentals of Biological Reactions; Physicochemical Processes; Flow and Contaminant Transport Modeling
	Groundwater Management and Contamination: Case Studies
Electives For Breadth	Field Techniques for Aquifer Evaluation
	Engineering Geology; Physical Geology; Low Temperature Geochemistry; Soil Chemistry; Soil Physics; Soil Mechanics; Aquatic Surface Chemistry
	Environmental Microbiology/Microbial Ecology; Biological and Physicochemical Unit Operations for Waste Treatment; Biochemical Engineering; Chemical Reactor Engineering; Petroleum Reservoir Engineering
	Organic Chemistry; Water Quality Analysis; Instrumental Analysis
	Environmental Toxicology; Risk Assessment and Public Decision Making; Statistics
	Environmental Law

Summary

Graduate-level opportunities in groundwater contamination and hazardous waste management have expanded dramatically in the last decade. Every indication is that this trend will continue, in view of the significant unsatisfied demand for experts trained in this field, the many unsolved questions and problems concerning subsurface behavior of contaminants, and the large expected expenditures by both the governmental and private sectors.

Coursework relevant to an understanding of groundwater contamination may be offered in a variety of departments or programs, including civil engineering, geological sciences, and agricultural engineering. A major goal for educators should be to integrate the training from these various disciplines so that today's groundwater professionals benefit from all these perspectives. Environmental engineering and science programs are particularly well-suited to orchestrate this synthesis. These programs are by nature interdisciplinary, integrating science and engineering for water quality protection and health risk reduction.

Universities differ in the extent to which their coursework emphasizes scientific and engineering fundamentals versus engineering practice, and it is appropriate that such a range exist. While it is clear that no single model exists for the "ideal" academic program in this area, we feel that a core program providing a blend of fundamentals and practical training offers the student the best of both approaches: a sense of how problems are addressed today, with basic skills to self-educate in order to understand new problems and new technologies in the future.

A core program should include a technical undergraduate degree followed by graduate-level coursework in groundwater hydrology, aquatic chemistry, biological and physico-chemical processes, and contaminant transport modeling as well as some exposure to case histories. Beyond this, certain electives should be offered to provide for advanced training in areas such as: scientific principles governing contaminant behavior, engineering processes for decontamination, principles of chemical behavior and analysis, and environmental policy and/or risk analysis.

The current need for groundwater and hazardous waste specialists outstrips the current capability of universities to supply such individuals; and there is a need for an expanded effort to develop existing and new technologies to meet our national needs in this area. The provision of appropriate governmental and industrial support via increased research and training grants is essential if we are to meet these needs. In the long run, such support repays itself many times over.

References

AEEP, Register of Environmental Engineering Graduate Programs, Sixth Edition, Compiled by William R. Knocke and Gary L. Amy, Association of Environmental Engineering Professors, January, 1989b.

Bouwer, H., U.S. Water Conservation Laboratory, Phoenix, AZ, personal communication, August, 1989.

Freeze, R.A. and J.A. Cherry, "What Has Gone Wrong?" Ground Water, Vol. 27, No. 4, p. 458, 1989.

Lehr, J.H., National Water Well Association, Dublin, OH, personal communication, August 1989.

Luthy, R.G., "Personnel and Research Shortages: Policy Recommendations for the Environmental Professions," Environment, Vol. 31, No. 3, p. 4, April 1989.

Luthy, R.G. and M.M. Benjamin, "Graduate Education in Environmental Engineering for Solving Groundwater Contamination Problems," to appear Water Environment and Technology, 1989.

Neuman, S.P., Department of Hydrology, University of Arizona, Tuscon, AZ, personal communication, May 1989.

INTEGRATION OF ENGINEERING STUDENTS INTO HYDRAULIC RESEARCH

Steven R. Abt[1], Member ASCE

Abstract

The Hydraulics Program at Colorado State University modified the traditional process of graduate education to integrate pre-engineering students and undergraduate engineering students into the research program. The program was continued for five years in which 58 students were employed. The results of the program are presented.

Introduction

The traditional means of conducting civil engineering research has emphasized the technical development and thinking processes of the graduate student, both M.S. and Ph.D. The graduate student program has included pursuit of a course of study, development and performance of the thesis or paper project, reporting project findings, and defense of the student's work. The research process facilitates the one to one working relationship between professor and student, and promotes the development of civil engineering/hydraulic technical skills.

Occasionally, an undergraduate student may be included in the research program to assist in labor oriented tasks. However, the undergraduate student's role is often nebulus and he/she is not accepted as an integral part of the research team.

The traditional graduate program of study has several limitations that provide opportunity to be explored and/or developed. For example, the undergraduate student is often ignored or by-passed as

[1]Professor and Assoc. Dept. Head for Res., Dept. of Civil Eng., Colorado State University, Fort Collins, CO 80523.

a contributor beyond laboratory role to the research program. Integration of undergraduates into the research program/project provides part-time employment, exposes the student to the technical, engineering aspects of a project not possible in the classroom, contributes to the team approach to problem solving and promotes recruitment for the graduate program.

A second aspect of the research program/project that is over-looked is the potential training ground that the typical M.S. or Ph.D. program can lend to personnel management and project fiscal responsibility. The graduate student often serves as the research project architect, laborer, and journeyman producer/analyzer. Upon program completion, the graduate student has not been directly involved with the managerial aspects including personnel supervision and counseling as well as technical direction.

A third aspect of the research program that is poorly developed is the potential recruitment of pre-engineering (high school) students into the research program, primarily during the summer semester. It is apparent that the traditional civil engineering research program can be planned and excuted to not only accomplish the technical development of the graduate student, but also influence and expose the undergraduate student and pre-engineering student to the civil engineering research experience.

Since 1985, the Hydraulics Program at Colorado State University has modified its graduate program experience to integrate undergraduate and pre-engineering students into the research program. The objective of this paper is to present aspects of this program.

Program Description

Modification of the graduate research program came about through necessity rather than through a well planned transition. The Principal Investigator (P.I.) received five contracts for conducting contract and grant activity in the hydraulics laboratory. Permanent, qualified personnel could not be hired and trained in a timely manner to support the research effort. Therefore, alternative resources, i.e. students, were sought to facilitate conductance of the research. The research program objectives were expanded to conduct the research for the sponsor, to provide research opportunity for graduate students, and to provide training opportunities for undergraduate and pre-engineering student labor force.

A research organizational structure was developed to allow the P.I. to direct and supervise the research efforts as shown in Fig. 1. A Ph.D. student served to assist the P.I. in routine personnel matters, in fiscal accounting and in project monitoring. The Ph.D. student also provided technical advise and/or assistance to each project graduate student while conducting his/her project. The individual research project was directed by an M.S. student. The M.S. students worked with the P.I. to define the project scope; to construct, instrument, and test each model facility; and to monitor project fiscal expenditures. The M.S. student was also responsible for data collection and data reduction. The M.S. student was acknowledged as a co-author on the final report(s).

Each project was allocated a staff to assist the graduate student in meeting the project objectives. The graduate student was responsible for staff training, accountability, and performance. The projects were each allocated at least one trainee at the sophomore or lower level. A minimum of two high school students served as members of the organizational staff each year.

The pre-engineering student and engineering student training tasks included IBM PC familiarization, hydraulic equipment, i.e. flow meters, piezometers, etc., familiarization, pump start-up and shut-down training, and data acquisition training. Students served as the primary labor force for construction, operation, cleaning and tear-down of a testing facility. In a majority of the cases, the undergraduate student was acknowledged as a co-author on the final report(s).

Since the original research structure was organized and implemented, a similar structure was continually used throughout the past five years. Although the initial need for the structure was no longer a factor, the structure was maintained to provide a long term evaluation.

Evaluation

The modifications made to the research and graduate programs enhanced the educational and training aspects of each of the student levels. During the five year program, 58 students; 2-Ph.D., 13-M.S., 12 seniors, 12-juniors, 9-sophomores, and 10-high school students, participated for one or more years. Four of the ten high school pre-engineering students entered a university as an engineering freshman. To date, two of the

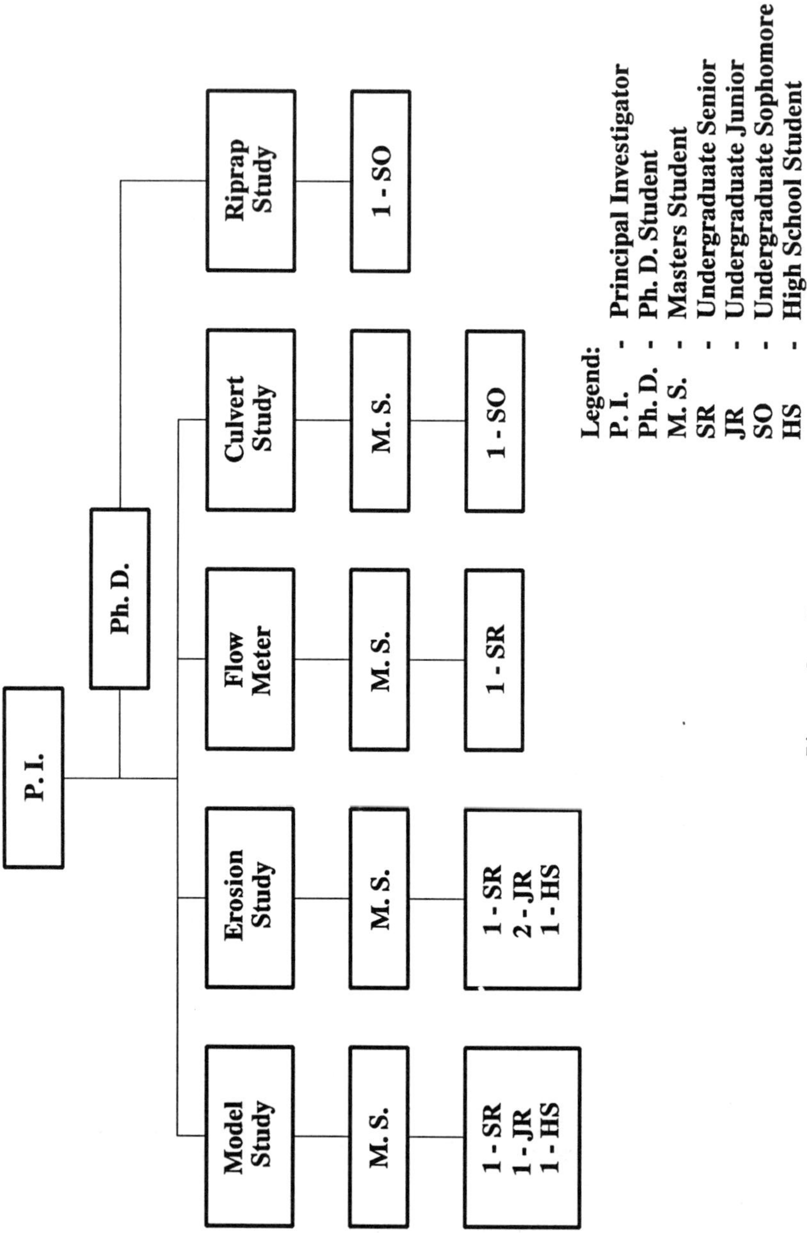

Figure 1. Typical structure

pre-engineering students have graduated with engineering degrees and two students remain in engineering curriculums. Also, five of the twelve seniors have entered graduate school immediately after completing the B.S. degree and one M.S. student continued to pursue the Ph.D.

Although the research program has been viewed to be successful from the P.I.'s view, there are advantages and disadvantages that must be weighed.

Advantages

- Graduate students were trained in both technical and managerial aspects of research,
- Pre-engineering students found employment and were exposed to the engineering process,
- Undergrauate students had hands-on applications of classroom principals, were employed, and were exposed to the graduate education process,
- The P.I. was released from many day-to-day, mundane managerial duties for other scholarly activities,
- All students were exposed to the team approach to problem solving,
- All members of a project staff became familiar with other projects as temporary staffing changes were occasionally warranted,
- Graduate students attained fulfillment by feeling in-charge of the project and by completing the project within budget and in a timely manner,
- Recruitment was enhanced at all levels.

Disadvantages

- The additional layer of 'management' dampened the close one-to-one student to professor relationship,
- The labor force, undergraduate and pre-engineering student staff, felt remote or removed from the P.I.,

- Projects required supplemental funds above project budgets to meet training and management requirements,

- The demands on the P.I.'s time often increased to properly monitor project progress and maintain graduate student contact,

- Care had to be taken to maintain a management hierarchy, or peck-order, in each project,

- Care had to be taken to expose pre-engineering students to as many engineering projects as possible to prevent stereotyping.

Project Financing

The technical projects conducted and managed by the graduate students were obtained through competitive contract and grant procedures from both the private and government sectors. In addition, pre-engineering programs administered through the U.S. Department of Agriculture and the National Science Foundation; undergraduate research programs administered through the National Science Foundation and Colorado State University College of Engineering; and minority and female recruitment programs administered by Colorado State University and the State of Colorado were found to be excellent sources of funds. The financing of pre-engineering students was limited only to the time the P.I. desired to expend in applying for resources.

Conclusions

The modification of a traditional graduate research program to allow the integration of pre-engineering and undergraduate engineering students into the research structure has resulted in several educational benefits. Graduate students have not only experienced the technical development they initially desired, but also gained valuable personnel and fiscal management experience. Undergraduates were exposed to hydraulic research, exposed to the graduate study process, and learned to work as part of an engineering team. Pre-engineering students were exposed to engineering and were mentored by the project M.S. and undergraduates. The success of the program is, in part, contingent upon obtaining funds beyond those necessary to conduct the research to cover additional training and management costs.

Spreadsheets in Geotechnical Engineering

Thomas F. Wolff[1], M., ASCE

Abstract

The slide rule, the mainframe computer, the electronic calculator, and now the personal computer have each, in their time and place, been the tool of choice for engineering computing. The premise of this paper is that spreadsheet software offers unique advantages for many problems in geotechnical engineering practice and education. Examples are cited and discussed.

Introduction

In this author's career, the "backbone" calculating machine has changed several times. "Backbone" calculating machine is herein loosely defined as that device on which most engineers rely most of the time to do most of their routine but complex calculations. The slide rule, the mainframe computer, the electronic calculator, and now the personal computer have each assumed this role for given times and problems.

In recent years, this author has found that the microcomputer spreadsheet has certain unique advantages for many classes of problems in both geotechnical engineering practice and education. For these problems, spreadsheet programs provide a level of power, flexibility, and user understanding generally not attainable with two common alternatives: hand solution with calculators or computer solution with special purpose FORTRAN or BASIC programs.

In geotechnical practice, there are many classes of problems where both the material parameters and the results being sought can be expressed as (and are) functions of depth. Examples include earth pressure problems, settlement problems, and pile foundation problems.

---

1. Assistant Professor of Civil Engineering, Michigan State University, East Lansing, MI, 44824

In geotechnical engineering education, it is difficult to foster what practitioners like to call judgment. There is often justification for more than one correct value of the parameters to be used in a calculation. Having little experience performing such calculations, a student may wonder how one can expect to obtain reasonable solutions.

Commercial Spreadsheet Software

Spreadsheets were originally developed to solve the classic accounting problem of balancing sums of rows and columns. Now in their third to fifth generation, spreadsheet programs such as 1-2-3 (TM), SuperCalc 5.0 (TM), Quattro (TM), Excel (TM) and others provide a broad range of mathematical functions and graphic capabilities. They can solve any problem that is solvable in principle with a calculator, but they afford the speed of a computer.

Speed is also provided by programs written in BASIC or FORTRAN. However, spreadsheets typically demand much less of the programmer or user regarding the format of input and output. Input, intermediate results, and output can all be displayed simultaneously, and changed instantly. Unlike BASIC or FORTRAN, a spreadsheet solution can easily be modified for a similar but different problem, or even adapted to suit individual user's preference, something quite dangerous to do in FORTRAN.

Spreadsheet Applications in Geotechnical Practice

Many geotecnical problems involve parameters and solutions that are functions of depth. Settlements are found by computing vertical strain as a function of depth and integrating the strain over the height of a layer. The side resistance of a pile involves integrating the skin resistance over the surface area of the pile.

Traditionally, these problems are modeled using one or a few layers, each assigned "design values" for the relevant parameters. In doing so, valuable information regarding how parameters vary with depth may be lost. Spreadsheets are capable of solving an equation at a very large number of thin layers, say 1 ft (0.3048 m) each. Using this approach, the parameters used in the analysis can readily be expressed as functions of depth: increasing, decreasing, or varying linearly between test data.

The modeling of a soil parameter as a function of depth is herein referred to as a *parameter function* (Wolff, 1989). This concept was recently used to make several *a priori* predictions of pile capacity (Wolff, 1989) where substantial test data were available. The method tested remarks by Kulhawy (1984), that problems

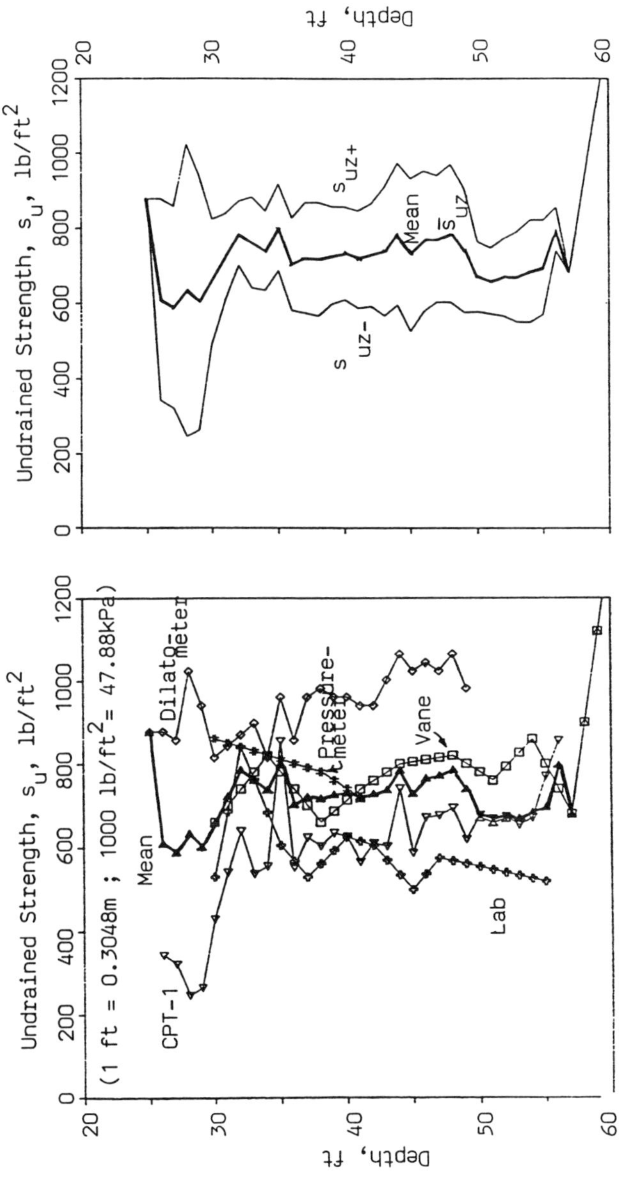

a. Derived from data

b. Mean plus and minus one standard deviation

Figure 1. Undrained Strength vs. Depth Functions

with static capacity predictions were often due to improper consideration of how parameters vary with depth. The predictions agreed quite well with observed capacities from load tests (Finno, et. al., 1989).

Parameters such as density, strength, and earth pressure coefficient were taken as their measured values at depths where tests were available, and the spreadsheet was used to linearly interpolate values at intermediate depths. Although such interpolation is perhaps fraught with assumptions, it is more rational and preserves more information than the common practice of arbitrarily assigning design values for parameters. Figure 1 shows an example of the undrained shear strength functions, their mean, and one-sigma bounds for the clay along the lower half of the pile. Once the parameters were expressed as discrete functions, several skin resistance formulas were put in the spreadsheet to obtain skin resistance functions. Only after developing these functions did the predictor's judgment significantly enter the analysis.

Spreadsheets and the parameter function concept allow one to carry much information through an analysis and apply judgments to calculated results that preserve the observed soil variability. The assignment of "design values" incorporates the judgments before the analysis and thus mathematical manipulations are made on the judgments as well as the data.

Similar approaches are applicable to other problems. For settlement on clay, the void ratio and compression index are commonly known at a few points and values vary. A spreadsheet solution allows these parameters to be modeled as functions of depth as are the vertical stress and the stress increase. Schmertmann's method (1970) for settlement of foundations on sand uses the concept of a parameter function and is readily amenable to spreadsheet solution. Lateral earth pressure problems have similar depth-dependent features.

Spreadsheet Applications in Geotechnical Education

On the education front, one needs to teach not only formulas and methods, but also to teach good design and begin to develop some insight regarding good engineering judgment. When a student asks why a friction angle of 32 degrees is chosen rather than 30 or 34 degrees, the pat answer is often "judgment." The next obvious question is "where does a student acquire this judgment ?"

The computer's capability to instantaneously produce solutions has led this writer to the observation that more could and should be done to teach problem modeling. For instance, foundation engineering texts provide solutions of the earth pressure increase for various surcharges behind a wall. Given the real problem of a train track behind a wall, what is the difference between modeling the surcharge as a single line load, two line loads, or a strip load ?

While good judgment takes a career to develop, spreadsheets permit students to obtain some cheap experience as to how variations in parameters and models affect solutions. In teaching senior level design courses, this writer has used pre-prepared spreadsheets with considerable success for all of the following problems:

1. Weight-volume diagrams
2. Lateral earth pressures
3. Surcharge effects on retaining walls
4. Rankine earth pressure coefficient
5. Coulomb earth pressure coefficient
6. Design of cantilever sheetpiles
7. Design of anchored sheetpiles (Wolff and Tipton, 1989)
8. Solution of dynamic pile driving equations and pile hammer selection (Hodapp, 1987)

Given a tool to quickly solve an equation once the concept has been taught and is understood, a student can obtain numerous solutions in the time formerly available for one. This allows problems such as the following:

1. Model a train above a wall as a single line load, two line loads, or a strip load and compare the difference.

2. Develop a set of footing width curves for a given soil and a range of column loads. Assess whether it is better to make footings wider or deeper.

3. Plot the value of the Coulomb earth pressure coefficient for a range of plausible values of the soil friction angle, wall friction angle and backfill angle and assess what parameters really affect the analysis.

4. Assess how uncertainty in the efficiency of a pile driving hammer and in the coefficient of restitution affect the calculated pile capacity.

5. Assess how the embedment depth of a sheet pile wall varies with the depth to the dredge line.

All of these are class problems that have been worked by Michigan State seniors in the past few years. These types of problems would be insurmountable with a calculator but are readily worked with pre-prepared spreadsheets. An example of the bearing capacity spreadsheet is shown in Figure 2.

Student response has been overwhelming. Course comments include statements such as: "we wish we had learned these techniques in our junior year," "we worked more complex problems than we ever worked but learned a lot more about design," and "I never realized problem

	B	C	D	E	F	G	H	I	J
1	BEARING CAPACITY				Ref: Das, p. 110--				
2	----------	----------	----------	----------	----------	----------			
3	Phi =	20.00	Degrees		Nc =	14.83		T. F. Wolff	
4		.35	Radians		Nq =	6.40		Mich State Univ	
5					Ngamma =	5.39		February, 1989	
6	B =	4.00							
7	L =	4.00							
8	----------	----------	----------	----------	----------	----------	q	Area	Q
9	c	Nc	Fcs						
10	1000.00	14.83	1.43				21234		
11									
12	gamma	D	Nq	Fqs					
13	100.00	3.00	6.40	1.36			2619		
14									
15		B	gamma	Ngamma	Fgs				
16	.50	4.00	110.00	5.39	.60		711		
17							--------		
18	For Educational Use Only					qu =	24564	16.00	393019
19						FS =	3.00		
20						qa =	8188	16.00	131006

Figure 2. Bearing Capacity Spreadsheet for Classroom Use.

modeling was as important as calculating the solution."

Summary and Conclusions

Although only another tool, the microcomputer spreadsheet is uniquely powerful for geotechnical engineering problems in practice and education. In the writer's opinion, it is often the tool of choice.

In practice, numerous geotechnical problems involve parameters that vary as a function of depth. Spreadsheets allow modeling in this fashion almost as easily as the assigning design values. In doing so, more information is preserved, and judgments can be based on the results rather than filtered through the analysis.

In education, judgment can begin to be learned by having the student explore the effects of variability in the soil parameters and in the models employed. Spreadsheets solutions allow this to be done in the time frame previously allotted to single solutions.

References

Finno, R. J., Cosamo, T., and Gitskin, B., (1989), "Results of Foundation Engineering Congress Pile Load Tests," Predicted and Observed Axial Behavior of Piles, ASCE Geotechnical Special Pub. No. 23, R.J. Finno, ed., June, 1989, pp. 338-355.

Hodapp, E. (1987), "DYNAPILE, A Dynamic Pile Equation Spreadsheet," unpublished undergraduate research report, Michigan State University.

Kulhawy, F., (1984), "Limiting Tip and Side Resistance: Fact or Fallacy," Symp. on Anal.and Design of Pile Foundations, ASCE, San Francisco, Oct., 1984, pp. 80-98.

Schmertmann, J., (1970), "Static Cone to Compute Settlement over Sand," JSMFD, ASCE, Vol. 96, No. SM3, pp. 1011-1043.

Wolff, T. F., (1989), "Pile Capacity Prediction Using Parameter Functions," Predicted and Observed Axial Behavior of Piles, ASCE Geotechnical Special Pub. No. 23, R.J. Finno, ed., June, 1989, pp. 96-106.

Wolff, T. F., and Tipton, R. W., "Spreadsheet Design of Anchored Sheetpiles," 7th National Conference on Microcomputers in Civil Engineering, Orlando, FL, Nov. 1989

Geological Education for Graduate Civil Engineers

Jerry D. Higgins,[1]
and
Gordon M. Matheson, MASCE[2]

Abstract

Many practicing engineers have recommended that geotechnical engineers need a strong background in geology so that they can effectively use geologic reasoning and techniques in geotechnical investigations. A survey of industry and government was completed to assess the level of geological knowledge that is desired by employers for graduating masters and doctoral level engineers. A survey of academic institutions was also made to assess what geologic background was typically being included in graduate-level geotechnical (civil) engineering programs. A comparison of the results of the surveys has been used to develop recommendations for the types of geologic background that are most desired by employers for a geotechnical engineer.

Introduction

Many references exist (Legget, 1979; Ter-Stepanian, 1985; etc.) indicating the importance of geological knowledge for geotechnical engineers. Traditionally, geotechnical engineers have dealt with subsurface investigations for construction of man-made structures. They, often in cooperation with an engineering geologist, design the investigation program, interpret the subsurface geologic conditions, collect and test samples representing the subsurface materials, and model the geologic conditions to predict stability and/or performance. All of these tasks deal with geologic materials and require the proper use of basic geologic reasoning, engineering geologic techniques, and an

[1]Associate Professor, Department of Geology and Geological Engineering, Colorado School of Mines, Golden, CO 80401.

[2]Senior Associate, Schnabel Engineering Associates, 274 Union Blvd., Suite 220, Lakewood, CO 80228.

understanding of the relationships between geologic implications and engineering behavior of the earth materials. Yet, many geotechnical engineering graduates do not possess the level of understanding of fundamental geologic principles and techniques required to accomplish these tasks or to thoughtfully consider the engineering geologist's opinion.

The objectives of this paper are to assess the desired level of geological education for graduate-level geotechnical (civil) engineers by employers, to compare that level to the perceived needs by academic institutions, and to make recommendations concerning geological curricula for graduate-level civil engineering programs.

Industry and Academic Institution Survey

To assess the desired level of geological education for graduate-level civil engineers by employers and compare that level to the perceived needs by academic institutions, two surveys were completed. The first survey contacted a broad cross section of industry. They were asked to specify the relative importance of various types of undergraduate and graduate geologic and engineering geologic course areas that they felt were important for graduate-level civil engineers. A total of twelve employers responded to the survey. These included three representatives in the consulting field that provide traditional geotechnical services, three representatives of the consulting field that provide environmental services, two representatives of major design firms that provide traditional geotechnical services, two representatives of design firms that provide environmental services, and two major government employers. The consulting firms contacted were all either regionally or nationally recognized.

The university surveys included contacting a broad spectrum of civil engineering programs with graduate geotechnical options. For this survey a total of 13 schools were contacted. Eight of the schools had either an engineering geology or geological engineering program on campus. The respondent was asked about the importance that their program placed on the same geologic course areas as asked of the industry representatives. Questions were also asked on the general relationship between the civil engineering and geology or geological engineering departments and how many credit hours of geologic related courses were typically taken by their graduate civil engineers. The survey included a list of geologic topics often offered in traditional geology, engineering geology and geological engineering programs. Table 1 presents the specific course areas included in the survey. The respondents were also asked what additional course areas not covered by these topics they felt were important.

The industry respondents were asked to rate each course area as either high, moderate or of low importance. The universities were asked to indicate whether a course area was considered to be an undergraduate deficiency, typically was taken for graduate credit, was required or recommended, in what department was it taught, if the course area was not available at that school, and to rank their opinion of the importance of the course area to graduate engineers.

It is recognized that the survey has by no means taken a large enough sample to adequately represent the full spectrum of opinion on the subject. It is felt, however, that sufficient data was collected to indicate the general attitudinal trends of both employers and academic institutions towards geologic education.

Results of Industry Survey

The results of the employer survey is presented in Table 1. The results of the survey were quantified by examining the ratio of the sum of the moderate and high interest responses over the moderate and low interest responses. If this ratio was greater than 1.0 then it was considered that a relatively high interest in the subject material was present. If the ratio was greater than 2.0 then it was considered that a very high interest in the subject material was present.

The results of the survey including all types of employers indicated undergraduate courses in geology for engineers, engineering geology, hydrogeology, and geological/geotechnical aspects of hazardous waste were of very high interest. Undergraduate courses in physical geology, structural geology, geomorphology, remote sensing and rock mechanics were of high interest. Historical geology, petrology/mineralogy, stratigraphy, field methods, engineering geophysics, and geochemistry were generally of little importance to employers.

At the graduate level, graduate courses in engineering geology, hydrogeology, ground-water modeling, slope stability and geological/geotechnical aspects of hazardous waste were of very high interest. Graduate courses in geomorphology, rock engineering/underground design/tunneling, and engineering geophysics were of high interest. Graduate level courses in remote sensing, structural geology and geochemistry were generally of little importance to employers.

For employers primarily interested in hazardous waste, an undergraduate level of understanding in geology for engineers, engineering geology, hydrogeology and geological/geotechnical aspects of hazardous waste management was almost universally given a very high priority. At the graduate level graduate courses in advanced engineering geology, hydrogeology, ground-water modeling, and geological/geotechnical aspects of hazardous waste were given very high priorities.

For employers primarily interested in the more traditional areas of geotechnical engineering, an undergraduate level of understanding in geology for engineers, engineering geology, hydrogeology and rock engineering were of greatest interest. At the graduate level, graduate courses in engineering geology,

TABLE 1
Results of Industry Survey
Geological Background for Graduate Civil Engineers

	Priority Ratio		
Undergraduate Course Areas	All Firms Contacted (12 Firms)	Traditional Geotechnical (6 Firms)	Environmental Geotechnical (6 Firms)
Physical Geology	1.6	1.0	3.0
Historical Geology	0.6	0.3	0.8
Petrology/mineralogy	0.7	0.7	0.7
Geology for Engineers	3.0	2.0	6.0
Structural Geology	1.4	0.8	1.2
Stratigraphy	1.0	0.8	1.3
Geomorphology	1.7	0.8	2.5
Engineering Geology	6.0	3.0	All High
Hydrogeology	2.2	1.3	6.0
Remote Sensing	1.1	0.6	1.5
Geologic Field Methods	1.0	1.0	1.0
Rock Engineering	1.4	1.3	1.5
Engineering Geophysics	1.0	0.8	1.2
Geotech/Geologic Aspects of Haz. Waste	2.0	0.8	6.0
Geochemistry	0.8	0.7	1.0
Graduate Course Areas			
Engineering Geology	3.0	2.0	6.0
Geomorphology	1.1	1.0	1.3
Hydrogeology	3.7	1.7	All High
Ground Water Modeling	2.0	1.0	6.0
Remote Sensing	0.6	0.6	0.6
Advanced Rock Engineering	1.3	1.3	1.3
Advanced Structural Geology	0.7	0.5	1.0
Tunnels/underground Design	1.2	2.0	0.8
Slope Stability	3.7	6.0	1.7
Engineering Geophysics	1.1	0.8	1.5
Geochemistry	1.0	0.8	1.0
Geotech/Geologic Aspects of Haz. Waste	2.5	1.0	All High

$$\text{Priority Ratio} = \frac{\text{Number of Moderate and High Responses}}{\text{Number of Moderate and Low Responses}}$$

hydrogeology, rock engineering/underground design, and slope stability were of most interest.

Results of University Survey

The results of the university survey were quantified in the same manner as the industry survey (Table 2). Undergraduate courses ranked at a very high level of importance by the universities were the introductory courses, geology for engineers and physical geology. No other undergraduate courses ranked as important. Engineering geology was ranked at a high to moderate level at six schools and was not offered at three schools.

Graduate courses considered to be of high priority included rock mechanics, slope stability, ground-water modeling, and geological/geotechnical aspects of hazardous waste. Even though these courses were ranked high, they are normally not required and often not recommended by the departments. The score for some courses was affected by the fact that some of the schools have developed program options such as hazardous waste management and some do not offer any courses on a particular subject. Several academic institutions that were surveyed offered specialty courses (in addition to those mentioned above) that geotechnical engineers took at least some of the time. These included seismology, geostatistics, geologic data analysis, unsaturated flow, clay mineralogy, and ground-water modeling.

Although a number of geological courses were ranked as having some level of importance by universities, the average credit hours in geological courses for a MS degree was 4.5 (range of 0-6) and for a Ph.D. was 6.7 (range of 0-15). The largest number of credit hours taken were typically at universities with engineering geology or geological engineering programs on campus. It appears that many students are not taking many of the courses ranked important by industry. These credit hours in some cases were taken in courses other than those listed as high priority by industry. The survey also indicated that the civil engineering faculty generally had significant interaction with geologic faculty at schools with engineering geology or geological engineering programs, but less interaction with faculty in traditional geology departments.

Discussion and Conclusions

The survey results indicate that there appears to be a gap between the background industry needs (or wants) in its new employees and what is included in most graduate programs. To compound the problem, new environmental regulations will cause the role of the geotechnical engineer to become even more diverse. Civil engineering graduate programs must continue to produce traditionally trained geotechnical engineers as well as those specializing in environmental issues. In general, the employer survey indicated that environmentally oriented geotechnical engineers require more geological background than traditional geotechnical engineers. It should be recognized that

TABLE 2
Results of University Survey

	Priority Ratio		
	School w/ EG/GE[1]	Schools w/o EG/GE[2]	Combined Ratio
Undergraduate Course Areas			
Physical Geology	-	-	4.0
Geology for Engineers			
Historical Geology	-	-	All Low
Petrology/mineralogy	0.1	All Low	0.1
Structural Geology	0.1	0.4	0.2
Stratigraphy	-	-	All Low
Geomorphology	0.5	0.8	0.6
Engineering Geology	0.8	0.5	0.7
Hydrogeology	0.3	1.0	0.5
Remote Sensing	0.3	All Low	0.2
Geologic Field Methods	0.1	All Low	0.1
Rock Mechanics	0.1	0.3	0.2
Engineering Geophysics	0.1	0.6	0.3
Geotech/Geologic Aspects of Haz. Waste	0.3	0.7	0.4
Geochemistry	All Low	0.3	0.1
Graduate Course Areas			
Engineering Geology	1.4	Not Offered	0.7
Geomorphology	0.4	All Low	0.2
Hydrogeology	0.6	1.0	0.7
Ground-Water Modeling	0.7	2.0	1.0
Remote Sensing	0.3	All Low	0.2
Advanced Rock Mechanics	1.8	0.3	1.0
Advanced Structural Geology	All Low	0.4	0.2
Tunnels/underground Design	1.0	0.5	0.9
Slope Stability	2.0	0.5	1.0
Engineering Geophysics	0.3	0.2	0.2
Geochemistry	0.3	0.2	0.2
Geotech/Geologic Aspects of Haz. Waste	2.0	0.5	1.1

$$\text{Priority Ratio} = \frac{\text{Number of Moderate and High Responses}}{\text{Number of Moderate and Low Responses}}$$

[1]Schools surveyed that have engineering geology or geological engineering programs on campus.
[2]Schools surveyed that have traditional geology programs only.

the current growth industry in geotechnical engineering is in the environmental field.

The survey of industry suggested a list of undergraduate and graduate courses that were considered important for a geotechnical engineer. Only about 24 credit hours are available in a typical masters level program for completion of core requirements and inclusion of electives. Of this, it is unreasonable to assume that more than six hours (in most cases) can be devoted to geology or geological engineering electives.

After consideration of industry responses and the constraints on available credits, the authors would like to suggest the inclusion of a specific geologic course content in graduate geotechnical engineering programs. It is recommended that graduate geotechnical engineers should be required to take or have taken 1) geology for engineers or physical geology, 2) undergraduate and/or graduate engineering geology, and 3) undergraduate and/or graduate hydrogeology at some point in their undergraduate or graduate programs. The engineering geology course should include elements of applied rock engineering, geological aspects of slope stability, and geotechnical aspects of hazardous waste disposal in addition to the normal subject matter. The hydrogeology course(s) should include well design and the basic applications of ground-water modeling. In addition to these core courses, a student planning to do traditional geotechnical work may want to take a course in rock engineering and/or geomorphology. A student specializing in environmental issues should take additional graduate courses in geological/geotechnical aspects of hazardous waste management and ground-water modeling. This would involve a commitment of approximately 6 to 12 credits depending on the previous undergraduate program of a particular student. Some of these courses, such as geology for engineers or physical geology, should be considered as undergraduate deficiencies. Students in a Ph.D. program could include additional courses.

At many institutions, some of the recommended courses are not offered or are inadequately taught for the civil engineer. At major universities where this is the case, it may be practical to develop the required course content in the civil engineering department by hiring an engineering geologist or geological engineer (possibly an adjunct), or by developing cooperative programs between the civil engineering and geology departments.

References

Legget, R.F., "Geology and Geotechnical Engineering," Journal of Geotechnical Engineering, ASCE, Vol. 105, No. GT3, March 1979, pp. 342-391.

Ter-Stepanian, G., "Geological Aspects of Geotechnical Engineering," Proceedings of the Eleventh International Conference on Soil Mechanics and Foundation Engineering, San Francisco, CA, August 1985, pp. 377-389.

Soil Mechanics Laboratory: A Revised Agenda

Edward F. Glynn[1], M. ASCE, William B. Fergusson[1]

Abstract

The paper discusses a new soil mechanics laboratory course currently being taught at Villanova University. The course is designed to enrich and broaden the students' understanding of geotechnical engineering rather than hone their laboratory skills.

Introduction

Soil mechanics laboratory courses have traditionally emphasized the technical aspects of laboratory testing. Although there is educational merit in developing the students' laboratory skills, most students will never be called upon to perform as laboratory technicians during their professional careers. However, most students will eventually be involved in civil engineering projects in which the designs are largely dictated by geological concerns. Therefore, a large portion of the time devoted to laboratory work could be better utilized in developing the students' understanding and appreciation of soil behavior and in emphasizing the role of geotechnical engineering in civil engineering design.

Course Philosophy

The goals of the revised soil mechanics laboratory are: (1) expose the students to laboratory tests which quantify engineering properties of soils; (2) introduce the students to soil identification, classification and site investigation; (3) demonstrate to the students that geotechnical engineering is one facet of the design process which typically begins with engineering geology and ends with soil-structure interaction; (4) emphasize

---

[1]Department of Civil Engineering, Villanova University, Villanova, PA 19085.

the empirical nature of geotechnical engineering by examining case histories of geotechnical engineering successes and failures; (5) develop the students' communication skills - through laboratory reports and an oral presentation in which each student presents a case study to the class.

The specific objectives of the laboratory are: (1) each student should learn to identify soils and to evaluate (at least qualitatively) their engineering properties; (2) the students will observe, and participate in, the operation of soil testing equipment and see how different soils respond to the same test; (3) the students are shown the intimate relation between soil mechanics and engineering geology through site investigation and evaluation; (4) the students will study and discuss the case histories of failures to benefit from the oversights of others.

The new course as currently constituted does not fulfill all the goals and objectives; however, it is a significant improvement over the traditional course. This paper focuses on two aspects of the new course: laboratory testing and case histories.

Course Constraints

The soil mechanics laboratory course at Villanova meets once a week for three hours. The semester consists of 13 weeks of classes plus a final examination period. Each section of the course typically has 12 to 18 students. The authors felt that each laboratory session should include a module of activities which the students can complete within 3 hours. Aside from preparing case histories, students should not have to devote outside time to the course.

Laboratory Testing

The traditional soil mechanics tests such as gradation analysis or consolidation test must be drastically altered to accommodate the new course format. A typical session might include a 30 minute introduction/discussion, a 60 minute experiment, a 30 minute report and 60 minutes of case histories. One hour is too short to run most traditional tests; however, it is enough to familiarize and enlighten the students if the experiment is creatively designed. For example, in the gradation exercise the students are divided into 6 groups. Each group is given a gradation curve and asked to blend a soil that meets the gradation. Raw material is available as dishes of sand

and gravel sorted by size: 3/4" to 3/8", 3/8" to #4, #4 to #10, etc. Once the groups blend their soils the 6 samples are placed around the laboratory. Each group receives a set of the six gradation curves and must match the soils with the gradations. The students seem attentive during the exercise and develop an appreciation for particle size distributions--perhaps better than the appreciation derived from running a single gradation analysis. The report consists of simple calculations and questions on test procedures and related topics. The reports are important tools in helping students correlate laboratory observations with lecture material. For example, the gradation report lists glacial landforms and asks which landforms might be good sources of clean sand and gravel.

There are some disadvantages to the abbreviated experiments. Some procedures such as consolidation testing do not lend themselves to an abridged format. The new experiments generally require much more preparation time on the instructor's part. Finally, the students do not get as much "hands on" experience with test equipment. The last problem is not particularly critical at Villanova because students may elect to take a second course in geotechnical engineering that includes a laboratory.

Case Histories

The case histories probably represent the single most important feature of the laboratory course. The cases were chosen using the following criteria: (1) the case history should have a definitive reference which includes figures and/or tables suitable for presentation to a group; (2) the case history should illustrate the use (or abuse) of some fundamental concept in geotechnical engineering; (3) the references should be technical enough to satiate (but not overwhelm) engineering students whose geotechnical backgrounds include one semester of engineering geology and a concurrent course in soil mechanics.

Developing the library of case histories was a challenge. The selections tended to focus on failure because failures are generally more enlightening and more interesting to students. Geotechnical engineering has many well documented failures. Many of the examples, however, involve failure modes or mechanisms which are too complex for the students to grasp with their limited backgrounds.

Appendix 2 lists some of the case histories used in the course. The complete inventory encompasses a broad selection of geotechnical topics including slope stability, settlement, lateral earth pressures, karst terrains and groundwater problems. The references used for the course are not necessarily the most recent or the most detailed account of the failure. For example, Kiersch's article (1983) on the Vaiont reservoir disaster serves as the primary reference for that case history. The Hendron and Patton report (1985) on Vaiont is too technical for the undergraduates. Not all of the case history involve spectacular failures. Some relatively obscure failures provide valuable insights into engineering practice--or the lack thereof. Kelly Barnes Lake Dam (Sowers, 1978) was a 40 foot high structure built with virtually no engineering design and no construction documentation. Thirty eight people died when the dam failed in 1977.

Each case history is summarized in the manner shown in Appendix 3. The Gros Ventre Slide is a typical example. The major reference is a contemporary account of the failure from a technical journal. The case history is updated with a modestly technical article from a recent periodical. The last line in the summary indicates that photographic slides of the failure are available.

Each student receives a complete set of summaries for the 23 case histories used in the course. The student uses the summaries to select a case history at the beginning of the semester. Note that the summaries do not include page numbers for the references. The students are encouraged to chose topics based on their interests, not on the perceived work load. The library call numbers are given for the truly ambitious students who wish to preview the topics.

The requirements for the presentations are outlined in Appendix 4 which is a copy of the instructions distributed to students. Each student receives a copy of the major reference for his/her project and must prepare a one-page written synopsis and a 15 minute oral presentation for the class. The synopsis and presentation should include: (1) a description of the problem; (2) a discussion of the geotechnical aspects of the problem; and (3) a summary of the remedial measures. The class has an opportunity to ask the speaker questions. The presentation concludes with a brief quiz on the topic.

The students prepare the presentations on their own time; however, all their efforts should be directed towards understanding the case history. No additional research is required. The major reference is an adequate source of information and it provides figures that can be used to make transparencies.

Conclusions

Villanova has tried to redirect the students' efforts in the new soil mechanics laboratory course. The course has been well received by the students and they seem to have a better appreciation of the engineering characteristics of soils and how the properties influence geotechnical design.

The course continues to evolve as new exercises are implemented and case histories are added.

Acknowledgement

This curriculum development project was funded by the Westinghouse Educational Foundation.

Appendix 1. References

Sowers, G.F. (1978). "Reconnaissance Report on the Failure of Kelly Barnes Lake Dam, Toccoa Falls, Georgia." Committee on Natural Disasters, National Academy of Science, Washington, D.C.

Hendron, A.J. and Patton, F.B. (1985). "The Vaiont Slide, A Geotechnical Analysis Based on New Geologic Observations of the Failure Surface." Technical Report GL-85-5, Department of the Army, Washington, D.C.

Kiersch, G.A. (1983). "The Vaiont Reservoir Disaster." in Focus on Environmental Geology, Oxford University Press, New York, N.Y.

Appendix 2. Case Histories (Partial Listing)

Kelly Barnes Lake Dam (Toccoa Falls, Georgia)
Vaiont Reservoir (Vaiont Valley, Italy)
Gros Ventre Slide (Teton National Forest, Wyoming)
Palace of Fine Arts (Mexico City)
Federal Courthouse (Philadelphia, Pennsylvania)
Anchored Bulkhead (Kake, Alaska)
Sinkhole Flooding (Bowling Green, Kentucky)
Love Canal (Niagara Falls, New York)
Great Salt Lake Railroad Embankment (Utah)

Appendix 3. Sample Summary Sheet

Case 200 Gros Ventre Slide

Subject: Slope Stability

Description: Landslide debris dammed valley; impounded water overtopped debris

Location: Teton National Forest, Wyoming

Date: 1925

Major Reference: "Landslide and Flood at Gros Ventre, Wyoming," William C. Alden, *Transactions of American Institute of Mining & Metallurgical Engineers*, Volume 76, 1928. (Falvey: TN1.A5)

Other References: "The Day the Mountain Came Down," *Off Belay The Mountain Magazine*, No. 37, February, 1978.

Comments: Slides Available

Appendix 4. Instructions for Presentations

Purpose

1. To expose the class to case histories of projects which involve geotechnical engineering problems.

2. To give each student an opportunity to research one history and to present his/her findings to the class.

Case Histories: Each student has a complete set of summary sheets for the 23 case histories used in CE 3901. The student can peruse the summaries before selecting a topic.

Synopsis Format

1. Each student must prepared a one-page synopsis of his/her case history. The synopsis should address the same items as the presentation. The entire synopsis (including figures) is limited to one side of an 8 1/2 x 11 sheet.

2. The synopsis must be available on the Thursday morning of the presentation. I will make copies and distribute them to the class just prior to your presentation.

Presentation Format

1. If necessary, I will provide a brief introduction to the topic.

2. The class will have time to read the synopsis.

3. The student will present his/her case to the class in 10 to 15 minutes. The presentation should include:

   a) A description of the problem.
   b) A discussion of the geotechnical aspect of the problem.
   c) A summary of the remedial measures.

4. The class will have an opportunity to ask questions.

5. The presentation will conclude with a short (2 or 3 questions) quiz administered by me.

Resources

1. Each student will have a copy of the "Major Reference" for his/her project. This Major Reference provides sufficient information for the presentation.

2. Students should feel free to discuss their project with me.

3. Students can and should take advantage of visual aids such as transparencies or slides. I will make the transparencies as long as you prepare the originals.

Grades

Your grade in CE 3901 will be determined as follows:

Laboratory exercises:	50%
Final Quiz:	10%
Synopsis:	15%
Oral Presentation:	15%
Case History quizzes	10%

TRANSPORTATION 2000- : A CURRICULUM CONCEPT

Venkataraman N.Shankar[1]
Student Member ASCE

Abstract

This paper addresses the very primordial basis on which transportation engineering as a field in itself has evolved. It deals with the issues that play an important role in the development of a transportation engineer. It also throws light on some of the needs of the hour that this field faces. A broad based evaluation of these needs has been done and suggestions put forward for the implementation of a curriculum, that would aid in a healthier and broad minded development of students into transportation engineers.

Introduction

The world is getting cluttered with people and buildings. Man in his blind craze for power and materialistic comfort has now landed himself in a predicament wherein many of his own creations are proving to be liabilities than assets. A classic example is the automobile. While it has brought greater flexibility, mobility as well as proved to be a reliable channel of economic growth, it has also made him its near captive user. Most of our daily activities are centered around it so much so that it is no longer a comfortable medium of travel. On the other hand its use has had a stressful effect on not only its riders but also on the environment and other components of the social system that derive their impetus to progress through the increased accessibility and mobility it offers. Such a situation which forseeably in the future could get exacerbated , coupled with the apparent fading of transportation awareness in the society warrants serious soul searching on the part of the transportation community including the up and coming student that is a prospective transportation engineer.

As some of us are aware , the issues currently at a premium in the above mentioned context and presumably very much so in the future are availability of fuel resources, both existing and alternative, atmospheric pollution, ecological equilibrium, urban sprawl, economic growth and global trade balance, international relations as well as societal attitudes and changing lifestyles. To realize that transportation has a major role to play in each of the above

1 Transportation Engineer, The Transpo Group, Inc., 14715, Bel-Red Road, Suite 100, Bellevue, WA 98007.

mentioned issues is the least we can do for the betterment of the future. There is considerable literature that has highlighted transportation's significance and diversity in its application to the real world situation. In (White,1988) the importance of transportation as a major contributor to the advancement of our economic welfare has been emphasized. In (Greene et al, 1988) the significance of our transportation energy consumption in the context of world energy market has been addressed. In (Lowry,1988) the balance between transportation costs and costs of compatibly configured land uses has been dwelt upon. His review of recent trends in human settlement in the United States gives us an insight into the planning process we could embark on in the future. In (Suhrbier&Deakin,1988) emphasis has been laid on the importance of environmental consideration in the transportation decision making process. In (Kitamura,1988) attention has been brought to the true picture that needs to be addressed in solving transportation problems and planning for the morrow. According to him and as is rarely realized, traffic congestion is only a symptom and the real problem lies in the life-style of the people. How lifestyle primordially affects travel behavior and hence accessibility as well as mobility is very important to us,if we have to take a step forward in our decision making process. Herein lies the challenge for the transportation community.

In essence, what is desirable for the future and what is required of us is the effort to deal with these afore mentioned issues at the grass root level and not continue to address them on a superficial and incremental basis as we have been doing. There are various forums in society where such an approach can be adopted on a coordinated , continued and comprehensive basis and in a way quite similar to the planning process that transportation engineers are aware of. However attention in this paper will be confined to one of these forums , primarily the educational system and to an extent certain institutions that have an interaction with it, for example, the industry

The University

The University, ever since its functional inception has been recognized as the center of academic excellence. More often than not the phrase 'academic excellence' has been misinterpreted and misrepresented as 'learning' and this reflects a sad state of affairs. The academic community is satisfying itself on this wrong notion. Learning is a process of a lifetime, not a four year or two year graduate course in which time the student assimilates incredible amount of information but not necessarily true knowledge. Learning is a benefit that translates information into useful knowledge and is derived more through experience than any alternative process. Along with experience it forms a virtuous circle with inter-related contributions to each other. While the University ideally should be a center of learning , time constraints obviously restrict that taking effect. What needs to be done to closely approximate this effect is a revamping of the educational system and the curriculum that imparts so called 'knowledge'.

The Curriculum

This forms the core of the educational process. It offers a product that is the assimilation of different thoughts. However noble it might seem in its definition, it has rarely lived upto it in its implementation, i.e, as a process through which all the useful information assimilated is handed down to the up and coming generation as a benefit of learning. A couple of features that identify this drawback are:

- Insufficient practical experience
- A grade oriented evaluation process

Insufficient practical experience addresses the fact that students are not given ample opportunities to translate their subject matter on the field. Laboratory experiments , to an extent satisfy this need but more needs to be done on this front. More often than not students have ready made clearly defined tasks and goals to be achieved in these experiments that they do not face a situation in which they have to wrestle with various options before defining the goals and objectives of the experiments themselves. Moreover the chance to operate within field imposed constraints doesn't arise too often at all. This ultimately reflects the fact that the curriculum fails to replicate a true real time professional process, which is what the student eventually fits into.

Grade oriented evaluation is a bane of the modern educational system. It derives quite as much importance in the selection process of students for advanced studies and also occasionally in the professional arena. The root cause can be traced back to the system tree itself. High school is visualized as the starting point of a responsible career upfront for the student but rarely used as an effective mean to that purpose. When students attend the University the realization that it is a different cup of tea hits them hard and too soon that they so often tend to associate this with a need to secure high grade point averages and not knowledge instead. For them excelling becomes a grade associated frenzy. While the system in all its earnestness should have nourished in them a desire to learn , it has only succeeded in providing a goal oriented four year process. Part of this obviously lies in the fact that there is a big jump from high school to University. It would be very desirable if this was modified into a smooth transition in responsibilities.

While the above mentioned features plague any curriculum because of their generic nature, transportation engineering continues to have its own plethora of drawbacks. It has ever since its definitional inception remained in the shadow of the traditional fields of civil engineering like structural , hydraulics and geotechnical engineering. These continue to draw more students sheerly by their 'beaten track' appeal and not as much by their ability to satisfy the true aptitudes of students. While the importance of these fields remains and deserves to be maintained, this young field in transportation needs to be nursed into significance. It is a burgeoning field that encompasses a wide range of applications in engineering, science and humanities. Through its diversity in applications and the versatility it offers in the development of the transportation engineer, it has come to occupy a regal place among a plethora of other young fields like CAE, AI etc. However the fact remains that it has not captured the attention of very many students as much as these fields have. Some of the drawbacks that could be attributed to this are:

- A lack of transportation awareness among students
- Lack of a strong computer based appeal which eventually relates to the job market
- Institutional constraints
- Lack of adequate industrial interaction

Transportation awareness as mentioned before lies in the student's understanding of the lifestyle of people and the overall process of landuse-transportation interaction. In order to cultivate this ,students need to be exposed to seminars and debates that involve key public issues of land use and transportation development as well as be afforded opportunities to work on projects that would address these aspects.

Computer based appeal is such a strong attraction for any student today because it is very much in consonance with the progress of the world into the computer age. Students evaluate a course by the opportunities it would give them , to use computers and eventually towards the realization of such skills on the professional front. While many transportation course offerings do accommodate this to an extent, not many of these, in the student's point of view are felt to be particularly advantageous on the professional front. This is partly because while other fields like CAE or AI would help the student develop skills on a more generic level like programming, transportation software is usually "off the counter" and more specific. More opportunities hence need to be thrown open for students to develop generic abilities while using computers since

these would help them relate their abilities to any situation that demands the use of the computer.
Institutional constraints address the lack of adequate qualified staff to impart transportation knowledge. As a field ,while it has not caught on among students, neither has it among faculty . Many civil engineering departments in the country boast of strong structural, hydraulics or environmental programs but not as many in transportation. It is the strong feeling of the author that this overwhelming zealousness in protecting certain fields is not desirable for the healthy future of civil engineering.
Industrial interaction with transportation programs at universities needs to be developed to a great extent. Students should be given the opportunity to work in industry , while attending school, on a cooperative basis. This is bound to reward them to an invaluable extent in terms of professional benefits and translation of theoretical expertise.

Recommendations

To summarize, the following recommendations have been drawn out on a near term and long term basis. The near term addresses the next five year period while the long term addresses its transition into the 21st century. The near term recommendations are:

- Introduce transportation courses that address lifestyle issues and travel behavior.

- Encourage interdepartmental course programs. For example, courses jointly offered by transportation departments with urban planning, architecture, economics, geography, policy planning and business administration need to be set up.

- Develop cooperative schemes for transportation students to work in industry.

- Lay more emphasis on development of generic abilities in computer use and applications in transportation engineering.

- Encourage student seminars in ASCE meetings of regional student chapters, wherein the chapter representatives could discuss ideas and regional issues that interest students.

- Encourage greater participation of students in public debates, and seminars addressed by professionals as well as in conferences.

- Encourage setting up of regional "OPEN HOUSES" in transportation engineering.

The long term recommendations are:

- Revamp the educational system, i.e, facilitate a gradual transition in student responsibilities from high school to University.

- De-emphasize grade point based performance and set up 'credit/no credit' courses that mandatorily involve industry interaction on the student's part.

- Deregulate faculty selection procedure in civil engineering departments and encourage addition of transportation faculty.

- Encourage "adjunct faculty" positions for reputed transportation professionals in the University.

- Restructure course offerings at secondary and high school level and lay more emphasis on transportation issues at a fundamental level.

References

1. White, Robert M: "Beyond the Millennium- Transportation and the Economy", Transportation Research Board, Special Report 220, 1988.
2. Greene, David L., Sperling, David and McNutt, Barry : "Transportation Energy to the Year 2020", Transportation Research Board, Special Report 220, 1988.
3. Lowry, Ira S: "Planning for Urban Sprawl", Transportation Research Board , Special Report 220, 1988.
4. Suhrbier, John and Deakin, Elizabeth A : "Environmental Considerations in a 2020 Transportation Plan: Constraints or Opportunities?", Transportation Research Board, Special Report 220, 1988.
5. Kitamura, Ryuichi : "Lifestyle and Travel Demand", Transportation Research Board, Special Report 220, 1988.

Attracting Students for the 21st Century

Harold Deutschman[1]

Abstract

In the next century, the United States will undergo a sharp change in demographics. The new workers entering the labor force will be predominantly women, blacks, and hispanics, all of whom are greatly under-represented in the engineering fields. This paper identifies the problem and offers programs to meet this new challenge in Civil Engineering.

According to figures from the U.S. Census Bureau, the United States is facing the challenge of changing demographics. Black and hispancis currently make up 25 percent of our school children; by the year 2000 they will make up 47 percent. In the years between 1985-2000, the U.S. Department of Labor estimates that there will be 25 million new workers, made up of the following: White Men 15%, White Women 42%, Non-White Men 7%, Non-White Women 13%, and Immigrant Men and Women 23%. At the same time, there are estimates of shortfalls of science and engineering professionals by the year 2000 by more than 550,000 workers. With the new workers entering the labor force being made up of primarily minorities and women, it is important to look at the current under-representation of these groups in science and engineering. For example, blacks make up 12 percent of the U.S. population and only 2 percent of all employed scientists and engineers. Similarly, hispanics comprise 9 percent of our population but only 2 percent of all employed scientists and engineers. Women make up 51 percent of the population, 45 percent of the workforce, but only 11 percent of all employed are scientists and engineers. The field of Civil Engineering is considered to be the area of engineering that reaches out to the public via issues of housing,

[1]Professor of Civil and Environmental Engineering, New Jersey Institute of Technology, Newark, N.J. 07102

transportation, environment, quality and quantity of water supply and in construction. With a mismatch of the composition of the practicing civil engineer to the general population, this may lead to a misunderstanding of issues and an improper setting of agendas for the future.

Along with the changing composition of the population of the United States (as we enter the 21st Century), and the under-representation of women and minorities in the Civil Engineering field, the problem of attracting new students to the field is the critical one. Students must be introduced to the potential of Civil Engineering at an early state of their school careers so that a proper sequence of math and science courses may be followed. In addition, many minorities and women are unfamiliar with the role of Civil Engineering in urban life, and generally would not even consider careers in this area. This problem, often combined with undeveloped curriculum and inadequate and negative counseling, severely limit the opportunities for these students in Civil Engineering. There is a need for the profession, through its societies, organizations, consultants and municipalities to reach out to these students and support programs at various levels to encourage minorities and women to enter the profession of Civil Engineering.

One such avenue of approach is through summer and academic year pre-college programs. In New Jersey, the State Board of Higher Education has identified a special pre-college program for junior high school students at the New Jersey Institute of Technology (NJIT) as a model for the state. This program, entitled Introduction to Urban Engineering has a primary focus on Civil Engineering covering such topics as urban planning, transportation engineering, energy systems, and environmental science, including instruction in computer programming and communications. A follow-up program in Urban Engineering is designed for high school students.

Modules of the Urban Engineering Programs

Transportation

This module emphasizes the relationships between transportation and city growth and development. The students learn the role that transportation has taken in shaping and developing the modern city. In addition, the transportation problems of the cities and potential solutions such as staggered work hours, expressway ramp metering, exclusive bus and car pool lanes and car pooling are discussed. The students apply mathematical equations to solve a typical ramp metering problem in which a number of entrance ramps are metered to enable the expressway to flow at a maximum capacity. To see, first hand, the principles of traffic control and surveillance and how it applies to other methods of transportation, trips are taken to the Newark Airport in Newark, New Jersey, and to the New Jersey Turnpike Surveillance Center in East Brunswick, New Jersey.

The students also learn how to use mathematical techniques for the optimum design location of shopping centers, hospitals, parks, and recreational centers and employment centers.

The student develop concepts for a new city, designed for the future, hopefully eliminating the mistakes of the past and integrating the transportation and land use functions of the city, so that the quality of life in the city is greatly improved (compared to current cities). The student teams also name their cities. The cities have such features as car free downtown areas, bus loops through each neighborhood and to the downtown areas and major attraction centers such as sport complexes, gambling centers, and recreational centers. The cities also provide for a transportation link to each of these centers.

Environmental Science

In the environmental science module, an awareness of the environmental problems facing future engineers is presented. Water supply, waste water treatment, and air pollution are tbe main topics covered in the lectures and discussions. Water supply topics include potential uses of water, availability of water supplies, methods of forecasting water demands and future water needs. Weste water treatment is discussed by stage of treatment (i.e., primary, secondary, and tertiary).

Urban Planning

The culmination of the Urban Engineering Program is the conceptual design of a new city of 10,000 residents on 193 acres of land in Secaucus, New Jersey. The students, applying their knowledge and background developed from the program in urban planning, transportation engineering, energy systems, and environmental planning, plan for a new town in the Hackensack Meadowlands area. With guidance and direction, the urban engineering students design for a full community, including parks, schools, shopping facilities, and mass transit.

Follow-Up Activities

Program directors, administrative and teaching staff maintain contact with students to provide assistance to the students in such matters as filing of applications to respective colleges and universities, providing letters of recommendation and general counseling, as well as any other services which the students may deem pertinent.

Students participating in the program have expressed a strong desire for academic year assistance. Students are encouraged to apply to our academic year pre-college programs and also invited to attend any other science of career workshops offered at NJIT.

Evaluation of Program

The student evaluation survey indicated the following:

1. Seventy percent of the students in the program felt that their summer experience affected their ideas about a possible career in science or engineering; many showing interest in Civil Engineering.

2. Ninety-five percent of the students would recommend the Urban Engineering Program to others.

3. Seventy-one percent of the students made a preliminary decision of a career goal as a result of the program.

It is recommended that the universities in urban areas and the profession jointly sponsor pre-college urban engineering programs such as the one illustrated at the New Jersey Institute of Technology. It is through such programs that we can attract a sufficient number of Civil Engineers to meet the needs of the 21st Century.

PREPARING THE TRANSPORTATION PROFESSIONAL FOR THE 21st CENTURY

by

Dr. Michael D. Meyer[1]
Associate Member
ASCE

ABSTRACT

The transportation engineering profession has undergone some major changes in the past ten years. The likely change over the next ten years will be even more profound. This paper examines the current status of transportation engineers in public sector agencies. Data from a recent state department of transportation survey of professional needs that will occur over the next ten years is used to portray the personnel problems that will be faced by one important public sector employer of transportation engineers. The paper then examines the likely engineering work environment in the 21st Century and the skills that will be needed by transportation engineers.

INTRODUCTION

State departments of transportation (DOT's) employ about 41,000 transportation professionals, more than are employed by any other type of public organization. About 30,000 of these professionals are civil engineers working in all facets of the planning, design, construction and maintenance of a state's transportation system. A major national study of the professional needs of transportation agencies conducted by the Transportation Research Board in 1985 concluded that staff retirements and changing professional skills would likely create a serious staff shortage over the subsequent five years [Transportation Research Board, 1985]. A national survey of state departments of transportation in 1988 conducted by the American Association of State Highway and Transportation Officials (AASHTO) confirmed that in some areas serious shortages of qualified staff did exist, but that in

1 Associate Professor, Georgia Institute of Technology Atlanta, GA.

others there did not seem to be a problem. The purpose of this paper is to summarize the information from these two national studies and to relate this information to the future needs of state DOT's, and by extension to other transportation agencies. In particular, the results of both studies are used to suggest skills and knowledge that will be needed by the successful transportation professional of the 21st Century.

TRANSPORTATION RESEARCH BOARD'S STUDY OF PROFESSIONAL NEEDS

The Transportation Research Board (TRB) conducted a study in 1985 of the nation's professional needs in the transportation sector. The results of this study are too extensive to repeat in this paper. However, several observations made in the study report merit special
attention in that they outline some of the major issues that state DOT's, in particular, are facing. First, the report notes that employment levels in state DOT's have generally been closely related to expenditures. During the late 1950's and early 1960's when the states were in the midst of constructing the Interstate Highway program, state DOT employment jumped by almost 100,000 employees. In the 1970's when construction levels tapered off and spending declined, the state DOT's experienced a net loss of almost 50,000 employees.

The importance of this spending--employment relationship is that beginning in the early 1980's, states once again began to spend more on their transportation systems. An increase in the federal gas tax and similar increases in numerous states provided the resources for a substantial expansion in the highway program in many areas of the country. It is likely that state highway programs will continue to grow in future years, if for no other reason than to keep pace with the increasing demand on transportation facilities (Federal Highway Administration estimates show an annual 2.5% increase in vehicle miles of traveled nationally to the year 2005), and to support expected growth in the economy (estimates vary from two to four percent annual growth in gross national product). It is likely therefore that state DOT's will continue to need new transportation engineers to replace those who retire and to add needed capabilities for managing the state's program.

The report also noted that, in many cases, the focus of the state's highway program was shifting. In 1954, 76 percent of the engineers in state highway agencies were involved in location studies, design and construction with only 5 percent in maintenance. In 1984, 66 percent of the engineers were involved in design and construction and more than 9 percent were involved in maintenance. Table 1 shows what state DOT officials felt would most likely be the changing emphasis in their highway programs over the following five years. This changing emphasis was reflected in these officials' estimates of required staff skills. In particular, computer programming and rehabilitation/recycling skills were identified as having more importance, with a substantial number of states also reporting increased needs for the more traditional design, construction, and safety skills. The skill needs least needed were in accounting, supervisory, drafting and surveying by technicians, and environmental impact analysis.

Many state DOT's use consultants as an extension of their own staff. According to the TRB study between 72 and 89 percent of the states contract for location studies, environmental analyses, geotechnical studies, right-of-way acquisition, and more than half the states contract for their construction inspection and materials testing. Because of the important role consultants play in state transportation highway programs, the TRB study also surveyed consulting engineering firms. Nearly 80 percent of the firms responding expected rehabilitation work to increase, and nearly 75 percent expected bridge and structure work to increase. The most important skill requirements anticipated by the consulting firms over the near future were computer operators, project management, construction inspectors, structural engineering specialists, computer programmers, highway designers, computer systems analysts, drafters, traffic engineers, and supervisory personnel.

In its conclusion, the study recommended that state DOT's take six steps to meet future professional needs: 1) revamp mid-level and management engineer training programs, 2) review internal certification practices, 3) exploit the potential of computers, 4) use more consultants, 5) monitor and adjust incentives to keep and motivate staff, and 6) increase the versatility and depth of technicians.

AASHTO'S STUDY ON PROFESSIONAL RECRUITMENT AND RETENTION

The American Association of State Highway and Transportation Officials, consisting of every state DOT in the U.S., conducted a survey of its members in mid-1980 to determine their experience with recruiting and training transportation professionals [AASHTO, 1988]. Forty two state DOT's responded to the survey along with a provincial highway ministry in Canada. Twenty two of the respondents reported that they expected serious problems recruiting civil engineers over the next five years, 19 felt that they would have problems finding traffic and construction engineers, 18 with design engineers, 17 with materials engineers and bridge/structures engineers, 13 with maintenance engineers, 10 with right-of-way engineers, 2 with planning engineers, and one with traffic engineering specialists. The survey also provides some interesting information on the current status of salaries for entry-level engineers. Entry-level salaries were similar in most of the states. The lowest was in Indiana at $16,749 to a high in Alaska at $28,236. After two years of work, the salaries improved in every state with the lowest in North Dakota at $19,656 and the highest in Alaska at $33,648. Fourest$30s000eported a salary over

---

TABLE 1--CHANGE IN HIGHWAY PROGRAM EMPHASIS

States Program	Percentage of Responding More	Less	Same
Rehabilitation	88	0	12
Bridges and Structures	70	2	28
Maintenance	68	2	30
Traffic Operations	46	0	54
Major Construction	24	34	42

Source: [Transportation Research Board, 1985]

Starting salaries for those having a masters degree varied across the country. Eleven states reported that the starting salaries for these engineers were the same as those having B.S degrees. Twenty one states reported a higher salary with the highest being Arizona at $35,636.

With regard to other modes, most states reported that starting salaries were similar to those for highway engineers. In some cases, however, there were differences. New Hampshire, for example, reported a starting salary of $41,352 in aviation, $36,757 in rail, and $21,742 for a safety expert. A similar salary pattern was found in Vermont, California, Connecticut, and Pennsylvania.

State agencies were also asked to list those strategies used to recruit college students to their agency. The following mechanisms were used most often.

Summer Employment: Thirty one states reported that they had a formal summer employment program for college students.

High School Contacts: Thirteen states said that they provide on an organized basis information to high school students and guidance counselors on transportation as a career.

College Contacts: Thirty states replied that they provide information on transportation careers to engineering students within their states. This information was provided through lectures, guidance counselors, or student professional society meetings.

On-Campus Activity: Eleven states reported that they have established a formal division of their agency on a college campus where students are hired, usually on a part-time basis, to do engineering work.

Student-Mentor Program: Five states said that they match professional career employees with engineering students to encourage a transportation career.

Pre-Career Development Program: Thirteen states replied that they enter into a formal arrangement with pre-selected students to provide them with work experiences during college in exchange for a career with the agency after graduation.

After Employment Development Program: Twenty eight states said that they have an organized program for entry-level employees to receive special training. Twenty states provided tuition reimbursement for post graduate studies; twenty seven rotated new employees through many different divisions; and twenty two agencies allowed the employee a choice in assignment after completion of the program.

The survey also asked state DOT's if they had a problem with retaining employees after they were hired.

Twenty one said they did not, and 17 states said 'yes'. The most often cited reasons for losing employees were: higher salaries and better opportunities in the private sector, desire to return to native states, and difficulty in passing EIT exams within a required time frame.

Finally, the survey asked state officials to identify specific recommendations for activities that would help recruitment and retention of engineers. Some of these recommendations included:

0 Encourage realistic pay scales
0 Distribute data relating to recruitment and retention
0 Sponsor "career" course for new hires
0 Cooperate with colleges to improve relevance of education
0 Encourage minority students
0 Provide incentives to junior engineers
0 Provide scholarships
0 Broaden scope of engineering education
0 Educate public about highway technology

The AASHTO survey provides some interesting information about the current status of transportation professionals in state DOT's. Many state agencies are having trouble attracting qualified engineers and have embarked on several initiatives to recruit and retain transportation professionals.

THE FUTURE TRANSPORTATION ENGINEER

The Transportation Research Board study and the AASHTO survey provide some interesting insights into the problems facing state agencies in attracting transportation professionals. However, both efforts focused on what was going to happen in the near term, that is, the next five to ten years. As we head into the 21st Century, it is appropriate to ask what will be the likely desired characteristics of a transportation engineer and what should we be doing now to prepare students for such a career? The following observations are based on the author's role as a state transportation official and active involvement in state and federal efforts to identify the possible futures associated with transportation and society.

"Futures" efforts, recently completed by several organizations, provide a useful point of departure for a discussion of what the future holds for the transportation profession. The Federal Highway Administration, AASHTO, American Public Transit Association, Highway Users Federation, the

Transportation Research Board and many others have prepared reports on the future characteristics of our society and where transportation fits or does not fit in this vision of the future (see, for example, [FHWA,1988; AASHTO,1988; TRB,1989; and APTA,1988]). In general, every report notes the importance of an efficient transportation system to the nation's economy and international competitiveness. In particular, congestion on the transportation systems is noted as a serious problem that must be addressed. In some cases, this means expanding the capacity of the transportation systems; in others, this means better using existing capacity through operational improvements or through new vehicle technology. In some cases, these reports also note that the transportation system often has a role in several environmental problems that will not likely disappear by the 21st Century. The most obvious example is the linkage between air quality and automobile pollution. Whether through alternative fuels or transportation control measures, transportation will have to receive major attention if the nation is serious about cleaning the air. Transportation, and transportation professionals, will thus continue to be an important area of concern as we head into the next century.

With this said, what are some of the areas in transportation engineering that will likely change or which will receive renewed emphasis in ten years and thus should provide some interest today on how to prepare students for a career? The following areas should not be considered all-inclusive; instead they are offered as a point of departure for a discussion on the appropriateness of current engineering curriculum for the future transportation professional.

Exposure to Design Concepts: Many metropolitan areas are experiencing serious levels of congestion. One of the ways of dealing with this problem will be to expand the highway network or to modify existing highways to handle more traffic. Undertaking such an effort will require engineers with a good understanding of geometric design concepts and of the design process. In many situations, any expansion of capacity will be tolerated only with substantial commitments from government agencies for mitigation of any environmental impacts caused by the project. The design process will thus be ever more sensitive to such issues.

Exposure to Operations Concepts: Just as some cities will build new highways to deal with the congestion phenomenon, many cities and towns will not be able to expand the highway system, either for lack of money or because of local opposition. In such cases, the two responses remaining to local officials is to manage the demand for transportation (through land use controls or some other means) or to improve the operational efficiency of the highway system. The latter will likely be tried first given the potentially controversial nature of managing land use. Traffic operations, and similarly safety improvements, will receive greater attention as we head to the 21st Century. Because of this, there will likely be a substantial increase in demand for engineers who are knowledgeable in traffic engineering and operations areas. Interest is such skills is already occurring. The Institute of Transportation Engineers, in responding to what it perceived as a tremendous need in the profession for these types of skills, is conducting with the Georgia Institute of Technology a Traffic Engineering Academy which is designed to provide entry-level engineers with basic knowledge in one area and a broad exposure to professional development issues during the one-week course. Three courses are to be offered during the first year of the Academy--signal operations and maintenance, geometric design of intersections, and site impact analysis and planning. Over 100 engineers from North America will be attending these courses, and will also be exposed to professional development talks on future technology, working in a "fishbowl" environment, how to communicate to non-technical individuals, and what it means to be a professional.

Rehabilitation and Maintenance: The survey of the state DOT officials discussed earlier identified rehabilitation and maintenance as key issues in the near future. It is likely that the rehabilitation and maintenance of infrastructure will continue to be a major professional and public concern well into the next century. Engineers will need to know why materials deteriorate and what can be done to slow down the deterioration process. In addition, the process of designing and replacing highways and bridges needs to be addressed somewhere in an engineer's education or training. A linkage to the operations area discussed above is the often difficult requirement to manage traffic on the construction or maintenance site while rehabilitation activities are underway. All of these issues require some special attention.

Materials and Pavement Design: The results of the Strategic Highway Research Program, the $150 million research effort to produce breakthroughs in materials and pavement design for highways, will have been put in place in most transportation agencies by the beginning of the next century. It is still too soon to determine what, if any, innovation will result for the practicing engineer, but it is likely that some change will occur in this area. It is incumbent on the profession, especially for educators, to introduce these changes into the next generation's engineering education as soon as they are adopted by the profession.

Geographic Information Systems: Geographic information systems provide a computer-based capability using relational data bases that could revolutionize how much of highway engineering is done. Such systems can be used for almost any aspect of the highway project development process--planning, design, construction, maintenance. As more regional areas become digitized, an increasing number of agencies will likely adopt an integrated data base approach to planning and engineering. The 1990 Census which will provide digitized, so-called TIGER files for all metropolitan areas, will speed up the process of such adoption. Using such a system, an agency could conduct all of its planning off of a single data base, and then using the same data base undertake project designs which could be directly related to any characteristic desired by the engineer (e.g., natural or man-made structures affected by each design). The adoption of geographic information systems could be as important a development as was the introduction of microcomputers into the transportation engineering process.

Computer-Aided Design: Computer-aided design has for many years been relegated to large workstations using fairly complex software. Major advances in software and their utilization on personal computers have provided impressive capability to the user of personal computers. Although there are many advantages to teaching students manual methods of design, it is likely that the transportation profession of the 21st Century will be heavily reliant on computer-aided design tools.

Artificial Intelligence and Expert Systems: Similar to computer-aided design, major strides are currently being made in applications of artificial intelligence and expert systems to engineering problems. These tools will be commonplace in the 21st Century and will greatly aid the engineer in his/her problem solving skills.

Technology: Great interest is being given today to future transportation technology. In particular, the so-called Intelligent Vehicle Highway System (IVHS) is receiving both governmental and corporate attention with regard to its potential impact on all facets of the transportation system. Benefits of this approach relate to improved safety, enhanced vehicle and system performance, reduced environmental impact, and maintaining a competitive edge in this highly technological area. It is likely that some form of technological innovation to either the vehicle or to how a facility is managed will be available by the early part of the 21st Century. However, it will not likely used in such a widespread manner or in a way that will greatly reduce the need for highway engineering. This technological frontier, however, does provide some potential motivation and excitement in attracting young engineers to the profession.

CONCLUSION

The next ten years until the 21st Century will see dramatic change in the way transportation engineers do their work. In particular, computer-aided tools which greatly enhance the productivity and capability of an engineer will become commonplace. The transportation engineer of the 21st Century will probably feel comfortable working with expert systems and computer-aided design, and some may be at the forefront of transportation technology in providing guideways and control systems that put "intelligence" into the vehicle and facilities. Even with these changes, it is also likely that the transportation engineer of the 21st Century will also need to know much of the basic knowledge that engineers today must know, in particular design and operations concepts. In addition, because of the massive national problem of rehabilitating and maintaining the transportation system, many engineers will need to have the skills necessary for undertaking these tasks.

What does this say about tomorrow's transportation engineering curriculum? At the undergraduate level it necessarily requires greater familiarity with computer-aided design and analysis tools. Such tools are already available and need to be put into the curriculum. However, the basic knowledge relating to design and operations needs to be retained and probably enhanced. It is also important to expose undergraduates to the possible technological futures associated with transportation. Such an exposure could provide some impetus to encouraging more students to become part of this profession.

At the graduate level, where there is a bit more flexibility in providing greater exposure to students, the likely future directions of transportation discussed above suggest some important initiatives. First, not only should graduate students be exposed to the many types of computer-aided tools available in the profession, they should be given the opportunity to use them. This will require acquisition and knowledge in use of typical software packages used today for design. Second, graduate students should be exposed in some detail to design and operations characteristics of transportation facilities. These are likely to be two important areas in transportation engineering's contribution to the future transportation system in this country. Third, students should be exposed to the use of geographic information systems both for planning purposes and engineering purposes. Finally, it is important that graduate students be exposed to technological innovation and what can realistically be expected. Many of these students could be the ones given the responsibility for implementing the concepts that today are only ideas.

REFERENCES

American Association of State Highway and Transportation Officials, **Transportation Professionals: Recruitment and Retention**, Washington D.C., 1988.

_________. **Keeping America Moving, The Bottom Line**, Washington D.C., 1989.

American Public Transit Association, **Transit 2000, Interim Report of the American Public Transit Association Transit 2000 Task Force**, Washington D.C., 1988.

Federal Highway Administration, The Future National Highway Program, 1991 and Beyond, Series of Working Papers, 1988.

Transportation Research Board, **Transportation Professionals: Future Needs and Opportunities**, National Academy of Sciences, Washington D.C., 1985.

Transportation Research Board, **A Look Ahead, Year 2020**, Special Report 220, National Academy of Sciences, Washington D.C., 1989.

Continuing Education via Satellite Television

John M. Mason, Jr.,[1] M. ASCE and
W. P. Kilareski,[2] M. ASCE

Abstract

This paper presents the attributes of using satellite television as an alternative media form of providing continuing education. A case study example reports on the effectiveness of using this media technique in the delivery of a statewide telecast for the Pennsylvania Department of Transportation.

Introduction

Career long education is advocated by various professional societies and is routinely cited as an identifiable need by both the public and private sector. The theme supporting this position is the desire to keep pace with advances by domestic and foreign competitors. The National Academy of Engineering (NAE) recently reported that "65% of engineers employed by small non-urban companies and almost 36% of engineers employed by large urban firms received no continuing education in a given year" (TR News, 1989). Additionally, NAE identified a heightened need for continuing education due to:

[1]Associate Professor of Civil Engineering and Director, Transportation Operations Program, Pennsylvania Transportation Institute, Pennsylvania State University, Research Building B, University Park, PA 16802-4797.

[2]Associate Professor of Civil Engineering and Director, Pavement Infrastructure Management Program, Pennsylvania Transportation Institute, Pennsylvania State University, Research Building B, University Park, PA 16802-4797.

1. the importance of constant upgrading of skills

2. rapid technological change/scientific advances

3. reductions in supply of new, young engineers

4. shortcomings in precollege mathematics and science training

5. the expanding field of opportunities for all engineering disciplines

Continuing education of civil engineers can be accomplished through numerous teaching/training techniques. Primary means of acquiring additional knowledge in a specialty area have been attending specialized short courses or enrolling in formal credit courses. Short courses usually take place within an organization, at a college or university conference center, or during professional society meetings. These short courses have proven to be technically effective and generally cost efficient. This type of continuing education activity provides unique interaction among the participants and an opportunity for immediate responses to questions raised by the students.

Supporting instructional materials for short courses have typically included a participants' notebook with the instructor using audio-visual materials in the form of slides, overhead transparencies, motion picture films, and videotape presentations. The use of supporting instructional materials naturally enhances the quality of the presentation by showing actual examples of the particular points or issues raised by the instructor.

Another teaching media form has been the use of televised broadcast lectures of an instructor from an off-site location (Greenberg and Biedenbach, 1987). This alternative has been used for many years on college campuses throughout the United States. It has been traditionally employed for college-level credit courses. The transmission and reception are both restricted and controlled from the production center. More recently, satellite transmissions have been used to conduct teleconferences throughout the United States. The Rural Transportation Assistance Program (RTAP) is one such example of successfully communicating new ideas, concepts, and techniques to other interested users.

Attributes of this type of instructional media technique are discussed in this paper.

Implementing Satellite Television

The following discussion highlights the advantages/disadvantages, the benefits/shortcomings, and the caveats/implications of using satellite television as a continuing education media technique. A case study example is provided along with an evaluation of the effectiveness of using satellite television as the presentation technique of a short course for state transportation engineers. The information presented in this paper provides guidance to civil engineering educators who are interested in implementing such an alternative media form of continuing education.

Background

In June 1988, the Pennsylvania Department of Transportation (PennDOT) expressed interest in providing continuing education opportunities for its traffic and highway engineering/design staffs throughout the state. PennDOT also indicated a desire to offer undergraduate/graduate credit to employees who were interested in pursuing or completing their respective degrees. Except for a few universities located in major urban areas, specific traffic/highway-related courses were not readily available to all PennDOT employees. The initial estimate of interested employees was 100; almost 500 individuals eventually attended the televised course.

The following issues had to be addressed to meet the needs of PennDOT:

* A large number of people should be reached simultaneously.

* Information must be provided to all locations.

* Topics may be very selective; audience may be pre-selected.

* Travel costs and related individual expenses must be minimized.

* Experienced employees must be provided with cross-training.

* Entry-level employees must be provided with practical information.

While closed-circuit television would be one media option, its transmission would be limited to a few specific locations currently on the Penn State system (the Penn State University System also includes a Commonwealth Campus network of undergraduate and graduate institutions). Another option was the use of the Penn State satellite "downlink" locations at 16 Continuing Education centers throughout the state. Their geographic locations would minimize travel for the PennDOT employees.

The satellite option satisfied the overall intent of the established criteria. Additionally, it provided scheduling flexibility, it would accommodate different class sizes, and audience interaction could be accommodated via two-way telephone links ("meet me bridge").

Technical Characteristics of Penn State Satellite Facilities

At the present time there are over 30 commercial satellites available for TV broadcast. These satellites are in geosynchronous orbit at a distance of 22,340 miles above the United States. The satellites are either C Band or Ku Band with up to 24 transponders for signal broadcast. Penn State uses the Ku Band and buys satellite time in bulk quantities. This allows for a bulk discount price of $380 to $450 per hour. Approximately 100 universities have uplink capabilities in the United States. Anyone with a satellite dish can be a downlink if they have the capability to monitor the signal (signals are unscrambled). All 19 Penn State branch campuses have downlink capability.

The Penn State satellite facilities are administered by the University's Commonwealth Education System, which supervises the University Division of Media Learning and Resources. Within this division is the University's public TV station, WPSX-TV. The broadcast section of WPSX handles routine public TV shows while the non-broadcast section handles the instructional portion of the TV unit. Specifically, the Center for Instructional Design and Interactive Video was responsible for the production of the satellite course.

At the University Park Campus there are three studios for instructional TV production. The studios are connected via fiber optic cable to the uplink facility. Each studio has a control room that monitors the production. A producer/director controls both audio and visual output. There are four remote control cameras in the studio. One of the cameras is used for slide or instructor overhead graphics. The entire studio is wired for sound from any part of the room. With this arrangement, the entire production can be done with one staff producer/director and the instructor.

Course Description and Presentation

The original intent of the overall program was to offer a full-semester 3-credit course via satellite. Because this would be the first televised course for both PennDOT highway design engineers and the Civil Engineering Department, it was decided that this initial effort should be conducted as a pilot short course. Students were able to register for one credit hour of Special Topics; a take-home written examination was given to those students who had formally registered. Approximately 17 percent of the participants elected to take the course for credit. Upon completion of the program, all participants received a certificate of 1.5 CEUs earned.

The purpose of the course was to introduce the American Association of State Highway and Transportation Officials (AASHTO) "Policy on Geometric Design of Streets and Highways" (1984 Green Book) to engineers not familiar with its content. Each participant received a loose-leaf notebook containing a set of summary notes keyed to each chapter of the Green Book and respective lectures. The session met for five Wednesdays from 3:00 to 6:00 p.m., March-April 1989, at the various Penn State campuses. The intended audience for this course consisted of "entry-level" highway engineers associated with the public sector (local, city, and state government) and the private sector (consultants, developers, and contractors). It was not intended for senior-level and experienced personnel. Lectures were given by three professional engineers who had practical design experience in addition to their university teaching and research activities. The program format was live video transmission and two-way audio via the University's "meet me bridge" phone system.

Evaluation of Program

Two formal evaluations were conducted to measure the effectiveness of the presentation and utility of the course content. A required "Instruction Evaluation Form," used for all Penn State University continuing education courses, was completed by each participant. To specifically address the course content and materials, a second list of 15 questions asked for a rating of usefulness of the notebook, visuals, examples, etc., and the likelihood of implementation of the knowledge presented in the course.

The overall results indicated that the course was effective in meeting its objectives, method of presentation, thoroughness of content, and instructors' speaking/teaching ability. Generally, over 80 percent of the respondents rated the instruction and content between average and excellent. Approximately 375 individuals submitted "complete" evaluation forms. Very little variation was found between lecturers or campus location or among the rating criteria. However, extreme ranges were noted in specific written comments regarding speaking style, clarity of explanations, usefulness of discussions, content of notes, and level of technical difficulty.

Several shortcomings were noted to benefit future efforts:

1. Carefully screen participants (not all participants were informed that the course would focus on entry-level engineers).

2. On-site instructors should be provided for after class discussions (students were encouraged to phone the instructors between weekly presentations).

3. High quality visuals were expected (35mm slides and prepared overheads were primarily used for lectures).

4. More telephones were needed at some sites (apprehensions had to be overcome to initiate a call).

Closing

The results of the evaluation process were very positive, and additional activities are envisioned. A list of over 30 civil engineering-related topics were identified by the course participants as potential topics for future televised courses. A few sample topics included: bridge design policies, traffic engineering, foundation design, pavement design, construction procedures, environmental issues, tort liability, and hydrology/hydraulics.

Although the case example represents a situation where the uplink transmission originated on a university campus and was downlinked to several Commonwealth-related campuses, it is just as possible to uplink from any location and to downlink to other preselected locations. Several large private industries and the military have used this technology to transmit formal and short course information to their employees. The benefit of the approach used in this case study is that it allows other local agencies and small consulting engineering firms to participate. Such organizations typically do not have large-size staffs or funds to justify the development and production of a particular course. However, through joint coordination and cooperation, local and private firms could easily participate in the cost of production.

References

Greenberg, J. S. and Biedenbach, J. M. Compendium of Users of Television in Engineering Education, Continuing Professional Development Division, ASEE, 1987.

TR News, Transportation Research Board, Washington, DC, 1989.

EDUCATION IN HIGHWAY SAFETY ENGINEERING

Janak S. Thakkar, P.E.*

SCOPE AND PURPOSE

The purpose of this paper is to examine the status of the Civil/ Transportation Engineering professionals in highway safety and show the need for adequate education in highway safety engineering. The paper identifies various courses in highway safety and suggests ways to attract prospective students to attend some or all the courses that may lead to a possible graduate level degree. The paper also describes some of the expected benefits resulting from education in highway safety engineering.

STATE OF HIGHWAY SAFETY

Highway safety has been identified as one of the two major goals of any transportation system, the other being mobility, i.e., efficient movement of traffic. However, many a times this safety aspect is not accorded its due consideration by the Civil/ Transportation Engineer. Often, emphasis is rather placed on mobility than on safety. Some professionals argue that mobility and safety cannot go hand in hand, i.e. mobility cannot be increased without sacrificing safety, and vice versa. It can, however, also be argued that these two goals need not be in conflict. For example, in congested urban areas, while improving capacity and hence mobility, we could also affect a reduction in accidents. Conversely, by reducing the potential for accidents, and hence improving safety, we could similarly reduce the accident-related interruptions in traffic and thus improve mobility on our highways.

At the present, the consideration of safety in highway planning, design, construction and operations appears to be rather minimal. Most states (with very few exceptions) select their highway improvement projects initially on the basis of structural and service (capacity) needs, and safety is not part of prioritizing and selection at that crucial stage. Again, at the design stage, safety is not so much on designer's mind. In the haste of finishing the project, a number of exceptions (not necessarily qualified) are made to save money as well as time (such as that required for acquiring right-of-way) which result in poor design from safety standpoint. In construction, the safety engineer's involvement is almost nil. Although Maintenance of Traffic requirements are presently being addressed, the process needs further conformance and emphasis so as to affect further reduction/ prevention in accidents in work zones. The highway built or improved with such cursory level of safety considerations at various stages, has built-in hazards.

There could be several reasons for safety not receiving its due consideration. First, the Civil/ Transportation Engineer, by his/her

----

*District Safety & Studies Engineer, Florida Department of Transportation, 1000 N.W. 111th Ave. Miami, Florida 33172

job description/statement of functional responsibilities, is not expected specifically to deal with the safety aspects of highway engineering. Second, he/she is not trained in the area of highway safety engineering, there being virtually no courses offered as part of Civil/Transportation Engineering curriculum. His/her understanding of the causation of motor-vehicle crashes and how to deal with them is, therefore, very much limited. Third, there appears to be a lack of awareness and concern on the part of transportation professional for consideration of safety. The thrust of the highway program is often to get the highway built as soon as possible, overriding many other considerations, including safety. Fourth, the complaints or comments generally received from the outside are aimed at pointing out potholes, broken or missing signs and other maintenance problems, but may not include specific safety/hazardous conditions. And finally, the political climate expects the professional to accomplish the roadway construction expeditiously, without diverting time and effort to build in safety. This lack of understanding, awareness and concern for safety is not surprising considering the limited exposure of the Civil/ Transportation Engineer to the safety-related courses at the learning institutions and the professional environment within which he/she operates.

LITERATURE REVIEW

Until the late 1960s/early 1970s, man-the driver was primarily considered responsible for "causing" an accident, and the role of the motor-vehicle and highway in the causation of accidents was either down-played or virtually ignored. The early literature, therefore, concentrated on educating the human operator of the vehicle or modifying/ improving his/her behavior when it was found to be affected by the use of alcohol, drugs or by other human characteristics that could contribute to an accident. A literature search on the subject of education in highway safety made through Transportation Research Information System (TRIS) of Transportation Research Board (TRB) also revealed some references that highlighted research in the manpower assessment/ development in the highway safety area and also in the area of technical training that may be useful in the occupations in highway safety field (1,2,3,4,5). Again, the focus of the research was mainly in the area of driver education, law enforcement, and rather minimally traffic engineering. The teaching of these topics was at a community college (Associate Degree) level.

In the early 1970s, the vehicle and the highway began receiving added emphasis. The U.S. D.O.T. through FHWA and TRB (for example, through its National Cooperative Highway Research Program) began studying relationship of safety aspects with various transportation elements, namely planning, highway design (geometrics, etc.), construction and maintenance and operations. Other organizations, such as ITE (Institute of Transportation Engineers), ASCE (American Society of Civil Engineers) and some universities (such as Texas A. & M.) joined in this effort. (The research publications by these organizations served the purpose of instructing engineers in the safety aspect of highway. These publications are too numerous to mention here,

Reference Nos. 6 and 7 are just examples). U.S. D.O.T. along with others also developed some courses, for example, Highway safety Engineering Studies (8), Functional Requirements of Highway Safety Features (9) and many others (10,11).

A search of the official literature revealed a letter from U.S. D.O.T.'s National Highway Safety Advisory Committee to the former Secretary of Transportation, Elizabeth Dole (12). This memorandum reflects the concerns of U.S. D.O.T. for the need for training in highway safety engineering as early as 1984. It states, "For many years, emphasis was placed on initially building a highway system,... Many times a road would be resurfaced, but a hazard would go uncorrected because highway personnel were not trained to recognize or properly correct or eliminate the hazard. This presented a "lost opportunity" to improve the road and provide a safer driving environment for the public. It also locked in the hazard for many more years... A systematic approach which would help bring about a uniform solution to the problem is a Highway Safety Engineering Training Initiative which is both timely and universal in need." This last statement says it all!

THE PROBLEM

The teaching of highway safety at the academic institutions is currently at a nonexistent level, barring a very few exceptions+. Some universities offer hardly one course on traffic safety. The Federal Highway Administration (FHWA) has offered, from time to time, short courses in traffic safety; such as Highway Safety Engineering Studies, Highway Safety Evaluation, Railroad-Highway Grade Crossing Safety. These short courses are limited in duration, with some courses lasting only one or two days. Because of the short duration of these courses, the coverage of various safety aspects is limited at best. With some exceptions, attendance in these courses is limited to those already involved in handling safety aspects. These courses generally reach certain Federal and State highway personnel, however, there appears to be a major breakdown when it comes to training for individuals at the local (county-city) level (12). This is not meant to say that these courses are not useful. However, because of the short duration of such courses and the lack of a comprehensive educational framework, emphasis is placed on a very few items and not in any consistent manner. Many times, it is assumed that the students have the necessary background in the field of safety or that they are knowledgeable in the safety area. Those who are not directly involved in the safety field normally may not understand what is being taught.

+Traffic Institute of Northwestern University, under FHWA sponsorship and direction, developed a "Highway Safety and Traffic Study" program. (Ref.13) It consisted of four courses: 1. Traffic Characteristics and Studies, 2. Traffic Operations, 3. Planning and Geometric Design of Highways, 4. Special Study in Highway Safety. This program was offered only twice in 1978 and was then discontinued.

This lack of coverage of safety aspects at the academic institutions is alarming, to say the least, considering that safety is one of the two major goals of the transportation system. Knowing the annual death toll on our highways in the 45,000 to 50,000 range, injuries in excess of 1.8 million and overall economic loss to the nation of over sixty billion dollars per year, greater attention must be focused on traffic safety. In recent years, we have seen health care costs increase quite dramatically. This means, future medical costs for the injuries resulting from highway accidents would be higher. There is a shift in the nation's demographics, in that the proportion of older population is growing ('Graying of America') for whom the health costs would be even higher. The fatality number is expected to increase in the next decade or two, almost doubling by the year 2000, according to one FHWA estimate. In addition, most public highway agencies are experiencing a large number of tort liability suits with increasing amount of settlements (some due to the "Deep Pocket Theory") against them as a result of loss of sovereign immunity in the last several years. The costs of neglecting safety which have been absorbed so far are going to become very visible in time of limited financial resources at each level of government. This is expected to put severe pressure on the political system to deal with the worsening of the accident problem on our nation's highways. We must prepare ourselves to deal with this increasingly costly and tragic situation before it is too late.

Prior to the 1950s, before we embarked on the building of the interstate system, the knowledge regarding highway transportation (design, construction, maintenance) was minimal compared to what we know today. Contributions to the success in highway design/ construction have resulted at least partly from the additional knowledge gained in the highway/ transportation engineering field. Many universities since late 1960s are offering graduate and undergraduate courses in Transportation Engineering. For these reasons, several universities have created a concentration of courses in Transportation Engineering, recognizing it as a separate discipline of Civil Engineering. A similar approach needs to be utilized for educating engineers in the area of highway safety.

If we are to increase our knowledge in the highway safety area and enhance the consideration of safety in design and operation, we must increase the number of safety-related courses taught at our engineering schools. Like Transportation Engineering, we may even initiate a new branch of civil engineering, namely Highway Safety Engineering, if not immediately, at least over the next several years. Examples of possible courses/ topics in this proposed field of study include: Highway Safety Engineering Studies/ Accident Analysis, Planning and Geometric Design, Highway Safety Management, etc. These courses of study should be pursued with safety as the underlying theme. Their basic premise should be that integrating safety into each aspect of highway transportation -- from planning through operation -- is not something that is merely desirable but it is a necessity to contain costs from the systemic point of view.

GRADUATE PROGRAM IN HIGHWAY SAFETY

The Traffic Institute of Northwestern University formulated some objectives for a FHWA-sponsored "Highway Safety and Traffic Study" program that was mainly oriented towards training engineers in highway safety (13). The general objectives of that program were as follows:

o To educate graduate level engineers, planners and safety managers in the area of highway safety engineering.

o To prepare students to conceive and manage projects, as well as assist in developing related policies

o To make students conscious of their responsibility for improving highway safety

o To impart knowledge regarding the latest technology and tools available to effectively manage safety programs and demonstrate the ability to use this knowledge and these tools.

In addition, specific objectives were also developed. Based on these general and specific objectives, the following courses are recommended:

1. Safety Design and Operational Practices for Streets and Highways
2. Highway Safety Engineering Studies (Accident Analysis)
3. Functional Requirements of Highway Safety Features
4. Accident Reconstruction
5. Human Factors in Highway Safety Engineering
6. Statistical Applications in Highway Safety Engineering
7. Traffic Characteristics
8. Traffic Operations
9. Safety in Highway Work Zones
10. Legal Aspects in Highway Safety/ Tort Liability
11. Highway Safety Improvement Program
12. Highway Safety Evaluation
13. Highway Safety Management
14. Special Topics in Highway Safety
    (a) Pedestrian and Bicyclist Safety
    (b) Skid-Accident Problem
    (c) Safety at Rail-Highway Grade Crossings
    (d) Land Development (Access Management) and Highway Safety
    (e) Incident Management for Safety
    (f) Maintenance Management for Safety
    (g) Changing Vehicle Mix and Highway Safety
    (h) Relationship of Design Standards to Highway safety

The Federal Government, through FHWA and NHTSA, having a nationwide perspective on the safety needs of the various States and having vast experience in the field of highway safety, can certainly help formulate the overall curriculum for the education in Highway Safety Engineering.

A brief description of these courses was developed but is not included here for the sake of brevity. If needed, the author will supply that.

Emphasis on Safety Research

Despite more than one-half century of modern road building, knowledge of the safety consequences of highway design decisions is limited (14). Research in the safety area is fragmented, uncoordinated, and more often the results are arrived at without proper application of rigorous statistical tests and appropriate methods. Some research results are based on safety experiences (gut feelings?) at two or three locations (statistically inappropriate sample sizes). Other research may be based on erroneous assumptions, such as normal distribution of data, etc. The education at the university level, under qualified supervision of faculty, could provide an excellent forum for performing very useful and needed research to increase the knowledge in the safety area. The students who may want to continue further study in highway safety engineering may be encouraged to conduct safety research under the guidance of university professors.

HOW TO ATTRACT PROSPECTIVE STUDENTS TO SAFETY AREA

Having established the recommended training curriculum for the study in Highway Safety Engineering, we must concentrate on how we can make this program sufficiently attractive for prospective students to attend some or all courses to complete the degree program.

A recent study by the ITE Technical Council Committee 2-32 (15), initially draws a rather bleak picture of declining enrollment in Civil Engineering schools, but it concludes by indicating that this situation could be turned around if we take certain measures in coming years. Since the study in Highway Safety Engineering could be considered a branch of Civil Engineering, at least two of its recommendations are applicable here. Those two recommendations along with others are listed below.

The environment in which engineers work, especially government, is perceived to be rather negative by the prospective student. This environment must be upgraded to allow wise long-term plans in favor of short-term benefits (and perhaps future heavy costs); sound technical solutions should be adopted over politically expedient decisions. The engineers should be provided with adequate resources, such as personal computers. Their advice and counsel on engineering matters should be sought and followed. This positive image of the employer would help eliminate the negative feeling for this people-oriented noble profession.

The beginning salary for a graduate Civil Engineer is rather low compared to that in other engineering disciplines. Again, in case of government agency, the salaries are based on established scales and all applicants are offered the same salary. This further dampens the competition among employers for the bright young engineers. The salaries should be upgraded and also be competitive.

Presently, the job of Safety Engineers is very much limited in scope. In Highway departments, they perform safety related activities which only include collection and analysis of accident data and proposal of some safety improvements. Their input is hardly sought by other areas such as Planning and Design. Their duties could be enlarged to allow active interaction between them and other areas of the department. Their responsibilities should be expanded to include providing thorough input with respect to safety considerations (accident histories, analysis and recommended countermeasures) in the project planning/ selection stage and also at the design stage. This enlarged role for the safety engineer further justifies elevation of the position of Safety Engineer.

The upper management must be supportive of Safety Engineer's opinions and recommendations. The management support lends further credibility and helps to get the recommendations implemented.

Also, the safety staff should be visible to management and within easy communication...(16). By virtue of specialized training and expertise in the safety matters, the safety group should be allowed to play active and significant role in the decision-making process at least in the safety related areas.

The Federal Government and also State Governments could encourage, if not require, the highway engineers to have additional education in Highway Safety Engineering and provide fellowships/ grants to deserving candidates to attend these courses.

There are over 27,000 officials in state and local agencies who are potential candidates and could benefit from the training in Highway Safety Engineering (12). These government officials supplemented by the new prospective students, can provide adequate enrollment in highway safety.

EXPECTED BENEFITS

The education in Highway Safety Engineering could provide several benefits to the nation and especially to the field of Civil/ Transportation Engineering. Listed below are some of those benefits.

o The education will bring awareness of the safety considerations in every walk of transportation. Their concern for safety considerations could begin right at the planning stage where initial selection of projects is taking place. The trained/ educted professional with assured support from top management will change the situation around by having safety deficiency (in terms of accident problem reflected as a high accident location) also considered in the final project selection process. The safety group with its experience and knowledge of the accident problem at a given location and subsequent analysis and determination can provide the necessary input from the safety standpoint in the final selection of the location for new improvement.

o Once the locations are selected, the safety group will provide input via review of the design plans at various stages of project development. Again the necessary analysis and evaluation, in which the Safety Engineer has expertise, will be prepared by him/her and the appropriate input will be provided at the design stage. With the backing of top management, the safety engineer will make sure that his/her recommendations are carried out which will improve the overall safety of the location being improved.

o Also, at the construction stage the safety engineer will be ready to make recommendations that will avoid oversight by the construction personnel especially when safety aspects are concerned.

o Throughout the various stages of project development and implementation, the safety engineer will help to build safety in the project. After the project is implemented, the Safety Engineer will monitor the accident situation at the location. While safety is built in the project right from the beginning, in rare cases he will propose further safety improvements if warranted by the accident situation.

o The building of safety into design and operation will reduce the need for safety improvements in the future as safety is already built in there. The engineers can build safety in the highway, if they are adequately educated in the Highway Safety Engineering field.

o The education in Highway Safety Engineering will greatly enhance the ability of the engineers to consider safety at all stages of road building and improvements. This enhanced consideration will very likely reduce the number of accidents and their severity.

LEFT TO THE ACADEMIC COMMUNITY

The views presented here are as perceived by the author. Obviously, the academic community involved in the developing, teaching of the courses and final certification of course completion, can provide major input as to which courses should be taught and how. Academicians can also determine the specific requirements for the degree. This paper probably has some limitations. For example, it may not include certain courses thought beneficial by the academicians. The conference forum would be well suited for discussion of these issues by the academic community. Even if the paper generates enough debate among the professionals in transportation and the academic community, it would have served its purpose.

REFERENCES

1. Marsden, W.E., Jr. and Sanders, R.C., Manpower Development in Traffic safety (Guidelines for Establishing Internship Programs) San Diego County Engineer Dept. Calif. 1973

2. Karim, A.S., Improving Highway Safety Manpower: Traffic Engineering Technician Project at Lansing Community College, Final Report, Phase I, American Assoc. of Junior Colleges, Wash. D.C. May

1972. (Also, by the same author, Phase II at Longview Comm. College and Comm. College of Denver, 1973).

3. Miller A.J., et al., Expansion of Vocational-Technical Programs to Accommodate Highway safety Manpower Requirements, Ohio State University, Center for Vocational and Technical Education; Columbus, Ohio, August 1971.

4. Daugherty, R.D., et al., Highway Safety Occupational Program Development Guide, Final Report, Ohio State University, Center for Vocational and Technical Education; Columbus, Ohio, July 1971.

5. Bishop, Richard and Sheehe, Gordon, The Role of the Community College in Developing Traffic Specialists and Technicians, American Assoc. of Junior Colleges, Wash., D.C. 1968.

6. FHWA, Motor Vehicle Accidents in Relation to Geometric and Traffic Features of Highway Intersections, U.S. D.O.T. Wash. D.C. 1976.

7. Transportation Research Board, Cost and Safety Effectiveness of Highway Design Elements, NCHRP Report No. 197, Wash., D.C. 1978.

8. FHWA, Highway safety Engineering Studies, Procedural Guide, U.S. D.O.T. June 1981.

9. FHWA, Functional Requirements of Highway Safety Features, U.S. D.O.T., 1984.

10. FHWA, Safety Design and Operational Practices for Streets and Highways, U.S. D.O.T. May 1980.

11. FHWA, Design and Operation of Work Zone Traffic Control,U.S. D.O.T. 1987.

12. National Highway Safety Advisory Committee, Highway Safety Engineering Training Initiative, A Letter to Secretary of Transportation, Elizabeth Dole, U.S. D.O.T., February 1984.

13. FHWA, Highway Safety and Traffic Study Program, prepared by Traffic Institute of Northwestern University, Evanston, Ill. 1978.

14. Transportation Research Board, Designing Safer Roads, Practices for Resurfacing, Restoration, and Rehabilitation, Special Report 214, National research Council, Wash. D.C., 1987.

15. Institute of Transportation Engineers, Attracting Students to a Professional Career in Transportation Engineering, Final Report of Technical Council Committee 2-32, 1989.

16. M. Horodniceanu and E.J. Cantilli, Transportation System Safety, Lexington Books, Lexington, Mass. 1978.

Transportation Student Recruitment Resources

L. David Shen* and Joan Al-Kazily**

Abstract

One-third of this nation's professional engineers in state, county, and city transportation agencies will retire over the next decade. This paper examines the reasons for this development and suggests the use of innovative audiovisual materials in transportation engineering education as a way to counter this problem by attracting students into the field.

Introduction

A recent National Science Fopundation study found that over the next five years, the college-age population will decline and the supply of engineers graduating from college will also decline, perhaps quite markedly. The future demand for engineers may have to be met by non-traditional sources which may result in a diminution of quality, threatening the long-term international competitiveness of the United States. This problem poses an especially serious threat to the future supply of transportation engineers because about one-third of this nation's professional engineers in state, county, and city transportation agencies will retire over the next decade.[1]

In 1976, about 47 percent of all enginerers were over age 40.[2] By 1986, that age group had increased to 59 percent. We are not attracting young people into the

---

*Assistant Professor, Department of Civil Engineering, Florida International University, The State University of Florida at Miami, Miami, FL 33199.

**Professor, Department of Civil Engineering, California State University, Sacramento, CA 95819.

profession. To complicate this already tight situation, the rate of retirement will be much greater than in past years. During the 1950s and 1960s, government agencies hired rapidly as the nation's highway programs expanded; those same professionals are now at retiring age.[1]

There are many explanations as how today's transportation engineer shortage problem was created. The three most important factors are as follows:

1) Inadequate training in math and science at the public school level,

2) Undeserved bad reputation given to transportation engineers by environmentalists and nature preservation groups, and

3) Inconspicuous status and inadequate compensation for transportation engineers.

It is not the intention of this paper to discuss means by which students can be attracted to engineering. The purpose of this paper is to discuss transportation student recruitment resource materials and their availability.

The Recruitment Problem

There are two ways to attract prospective students into transportation engineering: one is to attract more junior and senior high school students to all branches of engineering, to enlarge the pie, so to speak. One of the methods used in Florida to attract public school students' attention in engineering is to go to high school and elementary school. By talking to students about what engineering is so as to educate and motivate them to join the engineering profession. One of the programs starts as early as the fourth grade because it is important to reach young people early.

The second way of attracting students into transportation is to divert undergraduate enginnering students into the field. This strategy is being addressed here. Two primary factors identified by the ASCE Urban Transportation Division Committee on Education as major reasons why undergraduate engineering students are disinterested in choosing transportation engineering as their career:

1) Inadequate exposure. Most engineering schools require only one course in transportation engineering for Bachelor of Science in Civil Engineering (BSCE)

students to graduate. Compared with the early exposure given to other disciplines in civil engineering, it is not surprising that most BSCE students want to choose structure or water resources as their career.

2) Conventional teaching. Conventional teaching of transportation engineering course will not work because students cannot grasp the exciting nature of transportation engineering with chalk and blackboard in just one course. Therefore, innovative new teaching methods such as the use of audiovisual aids in teaching must be adopted in order to get their attention focused on transportation engineering.

There is very little that we can do to change the first factor. However, we definitely can make this single transportation engineering course more challenging and interesting. The use of audiovisual materials such as slides and videotapes is very effective in stimulating and attracting undergraduate engineering student's interest in transportation engineering.

Recruitment Resources

An Audiovisual Resource Guide for Transportation Engineering Education was prepared in recognition of the transportation educators and professionals need for information on current audiovisual resources which are useful references for transportation engineering education and practice. In addition to the educational value of such resources it is expected that by being exposed to these audiovisual materials, more undergraduate students will be attracted to the profession of transportation engineering.

The draft document[3] was compiled by Dr. L. David Shen of the Florida International University, as an activity sponsored by the ASCE Urban Transportation Division Committee on Education.[3] The audiovisual resource guide lists slides, videocassettes, and films, briefly describe their content, and also indicates the appropriate target audience and topical areas. The focus of the material contained in this guide is intended to be more within the domain of Civil Engineering applications to transportation, rather than the entire broad spectrum of transportation related disciplines.

To facilitate use of this resource guide each audiovisual resource is coded with a cross reference key. The numbers in the key reflect the target audience judged most appropriate. The letters in the cross reference key relate the materials to topical areas.

The cross referencing system for audiovisual materials is shown in Table 1.

Table 1. The Cross Referencing System for Audiovisual Materials

Reference	V	S	F	Total
1 - Undergraduate	18	31	32	81
2 - Graduate	18	31	32	81
3 - Reference	12	2	4	18
a - Transportation Planning	1	6	3	10
b - Highway Design	4	7	8	19
c - Pavement Design	10	4	3	17
d - Highway Safety		5	16	21
e - Mass Transit planning	7	1	1	9
f - Transportation System Management	2	1	3	6
g - Traffic Control	1	9	7	17

V = Videotape S = Slide F = Film

The audiovisual guide includes a brief description of each item. In some cases the strengths and/or weaknesses of the material are indicated. In order to expand on this information, which is standard as a guide for new users of the audiovisuial item, the Education Committee of ASCE Urban Transportation Division is following up the work by a survey of transportation educators. The goal of this survey is promotion of the exchange of information about transportation educators' experiences with the use of audiovisual materials inside and outside the classroom.

The survey will provide educators with the opportunity to inform the committee about audiovisual materials which they have found to be effective and about problems which they have experienced. The survey will also provide new users of audiovisual materials with information about their availability and with the opportunity to express their views on the subjects. The results of the survey will be used to improve the audiovisual resources guide for transportation engineering education by providing additional information and adding new items as appropriate.

Conclusion

With the engineering student recruitment problem, it is important for transportation engineers to help non-engineers, particularly young people, to gain a better understanding of the profession's struggles and rewards. It is clear that transportation engineering profession faces serious personnel and recruitment problem in the coming decade. National visibilty for the transportation engineering profession is needed. Transportation educators and agencies can make use of good audiovisual materials to intrigue our youth and challenge them to join us in the design and building of America's future.

Acknowledgements

The assistance of Dr. Yupo Chan, Chairman of Education Committee - ASCE Urban Transportation Division, is acknowledged and appreciated. The authors would like to thank all the members in the Education Committe for their support and encouragements. However, the statements and opinions expressed in this paper are the sole responsibility of the authors.

References

1. Hawkins, N.M. "Transportation Issues, 1985-1995: A University Persperctive." ITE Journal, Vol. 57, No. 1, January 1987, pp 41-44.
2. Kraft, W.H. "Hope for the Future." ITE Journal Vol. 59, No. 9, September 1989, pp 44-46.
3. Shen, L.D. Audiovisual Resource Guide for Transportation Engineering Education. ASCE Urban Transportation Division Committee on Education.

Appendix

Content of the Audiovisual Resource Guide

The Future of Airport Engineering

Lon-Li David Shen, A.M. ASCE

Abstract

The use of Automated People Mover (APM) systems has dramaticaly changed the way of airports being planned and designed in the United States during the last 20 years. This paper analyzes this new development's impact on the future of airport engineering. It is concluded that the trend of installing APM systems at major U.S. airports will be continuing well into the next century.

Introduction

The Airline Deregulation Act of 1978 had brought tremendous increase in air travel as the number of passengers increased 63 percent between 1978 and 1987, from 275 million to 447 million. The Fedearl Aviation Administration (FAA) expects the annual number of passengers using the nation's airports to grow more than 70 percent above 1986 levels in the next ten years.[1] In 1987, the top 16 airports in the U.S. accounted for nearly half of the air passengers in the country.[2] Since no major new airport is expected to open before 1993, the increased air passenger demands can only be met by enlarging the existing airports.

Expanding an airports is not a simple undertaking because an airport more closely resembles a city than a building. The design engineers must consider problems of infrastructure, roadway surfaces, pedestrian, vehicular and air traffic flow, and many other complexities. As the airport gets larger and larger to accommodate the increasing passenger traffic, it has created excessive

---

Assistant Professor, Department of Civil & Environmental Engineering, Florida International University, The State University of Florida at Miami, Miami, FL 33199.

walking distance for passengers at the nation's major airports. Passengers must move from terminals to aircraft, within the terminals themselves, or between terminals that may be miles apart. Delays and complications cause missed connections and irritation for passengers, and result in poor equipment utilization for airlines. Airport designed with speedy and comfortable ground transportation, on the other hand, pleases passengers, and boost airline and airport revenues by minimizing aircraft ground time.[3]

In recent years, an innovative Automated People Mover (APM) system has gained popularity in airport design in solving this ground transportation problem. With APM, new airport design and existing terminal modification which were impossible 20 years ago became regular. Therefore, the purpose of this paper is to analyze this new technology's impact on the future of airport engineering.

Airport APM Systems

The APM system is an automated, electric-powered vehicle operating singly or in multi-car trains on steel or concrete guideways. It can carry from 2,000 to 20,000 passengers per hour with headways as short as 60 seconds, offering convenience comparable to modern elevators. Ride quality is among the best of any transit system in the world. The APM vehicles travel at speed up to 35 miles per hour (mph), yet accelerate and decelerate smoothly and swiftly. They stop and start automatically and can operate in a demand mode during off-peak hours to minimize energy used.[3]

Since Tampa International Airport installed an APM system in 1971, similar systems have been built at airports in Seattle (1973), Dallas (1974), Atlanta (1980), Miami (1981), Houston (1981), and Las Vegas (1985).[4] In addition, ground was broken for the new APM system at Chicago O'Hare International Airport on September 10, 1987 and it is expected to open late 1991 or early 1992.[5] By adding an APM system to its current design, the Pittsburgh Airport's new midfield terminal is scheduled for completion in 1992. The $2.2 billion expansion program for New York's Kennedy Airport announced in 1986 also include a new APM system.[6] Many additional applications at the Nation's other major airports are currently being studied.

The airport APM's are extremely reliable. In addition, the operating and maintenance cost per passenger ride is extremely low. As of June 1987, more than 770 million airport passengers have used the systems at the eight airports, with an average system availability of 99.7 percent.[4] This kind of performance translates into satisfaction for passengers, higher equipment utilization for airlines, and greater efficiency for the entire airport complex.

At the present time, only five of the top 20 airports in the U.S. have operational APM systems in their terminals. For the remaining 15 major airports: two are under construction (Chicago and Pittsburgh); three are under design (New York JFK, Newark and Denver); and four are being proposed or studied (Los Angeles, San Francisco, New York LaGuardia and Boston). It is clear that the trend of installing APM systems at major U.S. airports will be continuing well into the next century.

Terminal Design before APM

The planning and design of airport terminal is the most important part of airport engineering. The function of an airport terminal is to effect the transfer of passengers and baggage from surface transportation to air transportation in a minimum time and with a minimum amount of confusion and discomfort. Prior to the introduction of APM system at airport in 1971, five passenger terminal designs each with varying degrees of centralization were commonly used.[7] Evaluation of these five basic terminal designs is shown in Table 1.

Table 1. Evaluation of Basic Terminal Designs.

Design	Level of Service	Design Economy	Flexibility	Security Control
Linear	Good	Poor	Very Good	Fair
Gate Arrival	Good	Poor	Very Good	Poor
Pier	Good	Very Good	Very Good	Good
Satellite	Fair	Good	Good	Good
Unit Terminal	Fair	Poor	Good	Poor
Transporter	Good	Good	Very Good	Fair

It is not uncommon that an airport utilizes more than one design configuration in its passenger terminal layout in order to reduce passenger walking distance. Most authorities agree that terminal walking distance of 600 feet is an ideal maximum and admit that any distance greater than 1,200 feet to 1,300 feet must be considered unacceptable.[8] Assuming an originating passenger steps off a limousine at the terminal entrance, goes directly to the ticket counter, then to the airline gate. Table 2 shows a typical ranges of maximum walking distance for inter-line passenger at airports with different terminal configurations. The figures reflect debarkation at the remote end of the farthest concourse with a walk directly to a gate at the remote end of the most distant concourse without stopping at a ticket counter.[8,9]

Table 2. Typical Maximum Walking Distance at Airports Before APM.

Design	Gates	**Inter-Line Passengers** Max Walking Distance (ft.)
Linear	5-20	600-2,000
Pier	16-136	1,500-7,500
Satellite	18-102	2,500-9,000
Unit	30-104	2,000-8,000
Transporter	40-50	1,000-1,300

Terminal Design after APM

The APM system can be built into any existing airport design at the time of airport expansion or modification in order to reduce excessive walking distance. This is the case in Seattle, Miami, Houston, and Las Vegas as they generally involve the construction of one or more new remote satellite, and the APM system is used to connect the satellite(s) with main terminal. In Seattle and Houston, the entire airport facilities are connected by the APM while in Miami and Las Vegas, only the new remote satellite is connected to the main terminal with APM.

New airport passenger terminal designed with APM system generally favors centralized operation because it utilizes airport resources more efficiently. The only exception is the Dallas - Ft. Worth International Airport, which is basically an old decentralized operation design conceptualized in the late 1960's. When a brand new airport passenger terminal complex is designed from scratch with APM in mind, three passenger terminal designs are generally used. These designs are as follows:

1. **Remote Satellites.** This design has a central terminal and several remote satellites which are connected by APM shuttle train either above or below the apron. This design usually takes full advantage of the APM system by having all ticket purchase, baggage check and reclaim facilities, and other main passenger services in the main terminal area only. Only the holding lounges and supplementary check-in facilities for passengers not carrying baggage are located in the satel-

lites.[10] Tampa and Orlando International Airports are examples of remote satellite design.

The design of one focal point (main terminal) with several remote satellites, is very simple and user-friendly for the first-time user. In addition, the design is extremely flexible and can be expanded with little difficulty. This design is efficient in its land use and, generally speaking has the shortest walking distance for passengers. However, for very high volume airports (annual passengers more than 40 millions), this design is inadequate. So far only the top four airports in the world have this kind of heavy passenger volumes. Therefore, the use of the remote satellites design sholud gain wide popularity in the future at the world's major airports.

2. **Remote Piers.** This is a new design based completely on the proven effectiveness of APM. The remote piers (concourses) are linked with central terminal via APM under the apron. The location of the terminal complex is in the middle of the airfield between the system of parallel runways. This facilitates the movement of aircraft between gates and runways, makes for short taxiing timed and alleviates congestion. This design allows efficient use of airport land as one of the heaviest air traffic airport in the world (Atlanta) was built in an area of less than 3,800 acres.

Parallel alignment of the piers (concourses) assures efficient use of apron space. One major problem with this design is that extremely long pier (concourse) may results in higher than desirable walking distance for transferring passengers. This is the case in Atlanta where the maximum walking distance for inter-line passengers - 2,000 feet, is the same as the length of one remote pier (concourse). This problem can be overcome by using a shorter piers (concourses) at greater numbers.

3. **Unit Terminal.** This decentralized design is the same as the previous unit terminal design before APM. The only difference is that APM system is built into the terminal to connect different unit buildings. Therefore, this provides a high level of interline ground connection service. Dallas-Ft. Worth International Airport is an example of this design. This design is not efficient in land use and requires an extraordinary amount of APM guideways to make it work. Subsequently, no new airport has adopted this design since Dallas-Fort Worth Airport opened in 1974.

Evaluation of terminal designs with APM systems is shown in Table 3. Typical walking distances at airports with APM systems are shown in Table 4. It is clear that APM eliminates one of the most aggravating facets of modern travel - the long, tiring, time consuming terminal trek from your car to the boarding gate. The maximum walking distance at the airport terminals designed with APM's are generally less than 1,300 feet. This is acceptable to most airport authorities.

Table 3. Evaluation of Terminal Designs with APM Systems.

Design	Level of Service	Design Economy	Flexibility	Security Control
Remote Satellites	Excellent	Good	VG	Excellent
Remote Piers	Excellent	VG	VG	Excellent
Unit Terminal	Good	Poor	Good	Poor

VG = Very Good

Table 4. Typical Walking Distances in Feet at Airports with APM Systems.

Terminal Design	No. of Gates	Originating Pax m.W.D.	Originating Pax M.W.D.	I.Pax M.W.D.
Remote Satellites	48-69	300	700	700
Remote Pier	142	400	1,300	2,000
Unit Terminal	101	300	700	1,200

Pax = Passenger I.Pax = Interlining Passenger
m.W.D. = Mimimum Walking Distance
M.W.D. = Maximum walking Distance

Summary and Conclusions

The main points that emerge from this paper are:

1. Existing airport designs generally require excessive walking distances for passengers at the nation's major airports.

2. New airport designs with APM eliminate one of the most aggravating facets of modern travel - the long, tiring, time consuming terminal trek from parked car to the boarding gate.

3. Remote satellites design is found to be better when the percentage of transfer passengers is lower and

the availability of land is not a problem.

4. Remote Piers design is found to be better when the percentage of transfer passengers is higher and the availability of land is a problem.

Acknowledgements

The assistance of Mr. Gary L. Brosch is acknowledged and appreciated. This research was supported in part by a grant from the Center for Urban Transportation Research (CUTR), University of South Florida.

References

1. Transportation Research Board, Measuring Airport Landside Capacity, Special Report 215, Wash., D.C. 1987.

2. FAA, Airport Activity Statistics, Washington, D.C., 1988.

3. Westinghouse Electric Corp., Automated People Mover and Regional Development, West Mifflin, PA, 1980.

4. Elliot, D., "Applications of Automated People Mover Systems at Airports," Automated People Movers: Engineering and Management in Major Activity Centers, pp. 384-398, ASCE, 1985.

5. Chicago Department of Aviation, Chicago O'hare International Airport Factsheet, 1988.

6. Maddison, Mandle and Godfrey, Jr. K, "Crunch Time for Airport Engineering," Civil Engineering, ASCE, April 1987.

7. Ashford, N., et al., The Design of Passenger Processing System for Airport Terminals, Report No.7407, Loughborough: Loughborough University of Technology, 1974.

8. Leigh Fisher Associates, Inc., Airline Airport Terminal Design Possibilities, San Francisco, CA, October 1963.

9. Shen, L. D., "Airport Planning and Design with New Technologies." American Society of Engineering Education Southeast Section 1989 Proceedings, pp. 20-24, Tuscaloosa, Alabama, April 2-4, 1989.

10. Asford, N. and Wright, P., Airport Engineering, Second Edition, John Wiley & Sons, Inc., 1984.

DEVELOPING CAREER GUIDANCE PROGRAMS FOR TRANSPORTATION ENGINEERING

Martin E. Lipinski, P.E., M. ASCE*

Abstract

Efforts are needed to overcome the shortage of qualified young people entering the transportation engineering profession. The positive aspects of choosing a career in transportation must be emphasized with effective programs which use available career guidance materials and the participation of educators and engineering practitioners. Personal contact and follow-up visits to perspective students are key elements in the overall recruiting plan.

The Problem

The transportation engineering field is facing a significant manpower shortage. Several reports document the problems facing state transportation departments (1)(2). Consulting engineering firms, the federal government, and city transportation departments indicate similar difficult in finding people to fill available positions.

The task of rebuilding our nation's infrastructure will increase the demand for transportation professionals. The magnitude of future mobility problems and the need for programs to address transportation deficiencies was documented in other reports (3)(4).

The supply of transportation engineers is provided by graduates of civil engineering programs. Enrollment statistics indicate that the number of students pursuing civil engineering degrees is not increasing and may not be adequate to meet projected demand. It is discouraging

*Professor of Civil Engineering, Memphis State University. Memphis, TN.

to note that enrollments in grades 9-12 will decline until 1990, resulting in a decrease in college and university enrollments. These trends indicate that civil engineering curricula will experience further declines.

The qualifications of students pursuing civil engineering degrees, as measured by high school grade point averages and standardized admission test scores, have been found to be below the qualifications of students selecting other engineering majors (5).

To meet anticipated demands it is apparent that efforts are needed to attract quality young people to careers in transportation engineering. Steps have already been taken by other professional organizations, such as the Institute of Transportation Engineers (ITE), which has prepared a report on attracting young people to transportation engineering careers (6) and has an on-going committee actively addressing the problem. Other professional organizations and some transportation agencies have undertaken similar projects.

Perceptions of Transportation Engineering

The principal causes of the shortfall in the supply of transportation professionals are the negative perceptions which resulted in the declining interest in civil and transportation engineering. These perceptions include: (1) a "low-tech" image of civil and transportation engineering, (2) poor starting salaries and limited opportunities for advancement, and (3) a diminishing public service ethic.

Addressing the Problem

Solving the manpower shortage requires an approach that will overcome the negative perceptions and create a positive image of the transportation engineering profession. The message must be communicated that jobs are available in transportation, that these jobs offer variety and challenge, that "high-tech" tools are used by the transportation engineers, that there are rewards from public service careers, and ownership opportunities exist for those who choose to work in the private sector. A comprehensive effort, involving educators, individual practicers, professional organizations, and employers is needed. This effort includes targeting different age groups and minority populations, developing promotional materials, formulating plans for distribution of information, and initiating and maintaining contact with young people.

Career Guidance Materials

Career guidance programs require high-quality promotional materials. Competition for the attention of young people is intense and professionally-produced materials are necessary to portray the challenges and rewards of transportation engineering careers. Several professional societies and associations have developed excellent brochures and audio-visual presentations. Videotapes are available from ASCE and The Institute of Transportation Engineers. The ITE videotape is aimed toward the undergraduate civil engineering student who has not selected a technical concentration. The ASCE videotape provides an overview of the civil engineering profession for a high school level audience. Slide-tape programs are also available from ASCE.

Professional societies that have brochures available include ASCE, ITE, The American Road and Transportation Builders Association (ARTBA), and The National Society of Professional Engineers (NSPE). Each piece has its own target audience ranging from junior high school to college age students.

Some transportation agencies and private consulting firms have recruiting materials which describe the types of jobs available and typical projects. Universities also produce brochures describing their overall engineering programs and the individual degree areas.

Producing new career guidance materials, especially videotapes or glossy brochures, is time-consuming and costly. While there may be unmet needs for new promotional materials targeted for specific audiences, existing sources should be reviewed before undertaking the development of new materials.

Comprehensive Program Development

The principal element in effective career guidance efforts is the individual contributions made by both educators and practitioners. Civil engineers must avail themselves of the opportunities to promote the profession. Simply handing out pamphlets or showing a videotape is not enough. Structured approaches should be developed which blend available materials, references to current or past local projects, individual experiences, and testimonials from the speakers about their career choice.

Increasing the number of transportation engineering graduates requires a pyramid approach that is based on establishing an interest in engineering among qualified grammar and junior high school students. Transportation engineers are produced from the pool of civil engineering graduates. Therefore, the next level of emphasis should focus on the challenges and opportunities in civil engineering. The primary audience for this message should be high school students. As a final step, specific transportation engineering career guidance programs should be developed for underclass civil engineering students.

Each segment of the civil engineering community, educators, employers, and professional organizations has a role to play in the overall career guidance effort. The following are examples of programs which have been used by each group to promote transportation engineering careers.

Educators Civil engineering professors, through their continual involvement with students, have many opportunities to recruit for the transportation engineering profession. Beginning with classroom contact, they can bring in examples of interesting projects they have worked on or are familiar with and the challenges which confront the profession. Serving as advisors to ASCE or other student groups, such as ITE, they can arrange field trips to local projects, arrange for practitioners to present programs related to their jobs or employment opportunities, organize student paper or design competitions centering on transportation topics, and assist students in finding summer, part-time, or permanent employment with transportation engineering firms.

Many universities have summer programs aimed at recruiting high ability or minority students. Professors can become involved in these programs using transportation projects to illustrate the application of engineering skills.

Engineering open house or career days on campus also provide opportunities to recruit students. Exhibits and demonstrations can be developed illustrating computer simulation of intersection traffic or the analysis of roadway design problems using CAD systems.

As members of professional societies, professors can also participate in career guidance programs organized by these groups.

Professional Societies Using the network of local sections and branches, professional societies can establish organized efforts in career guidance. Providing help to established programs, such as the Math Counts program administer by NSPE each spring or the competitions sponsored by JETS, is one way of generating interest in the engineering profession.

A local professional group can establish a speakers bureau to address high school students, scout groups, or other organizations. An outreach program which includes developing a list of contacts among school guidance counselors and their needs for both formal programs and career guidance materials can be beneficial. Appearances before university student groups are also valuable.

An effective speakers program requires preparation and training. The local organization should maintain a collection of brochures and audio/visual aids which can be used for programs with differing audiences. Individuals participating in the program should be given an outline script with examples of the topics that could be covered in a typical 45 minute presentation. An overview of the profession and its subareas, the relationship of transportation engineering to civil engineering, preparing for a career in transportation engineering, examples of transportation engineering assignment, and reasons for selecting a career in transportation are topics that should be covered in a general presentation. Personal experiences should be used liberally throughout the talk. The speaker should allow time for questions and answers, and the names and addresses of students attending the presentation should be taken. The organization should follow-up and contact the students that attended and give them a name they can call or write to for further information.

Local professional organizations could also provide schools and public libraries with career guidance literature.

Some national headquarters of professional organizations have developed summer job clearinghouses. Both ITE and ARTBA have such services.

At both the national and local levels, professional societies can develop scholarship programs. These may be designated for a specific university or for students from a particular community. These scholarships typically are awarded on the basis of interest in the field as demonstrated by a students' past academic record or through an essay competition.

Employers Employers have a major stake in career guidance activities as they are establishing a network of future employees. Providing summer intern and co-op work/study programs for college and senior high school students is an obvious project for employers. These programs should seek-out high ability and minority students. Job opportunities should provide the students with variety and challenging tasks, and not repetitious boring work.

Some companies have become involved with adopt-a-school programs in their local communities. These companies typically provide the schools with extra resources, such as science equipment. They also may include tutoring by employees and after school work opportunities.

Many governmental agencies have developed their own career guidance programs. Typical elements of these outreach efforts include displays for career days and visitations to high schools and universities.

Personal Contact

The most successful recruiting efforts will be those which involve one-on-one contact where transportation professionals seek out young people, speak with enthusiasm about opportunities in the field, and show interest, encouragement, and support for the students. Initial contacts could be made through formal speaking programs such as those described above. Contact should be maintained with perspective students with offers of more information or assistance to interested individuals.

Informal situations provide ideal opportunities to establish contact with young people. Family members, neighbors, young people in church groups and others should be encouraged to consider careers in transportation. Civil engineers can serve as counselors and role models in attracting quality individuals to the transportation engineering profession.

Conclusion

Effective career guidance programs requires the contributions of educators and practitioners. Universities, professional organizations, and employers must develop programs and commit resources if the shortage of qualified transportation professionals is to be overcome.

REFERENCES

1. Transportation Professionals: Future Needs and Opportunities, Transportation Research Board Special Report 207, Transportation Research Board, Washington, D.C., 1985.

2. Transportation Professionals: Recruitment and Retention, American Association of State Highway and Transportation Officials, Washington, D.C.,1988.

3. Fragile Foundations: A Report on America's Public Works, National Council for Public Works Improvement, Washington, D.C., 1988.

4. Beyond Gridlock, Advisory Committee on Highway Policy, 2020 Transportation Program, Highway Users Federation, Washington, D.C., 1988.

5. McDonough, James F., "The Qualifications of the Civil Engineering Student Entering College Compared to Other Engineering Students", Challenges to Civil Engineering Educators and Practitioners - Where Should We be Going?, American Society of Civil Engineers, New York, NY, 1985.

6. Attracting Students to a Career in Transportation Engineering, Report of Technical Council Committee 2-32, Institute of Transportation Engineers, Washington, D.C., 1989.

Integrated Surveying Engineering Education - New England

David L. Westerling, Member ASCE[1]

Abstract

In September of 1974 a white paper on surveying education was presented at the International Congress of Surveyors in Washington, D.C. This paper deals with the second recommendation of that white paper which called for an integrated regional policy for those aspects of surveying technology concerned with curriculum development, content and evaluation; faculty criteria; student enrollment and placement; and interface with the regional university program.

Introduction

The 1974 white paper on surveying education was endorsed by the six state societies of Land Surveyors, the New England Section of the American Congress on Surveying and Mapping (ACSM) and the Eighth National Surveying Teachers Conference. It contained five specific recommendations for an integrated regional survey education program. These were:

* that there be one professional degree program in surveying within the New England Region.

* that there be an intergrated regional policy for those aspects of surveying technology education concerned with curriculum development, content and evaluation; faculty criteria; student enrollment and placement; and interface with the regional university program.

* that a multiple entry approach towards land survey registration be formally recognized by all New England State Registration Boards.

[1] Assistant Professor, Department of Civil Engineering, Merrimack College, North Andover, Massachusetts 01845

* that a policy on continuing education requirements for surveyors in New England be formulated and officially adopted by the professional societies.

* that a common policy regarding professional practice requirements be adopted by the New England surveying societies.

This paper deals with the second recommendation on an integrated regional policy for surveying technology education and the intergration of this policy with the regional university program.

The Organization of a Regional Task Force

A task force to address a regional policy was formed at the Special Forum for Proposed Survey Technical Education. This forum was held at the annual meeting of the New England Section, ACSM at the Sheraton-Boxborough Inn and Conference Center, February 12-13, 1988. Since that time the task force has held regular meetings at Merrimack College, and has made significant strides towards a regional policy.

The task force has four subcommittees with specific functions. They are:

1. The Public Relations Committee - to investigate the methods which would increase public awareness of surveying engineering and which would increase the number of qualified students of surveying engineering.

2. The Practitioners' Needs Committee - to determine the expected needs of surveying and mapping companies for graduates. And to determine the qualifications of various classes or levels of graduates.

3. The Curriculum Development Committee - to identify existing programs in surveying technology and surveying engineering and to propose ideal curriculums to be used for planning academic programs.

4. The Surveing Engineering Education Integration Committee - to integrate the available and proposed surveying and mapping academic programs including certificate, two year Associate, four year Bachelor of Science, Masters Degree and Doctor of Philosophy Degree programs in surveying engineering.

An Integrated Regional Program

The integrated program must have three important components. First it must be structured such that students can progress to higher levels without undo hardships. Secondly, it must allow students to transfer between various educational institutions of higher learning. And third, it must allow for students from various disciplines to pursue careers in surveying engineering.

In a recent paper by Colcord (1988) a suggested educational model is proposed which relates various educational insitutions. A proposed model for New England is shown in Figure 1.

In order to enter any of the higher learning institutions a student would need a high school diploma with the appropriate preparations in math and science. A student could enter either the technical school, community college, liberal arts college, professional school, or university.

Currently the Accreditation Board for Engineering and Technology (ABET) has provisions for reviewing programs which lead to an Associate or Bachelor of Science Degree in Surveying Engineering. Those students not enrolled in these programs, such as those enrolled in certificate or Bachelor of Arts programs should not be penalized by uncertain curriculum requirements. The basic courses in surveying engineering such as introduction to surveying, plane surveying, photogrammetry, and others should have standard curricula. These courses, along with math, science and humanities electives could then be transferred through articulation agreements between schools. The formation of these articulation agreements should be promoted by providing generic agreements for use by academic administrators at both the accredited and non-accredited schools alike. This would allow for students to transfer credits between educational institutions with greater ease and with assurances that their coursework would not be lost.

The use of existing consortiums between colleges and universities in close geographical area could provide additional avenues of cooperation. Schools with accredited surveying engineering programs could enter into articulation agreements with consortium schools with each institution providing courses in their strong areas.

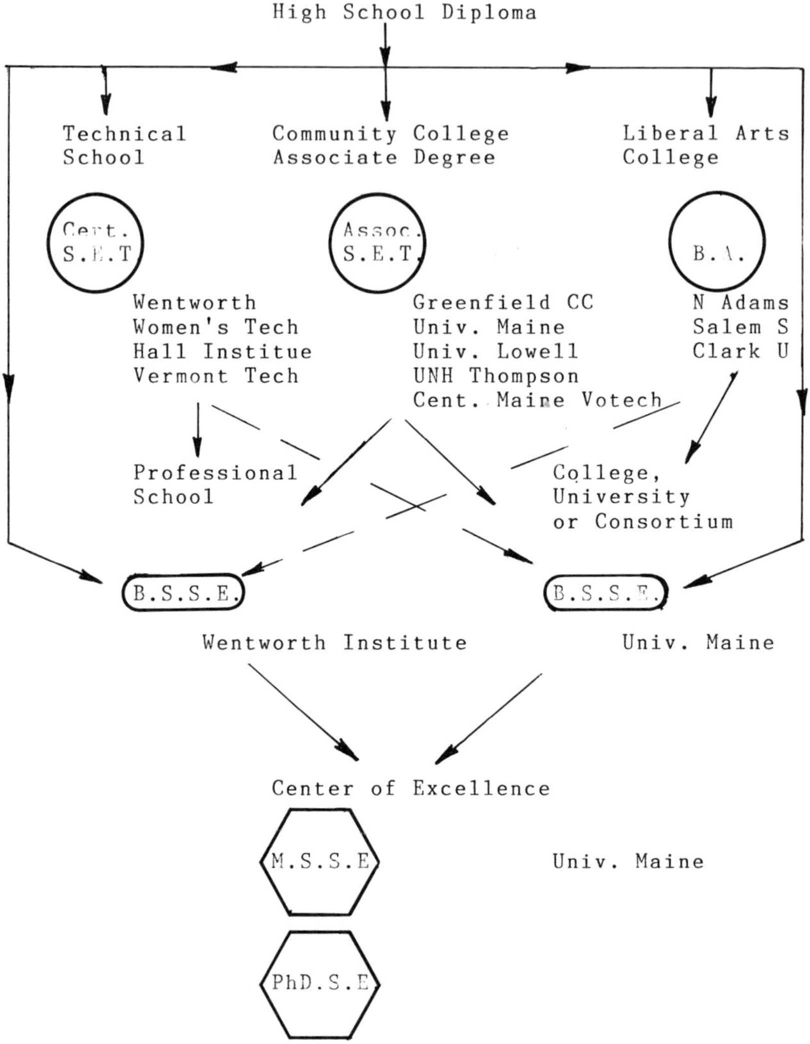

Figure 1. Suggested Educational Model for Surveying Engineering in New England
(Note - The institutions listed represent the best efforts of the Task Force in locating existing programs. Additional schools may be listed in the final Task Force report.)

With the advent of geographical information systems (GIS) and land information systems (LIS) it is apparent that many students are now being exposed to topics related to surveying engineering. These students are enrolled in geography and cartography courses and are persuing a Bachelor of Arts or Bachelor of Science Degree. Many are exposed to advanced digitized mapping equipment, resource information systems, and other cartographic specialties. These students may become involved in cadastral based systems which traditionally have been the land surveyors domain. It is essential that those students wishing to pursue these systems be allowed to enter an ABET accredited institution and begin a career path toward land surveying. The suggested educational model provides for this career path.

Education and Professional Registration

The suggested educational model must also be parallel with a multiple entry approach to professional registration as a land surveyor. If a student wishes to pursue a land surveying career he or she should have access to a multiple entry approach. This flexible approach should be stressed as an advantage in selecting surveying as a career.

For example, a student completes high school but due to lack of funding is unable to leave home to attend college. Instead, he or she attends a community college offering a two year Associate Degree in surveying engineering. This student is then immediately available to the profession and could begin working under a registered land surveyor on a traditional career path. This same student then might be forced to move due to family circumstances. He or she could then continue working as a technician or proceed toward a Bachelor of Science Degree in surveying engineering.

Conclusion

The suggested model for Surveying Engineering in New England is based upon Colcord's findings. It allows students flexibilty in entering the surveying career path and it provides options for students with differing academic abilities. It encourages institutions of higher learning to cooperate in the administration of surveying engineering programs through curriculum standardization and articulation agreements between schools. And it allows for a multiple entry approach, including the surveying experience entry, into the profession.

Appendix. References

Colcord, J. E. (1988). ASCE, "Surveying Engineering Education for the 21st Century", Journal of Surveying Engineering, Feb.

McLaughlin, John D. and Petersohn, Fritz K. (1974). "An Integrated Regional Approach to Surveying Education." XIV International Congress of Surveyors, Washington, D.C.,Sept.

Is Surveying Really Part of Civil Engineering?

Earl F. Burkholder[1] M.ASCE

Abstract

Surveying has been taught in departments of civil engineering and surveying practice has been associated with civil engineering for many years. In the past 30 years the number of surveying courses and credit hours required in most undergraduate civil engineering curricula has been significantly reduced and in some cases, eliminated entirely. This paper looks at the historical interaction between surveying and engineering and raises questions about impacts of the current trend in areas of accreditation, registration and professional practice.

Introduction

The terms "surveying" and "civil engineering" carry with them certain mental images which differ according to ones background and experience. Without attempting to define either one exclusively, it is noted that surveying as a measurement science has changed dramatically in the recent past through application of new technologies including; photogrammetry, remote sensing, electronic distance measurement, inertial positioning, satellite positioning, and the silicon chip (computers). Surveying as an information science (knowing what to measure as determined by evidence, historical fact and legal principle) has also come to be recognized for its contribution to development of a multipurpose cadastre and managing spatial attributes of a Land Information System (LIS). In general, the interaction between surveying and civil engineering has been in the measurement science arena and "land surveying" has been associated with land boundaries and property line surveying.

- - - - - - - - - -

1 Geodetic Engineer, Oregon Institute of Technology, 3201 Campus Drive, Klamth Falls, Oregon 97601-8801

A broad definition of surveying could cover both perspectives in which surveying falls under the broad umbrella of civil engineering. However, arguments were offered for separate status (registration) because specific differences are important to land surveyors who feel a civil engineer lacks sufficient knowledge and background in legal aspects to perform property surveys. Society, acting through state legislatures, laws and registration boards, accepted those arguments and separate registration is a fact in most states.

If surveying is truly under the umbrella of civil engineering, then the separate registration might have been a specialty certification (meaning a person would qualify for and receive engineering registration before becoming eligible for the added distinction of "land surveyor"). But, that is not the case as most (if not all) registration boards will grant a "land surveyor" license to one who would not otherwise qualify for engineering registration. The result is a double standard for registration. Does registration as an engineer qualify one to practice "land surveying"? Or, does registration as a surveyor qualify one to perform engineering surveys? These questions are too narrow because registration (even though widely used) is not a reliable indicator of professional competence. Of greater importance is the knowledge (education) and judgement (experience) one brings to a given task. As evidenced, in part, by persons registered in both catagories, there is a significant overlap between the professions and questions of academic preparation need to be discussed by both.

When discussing similarities of and differences between surveying and civil engineering, the issue is often clouded by a failure to distinguish the catgory in which comparisons are being made.

Research: Research activities which explore frontiers of knowledge and promote development of new technologies are conducted at universities, in both state and federal agencies and by private enterprise.

Management: Managers at all levels, both government and private, make informed decisions about allocation of technical, capital, and human resources to meet organizational missions or corporate (and business) objectives.

Application: Responsible professionals analyze existing technology and make judgements between alternatives in planning how given objectives or missions will be met.

Implementation: Persons knowledgeable about technical principles and concepts are responsible for getting the work done in a prudent manner.

The foregoing catagories are not meant to suggest that an individual is restricted to one or the other. In reality, some individuals may function in all catagories. Two points to be made are: 1) it is not fair to compare surveying in one catagory with engineering activities in another and, 2) educational curricula are often designed to prepare a person for one of the catagories; graduate school for research, business courses for management, engineering/surveying programs for applications and technology degrees for implementation. These catagories are neither complete nor exclusive.

Historical Interaction:

The historical interaction between surveying and civil engineering should be considered in trying to answer, "Should Surveying be Included in Civil Engineering Education?" Activities related to or called "surveying" have been part of civilization throughout recorded history. From construction of the Egyptian pyramids to Roman aquaducts to seagoing ships (and maps) to machines of the industrial age. From conquest of the seven seas and the uncharted continents to the fringes of space by landing on the moon, surveying and civil engineering professionals have worked together in a quest for a better life for all mankind. Specifically, surveyors and engineers both contributed in significant ways to settlement of the United States. Kreisle (1988) describes development of surveying engineering in the US, White (1983) chronicles the role of surveying in establishment of the U.S. Public Land System and Wilford (1981) traces the progress of the mappers (engineers, surveyors, cartographers and others) in exploring and charting vast territories of the western United States.

Since the early 1900's surveying has been part of the civil engineering curriculum in most universities in the USA. However, as other specialties in civil engineering (structures, transportation, planning, hydraulics, water supply & treatment, materials, geotechnical, environmental studies and others) have risen to prominence, other topics such as surveying have been reduced or eliminated entirely. Marks & Weeden (1985) did an extensive study of surveying course offering in civil engineering programs and found no program required more than two courses in surveying. Twenty three percent do not require any courses in surveying and ten percent don't offer any surveying instruction in the undergraduate curriculum. A follow-up study should be conducted.

With surveying instruction being squeezed out of civil engineering programs, the slack has been taken up by establishment of separate programs in surveying. In 1951 the Oregon Institute of Technology began offering a 2 year associate degree in surveying and in 1966 offered the first 4-year surveying program in the USA. Purdue University followed with a Land Surveying degree in 1971 and others have been established since then. Today, there are several very high quality university research based surveying programs in the USA offering both graduate and undergraduate degrees. There are also several professional level 4-year degree programs and numerous technical and associate degree programs available for the person who wishes to study surveying.

Definition of Issues:

Within the Surveying Engineering Division of ASCE, there is a Committee on Education and Professional Practice. At a committee meeting in April, 1989 the committee discussed the apparent decline of surveying instruction within civil engineering and identified five issues deserving study and consideration:

1. Should civil engineering programs offering courses in design and construction offer at least one course in surveying?

2. To what extent should surveying engineering (location, grades, drainage, etc) be included in transportation, site design, and facilities engineering programs?

3. Given the purpose of licensing is to protect public safety and welfare, is it necessary for an individual to be licensed as a surveyor in addition to his/her engineering registration in order to provide surveying engineering services?

4. At what point does specialty certification become necessary for one who has only an engineering registration or the person who has only a surveyor's license to provide services that are considered to overlap the other profession?

5. The licensing process should be flexible enough to accomodate an engineering or surveying degree from either profession as meeting minimum requirements for registration. Does it (or should it) work both ways?

It is hoped feedback from this forum on education and continuing development for the civil engineer will provide the basis for additional committee deliberation.

Areas of Impact

With due regard to the level (or catagory) of comparison and past interaction between surveying and civil engineerring, the complex issues raised by the Education and Professional Practice Committee need to be discussed in light of:

A. Accreditation - EAC, RAC and TAC

B. Registration - NCEE and individual state boards

C. Professional Practice - Overlap of professions

The Accreditation Board of Engineering & Technology (ABET) is the accrediting agency for civil (and other) engineering and surveying programs in the USA. Three accreditation commissions include:

EAC: Engineering Accreditation Commission
TAC: Technology Accreditation Commission
RAC: Related Accreditation Commission

The EAC accredits professional level engineering programs, including those surveying programs which meet EAC criteria. The TAC accredits engineering technology programs which includes some surveying programs, both at the associate and baccalaureate degree levels. The RAC was established in 1983 to accredit programs which are closely allied with engineering, but whose specific program content departs from traditional engineering programs. It was intended by those who established RAC criteria for surveying programs that the rigor would equal that of EAC progams.

EAC criteria do not require any surveying to be included in a civil engineering degree. TAC program criteria for both an associate and baccalaureate degree in civil engineering technology (submitted to ABET by ASCE) do require topics in surveying. The implication is that technical operations of surveying are important for technicians. Somehow, professionals either don't need it or are expected to get it on their own? Is (or has) ASCE going to abdicate its role in professional surveying? ASCE participates with the American Congress on Surveying & Mapping (ACSM) in submitting criteria for EAC Surveying and TAC Surveying Technology programs.

Additional evidence of the continuing growth and emergence of surveying as a separate profession is the name change approved August, 1989, at the annual meeting of the National Council of Engineering Examiners (NCEE). The new name is the National Council of Examiners for Engineers & Surveyors.

If surveying has its own (equal) accreditation criteria and enjoys equal stature in the registration process, then maybe civil engineers should abandon attempts to include surveying as part of civil engineering. However, in spite of the changes being made, some surveying engineers feel civil engineering should not abdicate its role in surveying.

NCEE is made up of registration boards (both engineering and surveying) from various states and composes and grades the examinations used by most boards as the basis for granting registration to both surveyors and engineers. Content for the surveyors exam is based upon a "task analysis" of activities of practicing surveyors and carefully analyzed by a hired psychometrician to assure validity of the testing process. Observations of the process are that many practicing surveyors do not have a college degree and perform many "surveying" tasks for which a college degree is not required. In addition, few states require a college degree of surveying registration applicants. Therefore, the testing process could be biased against improving the qualifications of those entering the surveying profession. A resolution passed at the 11th National Surveying Teachers' Conference (Fresno, CA, 1984) requested that NCEE base passing scores for the surveying exams on the performance of those examinees having a college degree. Unless and until that is done, the double standard for registration will be perpetuated.

In the area of professional practice, there is a significant difference between public perception of the surveyor and the civil engineer. Part of the perception is that many surveyors work directly for the lay public and most of their fees come directly from the consumer. Engineers tend to work for governmental or corporate clients. Added to that perception is the fact that surveyors are seen standing behind the tripod and the public fails to see or understand the planning and research which goes into surveying activities. The engineer is viewed as the person in a suit and hard hat with a roll of plans laid out on the hood of a pick-up on the construction site. Unwittingly, a surveying technician is compared to an engineering professional resulting in an inaccurate conclusion.

Admitting an element of truth in the aforementioned stereotype, it is not a correct picture. In an ASCE Workshop on Civil Engineering in the 21st Century, Dr. Wong (1988) described surveying and mapping activities beyond 2000 and pointed out that civil engineers have a vested interest in two areas of current research; real time surveying (positioning) systems and geographic data bases. Each of those surveying activities does and will have an impact on the scope of civil engineering activities. Civil engineers need significant training (more than a 2 credit hour course) in modern surveying methods and techniques.

Surveying as a measurement science is changing dramatically and those changes have a direct impact on civil engineering practice. Bro. B. Austin Barry (1987) focused specifically on that point in his presentation to the 12th National Surveying Teachers' Conference when he described "What Civil Engineers Should be Taught About Surveying." He was brave enough to suggest civil engineers should not waste time learning taping corrections and angles by repetition. Effort should be spent instead on three dimensional coordinate systems, EDM resections, stereoscopy, satellite positioning, computers and others.

Conclusion

In this writers opinion, there is a legitimate overlap of interest and practice between surveying and civil engieering, but two points need to be made: 1.) There is much more to surveying than a civil engineer can learn in one or two courses. The civil engineer who wants to "sign-off" on engineering surveys should obtain a surveying license.
2. The examination process for surveyors should be expanded to include modern measurement science at a level which promotes parity with other professionals.

References

1. Barry, Bro. B. Austin (1987); "What Civil Engineers Should be Taught About Surveying," Proceedings: 12th National Surveying Teachers' Conference, University of Wisconsin, Madison, Wisconsin.

2. Kreisle, William E. (1988); "History of Engineeing Surveying," Journal of Surveying Engineering, American Society of Civil Engineers, Volume 114, No. 3.

3. Marks, G. Warren & Harmer A. Weeden (1985); "Recent Issues in Surveying Education and Practice in the United States," Colloquium on Surveying & Mapping Education, Department of Surveying Engineering, University of New Brunswick, Fredericton, NB, Canada.

4. White, C. Albert (1983); "A History of the Rectangular Survey System," BLM, US Government Printing Office.

5. Wilford, John Noble (1981); "The Mapmakers," Vintage Books, New York, New York.

6. Wong, Kam W. (1988); "Surveying & Mapping Beyond 2000," ASCE, Journal of Professional Issues in Engineering, Volume 114, No. 3.

Surveying and Geodesy in the Exploration and Production of Oil and Gas

H.J. Krijnen[1]

Abstract

In the exploration and production of hydrocarbons, surveying and geodesy, or more generally termed geodetic engineering, constitute a support function. Shell Oil Company has recently made attempts to recruit geodetic engineers. These attempts have been rather unsuccessful. In addition it is noted that in the oil industry, geodetic engineering does not receive the emphasis it should.

Introduction

Exploration for and the production of hydrocarbons has an obvious spatial component. Exploration is conducted somewhere on the earth's surface, with the object to uncover the upper part of its internal structure in order to drill holes at locations with the highest chance of finding hydrocarbons. During exploration activities, positional information needs to be recorded. A drilling rig must be placed on the right location.

Positional information is a common thread through exploration and production, from the early geological reconnaissance to the construction of production facilities and pipelines. Surveying and geodesy should be a logical function in an oil company. It is not a core function, like for example geology or petrophysics, but it is definitely a support function.

In order to further define the role of surveying in the oil business, we may wish to clarify the terminology somewhat. Surveying, perceived as the trade of conducting and reducing measurements at a relatively small scale on land, would be too limited a description to indicate what is involved. Positioning in the oil business concerns a much wider spectrum of sciences and techniques. In the remainder of this article the term geodetic engineering

----

[1] Senior Geodesist, Shell Oil Company, Bellaire Research Center, P.O. Box 481, Houston, TX 77001.

shall be used.

Geodetic engineering activity

The geodetic engineering activity in an oil company can be subdivided into three main categories:

1. Onshore activities

When an exploration venture is started up in a new area, maps constitute one of the first requirements. If good maps do not exist, as may be the case in third world countries, mapping projects are undertaken. Today, often satellite images will suffice.

Before any further positioning work is carried out, it must be ascertained which datum and projection are to be used. Geodetic control must be recovered or established. Datum conversions may be necessary.

The most important exploration tool is reflection seismic. With this technique, traverses are laid out, geophones are planted and at regular intervals a dynamite charge is detonated or a vibroseis source is activated. All these positions need to be surveyed in. Such a traverse network along seismic lines must be tied into geodetic control on the proper datum. Depending on the situation, positional relations with land ownership boundaries may need to be established. The positional information thus obtained is stored in huge data bases.

After geologists and geophysicists have selected a drilling location, this position must be carefully staked out in the field. Again, property boundaries are often of great significance in this context. Surveys are normally required for site preparation and location building.

If a discovery is made, appraisal wells are drilled. If production is viable, facilities are designed and engineered, requiring further and more detailed surveys.

All these facilities and pipelines will be entered into an information data base. Photogrammetric mapping may be undertaken to provide, together with field surveys, the positional basis for such an information system.

2. Offshore activities

For exploration in an offshore environment, hydrographic charts are needed. These are normally obtained from appropriate authorities.

A radio navigation system needs to be set up. Base stations must be established and surveyed in, these days often by GPS. Datum issues are to be addressed.

The seismic data acquisition is conducted by vessels that tow streamers containing hydrophones. These streamers generally have a length of 3 to 4 kilometer. The seismic source is in most cases an air gun array, typically towed by the same vessel. Traverses are sailed, with the guns firing perhaps every 10 seconds or every 25 meters. Positioning of the vessel is obviously essential.

The introduction a decade ago of so called 3D surveys, where complete areas rather than individual traverses are covered, have dramatically changed the positioning requirements. Accuracies of a few meters are now needed. In addition, the need to exactly position the air guns and the streamers has led to the application of new positioning devices, such as streamer mounted compasses, acoustic ranging, scanning lasers, etc.

The activity sequence is roughly the same as onshore, but the survey techniques are different. Rig positioning is somewhat more complicated. Prior to drilling, the seabed must be surveyed using side scan sonar and shallow seismic technology. Underwater acoustics may be employed for emplacement of production platforms. Pipeline route surveys will precede pipe laying. Again, facilities and pipelines will be surveyed and entered into an information data base.

3. Coordinate data management

The most important asset (apart from people) of an exploration and production company is its information data base. This information, as stored and routinely accessed, will include geological, geophysical, petrophysical data, but also lease and land ownership information. All this information is related through positions.

The effectiveness of the data base is a function of (amongst other things) the integrity of the positional information. Coordinates will have a great variety of methods of acquisition: surveys, digitization, photogrammetric projects, satellite images, geological interpretation and mapping. They will come from a variety of sources: government, contractors, vendors, other oil companies. Coordinates will be on different datums and projections. They will have different accuracies. This data has to be managed properly if it is to truly serve as a reference for other data, based on which decisions are taken regarding lease sales, wells, operations and engineering projects.

A good overview of surveying as it relates to exploration can be found in (Spradley, 1985).

<u>Current geodetic engineering in the oil business</u>

There is no doubt that every oil company will, one way or another, deal with all of the geodetic engineering tasks as described above. The way this is done differs considerably from company to company. Some companies have fully developed geodetic departments. These departments are integrated in the organization and have operational responsibilities and authorities. Other companies may have a small group of specialists that internally provide technical advice on geodetic matters. Quite a few companies, however, have no or little in house expertise and rely on contractors or consultants, with geophysicists or engineers in a coordinating role. This may be logical approach for small operators, it is less understandable why a large organization would choose this option.

In this respect there is a clear difference between American and (at least some) European oil companies. The latter tend to place considerably more emphasis on an in house geodetic engineering capability. The reason for this is not entirely clear, but may have some historical background. In any case it would seem that many of the U.S. oil companies have not developed a tradition where geodetic engineering is a support function, to be managed by professionals.

This situation has had its impact on the way surveying and positioning is done in the exploration and production industry. As oil companies lacked the expertise, quality standards were not set, and for example seismic contractors had no incentive to employ trained surveyors. It is not long ago that seismic lines were surveyed by plane tables, while in other parts of the world more sophisticated equipment and procedures were used. The plane table was originally designed for small topographic surveys and is not really suited for seismic surveying. It had however one advantage: the operator needed little understanding of mathematics. Still today, the standards in offshore surveying in the Gulf of Mexico are lower than those in the North Sea. It is generally perceived that the coordinate data bases of many oil companies are in questionable shape.

The impact in business terms of suboptimal in house control over geodetic matters is not straightforward. It has undoubtedly never happened that an oil company went out of business because it had no geodetic department. However, there can be little doubt that overall effectiveness and efficiency of operations are not served by positioning and mapping problems. It has probably happened quite a few times that a well was positioned wrongly, or that some seismic survey yielded degraded data quality because of poor positioning. But in addition, lack of coordinate integrity causes many little inefficiencies that are difficult to measure. Each company will have to decide for itself how much inefficiency it can live with until it becomes worthwhile to increase the emphasis on geodetic engineering.

The profile of a geodetic engineer

Should an oil company wish to be serious about managing geodetic engineering properly, it will need some professionals in coordinating roles. No oil company with some sense will contemplate performing all surveying tasks itself, and develop its staff accordingly. Many jobs will be contracted out.

However, it is by far most effective if these geodetic projects are managed, from the clients perspective, by knowledgeable staff. Such staff will review requirements in consultation with ultimate coordinate users, design project specifications and manage project execution and quality control efforts. This could be considered as the operational and administrative component.

Geodetic staff will also be involved in management of coordinate data bases, the design of software for this activity, the development of data processing software and sometimes the design of new survey systems or procedures. This could be seen as the developmental component.

What type of person would be needed to assume the roles described above? Geodetic engineering is a support function. Inevitably, the group of people that will deal with geodetic engineering will be small. Therefore, the geodetic engineer working in an oil company should ideally be versatile and be able to cope with both components of the job.

He (or she) should be able to function in an operational environment and work under pressure. For example, he may need to take action if a seismic vessel is "down" for navigation, or if there are problems with the positioning of platforms offshore. He should be prepared to and feel comfortable with going to the field. He should have good interpersonal skills and be capable to deal with contractors, negotiate contracts and make arrangements with third parties as necessary. This requires a certain outgoing personality. It also requires a feel for the practicalities of field work and an ability to perform in a dynamic environment.

At the same time, however, he should be capable of doing some research and for example solve a datum transformation problem. It should not be a difficult task for him to translate such solutions into code in one of several computer languages. He must be familiar with e.g. least squares data reduction methods and feel comfortable when applying various filtering techniques.

This obviously requires a thorough theoretical knowledge of a variety of subjects: geodesy, satellite positioning (GPS), map projections, physical geodesy, statistics and adjustment theory, hydrography, computer science, cartography, GIS, photogrammetry, land surveying, geodetic astronomy. Of course, a good background in calculus, linear algebra, some physics and electronics is necessary.

To summarize, it requires someone with a good geodetic engineering background and who has preferably both operational and development/research type capabilities. Translated into academic requirements, this would mean possibly a Bachelors in geodetic engineering but ideally a Masters.

Geodetic engineering in Shell

In the Royal Dutch/Shell Group of Companies, geodetic engineering is an established support function. Each Operating Company of a certain size conducting exploration and production will have a geodetic department.

Worldwide, Shell employs some 100 geodetic engineers and surveyors in staff positions. Some 30 of these are international staff. They rotate in 2 to 4 year assignments between the various operating companies. Others are employed locally but may, however, be temporarily cross posted. Most international staff geodetic engineers are recruited from The Netherlands and The United Kingdom.

The limitations of the American market

Shell Oil Company is the U.S. branch of the Shell Group. Recently, Shell Oil Company has attempted to recruit geodetic engineers in the United States, with a view of strengthening its geodetic capability along the lines described above.

Universities were approached (Ohio, Purdue, Maine), advertisements were placed in newspapers and in professional journals. The contacts with the universities resulted in a few candidates, but no recruitment followed. It was remarkable that none of the candidates were native U.S. citizens.

Advertisements in newspapers were targeted at geodetic engineers with experience in offshore positioning. There are quite a few surveyors in the industry with substantial offshore experience, but they often lack the academic background that is felt is necessary to handle the entire spectrum of tasks as outlined above. This exercise, too, was unsuccessful.

Advertisements in professional journals were formulated in more general terms: experience preferred but not required. It had become clear by then that the market would be very limited and it was considered that e.g. offshore positioning could also be learned on the job. More than 70 applications resulted. The majority of these were field surveyors, many of whom no doubt are excellent land surveyors, but who would lack the academic background required. It was, again, remarkable that nearly all applicants with appropriate degrees had obtained them at foreign universities. There was only one American applicant with an MS in geodetic engineering and one with a BS. The only recruitment that resulted from this campaign, however, concerned a resident alien.

These efforts have shown that the supply side of geodetic engineers on the American market is not well represented. In discussions with professors at the universities, it was indeed learned that graduate students were either foreign or were sponsored by government agencies and hence committed.

For reasons that are not entirely clear, geodetic engineering in the sense described above does not seem to be a well developed discipline in the U.S.

Shell Oil Company has been very unsuccessful in its recruiting efforts and may have to revert to recruiting in Canada, where the situation is reportedly considerably better in this respect.

Conclusion

It has been found in recent recruiting efforts that there are virtually no geodetic engineers "on the market" of the type and caliber that would make them attractive to Shell Oil Company. For an as yet unclear reason there are very few private American students who complete an appropriate degree in geodesy, geodetic engineering or survey engineering.

Perhaps the reason is that, in the U.S., there is a rather rigid separation between practical surveying on the one hand and the academics of geodesy, map projections, statistics and the like on the other hand. Surveyors know little about geodesy, geodesists are primarily data analysts and have little affinity with the field. The field of geodetic engineering, which should integrate both the practical aspects as well as the academics, seems to be underdeveloped.

This correlates with the low emphasis that geodetic engineering receives in most if not all U.S. oil companies. This is unfortunate, as it has in general allowed low surveying standards to persist in the exploration industry.

Should this change at some stage, then undoubtedly many employment opportunities for geodetic engineers would develop. However, the numbers of students enrolled in the appropriate curricula would not be sufficient and would need to increase drastically.

Which comes first?

References

L.H. Spradley - Surveying and Navigation for Geophysical Exploration, International Human Resources Development Corporation, Boston, 1985.

SURVEYING IN THE CIVIL ENGINEERING CURRICULUM

J. L. Gattis[1], M. ASCE, and R. C. Knox[2], M. ASCE

Abstract

Educators have questioned the place of surveying in the undergraduate civil engineering curriculum. This paper examines the current status of surveying in undergraduate programs, the opinions of employers, and suggests alternatives to meet surveying education needs.

Introduction

In past times, surveying was an integral part of both civil engineering education and practice. Undergraduate catalogues from decades ago indicate that the civil engineering student took many semesters of surveying (Plympton, 1902). Even in recent years, some programs required a summer semester class which gave the student intense exposure to the use of surveying equipment. Civil engineering registration and surveying registration were one-and-the-same.

Today the situation is different. The demands for additional credit hours created by new education needs (e.g., computing skills) and by the desire for a "well rounded student" (i.e., one taking more liberal arts classes), coupled with the need to keep the total number of credit hours in a program from ballooning, led to an academic squeeze. Surveying classes often were consolidated or eliminated to accommodate these needs.

Surveying has developed as an occupation somewhat distinct from civil engineering. It is now common for states to register land surveyors separately from civil engineers. Many civil engineers had their last exposure to equipment when mechanical instruments were the norm, and lack familiarity with the new electronic surveying equipment.

Civil engineering departments are faced with the issue of what to do with surveying in the undergraduate program. Since the trend appears to be one of increasing distinction between the practice of surveying and the

----

[1,2] Assistant Professor, School of Civil Engineering and Environmental Science, The University of Oklahoma, 202 W. Boyd, Norman, OK 73019.

practice of civil engineering, does the college continue to teach surveying as traditionally taught? Now that electronic surveying equipment is dominant, does the department spend scarce funds on expensive surveying equipment which will be used a few times per year? With the divergence between surveying and civil engineering, can the department even expect to find a faculty member who can teach the classroom and field lab components of the traditional surveying class? Should the department respond by modifying the class material and teach an "Engineering Measurements" class, void of any experience with field equipment?

In response to these issues, some faculty at The University of Oklahoma sent questionnaires to civil engineering schools in the United States, asking about surveying education in their undergraduate civil engineering programs. Questionnaires were sent to firms or agencies which employed civil engineering graduates, asking what surveying knowledge they wanted the new employees to have. The group also investigated options which could reduce the cost of providing the students with exposure to the use of field equipment.

University Questionnaires

In order to determine the present status of surveying in the curricula of major universities, questionnaires were sent to civil engineering department heads and chairpersons. These letters were sent to Big 8 and Big 10 schools, as well as to schools in Texas, three to the west, and four to the east. The letters requested the following information.

1. Does your general civil engineering undergraduate program include required surveying classes? __ YES __NO

Please list the class title and number of credit hours of each required class; for each class, please furnish the number of hours per week in the classroom and the number of hours devoted to "field lab."

2. What instruments do the students in the required classes (not elective or advanced classes) use?

____ only electronic distance measuring devices (EDM);
____ mostly EDM, with some transit and chain;
____ a mixture of EDM and transit/chain;
____ mostly transit and chain, with some EDM;
____ entirely transit and chain.

A total of twenty-seven schools supplied information, for a 100% return rate. Twenty-four responded that their undergraduate civil engineering program included at least one surveying class; three programs did not require such a class.

Among those schools requiring surveying, the number of total credit, class, and laboratory hours varied widely. Table 1 presents the reported numbers of hours in the three categories. The more common practice is two hours of classroom (the practice at 11 schools), three hours of field lab (at 17 schools), and a total of three credit hours (at 13 schools).

TABLE 1. Surveying Hours in Civil Engineering Programs

Number of hours (1)	Number of schools having specified number of: class hours (2)	field lab hours (3)	total credit hours (4)
1	2	0	0
2	11	2	0
3	6	17	13
4	4	1	6
5 or more	1	4	5

Most of the departments reported that their required class gives the students some exposure to electronic surveying equipment, although the level of exposure varied among schools. Figure 1 summarizes the responses of those departments that require surveying.

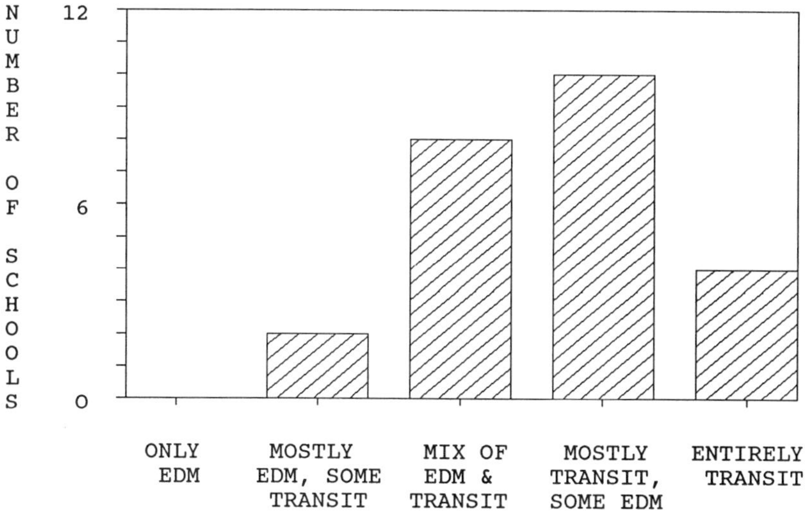

FIGURE 1. Surveying Equipment Used in Field Labs

A summary of the responses was sent to those who had returned the questionnaire. The summary was accompanied by a request for the department to furnish the names and addresses of employers who often hired their civil engineering graduates, two in the public sector, and two in the private sector.

Employer Questionnaire

Nine civil engineering departments furnished names and addresses of employers of their students. Fortunately, the nine were geographically dispersed over the country. A letter sent to these employers contained the following questions.

Q1. Is it important for the new employee to have studied fundamental surveying concepts by reading textbooks and listening to class lectures?

Very Important — Nice to have, but not imperative — Does not matter

Q2. Do you automatically assume that all students with B.S.C.E. degrees have had textbook and class lecture exposure to fundamental surveying concepts?

Yes — No

Q3. Is it important for the new employee to have had a field lab where he or she was exposed to the use of surveying equipment?

Very Important — Nice to have, but not imperative — Does not matter

Q4. Do you automatically assume that all students with B.S.C.E. degrees have had a field lab where they used surveying equipment?

Yes — No

Q5. When interviewing prospective employees who recently received their B.S.C.E. degrees, do you check to determine if their coursework included textbook and class lecture exposure to fundamental surveying concepts?

Yes — No

Q6. When interviewing prospective employees who recently received their B.S.C.E. degrees, do you check to determine if their coursework included actual field lab experience with surveying equipment?

Yes — No

Q7. Is it important that any surveying field labs include at least some exposure to the newer electronic surveying equipment (such as EDM)?

Very Important	Nice to have, but not imperative	Does not matter

The employers were also asked to categorize themselves. A total of 24 employers responded. The types of firms or agencies which responded are as follows:

Public works or hwy. dept.	General civil	Sanitary, or water resource related	Structural	Land develop., transport.	Other
9	7	2	0	3	3

Public agencies returned 11 questionnaires, and 13 came from private firms. A few wrote additional comments on their responses. Table 2 summarizes the responses.

TABLE 2. Responses from Employers

Type of agency or firm	response	Q1	Q2	Q3	Q4	Q5	Q6	Q7
Pub. Wk. or Hwy.	vi/yes	8	6	6	4	7	4	6
	nic/no	1	3	2	5	2	5	3
	dnm	0		0				0
Gen.l Civil	vi/yes	6	6	5	5	6	4	3
	nic/no	1	1	2	2	1	3	4
	dnm	0		0				0
Other groups combn'd.	vi/yes	6	5	5	4	4	4	2
	nic/no	2	3	3	4	4	4	6
	dnm	0		0				0
Totals	vi/yes	20	17	16	13	17	12	11
	nic/no	4	7	7	11	7	12	13
	dnm	0		0				0
	no response			1				

Note: vi/yes means "very important" or "yes"
nic/no means "nice to have ..." or "no"
dnm means "does not matter"

The responding employers strongly felt that new civil engineering graduates should have had exposure to surveying, both in the class and in the field (Q1 and Q3). The employers did not feel as strongly about the importance of undergraduate experience with electronic surveying equipment (Q7).

The proportion thinking the classroom experience was imperative (Q1) was greater than the proportion thinking that the lab was (Q3). More respondents assumed that students with B.S.C.E. degrees had surveying in the class (Q2) than assumed that students had a surveying lab (Q4). More checked to determine whether students had a surveying class (Q5) than checked to verify if students had a lab (Q6). The sample size was too small for significance, but the "sanitary/water" and "other" groups had higher proportions of "nice but not imperative" or "no" responses to questions 5, 6, and 7 .

Options for Accommodating Surveying Needs

The faculty group felt that surveying should remain in the undergraduate civil engineering program, and the students should be introduced to electronic measurement theory. Not only does current practice demand it, but the surveying texts now written give significant emphasis to electronic equipment.

It is desirable that the students also have hands on experience with electronic surveying equipment. When the labs only have the older mechanical equipment, students question why the field laboratory equipment does not correspond with the equipment described in the text. The students then question the validity of the lab experience. They express their dissatisfaction in the course and faculty evaluations. The faculty group felt that the major impediments to the use of electronic equipment in the labs were the costs of acquisition and upkeep, and the lack of faculty expertise with such equipment. The group suggested a number of options to address the surveying issue.

(1) Remove Surveying from the curriculum.

(2) Replace Surveying with a Measurements class, dropping the field laboratory which provides students experience with surveying equipment.

(3) Retain the Surveying class, and continue to use only mechanical equipment; thoroughly discuss electronic equipment in the lecture.

(4) Retain the Surveying class, and purchase expensive electronic equipment, either new or used.

(5) Retain the Surveying class. Rent the more expensive electronic equipment as needed for a few laboratory sessions, and use the older mechanical equipment for most of the lab sessions.

(6) Retain the Surveying class. Contract with a surveying firm to provide the students with exposure to electronic surveying equipment.

(7) Retain the Surveying class. Contract with a technical school that teaches surveying to provide the students with exposure to electronic surveying equipment.

(8) Retain the Surveying class. Purchase expensive electronic equipment, and rent it when not in lab use to recover the costs.

The faculty group concluded that the students need both a surveying class and field laboratory experience. Factors which influenced the faculty included the assessments by individual faculty concerning the skills which students will need when they are employed after graduation; the practice of the other universities as reported in the preceding "University Questionnaire" section; and the fact that a question requiring skills traditionally learned in a surveying class is a "standard" question on the national professional engineering exam.

Conclusion

Decision makers in civil engineering departments are faced with conflicting pressures regarding surveying in the undergraduate civil engineering program. The divergence of the civil engineering and surveying disciplines, the cost of modern surveying equipment for labs, and the lack of faculty with surveying experience are factors which tend to call for elimination of the traditional surveying class from the curriculum.

On the other hand, the questionnaire results from the twenty-seven universities indicate that most schools are including surveying in their programs. The employers who returned the questionnaires certainly want the graduates whom they hire to have had surveying. Surveying knowledge is tested on the professional engineering examination.

Even though land surveying may now be a separate discipline from civil engineering, the art of surveying crosses professional boundaries. Various aspects of surveying are utilized by the civil engineer working on environmental, geotechnical, hydrologic, or transportation projects. Beginning engineers need to know how to request and use surveying information. In light of the information gathered during this investigation, the authors concluded that surveying is a necessary curriculum component, even if its inclusion causes the department to hurdle financial and institutional barriers.

Appendix 1. References

Plympton, G. W. (1902). "How to become an engineer." D. Van Nostrand Company, New York, N.Y.

THE ROLE OF PRACTITIONERS IN ENGINEERING EDUCATION

Jeffrey B. Bradley[1], M. ASCE

Abstract

Several years ago the ASCE Engineering Management Division, feeling strongly that the experience of engineering practitioners was not being used as fully as it might be in civil engineering education, initiated a program to evaluate the existing level of practitioner utilization nationwide and to provide universities with "qualified" practitioner lecturers on design, management, and professional practice topics. This paper is the summary report of the Engineering Management Committee at the Individual Level (EMIL) which was delegated the project.

Introduction

"There is an unspoken 'ivy-covered Berlin wall' between academic culture and corporate culture. As a result substantial corporate resources are frequently ignored by universities, which come out short because of outmoded rules. Engineering and science faculties are increasingly threadbare. Universities cannot compete with corporate America, and not just in terms of salary. Many corporations offer better equipped research facilities, more research funding, more on-the-job freedom, and often more prestige. The talent hemorrhage out of academia will not be staunched. The only solution is for universities to bend-not-break traditions to allow for use of corporate talent as part-time teaching resources." (Baldwin, 1986) Statements like this, as well as a similar sentiment in the engineering management community led to the work discussed in the remainder of this paper.

The ASCE Engineering Management Committee at the Individual Level (EMIL), in pursuit of their charge to increase practitioner involvement in education, conducted a survey of Civil Engineering Department Chairmen to ascertain their interest in using practitioners in various classroom situations. EMIL also conducted a survey of ASCE Sections to determine if active programs existed within the Sections to provide civil engineering departments with "qualified" practitioner speakers. What follows is EMIL'S report on survey results, as well as a summary of where we go from here with the program.

---

[1] President, WEST Consultants, Inc., 2111 Palomar Airport Rd., Suite 180, Carlsbad, CA 92009-1419

Survey of CE Department Heads

A survey was conducted at three ASCE meetings of Civil Engineering Department Heads. These meetings are held annually throughout the country, and although it is not a complete survey (only 55 department chairs) it is a representative sample of their attitudes. The survey questionnaire results are reported below.

QUESTIONNAIRE
PRACTITIONER INVOLVEMENT IN EDUCATION
GOOD, BAD OR INDIFFERENT

The Engineering Management Division has been supporting a program encouraging increased involvement of practitioners in the classroom the past two years. Among other things, we have been attempting to develop speakers bureaus through local ASCE Sections that would provide a resource to local universities This questionnaire is intended to obtain feedback from the educational community as to the interest in this type of effort and will be used in determining future direction for the program.

Please circle the answer best reflecting your feelings. More than one may be appropriate on certain questions.

1. Is the utilization of practitioners in engineering education beneficial to the profession?

a) yes 53 b) no 2

2. Would a speakers bureau of high quality lecturers provided by local ASCE Sections be useful to CE department chairs and faculty?

a) yes 51 b) no 4

3. How would you utilize such speakers/lecturers?

a) to teach entire courses	12
b) guest lecturers in design courses	46
c) guest lecturers in seminars	45
d) speakers at student chapter meetings	49
e) other-please specify	5

4. What type of subjects would practitioners best be used to cover?

a) design courses	29
b) CE management	25
c) professional practice	42
d) case histories	44
e) other-please specify	3

5. Should ASCE actively support the use of practitioners in the classroom in a formalized manner. For example, should Sections be encouraged to form a current speakers bureau and have it written into their bylaws that said bureau be updated periodically?
a) yes 36 b) no 19

6. Would a set of class notes when developed by the Management Division for a course in engineering management be of interest/use to your department?
a) yes 37 b) no 18

7. Do you presently use practitioners in your department educational scheme, and in what fashion?

Seminar Speakers
ASCE Student Chapters
Guest Lecturers
Part-Time Instructors, Adjunct Faculty
Lectures in Design Courses
Professional Practice Courses
Field Trips
Specialized Conferences
Senior Project Evaluators/Consultants Capstone Design Course
On MS/PH.D. Committees
Lectures in Management and Professional Practice Courses

A preponderance of the respondees felt that use of practitioners in education is beneficial to the profession. They almost unanimously agreed that a speakers bureau of high quality lecturers provided by local ASCE Sections would be a significant resource to their departments. This is of importance as EMIL has attempted, with only moderate success, to stimulate local Sections to develop such programs. The response, by a two to one margin, to a similar question (number 5) was that ASCE should actively support the use of practitioners in a formalized way.

The survey showed that department chairs felt practitioners could provide significant input on professional practice, case histories, design courses and civil engineering management. The consensus also shows that optimal use of practitioner instructors would be as guest lecturers in seminars, design courses, and at student chapter meetings. Twenty percent also felt practitioners could be used to teach entire courses.

Respondees, by a two to one margin, felt that a set of class notes developed by the ASCE Engineering Management Division for a course in civil engineering management would be of interest to them. The Engineering Management Activities Committee (EMAC) is in the process of developing such a set of notes. EMAC has also set up a program to use practitioners in teaching portions of a class on engineering management at the University of California, Irvine.

Department Heads' responses showed that practitioners are presently used to some extent in their programs as lecturers in seminars, student chapter meetings, design courses, management and professional practice courses, and as part-time instructors or adjunct faculty. Present use of practitioners appears to center on one-on-one educator-practitioner relationships, the type of program, and on geography. Metropolitan universities, particularly those with night classes, use practitioners as adjunct faculty more frequently. The majority of responses did indicate that practitioners are not being utilized as fully as they might be in the educational process.

ASCE Section Survey

During the past several years EMIL sent letters to local Section Presidents, and wrote articles in ASCE News encouraging the formalization of programs at Section levels to increase the use and availability of practitioners in education by development of speakers bureaus and other similar activities. Engineering Management Division members were also encouraged to start programs or stimulate existing ones within their Sections. In order to determine the success of our effort, we conducted a phone survey of ASCE Sections. The response from this survey follows.

PRACTITIONERS IN EDUCATION-SECTION REPLIES

SECTION	RESPONSE
Montana	Montana State University has a 2 1/2 day non-stop seminar using mainly practitioners
Michigan	No formal program or knowledge of program
Colorado	Speakers Bureau of approximately 50, University of Colorado Denver and Colorado State University are interested in using. Director will take long term responsibility
Boston	University assistance program developed, local universities response negative
National Capitol	Minimal knowledge of program
New Hampshire	Interested, no action to date
Los Angeles	Developed speakers bureau, interaction with Loyola Marymont and CA State Long Beach
Hawaii	Committee established, speakers bureau, 9 lectures in graduate school
Tennessee	No action, not aware
Kansas City	Developed student support committee, speakers at Kansas City Campus U of Missouri, senior seminar
Kentucky	Nothing formal but active, University of Louisville mandatory co-op program & graduate course on technology & society which uses practitioners. U. of Kentucky 4-5 used in freshman class, senior seminar & 2 1/2 other classes sole responsibility of

	practitioners
Wyoming	Not aware
North Carolina	Program being formalized
Wisconsin	Universities contacted but only interest through student chapters, suggested use of quality manual as lecture material
Mohawk-Hudson	Not aware
Alaska	Committee chair on student members delegated responsibility, no further action
Illinois	Speakers bureau (14), committee formed under management group, 3 schools
Inland Empire	Speakers bureau available, U of Idaho has formal program
Texas	Minimal action
St. Louis	Washington University & University of Missouri as Rolla contacted, Paul Munger at U of M shown interest but not further action
Iowa	Committee, Iowa State Student Chapter, speakers, sophomores two classes with 2 hour lectures on ethics, professional liability and registration field trips to CE Projects, CE296 4-6 speakers on consulting, CE295-City Engineer, County Engr., State & Federal, class on government practice
Cincinnati	U of Cincinnati large co-op program but nothing formalized
Maryland	Minimal interest
Central Penn	Formalized with workshops, student chapter meetings and a mentor on campus retired, no action
Arizona	No action
Columbia	Speakers bureau at University of Walla Walla and Community College
Nevada	No action, unaware
Oregon	No action, unaware
San Diego	No action, unaware
Seattle	No action, unaware
Southern Idaho	No action, unaware
Cleveland	No action, interested
Akron	Nothing formal
Dayton	No action, want information
Lehigh Valley	No action, unaware
Philadelphia	Informal program, unaware
Pittsburgh	No action minimal, interest
New Jersey	No action, unaware
Metropolitan	Unaware, local program, enthusiastic
Buffalo	Vaguely aware, actively supports program at University of Buffalo
Ithaca	Unaware, some interest
Delaware	Unaware, limited informal program, senior Design Project

Maryland	No action, somewhat aware
Rochester	Unaware, small informal involvement
Syracuse	Unaware, active use of adjunct faculty, very interested
Connecticut	Not aware, adjunct faculty used
Vermont	Vaguely aware, very interested
Maine	Program with practitioners in Continuing Education and EIT Reviews
Rhode Island	Unaware, adjunct faculty used
Oklahoma	No action to date
Quad Cities	No action at Universities/only high schools

Number of Sections Contacted 48 Percent Response 60 %

The majority of Sections were unaware of any effort within ASCE to increase the use of practitioners in education. This can be attributed to the massive quantity of paper sent to Section Presidents, some of which gets "round filed", the discontinuity from year to year as Section Boards change members, and to the fact that ASCE has not as of this writing made this a high priority item within the Society. Some Sections do have very active programs. Two such examples are the Hawaii and Colorado Section.

The Hawaii Section has developed a three pronged approach to explore programs to promote student participation at technical and professional meetings, explore programs at local universities to enhance interaction between members in practice and civil engineering students, and develop a mentor program for younger members. They have proposed that a "plan-design-build" projects class or seminar be made available to the University of Hawaii as a regular course, and also proposed a one credit seminar on professional practice to provide students a broader understanding of what is required in the practice of Civil Engineering. Practitioners would be heavily relied upon in both courses.

The Colorado Section developed a speakers bureau of approximately fifty "highly qualified" practitioners. These people were selected by a Section Task Committee which was composed of the Dean of Engineering at the University of Colorado at Denver, the past department head at Colorado State University, the past ASCE Colorado Section President, and another Section representative. It was the intent of this Task Committee to select speakers with interest in the program, who had strong technical or professional practice background and who were good speakers. Both Colorado State University and the University of Colorado at Denver have indicated interest in using the speakers bureau. The Section designated that a member of the Board would be in charge of the program. This provides for perpetuation of the program in their Bylaws including updates of the speakers bureau. There is no doubt that the membership of university faculty on the Board of Directors helped in the implementation of this program, and it is highly recommended that university department heads, as well as other faculty, be involved in the development of any such program.

Conclusions-Where Do We Go Now

There **is** a need for practitioners in civil engineering education. They can complement full time faculty, and provide their experience of professional practice, management and design to the student. There is a consensus between practice and education that increased involvement of practitioners in education is important to the profession. The process of making this happen appears to be our major stumbling block. Problems do exist, including faculty resistance to adding courses to their curriculum with the present limitations of the existing four year education. As each university and Section has differing constraints, any program must be tailored to individual systems and geography. It is clear from past experience that a cooperative effort between education and practice is necessary to increase utilization of high quality practitioners in the classroom is necessary.

The Engineering Management Division has proposed to ASCE formation of a Task/Steering Committee on the Practitioner Educator Partnership. The proposed committee would be composed of members from the Engineering Management Division, Educational Activities Committee and ABET. The committee would initially develop program goals at a national level. Eventual outputs of the committee and program might well include the following:

1) formalization of the program in the ASCE Strategic Plan;
2) a statement of program goals to be included in Section Handbooks;
3) development of speakers bureaus and other similar programs at the Section level;
4) provide input for ABET accreditation;
5) and implementation of trial programs at selected universities.

An engineering educational crisis is predicted within the U.S. in this century if present trends continue. Certainly that crisis is not solely related to the topic of this paper, however, one partial solution lies in our use of an existing resource, the practitioner, in civil engineering education. My challenge to both practice and education is that we do so.

References

Baldwin, Lionel (1986), "Universities Must Use Part-time Corporate Teachers to Survive, Says NTU Head," Engineering Times.

SYNERGIST FOR THE NINETIES: THE CIVIL ENGINEERING MANAGER

Damodara U. Kini[1]
M.ASCE

Abstract

The next decade calls for a new approach to our handling of engineering and construction projects. They will need to be managed to maximize benefits while minimizing any long term adverse effects. Civil engineers are ideally suited to play this leadership role and ASCE can facilitate their transformation into managers who produce synergy.

Introduction

The pressing need of the nineties will be for technical managers who have a global perspective in looking at the interaction of the various issues facing mankind, and who are capable of developing the means of meeting these challenges with minimum sociological and environmental impact. These managers will be leaders with the ability to bring a variety of experts together for the common good and create a result that is greater than the sum of the individual parts. In other words, be the catalysts to promote synergism.

The challenge of the nineties is to ensure that our children and grandchildren will benefit from our engineering accomplishments and will not have to deal with any adverse consequences of such work. Every future engineering project will be like a jigsaw puzzle in which all the pieces fit together perfectly, not just in one position, but in several different positions to come up with the same picture. Further, regardless of which position they are in, the final picture will fit into the same frame without having to stretch or strain the frame. The individual pieces are the products of the various engineering disciplines such as civil engineering, mechanical engineering, electrical engineering and chemical engineering. The frame is the earth with its

----

[1]Materials Manager, United Engineers & Constructors. Stearns-Roger Division. 700 S.Ash, Denver, CO 80222.

environment. The final picture is the end product of the engineering effort, which could be a dam, a power plant, a chemical processing plant or even a space station. The objective is to ensure that this engineering project does not stress or strain planet earth in any way.

To be able to execute engineering projects in this manner the traditional engineering disciplines will need to be supported by personnel in other areas of expertise such as biology, chemistry, medicine and meteorology. The formidable task of molding this diverse group of experts into a whole will be up to our engineering manager of the nineties. The manager must have the ability to direct these experts to explore different paths to meet the end objective, and must be able to choose the one path that is likely to have the least future detrimental effects. Further, the manager should have the capacity to fuse the talent at his disposal into an end product which is greater than the sum of the individual parts. That is, the individual parts must fit together in a way to produce synergy.

Past Problems

The main challenge for the managers of the nineties will be not only to undo the sociological and environmental problems created as a result of our actions in the past, but also to ensure that these mistakes are not repeated in the future. In the past, if a dam was required to meet water needs, we built it. If power was in short supply, we built new power plants. Waste products of all types were dumped at the nearest open land area. Science and engineering had not yet begun to understand the long term consequences of these actions. The ecological impact of dams, acid rain due to sulphur dioxide emissions from power plants and groundwater pollution from land fills had not been understood yet. The focus was on fulfilling short term needs without much attention being paid to future impact of current actions. Now that our knowledge of long term effects is vastly improved, these managers will be called upon to examine and prove that projects implemented for our benefit now will not turn out to be hazards to our descendents. They must have the analytical ability of weighing available alternatives to satisfy immediate requirements while minimizing future problems. Their technical knowledge will have to be supplemented with communicative skills in order to show the public the pros and cons of any public work.

Challenges to Come

To better illustrate what the future will have in store

for the managers of the next decade, let us consider the complex problem of disposal of the tons of garbage that even a medium size city generates. Landfills are an unsatisfactory way of disposing of the garbage not only because open land is becoming scarce, but also because of the hazards of methane gas generated at the landfills. There is also the possibility that hazardous substances in the garbage could contaminate the ground water. Thus, incineration of the garbage has become an acceptable alternative, especially where the steam could be used to generate electricity or used for industrial purposes. Such cogeneration plants have their own challenges to contend with.

To begin with, although the location should ideally be close to the metropolitan area that the plant will serve, city dwellers rarely want an industrial plant as a neighbor. The emissions from the plant have to be free of substances harmful to the health of the local residents. The noise from the plant has to be minimized, and traffic around it should be routed in such a way as to minimize disruptions to the normal flow. Most important, the local citizens have to be informed and convinced of the need for the plant. All of this will take the efforts of a variety of experts working together towards a common goal.

To discuss the functions of only a few of these experts, mechanical process engineers will need to analyze the refuse and devise means of separating materials that do not burn easily. Environmental engineers must determine the maximum levels of emissions from the chimney while electrical engineers provide the control systems for the plant. Civil engineers will determine the grading of the site and structural engineers must ensure that the building will support all the equipment needed. Procurement personnel must ensure that the materials of the required quality are obtained, processed or fabricated and delivered to the site at the time needed. Construction specialists will then put this all together into an operating plant. To make this work, it is essential that the entire effort be managed in such a manner that all the diverse expertise is focused towards the common goal of building a workable garbage incineration plant. The manager has to be one who has the ability to grasp the essentials of the process, coordinate the various experts to dovetail into each other's efforts, minimize conflict, control costs and direct the entire effort to maximum efficiency. He has to generate synergy.

The next decade will see more exploration of space with

the ultimate goal of setting up stations in space, or on some of the planets, for research and possible travel beyond the solar system. Innovative approaches will be needed to design structures that stand up to the rigors of space while supporting a habitable environment. Stations in space or on other planets will require new materials, new methods of design and techniques of construction unheard of today. It will need the efforts of propulsion engineers who can design engines that can perform at speeds only imagined today and specialists in energy generation who will be called upon to provide a nearly inexhaustible source of energy. Creative medical professionals will have to device means of avoiding the decay of human muscle during years of space travel at zero gravity. It will require innovative foresight from psychologists to provide means of keeping the human mind sharp during long voyages through the emptiness of space. For this type of effort to be successful, will require leaders who have a thorough grounding in the physical sciences, can cut through the technical jargon to get to the heart of a problem, and point the way to a solution.

Qualifications and Skills Required

The key point of this foregoing discussion is that the challenges facing us in the nineties and beyond need a special breed of leaders. Such leaders will not be experts in a narrow area of expertise but rather will have a solid base of general scientific, engineering expertise. This knowledge combined with skills such as interpersonal communication skills, an understanding of human motivation, awareness of international affairs, and an understanding of contractual relationships will enable them to combine the energy of their subordinates into a force which is greater than the sum of the individual parts.These synergists will require a mix of knowledge and skills to manage the projects of the nineties which will be somewhat different from those that are required today:

1) A good understanding of engineering basics, fundamentals of physics and chemistry, and their practical applications.
2) Excellent analytical skills and the ability for logical thinking.
3) A working knowledge of mathematics and its applications.
4) Good verbal and written communication skills.
5) A good grasp of contracts and business law.
6) Familiarity with world politics and international relations.

7) Thorough understanding of motivational and management theory.
8) Negotiating skills.

In light of the discussion above on the challenges of the next decade, the need for this type of background and expertise becomes obvious. Rather than experts who perform a microscopic examination of selected small areas in science, we will need those who may be called generalists and who have the ability, due to their breadth of knowledge, to foresee future consequences of current actions. This means that these managers must be able to understand the functions performed by the specialists and through logical reasoning ask the right questions in order to forecast the results of such functions. A good understanding of engineering basics such as mechanics as well as a similar grounding in the fundamentals of physics and chemistry are essential for our managers of the future to control and focus the expertise of their specialist subordinates. Similarly, a working knowledge of mathematics provides them with a powerful tool that can be used to analyze not only engineering problems, but also financial reports. The ability to grasp financial and cost data will be critical to control costs.

In addition to these basic qualifications, the need for the other characteristics listed above falls into place when we realize that it will take teams of people with a variety of expertise to meet the global challenges of the nineties. Good communication skills will be essential if the teams are to act as cohesive units and also for teams within separate organizations to interact effectively with each other. The organizational entities may be in one country or more likely be located in different countries. In either case, the relationship between organizations will need to be governed by mutually acceptable contractual documents. Our managers of the future must be able to understand how these contracts are formulated and implemented. Since the challenges of the nineties will be global in nature, the methods of handling such challenges will transcend international boundaries. A general knowledge of international political issues of the day will be an important asset to the managers. The performance of any manager is measured by how well the manager can motivate subordinates to maximize their effectiveness in the functions that they perform. If our managers are to be synergists, they must sharpen their motivational skills and understanding of management theory in order to get the best out of their personnel. An effective manager will excel in using communication and negotiating skills to convince others to the manager's point of view and to channel their energies towards a goal which, at first, only the manager can visualize.

Civil Engineers Ideally Suited to be Synergistic Managers

Reviewing the characteristics for the synergistic managers discussed above, it is important to note that the first three qualifications are essential and that these three can only be obtained through a formal education. The remaining five skills can be acquired through experience, educational courses or a combination of both. Civil engineers are ideally suited for this role of synergistic managers since an undergraduate degree in civil engineering imparts an excellent understanding of the three essentials defined above, which are, engineering basics, fundamentals of physics and chemistry, and mathematics. The very nature of these subjects makes it necessary for the students to develop analytical skills and logical reasoning. Those who go to graduate school generally tend to specialize in structural engineering, geotechnical, engineering mechanics, environmental engineering or water resources. Each of these areas of specialization requires prowess in mathematics, statics, dynamics, and fluid mechanics. Environmental engineering may require a knowledge of biology and chemistry. No other branch of engineering provides this breadth of knowledge. However, the other five characteristics required for our managers of the future are usually not included in the undergraduate or graduate curriculum.

ASCE's Role

One of the challenges of civil engineering education for the next decade is to device ways and means of including these five skills among the core subjects required for an undergraduate degree in civil engineering. ASCE should take the lead to meet the challenge through the following steps :

1) Work with the engineering schools to develop courses of study in the areas of communications, contract law, international relations, management theory and negotiations. In many instances, basic courses in several of these subjects may already exist in departments other than engineering. ASCE's role will be to identify specific area of emphasis that would be particulary useful in practice and include these in the core subjects for the undergraduate degree.
2) Encourage utilizing the skills of practicing civil engineers who are experts in these subjects, but who may

not have doctoral degrees. In the subjects under discussion, two to three years of experience is worth much more than a doctorate. For instance, there is no substitute for experience in Materials Management for an instructor in contracts and negotiations. Only a person who has performed negotiations, formulated appropriate language for contracts, and administered such contracts to a successful conclusion will have an insight into the nuances of these topics. ASCE can develop and furnish to the universities a listing of members interested in teaching these subjects.

The second challenge to ASCE is to develop a plan for imparting the same five characteristics to practicing engineers. For many of the practicing civil engineers who already possess some of these non-engineering skills, the need will be to develop a proficiency in the remaining skills. Following are some of the actions that ASCE will need to take to service the practicing engineers:

1) Sponsor seminars in these topics. Once again, utilize the services of members who are ackowledged experts in these areas. ASCE can define the type and length of experience required for members interested in presenting these seminars, obtain resumes and select the best qualified.

2)Define a minimum number of continuing education units that an engineer will need per year in order to stay current in each of these subjects, which would be critical for our synergists for the nineties.

3) Organize combined meetings or other types of liaison with organizations with expertise in some of these areas. For instance, this type of relationship with the National Association of Purchasing Managers (NAPM), and the National Contract Management Association (NCMA) will allow the ASCE to draw on the resources of these organizations in materials management and contracts.

Conclusion

These actions by the ASCE will ensure two things. First, practicing civil engineers will gain the skills necessary for them to develop into synergistic leaders with the vision to device sociologically and environmentally accceptable solutions to the problems of the early part of the next decade. Second, the next generation of civil engineers will graduate out of school already equipped with the expertise to become the synergists for the latter part of the next decade.

QUALITY MANAGEMENT AND THE CIVIL ENGINEER

By Charles H. Samson,[1] Fellow, ASCE

Abstract

Some of the key elements of the quality management approach are discussed, with particular reference to the work of W. Edwards Deming and Joseph M. Juran. For many years the primary application of quality management principles occurred in Japan in the area of manufacturing. Now quality management is finding wide acceptance in American industry, and is being adopted in non-manufacturing areas as well as manufacturing. The writer concludes that quality management has many implications for the civil engineer, and should be considered an essential element in civil engineering education.

Introduction

Juran (1969) refers to "life behind the quality dikes":

> Ours is the century in which, for the first time in human history, great masses of human beings have placed their safety, health and even their daily well-being behind numerous protective dikes of quality control. Just look at a few of these dikes . . .
>
> The very continuity of our daily lives is built around the continuity of numerous vital services: power, transport, communication, water, waste removal and many others. We have structured our society on the premise that these services will continue without interruption. A major power failure paralyzes the lives of millions of people.

---

[1]Prof. and Assoc. Head, Dept. of Civil Engineering, Texas A&M Univ., College Station, TX 77843

> What is new is the decision of the human being to risk his safety, health and daily well-being on the integrity of the dikes of quality control. (This decision has already gone past the point of no return.) Having made the decision and now experiencing life behind those dikes, the human being has discovered that the dikes are suspect. He is in no mood to be told that we have always had failures. Instead, he is taking the position that the failures had jolly well better stop, and he is enlisting quite an array of support to make his point.

It is the writer's opinion that this observation by Juran has even greater meaning today, and further, that the civil engineer has a special responsibility in maintaining the "quality dikes." The work of the civil engineer has historically been closely involved with the health, safety, and well-being of society. Concern with this obligation has necessarily involved consideration of quality. In ASCE this concern has led to addressing quality through the "Manual of Professional Practice for Quality in the Constructed Project" (ASCE 1988).

Quality

Quality has been defined in many ways, creating considerable confusion as to whether quality costs more or less, how to measure it, and how to achieve it.

Juran (1989) suggests "fitness for use" as a short definition of quality, and distinguishes between quality as "freedom from deficiencies" and quality as a measure of "product features."

Deming (1982) states: "Quality should be aimed at the needs of the consumer, present and future." He also states: "Quality begins with the intent, which is fixed by management. The intent must be translated by engineers and others into plans, specifications, tests, production."

Garvin (1988) provides an extended discussion of definitions of quality proposed by various writers.

Historical Development

"Scientific management" refers to the management approach developed by Frederick Taylor around the turn of the century. This has been the primary basis for the management of quality until the past few years in the United States. (See Juran 1973 for an expanded discussion of the Taylor system.) Although quality control was pioneered in this country, with special attention during

World War II, its evolution into "quality management," as the term is now used, developed largely in Japan.

The adoption by the Japanese grew out of seminars that were part of General MacArthur's effort to revitalize Japanese industry in the latter 1940s, combined with visits to Japan by Deming beginning in 1950 and by Juran in 1954 to offer seminars on quality. Following these events, the quality of Japanese products improved markedly.

In the United States, as competition in many areas has expanded to a global arena, much discussion has occurred concerning the need for a new approach to managing quality. While it is difficult to identify a specific turning point, during the past decade interest in this subject appears to have increased rapidly with a large number of major corporations moving toward adoption of many of the quality management principles.

Quality Management

As noted earlier, quality management as considered here is based primarily on the work of Deming and Juran. Only a few key elements of their work will be presented.

Deming's 14 Points. - Deming (1982) states 14 points that he believes are essential to the enhancement of quality in this country. With reference to these points, he comments:

> The 14 points are the basis for transformation of American industry. It will not suffice merely to solve problems, big or little. Adoption and action on the 14 points are a signal that the management intend to stay in business and aim to protect investors and jobs. Such a system formed the basis for lessons for top management in Japan in 1950 and in subsequent years . . .
>
> The 14 points apply anywhere, to small organizations as well as to large ones, to the service industry as well as to manufacturing. They apply to a division within a company.

Various writers expand on the interpretation and implementation of the 14 points. (See, for example, Deming 1982, Scherkenbach 1986, Gitlow and Gitlow 1987, and Walton 1986.)

Juran Trilogy® Model. - This model (Juran 1989) is shown in Fig. 1. The primary components of the model are quality planning, quality control, and quality improvement.

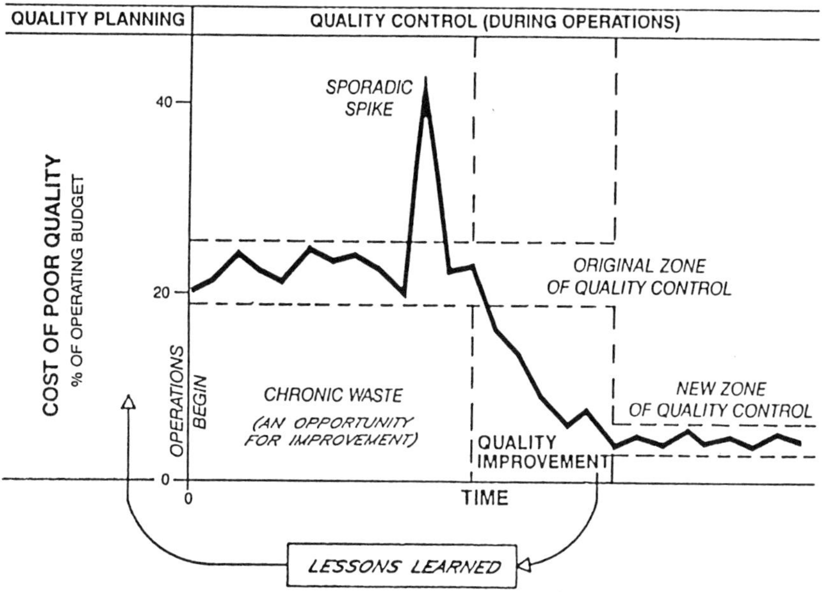

Figure 1. The Juran Trilogy®
(Reproduced with permission of Juran Institute, Inc.)

Juran lists the essential features of each component as follows:

Quality Planning
Determine who the customers are
Determine the needs of the customers
Develop product features that respond to customers' needs
Develop processes able to produce the product features
Transfer the plans to the operating forces

Quality Control
Evaluate actual performance
Compare actual performance to product goals
Act on the difference

Quality Improvement
Establish the infrastructure
Identify the improvement projects
Establish project teams
Provide the teams with resources, training, and motivation to:

Diagnose the causes
Stimulate remedies
Establish controls to hold the gains

Juran (1989) also emphasizes that quality must be a primary consideration as every activity (subsystem) in a process (system) fills various roles: customer, processor, and supplier:

> As a customer - The processor team acquires various kinds of inputs that are used in carrying out the process. The processor team is a customer of those who provide the inputs.
>
> As a processor - The team carries out various managerial and technological activities to produce its products.
>
> As a supplier - The processor team supplies its products to its customers.

Quality "Tools." - A variety of tools are used in the application of quality management. Although it is not feasible here to discuss these in detail, they would include cause-and-effect diagrams (also called "fishbone" or Ishikawa diagrams), flowcharts, Pareto charts, run charts, histograms, control charts, and scatter diagrams. (See for example, Wadsworth, Stephens, and Godfrey 1986; Ishikawa 1986; Gitlow, Gitlow, Oppenheim, and Oppenheim 1989.)

Figure 1 is a form of control chart. If the ordinate represents cost of defects or nonconformities ("up is bad") and the abscissa represents time, then performance of a certain activity can be tracked. Through appropriate statistical techniques, upper and lower control limits (horizontal dashed lines) can be constructed. Variations within these control limits are normally treated as "common," or "random," causes. Spikes outside the control limits are "special," or "assignable," causes, and represent the "firefighting" in which managers typically spend a significant portion of their time, while the greatest opportunities for improvement lie in reducing the level of average nonconformities and reducing the variability. Deming (1982) estimates that 94% of improvement opportunities belong to the system (the responsibility of management), with 6% related to special causes. Tribus (1988) states: "Workers work <u>in</u> the system; management should work <u>on</u> the system."

Conclusions

1. Although much of the application of quality management has occurred in the manufacturing industry, there is considerable evidence that the approach can be applied with comparable benefits in civil engineering activities.

2. There is a need for collaboration between those knowledgeable in quality management application and those who perform civil engineering functions to develop new techniques with direct applicability to these functions.

3. There is a need to facilitate the use of quality management by civil engineers though a networking and documentation of knowledge gained from experience.

4. The integration of quality management and systems engineering is of great importance to civil engineers. Civil engineering education should move quickly to provide needed leadership.

Appendix. References

ASCE (1988). Manual of professional practice for quality in the constructed project, ASCE, New York, NY.

Deming, W. E. (1982). Out of the crisis. MIT Press, Cambridge, MA.

Garvin, D. A. (1988). Managing quality. The Free Press, New York, NY.

Gitlow, H., and Gitlow, S. (1987). The Deming guide to quality and competitive position. Prentice-Hall, Inc., Englewood Cliffs, NJ.

Gitlow, H., Gitlow, S., Oppenheim, A., and Oppenheim, R. (1989). Tools and methods for the improvement of quality. Richard C. Irwin, Inc., Homewood, IL.

Ishikawa, K. (1986). Guide to quality control. Asian Productivity Organization, Tokyo, Japan.

Juran, J. M. (1969). "Mobilizing for the 1970's," Quality Progress, March issue, 8-17.

Juran, J. M. (1973). "The Taylor system and quality control." Quality Progress (1973). American Society for quality Control, May through December, 1973.

Juran, J. M. (1989). Juran on leadership for quality. The Free Press, New York, NY.

Scherkenbach, W. (1986). The Deming route to quality and productivity: road maps and roadblocks. Ceepress Books, Washington, DC.

Tribus, M. (1988). Quality first. National Society of Professional Engineers, Alexandria, VA, 32.

Wadsworth, H. M., Stephens, K. S., and Godfrey, A. B. (1986). Modern methods for quality control and improvement. John Wiley & Sons, New York, NY.

Walton, M. (1986). The Deming management method. Dodd, Mead & Company, New York, NY.

PROJECT MANAGEMENT AS A PROFESSION--GOAL OF PMI

Henry Padgham[1]
Member ASCE

ABSTRACT

This paper describes the project management professionalism program developed by the Project Management Institute. The program is designed to improve the quality of project management and provide a recognized standard of practice for this field of endeavor. These services should be of particular importance to the civil engineer project manager practitioner and to related educational programs.

PROFESSIONALISM OBJECTIVES

The Project Management Institute (PMI) has in place a 5-year planning process, and the principal goal of the current plan is to move the practice of the field of project management into the arena of professional recognition as soon as possible. The reason is that improvement in management of projects is essential to obtaining better project performance and quality, thereby better serving the public need.

While the PMI membership is multi-industry, it is, in fact, dominated by engineers and the focus of their interest in the organization is predominantly the advancement of the state of the art of management of the constructed project.

To accomplish its goal PMI's focus is on education and certification of the Project Manager practitioner and hence also the accreditation

---

[1]Chairman of the Board, Project Management Institute; Vice President, CH2M HILL; 806 West Glen Oaks Lane, Mequon, Wisconsin, 53092

of educational curricula. To this end, its 5-year plan includes the following salient features:

- o Preparation of broad based educational material that will provide the means for teaching and learning the fundamentals of good project management as embodied in the Project Management Body of Knowledge (PMBOK) and associated standards to improve professional knowledge, skills, and attitudes.

- o Expanded services in accreditation and reaccreditation of graduate level university degree programs in project management. These types of programs are expanding and PMI is at the forefront of being able to provide this service. This is in the interest of assuring high quality programs with the educational content such that its graduates can truly help improve the state-of-the-art of project management.

- o Expanded opportunity for project managers to become Project Management Professionals (PMP) through the project manager certification (testing) processes pioneered by PMI in 1982. This program is the cornerstone for establishing project management as a professional field and includes observation of a professional code of ethics as part of the program.

- o Development of programs to promote standardization and quality in the proliferation of privately and publicly sponsored training courses in the field of project management.

- o Stimulation of research to enhance and advance the state-of-the-art in project management concepts, processes, and techniques.

PMI strongly believes that appropriate project management course work needs to be a part of our engineering curricula, both graduate and undergraduate. Further the institute is uniquely qualified to assist in this regard. It has already set the stage to provide this support and welcomes the challenge to be of service in promoting quality in the education of the practitioner in this emerging vital profession in which the civil engineer currently plays, and will increasingly play, a predominant role.

PMI would welcome and encourage the establishment of a joint ASCE/PMI project management educational committee to work on implementation of specific curricula recommendations and accreditation recommendations in the field of management of constructed projects. An important part of this work should embody a review of the PMBOK for jointly recommending its adoption as the frame work for project management education and training.

Exhibit A shows the PMBOK. This was developed over several years of intensive effort by PMI members and now serves as the basis for the PMI certification testing program and for the university accreditation program. It is increasingly being adopted by numerous organizations providing project management instructional services. PMI believes it could become the standard for all educational curricula involving project management course work.

Exhibit B is an excerpt from the PMI accreditation program policies, procedures, and standards publication. As of this writing two schools have this accreditation. There are several graduate university programs in engineering schools that presently exist without accreditation of this type. PMI would encourage ASCE in joining with it in a joint promotional effort to encourage these schools to obtain this accreditation.

Exhibit C is a description of the PMI certification program. It is believed that the quality of the management of projects would be significantly improved with increased emphasis on PM practitioners becoming PMPs. The purpose of the PMP test and subsequent certification is to help assure that the PM practitioners are at least knowledgeable of the basic principles for the successful management of projects.

The PMI was established in 1969 and presently has a membership of 6,200. A significant percentage of the membership belongs to other professional or technical organizations like ASCE. Membership growth is expected to increase substantially with increased marketing emphasis on what PMI has to offer to those individuals, whether in the academic, government, or private sectors and who are becoming increasingly involved with PM of our constructed projects. These projects are becoming more complex and increasingly require expert management from a limited pool of human resources.

EXHIBIT A

PROJECT MANAGEMENT INSTITUTE
PROJECT MANAGEMENT BODY OF KNOWLEDGE

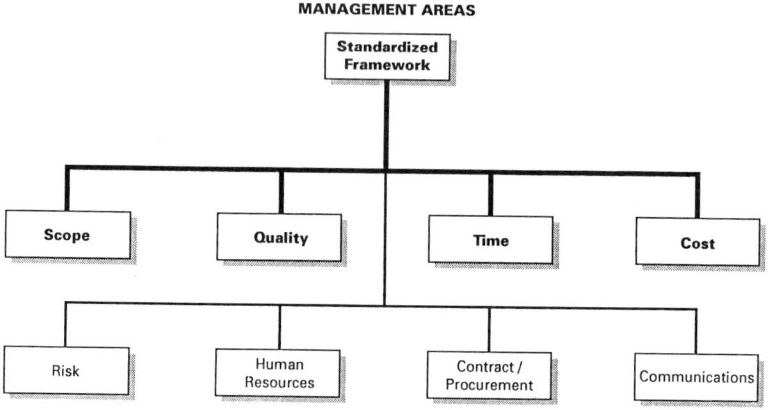

EXHIBIT B
EXCERPTED FROM
PROJECT MANAGEMENT INSTITUTE ACCREDITATION POLICIES, PROCEDURES, AND STANDARDS

PREFACE

This publication describes the objectives, policies, procedures, and standards governing masters accreditation of degree-granting programs in project management by the Project Management Institute (PMI). This publication is made available to all interested parties.

Persons serving as elected or appointed officers of the Institute, as well as a description of the full range of programs and activities undertaken by PMI in addition to accreditation, are published in the PMI Annual Report. PMI publishes the *Project Management Journal*, a refereed professional journal designed to provide a medium of exchange and education among project management professionals and other interested parties. Other specialized project management educational documents are published by the Institute. PMI also conducts an international examination and certification program for project management professionals.

CONTENTS

THE PROJECT MANAGEMENT INSTITUTE ACCREDITATION POLICIES AND PROCEDURES

The need for educational programs in project management has evolved to meet the contemporary needs of business and government organizations for qualified professionals to enter the project management field. The accreditation process seeks to review the content and process of project management education in graduate and professional schools to assure that students who study project management will have the educational background to effectively serve modern society within this field. Therefore, those institutions which have made the commitment to meet and maintain the required level of quality for accreditation of their project management degree programs represent sound choices for prospective students and for those persons responsible for recruiting students with professional preparation for positions within the project management profession.

The Project Management Institute's Board of Directors administers its accreditation function through the PMI Accreditation Committee. The Board of Directors is responsible for accreditation standards, procedures, and decisions, as well as the appointment of the Chairperson and members of the Committee to carry out the accreditation process. The membership of the PMI Accreditation Committee includes not only representatives of accredited project management educational programs, but also representatives of noncollegiate organizations reflecting the diverse interests and viewpoints of business, industry, and allied professions. Thus, the accreditation activities of the Institute are strengthened by the contribution of representatives possessing a wide variety of educational and professional backgrounds as well as a common desire to establish and improve graduate education for project management.

PMI's accreditation interests lie exclusively within the field of project management, and the Institute neither desires, intends, nor is prepared to review or accredit institutions of higher learning or portions of those institutions on a broader scale. Graduate education programs for project management generally fall within the structure of either business or engineering graduate schools. The U.S. national accrediting agencies for such schools are either the American Assembly of Collegiate Schools of Business (AACSB) or the Accreditation Board of Engineering and Technology (ABET). The AACSB and ABET also conduct activities in Canada and Countries other than the U.S. may have their own national accreditation agencies. **The Project Management Institute strongly supports the activities of these accreditation agencies, and none of PMI's activities are intended to impede or conflict with their standards or procedures. Should any conflict be discovered between the standards and procedures defined in this publication and those of AACSB (for business schools), ABET (for engineering schools), or other independent national or international accrediting agencies, then the standards and procedures of their agencies will apply as appropriate.**

EXHIBIT C
PMI INTERNATIONAL CERTIFICATION PROGRAM: CERTIFYING PROJECT MANAGEMENT PROFESSIONALS

PURPOSE AND PHILOSOPHY

The basic purposes of the certification program are to enhance professionalism in project management and to provide project management personnel with a structured method to aid their personal professional development. It is a program which PMI offers to both members and nonmembers of the Institute. The program is designed for individuals working in project management who seek to enhance their knowledge, skills, and visibility as project management professionals. You do not have to be a PMI member to apply for certification and it is not anticipated that all members will apply. It is an individual decision based on a desire for professional development and enhancement.

The certification program is structured around the generally accepted characteristics of a profession in terms of a United States and an international context. There are five major characteristics which must be satisfied for a quality certification program. These include: a common body of knowledge; working experience in the project management field; a service motive; a code of ethics; and an authenticating body.

The authenticating body for our program is the Project Management Institute, as a legally constituted international organization.

The Project Management Body of Knowledge comprises eight areas of concentration: Scope Management; Quality Management; Time Management; Cost Management; Risk Management; Human Resource Management; Contract/Procurement Management; and Communication Management. It is based on a report published in the August 1986 special issue of the *Project Management Journal* and approved by the PMI Board in 1987.

The Code of Ethics is based on a report which appeared in the December 1982 *Project Management Quarterly*. Experience requirements encompass work as either a project manager or in project management-related jobs.

The service motive can be demonstrated by service to PMI or any similar professional organization, such as the American Association of Cost Engineers, the National Contract Management Association, the Institute of Industrial engineers or other professional society.

Essentially, these criteria require that an individual meet eligibility requirements in the areas of education, experience, service and ethics. If so qualified, then the person takes an examination to demonstrate an acceptable mastery of the approved body of knowledge. A critical consideration is that individuals wishing to be certified are assumed to bring *basic industry knowledge* to the program and, therefore, should be tested only in the project management area.

Who Is Responsible for Materials?

Edwin R. Dunstan, Jr. [1]

Abstract

Can an engineer remain expert in everything? We are in an "Age of Materials." Should a structural engineer remain expert in all materials, some materials, or does it matter?

Introduction

Historically, an engineer, an individual, is responsible for all of the structural aspects of a project. However, we have now entered into an "Age of Materials," an "Era of Advanced Materials" (Langer 1989). In the past an engineer could know it all. He could prescribe how each material was to be produced and how it was to be used. It is now impossible for the responsible engineer to remain an expert in everything. What should he know? What are the minimum requirements?

An Age of Materials

Periods of history have been identified with a particular material such as the Stone Age, the Bronze Age, the Iron Age, etc. Continuing progress in the use of many different materials has led many to call the recent years the "Age of Materials." Economic growth and cultural growth has been measured by consumption of a basic material. With this thought in mind, this writer would like to suggest that since the invention of portland cement in 1824, we have been in the Concrete Age.

How many construction projects can you identify that contain no concrete? There is currently 250 million cubic yards of ready-mixed concrete placed each year in the USA. Concrete products such a block, pipe, precast and other uses add another 100 million for a total of 350 million cubic yards. How much does the average graduate engineer know about concrete? How much does the average practicing engineer know about concrete? "It's gray, it gets hard, and it cracks"?

---

[1] Engineering Manager, Fly Ash Division, Florida Mining & Materials, P. O. Box 31965, Tampa, Florida 33631.

Why the emphasis on this most basic of materials? What would the engineer do without materials? Are not designs based on the materials that are to be used? I wonder about the Bible Genesis account of the builder's of the Tower of babel and their goal of reaching the heavens. When they chose asphaltic concrete (bitumen and brick) did they understand more than the basics, "It's black, it gets hard, and it cracks"? When we build a modern tower to reach the heavens, that is scrape the sky (a skyscraper) do we know more? As a minimum, what should we know?

With a basic understanding of mathematics (geometry and trigonometry) the ancient Mesopotamians built Ziggurats which progressed to the building of pyramids by the Egyptians. Their understanding of mathematics was far ahead of their understanding of materials. There is evidence of materials failures in Ziggurats which moved the ancients to use stone (the Stone Age). The bent pyramid of Snerfu indicates a design change. The slope of the pyramid at the base is very steep. Part way up the pyramid the slope is changed to a smaller angle. It is possible this change is due to cracks that formed in the interior of the pyramid and the angle was changed to reduce the load. It is likely that this trial and error knowledge, helped Snerfu's son Cheops put it all together producing "the great pyramid" with a 755 foot square base and a height of 481 feet.

Using the current annual concrete production in this country, each state could build two great pyramids, as could Washington, D.C. Those wishing to go through the mathematical volume calculations will find that there is still enough concrete left over to build two more for a total of 104 great pyramids.

I wonder what would happen if the U.S. Government commissioned two great pyramids for each state, and each state had the State DOT, with the assistance of the State University, design a pyramid and write the concrete specifications? What if, the second pyramid for each state was open for bid and various private companies or consortiums designed the second pyramid? What guide specification would be used? Would the guide specification be one written by the U.S. Government for the pyramids to be built in Washington, D.C.? It is possible that a pyramid for Florida could be designed by an engineer from Detroit, or New York, or Boston. A pyramid to be located in Southern California might be designed by an engineer from San Francisco. The Alaska pyramid might be designed by an engineer from Arizona, etc.

Even if everyone used the same specification, would there be any similarities in the specifications? What differences would be found? If an expert panel was formed to judge the designs and the specifications, who would win? Would they judge the designs and specifications by performance, or prescription? The designs provided by Universities might reflect lack of practical experience. The designs from private firms might reflect the thoughts of strong individuals with singular views or prejudicial views of various design procedures or prescriptions.

To minimize differences the concrete materials industry has long recognized the need for education. The industry has a Standard Practice ASTM C1077, "Laboratories Testing Concrete and Concrete Aggregates for Use in Construction and Criteria for Laboratory Evaluation." The practice identifies and defines duties, responsibilities and minimum technical requirements of testing lab personnel and equipment.

The need for a quality standard was demonstrated on a recent project in Florida. Tested concrete strengths from two different laboratories were 1000 lb/in^2 apart. In one case the concrete met the strength requirement and in the second case it failed. What guidance does this provide the structural engineer?

The concrete industry, through ACI, also trains and certifies concrete inspectors. Work is now underway in Florida to evaluate and certify ready-mix concrete plants and personnel.

The concrete industry has recognized a need to police its own members by education and requiring minimum standards. I'm sure most in the concrete materials industry congratulate ASCE for their efforts in producing a manual of quality. (Korman 1989).

Is there a similar need to certify engineering firms, contractors, etc. for quality? The manual of quality was launched after the collapse of the walkway at the Hyatt Regency in Kansas City. A major goal was to clarify responsibilities and to prevent legal claims and collapses. Some feel the manual of quality will only become an educational tool. The initial idea was for the manual to become a standard, a code.

The concrete industry is beginning to recognize benefits from training, education, certification, codes, and standards. Could this benefit also extend to engineering firms? How about, "ASTM____Standard Practice for Engineering Firms Designing Structural Buildings, Criteria for Evaluation"?

Does this standard scare anyone? It would include only

minimum requirements. James W. Poirot, the leader of the ASCE quality manual project, adds "Those that are dissatisfied might be - because they want to operate contrary to the manual." (Korman 1989)

I wonder how many people agree with a few critics of the medical profession. These critics argue that the profession overly protects each other and there does not appear to be a real policing of the profession. How many of us would allow our family medical doctor, a good doctor, to operate on our kid? We would challenge or question his surgical ability even though he is an M.D. We would prefer a trained and experienced surgeon who far exceeds the minimum standards. If we expect this of others, how about ourselves? In our own branch of engineering, can we identify the minimum standards? Even if we do not develop a code or standard specification (ASTM_____), we must identify the minimums and redefine responsibilities.

What are the minimums? Should the structural engineer know more than, "It's gray, it gets hard, and it cracks." Should he be expert in steel, wood, masonry, concrete, paint, grout, soils, welding, etc? Or does the engineer only need to specify performance, such as 4000 lb/in^2, etc. It would then be the responsibility of the supplier. What are the qualifications of the supplier? Does he then have to have a responsible professional engineer? Or should the structural engineering firm be required to have a materials expert on staff? How do we certify or determine that an engineer is "expert" or meets a minimum requirement?

The current system for specifying concrete needs improvement and I am sure that is true of other materials. Materials engineers with ready-mix companies spend a lot of time suggesting specification changes or dealing with impossible specifications. In this "Age of Advanced Materials" a new definition of responsibility for materials is needed. Even concrete, once thought of as a basic material, is not basic any longer. We now have third and fourth generation admixtures, polypropylene fibers, artificial aggregates, super-this and super-that. How will we handle future "Advanced Materials"?

Who is, or who should be, responsible for materials?

References

Korman, Richard, "Endorsements Elude ASCE," Engineering News Record Magazine, October 5, 1989, page 14.

Langer, Edward L., "Advanced Materials in a Changing World," ASTM Standardization News, October 1989, pages 32-34.

The Role of Computing in the Civil Engineering Curriculum

Kincho H. Law, Stanford University , Stanford, CA
William J. Rasdorf, North Carolina State University , Raleigh, NC
Mohammad Karamouz, Pratt Institute , Brooklyn, NY
and
Osama Y. Abudayyeh, North Carolina State Univ.
Raleigh, NC

ABSTRACT

The effective use of computers in engineering processes and applications is recognized by many as the key to increased individual, company, and national productivity. This paper presents the results of a survey that was conducted by the ASCE Task Committee on Computing Education to assess the computing needs in the civil engineering profession and to assess the current computing curriculum in civil engineering education.

The survey results presented in this paper can be used as a guide by educators who wish to enhance the computing aspects of their engineering curricula and by engineering firms that are evaluating the educational profile of new graduates. The goal is to enable students to possess, in addition to their engineering skills, both the ability to evaluate and use production software and the ability to organize and supervise the development of software.

1 Introduction

In the future, an integrated combination of computer-aided analysis and design tools and techniques will be developed for all types of engineering design problems. This will require applying computer science principles and practices to a variety of engineering systems in order to determine their response to external influences. The implications of this requirement for the academic community are clear: we must prepare our students to use computer methods and applications as a part of their fundamental engineering education [Rasdorf84, Rasdorf87, Rehak83].

Considering the above observations, the American Society of Civil Engineers' Education Committee proposed and started a study on computing education requirements for the civil engineering curriculum. One of the tasks in this study was to assess the current teaching practices and course offerings in civil engineering departments of various universities and to assess the educational requirements, relative to computing, of professional engineering firms. The task committee designed a survey and send it to both academicians and practitioners to solicit responses. The responses to the survey have been accumulated and analyzed. This paper presents and discusses those results.

The paper is organized as follows: section two provides a discussion of the history of the ASCE Education Task Committee on Computing Curriculum and on the survey conducted; section three lists the results of the survey in tabular and graphical forms; section four provides a commentary on the results; and, section five provides a summary.

2 Background

Adequate computer resources have been a subject of concern in engineering education. In 1986, the education committee of the Technical Council on Computer Practices conducted a survey to determine the availability of computing resources in civil engineering departments and to determine the attitude of faculty towards computing in civil engineering education. The committee received over 93 responses from the civil engineering departments and 550 responses from faculty across the country.

The survey indicated that while almost all universities require a programming course, most civil engineering departments do not require, nor do they have, any additional computing courses beyond the first programming language course. The faculty responding to the survey strongly agreed that some type of a computing technology course should be a fundamental course within a civil engineering degree granting program. Furthermore, the faculty strongly agreed that the civil engineering curriculum should allocate course time to train students to use computers even if the result is an increase of 3-6 hours in the overall number of courses required to complete the BSCE degree. Three specific areas were identified by the task committee that civil engineering students need to be exposed to:

1. The technology of computers - how they work and how to program them.
2. Computers as problem solving tools - how to use spreadsheets, graphics, CAD, databases, etc. to solve engineering problems.
3. Computers as engineering simulators - how to use software to illustrate and evaluate a variety of solutions or alternatives.

Because there is a lack of consistency in computing education to address these areas, there is clearly a need for a comprehensive review of the computing aspects of the civil engineering curriculum.

The formation of a task committee under the ASCE Education Committee for assessing the status of the civil engineering curriculum with respect to computing was first initiated in 1987. A proposal was written, submitted, and approved by the Technical Council on Computer Practices of ASCE. The proposal called for developing guidelines and recommendations for a computing curriculum in civil engineering education. The activities involved in the effort included:

1. Assessing the current computing curriculum in civil engineering education by sending out surveys to academicians that are teaching computing courses in different schools across the nation;
2. Assessing the computing needs of the civil engineering profession by sending out surveys to practitioners in both the public and private sectors in different civil engineering firms across the nation; and

3. Developing recommendations and suggesting formal guidelines for a modified civil engineering computing curriculum.

The proposal called for completion of these activities in a two-year period. This paper addresses the first two of these three issues.

The task committee members, represented by the authors of this paper, approved two mailing lists: one for the professional candidates and one for the academic candidates; sent the questionnaire to each candidate; collected the responses; and, tabulated and analyzed the results that are presented in this paper.

3 Survey Results

The results of the academic survey are tabulated in Tables 1 and 2. They are also displayed in Figures 1 and 2. The results of the professional survey are tabulated in Tables 3 and 4. They are also displayed in Figures 3 and 4. In the tables, the number of responses shown indicates how many survey responses, out of the total number of responses (given in the table header), selected the item shown. Also, the percentages shown represent the ratio of the number of responses to the total number of responses (given in the table header).

Figure 1 is a bar chart of the prioritization of the different subject areas in civil engineering that were provided by the participants of the academic survey. Table 1 lists the titles of the subject areas displayed in Figure 1. Figure 2 is a bar chart of the number of computing courses offered at the universities represented by the academic survey participants. Table 2 lists the actual number of computing courses displayed in Figure 2. The "no course" percentage (36%) item in Figure 2 and Table 2 includes both those institutions where there is no computing course offering (30%) and those institutions where there are future planned courses (6%).

Figure 3 is a bar chart of the prioritization of the different subject areas in civil engineering that were provided by the participants of the professional survey. Table 3 lists the titles of the subject areas displayed in Figure 3. Figure 4 is a bar chart of the civil engineering computing application areas that were provided by the participants of the professional survey. Table 4 lists the titles of the application areas displayed in Figure 4.

The survey has determined that there is a general consensus among academicians and practioners with respect to the need for an increased emphasis on computing in existing civil engineering courses. This conclusion is based on such an indication from 42% of the responses from the academic survey and 28% of the responses from the professional survey, where these percentages were the highest on both surveys.

The second highest percentage on the academic survey was different than that of the professional survey. While academicians believe that acquiring good software and hardware and maintaining a balanced computing education curriculum that puts equal emphasis on computing and basic understanding of the underlying civil engineering principles; practioners believe that it is more important to emphasize the use of personal computer based software packages that relate to civil engineering as well as to teach students how to judge and evaluate computer program results.

The third highest score on the academic survey emphasized that there is a need for

Table 1: Academic Survey Subject Results (85 Responses).

Rank	Subject	Responses	%
1	Spreadsheet	63	74
2	CADD	57	67
3	Programming	55	64
4	Graphics	46	54
5	Databases	27	32
6	Wordprocessing	23	27
7	Software Packages	22	26
8	Expert Systems	15	18
9	Software Engineering	14	17
10	Geometric Modeling	12	14
11	Numerical Analysis	11	13
12	Operating Systems	11	13
13	Networking	3	4
14	Finite Element Methods	2	2
15	Simulation	2	2

Table 2: Academic Survey Course Offering Results (85 Responses).

Course Offering	Responses	%
no courses	31	36
1 course	20	24
2 courses	18	21
3 courses	12	14
4 or more	4	5

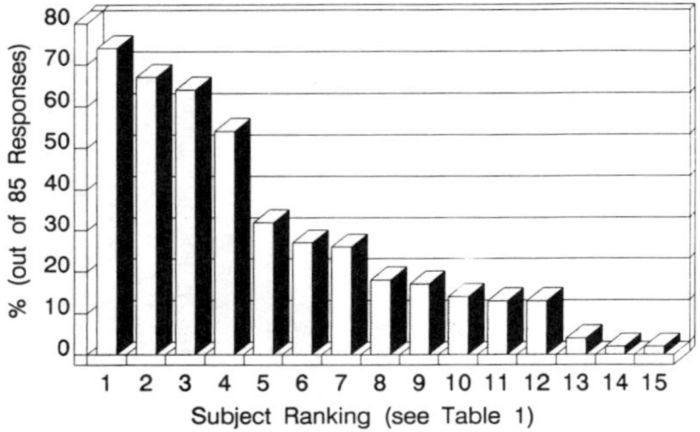

Figure 1: Academic Survey Subject Results.

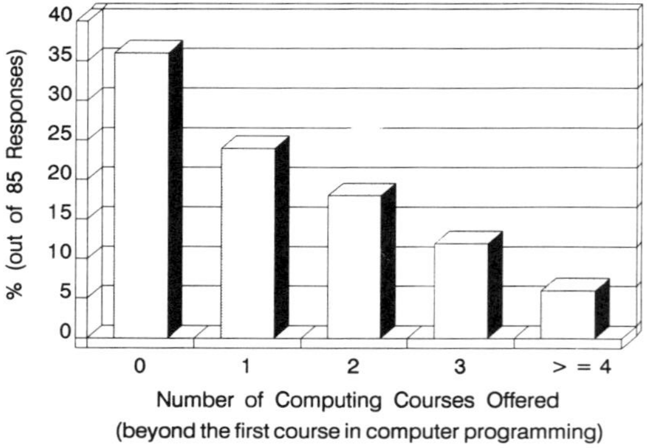

Figure 2: Academic Survey Course Offering Results.

Table 3: Professional Survey Subject Results (86 Responses).

Rank	Subject	Responses	%
1	CADD	59	69
2	Spreadsheets	59	69
3	Databases	47	55
4	Programming	45	52
5	Graphics	32	37
6	Wordprocessing	28	33
7	Expert Systems	16	19
8	Software Engineering	15	17
9	Software Packages	15	17
10	Operating Systems	13	15
11	Communications	8	9
12	Finite Element Methods	5	6
13	Numerical Methods	3	3
14	Hardware	3	3

Table 4: The Professional Survey Application Area Results (86 Responses).

Rank	Application Area	Responses	%
1	Design	72	84
2	Management	62	72
3	Analysis	58	67
4	Drafting	55	64
5	Simulation	20	23
6	Optimization	14	16
7	Data Management	12	14
8	Report Writing	7	8
9	Modeling	6	7

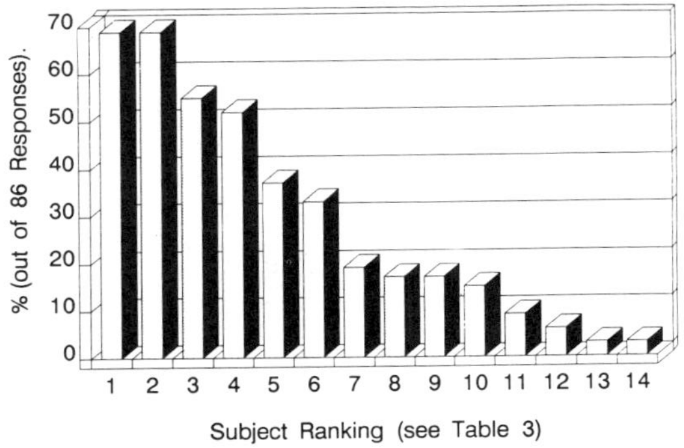

Figure 3: Professional Survey Subject Results.

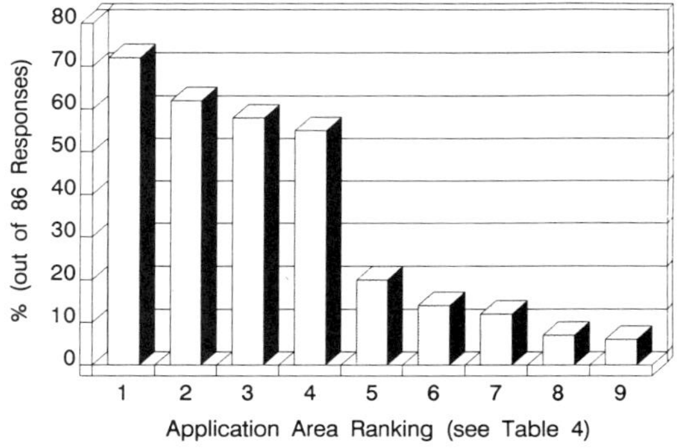

Figure 4: Professional Survey Application Area Results.

educating faculty members on computing concepts, techniques, and utilization. The third highest score on the professional survey was given to the fact that students must learn when it is appropriate to use computers and when it is appropriate to use other problem solving tools.

4 Survey Commentary

This section discusses the results of both surveys, list the conclusions, and suggests general guideline for schools that intend to develop a new computing course or courses.

From Figures 1 and 3 and Tables 1 and 3 one can observe that the following subjects were the top highest five subjects that both academicians and practioners believed to be the most promising and awarding to the civil engineering profession: spreadsheets, CADD (Computer Aided Design and Drafting), programming, graphics, and databases.

From the above list one may conclude that both academicians and practioners realize the need for adopting computing concepts and technologies and integrating them into the civil engineering educational program. It was encouraging to note that both surveys were in almost total agreement on the prioritization of the above list which may indicate that these subjects are in fact emerging as substantial prerequisites for civil engineering practice and research.

From Figure 2 and Table 2 one can observe that civil engineering departments clearly realize the need for adopting courses to teach computing concepts and techniques. 64% of the schools represented in the survey offered at least one computing course and additional 6% are planning to offer one in the near future.

From Table 4 and Figure 4 one can conclude that practioners overwhelmingly favor (76% of the participants) the design, management, analysis, and drafting application areas to the other areas. This should convey a message to academicians about which existing courses should receive the most attention and provide the largest professional payoff relative to computing.

The following guidelines are based on the comments provided by the survey respondees as well as the observations made in this section. These guidelines are not intended to be complete or exhaustive but they do represent somewhat of a consensus distillation of the survey results. Participants generally suggest that colleges and universities should:

1. Emphasize computing in existing civil engineering courses by requiring the use of computers in junior and senior level courses.

2. Acquire and maintain quality hardware and software.

3. Maintain a balanced computing education curriculum that gives increased emphasis on computing concepts as well as on civil engineering principles.

4. Organize workshops for educating faculty members on computing concepts and computer use.

5. Teach students how to use computing concepts effectively and how to judge and evaluate computer program results. The best way to accomplish this is to design

homework problems that require students to interact with the computer rather than to use it as a black box.

5 Summary

This paper presented the results of a survey that was conducted by the ASCE Task Committee on Computing Education to assess the role of computing in the civil engineering curriculum. The survey sought to determine the depth of course offering in civil engineering departments of various universities, to assess the computing education specifications of educators, and to assess the computing educational requirements, relative to computing, of professional engineering firms. The motivation for the survey and the origins of the ASCE Education Task Committee were described. The results of the survey were provided and discussed. Conclusions were drawn based on the accumulated results and a commentary on the results was provided.

6 Acknowledgment

The authors would like to express their gratitude to everyone who participated in this survey for their cooperation and subjective comments. In addition, ASCE is acknowledged for supporting the task committee in its charge, for providing the resources to conduct the survey, and for providing the forum to convey the results to the civil engineering community. The former task committee on computing resources was chaired by professor Darrell Fontane of Colorado State University, whose encouragement and suggestions were very helpful in completing the study described here.

References

[Rasdorf87] Rasdorf, W.J. and Storaasli, O.O., "Educational Fundamentals of Computer-Aided Design," International Journal of Applied Engineering Education, Volume 3, Number 3, Pages 247-254, 1987.

[Rasdorf84] Rasdorf, W.J., "Computer Programming in the Civil Engineering Curriculum," Journal of Professional Issues in Engineering, American Society of Civil Engineers, Volume 3, Number 4, Pages 141-148, October 1984.

[Rehak83] Rehak, D.R., "The Use of Powerful Personal Computers in Engineering Education," Proceedings of the American Society of Engineering Education Annual Convention, American Society of Engineering Education, Volume 1, Pages 165-169, June 1983.

The Use of an Expert System for Teaching Civil Engineering Systems

Slobodan P. Simonovic[1], M.ASCE, and Brian A. Kapalka[2]

Abstract

During the past decade computers have become a vital component of our world. Recently, the computer is taking on an educational role. There are many "toys" on the market to help a child learn to spell or add. Why not have a "toy" to help students learn more serious subjects like Calculus, Chemistry, or English? It is with this goal in mind that the software package described herein was developed. It was developed using a branch of Artificial Intelligence (AI) known as Expert Systems (ES).

Introduction

Civil Engineering Systems is a second year engineering course at the University of Manitoba aimed specifically at exposing students to systems analysis concepts and procedures. The basic motivation for such a course is that systems analysis, in its general sense, is as fundamental a component of engineering education as mathematics and statistics (Goulter and Simonovic, 1988). Systems is the means by which engineering problems can be reconstructed following the decomposition of the original problem for the detailed analysis. The approach used in teaching the Civil Engineering Systems course at the University of Manitoba is analytical or "techniques" oriented, providing students with the insight in a set of operations research techniques and their application in Civil Engineering.

Dramatic changes are under way in the Civil Engineering workplace as a result of society development, automation, fast growth of scientific and technological knowledge, increasing complexity of physical systems, and expansion of technical occupations (Simonovic, 1989). It seems that AI will have a strong impact on the future of Engineering, specially through the ES technology known also as knowledge-based engineering. An Expert System is a computer program which incorporates expertise in a specific domain in order to provide knowledgeable advice (Rolston, 1988). Expert Systems are beginning to be used in Civil Engineering research, but are little used in Civil Engineering education. There are a number of Expert Systems characteristics that look particularly promising, but thus far have not been applied to teaching. Perhaps the most intriguing and powerful characteristic of ESs, and that which distinguishes them from traditional computer applications, is their ability to explain the reasoning process used in reaching a conclusion. The explanation typically consists of an identification of the steps in the reasoning process and a justification for each step. Because the ESs structure of rules, inference engine, and learning by experience have parallels in research on engineering education, the engineering classroom appears to be a logical place for the future application and development of ES models.

---

[1]Associate Professor, Department of Civil Engineering, The University of Manitoba, Winnipeg, Manitoba, R3T 2N2. CANADA.

[2]Research Assistant, Department of Civil Engineering, The University of Manitoba, Winnipeg, Manitoba, R3T 2N2. CANADA.

This paper illustrates an effort in using ES technology to help students learn Civil Engineering Systems. The system named SYSES has been developed at the University of Manitoba to be used on the PC environment to aid students in learning different optimization techniques.

Issues in SYSES Development

SYSES was developed using VP-Expert, a rule-based ES development tool (Paperback Software, 1987). VP-Expert offers a combination of powerful features important for the application discussed in this paper. The tool is menu driven and very friendly. All the instructions are stored in the help menu which may assist directly in the development process. In VP-Expert the knowledge base files contain three basic elements: a) the ACTIONS block; b) rules; and c) statements. A fourth element, clauses, are contained within the ACTIONS block and rules of the knowledge base. The ACTIONS block tells the inference engine what needs to be found out, and in what order. Rules, stated as IF/THEN propositions, contain the actual knowledge or expertise. Statements generally contain information pertinent to the consultation itself. Unlike statements, clauses are not independent in the knowledge base. They always occur as part of the ACTIONS block or the conclusion of a rule. The special characteristics of VP-Expert investigated and used for the presented application include: a) selection capability; b) support capability; c) integration capability; and d) graphical capability (Simonovic, 1989a).

Selection capability. The selection of an appropriate optimization technique depends on several factors which will be discussed in the next section. The knowledge has been stated in the form of IF/THEN propositions. The FIND clause is used to identify variables whose value is needed to successfully complete consultation. Whenever a FIND clause is executed, the inference engine attempts to find a value for the given variable. To make the consultation more convenient, an EXECUTE statement has been used. This statement causes a consultation to begin automatically upon the execution of the Main Menu CONSULT command. In SYSES, the WHILEKNOWN clause has been used very often. This clause creates a loop which continues to be executed as long as the variable has a value assigned to it. To perform forward chaining the WHENEVER clause is used in VP-Expert. This is the most widely used statement in SYSES. The WHENEVER statement is executed every time the variable is assigned a value which causes the IF condition to be true.

Support capability. Application of ES technology in teaching a Civil Engineering Systems course relies heavily on the support of external optimization programs. The VP-Expert capability regarding communication with other programs has been studied. The CALL, BCALL and CCALL clauses let VP-Expert run external programs. These clauses suspend a VP-Expert consultation and load an executable (.EXE) program file, a DOS batch (.BAT) file, and a command (.COM) file respectively. When the execution of the program is finished, the consultation resumes where it left off.

Integration capability. The modular structure of the SYSES requires the different knowledge bases to be linked together. In VP-Expert, the CHAIN, LOADFACTS, and SAVEFACTS clauses link or chain knowledge bases together. The basic steps required in their use are: a) use SAVEFACTS to save the known values from the current module; b) issue a CHAIN clause naming a new knowledge base (module). The old knowledge base will be closed and this file will take its place in the current consultation; c) use LOADFACTS to load the values saved by SAVEFACTS into the new knowledge base.

Graphical capability. Anytime the programmer wishes to use a graphics screen, he/she must include a GMODE clause. Some clauses can only be used following a GMODE statement. These include RECTANGLE, FILL, GLOCATE, GDISPLAY, GCLS, FORMFIELD, and LBUTTON. Only some of these clauses will be described in this paper. For others, the reader is directed to consult the VP-Expert Manual. The RECTANGLE clause draws a rectangle on the screen at the location determined by the provided coordinates of the upper left and bottom right corners. The rectangle is drawn in whichever colour was last specified by a GCOLOR clause. The FILL clause fills an area of the screen with the colour last specified in a GCOLOR clause. A FORMFIELD is often referred to as a "smart form". A rectangular box or formfield is displayed any time the named variable is assigned a value. Pressing the mouse while pointing to the FORMFIELD enables the user to input a new value for the variable. The LBUTTON can be thought of as a formfield with predefined values (yes or no). If the user presses the button with the mouse, the variable is assigned the opposite value to what it was assigned.

Description of SYSES

The purpose of SYSES is to help a student or non-expert learn about different optimization techniques and understand which types of problems each method can be used to solve. Knowledge about 18 optimization models have been incorporated into SYSES. The algorithms are taken from Baldiserra et al. (1983) and Kuester and Mize (1973) and modified to run on a PC and to produce a more understandable output. In some cases, such as for the Simplex method, the programs were further changed to allow for more complex problems. The methods incorporated in the SYSES are listed below with the names of their respective knowledge bases, which will be described later:

Linear programming methods	
Simplex method	- SIMPLEX.KBS
Revised Simplex Method	- REVSIM.KBS
Integer Programming	- INTEGR.KBS
Assignment Method	- ASSIGN.KBS
Transportation Method	- TRANS.KBS
Nonlinear programming method	
Quadratic Programming	- QUAD.KBS
Fibonacci Search	- FIBON.KBS
Golden Search	- GOLDEN.KBS
Bisection Method	- BISECT.KBS
Pattern Search	- PATTERN.KBS
Gradient Search	- GRADIENT.KBS
Zoutendijk's Method	- ZOUTEN.KBS
Fletcher-Powell Method	- FLETCHER.KBS
SUMT Method	- SUMT.KBS
Dynamic programming methods	
Dynamic Programming	- DYNAMIC.KBS
Network methods	
Maximum Flow Method	- MAXFLOW.KBS
PERT Method	- PERT.KBS
CPM Method	- CPM.KBS

Simplex Method. This method is used to find the maximum or minimum of a linear function which is subject to linear constraints. SIMPLEX.EXE has been modified to accept constraints of all types (<, >, =). The limits on the problem size are: i) the maximum number of variables is 20; and ii) the total number of constraints is 20.

Revised Simplex Method. The Revised Simplex solves the same type of problem as the Simplex method solves. However, when the number of variables becomes large, the Revised Simplex method is much more efficient.

Integer Programming. This method is used to solve a linear programming problem whose variables are further constrained to be integers.

Assignment Method. The Assignment problem is a special kind of transportation problem in which only one unit of resource is available at each source and one unit of resource is required at each destination. In this type of problem, the number of origins must equal the number of destinations.

Transportation Method. In this type of problem, the objective is to assign quantities of a single commodity to various origins and destinations in such a way as to minimize transportation costs or to maximize profits.

Quadratic Programming. This method is used to locate the optimum of any unconstrained multiple variable-function written in quadratic form.

Fibonacci Search. In general this method can be used to find the optimum of any single variable unimodal function over a specified interval. The program FIBON.EXE is restricted to polynomial functions with 15 terms or less.

Golden Search. This technique is applied to the same type of problem as the Fibonacci method. However, when it is known in advance that a large number of iterations will be required, the Golden Search method provides a more efficient algorithm.

Bisection Method. This method can be applied to locate the optimum of any single variable function whose derivative exists over the given interval. The program calculates the derivative of the function alone.

Pattern Search. The Pattern Search is a widely used search for the optimum of an unconstrained multi-variable function. This method does not require the function to have first order partial derivatives over the specified interval.

Gradient Search. This search is used to find the optimum of a multi-variable unconstrained function which can be either linear or non-linear.

Zoutendijk's Method. This method of feasible directions can be applied to find the optimum of a multi-variable, non-linear function which is subject to linear equality constraints and non-linear inequality constraints.

Fletcher-Powell Method. This technique can solve unconstrained multi-variable functions, whether linear or non-linear. The partial derivatives of the function must exist over the interval in question.

SUMT Method. The sequential unconstrained minimization technique, SUMT, can be used to solve any constrained, multi-variable function regardless of whether or not the function and/or constraints are non-linear. Using penalty functions the method attempts to eliminate the constraints and solve the problem as unconstrained.

Dynamic Programming. This technique is used to find the optimum of constrained single or multi-variable function whether linear or non-linear. The only requirement of the method is the function to be separable. This method is well suited for sequential decision making problems. The program DYNAMIC.EXE is restricted to nine predefined types of recursive equations.

Maximum Flow Method. This technique is concerned with the maximum possible flow through a network, given the maximum flow through each branch.

PERT Method. The Program Evaluation and Review Technique algorithm determines the critical path of a project when only one completion time and cost are given.

CPM Method. The Critical Path Method is a project management technique which is used to plan and schedule a project defined as a series of tasks which must be performed in a certain order and at some cost.

SYSES is a compilation of 19 knowledge bases linked together as shown in Figure 1. The main knowledge base, OPTIMIZE.KBS, contains the expertise of the system and selects the appropriate optimization technique for the given problem. The other 18 knowledge bases, listed above, are used for data preparation and calling executable programs. OPTIMIZE.KBS is the "brains" of the SYSES. Using a backward chaining technique (starting with the goal and moving backwards to find the conditions for satisfying that goal), OPTIMIZE.KBS prompts the student for information about the problem in the following "simplified" manner. OPTIMIZE.KBS begins by asking the student questions such as "Is the variable time dependent?" in an effort to determine whether or not the problem is dynamic. If it is not dynamic, the knowledge base attempts to classify the problem as a network problem. The student must answer questions along the lines of "Are you trying to calculate flow through a network of nodes?" and "Are you interested in a project's completion time?". If OPTIMIZE.KBS decides that the network methods are appropriate, it asks more specific questions about the problem in order to choose the one model that is bested suited for the given problem. Otherwise, OPTIMIZE.KBS asks a series of questions designed to indicate if the problem is in the "special" LP class. These include "supply and demand", "commodity assignment", and "transportation cost" questions. Again, a more detailed inquiry is performed should the knowledge base conclude that the problem falls into this class of problems. Barring that, OPTIMIZE.KBS prompts the student for information on the objective function and constraints. If the problem contains a linear objective function and linear constraints, questions pertaining to the number and type of variables are posed and OPTIMIZE.KBS selects the appropriate linear programming method. If the problem contains some nonlinear aspects, OPTIMIZE.KBS eliminates all methods other than the nonlinear programming methods and further reduces the number of possible methods by asking questions such as "How many variables are in the problem?", and "Are the number of iterations required known beforehand?". If the problem contains constraints, the student must give information on both the equality and inequality constraints. At this point, OPTIMIZE.KBS has enough data to make an "informed" decision and can choose one of the techniques contained in SYSES.

Once the method has been selected by OPTIMIZE.KBS, a "chain" connects OPTIMIZE.KBS with the appropriate knowledge base. Each optimization technique has its own corresponding knowledge base. Although the required input differs for each technique, the knowledge bases all follow a general pattern as shown in Figure 1 for Simplex method. Once "chained" to, the knowledge base reads in a text file describing the method. This is followed by a screen which lists the program's limitations (eg. the maximum number of variables) and then the data entry screens. In the case of problems composed of an objective function and constraints, the coefficients of both are entered into the formfields. If it is an allocation type problem, the knowledge base provides the user with formfields in which to enter supplies, demands, and associated costs. In all cases, the knowledge bases entry a simple task. Once the data is entered it is "shipped" to an input file. The user is presented with a choice of output formats (final solution, input and final solution, or input, final solution, and intermediate steps). By way of a CALL clause the executable program reads in the data from the input file. Next a BCALL clause is executed to delete the input file so that the data from future problems will be written at the beginning of the input file. The solution is printed to the screen using a RECEIVE clause

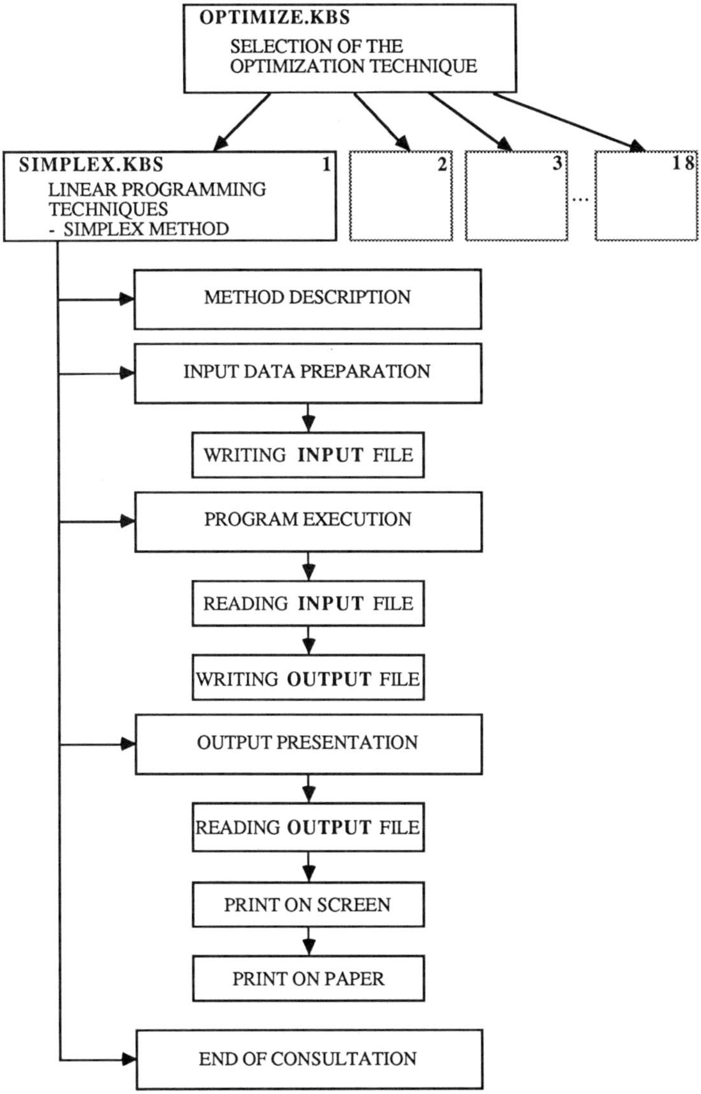

Figure 1. The structure of SYSES

coupled with a line counter that causes a pause after every 20 lines. Finally, the knowledge bases give the user the following print options:

(i) print solution
(ii) print introductory text
(iii) print solution and introductory text
(iv) do not print anything.

If a print-out is desired, the knowledge bases use another BCALL to send the appropriate files to the printer. The consultation finishes with VP-Expert's Main Menu.

Conclusions

Some of the characteristics of ESs are very appealing for the application of this technology in teaching engineering courses. An attempt to support teaching Civil Engineering Systems at the University of Manitoba has been presented in the paper. The course is techniques-oriented which allows straightforward use of ES technology. A system named SYSES has been developed to help students select, learn and understand the use of different optimization techniques. SYSES has been developed for PC computers and contains 18 optimization techniques linked together within 19 knowledge bases. The development tool VP-Expert has been used to develop SYSES. Flexibility of the tool, and characteristics briefly discussed in the paper were found applicable for SYSES's development. System is in the verification process during this academic year and will be fully implemented next year.

Acknowledgement

In the present form this research has been supported by an operating grant from the Natural Sciences and Engineering Research Council of Canada. This help is gratefully acknowledged.

Appendix I. - References

Baldiserra, C., Ceri, S., and Colorni, A. (1983). *Metodi di Optimizzazione e Programmi di Calcolo*, CLUP, Milano, (in Italian).

Goulter, I., and Simonovic, S.P. (1988). "The need for systems in first year engineering", *Proceedings*, Sixth Canadian Conference on Engineering Education, The University of Manitoba, 422-440.

Kuester, J.L. and Mize, J.H. (1973). *Optimization Techniques with Fortran*, McGraw-Hill Book Company, New York.

Paperback Software (1987). *VP-Expert rule-based expert system development tool*, Users Manual, 218 p.

Rolston, D.W. (1988). *Artificial intelligence and expert systems development*, McGraw-Hill Book Company, New York.

Simonovic, S.P. (1989). "Challenges of the changing profession", *Proceedings*, 4th Canadian Seminar on Systems Theory for the Civil Engineering, The University of Manitoba, 20-28.

Simonovic, S.P. (1989a). *Expert Systems for Water Resources Engineers*, Short Course, Department of Civil Engineering, The University of Manitoba, 100 p.

Educating Computer Literate Engineers?

Michelle Crull, PhD[1]
Associate Member, ASCE

Abstract

With the increasing use of computer aided engineering in the workplace, the engineering education process must be examined to determine if engineering graduates are capable of the demands placed on them by such use. Some general problems with the existing educational system are discussed. The civil engineering curriculum at The University of Alabama in Huntsville (UAH), in particular, is examined. Some of the problems encountered by the author while incorporating computer usage in the classroom are discussed and some possible solutions are presented. Industrial support of educational efforts to include computers in the classroom is also discussed.

Introduction

The use of computer aided engineering has increased sharply in the past several years. Education of civil engineers must change to reflect this increased use of computers. A traditional curriculum with one course in computer programming and limited use of computers in other courses may not adequately prepare students for careers in which computers are used extensively. Educators and industry must work together to update the civil engineering educational process.

As an example, the civil engineering curriculum at The University of Alabama in Huntsville (UAH) is examined and some of the problems discussed. Some changes are proposed and some possible solutions suggested. The cooperation that educators need from industry is outlined.

---

[1]Assistant Professor of Civil Engineering, University of Alabama in Huntsville, Huntsville, AL 35899.

A Civil Engineering Curriculum

Traditionally, civil engineering students have been required to complete one course in computer programming. Other computer experience has been gained in a "hit or miss" fashion with the individual teachers deciding how much and what kind of computer usage will be included in their courses. Many colleges have incorporated computer aided drafting into their curriculums either in place of or in addition to standard drafting by hand. Also, the use of word processing and spread sheets has become required for technical writing classes and lab reports. However, while these are powerful tools, it is still necessary to incorporate computers in the solution of analysis and design problems and this is where major changes are required.

Students resist using computers for many reasons including a fear or dislike of computers, inconvenience of access, and inadequate computer facilities. This resistance by the students makes it harder for teachers to incorporate computers into their courses. Also, developing problems which lend themselves to computer solutions is time consuming and heavily dependent on the software available.

It is easy to say "make the students use the computer in each class", but somewhat harder to implement this policy. All too often students take computer programming as freshmen and then are rarely required to use the computers until years later when they are asked to do some programming in a course such as indeterminate structural analysis. By this time, the students have forgotten whatever computer language they have been taught and the teacher must either spend time reteaching them computer programming or leave the students to learn it on their own. Neither solution is really acceptable and so all too often the teacher leaves out any computer programming exercises.

For example, UAH civil engineering students are required to take one course in computer methods in engineering which includes solution of engineering problems using a digital computer and practice in solving engineering problems on a mainframe computer using FORTRAN. This course is usually taken in the third term of the student's freshman year. Until the second term of the junior year the students are occupied fulfilling math, science, humanities, and core course requirements. When they take structural analysis I as juniors they have frequently forgotten how to program.

While teaching structural analysis II last spring, the author conducted an informal survey before asking the students to do any computer programming. The students were asked to write a short program to read a list of ten numbers, compute the sum of these numbers, and print the numbers and their sum. In a class of ten students, only two could write a program which might work and only half of them knew enough programming to even attempt the problem. Since the topics covered in the course required some computer programming, it was necessary to spend some time reviewing FORTRAN and this eventually used up approximately ten hours of class time which could have been better spent teaching structural analysis. However, the students were required to write several structural analysis programs during the term, including an analysis program which they were required to present as if they were trying to sell it for a term project. The students were allowed to use programs they found in textbooks as long as they could fully explain the algorithms used by the program. Only two of the students elected to use programs found in a textbook and they spent a great deal of time studying the program in order to draw a flow chart. All of the students presented programs which were easy to use, well documented, and gave reliable results. The results of requiring the students to write structural analysis programs were good, but the drawback was that the author spent a great deal of time with the students on an individual basis trying to debug programs and with a larger class this just wouldn't have been possible.

The author ran into similar problems while teaching numerical methods and computations and statics last summer. Numerical methods students had been expected to use "canned" programs and possibly make adaptations to these programs. As the civil engineering faculty started teaching this course they decided that this did not adequately teach the students the topics and started requiring extensive computer programming. The students complained vigorously, and the faculty have received some of their worst evaluations by the students as a result. Since these evaluations are used in promotion and tenure decisions, there is little incentive for the faculty to continue to require this computer programming.

As a result of these experiences, the author conducted a more widespread survey over the entire engineering school. Students from statics, dynamics, mechanics of materials, and numerical methods were polled. The survey was completed by 148 students of which 19 were able to write a program which would solve the problem described previously and 39 students did not remember enough programming to even attempt it. The students polled were from every engineering discipline offered at UAH. This indicates that the problem is not restricted to civil engineering alone.

"Canned" programs introduce a problem of their own. Much of the time, programs which are very useful in practice are not really suitable for educational purposes. Many of these programs are prohibitively expensive and, due to the competitiveness of the computer software industry, many software manufacturers will not reveal enough details about what the programs are doing. Too much of the time the tendency is to use such programs without sufficient knowledge of the methods used and the limitations of the programs. Then the student takes the results of the program and assumes that they must be correct because they came from the computer. This "black box" approach can be extremely dangerous, especially when it is carried over into the working world where mistakes can cost lives.

Changes in the Educational Process

Some schools have addressed the problem of increasing student computing experience by introducing new courses such as civil engineering computer applications (Fenves, et al., 1988). However, as all educators know, due to time constraints it is difficult to fit a new course into the curriculum. It is proposed that computer education may be accomplished within existing coursework and a course specifically for computer usage in civil engineering, while advantageous, is not necessary.

First, a computer programming course should be made a prerequisite for all other engineering courses. Otherwise it is impossible to require the students to do any programming in these other courses. Next, starting with the most basic engineering courses (usually statics and dynamics), students should be required to use the computer to solve certain representative problems in each engineering course. Programming by the students can increase their understanding of the engineering principles as well as help them to gain computer experience. "Canned" programs are also useful in that students can solve a variety of problems quickly or vary certain parameters and study the effects of such variations. At least one programming assignment in each course insures that students do not forget how to program due to lag time.

The statics and dynamics text used at UAH is _Vector Mechanics for Engineers_ (Beer and Johnston, 1988). The latest edition of this text includes computer programs on diskettes as well as problems which have been designed for computer solution in each chapter. In teaching statics, the author has included one programming exercise from each chapter. Although the students complained initially, the quality of their programming improved over the term. By the end of the term, not only were the students concerned with obtaining correct solutions, but they were attempting to write programs so that the solutions were presented clearly. Also, many of these students participated in the survey taken the next term and they were all able to write the computer program to read and write a series of numbers and calculate the sum. The main problem encountered in requiring this programming was with the non-traditional and part-time students. Many of these students had taken their computer programming class years before taking statics and had forgotten how to program. The inclusion of suitable computer problems in the textbook was extremely helpful, relieving the teacher of the time-consuming task of designing appropriate problems. The authors of textbooks should keep this in mind and try to include suitable computer problems for each topic as they write or revise their textbooks.

It is especially difficult to incorporate computers into design courses while at the same time these are the courses that might benefit the most by including computer usage. While using available software allows students to study a number of alternate designs quickly and easily, the students are using a "black box" approach and are learning nothing about using the algorithms for design taught in class. On the other hand, programming for design problems can be too tedious and time consuming to be practical. For example, design of steel structures requires an immense data base. One solution may be to have the students use "canned" programs to obtain the actual designs and, in addition, to require them to flow chart the design procedure to demonstrate their knowledge of the algorithms involved. Another solution may be to have the students program design algorithms as a term project with the students adding to the program as new topics are introduced. Such a project has been used by Professor Shoemaker at Auburn University for a course in prestressed concrete design with very good results (Shoemaker, 1988).

Cooperation from Industry

Engineering education has always depended heavily on industrial support for guidance in what is required of graduating engineers as well as for support of equipment and funds. Both kinds of support are necessary to solve this problem as well. Practicing engineers know how often they use computers and what kind of computer aided engineering is used most. Also, these engineers know some of the pitfalls of using computer aided engineering. In addition, if up-to-date computers and computer software is not available to the students, they cannot gain the necessary computer experience.

Most computer software manufacturers have recognized the advantages of supplying low cost software to educational institutions. Obviously, if an engineer has used some software successfully as a student, he is going to be more likely to recommend that software's use as a practitioner. Many computer hardware companies give educational discounts on hardware as well. However, low cost is not the only thing that educational institutions require of computer hardware and software.

More than practicing engineers, educational institutions require hardy machines. Because of the tremendous hours and number of users to which computers at schools are subjected, these machines must be able to withstand abuses for which they were never intended. It must be realized that schools cannot afford to replace computers every couple of years.

Computer software should come with technical support. If a student or a teacher has a question about a software package, he should be able to obtain assistance from the manufacturer. The technical support needed by education institutions and industry are not always the same. Both practicing engineers and students need to know what are the limitations of the programs. Computer software manufacturers do not want to admit that their programs have limitations, but obviously no one program can do everything, they all have limitations. However, except as they pertain to the limitations, practitioners rarely care about the exact algorithms used by a software package. Students and teachers, however, need to know what design and analysis procedures are used by the programs. Also the hardware requirements for each software package should be made very clear.

Conclusions

Traditional civil engineering curriculums often do not meet the needs for educating computer literate engineers. However, time constraints may prohibit the introduction of new courses into the curriculum. Using the existing coursework, students' computer experience may be increased by a change in attitude. Incorporating the use of computers into each engineering class reinforces the knowledge gained in a computer programming course and helps alleviate students' computer phobia. Where "canned" programs are used, care must be taken to avoid a "black box" approach. While educators must be instrumental in bringing about the changes necessary to educate computer literate engineers, industry must cooperate. It is impossible to expect students to use computers if there are insufficient or out-dated machines available and educational institutions cannot afford to replace all their machines every couple of years. Educators must take the time and make the effort to develop methods for incorporating computers into the classroom. Industry must recognize that the technical support required by educators is not quite the same as that required by practicing engineers. They must be willing to give educators such technical support as well as financial support. With such cooperation between educators and industry, civil engineering graduates can be truly computer literate.

References

Beer and Johnston, Vector Mechanics for Engineers, fifth edition, McGraw Hill, New York, 1988.

Fenves, Hendrickson, Maher, Rehak, and Thewalt, "Two Undergraduate Courses in Computer-Aided Engineering", Engineering Education, November 1988, pp. 122-126.

Shoemaker, W.L., "Integrating Microcomputers with Design Courses", Engineering Education, November 1988, pp. 126-127.

Interactive Computer Graphics in CE Education

Leroy G. Holub[1], Member, ASCE and Darryl J. Holman[2]

Abstract

The usefulness and need for computer applications in civil engineering education are generally recognized. The degree of utilization and the nature of the applications remain areas subject to a range of opinions. However, the framework for discussing these areas evolves with changes in computer technology, the expanding role of computers in the profession, and research in information transfer. This paper discusses interactive graphics software in the context of these considerations. Arguments for their use are discussed and examples are presented.

Introduction

Over the past decade, the power and availability of computers have increased substantially. Especially evident is the increased access to personal workstations with a combination of high speed processors and high resolution graphics displays. Sophisticated graphics-oriented engineering software is increasingly available. This combination of hardware and software not only impacts the practice of the profession, but also provides opportunities for new and broadened teaching and learning strategies.

----

[1] Faculty Associate, Department of Engineering Professional Development, 432 N. Lake Street, Univ. of Wisconsin, Madison, Madison, WI 53706.

[2] Computer Specialist, Department of Engineering Professional Development, 432 N. Lake Street, Univ. of Wisconsin, Madison, Madison, WI 53706.

The Computer Learning Environment: A Synopsis

Interactive computer graphics has played a role in engineering education for many years. A sample of past applications of this type of learning environment yields remarkable similarities.

Wozny (1978) described the interactive computer center at Rensselaer Polytechnic Institute as a major commitment to the use of computer graphics in engineering education. He found that "interactive computer graphics gives the students deeper insights into fundamental concepts and fuller understanding and appreciation of the role of computers in creative design." and that "[c]omputer graphics is effective in engineering education because it emphasizes intuition rather than exact calculations."

Kamel (1978), in describing a graphics-oriented structural analysis package, concluded that this technology "helps the student develop a feel for the finite element method....helps to reinforce and challenge the physical intuition of the student with regard to complex structural behavior....graphics mode of interaction is a superior debugging tool....the student can solve realistic problems of practical interest...." He pointed out the motivational advantage of having students solve realistic problems of practical interest. He also discussed student use of programs in a cookbook fashion without an understanding of the fundamental principles on which they are based. Kamel's experience was that poor understanding resulted in inferior modeling, which forced a return to fundamentals.

Using a structural design example from engineering practice, Wilson and Lansberry (1976) found that "[i]nteractive computer graphics...permits an overall unification of the design process so that the designer will be able to examine alternative schemes in greater depth and achieve a greater feeling for the response of a physical system" and that "[t]he use of interactive graphics as an aid in model generation and debugging is significant."

Holub (1983) discussed the use of student design projects and professional software in civil engineering education. He stressed the need for students to create solutions for practical engineering problems that require the evaluation of their proposed design and design conditions. This can readily be accomplished by the use of professional software and has a positive impact on students' creative problem solving skills.

The above examples refer to complex systems and applications in upper level design courses. Graphics applications are also useful in illustrating basic concepts in mathematics and engineering science. Examples of applications to fundamental engineering concepts include moment, shear, and deflection diagrams of structural elements, and for fluid systems, velocity distributions and energy profiles.

Demana and Waits (1987), in an application for precalculus students, concluded that "microcomputers [utilizing interactive graphics] encourage abstraction, because students are able to investigate many special cases quickly, gaining the necessary evidence needed to make generalizations...permitting students to access problems which are more interesting, realistic, and broader in scope than those found in standard textbooks."

These authors agree in many ways, but their statements are largely perceptions rather than quantitative analyses. Some engineering educators are unlikely to be swayed by arguments based on intangibles such as "intuitive feel." Two additional points will be developed based on the continuation of a technological transition in civil engineering.

The first point is that the current technological revolution supports more fully a design-oriented engineering curriculum. Indeed, the requirement of educating functional engineers will likely necessitate the use of these technologies.

The Computer Revolution

It is clear that we are undergoing a revolution in computer technology. Computer processor power is roughly measured in millions of instructions per second (MIPS). According to notable computer expert Bill Joy, a simple predictive formula is MIPS = $2^{Year-1984}$ (Greenberg, 1988). This relation is likely to overestimate computing power for a long time horizon, but consider the near term. Nineteen ninety is the year of 64 MIPS processors - a quantity beyond belief a few years ago. Even microprocessor chips, which are slow compared to the fastest processors, will be rated at 30 MIPS by the end of 1990. In five years, Joy's formula predicts an unbelievable 2 billion instructions per second (BIPS or is that kMIPS?)!

An equally important aspect of this revolution is the shift in computer accessibility. The redistribution of computing power into the hands of scientists and engineers (and everyone else) began with the introduction of Digital Equipment Corporation's PDP-8 in 1964 (Evans and Bernard, 1987). The PDP-8 was significant because it was affordable at a mere $16,200, and because it allowed for easy interfacing to special equipment. No longer was the engineer's computing restricted to batch input and line printer output, but special graphics output devices and a wide range of input devices could be readily employed. Today, computers many times more powerful are available for under $2,000.

Software improvements, although impressive, have not tracked the exponential progress of hardware. In part, this is because hardware power is related to the physical size of two-dimensional arrays etched into silicon. Putting 300 solid-state switches along an axis that was used for 200 will more than double the number of switches. Software, on the other hand, doesn't lend itself to the same kind of large scale increases in power; there are very few obvious algorithms that lead to major improvements in speed. The improvements in software over the last few years largely result from greater processing speed and more memory. Even if the analytical portion of the software remains unchanged during the next five years, the improvements in microcomputer power will make it feasible to use software that only runs on supercomputers today. It is chiefly the hardware improvements that are responsible for the increased power of interactive computer graphics.

Another significant development in software over the last decade is the standardization of environments from which the software is used. In combination with the large number of these computer systems in use, development of commercial software is feasible; a software company can profitability sell a $500 software package for a specialized area of engineering. The result is a proliferation of engineering software.

These changes are not without problems. Extensive computer power encourages the production of immense quantities of data, much of which must be analyzed by the engineer. This leads to the second point; research and development in interactive graphics make the management of these data feasible.

Research in Interactive Graphics

The primary research communities for interactive graphics come from statistics, in a specialization called Statistical Graphics, and computer science, in a field called Visualization in Scientific Computing (ViSC). Statistical Graphics is an established field with publications going back to a study of pie charts versus bar charts by Eells in 1926. Statisticians do not focus their research specifically on interactive graphical methods, but include it as an extension.

The ViSC research is focused on dynamic graphics techniques. This is a formalized subdiscipline within the computer science community. The summary from a Panel Report on VISC defined the emerging discipline as "a major computer-based field, with a body of problems, a commonalty of tools and terminology, boundaries, and a cohort of trained personnel. As a tool for applying computers to science, it offers a way to see the unseen. As a technology, Visualization in Scientific Computing promises radical improvements in the human/computer interface...." (Frenkel, 1988).

Research in computer graphics can be categorized into three topics: graphical methods, computing, and graphical perception (Cleveland 1987). Graphical methods refers to the study of the ways quantitative information can be shown on a graph. The computing research is in the domain of computer scientists and typically involves algorithm design and building interfaces. Cleveland and McGill (1985) write "What has been missing, until recently...is the study of graphs and the human visual system.... Informed decisions about how to encode data can be achieved only through an understanding of this visual decoding process which we call graphical perception."

Examples

Two examples of the application of interactive graphics in the study of hydraulic engineering are presented to further illustrate the benefits discussed above. These applications are used in both college credit courses and continuing education programs for practicing engineers.

The first example comes from the analysis and

evaluation of municipal water distribution systems. Software is used to determine the adequacy of a system in providing present and future demands. The results of a system analysis can be quickly displayed in summary tables or graphical form. If the system is shown to be inadequate, then improvements can be designed, incorporated into the computer model, and tested. Assumptions regarding demands and system characteristics can be studied, for example, the increase in pipe roughness with age. Dynamic simulation of system response over an extended time period is also possible but produces even larger quantities of output data. The only feasible way to assimilate this information is in graphical form. Especially for the undergraduate, the key is to increase their understanding of system response by completing a comprehensive system evaluation. This reinforces understanding of hydraulic principles, develops an intuitive feel for fluid energy relationships, and provides experience in the design and selection of system improvements.

The other example is based on fluid system transient analysis. Karney and McInnis (1989) have pointed out the importance of transient conditions in distribution networks and the need for careful consideration of how a system is being modeled. Comprehensive analysis of realistic representations of actual systems is required to determine the critical conditions for design. This results in the need to "think fundamentally and analyze comprehensively." Karney and his colleagues have developed analysis and graphics software for presenting results in two and three dimensions. Graphics representations are essential for interpreting these results. Again, fundamental principles are emphasized and an understanding of system response is enhanced through the use of comprehensive problems solved with interactive graphics software.

Closing Thoughts

Computer driven technology is changing the face of engineering practice. Educators must, in turn, respond to the needs of the profession by incorporating these tools more fully into the curriculum. The debate has shifted from the question of whether this technology should be incorporated into the educational process; now, the question is whether schools can produce competent engineers without its inclusion.

Appendix. References

Cleveland, W. S. (1987). "Research in statistical graphics", J. Am. Stat. Assoc. 82(398):419-423.

Cleveland, W. S., and McGill, R. (1985). "Graphical perception and graphical methods for analyzing scientific data." Science, 229:828-833.

Demana, F., and Waits, B. (1987). "Enhancing problem solving skills in mathematics through microcomputers." Collegiate Microcomputer, 5(1):72-75.

Eells, W. C. (1926). "The relative merits of circles and bars for representing component parts." J. Am. Stat. Assoc. 21(154):119-132.

Evans, M., and Bernard, H. R. (1987). "Word processing, office drudgery, and the microcomputer revolution.", in H. R. Bernard and P. Pelto eds. Technology and Social Change, 2nd ed. Waveland Press, Prospect Heights, Illinois, 330-358.

Frenkel, K. A. (1988). "The art and science of visualizing data." Communications of the ACM, 31(2):111-121.

Greenberg, D. P. (1988). "Coons award lecture." Communications of the ACM, 31(2):123-129.

Holub, L. G. (1983). "The use of professional programs in civil engineering education." Civil Engr. Edu. American Society Engineering Education, 5(1):8-12.

Kamel, H. A. (1978). "Interactive computer graphics as a tool in teaching structural analysis." J. Struct. Div., ASCE, 104(8):1299-1314.

Karney, B. W., and McInnis, D. (1989). Transient Analysis of Water Distribution Systems, Publication 89-01, University of Toronto, Toronto, Canada.

Wilson, J. L., and Lansberry, C. R. (1976). "Interactive computer graphics for computer aided design in civil engineering." Proc. 3rd Annual Conference on Computer Graphics, Interaction, and Image Processing, ACM, 89-96.

Wozny, M. J. (1978). "Interactive computer graphics for engineering education." Professional Engineer, 48(6):14-18.

Video/Computer Integration for Education

Donna C. Nelson,[1] Associate Member, ASCE
Jorge A. Vanegas,[2] Associate Member, ASCE

Abstract

Classroom coverage of topics within Civil Engineering that involve dynamic behavior over variable periods of time create challenging situations for instructors who are limited to traditional classroom presentation methods. The use of video technology in the classroom is a feasible alternative to enhance the visualization of dynamic events. A second alternative is the use of existing computer technologies for graphical two- and three-dimensional representation or modelling of real-world events. Individually, both technologies present serious limitations such as cost, complexity, time investment, that preclude a more widespread classroom use. Integrated video and computer technologies can overcome some of these limitations. This paper describes the results of a research project on the use of integrated video/computer technology for Civil Engineering education, and implementation of educational support material for two courses.

Introduction

This paper describes the results of a research project at Purdue University that investigated the use of integrated video/computer technology for Civil Engineering education as a solution to overcome some of the limitations of traditional presentation methods in classrooms. The prototype *Integrated Video/Computer System* (**IVCS**) developed during Phase I of the project provides a conceptual framework for development of applications that use different types of image displays to enhance the student's visualization and understanding of dynamic engineering processes in Transportation Engineering and Construction Engineering.

Classroom coverage of topics within Civil Engineering that involve dynamic behavior over variable periods of time create challenging situations for instructors who are limited to traditional classroom presentation methods. For example, traditional methods of instruction are not conducive to the easy visualization of topics such as traffic flow models or construction methods like concrete placement operations. To enhance visualization of topics such as these, instructors may use field trips, slides or

[1]Assistant Professor, School of Civil Engineering, Purdue University, West Lafayette, Indiana 47907

[2]Assistant Professor, School of Civil Engineering, Purdue University, West Lafayette, Indiana 47907

photographs to supplement classroom presentations, but there is a limit on how many field trips can be done in a semester, and still images do not convey movement.

The use of existing portable and economic camcorders (i.e., video-recorder cameras) enhances visualization, but at the same, presents some serious obstacles. With camcorders, an instructor can record real-time events and then present them in class on a TV monitor. However, many events occur over long periods of time and presenting them in real-time is difficult within the time constraints of a class. Furthermore, when many topics need to be covered, the instructor has to edit the material extensively. Editing videotaped events is a complex process because access and retrieval of specific images is cumbersome and difficult. Finally, simply showing a videotape in class does not contribute to the education of the engineering student and can become distracting or may even be uninteresting.

Another alternative to enhance the visualization of dynamic events is the use of existing computer technologies. Generic programs are available for graphical two- and three-dimensional representation or modelling of real-world events, either in static (i.e., still images) or dynamic (i.e., animated images) form. Other programs allow massive storage and easy manipulation of graphical, textual or numerical information. However, computer technologies also have limitations: developing applications can be very time consuming, especially if they require extensive programming; high-end image processing requires specialized knowledge and skills; and some of the hardware and software requirements are still very expensive.

System Description

The system uses a Macintosh II computer with a 40 Megabyte hard-disk, 5 Megabytes RAM and a high resolution RGB color monitor and is driven by the Macintosh HyperCard development environment. The overall system, shown in Figure 1, contains two basic modules.

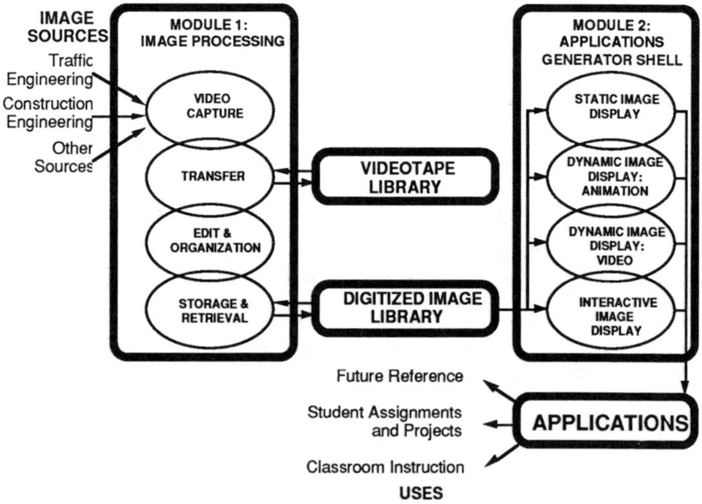

Figure 1.- Principal Components of IVCS

The first module, the *Image Processing Module* (**IPM**), is used to collect digitized images of real operations or processes, edit and organize them, and finally, store them in an image library which then can be accessed easily for use in customized applications. The second module, the *Applications Generator Shell* (**AGS**), uses the images processed in IPM to create different types of image display options for specific applications.

The main inputs to **IVCS** are images generated by different sources. For the scope of the system described in this paper, the main two sources were Transportation Engineering and Construction Engineering phenomena or operations. Future expansions of the system will include other disciplines within Civil Engineering. The main outputs of **IVCS** are specific applications used for classroom instruction, for student assignments and projects, and for future reference.

Image Processing Module

The principal components of **IPM** are shown in Figure 2. This module contains four distinct types of processes: 1) capturing images of real operations or processes in both real-time or time-lapse video form; 2) transferring these video images from tape to digitized images in the computer; 3) editing and organizing the digitized images; and 4) storing and retrieving the digitized images.

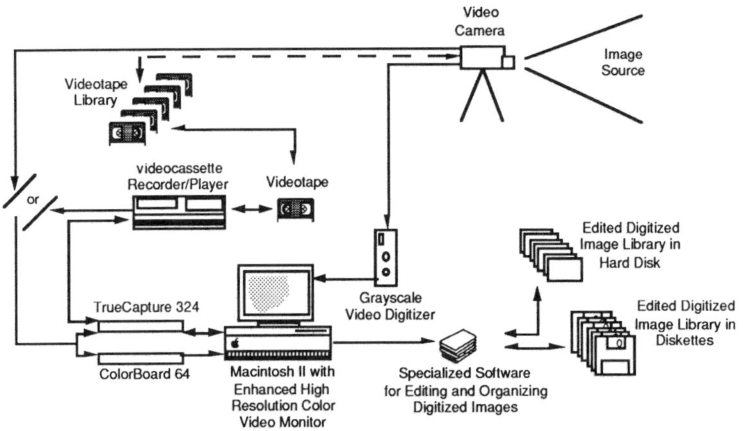

Figure 2.- Principal Components of IPM

Video Capture.- Any videotape can be used as a source of images. However, the images used in this project were recorded directly using a professional/industrial Panasonic AG-170 VHS camcorder (1/2" CCD, 270,000 pixels, EIA Standard, NTSC color signal). The built-in features of the camcorder allow the user to record high resolution images, and in addition, a character generator permits adding documentation directly to the images filmed.

While it is more expensive, a camera with color capabilities may be important for some applications (e.g., filming a color-coded piping system in a process plant). Also, since there are many operations that require recording from two different angles to capture the processes more effectively (e.g., filming the traffic activity in a busy intersection), two camcorders are recommended.

Transfer.- To successfully digitize the images and display them in high resolution form, two different approaches may be used, depending on whether the image desired is in grayscale or in color.

To transfer 16 grays or 256 grays images, a Pixelogic ProViz™ Grayscale Video Digitizer is used. This type of digitizer is an external peripheral device that connects to the SCSI port of the computer. The set-up requires a direct connection between the camcorder or a video-cassette recorder/player and the computer using special cables. The ProViz™ software allows continuous display of the scanned image in the Macintosh II window, interactive focus, triggering to specific image capture, multiple takes using timed intervals, image enhancement and adjustment, and finally, saving and restoring saved images. In addition, a special software interface, HyperVision™, allows to transfer images and incorporate them directly to HyperCard. The images are saved in a variety of formats including PICT, Raster Image File Format (RIFF), Tagged Image File Format (TIFF), and Encapsulated Postcript (EPSF). The size of the image file depends on the grayscale type with which it was transferred. The size of a typical image transferred in 16-grays is 50-175K, while a 256-grays image requires 200-400K.

To transfer 8-bit or 24-bit images in true color, a special internal board is used. The RasterOps™ TrueCapture 324 Board allows to capture from either a separate RGB or a composite NTSC/PAL video source. In addition to capturing the image, the TrueCapture software allows the same options as the ProViz™ software with the exception of multiple takes. The 324 board also has an NTSC/Genlock Option that generates NTSC video for sending screen images to a VCR. The images are saved in PICT format. The size of the image file depends on the color type with which it was transferred. The size of a typical image transferred in 8-bit color is 100-150K. When the image is transferred in 24-bit, the size increases substantially to 500-900K.

A RasterOps™ ColorBoard 64 enhances the display resolution of the Macintosh II Monitor to a maximum resolution of 640 by 480 pixels, each 24-bits deep, and the possibility of displaying 786,432 colors from a palette containing 16.7 million colors. The result is a digitized image with photograph likeness in either 8-bit or 24-bit color.

Edit and Organize.- The transfer process results in digitized images that can occupy substantial storage space and which by themselves, are of little value. Thus, some time has to be devoted to editing and organizing the images.

Partial editing can be done during the transfer process by modifying the display window (i.e., the rectangle that is showing the video image) to transfer only the relevant portions of images, or by adjusting the color settings to increase or decrease the contrast and the level of brightness or intensity of the image colors. After the image is digitized, the user can also crop (i.e., cut) the image to eliminate any additional unwanted sections, and in some cases, add special effects to the image (e.g., creating negative or inverted images). A more sophisticated type of editing uses specialized software for image enhancement. These type of programs function as computerized darkrooms to enhance, retouch, or compose digitized images.

Organizing the images is a totally different process. It begins with an initial plan or structure of what needs to be recorded before any video images are actually recorded. The contents of each tape need to be documented well, preferable as the recording is being done. This preliminary organization will reduce significantly the time used to search and find an image to digitize. The next stage involves creating and cataloguing special image files. Initially, a HyperCard stack was developed to assist in the process of organizing the images. This application enables the user to attach keywords, and written or sound explanations to each image. However, images can also be organized in files created with any graphics or presentation software that accepts digitized images (e.g., MacDraw II or PowerPoint™).

Storage and Retrieval.- The two types of storage in **IVCS** are: 1) raw images stored in videotapes to be used as needed; and 2) digitized images stored in a computer-based media (e.g., hard disk or diskettes) to be retrieved easily as needed.

The HyperCard stack used for editing also serves as the initial database for storing the images, enabling the user to search and sort quickly using keywords. Its only limitation is available disk space. Image files created with other programs remain as individual files that are accessed as needed.

Applications Generator Shell

AGS was developed using HyperCard. Its principal components are shown in Figure 3. **AGS** provides a generic framework for the development of four types of customized image display applications: 1) static image displays; 2) dynamic animated image displays; 3) dynamic video image displays; and 4) interactive image displays. Each specific type of display application is developed using a specialized HyperCard stack. These display options are used to complement and enhance classroom instruction, for student assignments and projects, and for future reference. Phase I focused on the first 2.

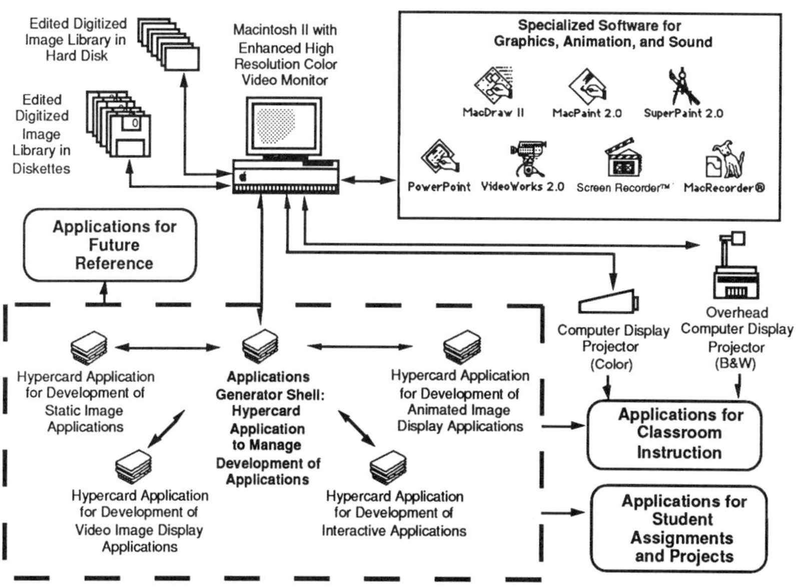

Figure 3.- Principal Components of AGS

For applications to complement or support classroom instruction, two types of devices may be used: 1) an overhead computer display projector such as the Kodak Datashow HR/M, which provides real-time B&W projection; and 2) a color projector such as the BARCO 3, which provides real-time color projection.

HyperCard Stacks in Phase I.- The static image display stack assists the user in developing applications that are analogous to a slide show. With this stack, the user creates individual cards using a combination of digitized images and computer-generated text and graphics, and relates them to each other using the linking capabilities of the HyperTalk programming language used in HyperCard. The cards can then be displayed one after the other using either automatic or manual mode.

The dynamic animated image display stack assists the user in developing applications that are analogous to cartoons animated over "real" backgrounds. With this stack, the user combines digitized images that serve as backgrounds with computer-generated animated text and graphics created with specialized software. The animated sequences can then be related to each other using HyperTalk, and displayed one after the other using either automatic or manual mode.

Sample Applications.- In Phase I, **IVCS** is being implemented and tested in one Transportation and one Construction Engineering courses. In Transportation Engineering, current efforts include 1) the development of an on-line guide to the organization and implementation of traffic volume studies; 2) a presentation of weight-in-motion technology; and 3) an instructional module on pavement deterioration and rehabilitation. In Construction Engineering, current efforts include 1) a presentation of a concrete placement or a steel erection operation; 2) the development of an on-line guide to mechanical systems in buildings; and 3) the development of an on-line guide to electrical systems in buildings.

Future HyperCard Stacks.- Future extensions of **IVCS** will incorporate more advanced computer technologies. For dynamic video image displays, currently available boards and peripherals can turn a Macintosh workstation into a full-fledged video editing and projection station. Video clips may be shown either in real-time or in time-lapse form directly in a display window within the computer monitor. Furthermore, the stack may be used to drive laser disk or video-cassette players.

Additionally, with the increasingly powerful analytical and computational software available for Macintosh computers, interactive image displays will become an important tool to strengthen the integrative skills of students for the analysis, synthesis, and contextual understanding of engineering problems. Using the fourth stack, the user can integrate image displays with analytical capabilities (e.g., computational tools or expert systems), and can modify and simulate engineering phenomena, processes, or operations.

Obstacles and Limitations

Efforts in integrating video and computer technologies to support Civil Engineering education are promising but unfortunately, there are several obstacles and limitations that still preclude a more widespread use:

- The cost of the specialized hardware and software required for development and implementation of applications is still high.
- Development of applications can be very time-consuming without support from specialists and programmers, despite the simplified programming environments,

such as HyperCard, that are currently available. The need for these additional personnel adds to the development costs.

- The applications development process requires extensive planning to obtain meaningful results. Planning avoids unmanageable scopes or premature obsolescence of the applications developed, especially with the wide range of choices and increasingly powerful software and hardware capabilities available.
- The quality of the applications developed is directly proportional to the quality of the images captured and the amount of editing done. Thus, there is an often overlooked, and many times high, cost associated with collecting data.
- There is a potential threat of emphasizing the tools rather than the content of the application; applications need to be designed to go beyond being an attractive way of presenting information.

Benefits and Opportunities

Despite the obstacles and limitations, the integration of video and computer technologies creates a powerful tool that will enhance student visualization of dynamic behavior of engineering phenomena and processes over variable periods of time. Some of the most important benefits and opportunities are:

- The integration of "real" processes and phenomena overcomes the limitations of static or schematic representations, and places learning in context.
- Better visualization tools permit a deeper exploration of certain topics in engineering, and enhance the communication between the instructor and the students.
- Pre-defined lessons or demonstrations developed by the instructor may be distributed to students for review and consultation.
- With a generic methodology, such as the one developed in this project, applications can be developed easily and always be customized to satisfy unique or specific needs and requirements.
- The cost of developing applications decreases over time. Once produced, displays and presentations can be easily modified, updated and improved without the need to start from the beginning.
- Visual displays can be linked eventually to powerful analysis tools, such as expert systems, to integrate the processes studied with the analysis being done.

Acknowledgements

The project described in this paper was primarily funded by the Purdue Engineering Dean's Club at Purdue University under the Creative Undergraduate Instruction Projects Grant. The computer workstation used in the project forms part of the Computer Research Laboratory of the Division of Construction Engineering and Management's at Purdue University. This support is gratefully acknowledged.

ADVANTAGES OF COMPUTER BASED CONTINUING EDUCATION

Jose D. Mitrani P.E.[1]
Zohar Herbsman Ph.D.[2]

ABSTRACT

Continuing education programs for practicing engineering professionals are faced with the dilemma of reconciling the rigorous and demanding exposition of knowledge required by rapidly evolving technology with severe time and logistical constraints of the practitioners the programs are intended for. This dilemma frequently has undesirable ramifications. For instance topics may be diluted as to both their breadth and depth. Another consequence may be a reduced expectation of the participant's ability to modify his/her practice to implement and to apply the material learned. The authors have done considerable experimentation in the use of microcomputer aided continuing education for professionals. They have succeeded in presenting highly technical workshops to audiences with a broad range of backgrounds, education and experience. This paper describes their philosophy and methodology of microcomputer aided instruction.

INTRODUCTION

Rapid advances in technology have made it necessary for the civil engineering profession to promote continuing education and professional development for its members. It has even been suggested that continuing

---

[1]Associate Professor, Florida International University, Miami, Florida. Member ASCE.

[2]Professor, University of Florida, Gainesville, Florida. Member ASCE.

education should be a requirement for renewal of professional licensing, and steps towards implementing such a requirement have been taken by various state legislatures.

Existing continuing education programs for practicing civil engineers are faced with the dilemma of reconciling the rigorous and demanding exposition of knowledge required by rapidly evolving technology with the severe time and logistical constraints of the practitioners the programs are intended for. As a consequence, traditional continuing education programs frequently have undesirable results. For example, presentation of topics may have to be diluted as to both breadth and depth, and the programs do not include opportunities to investigate case studies that involve actual applications of the new technology. The short term result is an inability of the participants to effectively and efficiently implement the new technology in their practice. In the long term the consequences are more drastic. There is an inevitable erosion in the practitioners' perception of the value and the relevance of the concept of continuing education.

TRADITIONAL CONTINUING EDUCATION ALTERNATIVES

Civil Engineers desiring continuing education have traditionally been offered three alternatives:

1) Enrollment in full term courses at institutions of higher learning, where available.
2) Correspondence courses.
3) Attendance at meetings and conferences.

Participation in any of these involves a compromise between the activity and the participants' ongoing professional and/or employment obligations. It also must consider the cost of the activity as well as the cost of being absent from one's practice.

ENROLLMENT AT INSTITUTIONS OF HIGHER LEARNING

Enrollment in full term courses at an institution of higher learning (if available) is one possible alternative. This involves a relatively long term commitment to attend classes that must be scheduled around the full time work obligations of the participant. The quality and relevance of the material is greatly dependent upon the resources of the institution and the makeup and level of the students that comprise the class. The length and logistics of this alternative make it unpopular to practitioners who are unable or reluctant to make the commitment necessary to complete the course of study for its full length. If a significant number of Continuing Education Units are needed, (i.e. for license renewal), an engineer may have to take several courses over many academic terms in order to obtain them.

CORRESPONDENCE COURSES

A second traditional continuing education alternative is to take correspondence courses. The quality of these varies, and they have many shortcomings. The user generally is unable to get help on the subject and feedback is limited to comments on graded tests, if any. There is no control over the learning environment and, although the courses are usually quite structured, a user interested in simply complying with the testing requirements and getting C.E.U.'s may devote himself entirely to working the tests successfully and not to learning the material. As such, the overall effectiveness of this method is questionable.

ATTENDANCE AT CONFERENCES AND MEETINGS

Attendance at specialty conferences and meetings is also a traditional method of achieving continuing education. Undoubtedly an engineer may obtain exposure to current developments in his field by attending these. Conferences and meetings, however, are generally relatively loosely structured. The cohesiveness and continuity of the papers presented are limited to the general session topic under which they appear. The time limitation on the presentations, as well as the large audience settings, are likely to inhibit a participant from asking detailed questions. The participant, in order to effectively implement knowledge obtained, must spend considerable time studying and interpreting conference proceedings and doing additional research on the subject. Apart from an overall exposure to developments in the field, attendance at conferences and meetings may be considered only a superficial exposure to the state-of-the-art.

COMPUTER BASED CONTINUING EDUCATION

The authors have considerable experience in the presentation of continuing education and professional development workshops for practicing engineers in construction. They have presented workshops under the following conditions:

1) Lecture only. No computers used.
2) Lectures with computers used for demonstration only.
3) Lectures followed by hands on computer use by each of the participants.

Surveys of workshop registrants revealed that the highest degree of participant satisfaction has been achieved in those workshops where there was a one to one correspondence between attendees and computers, and where the partticipants were computer instructed by hands on participation during parts of the workshop. Those surveyed repeatedly commented that the most beneficial consequence of the workshop was their ability to quickly grasp the essence of the subject discussed. This was attributed to the opportunity to work case studies during the workshop and get immediate feedback from the lecturers and workshop assistants. Also rated as very significant was the participants' ability to quickly and effortlessly implement much of what they learned into their own practice.

RECOMMENDATIONS FOR COMPUTER WORKSHOPS

Following are details of the hands on workshops which the authors consider essential for high workshop effectiveness:

1) Lectures and hands on sessions should be held in the same room for minimum disruption and ease of flow from lecture to hands on sessions and back.

2) Each participant should have his/her own computer.

3) There should be a low (e.g., 5 to 1) ratio between registrants and lab assistants/lecturers.

4) All the lab assistants/lecturers should be intimately familiar not only with the

applications software presented, but also with basic operating system functions and with hardware fundamentals.

5) Workshop lecturers should be able to work computer problems concurrently with the participants using a computer attached to projection equipment, so that participants can follow the lecturer's work.

6) Whenever possible, course software (courseware) that utilizes generic programs (such as spreadsheets and word processors) should be prepared on public domain shells and programs that are compatible with industry standard software such as Lotus 1-2-3 or WordStar. This way participants may be permitted to take the programs and the class models with them when they leave, and will be able to immediately apply them to their own practice.

7) The courseware should, whenever possible, have built into it context sensitive help screens to help the user when stuck. The software itself should be as user friendly as possible.

8) Reference material and detailed examples should be provided the participants for use during and after the workshop.

ADVANTAGES OF COMPUTER BASED INSTRUCTION

Using hands on computer instruction coupled with traditional lecturing, the authors have succeeded in presenting highly technical workshops to audiences with a broad range of backgrounds, education and experience. This is a result of the many advantages associated with computer based instruction. The advantages are as follows:

1) Participants have an opportunity to practice and to reinforce the lecture material during the hands on sessions by working on carefully prepared case studies. Since the computational aspects of the problems are all handled by the computer the participant may concentrate on understanding the more significant elements of the theory and the limitations of the methods presented.

2) When properly prepared, the courseware provides immediate feedback to the user working problems.

This rapid response is very conducive to effective learning of the material.

3) The hands on sessions allow the participants to work case studies at their own pace. A slow learner may repeat the problems several times without affecting the pace of everyone else, while a fast learner may advance to more complex problems relatively quickly.

4) The low lecturer/lab assistant ratio permits a participant to get immediate help on both the subject and the computer application. This is very important in assuring effective learning of the material. This ability to closely interact personally with the participants also eliminates the problems large audiences invariably have with asking questions. The participants generally feel much less inhibited in seeking help in this setting.

5) Computer based hands on instruction makes the participant take positive action to learn. Working problems immediately after lectures establishes immediate reinforcement of the material learned.

6) The participants can take the courseware with them after the workshop. They may repeat them as necessary to reinforce the learning, may use them as reference, and in some cases may apply them to their practice immediately.

7) Hands on courseware is relatively easy to modify and to update as frequently as the need arises. Updates can be mailed to participants long after they attended the workshop.

CONCLUSION

The use of computer based instruction for continuing education can be an effective means of eliminating problems associated with its traditional delivery. Through its use participants can be provided with a meaningful educational experience to enhance their professional knowledge and to enable them to upgrade their practice to the state-of-the-art.

Only Microcomputers for Undergraduate Education ?

M. U. Hosain[1], M.ASCE

Abstract

Until recently, civil engineering undergraduate courses requiring computer usage were oriented toward a mainframe or a minicomputer system. Many engineering institutions in North America have now switched to microcomputers for undergraduate teaching. Even in universities where mainframe and minicomputer facilities are available, there has been a trend to place more emphasis on microcomputers at the exclusion of minicomputer and mainframe experience. Should microcomputers be used exclusively for undergraduate engineering education ? The purpose of this paper is to generate a discusion and find an answer to this very important question.

The inherent graphics capabilities of microcomputers make their utilization particularly suitable for some civil engineering courses and the usefulness of electronic spreadsheet is remarkable for others. The unrestricted availability of microcomputers for wordprocessing, drawing, drafting and computing works tends to make the students competent in computer-related skills. The author, however, does not recommend a total dedication to microcomputers at the exclusion of minicomputer and mainframe experience.

Introduction

Today's engineering educator is faced with the challenge to preserve the traditional emphasis on the teaching of engineering fundamentals, while adequately preparing the students for a career in a

[1]Professor, Civil Engineering Dept., University of Saskatchewan, Saskatoon, Canada, S7N 0W0.

modern workplace. The graduates must be able to step into a wide range of computer oriented careers but, above all, they must have the basic attributes to be able to enter into new areas in the near future. There is no doubt, therefore, that teaching of computer related subjects must be preceeded by a thorough grounding of engineering basics. This paper will concentrate on the computer training aspects of engineering education. Until recently, civil engineering undergraduate courses requiring computer usuage were oriented towards a mainframe or a minicomputer system. Now, many engineering institutions in North America have switched to microcomputers for undergraduate teaching. In some universities, a shift toward microcomputers has been dictated by charge policies for the use of large computers and in others microcomputers are been utilized for undergraduate teaching so that existing mainframe and /or minicomputer facilities can be devoted exclusively to graduate studies and research. Even in institutions where such restrictive measures have not been necessary, there has been a trend to place more emphasis on microcomputers at the exclusion of minicomputer and mainframe experience.

Due to this new development, some engineering educators are faced with a dilemma in preparing course material. They are wondering if well planned courses based on the use of mini and mainframe computers be dismantled to accomodate the use of microcomputers. Should these courses be modified just a little and microcomputers be used exclusively in some other courses ? If so, in which ones ? The problem is compounded by the fact that some senior members of the faculty have adhered to a traditional approach of teaching and have largely ignored computer usage on mainframes, minis, or micros. The responsibility of ensuring student familiarity with computer aided analysis and design is left on the shoulders of a few professors. The computer technology is changing so fast, it is difficult to stay abreast with the latest developments of only microcomputers let alone minis and mainframes. Is specialization the answer ? The objectives of this paper are to share similar experiences, generate a discussion on these issues, and find some answers.

Computers in Undergraduate Teaching

Microcomputers are excellent tools for teaching some engineering courses because of their inherent graphics capabilities (O'Leary et al. 1989). Proceedings of any recent computer related conference would more than likely include several articles on this topic.

The electronic spreadsheet is particularly suitable for solving routine problems in practically all areas of civil engineering(Grauer and Sugrue 1989). The Apple Macintosh with its mouse technology and icons has turned computational work into an enjoyable task. The availability of excellent business and engineering software packages puts micro-computers head and shoulders above mini and mainframe computers for certain tasks. As the cost of software development can be passed on to a large number of users, the price for most of the programs is quite reasonable.

If microcomputers are so good why not use microcomputers exclusively for undergraduate teaching ? If an educational institution does not have mini or mainframe computer facilities, this would be an acceptable but not necessarily the most desirable option. The author does not share the opinion that the use of microcomputers will become the only mode of computing in the very near future. Based on his own experience, he feels that a strong case can be made against the exclusive use of any one type of computers.

At the University of Saskatchewan in Canada, as in many other universities in North America, an enormous number of computer programs have been developed over the years by civil engineering graduate students and faculty. Some of these program were written specifically for use by undergraduate students. Fully debugged and tested these programs have been very useful in teaching undergraduate courses. Originally operational on IBM mainframes and DEC minicomputers, these programs now run on the campus VAXcluster consisting of two VAX 6330s and two VAX 6320s. The author can not see any reason why these programs should not be used for instructional purpose.

In the early 1980s, when stand-alone microcomputers were first acquired for use by undergraduate students, licensing problems associated with volume purchase of microcomputer software favored the use of large computers in an education institution. It was more economical to lease a single package for the mainframe computer than to purchase a large number of less powerful microcomputer version of the same package(Univ. of Saskatchewan 1984). It was also generally agreed that it would be cheaper to maintain a small number of large systems rather than a large number of small systems.

Unlike business type programs, engineering analysis programs for microcomputer initially were relatively expensive. The inexpensive ones were released without an acceptable level of validation. In the undergraduate microcomputer laboratory, software piracy surfaced as a major problem requiring constant vigilance on the part of the instructors. Moreover, a listing of the actual source program was not made available for proprietary software and thus it was considered somewhat unsuitable for undergraduate teaching where an understanding of how the program derives its answer to a problem is more important than the end results. Therefore, it was necessary to develop in-house programs for microcomputers. Usually these were downloaded versions of in-house programs already in operation on Vax11/780 minicomputers. Since the original programs included graphics routines and operated in interactive mode, the microcomputer versions did not have any special advantage. Moreover, it was possible to run VAX programs from terminals placed inside the classrooms during problem sessions, thus nullifying the special advantage provided by a portable stand-alone microcomputer.

At present, civil engineering students also have access to a microcomputer laboratory, which includes 25 PC-compatible machines connected by a local area network and served by an AT&T 3B2/400 superminicomputer fileserver. A host of popular software packages is available for use by undergraduate students. The local area network provides adequate security and protection against copyright violations. Downloaded versions of in-house VAX programs are also available in this system. For a small size problem, there is no difference as to which version is used. However, for a large size finite element analysis problem, the microcomputer version takes too long to execute and the advantages of a minicomputer over a microcomputer become apparent.

Many large programs are only operational on the VAXcluster. For example, students are required to utilize computer aided design and analysis packages, both in-house as well as commercial packages such as ANSYS, RESST, etc. In some courses, students are required to use solid modelling and design drafting packages on a Tektronix workstation located in the Engineering Building but run on a CYBER 180/830 mainframe housed in a Provincial Government Training Centre downtown. The Engineering Computer Centre also features a DECMicroVAX3600, a number of SUN 4/10 and SUN 3/50 workstations, and two DECstation 2100 workstations. These workstations are being used for demonstration purposes but can also be used by undergraduate students for their undergraduate thesis projects.

This wide variety of machines allows the students to experience.

the "feel" of various computer systems, as well as the associated software, and gives them some idea of the ability at each level of the computer hierarchy. This will give them a better chance of becoming quickly familiar with new machines when confronted with them in their careers. Since the main objective of an engineering education is to instill in our students the basic scientific principles, it really does not matter what type of computer is used to accomplish this task. However, through the proper selection of a computer system or systems it may also be possible to give the students some marketable computer skills.

Computer Usage in the Modern Workplace

Present Situation: Upon graduation a student may find employment in a small consulting engineering firm where all computing needs are probably met by a number of stand-alone microcomputers and a host of commercially available program packages. A number of structural engineering firms have found it necessary to have occasional access to a more powerful computer for some of their complex analysis work involving finite element method.

Many corporations and engineering firms have opted for the so-called "distributed data processing system" which consists of a number of microcomputers networked together with a central mainframe or minicomputer. Normally, jobs requiring the use of a large CPU and access to corporate databases are transferred to the central computer. After the program is executed, the files are transferred to the microcomputer and stored on the microcomputer's diskette and printed on a local printer. The newly introduced IBM AS/400 System is also becoming very popular.

Many large engineering firms have acquired in-house supermini computers such as VAX11/780s, IBM 4361s, and powerful design-drafting packages such as Intergraph. These companies have recognized the convenience of a common shared database for faster design, analysis and drafting processes. Other engineering firms have opted for powerful workstations such as SUN 4/10s and DECstation 2100s.

Mainframe computers, such as IBM 4381, IBM 3090, etc. are in widespread use in governmental and municipal agencies, universities, research institutions and large corporations. Famous for its minicomputers, the Digital Equipment Corporation will soon release a mainframe computer named VAX 9000. Use of mainframes by consulting engineering firms has been mainly through service bureaus. A survey conducted by the Civil Engineering magazine(ASCE 1985) revealed that 10% of the consulting engineering firms in USA used mainframes in 1985. The corresponding figures for minicomputers and microcomputers were 33% and 84% respectively. Only 25% reported CADD usage.

American military has been the principal user of the so-called "supercomputers". These giants are also being used in many other fields including seismology, mining, gas & oil explorations, weather forecasting, computer animation and filmmaking. The Boeing Computer Services has recently installed a CRAY X-MP/24 in their Seattle Data Centre. Many American universities are also making use of these powerful machines. And closer to home, the University of Calgary has acquired a CDC Cyber 205, and a CRAY X-MP/24 has recently been installed at the University of Toronto, Toronto.

It appears that there is a wide spectrum of computer usage in the real world and that no one type of computer has a virtual monopoly. A well-rounded training involving student exposure to different types of computers appears to be the logical choice at least for the time being. This familiarization will also provide the background that the students require in all computer aided schemes, so that they will not have to endure as extensive a re-learning experience when thay encounter a new system.

Future Trends: The recent breakthroughs incorporated in the NeXT Computer System will likely change the way computers are used in the 1990s. Because of these new developments, there is a tendency to assume that microcomputers will soon put an end to the mainframe industry. This is not likely to occur since the technology which created the microcomputer industry will also facilitate the production of cheaper mainframes exceeding the capacities of today's supercomputers(Torrero 1985).

In order to gain business advantage, more and more engineering firms will seek sophisticated CAD/CAM systems. General purpose superminis or mainframes will possibly continue to host such systems. Applications in pattern recognition and knowledge-based systems will require vast amount of computing power. Widespread use of massive databases by large corporations, governmental and municipal agencies, research needs in earthquake engineering and artificial intelligence techniques, the American Star Wars program, etc. will continue to support the development of supercomputers with unprecedented capacities. This development will be supplemented by newer technologies such as vectorization and parallel processing(Noor 1987). Together with the development of supercomputers, artificial intellgence techniques will be used extensively in automating many process planning tasks. American manufacturers will face steep competitions from Japanese firms such as Hitachi and Fujitsu. And, there are rumours that the Japanese Fifth Generation Computer Project, announced in 1982, will be completed by the early 1990s.

Conclusions

1. The rapid development of computer technology will place new demands on the future generation of engineering students and educators. Extensive training in computational sciences will be necessary for students in all branches of engineering.

2. It appears that there is a wide spectrum of computer usage in the real world and that no one type of computer has a virtual monopoly. A well-rounded training involving student exposure to different types of computers appears to be the logical choice, for the time being.

3. At a time when Japanese Engineers are threatening to develop the so-called "Fifth Generation Computers", an undergarduate training solely based on microcomputers does not appear to be the right choice.

4. A strong case can be made against the exclusive use of any one type of computers for undergraduate education.

References

ASCE 1985. Engineers' Wish List: More CADD, Civil Engineering, Vol. 55, No.10, pp.22.

Grauer, R.T., and Sugrue, P.K. 1989. Microcomputer Applications. McGraw-Hill Book Company, New York.

Noor, A.K.(Editor) 1987. Parallel Computations and Their Impact on Mechanics. ASME, New York.

O'Leary, T.J., Williams, B.K., and O'Leary, L.I. 1989. Microcomputing. Annual Edition, McGraw-Hill Book Company, New York.

Torrero, E.A.(Editor) 1985. Next-Generation Computers. Spectrum Series. The Institute of Electrical and Electronics Engineers, Inc., New York.

University of Saskatchewan 1984. A plan for Excellence. University of Saskatchewan's Plan for Future Academic Computer Facilities.

MOVING FROM PERSONAL COMPUTERS TO WORKSTATIONS IN UNDERGRADUATE CIVIL ENGINEERING EDUCATION

by

James O'Shaughnessy[1], Guillermo Salazar[2], Frederic L.Hart[3], and P.Jayachandran[4]

INTRODUCTION

Civil Engineering problems comprise a wide variety of planning, analysis, design, construction and maintenance activities. Due to the complex nature of these engineering components, computers have become valuable tools in the support of today's Civil Engineering profession. As part of the undergraduate educational process, colleges and universities .attempt to develop highly qualified graduates who are proficient in the use of computers in the engineering workplace. To insure the development of such skills, the WPI Civil Engineering Department has required students to use computer applications throughout the educational program. During the past decade educational computer use has increased drastically in both the mainframe and microcomputer environment.

Currently the Department has developed more than 100 software programs which are available for use in its coursework and support student projects. The Department has also identified the need for an integrated computer network including workstations as a method of implementing of its educational goals. The National Science Foundation has recently awarded a grant to help in establishing this workstation network. The workstations are Unix based and graphics oriented to introduce students to state-of-the-art computer aided analysis, design, simulation and mapping and construction.

OBJECTIVES

The specific objectives of this project are:

1. To establish an Integrated Computer Insructional Classroom and Laboratory using a UNIX based workstation network.

---

1. M. ASCE., Prof and Head, Dept. of Civil Eng.,Worcester Polytechnic Inst.
2. M. ASCE., Asst. Prof , Dept. of Civil Eng.,Worcester Polytechnic Inst.
3. M. ASCE., Asoc. Prof , Dept. of Civil Eng.,Worcester Polytechnic Inst.
4. M. ASCE., Asoc. Prof , Dept. of Civil Eng.,Worcester Polytechnic Inst.
Worcester, MA.

2. To develop and transport existing software for analysis and design from mainframe and PC's to workstations.
3. To develop graphic utilities software with pre and post-processing capabilities in structural analysis and design in order to help specify input-output information, display results and scan output for maxima and minima for design and optimum designs to be selected in several phases.
4. To simulate the engineering processes, construction processes and sequences used to arrive at optimum strategies for studying parametric variations of systems behavior. These results can then be used to optimize material, labor, and cost; also evaluate feasibility studies for various alternatives in systems for these processes. In addition, surveying and mapping software will be developed to teach design and drawing skills in topographical and environmental planning.

MODERNIZED CURRICULUM

One of the major objectives of the Civil Engineering Department at WPI is to develop highly qualified graduates who are proficient in the use of computers. The CE Department, therefore, has traditionally used a computer aided instructional approach in their courses and has developed over 100 software programs for use in courses and projects. In addition, outside software developments in areas such as CAD, expert systems, artificial intelligence and graphical displays have dramatically altered standard engineering practices.

The establishment of a comprehensive Computer Instructional Laboratory using a UNIX based workstation network is a most effective method of providing this type of modernized curriculum to a large number of students. With this type of computer system, files (both specific and general application) may be accessed by users throughout the campus. Because access privileges may be assigned to individual users, both executable files and data-base files may be assigned to selected project or classroom groups. The end result is a highly efficient exposure of high technology to undergraduate students.

HANDS ON LABORATORY EXPERIENCE

There is no substitute in the educational process for actual hands-on experience. A computer network is capable of providing that type of experience simultaneously to a large number of users. As networks have evolved the UNIX operating system is becoming the standard in most academic settings. In addition its acceptance is growing throughout the engineering, scientific and business communities. The use of computers as the major tool in engineering analysis and design is a definite reality for the 1990's. Problem solutions and tutorials are growing in use throughout most curriculum and the computer allows each student to learn new applications at individual rates.
In addition to classroom applications the aid of computers in laboratory data collection, data analysis, and report preparation is becoming the standard method for laboratory instruction. The computer is accepted as the integrated workplace in engineering applications.

The Engineering discipline, and particularly the Civil Engineering Discipline, is actually an integration of many technologies and skills. Modern engineers, and therefore engineering students who will become tomorrow's engineers, should be able to define the over-all problem and to call upon a variety of skills to help solve the problem. The use of sophisticated application software (such as an auto CAD/data-base package) helps the engineer to breakaway from the traditional time consuming detailed computational tasks that is often associated with an engineering problem. The total engineering system (various technical disciplines and social issues) becomes less transparent and therefore a significant part of the evaluation process. The workstation environment provides the required computatinal power and graphics support to speed up the educational process.

SELECTION OF SOFTWARE AND HARDWARE

Over the last ten years the Civil Engineering Department has developed a larger number of computer programs on the mainframe computer. A concerted effort was initiated in the last four years to port a significant part of this software to the personal computers so students could use them in the classroom and in solving their homework assignments. This development effort has taken the form of undergraduate Major Qualifying Projects, Master Thesis, and use of department budget allocations for purchase of third party software. Although PC's were very instrumental in achieving these objectives, it was clear that their limitations in terms of memory and processing speed prevented full utilization of the program capabilities available in mainframe computers.

Introduction of current workstation applications requires a long leaning curve for each new application. Also each software program has a cost which is usually 10 times the cost of PC compatiability software. In addition many software applications are not available for use on the more powerful machines. Table 1 reviews the status of a number programs the Department will eventually use throughout the curricula. These programs include both WPI and third party software.

TABLE 1: SUMMARY OF HARDWARE AND SOFTWARE EVALUATION

SOFTWARE	386 & 68020	RISC	68030	286 MS-DOS
SOCRATES	No	No	No	No
CLNET (WPI)	Yes	No	No	Yes
ANSWERS (WPI)	Yes	No	No	Yes
SIMSUPER (WPI)	Yes	No	No	Yes
CAD	Yes	Yes	Yes	Yes
MATH & STATISTICS	Yes	Yes	Yes	Yes
SAP4 (WPI)	No	No	No	Yes
Drain 2D	No	No	No	Yes

After review of the status of applicable software for education application at the undergraduate level It became clear that software would not be the major factor in equipment purchase. Following this analysis it was determined that the equipment that was to be acquired must include the following features: graphics, color display, multitasking, UNIX operating system, and X-windows.

Most engineering problems require some kind of graphic input and output of the problem. Therefore, an essential feature of computers is the capability to generate graphic display of the results of a computational analysis. Mainframe and PC's are somewhat restricted in this respect. With the availability of powerful speed and memory of the workstations, graphical manipulations have become fairly easy to achieve. Additional features such as UNIX operating system which permits multitasking, and X-Windows which allows graphical user interfaces (GUI) and efficient network communications permit both numerical computations and graphics to be processed and displayed. Also, academic institutions have limited resources for software acquisition. There is some software that is available in the public domain but it is often limited in its user interface capabilities and hence cumbersome. In order to provide a clear visualization of results of design, simulation, and analysis, color display is extremely useful, especially in three dimensional modeling such as drawing and mapping.

The above summarizes the needed features for the computational capabilities in the development of the Civil Engineering Computing Laboratory (CECL). Together with these features there is a need to incorporate smaller workstations into the network which can share data and files and still have color display capabilities. This feature permits an efficient allocation of resources so smaller processes such as spreadsheet analysis, word processing, etc. can be done locally; larger processes can be done simultaneously in the workstations.

The current workstation technology offers the following processors: Motorola M68030, Intel 80386, MIPS 2000 (RISC processor). The first two offer speeds in the range of 4 to 5 Mips whereas the RISC processor offers speeds in the range of 12 to 15 Mips. Workstations designed around these processors are: Apollo Workstations, Sun 3/60, Sun 386i, Sun Sparcstation, DEC systems 2100 and 3100. All of these machines use different variations of the UNIX operating system, AT&T or BSD versions. The price of these machines are in the range of $15,000 to $30,000.

To combine the powerful capabilities of the above workstations with an efficient local processor it was found that the MacIntosh IIcx together with Mac-X and AUX operating system was a new option. The final configuration for CECL was DEC-3100 and MacIntosh IIcx with Ethernet as a network. This network will also connect the WPI mainframe computer, an ENCORE Multimax UNIX-based parallel computer, as well as the PC's (See Figure 1). This selection was based on a combination of price and performance ratio and the convenience of campus wide service. This choice also allowed some freedom to buy additional software for word processing, spreadsheet, and graphics. Figure 2 shows the intergration of all of the computers used in the Department of Civil Engineering.

The DEC system 3100 uses Ultrix and DEC-Windows which is the enhanced implementation of X-Windows. The MacIntosh IIcx uses Mac-OS and Mac-X to communicate with the DEC system. It also uses AUX for local UNIX operations. Digital Equipment Corporation will provide support to CECL to port software developed at WPI in the mainframe. This effort will include porting programs such as ANSWERS, CLNET, SIMSUPER5 and others, with an emphasis on developing

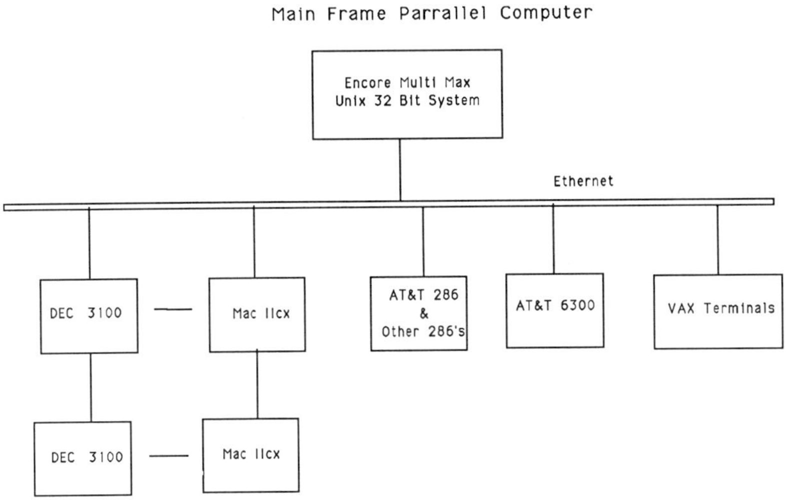

Figure 1 Civil Engineering
Work Station Computer Utilization

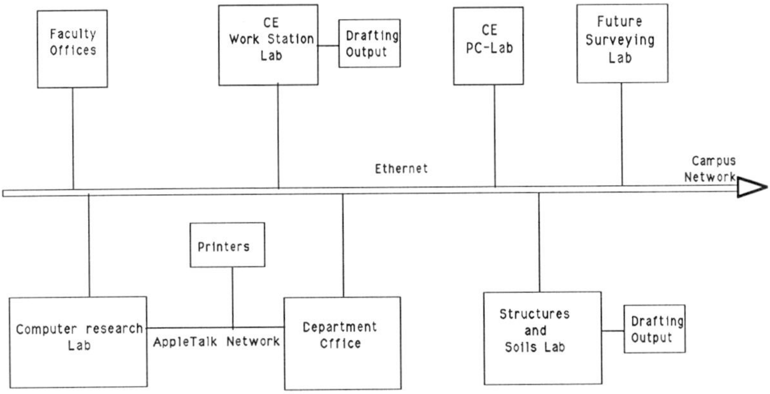

Figure 2 Civil Engineering
Work Station Computer Location

input/output graphic interfaces.

IMPACT ON CURRICULUM

The Department's approach to achieve computer literacy has been to develop a series of computer applications within many courses rather than one specific C.E. computer course. The vast majority of our courses (24 out of 32) have computer applications within the course curriculum. Our philosophy has been to allow the student to matriculate through both breadth and depth areas and use computer applications to enhance engineering solutions.

Irrespective of the sub-area of Civil Engineering students choose, they achieve a required degree of computer literacy by the time of graduation. With the advent of the workstation in the curriculum even more difficult engineering analysis and design problems can be introduced without detracting from engineering solution.

Initially these advance problems will be introduced in the senior level courses in the areas of environmental engineering, structural / geotechnical engineering and construction project management. In addition, in the basic structural analysis sequence, CE2000, 2001 and 2002 programs such as MacBeam, Socrates and other analytical packages will be integrated into the problem solving environment. Concurrent with third party software acquisition will be the development of in-house programs and graphic output. In order to maximize computer use at all levels all of the MS DOS based PC compatible computers will be fitted with software that emulates a mouse driven menu prompted window environment. We are implementing the graphical user interface (GUI) on all PC'S and workstations in the Department. The use of files between the Macintosh environment and the DEC environment will initially be via Mac-X. In the near future we will evaluate the possibility of using a AUX as the Macintosh DEC connectivity mode.

With the development of new UNIX based software as well as an expanded investment in third party applications the Civil Engineering Department expects to have approximately 300 computer applications available for student use. These applications will support both courses and project efforts. Finally, as more and more texts are coming into the market which include particular computer applications, we expect that the typical Civil Engineering curriculum of the 1990's will have over 1,000 specific programs available to support classroom activities.

COMPUTERS IN STRUCTURAL ANALYSIS

R.L. Sack, Fellow, ASCE

Abstract

The invention of the transistor in 1947 and the silicon chip in 1959 were pivotal discoveries that accelerated the development of the digital computer. By embracing this new computing technology, the engineers of the 1940's and 50's completely changed structural analysis. The microcomputer of today compels educators of structural engineering to make pedagogical choices. One strategy for incorporating computing in courses is to require students to write source-code programs; another tactic is to use existing programs. The divisions between hand-calculation methods and computer methods are becoming fuzzy. Probably, the course of the early 21st century will simply be titled "structural analysis" because computer methods will pervade all topics. However, today we must grapple with the issues of how we effectively incorporate computing into structural courses.

Introduction

How do civil engineering educators prepare students to be successful structural engineers? The revolution in structural analysis was started four decades ago by the demands of the airframe industry, and the digital computer stimulated the movement. The mini- and microcomputers of today are forcing us to incorporate computing into structural analysis courses, but there is no consensus among educators how this should be done. During the past five years, 17 new textbooks or editions on basic structural analysis have been published, with the primary

---

[1]Dir., School of Civ. Engr., Univ. of Oklahoma, Norman, OK 73019

emphasis on classical structural analysis. All of these texts treat matrix structural analysis, but notation, emphasis, and philosophy vary widely. Other books have emerged that contain little theoretical background and are filled with source listings and diskettes ready for execution. The alarming fact for students and practicing engineers is that educators are sharply divided over the best way to integrate computing into structural analysis courses.

Many engineering colleges incorporate computing into the general requirements through the freshman course on introduction to engineering. A remarkable success story is the curriculum at Arizona State University, wherein a two-course sequence, with 3 hours per semester, introduces students to several computer languages, spreadsheets, equation solvers, databases, and computer graphics (Beakley, et al, 1986).

This article provides a historical perspective of both computing and structural analysis and describes choices that must be made by civil engineering educators. Personal anecdotal experience provides insight into options for incorporating computing into the structural analysis curriculum. We also note developments in artificial intelligence that hold promise to enhance structural analysis courses.

The Evolution of Computing

The first electronic digital computer, ENIAC (Electronic Numerical Integrator and Computer), was completed in 1946 at the University of Pennsylvania and marked the beginning of the first generation of computers (1946-1959). The machines of this era employed vacuum tubes, with magnetic tape or magnetic drum peripheral storage. Main memory cycle times for computers of the early first generation (1946-1953) were 40-40,000 μs, while those of late first generation machines (1953-1959) decreased to 10-20 μs. The IBM 702 is a product of the early first generation, while the IBM 650 and UNIVAC 80 and 90 exemplify the late first generation.

The transistor was discovered in 1947 and served as the component technology for the second generation (1959-1964). Main memory cycle times during this time were 2-10 μs, and laser and magnetic bubbles were added as peripheral storage devices. The IBM 7040, CDC 1604, and UNIVAC 1107 are representative of large scientific computers of the second generation.

Integrated circuits and large-scale integration (LSI) were

used in the third-generation computers (1964-1975). Main memory cycle times were 0.5-2 μs in the early third generation (1964-1969), and reduced to 0.020-1 μs during the late third generation (1969-1975). The IBM 360 series, UNIVAC 1108, and CDC 6600 appeared during the early third generation. The IBM 370 (System 3) and the CDC 7600, were produced during the late third generation; minicomputers were also introduced.

Many innovations have appeared since 1975. Mainframe computers of note are the IBM 3000 and 4300 series, and Control Data introduced the Cyber 170 and 200 series. The Amdahl Corporation emerged with their powerful V/7 and V/8 machines, and Cray Research offered their supercomputers. The minicomputer business grew in the 1970's with Digital Equipment Corporation dominating the market.

In the late 1970's the microcomputer explosion was made possible by LSI technology. IBM's early PC and XT machines had cycle frequencies of 4.77 MHz; these were improved to 6-8 MHz in the IBM AT. IBM PS/2 models operate between 10 and 25 MHZ, and the 386 chip has clock speeds of 16, 25, and 33 MHz. Integer addition on the IBM PC requires approximately 3 μs, but the IBM AT needs only 1 μs. Floating point arithmetic with an 8087 coprocessor is executed in 30 μs for multiplication, 45 μs for division, and 20 μs for addition. A modest configuration of an IBM PC has 640 kbytes of RAM and a hard disk of 20 Mbytes.

Classical Structural Analysis

Most undergraduate civil engineering curricula contain a one-semester introduction to structural analysis. Typical coverage includes: influence lines; calculation of deflections using the governing equations (e.g., the moment-area and the conjugate-beam methods) plus energy methods (e.g., complementary virtual work and Castigliano's Theorem, Part II); and analysis of statically indeterminate structures using classical compatibility methods (e.g., consistent displacements, least work, and the three-moment equation) and also classical equilibrium methods (e.g., slope-deflection method and moment distribution).

Students solving structural analysis problems on the calculator are commonly overwhelmed by the arithmetic and have difficulty distinguishing basic concepts. For example, after finishing a solution by moment distribution, if the student finds a mistake in the initial steps (e.g., the distribution factors or fixed-end moments), s/he must perform the calculations again (and usually again) before successful completion. Or the student using complementary virtual work typically becomes entangled in the task of integration to obtain the deflection of a beam.

The frustrations of using calculators to solve classical structural analysis exercises can be removed by using equation solvers, spreadsheets, and computer graphics on microcomputers. Many textbooks have cast moment distribution as a multi-columnar array process. The student familiar with a spreadsheet will immediately solve moment distribution exercises on a PC. Furthermore, equation solvers relieve the student of being sidetracked to evaluate integrals, obtain solutions of differential equations, or plot results. Using PC's, the method of complementary virtual work does not degrade into a boring arithmetic task; the student gives an expression for the complementary strain energy density and obtains the displacement computed by the equation solver. Computer hardware and software allow students to engage in "what if" exercises by changing member stiffnesses, loads, span lengths, and material properties.

Matrix Structural Analysis--Historical Context

The airplane and digital computer are responsible for revolutionizing structural analysis. In the 1940's and 1950's structural engineers were confronted with two highly statically indeterminate systems: the swept-wing and delta-wing aircraft. The governing equations were cast ab initio in matrix format, but this approach required solution of large sets of simultaneous linear algebraic equations. Fortuitously, the emergence and developments of the digital computer in the late 1940's and 1950's gave impetus to the structural analysis revolution. Trusses, beams, and frames were initially investigated, but in the mid-1950's a group at The Boeing Company demonstrated the procedure could be extended to continua. Common usage now dictates that matrix structural analysis (MSA) designates investigations of structures composed of articulated or discrete components, while the expression, finite element method, denotes analysis of conintua.

Matrix Structural Analysis in the Curriculum

Most civil engineering curricula contain a one-semester offering in MSA that the student takes after completing classical structural analysis. The course in MSA typically employs equilibrium methods to study structures composed of discrete components; the equations are cast in matrix form. The students are usually seniors or beginning graduate students and are nominally acquainted with computing. The course materials and audience make a perfect blend for a computer intensive orientation.

There are three approaches for integrating computing into MSA.

a) Students write their own software in one of the standard computer languages (e.g., FORTRAN, Basic, or C).

b) Students use programs such as spreadsheets, equation solvers, or matrix intrepretive systems.

c) Students use automated structural analysis programs.

In the first approach, students must write program modules that: compute element stiffness matrices; assemble the global stiffness matrix; solve simultaneous linear algebraic equations; obtain element forces. Formulating pre- and post-processors may be the obligation of instructor or students. Writing a program is a good gauge of ones understanding of the process being modeled, but this approach seems inefficient for a one-semester introductory course. The enormous time requirements of program writing could be invested in learning the concepts of matrix structural analysis.

Computer Assisted Learning of Structural Analysis (CAL86) (Wilson, 1986) is an excellent program for implementing the second approach. This software requires the user to load the basic information and write the appropriate code in the CAL86 metalanguage to attain the desired outcome. The program operates in an interactive mode or it can read information from a batch data file. Using a restricted set of commands, CAL86 will operate as a matrix interpretive system only. Expanded logic allows the user to activate limited commands of the stiffness method, and dynamic response analysis is also possible.

The author has used versions of CAL since 1970 to teach MSA in industry and at several universities; the reception by students is universal acceptance. The first assignment typically consists of performing standard matrix manipulations using only the CAL86 matrix commands. Subsequently, the students are required to solve several trusses by MSA using only the matrix commands. After introducing the direct stiffness commands and assigning several exercises, we urge the students to use a mix of commands of their choosing. CAL86 allows the students to investigate structures sufficiently large and complex to reveal the power of MSA; a space truss with 75 degrees of freedom presents no trauma to the students.

The third approach for incorporating computing into MSA is pedagogically unsound. The user of automated structural analysis programs simply furnishes nodal coordinates, element connectivity, element properties, applied loads, choose the type of analysis (e.g., static, dynamic, linear, or nonlinear) and selects output options. Asking a student to believe the output from a blackbox program is tantamount to destroying the spirit of intellectual inquiry. Students should learn to suspect any computer output and accept it only after they are confident of the logic and details within the program.

Artificial Intelligence and Structural Analysis

Artificial intelligence (AI) embraces the subfields of robotics, automated vision, automated problem solving, knowledge-based expert systems (KBES), and natural language processing. KBES programs can provide information to structural engineers on many topics such as operation of finite element analysis systems and seismic damage analysis

KBES can also be useful in teaching structural analysis. Sack and Schroeder (1987) developed computer software that generates an unlimited number exercises, and their solutions, for two-dimensional beams, frames, and trusses. Teachers may wish to remove the solve option and distribute the problem generation portion to students. We also envision that an allied program could be written to analyze homework, check it for solution technique and completeness, and assign a grade.

The program PROUST (PROgram Understander for STudents) provides encouragement in the area of intelligent tutoring systems (ITS) (Johnson and Soloway, 1985). PROUST is a knowledge-based system that finds nonsyntactic bugs in Pascal programs, determines how the bug can be corrected, and suggests why the bug arose. Program oriented courseware has also been developed, and tested for structural engineering courses (Lundgren, 1986). The student requests a menu of programs, selects and solves the problem, enters selected computed values, and requests grading.

Discussion

Matrix structural analysis has come to fruition since the 1940's, but its roots are in classical structural mechanics. This revolution in structural analysis was stimulated by the demands of the airframe industry and reinforced by the emergence of the digital computer in 1946. Subsequent improvements in computer technology, occurring in three distinct generations, provided additional stimulus. Interactive computing is now available to students at a reasonable cost because of the mini- and microcomputer that have emerged since the late 1970's. Over the past 25 years, computational speed has increased by a factor of 200, while the cost, energy consumption, and size of computers of comparable power have decreased by a factor of 10,000.

The advances in structural analysis and computer technology are forcing teachers to make pedagogical choices. We have three distinct options for incorporating computers into structural analysis courses by: writing source code; using spreadsheets, equation solvers, and matrix interpretive systems; or using automated structural analysis programs. No concensus has been reached on the optimum approach; each strategy has its vociferous advocates and detractors. Perhaps in a few years we will not be required to distinguish computer-oriented structural analysis from classical topics because computing will pervade all topics in the field. In the present transitional period, we have the unique opportunity to mold the alliance betweem computing and structural engineering in the educational arena.

References

Beakley, G.C., Evans, D.L., and Bowers, D.H. (1986). "Freshman Engineering for the 21st Century Engineer," ASEE 1986 Annual Conf. Proc., Cincinnati, Ohio, pp. 1126-1136.

Johnson, W.L. and Soloway, E. (1985). "Proust," Byte, vol. 10, no. 4, McGraw-Hill, New York, pp. 179-184.

Lundgren, H. and Sims, J. (1986). "A Problem Oriented Courseware Interface," ASCE, Proceedings of the Ninth Conference on Electronic Computation, Ed. K.M. Will, Birmingham, Ala., pp. 566-569.

Sack, R.L. and Schroeder, R.D. (1987). "PANDORA'S BOX: An Automatic Generator of Structures," The Application of Artificial Intelligence Techniques to Civil and Structural Engineering, Ed. B.H.V. Topping, Civil-Comp Press, Edinburgh, pp. 145-152.

Wilson, E.L. (1986). CAL86 Computer Assisted Learning of Structural Analysis and the CAL/SAP Development System, Report No. UCB/SESM-86/05, Berkeley, CA.

STUDENTS - THE FUTURE OF CIVIL ENGINEERING

by: John P. Gnaedinger[1]
Harry F. Tobin[2]
Gregory Kazarian[3]
Joel Getzendanner[4]

Abstract

ASCE has been active for many years in attempting to attract bright high school students into the profession. This is usually attempted by providing speakers at Career Days for high school seniors, and/or interviewers to meet with honor students. Many others are doing likewise, focusing on the bright seniors.

The day has arrived, however, when this system is in need of change, for a number of reasons, particularly because of the high drop-out rate among minorities. Minority scholarships have been going begging, with fewer and fewer minorities filing, much less qualifying.

Civil engineers can blame it on high schools, who blame it on parents, or on funding. However, twenty-five years of increases in teachers salaries have had a negative correlation with student achievement in those years. Yet public clamor is for more federal dollars from our bankrupt government.

"Tis better to light a candle than to curse the darkness!"

The Chicago Careers for Youth program has been developed by the senior author, and is being implemented at this time in Lake County, Illinois since 1984, and in one form or another since 1976 in Chicago.

----

[1]Chairman, STS Consultants, Ltd., Chicago, F. ASCE; [2]Director of Vocational Training, Chicago Public Schools; [3]Attorney-at-Law, Pedersen & Haupt, Chicago; [4]Economics Group, The Joyce Foundation, Chicago. IL

It is proposed at this ASCE Conference on Education, with the hope that ASCE will not only participate in the Built Environment Cluster, but in improving it, and in working with other cluster groups to start programs in major cities in the U.S.A., and later in rural communities.

Introduction

Educational reform at the primary and secondary school levels is clearly one of this nation's highest priorities, and the challenges are substantial. We must prepare all of our students to be active participants in the economy - training a select few is not enough. The historic distinction between "vocational" and "academic" training is no longer meaningful, each must be fully integrated into the other. All this must be accomplished in the face of budgetary constraints. Every effort must be made to reallocate existing resources before new revenue sources are requested. Unfortunately, our public schools seem to be opposed to these requirements.

Yet, in the school system once dubbed "the worst in the nation", a remarkable effort is in its early stages. Corporate and civic leaders are joining with state and local educators to transform the way that the Chicago Public Schools prepare students for their working lives. While many changes in Chicago and elsewhere have required changes in laws and a great deal of political squabbling, Chicago Careers for Youth (CCFY) is operating on a small grant from a local foundation, the dedicated volunteer effort of hundreds of business people, and the complete cooperation of the school administration. It is a public/private partnership of the most ideal form, one in which the role models from the business world interact directly with all students in grades six through twelve.

Notable Features of Model

1) CCFY appears to be the only major effort to allow "vocational" instruction to mirror the structure of the economy; that is, organize it along sectorial lines. Most others organize around tasks (e.g., engine repair), occupational types (e.g., business management), or academic skills (e.g., mathematics).

2) CCFY is not a "magnet school" program. All high schools will develop at least one cluster focus. This turns the notion of "selectivity" on its head. Schools won't get to select only the best students. Students will select the schools that have programs

that match their skills and interests. There will still be selective programs, but they will play a subordinate role to preparing the entire student population.

3) CCFY intentionally blurs lines between vocational and academic education. Tracking is a put-down to vocational students, who are thereby deprived of the opportunities to achieve, and to develop a sense of self-worth. The economy is demanding a mix of skills for all workers -- top to bottom -- the educational system currently teaches the skills separately.

4) CCFY enhances the role between businesses and educators. Businesses will help make sure instruction is reasonably in line with economic realities, but professional educators will continue to teach and make sure that the broader goals of education are also protected.

The Principals

Chicago Careers for Youth is based on the following principles:

1) Career preparation must reach all of the students within the system.

2) The program should utilize volunteers from business, the professions and industry to the greatest possible degree, and be structured in such a way as to maximize the potential of the volunteers acting as role models for the students.

3) Career options are organized using "clusters" of industries defined by the human needs they satisfy, (e.g., the "Hospitality Cluster" includes restaurants, hotels, etc.).

   a) No single industry should dominate a cluster.
   b) All occupation levels are represented in the clusters, (e.g., both managers and maids).

4) All activity is organized for the students' benefit, so that each might achieve his/her highest potential. The fact that the businesses benefit from the improved preparation of the students is intentional, but should never be the driving force. Each student, hopefully, will see himself/herself as a potential entrepreneur.

The Model

All activity is organized around 16 clusters of industry that were developed for the Illinois Department of Education:

Agriculture	Health Care
Arts, Culture & Religion	Hospitality
Built Environment	Insurance
Communications	Manufacturing
Education	Natural Resources
Energy	Personal/Bus. Services
Finance	Retail/Wholesale
Government	Transportation

For each cluster, an advisory panel is formed that is made up of representatives from the relevant industries. This panel recruits volunteers to visit the schools, work on curriculum issues, and develop policy for school/industry interaction.

The school-based activity is thought of as having three phases:

1) Career Awareness -- In the sixth grade, students will be visited by "role models" from half of the clusters and will view videotapes describing the range of occupations that are available. Teachers will have instructional materials available for follow-up. In the seventh grade, students will be visited by volunteers from the remaining eight clusters. Chicago Careers for Youth will develop modules that will be approximate for use in the early grades. The eight programs should optimally be scheduled to best suit the classwork, such as Finance during the math class, etc. The awareness activities can be so presented to the students that they understand the relevance of academic courses.

2) Career Exploration -- At the end of seventh grade, each student will select the cluster of greatest interest, and immediately be entered into a computerized network for that cluster. This will permit information on high school, community college, trade school, and university education to be sent to the student's home; as well as information on interesting accomplishments and opportunities in the cluster, and on the multitude of job opportunities within the cluster.

   In eighth grade, half-day programs will be scheduled for those students selecting each of the sixteen clusters. At these programs, perhaps ten role models from the cluster will discuss their jobs, whatever the level (e.g., from ambulance drivers to surgeons). Exhibits will be on hand regarding human needs served by jobs in the cluster, emphasizing exciting and new activities and accomplishments. Creative and entrepreneurial aspects will be emphasized for each cluster, and job and site visits will be scheduled. Eighth graders will select a high school which will focus on their cluster of choice.

3) Career Development - The High School Phase -- Three to nine high schools are being identified to focus on each cluster, with more than one cluster being presented at most high schools. Role models will continue to interact with the students in the career related classes, perhaps monthly. Each student will have a one term course which will focus on the various industries in the cluster, entrepreneural opportunities, the relationship between quality and achievement, and lifestyle and success.

Networking with the role models and fellow students will lead to part-time and summer jobs, shadowing, mentorships, and opportunities for post-secondary education and jobs.

The Expected Results

The program will:

1) Bring new resources to the educational system through new, more practical relationships with the business community.

2) Make it easier to develop curriculum that stresses proficiency in communicating ideas and working in teams, an increasing requirement in the workplace.

3) Naturally teaches students about the structure and operation of the economic system.

4) Easily accommodates entrepreneural curiosity by introducing all students to entrepreneurship as the basic process of economic development and individual self-fulfillment.

5) Provide means for businesses to participate constructively in curricular discussions without allowing schools to become "corporate training centers".

6) Allow students more real choice in pursuing interests.

7) Provide the context and facilitate the retraining of teachers.

8) Provide students with continuous information regarding educations and employment within the clusters.

9) Provide contacts into the job network not present in many communities.

Future Opportunities

The model is not closed ended. It's only limitations are the ingenuity of the human mind, the commitment of its operators to the well-being of the students, and to the satisfying of human needs.

Conclusions

This paper deals with the entire economic system, represented by the sixteen clusters. Thus it is an introduction to win-win economics, not the traditional supply and demand model that has trouble relating to many aspects of our present economic system, such as win-lose activities in currency and commodity trading, gambling, program trading, options trading, leveraged buyouts, hostile takeovers, and implied stock price manipulation, and others.

With a primary ASCE focus on the Built Environment, and secondly, Transportation, interactions with students by Universities in these disciplines can easily be implemented, using the computer network system for each of the sixteen clusters, and city-wide cable TV already in Chicago schools. The program starts with the role models talking to 100% of the 6th and 7th graders. Video tapes are available, but can be improved.

Active ASCE participation and even leadership certainly fro the Built Environment and Transportation Clusters is hoped for. Information will be sent on request.

"Your expectation of youngsters largely determine how they behave. They will never let you down. Treat them with high expectations, treat them with respect, and they will perform. Show them you expect them to be thugs and they will never disappoint you."

" The only true service to God is Service to man."

"Give a man a fish and he will live for a day.
Teach a man to fish and he will live for a lifetime."

"The only truly happy people are those helping others."

"Freedom With Responsibility" is a better characterization of the Win-Win "American Way" than "Freedom" alone, because it respects the Freedom of others.

"Arise! Awake! And stop not until the goal is reached."
Vivekananda

"Goal people are winners! Process people are losers! Set your goals as high as you can, never lose sight of them"

"It is not who is right, but what is right that is important."

"Reform is the realization of the potential of every youngster in the Chicago Public Schools"
Joe Reed LQE

"The fault dear Brutus, lies not in our stars, but in ourselves, that we are underlings"

"Pity the man who knows the price of everything, and the value of nothing.

"Ad astra per aspera!" (To the stars through difficulties.)

"God help those who help themselves."
Vivekananda

"Luck may sometimes help, Work helps always"
Indian Wisdom

"There is no limit to the goals you can attain,
The success you can achieve —
Your possibilities are as endless as your dreams."
D. Crowe

"The smile that you send out returns to you"

"A pure and strong will is all—powerful."

"What we have done will not be lost to all eternity. Everything ripens at its time and becomes fruit at its hour."

"Get up and set your shoulder to the wheel
—how long is this life for?
As you come into this world, leave some mark behind.
Otherwise, where is the difference between you and
the trees or the stones? — they too come into existence,
decay and die."

"True education makes one humble.
Unlike wealth, knowledge grows with sharing and brings happiness and fulfilment in human life."
Sanskrit Sloka

"Everything but thy knowledge may get stolen."

"God respects me when I work; but he loves me when I sing.
Tagore

"Never compare between a king and a wise man.
A king is known to his subjects only, wheras the wise man is respected by the whole world."
Indian Wisdom

"If you plan for a year, sow paddy.
If you plan for a decade, plant a tree.
But if you plan for a lifetime, Educate yourself."

"We must face the truth: the quality of the early school years is critical to later academic and career choice. When those of us in higher education discuss the problem of underrepresentation of minorities in the quantitative disciplines, we must raise our sights to the elementary school levels and earlier years: we must look to the home; we must examine the role models we provide; the encouragement we give and the performance we expect."
Dr. Stanley Ikenberry Pres. U.of Ill.

SIXTEEN CLUSTER PARTNERS FROM WORLD OF WORK

1. Provide Role Models for sixteen cluster awareness programs at every elementary school, sixth and seventh grades.

2. Provide Career Days in eight grade for each cluster, for only students have selected each cluster — Career Exploration. Follow career days with plant, office and site visits. Provide information on high schools focusing on each cluster, on community college and trade schools, and on university sutdies and degrees.

3. At High School focusing on their cluster, continue to provide role models monthly, assist as requested in curriculum and staffing help provide part—time, summer and on graduation, full—time jobs (high school or college grads) assist where possible with scholarships mentoring and summer training opportunities for faculty of high schools. Maintain network communications thru high school with students at their homes, to maximize parent involvement.

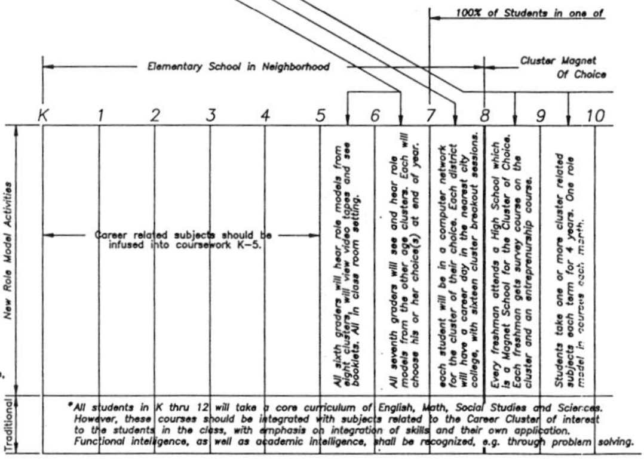

• Note that the proposed curriculum combines the preoccupations of those who urge "back to the basics", and those who favor tracking such as in the typical "vocational or college bound" tracks. It is relevant that the words "Career" and "Curriculum" have the same roots in Latin.

CHICAGO CAREERS FOR YOUTH

VISION OF THE FUTURE —

Chicago Board of Education Partner

1. Provide information to Cluster partner on Principals, Counsellors, Schools, Addresses and Phone Numbers, to permit scheduling of programs in 6th and 7th grades awareness programs.

2. Director of Vocational Education select Magnet High Schools for each Cluster, after discussions with Principals and key vocational teachers.

3. Appropriate CPS Director set up computer network, or modify existing student data base, to show Career Cluster selected by each and every seventh grade student, including career orientated courses taken each year, as well as academic courses. Information on career related high school and post—secondary educational opportunities to be sent to every eight grader, through high school at the student's home to involve the parents. Information on jobs, and exciting activities in each cluster, also to be sent.

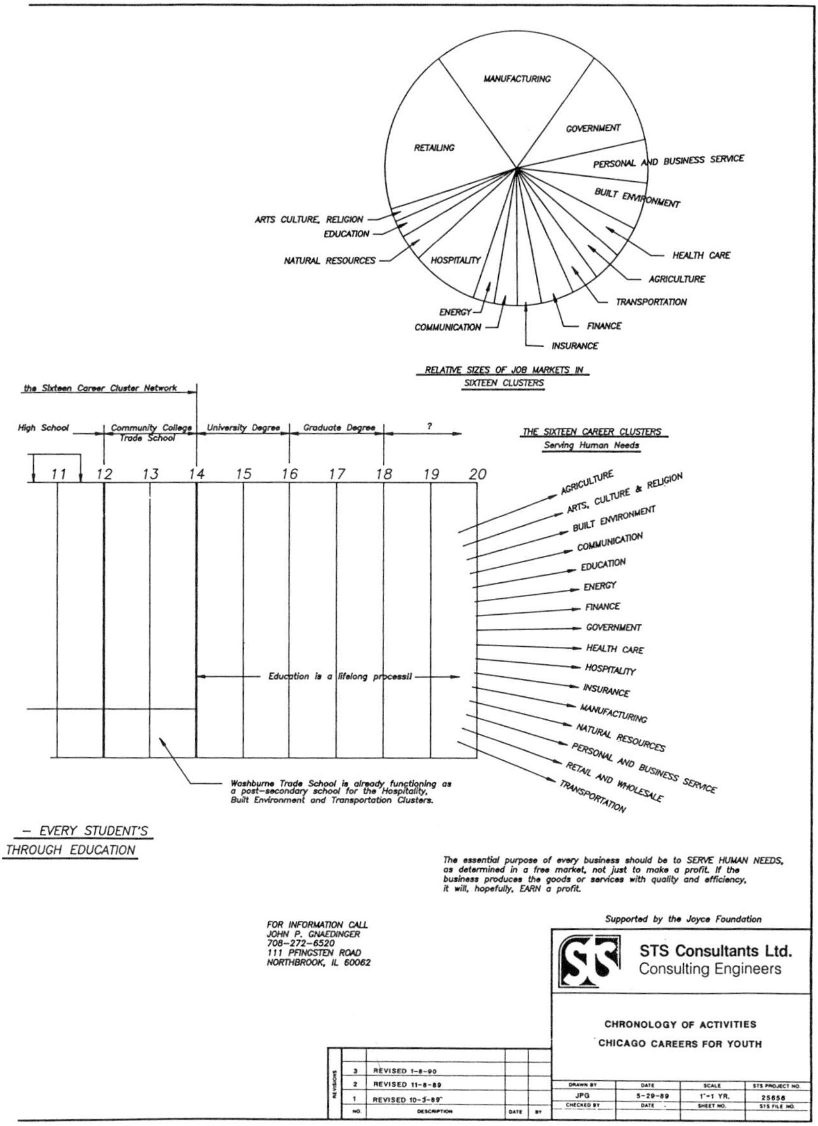
MANUFACTURING
GOVERNMENT
RETAILING
PERSONAL AND BUSINESS SERVICE
BUILT ENVIRONMENT
ARTS CULTURE, RELIGION
EDUCATION
NATURAL RESOURCES
HOSPITALITY
HEALTH CARE
AGRICULTURE
TRANSPORTATION
ENERGY
COMMUNICATION
FINANCE
INSURANCE
RELATIVE SIZES OF JOB MARKETS IN SIXTEEN CLUSTERS
the Sixteen Career Cluster Network
High School
Community College Trade School
University Degree
Graduate Degree
?
11 12 13 14 15 16 17 18 19 20
THE SIXTEEN CAREER CLUSTERS
Serving Human Needs
AGRICULTURE
ARTS, CULTURE & RELIGION
BUILT ENVIRONMENT
COMMUNICATION
EDUCATION
ENERGY
FINANCE
GOVERNMENT
HEALTH CARE
HOSPITALITY
INSURANCE
MANUFACTURING
NATURAL RESOURCES
PERSONAL AND BUSINESS SERVICE
RETAIL AND WHOLESALE
TRANSPORTATION
Education is a lifelong process!!
Washburne Trade School is already functioning as a post-secondary school for the Hospitality, Built Environment and Transportation Clusters.
– EVERY STUDENT'S THROUGH EDUCATION
The essential purpose of every business should be to SERVE HUMAN NEEDS, as determined in a free market, not just to make a profit. If the business produces the goods or services with quality and efficiency, it will, hopefully, EARN a profit.
Supported by the Joyce Foundation
FOR INFORMATION CALL
JOHN P. GNAEDINGER
708-272-6520
111 PFINGSTEN ROAD
NORTHBROOK, IL 60062
STS Consultants Ltd.
Consulting Engineers
CHRONOLOGY OF ACTIVITIES
CHICAGO CAREERS FOR YOUTH
REVISIONS
3 REVISED 1-8-90
2 REVISED 11-8-89
1 REVISED 10-3-89
NO. DESCRIPTION DATE BY
DRAWN BY JPG
DATE 5-29-89
SCALE 1"=1 YR.
STS PROJECT NO. 25658
CHECKED BY
DATE
SHEET NO.
STS FILE NO.

Developing a New Agenda for Training of Civil Engineers

Harold Deutschman[1]

Abstract

Approaching the 21st Century, we are in the midst of a crises in the use of our transporation facilities. Problems of congestion, pollution, and delays have become commonplace. Urban cities have lost population and jobs to suburban sprawl. This paper focuses on developing a new educational agenda for the training of Civil Engineers to reverse the above trend and to develop more effective transportation/land use systems.

Introduction

In New Jersey, state figures of population and employment indicate a significant growth in the suburbs accompanied by a decline of the cities. For example, from 1960 to 1985, the state's employment grew more than 70 percent with the cities losing more than 35 percent of their employment base. With the state's population growth of 5.5 percent increase from 1970 to 1985, the cities lost 13 percent of its population during this time period. With the shift of population and jobs, there has been a dramatic shift from mass transit to the auto for the journey to work. It is now estimated that 60 percent of New Jersey's highways are currently either at capacity or are exceeding capacity. Furthermore, the shift of population and employment have meant an inefficient use of resources. For example, for every shift of 100 jobs from the cities to the suburbs, there is an additional cost of 25,000 gallons of gasoline consumed due to the shift from public transit to the auto for the commutation. According to the 1980 Census Journey to Work, only <u>2 percent</u> of all suburban to suburban work trips utilize mass transit. This figure is on the

---

[1] Professor of Civil and Environmental Engineering, New Jersey Institute of Technology, Newark, N.J. 07102

national level. For New Jersey in 1980, 82 percent of all workers used auto for their journey to work. An obvious by product of the dependence of the auto for the work trip in New Jersey is the increased noise and air pollution in the heavily travelled corridors.

In a 1987 study "Growth Without Chaos," commissioned by the New Jersey Conservation Foundation, the following were among the findings of the report

- o Scattered development is a fact in New Jersey, but undeveloped areas could benefit from better planning;
- o scattered development is almost impossible to serve with adequate transportation facilities;
- o officials, citizen leaders, and even professional planners lack a vision to enhance the quality of life;
- o attempts to augment the tax base often outweigh other issues, such as livability, conservation and efficiency;
- o the development industry drives the planning and growth management system.

A new image of development for the state would involve "more dense and more focused development, with a sense of place and centrality, as well as open spaces used solely for agriculture, forestry, recreation and conservation," the study said.

"The market, rather than the public interest, is the predominant force operating to schedule development in an orderly manner," the report concluded.

The first statewide land use master plan for New Jersey was developed by the State Planning Commission in May 1987 and included the following observations and conclusions:

"Economically distressed cities, rapidly disappearing farmland and open spaces and increasing taxes to subsidize new public facilities are evident"planners observe

"The burden has fallen most heavily upon our now distressed cities and, through the degradation of our natural resources, upon future generations of New Jerseyans."

Looking at trends, the planners found that "rather than locating in concentrated nodes of development served by efficient forms of mass transit, sewerage, water and police systems, these developments are evolving in a low-density sprawl fashion requiring large investments of public tax dollars to construct smaller, relatively ineffficinet service systems."

Sprawl dependent on the automobile has created a situation in which the entire state is in violation of national ozone standards, planners report.

The strategy set forth in the draft plan is designed to:

- o Revitilize the state's urban centers and surrounding declining surburbs.
- o Concentrate development in "nodal centers" (in and around major transportation hubs) along growth corridors and in other separate towns.

Potential for Change

In a study by James Redeker[2] of the industrial migration patterns of New Jersey business over the 1970-1980 ten year period, 35 percent of the firms relocated at least once, and 9 percent of the firms were newly organized entities. This movement, unfortunately was primarily from the central city to the suburbs, corresponding with the decay of the central cities and the increasing accessibility of suburban locations. However, there is room for a change back to the urban areas, with tax incentives, new visions of improved mass transit, revitalization of the cities and an overall plan for growth and change.

New Agenda for Training of Planners

America is changing in a dramatic fashion, particularly in the composition of its young people. By the year 2000, 47 percent of the schoolchildren will be Black and Hispanics.[3] By 2010, one in every three 18 year olds will be Black or Hispanic. These newcomers have traditionally lived in the urban centers. They should be committed to and play a role in the revitalization of the cities as Civil Engineers. However, Blacks and Hispanics have been to this point under-represented in

[2] James Redeker, "Evaluating the Energy Consequence of Industrial Relocation in the State of New Jersey, unpub-unpublished Master's Thesis, 1980.

[3] U.S. Department of Labor

the fields of science and engineering. For example, Hispanics constitute 2 percent of all employed scientists and engineers, and hold 2 percent of all Bachelor's degrees. The U.S. Hispanic population is the fastest growing minority group and currently comprise 9 percent of our population.[4] This represents a mis-match of a factor of 9/2 = 4.5.

For Blacks, this mis-match is even greater. Blacks are 12 percent of the U.S. population but represent only 2 percent of all employed scientists and engineers. They earn 4 percent of all baccalaureates.[5] This mis-match on the employment level is 6.0.

New Agenda

The Civil Engineering profession must take a leadership position in the development of Civil Engineers, committed to the revitalization and development of the urban areas. This agenda must include the following:

1. Recognizing the need to reach out for the Black and Hispanic population at a very early stage to encourage them to plan for a career in Civil Engineering; to show them the excitement and importance of such a career.

2. To develop and fund pre-college courses starting at the junior high school level and below to introduce the new members of a changing America to the world of Civil Engineering and the need to include math and science in their high school curricula.

3. To send Civil Engineers from government and industry on loan to the urban school systems for guest lectures in math classes to introduce the mathematical relationships in developing solutions to transportation/city growth problems. These could be developed in structuring simultaneous equations for an expressway ramp metering problem, or determining the economics of car pooling, or exclusive bus and carpool lanes, for developing the energy demand differences of land use patterns. Furthermore, an introduction to applications of computer programming is needed in problem solving.

4. To offer special scholarships to minority students who will agree to work after graduation as Civil

---

[4] U.S. Department of Labor

[5] National Science Foundation

Engineers in municipal government in the inner cities.

5. To have minority Civil Engineers serve as role models, and be offered incentives to create the time and resources needed to meet with inner city students to counsel and direct these students into a Civil Engineering career.

6. The training of Junior High School and High School Counselors to develop career development strategies whereby students will:

a) increase their knowledge and develop an awareness of the Civil Engineering profession;
b) gain an understanding of the career decision-making process, and;
c) learn how to conduct a self-assessment as an important step in setting realistic career goals.

In developing a new agenda for the training of Civil Engineers in the 21st Century, one must identify the new faces of a changing America, and offer them the opportunities, challenge, and commitment to play a role in revitalizing our cities and our urban areas.

Career Education and the Engineering Team

Kurt H. Schuler, CET[1]

Abstract

The importance of a sound education in both the fundamentals of any engineering discipline and continuing education in the field along with the ability to apply that education are essential to the successful completion of any engineering endeavor. This paper discusses the premise that engineering is not a singular entity but rather a team effort comprised of several key players, and that without the emphasis of core preparatory education beginning at the elementary school level, the student is either relegated to mediocrity or failure. Although this conference stresses Civil Engineering, the premise is multi-disciplined.

Introduction

Many students, and to a large degree the general public, are aware of the basic term "Engineer" but are unaware of what specifically an engineer does or the preparatory and continuing education required to maintain proficiency as the state-of-the-art changes. Most are unaware that depending on the area of specialization, state licensing as a Professional Engineer may be required for the individual to practice. Without this basic knowledge they also cannot comprehend that with a team of skilled individuals any engineering program or development can be conceived, designed, proto-typed, tested, constructed, and implemented or marketed.

The Engineering Team

The Engineering Team is comprised of the Scientist, Engineer, Engineering Technologist, Engineering Technician, and Craftsman. Either all, or specific members, comprise a Team for the successful completion of any given engineering task, project, or program. The educational and skill

----

[1]Secretary-Treasurer, American Society of Certified Engineering Technicians, 3202 Donegal Road, El Paso, TX 79925.

levels of the various members differ, but the total cohesive combination of their talents makes the overall project a success.

A basic example of the functional roles of all Team Members is in a Research and Developement (R&D) effort. The Scientist would be the conceptualizer. The Engineer would develop the concept, and through design and application of engineering principles, transform it into something which would be a tangible end product. The Engineering Technologist and Engineering Technician would assist in taking the design through the drafting, proto-typing, testing, and production process. The Craftsman would perform the skilled level work required to build the final product. Outside the R&D environment, the Scientist does not play as prominent a role and the Engineer tends to function as both the conceptualizer and designer.

The educational requirements of the Team Members form a somewhat logical breakout for definition of complexity of duties. The Scientist has attained an engineering education to at least Graduate or Doctoral level. The Engineer has attained at least a Bachelor of Science Degree in Engineering and quite often a Graduate Degree. The Engineering Technologist has attained at least a Bachelor of Science Degree in Engineering Technology. The Engineering Technician has attained at least an Associate in Applied Sciences Degree in Engineering Technology. The Craftsman normally has completed a Trade School or has served an apprenticeship.

Student Opportunities

The purpose of detailing the Engineering Team concept was to illustrate the broad spectrum of opportunities available to students desiring a career in the engineering area. Although many express interest in becoming Scientists or Engineers, a number of factors may contribute to the non-fulfillment of this desire. The Team concept shows that there are other options they can take and still be involved and contribute to the overall engineering effort. It also indicates entry at one level, the ability to remain at that level if it satisfies the personal need, or continue with education and advance to the next level. The key is flexibility.

Student Literacy

A multitude of student scholastic aptitude testing programs have been conducted by various federal, state, and local groups to determine scholastic aptitude. The results have been published in newspapers, magazines, and various professional and academic journals. Results of these tests have shown that by the time most students leave elementary school, many have little more than the most basic skills in math and science. The same is true, possibly to a lesser degree, in the basic oral and written communicative skills. When they enter high school, this lack of preparation forces most into a continuation of rudimentary courses and a series of "Basket Weaving" courses needed to satisfy the minimum requirements for a diploma.

Scholastic aptitude testing at the various levels appears to be more of a rote rather than a comprehension method of testing. It forces teachers to "Teach the Test" so that a given school can maintain a statistical quota and standing which, in effect, measures the student's capability to memorize. The net result is development of a generation of functional illiterates.

The Elementary and Middle School System

There is a basic weakness in the educational system which prepares teachers. Most pursue a Liberal Arts Degree program with a limited curriculum in the math and science areas. Most go directly from college into the classroom with no other work experience. In many states the teacher certification process requires little more than an understanding of addition, subtraction, multiplication, and division as a demonstration of math skills. If one takes this as a starting point and then projects that a percentage will continue in Graduate school in order to receive advanced degrees in Educational Administration, the root of the problem begins to surface. Through no direct fault of these educators, most of whom are extremely dedicated individuals, how can they begin to understand the heavy reliance on advanced math and science required by potential students of engineering areas? At what point in the students educational development is the desire for an engineering related career recognized and the proper preparatory courses started to assure the students success?

The above area is one in which I have had personal experience when discussing preparation for a career in the engineering environment. Students having mechanical aptitude and taking vocational training programs in high school are viewed by a large number of teachers and guidance counselors (Liberal Arts Background) as not being suitable for a continuing education at the collegiate level, notwithstanding that plumbers or electricians have greater earning potential than elementary and middle school teachers. The educators do not comprehend that in most areas of vocational training, students are exposed to the basic engineering theory, mathematical problem solving, and hands-on experience which allows them the chance to succeed in an engineering related continuing education program.

Changes Needed in the Educational System

Since the entire engineering spectrum is based upon mathematics and physical science, these are the basic foundations required by all Team members. As technology advances, these basic foundations may be required by all students to successfully compete in the future high-tech world. For the immediate future a need exists to identify potential engineering spectrum students early on, and to develop the required progressional courses in these areas. This would enable the students to prepare for their college years. In addition, greater emphasis needs to be placed on the importance of English in the oral, written, and reading comprehension communications skills. Extensive knowledge about a subject is an attribute, but if the proponent can neither understandably write or talk about it, and the recipient cannot understand or read and comprehend the information, knowledge is not passed. At some point in time, prior to college, these students must also be made aware of their eventual need of knowledge in economics, engineering ethics, business law, and the numerous other material that they will need if they are to engage in the field, especially in private practice.

At what point in time during a student's elementary and middle school years does this identification of potential engineering students and the reeducation process begin? The total answer to this question will take research, time, money, and a major rework of our current educational process.

An innovative program was commissioned by the Governor of Virginia in 1984. It was developed by the Virginia Peninsula Local Education Agencies in conjunction with the Virginia Department of Education and the Virginia Community College System. The state's pilot center for this program entitled 2 + 2 MASTER TECHNICIAN PROGRAM[1] is, the Thomas Nelson Community College. The Master Technician developed by this program is an engineering technician skilled in more than one subfield.

The concept of this program is to determine which students in grades 9 and 10 would be interested in pursuing a career as an engineering technician. Once identified, these students enter the program in grade 11. In addition to their normal studies, they take courses in mathematics, principles of technology or physics, laboratory science, electronics, and materials and processes or chemistry in preparation for the next phase of their education. Upon graduating from high school, they enter the Community College fully prepared to successfully complete the two year Associate Degree program in Electronic/ Electromechanical Engineering Technology. The first class is scheduled for graduation in 1989.

The program was developed based on industrial needs in the state. Since area business and industry representatives provide input to the program, it can be changed or expanded to meet the ever changing technological needs. The input to, and support of the program by the business sector then provides the graduate of the program the next opportunity - a source of possible immediate employment in their chosen field. Once in the field, the former student can remain as a technician or continue with their education and advance up the Engineering Team ladder. Figure 1. is an illustration of the educational ladder in the engineering area.

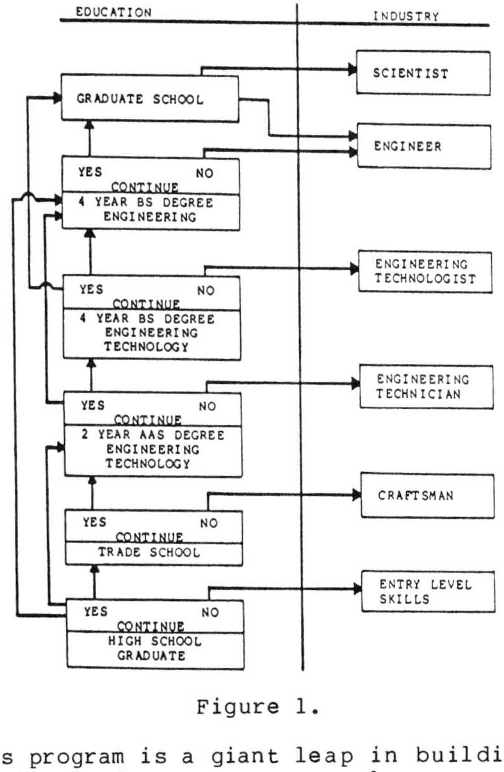

Figure 1.

This program is a giant leap in building a strong technological base for our country and can serve as a model for other states and communities. It is not the total answer for revision of the educational needs of students wishing to pursue an education leading to a career on the Engineering Team, but it can serve as a model on which to build the ultimate program. The development of this program started at the state level and had sufficient political push to see it carried through. Similar state or federal emphasis would be needed for the ultimate program.

Conclusion

Restructure of the educational system will be a long term effort. A major source of support for this effort involves education of the parents of potential students. They have the greatest concern for the future of their children. They are the individuals who pay the school tax and comprise the school boards and Parent Teacher Associations. Through a comprehensive parent education program, which stresses the importance of the early recognition of potential Engineering Team members and establishes an educational system which will assure their success, the required changes are possible.

Most Professional Societies, many of which have engineering educators as members, have their own programs and materials which are available to individuals who ask for them. Many sponsor local programs and there are several national math and science competition programs which draw a degree of interest to the problem while they are being presented. Unfortunately this material and these programs get recognition from only the small group of individuals who already have a knowledge of the engineering spectrum. They do not reach the majority of parents, students, educators and guidance counselors in the elementary and middle schools, and legislators who need to be educated in the preparatory needs of pre-engineering students and have the ability to make the required changes.

The promotional material most Societies produce and distribute is geared to their specific engineering disciplines which is an ideal supplement for use with a basic program. Some Societies and engineering educators provide career preparatory and guidance information which is also helpful. The net result is that many groups are spending time and money to provide programs which are in effect overlapping splinter groups, but splinter groups with no common direction. What is needed is a core group to define the program and unite the splinter groups under a total and comprehensive effort.

Since this is an ASCE sponsored conference with one of the focuses being education, a suggestion is to designate a small committee to establish contact with the various Professional Societies and start a united effort to develop the core course outline required by students in the elementary and middle school systems. Once the core course outline is developed, existing Society material can be reviewed and appended to provide a greater understanding of the major engineering areas and associated specialty subfields. The completed material should be in loose leaf format to permit easy update and customizing. This is the preliminary organizational process to develop an end product for use in the educational campaign.

The next phase is to use this material in a lobbying effort of parents and legislators in order to thoroughly acquaint them with a solution to the well publicized problem: This country is now lagging in both technology and product quality!

I have oversimplified the conclusion, but am aware that this is a time consuming and costly solution. It will take legislative action and funding along with continual monitoring and change to accomplish. I feel we are already behind the times but if something is not started, twenty years from now we will still be asking what can be done to provide career education to potential students of engineering related programs.

Appendix I. Reference

1. 2 + 2 MASTER TECHNICIAN PROGRAM, Thomas Nelson Community College, 99 Thomas Nelson Road, P.O. Box 9407, Hampton, VA 23670. Contact Debra Morrison Eysmans (804) 825-2271 or Donna D'Amelio (804) 835-2774.

Civil Engineering Education for the Handicapped

Erdogan M. Sener[1],Member, ASCE

Abstract

This paper brings forth the underrepresentation of the handicapped in civil engineering (CE) education and careers as an urgent concern that needs to be addressed in the near term. A design and implementation framework to make the CE programs more accessible to the disabled using opportunities that currently exist as a result of developments in personal computer hardware and software, adaptive devices for the handicapped, and computer-aided-instruction/learning is presented. The paper aims to provide a general model for approaching the problem of education of the disabled in CE programs, devising plausible techniques, and implementation. It does not aim to bring forward concrete methodologies or solutions to the problem since the specifics of the undertaking will depend on the type of disabilities to be addressed in the respective CE program. The model can be modified and revised as appropriate to accommodate a broader spectrum of disabilities or to improve the methods of implementation.

Introduction

Even though much progress has been achieved in ensuring a respectable quality of living for the handicapped, their education and training for gainful employment still remain as areas of grave concern. Lucas (1989), states that the unemployment rate among the working age (16-64) disabled in this country is about 70% though it is only about 5.2% for the whole nation. The main reason for this is given as the lack of proper education and training on the part of the handicapped.

Inadequate education and training of the disabled is not due to unavailability of funding. It is estimated that (Lucas, 1989) by 1990's more than 100 billion will be spent for diverse programs for the handicapped. The problem is that a very small percentage of this money (presently about 2.5%) goes into education because of mainstream education

1 Assistant Professor, Purdue University School of Engineering and Technology at Indianapolis, Indiana University-Purdue University at Indianapolis, 799 West Michigan Street, Indianapolis, IN 46202.

programs being still not very easily accessible. The percentage of the handicapped in post-secondary education now stands at about 10% of the total enrollment, up from 2.6% in 1976 (National Center for Education Statistics, 1989). However, the percentage of enrollment in engineering programs is only about 1%. Individual data for CE programs is not available. If the percentage at this institution is any indication (about 0.5%), the disabled are highly underrepresented in civil engineering education.

Contemporary advances in computer hardware, software and adaptive devices for the handicapped have brought forward the opportunity to consider (or reconsider) how to educate the handicapped in CE programs. It is my contention that the opportunity for increasing the number of disabled people in civil engineering education was never better. It is ironic that this coincides with a period in which engineering enrollment in our higher education institutions is declining. The effects this decrease will have on the technical capacity and competitiveness of the nation is increasingly under discussion. Amidst all this the disabled stand out as an untapped resource pool.

Advances in Personal Computer Accessibility for the Disabled

The framework of the methods for accommodating the disabled in civil engineering programs should rely mainly on the use of personal computers by the disabled for carrying out all the work that is required in an engineering curriculum.

Computers are increasingly accessible to the handicapped. This accessibility has three main parts to it:

(1) Hardware accessibility: Accessibility of computers is evaluated in terms of the transparency and redundancy of the hardware (Bowe,1987; Roessler,1986). Transparency refers to designing of the parts of a personal computer to work together without the need for artificial aids. Consequently, the computer does not distinguish between different data input systems. It processes data regardless of whether they are entered through a standard keyboard, mouse, touchscreen, joystick, switches or any other type of interface. There is no need for emulators between the input interface and the central processing unit to emulate a certain input device. Redundancy, on the other hand, is designing hardware (and software) to provide output simultaneously in different ways. It is the capability for multiple modes of communication for input and output such as speech recognition, screen, speech synthesis, keyboard, print, braille, and so on.

(2) Software accessibility: Accessibility limitations in relation to software have traditionally been due to two main problems:

(a) Incompatibilities and operational conflicts between emulators and adaptive devices being used by the disabled and the copy protection features of the software.

(b) The above problem getting magnified due to different tasks like wordprocessing, database management, spreadsheet applications and graphics necessitating separate software and a new access in each case.

Transparent and redundant hardware and software have minimized such problems. Integrated software combining word, data, and spreadsheet processing into one package and allowing communication and graphics operations have eliminated the need to access various different software.

(3) Adaptive devices: Any device which help the handicapped individual access a computer system from the perspectives of computer hardware, software and physical access are given the name of adaptive device. These range all the way from a simple page turner to a computer desk and to voice recognition systems which make it possible to operate a computer by voice commands alone. Quite an impressive number and kinds of adaptive devices currently exist to help the disabled (Burkhead, 1986; Brandenburg & Wanderheiden, 1987; Houston & Cress, 1986) and research and development in this area is continuing.

Analysis of the Curriculum

The process of accommodation of the handicapped in CE education should start with a detailed analysis of each course in the curriculum. This should be done with the aim of determining how personal computers and computer-aided-instruction/learning technologies should be combined and used to arrive at the best configuration of accommodations for the disabled student and the disability in question.

Each course in the curriculum is analyzed in different respects. An overview of this is given below for departmental courses. The approach is not any different for the non-departmental courses; the arguments and the implementations apply equally well. The analysis should concentrate on:

(1) Whether the specific course is a lecture, problem solving, drafting, design or a laboratory type of course.

(2) Present method of instruction.

(3) Instructional aids being used for the course (audio tapes, video tapes, overhead projectors, slide projectors, computer-aided presentation systems, etc.).

(4) The main purpose behind the course and its specific contribution to and the individual value in the overall education of the CE graduate.

- Lecture courses:

For handicapped accessibility, the method of presentation is the most important aspect in these type of courses. Depending on the type of disability the presentation may be seen but not heard, heard but not seen, seen and heard but no notes can be taken or variations thereof. For accessibility, presentation methods may be changed as appropriate making use of a vast array of instructional aids currently available to accommodate the specific handicap(s). Students may be given audio, video, or computer diskettes of the presentation for outside class viewing, listening or reviewing.

- Problem-solving courses:

In these types of classes the approach taken in lecture type of classes must be supplemented with drill and practice type of computer-aided-instruction/learning (CAI/L) software with appropriate technical format for different types of disabilities.

- Drafting courses:

These courses are perhaps the easiest to handle in view of the present commercial computer-aided-drafting (CAD) and graphics software packages available and their accessibility by the disabled. Some popular CAD and graphics software can nowadays be run by voice recognition (Kramer, 1988; Burnette, 1988). Capabilities of the software range from simple two-dimensional drafting to three-dimensional solid modelling. Research has shown that (Resetaris, 1989) there is no difference in learning and resultant competency in drafting depending on whether the material is taught and practiced manually or using CAD. Work in this area is well under way for increasing the accessibility for the disabled.

- Design Courses:

There can be a broad spectrum of courses in this group. Some can be handled by means of commercially or academically available computer-aided-drafting and design (CADD) software packages which are continuously becoming more powerful and user friendly. Others will necessitate finite element method (FEM) software integrated with various databases which instantly supply material information such as sectional properties of structural members, rebar areas, allowable stresses and so on. A number of courses in this group may necessitate development of special computer programs and CAI/L material for accommodation of the handicapped.

- Laboratory Courses:

These type of courses may be the most difficult to handle in the CE curriculum depending on the type of laboratory and hands-on work that needs to be undertaken by the student individually.

Simulation of some of the laboratory work using already available CAI/L material and developing others is one way. Another way is making use of the fact that in some of today's CE laboratories data collection is almost totally done by computers even though a manual setup may be involved at the beginning of the experiment (Smith, 1988). In other laboratory classes like surveying, for example, even though adaptive devices can be designed to accommodate certain disabilities it may not be worth the effort. It may be sufficient just to have the student observe the process or be told about it and then have him/her write a computer program to do the calculations to ensure that the concepts and the methodology have been understood. I do not think this kind of an approach can be considered making a sacrifice from the real purpose of such laboratory classes. For other laboratory work like geotechnical for example, a combination of the above methods may be employed as appropriate.

Implementation

The above type of analysis will lead to decisions on the type of accommodations that will be needed for each single course. It should be noted that this is something that needs to be done on a case by case basis depending on the type of disability. The accommodations that will be required by the orthopedically impaired will be very different from those that will be required by the visually impaired. So both for practical and economical reasons some decisions need to be made as to the type of disabilities that can be accommodated at least initially. One should also accept the fact that it may not be at all possible or practical to try to accommodate all types of disabilities in a CE program.

The accommodations to be implemented manifest themselves mainly in the form of:

(1) Changes to method of instruction.

(2) Changes in instructional aids and/or inclusion of other instructional aids.

(3) Provision of special study aids to the disabled student.

(4) Changes in expectations from disabled students without sacrificing from the quality and main purpose of the course.

Computer-aided-instruction/learning (CAI/L) software can play a very significant role in the above described accommodations. Currently CAI/L software fall into three basic categories:

(1) Presentation, demonstration and electronic blackboard type of software: These enable quite a variety in instructional techniques. Common examples are the authoring systems, simple presentation programs, interactive video and others. These type of CAI/L material will be useful in all types of courses.

(2) Drill and practice type of software: This type of CAI/L software is especially suitable for problem-solving courses. Tutorials, tests, and study guides can be prepared using readily available authoring systems, hypercard, interactive video, different programming languages and other courseware.

(3) Computer-aided-drafting/design and graphics software: Highly capable commercial CADD software is available nowadays for several facets of civil engineering work involving drafting and design. Finite element analysis programs with graphics capabilities are able to handle structural design part of the civil engineering work. Combined, CADD and FEM software can cover all facets of CE work involving analysis, drafting and design.

(4) Simulation software: Several types of commercial software and courseware currently exist for simulating different laboratories in a CE curriculum. The same holds true for non-departmental courses in physics, chemistry, mathematics and others. Other CAI/L material can be prepared using programming languages and authoring languages enabling simulation, hypercard and so on.

Other possibilities

Whatever may be the success in making computers more accessible to the handicapped, the mobility of the handicapped still remains a big obstacle to their education in several cases. Off-campus credit programs may provide access for some students to CE education. As far as technology goes, several options exist for implementation of distance education through electronic bulletin boards, electronic classrooms, satellite delivered instruction relying on computer and communication networks (Czeck, 1989; Holt, 1989; Scigliano et al.,1989). Telecommuting, that is, using a modem-equipped computer in the home to receive, process and send information back to the place of study or employment (Bowe, 1987) holds a great promise in the CE education and careers area.

Conclusions

As a result of advances in personal computer hardware, software and adaptive devices, computers have become increasingly accessible to the handicapped. These advances, combined with CAI/L and other contemporary instructional media, have created significant opportunities for educating the handicapped in civil engineering programs. Diverse possibilities currently exist for accommodating individuals with various disabilities in mainstream civil engineering curricula.

What is needed at this point is for implementation to be started at various CE institutions so that data, experience and CAI/L material for furthering and improving the undertaking can be obtained. It is important that the handicapped not be seen as a problem in this undertaking but as resources in solution of the problem.

Even though the education process will be greatly facilitated through the use of computers and CAI/L, this should not be seen as a panacea for the problems of the handicapped. Computers should be viewed as another prosthetic device in this whole endeavour.

The need to prepare, acquire and use appropriate CAI/L material and instructional media is a requisite for success. This is one area where American Society of Civil Engineers (ASCE) organization could play a vital role by stimulating development of handicapped oriented software and CAI/L materials and acting as a distribution center for the same.

Research is needed in the area of implementation of education of the handicapped in CE programs and in the area of techniques for development of appropriate computer based education materials for the disabled.

All these very challenging and equally rewarding tasks await the urgent attention of civil engineering faculty and other members of the profession.

REFERENCES

Bowe, F. (1987). Making Computers Accessible to Disabled People. Technology Review. Vol.90, No.1, pp.52-72.

Bowe, F. & Little N. (1984). Computer Accessibility: A Study. Rehabilitation Literature. Vol. 45, No. 9-10. pp.289-291.

Brandenburg, A. S. & Vanderheiden, C. G. (1987). Communication, Control and Computer Access for Disabled and Elderly Individuals: Hardware and Software. College-Hill Press, Boston, 1987.

Burkhead, E.J., Sampson, P., McMahon, J. & Brian, T. (1986). The Liberation of Disabled Persons in a Technological Society: Access to Computer Technology. Rehabilitation Literature. Vol. 47, No. 7-8, pp.162-168.

Burnette, J. (1988). Computer Workstation for the Physically Disabled (Videotape), School of Architecture, Washington State University, 1988.

Czeck, D. (1989). Fax, TV and the Remote Classroom. Technical Horizons in Education Journal. Vol.16, No.8, pp.69-72.

Holt, L. S. (1989). The Future According to Pogo. Technical Horizons in Education Journal. Vol.16, No.6, pp.58-52.

Houston, K. and Cress, C. (1986). Disabled Access to Technological Advances. Proceedings of the Conference on Training and Technology for the Disabled, Milwaukee, Wisconsin, March 10-12, 1986, pp.29-31.

Kramer, K. (1988). Computer Accessibility For Students With Disabilities. Academic Computing. Vol.3, No.2 .pp.26-39.

Lucas, T. (1989, June 25). Invisible Barriers Block Access to Jobs-Getting Right Training Tough for Disabled. The Indianapolis Star. pp.H1,H6.

National Center for Education Statistics, U.S. Department of Education. (1989). Profile of Handicapped Students in Postsecondary Education, 1987. Survey Report. pp.14-16.

Resetaris, J. P. TRAD or CAD? (1989). A Comparison. Technical Horizons in Education Journal. Vol.16, No.7, pp.93-96.

Roessler, R. T. (1986). Technology Utilization in Rehabilitation. Rehabilitation Literature. Vol. 47, No. 7-8, pp.170-173.

Scigliano, A. J., Joslyn, L. D., & Levin, J. (1989). The Non-School Learning Environment: ECR. Technical Horizons in Education Journal, Vol.16, No.6, pp. 63-67.

Smith, W. S. (1988). Enhancement of Soil Testing Courses using Microcomputers. Civil Engineering Education. Vol.X, No.2, pp.1-12.

Why Do Research With Undergraduates?

Donald E. Milks, P.E., MASCE(1) and
William E. Brewer, P.E., MASCE(2)

Abstract

Reports on the use of undergraduates for applied research projects. Paper endorses the involvement of applied research on the undergraduate level because of the positive effect it has on participating students, faculty, and non-participating students and faculty. Actual cases studies presented with anticipated results.

Introduction

Research activities with undergraduates provides opportunities for both faculty and undergraduates. Faculty members can incorporate related research activities such as; literature search, data collection, statistics, synthesizing, and communication skills into their teaching. The student has an opportunity to further understand problem solving techniques and how they relate to academic training.

In many cases, this will be the only opportunity for students to participate in research since the majority do not go onto graduate school. This exposure to research has and will change the student's career goals. Students become aware that research is a transfer of knowledge from one generation to another. They see how the technical base is built through published papers and the need for good communication skills. They also participate in project meetings and review sessions with granting agencies.

Students for research projects are selected from those student exhibiting good scholastic records. Through an interview process students are selected by the faculty member conducting the research, the Principal Investigator (PI). The selected students must be capable of following directions without a considerable amount of supervision.

---

(1) Prof. Civil Engineering, Ohio Northern University, Ada, Ohio.
(2) Adjunct Prof. Engineering Research, Ohio Northern University, Ada, Ohio and President, Brewer & Associates, Toledo, Ohio.

Students also become aware of the demands that good research requires and the price (time required) for thoroughness.

Applied Research Versus Theoretical

Applied research primarily deals with findings solutions to immediate problems. Research with undergraduate students is applied research. This research consists of both laboratory and field practices. The majority of the research projects are developed either through industry contacts or through faculty members awareness of industry needs. Theoretical research is beyond the education of undergraduate students and therefore not considered for research activities at Ohio Northern University. It should be noted that students participating in applied research activities could easily move along to theoretical research in graduate programs.

Need For Industry - Applied Research Answers

Industry is constantly looking for "better mouse traps". Applied research, especially with undergraduates, provides industry with fresh minds with newly acquired tools for problem solving; social, economical or technical. The majority of the undergraduate research participants have <u>not</u> been "turned off" by "it can't be done" and "why try" attitudes of their elders. It appears that applied research provides student growth which is not usually provided in the traditional classroom setting. This student growth needs to be monitored to successfully make this claim.

Currently industry has not considered research with undergraduates. Frequently a <u>BIG</u> name school is contacted for problem solving needs. What is unique about the undergraduate resrarch activities at ONU is that some faculty members, because of the applied research activities at ONU, have solicited industry about problems that needed to be addressed. These problems relate to; continuing education, management, safety, construction methods and materials and others.

Utilization of Existing Equipment

The university has educational equipment that now can also be used for applied research. This expensive equipment, if used only for educational purposes, could be used only two or three weeks a semester. With applied research activities, this equipment can be utilized every week.

The utilization of the equipment also affords the student more experience with the equipment than provided in the academic program.

New Equipment Availability

The cost of equipment continues to rise. Research activities help off set this continued rise in equipment costs. Many applied research projects offer the opportunity to include equipment that can be used for educational purposes after or even during the research activity. This benefits all participants; the university, research students, future students, project sponsors, and equipment manufacturers.

Sometimes the project sponsor will retain the testing equipment purchased for the project. In several cases, the sponsor has, upon the completion of the project, donated the equipment to the university.

Types of Applied Research

There are many types of research that are considered applied research. A brief discussion of each type follows:

Design

Applied research in design deals with new design techniques for buildings, pavements, sewers and bridges. The existing culvert research project is an example of this type of applied research.

Materials

Material research assist industry in the enhancement of existing materials and the development of new materials. The construction industry in particular offers many opportunities for material research.

Education

The methods used in the education process of undergraduates, graduates, and professionals provides research opportunities. What are the benefits of a new material discovery if the design professionals are not aware of the material?

Construction

Research in construction needs to be updated. This industry is on the edge of major changes in its approach to; education, construction practices, and research. More and more schools are developing construction programs. Changes in technology have made it impossible for the traditional Civil Engineering program to cover the many aspects of construction. The majority of the Civil Engineering programs have opted for design and construction. Those who have implemented construction have become aware of the construction industry's need for applied research.

Computers

Research in computer applications needs to be extended. Research projects, referenced in this paper, would not have been possible without the computer. More development is needed in the use of spreadsheets and the comprehension of artificial intelligence (AI) and expert systems. These items hold great research potential for all aspects of Civil Engineering.

Human Relations

"People Research" needs to be considered for Civil Engineers. How to "get along" with the boss and associates needs to addressed at all levels of the Civil Engineering world. Has our world become a place where no one likes; the president, the dean, the department chairperson, co-workers, assistants, the boss, themselves...and so on? If the answer to just one of these is yes, then research dealing with people is necessary!

Case Studies

To help illustrate how undergraduates can participate in applied research, several case studies are presented. These case studies deal with two different types of applied research. One is related to material development and construction while the other is computer database oriented. Both projects are being funded through the Ohio Department of Transportation (ODOT).

Controlled Low Strength Materials (CLSM)

The material development and construction research is titled; "Design and Construction of Small Span Bridges and Culverts Using Controlled Low Strength Materials (CLSM)". Its major objective is the development of design information for the use of CLSM with metal pipes.

The undergraduates participated in the material testing phase by performing laboratory tests on selected materials. These tests consisted of; flow, gradation, angularity, mixture proportioning, strength, and stress-strain. The selected materials represented 100 different industrial "waste" products from all parts of the United States.

During the construction and field testing operations the students will be monitoring the pressure cells that have been placed in the CLSM backfill around the metal pipe arch. The construction site is approximately 50 miles from campus. The students will also participate in the analysis of collected field data that will be incorprated into the final research report.

Another aspect of this research is the participation of the students in ODOT review meetings. Students are encourage to attend these sessions to obtain experience in professional communications.

Truck Accident Investigation

The computer database research is titled; "Truck Accident Investigation". Its major objective is to inventory truck related accidents and to identify the parameters which are statistically significant in truck accidents.

Students have participated in this research in several ways:

- conducted literature searches to determine what had been done by other agencies relative to truck accidents.

- analyzed computer data tapes provided by ODOT.

- developed a questionnaire for interviewing truck drivers.

- surveyed and interviewed truck drivers at nearby truck stops and rest areas and tabulated the survey results.

Anticipated Results

The results of applied research, at an independent school, can affect many individuals, agencies, and organizations. For example, the two cases presented in this paper have affected the following.

- students involved in projects

Nine students from three different graduating classes were involved in these two projects. The use of different levels of students permitted good carry over from year to year. The effect that this type of activity has on the student is best explained by quoting, in part, directly from a letter received from one of these students.

"Working on the research project helped me learn several things besides just testing procedures. Learning to budget my time to be able to keep up with classes was important. I gained experience in working with others and sharing responsibilities. Organization, which is extremely important in engineering, was also necessary in research. It was needed to keep track of data obtained and of what other students were testing. I also learned to have appreciation of laboratory work. By neing familiar with lab testing procedures, I, as an engineer, will be better able to understand work that I need to have done by a lab technician in the future."

- students not involved in projects

Students have come to the Principal Investigators and requested to be involved. Requests have been accommodated if funds were available and the student showed academic promise. Students not involved were aware of the educational benefits of applied research. It was an extension beyond the traditional classroom setting.

- faculty members involved

Five faculty members have been involved with these projects. In some cases, the research results directly affected their classroom teaching. In other cases the research did not relate directly to a class situation, but the research did broaden their outlook. In either case, their teaching was improved.

- faculty members not involved

Research awareness has been generated at the deparment, college, and university levels. Faculty members not directly involved are now exploring possible research opportunities within their specific discipline. Interest has also been generated with possible cross discipline research...engineering and business...engineering and education...and so on.

- industry

Industry related results, during the research stage, are not as apparent as they will be in the future. The culvert project will yield a number of industry benefits; lighter gage steel, reduced construction costs, more reliable structues, use of waste materials, and improved construction techniques. Most industries are slow to respond to research results unless they can see an immediate profit. It is the investigators opinion that only through applied research can future profit increases be realized.

- university administration

The administration at Ohio Northern University has been supportive of applied research. Direct benefits to the administration has been the recruitment of students and faculty through the publicity generated by research. This publicity also helps with the solicited of other grants and endowment funds.

"Issues to Improve the Status of Minority Students in the Civil Engineering Profession"

Judy A. Perkins*
Associate Member ASCE

ABSTRACT

Despite a steady increase in the number of minorities receiving undergraduate degrees in engineering and science fields, underrepresentation of Blacks, Hispanics, and American Indians remains a serious problem. These groups receive science and engineering degrees at less than half the rate of whites, reflecting both a lower level of participation in higher education and a lower rate of selection of engineering and science-related majors (Kimmel, Martino, and Tomkins, 1988).

Similarly, if one was to disaggregate the engineering and science fields further, the statistics would reveal that minority enrollment and degree recipients in the civil engineering field are substantially less than the minority enrollment and degree recipients of other engineering fields (e.g., electrical/electronic engineering, mechanical engineering, etc.). This not only reiterate the existence of a problem but raises an awareness of the trends displayed by minorities in this field.

The purpose of this paper is to address those issues which may assist the Civil Engineering profession in recruiting and retaining minorities in the next decade.

INTRODUCTION

To recruit a work force that truly represents all facets of society will require fundamental changes in the way the civil engineering profession interfaces with minority students. A strategy to increase the supply and quality of young minority civil engineers must be based on an understanding of the unique problems fostered by a dra-

---

*Assistant Professor, Civil Engineering Dept., Southern University, Baton Rouge, LA 70813. Doctoral Student, School of Civil Engineering, Georgia Institute of Technology, Atlanta, GA 30334

matic change in the ethnic and racial composition to the workforce. Such strategies should include altering the pattern by which potential minority students are identified, recruited, supported, retained, and graduated (Adams, 1988) from degree programs in civil engineering.

If we are to bring more minority students into the ranks (Office of Technology Assessment, 1989) civil engineers, promising education and recruitment programs must be adopted even if they are unproven. The need to revise our methods and models of recruitment, clarify the image of the (OTA, 1989) "civil engineer" and rethink the notions of "professional calling" as it relates to the accessibility of the (OTA, 1989) civil engineering career requires constant attention.

PROGRESS TO DATE

Over the past decade, the civil engineering profession and the American Society of Civil Engineers have made substantial progress toward increasing the representation of minorities and women in the work force. In 1979, minorities represented 7.4 percent and women 2.5 percent of the total permanent work force of almost 162,000 civil engineers (U.S. Bureau of the Census, 1980). In 1988, minorities and women represented 10.0 percent and 6.0 percent, respectively, of a work force of almost 218,000 civil engineers (U.S. Bureau of the Census, 1987). In 1986, the number of graduating Hispanic and Asian-Pacific students went up 13.3% and 2.3% respectively. But Blacks dropped 9.6% to 168 graduates, and there were 14 new American Indian BSCE's, down two from the proceeding year (Basta, 1987). There has also been a noticeable upward movement of minorities and women into various managerial positions. In 1989, James E. Davis became American Society of Civil Engineers' first black assistant director at the Society's New York head-quarters location. Also, during this same year, ASCE planned to expand their sponsor/co-sponsorship of regional Summer Institutes to introduce Minorities to Civil Engineering to 13 university locations (Berson, 1989), a 62% increase over 1988 figures.

FUTURE AIMS

Recruitment and retention are the most direct routes to achieving the goal of increasing underrepresented minority groups in all specialties of the Civil Engineering's broad spectrum of opportunities. However, for the civil engineering profession to meet its goals for the rest of the century and beyond will require a much broader focus than recruitment and retention. The action items discussed below will enhance the rate at which the civil engineering profession can reach its goal. These action items include: (1) identification and recruitment of students, (2) financial support for students, (3) nurturing/retention of students, and (4) enhancing public image/awareness.

A brief description of each of the action items along with some general recommendations to accomplish these actions will follow. These recommendations are by no means all inclusive. Creativity and innovation in expanding these approaches in many different ways can only help to better achieve greater minority representation in the profession.

Action Item #1

AIM: Intensify efforts to identify and recruit minority students into the Civil Engineering profession.

RECOMMENDATIONS: 1) A recent study confirmed that "by the seventh or eighth grade the die is cast" (Massachusetts Department of Education study was quoted in the New York Times, 1986). Therefore the time to motivate and excite students about science and technology is during the elementary and middle school years (Kimmel, 1987). Encourage the ASCE professional section chapters in each state to adopt a number of students from disadvantaged elementary, junior high, and/or high schools based on their aptitude in mathematics and a desire to attend college (Debley, 1987). Provide activities that include science projects, tutoring workshops on critical thinking and test-taking strategies (Debley, 1987), and field trips to civil engineering firms; 2) Expand the existing pre-college summer institute programs to incorporate junior colleges with large minority enrollment. This can serve as a beacon for partnerships between two- and four-year colleges to increase the number of minority students earning baccalaureate degrees in civil engineering; 3) Initiate a program which seeks to expand the recruiting network to non-traditional institutions and incorporate women functioning as civil engineers as often as possible, so that young women have access to experienced women engineers; 4) Establish a program that enables experienced minority civil engineers from the industrial sector to undertake a 1-year visiting professorship at a host institution, a university, a four- or two-year college which have a large minority population. This will increase the visibility of minority civil engineers and encourage other minority students to pursue careers in the profession; and 5) Develop and disseminate material on corporate life and organizational culture----from a minority student point of view----and include photographs of minority employees as part of the organization. Provide these to local counseling centers at elementary, junior-high and senior-high schools and universities.

Action Item #2

AIM: To develop adequate financial resources to support minority students before and after enrolling in college.

RECOMMENDATIONS: 1) Establish a substantial number of Civil Engineering scholarships for incoming freshmen as well as underclassmen based on the minority enrollment at every accredited civil engineering program in each state. This vehicle wil enable existin minority students with financial difficulties to remain in school and encourage incoming students to select civil engineering as their major; and 2) Emphasize industrial support by challenging civil engineering firms to create advisory councils of industrial representatives to network with alumni of the various institutions to generate funds that can be useful for these scholarships.

Action Item #3

AIM: To increase the overall retention rate of minority students majoring in and planning to major in a civil engineering degree program.

RECOMMENDATIONS: 1) Conduct workshops led by civil engineers to train secondary school counselors, higher education counselors and a number of civil engineering agencies and organizations interested in the problems of minority student retention, on how to organize their resources and provide support to each other which is necessary to stimulate and develop student interest in the various civil engineering disciplines; 2) Provide a forum for universities and colleges under the auspices of the ASCE section chapters and local university student chapters in each state to share ideas on the problems concerning advisement of disadvantaged youths in their pursuit of a civil engineering career, and the solutions that they have attempted to implement. This gives the schools and the professional society the opportunity to alter existing methods and/or develop new ones to increase and retain the number of minorities entering civil engineering and those currently enrolled in a civil engineering program respectively; and 3) Provide workshops at the regional and national conferences on how to better facilitate the transition of minority students from high school to college and to assist school districts and universities in developing curricula that prepare students for civil engineering careers.

Action Item #4

AIM: To have a minority public that is aware of and knowledgeable about the civil engineering programs, benefits and professional opportunities.

RECOMMENDATIONS: 1) Expand the publicity of the civil engineering profession by providing advertisement (articles that communicate the profession's philosophy of "strengthening through diversification") to the National Society of Black Engineers Journal, Society of Women Engineers Journal, and the Journal of the National Technical Association, etc.; 2) Produce a video spot on the civil engineering profession which can be aired on network and minority TV channels; 3) Encourage the ASCE section chapters and local student chapters in each state to offer workshops on state-of-the-art computer applications to elementary, middle, and high school minority science and engineering teachers; and 4) Sponsor and participate in minority community service projects to increase the visibility of the civil engineering profession in the community. Also, establish ties with other non-traditional community service organizations to promote the image of a "peole serving profession."

CONCLUSIONS

Our goal of having a profession that better reflects the nation's diversity can only become a reality with vigorous application of these recommendations. The above action items are required not only to ensure an adequate supply of quality civil engineers which will be needed if this profession desires to retain its technical leadership in the world, but also will allow for equal opportunities for all our citizens.

REFERENCES

1. Adams, H.G. "Advanced Degrees for Minority Students in Engineering." ASEE, vol. 78, no. 8, 1988, pp. 775-777.

2. Basta, N. "Job Prospects for Civil Engineers." Graduating Engineering-Minorities Issue, October, 1987, p. 166.

3. Benson, N. "ASCE Committee on Minority Programs." ASCE-Field Services Memorandum, 1989.

4. Debley, T. "Berkeley Chapter Programs-Black Graduate Engineering and Science Students Take On Sixth Graders." NSBE Journal, vol. 2, no. 5, 1987, p. 12.

5. Kimmel, H., Martino, N., and Tomkins, R. "An Approach to Increasing the Representation of Minorities in Engineering and Science." ASEE, vol. 78, no. 11, 1988, p. 186.

6. Kimmel, H. "An Engineering-Science Outreach Effort for Minority Elementary Grade Students." ASEE, vol. 78, no. 1, 1987, p. 64.

7. New York Times. "Massachusetts Department of Education Study." 1986.

8. U.S. Bureau of the Census. "Statistical Abstract of the United States: 1988 (108th edition)." Washington, D.C., 1987.

9. U.S. Bureau of the Census. "Statistical Abstract of the United States: 1979 (101st ediidtion)." Washington, D.C., 1980.

10. U.S. Congress-Office of Technology Assessment. "Higher Education for Science and Engineering-A Background Paper." OTA-BP-SET-52 (Washington, DC: U.S. Government Printing Office, March 1989).

THE ROLE OF STUDENT CLUBS WITHIN ASCE

Milan E. Degyansky, M.ASCE[1]

Abstract

ASCE sponsors both clubs and chapters at colleges. Most other major engineering organizations have only chapters. This paper explores the differences between clubs and chapters. Observations/suggestions offered are aimed at strengthening ASCE's student programs. The professional development of students in civil programs is one of the principal thrusts of these suggestions. Examples of Section/Branch involvement will be taken from the Georgia Section and its branches which sponsor one chapter and three clubs.

INTRODUCTION

While participating in "Setting the Agenda" in education and continuing development for the civil engineer, why not direct some effort toward ASCE's student programs? The vitality of the civil engineering profession is directly dependent upon attracting, invigorating and retaining competent and committed individuals. Many ASCE members had their first exposure to ASCE while attending college. If the role of civil engineering and of ASCE in the modern world is to be maintained or more importantly improved, educators, practitioners, and the Society must do a better job of providing students the opportunity to begin their professional development while in college. Offering civil engineers, technologists and technicians a positive and nurturing relationship at the start of their professional careers seems not only wise but also necessary if these individuals are to be effective and con-

---

[1]Professor, Civil Engineering Technology, Georgia Southern College, Landrum Box 8045, Statesboro, GA 30460.

tributing members of the profession and society as a whole. With this as a backdrop, an examination of the role that Student Clubs play within ASCE reveals some areas where the profession and ASCE have an opportunity to be more active in developing this element of the whole organizational structure of civil engineering.

CLUBS AND CHAPTERS

According to the 1989 ASCE Official Register, Clubs "are intended to stimulate student interest in civil engineering and to provide Society services to undergraduate students." Student Chapters have their objective listed as "to help students prepare themselves for entry into the Civil Engineering profession and society." It is worth noting at this point that several major engineering organizations, ASME, IEEE and IIE, charter student organizations only as chapters and do not have the club distinction as ASCE does.

Student Clubs were introduced in 1970 by ASCE. Student Chapters were authorized in 1920. The Student Line, Fall 1989, lists membership at 25 clubs and 214 chapters. The major distinction between clubs and chapters is that chapters may only operate at schools which have a four-year EAC/ABET accredited engineering program related to Civil Engineering and leading to a bachelor degree. Since the introduction of clubs, several clubs have become chapters after their engineering program achieved EAC/ABET accreditation. However, numerous clubs do not have the option to become chapters because of the nature of their programs and/or lack of opportunity to change their programs to an accreditable four-year engineering program.

An analysis of the ASCE Constitution, By Laws, Rules of Policy and Procedures, Student Club Handbook, and flyers sent to Student Clubs reveals that the following services, competitions, handbooks, and society definitions are available for clubs and chapters.

STUDENT CLUBS

- Rules of Policy and Procedures Article XI.6
  - Advisor may be any ASCE member in area
  - No procedure given for appointment

- Student Club Handbook

- Construction Engineering Scholarship competition
- NIOSH Safety Student Technical Paper competition

STUDENT CHAPTERS

- Constitution, Article II.5 and IX.3
  Student membership and Student Chapter designation
- By Laws, Article XII.12-18
  Advisor must be a member of the faculty
  Two Section members serve as contact members
  District Director forwards Advisor and Contact Members to ASCE
- Rules of Policy and Procedure, Article XI.4.
- Student Chapter Handbook
- Student Chapter Faculty Advisor Handbook
- Student Chapter Contact Member Handbook
- Several different chapter awards
- Chapter review and rating sheet
- Daniel W. Mead Paper Contest and several scholarships.
- Construction Engineering Scholarship competition
- NIOSH Safety Student Technical Paper competition
- Workshops for student chapter leaders
- Brochure, ASCE STUDENT CHAPTER

The wide difference between clubs and chapters, starting with the lack of club definition or authorization in the Constitution or By Laws and continuing on with the limited number of services, handbooks, and competitions, should make an observer wonder about the role or place of clubs within ASCE. However, if the changes in the Rules of Policy and Procedures and the listing of Student Clubs in the Official Register are noted for the years 1985 thru 1989, it becomes clear that some effort has been made to draw the club into more active participation within events sponsored by ASCE. The changes in the ASCE Official Register were:

1985 - At the listing of the Student Clubs both two-year and four-year undergraduate programs are stated as qualifying for the establishment of clubs.

1986 - The 1985 listing of the Student Clubs shows that only four-year undergraduate programs qualify for clubs. This is the first year that clubs have the opportunity to participate in the Spring Student Regional Conferences. This opportunity is dependent upon the host chapter inviting the clubs to participate.

1987 - This is the first year that clubs are included in the listing of Organized Student Chapter Regional Conference Groups. An asterisk is used to designate clubs. Incorporated into the introduction at the beginning of the listing is a statement encouraging the host chapter to invite the Student Clubs within their regions to the Spring Student Regional Conference.

1988 - The statement at the beginning of Organized Student Chapter Regional Conference Groups now states that Chapters and Clubs assemble for Spring Conferences.

1989 - No changes from the 1988 Official Register.

At no time during these various changes were the Rules of Policy and Procedure changed to more clearly define or improve the services, structure or operation of Student Clubs. The Constitution and By Laws were not amended to designate Student Clubs as a Local Organization. However, the proposed Amendments to the ASCE Constitution listed on page 3 of the September 1989 issue of ASCE NEWS would make members of Student Clubs eligible for Student Member status.

Student membership would go a long way in providing club members with many opportunities within ASCE that are not presently available to them. This change would entitle club members to receive publications and to be eligible for insurance and other benefits such as the following: At the most recent New Orleans convention the forms required an ASCE membership ID# to register at student member rates. This automatically excluded all club members from registering at member rates, since they presently do not qualify for student membership. The ASCE JOB SEARCH HANDBOOK FOR THE GRADUATING CIVIL ENGINEER is available free only to Student Members.

What could/should be the role of Clubs within ASCE? As the world becomes more complex and the need to specialize becomes increasingly important in all aspects of education and employment, the need to keep the lines of communication open and the flow of information ongoing would appear to be a driving force in structuring any organization that has education and professionalism as two of its tenets. Also, as society becomes more driven by liability, certification, and litigation, it behooves any organization of stature, and ASCE should be of such stature, to be aware that its stamp of approval or membership criteria serve society as a measure of value or worth. In support of these suppositions the following recommendations are offered.

To support the education and professionalism effort of ASCE, the membership of ASCE needs to amend the Constitution and the By Laws to include Student Clubs and to make Student Club members eligible for Student Member status. The Board of Direction of ASCE needs to make the Rules of Policy and Procedure for Student Clubs parallel those of the Student Chapter relative to faculty advisor, Contact Members, recognition, etc. The Board of

Direction of ASCE needs to expand the services and competitions available, such as handbooks, scholarship competition, workshops for student leaders, brochures, etc., to the Student Clubs and their members.

In support of the certification/stamp of approval aspect of ASCE's designations, the Club/Chapter distinction based upon ABET accreditation and type of academic program should be retained. With numerous two-year programs continuing to be started and expanded, eligibility for a club should be permitted at schools with TAC/ABET accredited associate degree programs. To better understand and support the technician/technologist programs, ASCE needs to work at strengthening the Contact Member part of the Club structure. This may increase the availability of practitioners for ABET accreditors of two and four-year technology programs. ASCE needs to retain the Associate and Affiliate member categories to provide Club members an avenue for continuing their ASCE membership, once in the work force. Some club members may not be permitted to become registered as engineers-in-training or professional engineers, since the various state registration boards have many different formats for evaluating technology graduates.

Professionalism should be fostered by promoting more interaction between practitioners, clubs, and chapters. Expanding the competitions for club awards would be a good start. Clubs presently submit annual reports but receive no feedback or evaluation of their activities. The opportunity to enter the Daniel W. Mead paper competition could only raise the level of awareness of professional ethics among club members. Contact members for student clubs would provide club members additional insights into professional practice. What an excellent way to "stimulate student interest in civil engineering."

The Georgia Section currently sponsors one student chapter, Georgia Tech, and three student clubs, Southern College of Technology, Savannah State College, and Georgia Southern College. The last two are sponsored by the Savannah Branch. The Section sponsors a yearly outstanding student member award at each institution. The Section, through the Georgia Engineering Foundation, awards a scholarship, presently $2000 per year, to a promising upper classman who is a club or chapter member. The Savannah Branch invites as its guest a club member from each of its two clubs to the monthly branch meeting. The Section currently has the Georgia Tech Chapter run the June meeting. It is hoped that the clubs can be given a similar opportunity in the future. The Section budgets money each year to help the clubs and chapter meet some of its operating or Spring Conference expenses. In the past the section has covered some of the expenses for club or chapter members who are presenting a paper at an ASCE convention/conference.

CONCLUSION

ASCE has set a goal of 200,000 members by the year 2000. ASCE needs to strengthen its student programs and to cultivate this part of the pool of potential members. By including the student clubs in this cultivation, ASCE has the opportunity to develop members for the long term. In addition, ASCE can capitalize on its intention "to stimulate student interest in civil engineering." Additionally, ASCE may be successful in preparing students for "entry into the Civil Engineering profession and society." A commitment to student programs is a wise investment. Students and graduates are a renewable resource.

APPENDIX-REFERENCES

ASCE NEWS (1989), ASCE, September 1989, 14(9), 3.

OFFICIAL REGISTER (1989), ASCE.

OFFICIAL REGISTER (1988), ASCE.

OFFICIAL REGISTER (1987), ASCE.

OFFICIAL REGISTER (1986), ASCE.

OFFICIAL REGISTER (1985), ASCE.

STUDENT CLUB HANDBOOK (1989), ASCE, July 1989.

THE STUDENT LINE (1989), ASCE, Fall 1989, 3(1).

Activity of the ASCE
Committee on Career Guidance

Kenneth D. Munski, Member, ASCE[1]

Abstract

The ASCE Committee on Career Guidance has prepared and cataloged a variety of guidance materials to aid engineers in preparing and presenting effective guidance programs to prospective engineering students. The materials currently available are described along with committee goals for continual updating of the materials. Guidance materials descriptions are based upon the "ASCE HANDBOOK FOR CAREER GUIDANCE" recently published by the Committee on Career Guidance. The guidance materials are intended to improve the quality of society guidance activities, as well as to reduce the time commitment of individual presenters for background research. The guidance materials are available through section and branch officers, as well as through ASCE Headquarters.

Introduction

Attracting and educating an adequate number of talented people to meet the nation's requirement for civil engineers in the upcoming years poses an ongoing challenge to the engineering community. Substantial declines in civil engineering enrollment beginning in the mid 1980's, coupled with agressive plans for accelerating improvements to America's infrastructure in the 1990's, foretell an upcoming shortage of entry level Civil Engineers. One crucial step toward mitigating the probable upcoming gap between civil engineer supply and demand is for today's practicing civil engineering community to

----

[1]Division Manager, Chen-Northern, Inc, Box 951, Great Falls, MT 59403. Mr. Munski is currently chairman of the ASCE Committee on Career Guidance.

become actively involved in locating and recruiting tomorrow's entering professionals. Preliminary data indicates some enrollment improvement within selected civil engineering disciplines beginning in 1988, but overall, substantial shortfalls are still projected.

To grasp the nature of the perceived upcoming shortage of Civil Engineers, it is necessary to consider the driving demographic trends. United States Labor Department statistics indicate that by the year 2000, the total U.S. labor force will be growing more slowly than at any time since the 1930's. The number of young workers, aged 16 to 24, will actually <u>drop</u> by nearly 2 million, or about 8%. Within the engineering professions, which have traditionally been dominated by white males, the new demographics spell even greater reductions in potential recruits, if we continue to rely on the traditional gender and ethnic backgrounds. In 1985, American born, white males made up 47% of the total labor force; whereas by the year 2000, this group will make up only 15% of the new entrants into the total labor force. In 1985, women and non-white ethnic groups made up only very minor percentages of engineers. The engineering professions of 1985 were about 4% women, 2% blacks and 5% Hispanics. Considering that these three groups, which make up only about 11% of the 1986 engineering communities, will comprise most of the new entrants into the total labor force of the year 2000, it becomes obvious we must attract large numbers of women and today's minority ethnic groups if we hope to fill our ranks in the future.

It is the objective of the Committee on Career Guidance to catalog the currently available guidance materials and to aid practicing engineers in using the materials as part of effective guidance presentations. By employing the nationwide resources of the Society for central production and distribution of guidance materials, it is hoped that our guidance presentations will be highly effective, yet require only a minor time commitment from the presenter for background research and program preparation. Ready access to up-to-date guidance materials and ideas gained from prior experience of other engineers with guidance presentations should encourage a significant portion of our 100,000 member Society in becoming "recruiters" for our future professionals.

It must be emphasized that the goal of these guidance efforts is not to simply forever attract higher numbers of incoming students. But rather, our goal is to realistically present the challenges and rewards of civil engineering, along with the best available job market data and projections. With such presentations, we will encourage some of tomorrow's most talented and productive citizens to consider civil engineering as a career, helping to balance the number of entering level engineers with demand.

Overview of Available Guidance Materials

A variety of high quality guidance materials have been developed through national efforts of ASCE, as well as by other organizations. The Committee on Career Guidance is continually developing new materials, and distributing materials produced by others. Materials believed to be appropriate as part of the Society's guidance efforts are made available to all ASCE Sections and Branches (as well as individual members) through national headquarters. Central to the guidance materials and their use in the "ASCE HANDBOOK FOR CAREER GUIDANCE" which describes the available items in detail. The Committee on Career Guidance has developed a series of guidance materials entitled "Is Civil Engineering for You?" to help engineers give effective guidance presentations to potential recruits into the profession. The materials are targeted primarily toward high school students, but are also useful for junior high school and incoming college students. Four different formats are available:

1. Fourteen Minute Video
   A VHS format video describes the various subdisciplines within civil engineering and shows examples of representative projects. Several engineers describe in their own words why they have found civil engineering a challenging and rewarding profession.

2. Sixteen Minute Narrated Slide Show
   This is the 35mm slide version of the 14 minute video described above. One hundred and twenty five slides and an audio cassette with a corresponding script present many of the pictures and portions of the sound track used in the video. It allows the presenter to insert local slides for the audience to more easily relate to.

3. <u>Fourteen Page Brochure</u>
This brochure describes the work done by civil engineers and answers several questions that an interested high school student might ask about the profession.

4. <u>Four Section Flyer</u>
The flyer is an abbreviated version of the brochure described above, and includes a postage paid, tear off post card that may be sent in to request a copy of the fourteen page brochure.

A wide variety of guidance materials developed independently of CCG are also available through ASCE. Depending on the audience, the presenter may wish to use some of the materials described as follows:

1. <u>Would You Like to be an Engineer?</u>
Coloring book. For kindergarten through 5th grade students, it was prepared by the ASCE Committee on Minority Programs (COMP) and describes a civil engineer's role and accomplishments in an urban setting. Guidelines are provided to help teachers implement the coloring book into classroom activities.

2. <u>Dream Your Own Dream</u>
Video tape. The ten minute video was also prepared by the Committee on Minority Programs for grades 8 through 11 using an amusement park setting to present an overview of civil engineering and its different disciplines. The video emphasizes women and minorities and describes the need to take math and science classes in high school.

3. <u>Subject: Civil Engineering</u>
Video tape. The tape is an 8-1/2 minute, fast paced, overview of careers in civil engineering. The tape was not specifically designed for career guidance, but is applicable to college students, parents, guidance counselors and the general public.

4. <u>Agents of America Program</u>
Slide show. The presentation of 80 slides was not specifically prepared for career guidance but is useful in talking to a general audience about what civil engineers have done in the past. It provides an opening to describe future careers in civil engineering.

5. <u>The Ballad of a Civil Engineer</u>
   Slide show. The presentation is a 140 slide narrated show describing the many civil engineering disciplines.

Several other organizations also make engineering career information available. One of the most noteworthy is JETS (Junior Engineering Technical Society), a national non-profit organization whose goal is to encourage interest in engineering, technology and science through junior high school and high school clubs, academic competitions, and other programs. There are over 25,000 JETS members in 42 states. Information available from JETS consists of flyers on the various engineering disciplines, including civil engineering. The flyers are helpful to students, parents and counselors in understanding the difference between the various engineering fields.

In addition to samples, descriptions and availability information regarding the guidance items described above, the "ASCE HANDBOOK FOR CAREER GUIDANCE" contains the following:

1. ABET List of Accredited Institutions
2. ASCE Audio-Visual Brochure
3. "Quality of Life" Posters and Pamphlets
4. Recent Salary Survey
5. Scholarship and Fellowship Brochure
6. Student Line ASCE Newsletter
7. Supply Order Form
8. TEAMS Sponsor Guide

The handbook also provides suggestions for guidance programs using the Mississippi Section program as an example. It is hoped that more example programs from other Sections and Branches will be added as experience is gained with career guidance activities. A list of members with presentation experience who have agreed to share ideas, answer questions, and offer advice is also included.

Conclusion

Additional guidance materials are continually being developed by the Committee on Career Guidance. Information regarding new materials can be obtained from ASCE Headquarters or from any member of CCG. As the new guidance materials are put into use, CCG is encouraging feedback for improvements to future materials and for improvements to recommended presentation methods.

BECOMING A MASTER ENGINEERING STUDENT

Gail Utley[1]

ABSTRACT

In the Fall of 1988, and again in the Fall of 1989, the Speed Scientific School of the University of Louisville offered a credited orientation course for freshman engineering students. Where data is available, it has been demonstrated that students who take these classes tend to be 'persisters,' achieving success on a number of measures. Such a course has proven valuable for high-risk student populations such as ethnic minorities, as well as the very most qualified students. Our curricular components include academic information and policy, study skills, library usage, career planning, problem solving strategies, use of university computing facilities, goal setting and time management, and active use of campus resources, with emphasis on 'departmental exploration,' to discover more about the chosen major field. This paper outlines the structure, goals and curriculum of our 'Master Engineering Student' class.

The University of Louisville

The University of Louisville, founded in 1798, became a part of the Commonwealth's system of higher education in 1970. It is situated in the major urban area of Kentucky, and thus serves its community through close ties with the City of Louisville and Jefferson County. Nearly three-fourths of the University's 20,000+ undergraduate and graduate student population are commuters.

The Speed Scientific School

The Speed Scientific School is one of eleven degree-granting academic units within the University of Louisville, offering a Master of Engineering degree in six areas of specialization. The total enrollment of

---

[1]Assistant Director, General Engineering Studies, University of Louisville Speed Scientific School, Rm. 213 J.B. Speed Bldg., Louisville, KY 40241

Speed numbers approximately 2,100 students. Approximately 700 of these are freshmen and sophomores, with 80%+ of these under age 20.

The Speed Scientific School attracts the area's most qualified high school graduates, and has more stringent admissions requirements, with a full 67% of the 1989 freshman class having an ACT composite of 22 or better.

Course Rationale

Even given the high potential of Speed students, our attrition rate is much too high. Most recent attempts to assess the 'dropout' rate show that we are losing a fair number of students for reasons other than academic difficulties. Data also shows that 38% of attrition is occurring during the first year (national data suggest that most universities lose a disproportionate number of students before Christmas). It seems evident that our traditional two-day intensive orientation sessions, held during the summer prior to the beginning of classes, are not sufficient to prepare students for the rigors of life at Speed School.

Nationwide research efforts into student retention initiatives suggest that an 'extended orientation' experience is needed in order to provide students with 'critical connectors' during the first few weeks of their collegiate career. Inspired by this reality, the Speed School faculty and staff joined in a campus-wide effort to reduce student attrition by offering a freshman orientation course.

Course Goals and Objectives

The guiding purpose of a freshman seminar course is to help students feel more comfortable in the transition from high school to the university environment. Activities must be designed to help students feel more secure and competent during this period of adjustment. Objectives designed to help students reach these goals include:

(a) active use of various on-campus academic and non-academic resources, personnel and facilities,
(b) development of student peer network relationships and use of faculty mentoring,
(c) improvement in academic and study skills,
(d) commitment to the value of higher education, and to the student's particular course of study,
(e) Exploration of the relationship between academic preparation and chosen career, and
(f) Providing general support and encouragement for all aspects of the transition from high school to college.

Instructional Staff

In the fall of 1987, a committee was appointed within the Speed Scientific School to develop a pilot program for a freshman orientation course. This committee felt that the instructional staff should include a combination of faculty, professional staff, and upper level students. The General Engineering Studies office, as the advising home for all freshman and sophomore engineering students, was appointed to coordinate this program. This instructional combination puts students in touch with a variety of resources and differing points of view.

Format of the Course

The Speed planning committee recommended a credited orientation course, to be conducted during the first few weeks of the fall semester. The orientation class is thus offered during the first five weeks of the fall semester, with classes meeting twice weekly for 75 minutes, for a total of 15 class hours.

Although other academic units of the university instituted a mandatory orientation class, the Speed committee chose to have the pilot program be optional, with possible expansion after a period of evaluation.

A wide variety of teaching strategies and learning activities are employed to accomplish the objectives of the course. They include assigned readings, lectures, group discussions and projects, quizzes, guest lectures, interviews and surveys, and both written and oral presentations. One key activity involves having students keep a weekly journal, so that the instructional staff can monitor student progress and address problems early. Special emphasis is placed on student participation, stressing that this course does not call for a passive learning experience.

Required Text

There are a myriad of publications available to provide students with an excellent resource and guide for their college adjustment. We were fortunate enough to find one very specific to the concerns of engineering students. "Academic Gamesmanship: Becoming a 'Master' Engineering Student" , published by NACME (National Action Council for Minorities in Engineering), is a clear and concise guide, providing an excellent overview of the issues and concerns facing engineering students. In addition to this booklet, we assign reading from our Speed School catalog and Academic Advising Handbook.

Selection of Students

Because attendance was not required, students were identified and selected for the orientation course by the advising staff of the General Engineering Studies office during summer orientation sessions. A limit of 25 students was set, with one section offered in 1988, and three sections in 1989. A concerted effort was made to select a diverse student population. Thus we included those at the top of their graduating class, as well as those with a more shaky preparation, those with both top and borderline ACT scores, commuters as well as dorm residents. A special effort was made to include minority students (both ethnic and female) and students who might be at risk because of the number of hours they work off-campus. In 1989, we were able to accomodate more students who simply expressed a desire to take the class.

Grading

As a freshman seminar course that is primarily non-academic in nature, with emphasis on such topics as college procedures and resources, study skills, goal setting, self-assessment, and other adjustment issues, we have chosen to grade the course on a pass-fail basis. This method of grading indicates the process nature of becoming oriented, while making it clear that it is possible to not pass the class without adequate involvement in the process.

Course Content

The Speed School freshman orientation class differed from others offered on campus in that it was major-specific. Therefore, course content was specifically tailored to cover areas particularly salient to engineering students. Keeping in mind the goals and objectives set forth by the Speed committee, ten modules of instruction were developed to be presented in the ten class meetings. The following list provides a brief overview of each area:

(1) Study Skills - To help narrow the focus of the many topics which might be addressed in this area, we evaluated a 'Needs Assessment Survey', mailed out to each new Speed student during the summer prior to the beginning of classes. In this survey, students have an opportunity to express concerns about such issues as test anxiety, reading of technical material, anticipated tutoring needs, time management, notetaking, etc. We then arranged for a professional staff member of the university counseling center to tailor a workshop session to the needs of each individual class. In addition to using the survey results, we polled the class once again, to be certain these topics

are still of the highest concern after the actual beginning of classes.

(2) Learning Styles Inventory - There is growing evidence that many difficulties in college result from mismatch in a professor's teaching style and a student's preferred learning style. To provide insight into this problem, we selected the Learning Styles Inventory, developed by David A. Kolb in 1970. Following completion of the inventory in class, students were provided with a brief overview of the development and uses of this assessment. Students were then asked to participate in class discussion on how this information might be helpful to them.

(3) Campus Resources - Includes knowing where to go for what. An upper level student assistant serves as a referral source. Students are asked to find and bring to class a schedule of study skills workshops, career planning services, counseling center services, tutoring and supplemental instruction schedules, etc. Students can earn extra credit by taking advantage of one or more supportive activities.

(4) Campus Life Experience - Students may earn points for attending campus social events, or for joining any student organization. The student assistant makes sure that the class knows about student chapters of the professional engineering societies.

(5) Introduction to Computing and Library facilities Library and computing center staff members conduct tours of the engineering library and computing facilities. Students are given assignments to assure they can use these facilities at least well enough to accomplish assignments in other classes.

(6) Time Management and Goal Setting - Extra time and effort is devoted to time management and goal setting, beyond the study skills presentation. Because the engineering students pursue a much more rigorous academic program than some of their peers, time management becomes an immediate problem for them. Many students were successful in high school with very little effort applied in this area. In addition to time management, the high level of

commitment required in this program makes goal setting essential. Students are assisted with organizing long range goals into a series of short-range objectives that are relevant and attainable in the here and now.

(7) Departmental Exploration - Selection of a major seems to be the one thing that most prospective engineering students are unsure about. In the student surveys mentioned above, a very large percentage of students indicate they really know very little about the differences in the various engineering specialties. Thus, we have each student conduct interviews, involving both faculty and upper-level students, within their chosen major department. Each major group is responsible for both written and oral presentations.

(8) Career Planning and Cooperative Education Speed students will complete cooperative internships as a mandatory part of their degree requirements. There are few things that could interest a new freshman more than this opportunity to earn money while gaining experience. Guests from the university's Career Planning Office and the Speed Cooperative Placement Office make presentations to the class, involving students in activities to enhance knowledge in these areas. A panel of graduate students also attends one class, to give their own unique perspective on the coop experience.

(9) Becoming an Educated Person/Critical Thinking Skills - Considerable group discussion is encouraged to get students thinking about what it means to be an educated person in our modern society. Sometimes engineering students, not to mention faculty, have different ideas about the practical value of the subjects they must study (especially the dreaded 'humanities-social science' electives). Especially important to these students is the value of critical thinking. Students are thus exposed to critical thinking skills which can be immediately applied to the courses they are taking.

(10) Academic Procedures and Policies/Academic Advising - It is hard to underestimate the importance of a student's understanding of the basic 'rules and regs' of their institution.

Members of the Speed advising staff make sure that students examine their school catalog and advising handbooks very closely (required reading for the class). Special emphasis is placed on the role of advisors and mentors - what the student can expect, as well as what is expected of the student.

Course Evaluation

Students are asked to complete two separate course evaluations. One is a standard evaluation used for all university courses. The other is more specific to this class, and asks students to rate specific course content as well as instructional style.

The history of this course is too brief to assess specific student outcomes. However, each student participant will be followed on a computerized 'tracking' system. Plans for future research involve pairing each participant with an engineering student who did not take the class. The pairs would be matched on several criteria, such as ACT scores, high school GPA, commuter status, and so forth. We hope to find a higher degree of retention as well as a higher level of academic success among the orientation class participants.

Conclusions

Initial student evaluations are quite positive. The students perceive substantial benefits involved in their participation. They rated instructional staff as well as course content highly. Having the students keep journals provided the instructional staff with additional evaluations of an anecdotal nature. One of the more positive side effects was that students identified with a socially cohesive group early in the semester. Many indicated that this was very helpful in relieving anxiety and developing competence.

Summary

More and more colleges and universities are turning to the freshman seminar class to improve retention. Today's student faces a more complex period of adjustment upon entering college. The modern college student is more practically oriented than counterparts of the past. Their educational experience must be linked to the real demands of the world of work. The freshman seminar class both eases this complex transition and establishes a critical link to the real world.

Relative Performance of CE and ME Students

by William H. Highter[1] and
Samuel P. Clemence[2], Fellows, ASCE

Abstract

Civil Engineering enrollments and undergraduate degrees conferred have decreased in recent years. The performance of Civil Engineering (CE) students on the Fundamentals of Engineering Examination (FE) has been disappointing as well. This paper investigates the performance of Civil Engineering and Mechanical Engineering (ME) students in common courses -- statics, strength (mechanics) of materials, dynamics, and thermodynamics at two private and one public university and compares these results with the performance of CE and ME students in the same subjects on the FE examination. A lumped comparison of CEs and MEs from all three universities indicate that there is statistically no significant difference in the means of the class averages of CEs and MEs in statics. The same is true for dynamics. However, the class averages of ME students in strength of materials and in thermodynamics were significantly greater than those of CEs taking the same courses. Data from the results of FE examinations for the past few years indicate that MEs and CEs have about the same performance in statics and strength of materials problems on the exam but that MEs on the average solve about 2 more problems out of 16 correct than CEs in dynamics and about 2-1/2 more problems out of 14 correct in thermodynamics. Data indicate that if CEs performed as well as MEs in these four subject areas, about 70% of CEs would pass the FE exam rather than the approximately 63% who do so at this time.

Introduction

During the past eight years national undergraduate enrollments in Civil Engineering have declined (Table

----

[1]Prof. and Head, Dept. of Civil Engrg., Univ. of Massachusetts, Amherst, MA 01003.

[2]Prof. and Chairman, Dept. of Civil and Environmental Engrg., Syracuse Univ., Syracuse, NY 13244.

1). The peak enrollment was experienced in 1981 with 44,404 undergraduate students in Civil Engineering. Mechanical Engineering experienced increases in enrollment until 1983, and then enrollments declined. In 1988 enrollments in Civil Engineering were approximately half those in Mechanical Engineering as were the number of degrees conferred (Table 1).

TABLE 1 UNDERGRADUATE ENROLLMENTS AND DEGREES CONFERRED IN CE and ME, 1981-1988.
From Day, 1989a and 1989b, and Ellis, 1989.

Year	CE Enroll	CE (Degrees)	ME Enroll	ME (Degrees)
1981	44,404	(10,547)	69,608	(13,462)
1982	41,368	(10,330)	71,388	(14,178)
1983	38,857	(10,484)	71,459	(16,484)
1984	35,403	(9,877)	69,163	(17,214)
1985	34,547	(9,468)	66,738	(17,152)
1986	32,748	(8,798)	67,391	(16,706)
1987	32,577	(8,388)	65,952	(16,056)
1988	32,620	(7,714)	63,971	(15,610)
1989	NA	(7,694)	NA	(15,374)

There is a widespread belief among educators that there is a quality as well as a quantity issue that favors ME students with respect to those enrolled in CE. The quality issue is based on several factors including a) Civil Engineering admission standards as compared to Mechanical Engineering standards, b) academic performance as measured by overall grade point averages, and c) results of student performance on the Fundamentals of Engineering Examination.

Table 2 gives the performance of examinees on the Fundamentals of Engineering exam. Overall, in the past eight years only about 63% of the Civil Engineers taking the exam received a passing grade; about 73% of all examinees from programs accredited by the Engineering Accreditation Commission of the Accreditation Board for Engineering and Technology passed the exam; and 81% of the Mechanical Engineers sitting for the exam were successful. The data in Tables 1 and 2 suggest that nearly all of the Civil Engineers who received BS degrees took the exam whereas about 2/3 of those receiving degrees in Mechanical Engineering did so.

To evaluate the validity of the quality issue, a comparison of academic performance data for Civil Engineering and Mechanical Engineering students has been

TABLE 2 PERFORMANCE OF CEs AND MEs
ON FUNDAMENTALS OF ENGINEERING EXAMINATION
(Examinees from ABET/ECE Programs).
From LeFevre, 1989.

Year	Number of Examinees and (Per Cent Passing) All Programs	CE	ME
1981	33,911(74.5%)	10,046(66.2%)	9,219(83.4%)
1982	33,946(74.3%)	9,775(68.0%)	9,854(80.1%)
1983	34,695(74.4%)	9,627(65.0%)	10,163(82.4%)
1984	34,128(75.6%)	9,981(64.6%)	9,949(84.0%)
1985	35,559(71.7%)	10,262(59.6%)	10,135(78.9%)
1986	35,750(73.0%)	10,131(61.7%)	10,006(80.2%)
1987	35,030(70.7%)	10,362(57.1%)	9,462(78.3%)
1988(April)	21,227(72.7%)	6,177(59.3%)	5,914(81.2%)
Total (%)	264,246(73.4%)	76,361(62.8%)	74,702(81.0%)

developed for three universities. A public university (University of Massachusetts at Amherst, UMass) and two private universities (Clarkson University, CU, and Syracuse University) were included in this study. The data used for a basis of comparison included Scholastic Aptitude Test (SAT) scores; grades received in the following subjects that are common to both Civil and Mechanical Engineering students: statics, dynamics, strength of materials, and thermodynamics; and performance on the FE examination. The study was restricted to Civil and Mechanical Engineers since these majors have substantial overlap in required engineering courses. Academic data were obtained for students graduating in the 1988-89 academic year.

The curricula of the three universities contain required courses in statics, strength of materials, and thermodynamics. Dynamics is required of CE and ME students at UMass and Syracuse.

Analysis and Results

For each course (statics, strength of materials, dynamics, and thermodynamics) a student's t test of the means of the averages for CE and ME students in each course at each university was carried out. The null hypothesis was that the means of the CE average score in a course and ME average score in the same course were equal. The probability chosen for a Type I error i.e. the error made if a true hypothesis is rejected was $\alpha = 0.05$. The number of subjects in each population varied with the course as shown in Table 3. Results of the statistical analysis are summarized in Table 4. No significant difference was found in the means of the grades of CE students and ME students

TABLE 3 SAMPLE SIZES

Subject	CU		UMass		Syracuse		Combined Sample	
	CE	ME	CE	ME	CE	ME	CE	ME
Statics	62	109	42	116	13	28	117	253
Str of Mat	67	126	44	121	17	34	128	281
Dynamics	Not Req'd		45	123	17	29	62	152
Thermo	74	152	47	122	16	34	137	308
SAT Math	No Data		30	97	16	26	46	123
SAT Verbal	No Data		30	97	16	26	46	123

taking statics. The same was true for strength of materials at UMass and Syracuse, but at Clarkson, the ME average was statistically significantly higher than the CE average. Clarkson University does not require the student to take dynamics. The means of MEs and CEs at UMass and at Syracuse were not significantly different in this course. Mechanical Engineering students at Clarkson received significantly higher grades in thermodynamics than the Civil Engineering students did. However, there was no statistically significant difference in the means at either UMass or at Syracuse in thermodynamics. There was no significant difference in the means of the SAT math score between ME and CE students at Syracuse. However, the ME students at UMass had a statistically significant higher average SAT math score than did the CE students. There was no differ-

TABLE 4 RESULTS OF STATISTICAL ANALYSIS OF PERFORMANCE OF CE and ME STUDENTS IN COMMON COURSES (TEST OF MEANS, $\alpha = 0.05$)

	University			
	CU	UMass	Syracuse	All CEs & MEs
Statics	A	A	A	A
Str of Mat	R(ME>CE)	A	A	R(ME>CE)
Dynamics	*	A	A	A**
Thermo.	R(ME>CE)	A	A	R(ME>CE)
SAT Math	No Data	R(ME>CE)	A	R(ME>CE)**
SAT Verbal	No Data	A	A	A**

A - Accept the hypothesis that CE and ME grade means are equal.
R - Reject the hypothesis that CE and ME grade means are equal.

* Dynamics is not required at Clarkson University.
** UMass and Syracuse data only.

ence in the SAT verbal scores between MEs and CEs at either UMass or Syracuse. Data were not available from Clarkson on the Scholastic Aptitude Test for the students in the sample. When all CEs from Clarkson, UMass, and Syracuse were combined into one sample and compared to a similar ME sample, it was found that there was no significant difference between the means in statics, dynamics, or in the SAT verbal scores between MEs and CEs. However, MEs had significantly higher averages in strength of materials, thermodynamics, and had significantly higher SAT math scores.

Discussion

That the ME students had a higher average than CEs in strength of materials at one of the universities was a mild surprise to the authors. This was expected for the thermodynamics course and yet at UMass and Syracuse MEs and CEs did not have statistically significant different grade point averages. It is interesting to compare the results of the study on these three universities with results of all Civil Engineering and all Mechanical Engineering students who took the FE exam in the last three years. These data are summarized in Table 5. Results show that, generally, ME students performed slightly better than CE students in statics. ME students, on the average, correctly answered 8.2 out of 13 statics problems while CE students responded correctly to 7.8 questions. It is not known whether the difference is statistically significant because data for making a test on the means were not available. Table 5 shows similar results for strength of materials. Again the ME students did slightly better than the CE students getting 7.2 out of 13 problems correct compared to the CE students' performance of 6.8 correct problems. In dynamics the difference between the performance of ME students and CE students is much larger. ME students answered nearly 9 out of 16 problems correctly while CE students answered a little more than 7 problems correctly. As expected, the ME students performed better than CEs in solving thermodynamics problems. They answered 7.8 out of 14 problems correctly while CE students answered only 5.2 problems correctly on the average.

It is interesting to examine the differences between the results in Table 4 for the three universities under study and the national figures represented in Table 5. CE and ME students performed about the same in dynamics in the classroom yet did less well on the FE exam. This may support the contention of the ASCE Task Committee, which studied the Fundamentals of Engineering Examination in 1988, that Engineering Science

TABLE 5 CE and ME PERFORMANCE FOR MORNING SESSION OF FUNDAMENTALS OF ENGINEERING EXAM (From National Council of Engineering Examiners, 1986, 1987, and 1988).

		CE Students		ME Students	
Course	No. of Probs.	% Correct	No. of Correct Probs.	% Correct	No. of Correct Probs.
Statics	13	59.8	7.8	63.2	8.2
S of M	13	48.9	6.8	51.5	7.2
Dynamics	16	45.2	7.1	57.0	8.9
Thermo	14	36.9	5.2	56.0	7.8

Note: Data compiled and weighted by number of examinees from the Oct. 1985, 1986, 1987, and April 1988 exams.

courses receive continuing emphasis in disciplines other than Civil Engineering. Given that many Mechanical Engineering curricula require two courses in thermodynamics while many Civil Engineering curricula do not require thermodynamics at all, it is not surprising to note that MEs are much more successful in solving thermodynamics problems on the FE exam than are Civil Engineering students.

Summary and Conclusion

1. Based on statistical analyses of grades from three northeast universities, Civil Engineering and Mechanical Engineering students perform equally well in statics courses. This performance is also reflected in CE and ME scores on the FE examination where ME students perform only slightly better than CE students in statics.

2. Mechanical Engineering students perform significantly better than Civil Engineering students in the classroom in strength of materials. However, this is not reflected in the FE exam where MEs on the average solved only 0.4 more problems correctly out of 13 than CEs did.

3. At UMass and Syracuse the means of the grades of CEs and MEs in dynamics were not significantly different. However, MEs performed better on the FE exam than CEs getting, on the average, nearly 2 more problems correct out of 16.

4. As a group the MEs at Clarkson, Syracuse, and UMass did better in thermodynamics than their CE counterparts; this was also true of the FE exam.

5. The percentage of Civil Engineering students taking the FE examination is much higher than the Mechanical Engineering students. Almost all the Civil Engineering students take the examination while only about two-thirds of the Mechanical Engineers did. It is possible that only the more academically successful and professionally motivated ME students attempt the FE examination.

Acknowledgement

The authors would like to thank Drs. Norbert L. Ackermann and Kenneth D. Willmert of Clarkson University who compiled the Clarkson data for this study.

References

1. ASCE Task Committee on the Fundamentals of Engineering Examination Review, "Fundamentals of Engineering (FE) Examination Review, 1986," Journal of Professional Issues in Engineering, ASCE, Vol. 114, No. 2, April 1988, pp. 173-191.

2. Day, Charles, "Engineering and Technology Degrees, 1988," Source: Engineering Manpower Commission, American Association of Engineering Societies, Washington, DC, 1989a.

3. Day, Charles, "Engineering and Technology Enrollments, Fall 1988," Source: Engineering Manpower Commission, American Association of Engineering Societies, Washington, DC, 1989b.

4. Ellis, Richard, Engineering Manpower Commission, American Association of Engineering Societies, Washington, DC, Personal Communication, 17 October 1989.

5. LeFevre, E. Walter, Compilation of "Candidate Performance Summary for October 19__ and April 19__, Fundamentals of Engineering Examinations," Source: National Council of Engineering Examiners, Clemson, South Carolina, 1981, 1982, 1983, 1984, 1985, 1986, 1987, and 1988.

THE CIVIL ENGINEERING SHORTAGE: REALITY OR MYTH?

By John A. Alexander[1], Member ASCE

Abstract

This study used the numbers and starting salaries of bachelor-level graduates to evaluate the anticipated shortage of civil engineers. Long term trends, demographic projections, and the most recent data on numbers of graduates and their starting salaries were considered. The study found that, contrary to the present conventional wisdom, there is no valid evidence to support the forecast of a future shortage of civil engineers, and that we are now in a period of oversupply. Additionally, it is argued that recruiting efforts will indirectly lower the quality and stature of the civil engineering profession.

Introduction

The civil engineering profession is currently highly concerned that the future need for civil engineers will far exceed the supply and that the subsequent shortage of civil engineering expertise will become a national problem. The goal of this study was to assess the need for civil engineers in the marketplace and was conducted as part of the work of ASCE's Committee on Career Guidance.

Method of Study

Most previous papers addressing the question of need for civil engineers in the marketplace have used

----

[1]Prof. and Chair, Dept. of Civ. Engrg., Univ of Maine, Orono ME 04469

the term "demand" interchangeably with the term "need". Estimates of need/demand have been derived by either trying to predict future economic trends and extropalating from the number of engineers employed at the time of the prediction, or by surveys that ask employers to estimate how many engineering positions they have open, or will have available at some time in the future. It was found that these approaches have usually resulted in overestimating, sometimes grossly, the future need for engineers.

This study is based on the established economic principle that, in a free market, the most significant indicator of changes in the relative scarcity (shortage or oversupply) of a commodity is a change in the market price of that commodity.

It is meaningless and misleading to speak of a shortage of civil engineers in a free market unless increasing salary levels validate the claimed shortage. An increase in demand is not created by an employer saying she would like to hire additional engineers. Demand is only created if the collective market is willing to put its money where its mouth is. Anything short of this is only wishful thinking.

To gain historical perspective on the present situation, statistics for both the award of BSCE degrees and average starting salaries were examined for the 1955 to 1988 period. A search was also made for forecasts and predictions of the future need for civil engineers during this period. Particular attention was focused on the most recent data on BSCE graduation rates, trends in entry level salaries, current civil engineering enrollment, and demographic trends.

Findings

The award of BSCEs has grown from 5200 for the ten year average of the 1950s, to 7700 at present. This represents an average growth rate of less than 1% per year. Year-to-year growth has been very uneven with both increases and decreases during the three previous decades that were much sharper than the decrease in the 1980s.

Between 1955 and 1988, starting salaries for civil engineers increased a total of 7% (0.2% per year) in constant dollars (Professional Income for Engineers 1956), (Statistical Abstract of US 1989). Average income for all full time employees increased 55% during the same time period (Handbook of Labor Statistics 1989).

Instead of finding evidence of past shortages of entering civil engineers, the reduction of BSCE salaries relative to other jobs indicates a substantial oversupply for most of the last 33 years. We share this unenviable distinction with other engineers. However, civil engineering's record is the worst of the four major engineering disciplines. In 1976 the other three disciplines enjoyed starting salaries 9% above the average for civil engineers (Prospects for Engineering Graduates 1976). By 1988 that advantage had doubled to 18% (Statistical Abstract of US 1989). It is not surprising that more bright students now seem to be choosing other engineering majors.

We are 10 years into the much heralded decline in the number of 18 year olds, and the number of BSCE's awarded has declined 20 % over the last 6 years. In spite of this declining production in a period of economic prosperity, starting salaries have decreased in constant dollars since 1980. The latest data show a 1.0 % decrease between 1987 and 1988, indicating that more than enough civil engineers are presently available (Statistical Abstract of US 1989).

The Civil Engineering Shortage is an Illusion

Although this study found no evidence of a national shortage of civil engineers in the last 33 years, it found several extravagent forecasts of the growth in engineering jobs. Some of these could be characterized as dire predictions of future shortages that never happened. These past forecasts of shortages sound very much like the ones we are now hearing.

The inescapable conclusion seems to be that the civil engineering profession has a collective and irresistible proclivity for forecasting national shortages of civil engineers even though none of us has ever seen one.

A cynical observer might attribute this urge to predict shortages that never occur, and an obsession with recruiting, to the self-interest of engineering employers who want a ready supply of cheap labor, and from engineering deans and chairmen, whose welfare within the university political system depends partly on maintaining or increasing student enrollment.

Although both engineering employers and engineering faculty have a vested interest in seeing that a growing number of people enter the field, and even though human frailty sometimes leads us to less than noble actions, this doesn't appear to be the major cause of this phenomenon. Most engineers with a strong interest in recruiting seem to have a genuine concern that more young people get a chance to enjoy the satisfactions of a civil engineering career. This seems to be a natural outgrowth of their love for the profession and their natural optimism about the future.

However, if past forecasts of shortages, coupled with recruiting, have contributed to a perpetual state of oversupply, these projections and recruiting efforts may have played a major role in the decline of both the quality of civil engineering students and the stature of the civil engineering profession.

The Quality Problem is Real

Many civil engineers are concerned that the quality of civil engineering students, the state of civil engineering practice, and the stature of the profession are not what they should be and are declining. Problems cited include

1. Low academic ability of civil engineering students compared to other engineering students,
2. Low passing rates on the fundamentals of engineering exam,
3. The weak position of civil engineers in resisting the spread of price competition for engineering services,
4. Declining professionalism, as more and more civils engineers view themselves as employees who are too financially insecure to rock the boat over professional ethics.

Rationalizations have been offered for all of these, but the profession remains concerned and has initiated a variety of actions in recent years, ranging from the formation of committees and task forces to study the phenomena, to sponsoring floats in the Rose Bowl Parade.

Several papers and reports have recommended action. Oddly enough, the most commonly recommended cure for the quality-related ills of the profession is to improve public relations and to recruit more diligently (Ruess and Vogel 1989).

There is no reason to expect that recruiting alone can raise the quality of entering students. In fact, it is likely to lower that quality. If recruiting is effective in attracting more people into the profession, the resulting lower starting salaries, lower financial security for practicing engineers, and ultimately lower stature for the profession will deter entry of the exceptionally able young people whom the profession needs to help reverse the trends implied above. These young people have the ability and self-confidence to choose the profession that promises the most challenge, financial reward, and stature. Civil engineering will offer less and less of these enticements if the oversupply of engineers continues.

Conclusions

The consistant inaccuracy of our past predictions should make us somewhat humble about our ability to forecast the future scarcity or oversupply of civil engineers. However, on the basis of the findings of this study, especially our recent experience with falling salaries several years into the dreaded demographic problem, it is reasonable to predict that the market forces that have worked in the past will continue to work and will prevent any substantial future shortage of civil engineers.

Further, if the vigorous recruiting efforts now underway are effective, it is reasonable to assume that they will distort the market process by increasing the number of civil engineers beyond what the market calls for. This will result in lower starting salaries, difficulty in attracting high quality students, and the continuing decline in stature of the profession.

Recommendations

The findings of this study lead to conclusions that are not in agreement with the current conventional wisdom of the engineering profession. However, the recent trends in civil engineering starting salaries, the consistant inaccuracies in our past forecasts of shortages, and the present decline of the stature of the civil engineering profession call into question the assumptions that better public relations efforts and more effective recruiting are appropriate strategies.

A detailed set of recommendations would require more space than is available for this paper. However,the following general recommendations are strongly supported by the findings of this study.

1. We should immediately discontinue all general recruiting efforts. Recruitment focused only on high quality prospects, or directed toward attracting women or minorities into civil

engineering, may still be justified, but should present a totally honest picture of the profession including the economic and professional disadvantages.

2. We should raise the standards for civil engineering both in engineering education and engineering practice. This is not a simple or easily achievable goal, but ultimately, civil engineers have the power to control their profession. There will be little progress until we stop deceiving ourselves that the easy approach of better public relations and recruiting can solve our problems.

In a 1976 paper, Milton Alpern forcefully argued that the continued predictions of engineering shortages and recruiting of engineers would perpetuate the oversuppply of engineers and the decline in "any sense of economic security for engineers", and that this in turn would make "real professionalism a practical impossibility" (Alpern 1976). Alpern's contentions were well supported in his paper but were counter to the prevailing conventional wisdom of 1976. The idea persisted that better public relations and more effective recruiting were needed. I suggest that every engineer read Alpern's paper and then take another look at the findings of this study. How long can a profession that prides itself on logical thinking continue on this path?

Appendix-References

Alpern, M., "Quantity Instead of Quality: A Sabotage of Engineering and Its Education By 'Engineer Shortage' Propaganda?", J. Prof. Issues Engrg., ASCE, 102 (E14), 457-469.

Handbook of Labor Statistics, US Bureau of Labor Statistics, 1967, 167.

Handbook of Labor Statistics, US Bureau of Labor Statistics, 1989, Table 41.

Professional Income of Engineers, Engrg. Manpower Commission, 1956, 25.

Prospects for Engineering Graduates, Engrg. Manpower Commission, 1976, 13.

Reuss, M.C., and Vogel, R.M., "Attracting Today's Youth to Civil Engineering", J. Prof. Issues Engrg., ASCE, 115(4),363-368.

Statistical Abstracts of US, US Dept. of Commerce, 1989,156.

THE FIRST JOB--ELATION OR FRUSTRATION

Gordon L. Dugan[1], M. ASCE

Abstract

The first professional position of newly graduated civil engineers is typically an eye opener. No longer are they exclusively associated with their fellow student peers who are primarily striving for one principal goal, the elusive diploma, or with professors who specialize in their particular area of expertise. Relocation to an unfamiliar geographic area is frequently necessary for the first position. Nontechnical problems appear with the new position, such as filling out time cards, selecting insurance plans, and working with seemingly incompatible fellow employees and staff. The time required for the recent graduate to adapt to his/her first position is dependent on numerous factors. Nevertheless the transition period from academia to the so-called "real world situation" can be quite traumatic to many individuals. To help alleviate the frustration encountered by many recent graduates, a 1-credit flexible interaction seminar is suggested which allows a student to complete the seminar by accumulating a given number of points during the semester/term by various activities that range from attending seminars to working for a professional civil engineering organization.

[1]Professor, Department of Civil Engineering, University of Hawaii at Manoa, 2540 Dole Street, Honolulu, HI 96822

Introduction

The first professional position for newly graduated civil engineers is typically an eye opener. No longer are they exclusively associated with their fellow student peers who are primarily striving for one principal goal, the elusive diploma, or with professors who specialize in their particular area of expertise. Gone is the selection of the concentration and amount of their workload, which relates to the type and number of courses enrolled in, albeit that students may put in longer hours during the school week than their professional civil engineer counterparts.

The theoretical/empirical problem-solving approach of student life is replaced by the practical expedient method, which is mired in company/institution policy, political decisions, and the "we have always done it this way" attitude. However, many professional positions are closer to the "cutting edge of new technology" as they relate to projected monetary returns, and the employer's ability to respond to the purchase of new time saving devices, such as the escalating availability of new computers and computer programs.

Relocation to an unfamiliar geographic area is frequently necessary for the first position, while the 40-hour week on the job at set starting and quitting times replaces the concept of when, where, and what to study. Nontechnical problems appear with the new position, such as filling out time cards, selecting insurance plans, working with seemingly incompatible fellow employees, and staff who have been with their employers for possibly a decade or two. In addition, recognition of the recent graduate's ideas and work is probably as important to the individual as the actual compensation they receive.

Influencing Factors

The time required for the recent graduate to adapt to his/her first position is dependent on numerous factors, none the least of which are involved with personality traits, family background and economic status, social involvements, and previous work experience. The environmental setting at the new work place, personality and helpfulness of supervisors and co-workers, employers' work ethic, and challenge of the assignments are all a function of the time required for the transition period. The financial compensation received, location of the job assignment, and reputation of the employer generally have the greatest influence on the choice of the first position after graduation, if a choice of desirable positions are available. However, during recessive economic times a graduate may find it necessary to accept the first reasonable position that is offered, with the intention of possibly finding a new position at a later date. Fortunately, recessive conditions have been relatively rare in terms of hiring civil engineers during the past three decades. During periods of economic growth the civil engineering graduate typically has multiple choices if he/she combs the market and is willing to relocate.

Some new positions have definite training programs and/or a relatively set schedule of alternate job assignments, while others use the "on the job training" that approaches an apprentice type program. The former may be more effective in easing the transition period; however, formal training programs are typically limited to employers with relatively large professional staffs. Regardless of the quality of the employers' training program, if any, the exposure to the so-called "real world situation" of the first position after graduation can undoubtedly be eased by various forms of pregraduation exposure.

Pregraduation Preparation

The apparent most expedient approach to any problem involving the transformation of information to students is to add another course. However, civil engineering undergraduate curriculums are typically already saturated to the point that if a new course is added another will have to be eliminated, if the student intends to graduate in a four-year time period. Complicating the situation are the required basic university/college course requirements and the array of courses necessary for program accreditation.

In terms of establishing a program to attempt to ease the transition period from academic life to professional work involvement it would appear appropriate to operate on a seminar-type mode, with perhaps a 1-hr credit being given for completing/participating in the program, in lieu of a strictly voluntary program. Such a program, by necessity, should remain somewhat informal and flexible, inasmuch as there is no definite method for analyzing or measuring the magnitude and time span of the potential frustration in the first or subsequent positions. Some graduates may essentially have no or very little transition difficulties, for others it may be a near traumatic experience. Graduates who have worked for one or more employers for significant periods of time, including summer and/or part-time employment, would be expected to be included in the former, while someone who went from high school to university/college without outside work experience would be expected to be a potential candidate for the latter.

The suggested 1-hr "interaction" seminar could be structured around utilizing outside individuals for speakers, panelists, and members of discussion groups on a weekly basis, coupled with field trips to several major significant engineering projects and facilities, and involvement with professional engineering organizations. The completion of the seminar could be established along a minimum number of points earned in certain established categories. The minimum number of points and categories could be adjusted by the particular situation at the given educational institution. For example, a wider variety and number of outside professional civil engineer participants and engineering facilities and projects would be expected to be located near a larger metropolitan area setting than in "smaller university/college towns."

The concept of the interaction seminar is not only to encourage interaction with outside professional engineers, but to become knowledgeable in the interworkings of various professional organizations. The latter could cover mundane items of the work place such as, health insurance coverage, sick leave benefits, retirement options, potential annual or periodic bonuses or investment options -- in the case of private organizations, work hours, vacation time, overtime policy, compensation and time-off for additional education, transportation and/or parking options at the work place, just to mention a few.

The so-called mundane items could be covered in one or more of the seminars, meetings or when a panel or discussion group has been scheduled. Members of the panel and/or discussion group should potentially involve a wide variety of professional staff, including not only engineering managers and senior engineers, but recent civil engineering graduates, administrators, technicians, and secretaries. Such a group could address a fuller spectrum of activities at the work place, besides the more exciting conceptual engineering designs that a student has been exposed to during his/her undergraduate studies.

Being actively involved in student professional organizations, notably, the American Society of Civil Engineers, provides interchange with the professional engineering community. Where practical attendance at local, state, regional, national, and even international meetings, conferences, and conventions is an excellent opportunity to become acquainted with a wide variety of engineers and other professionals. Part-time and summer employment provides first hand experience of what to expect upon graduation, but these opportunities, at least for part-time employment, can be quite limited in smaller communities. In addition, students with heavy course loads may not be inclined to accept part-time employment.

Work-study programs (alternately working one term, attending school the next term) are available, particularly through selected governmental organizations, however, this usually approximately doubles the length of time required for graduation. The various working programs, whether it be part-time, summer, or work-study not only gives the student experience, but also provides an opportunity for the employer to evaluate the student for potential future employment. Personal finances has a major influence in securing work during and between school terms. The interaction seminar could also serve as a vehicle for providing information concerning outside work opportunities prior to graduation.

Points could be assigned for the aforementioned work experiences, as well as for attendance to organized engineering functions. These points when placed in established categories

TABLE 1

Proposed Professional Interaction Seminar

Category Participation	Points per/event	Points Minimum Required Per Category
A. Weekly seminars	10	100
B. Organized engineering field trips	20	40
C. Professional engineering organizations		
1. Active student member	50	50
a. Elected officer	25	
2. Attend professional meetings		
a. 1- to 2-hr session	10	
b. All day session	20	
3. School-related engineering projects	(variable)	
D. Work for private or governmental engineering	5/hr	100

could help fulfill the requirements for the interaction seminar. While attendance at a minimum number of weekly seminar sessions would be expected, the student should be able to alternate other activities at his/her discretion in order to obtain the minimum number of points in each of the established categories, as well as the total required to obtain credit for the seminar course. It is conceived that each educational institution's program would have to tailor the interaction seminar according to their own particular situation. The seminar could be a one, two or more term requirement. To be effective the student shouldn't wait until his/her last term to enroll in the seminar, rather it would seem more logical to encourage the student to take the seminar during the junior or first part of the senior year.

An example of a proposed interaction seminar course is presented in Table 1. The assumption utilized for establishing the seminar was that the educational institution was located in a moderate-sized community where access to professional engineering organizations is readily available. For such a scenario it is suggested that the student earn a total of 400 points to complete the seminar requirements, which could be on a credit/no credit basis. To fulfill the minimum number of points in each category a student would accumulate 290 points. Thus, an additional 110 points would be required from one or more of the four categories. Under this scenario students with part-time employment at civil engineering organizations would only need 190 points, provided the student has worked at least 42 hours (210 points) during the term, and/or possibly during the preceding term, summer or vacation times. It is assumed that part-time civil engineering employment would lessen the impact and time required for the transition period after graduation. The proposed points required per category would also encourage students to become active in the student professional organizations.

Overall the interactive seminar should remain flexible and innovative and tailored to the individual needs of the particular educational institution involved.

SUMMER INTERNSHIPS AS CURRICULA AUGMENTATION

Robert C. Williams[1], P.E., M. ASCE
C. Harold Emmett[2], P.E., M. ASCE

ABSTRACT

Although professional work experience has long been recognized as a valued learning tool for the engineering student, the emphasis of student-employer programs has often been placed on personnel economics (for the student) or supplemental labor (for the company). These can be meritorious reasons for hiring students, but the major emphasis for summer employment should be on educating the student. The Agency for Toxic Substances and Disease Registry (ATSDR) has developed a successful summer internship program that focuses on the student. The experiences gained during development of the ATSDR program are beneficial to those contemplating similar summer programs. Furthermore, recognizing how summer internships benefit the employer, the student and the student's academic institution is critical to the discussion of the merits of summer internships as curricula augmentation.

INTRODUCTION

Actual work experience has always been particularly valuable to engineering students. The experience of doing meaningful work that is related to a student's courses gives a sense of reality to the classroom work. This extracurricular work experience results in more interest in the course work and higher achievement.

The Agency for Toxic Substances and Disease Registry (ATSDR) became interested in summer internships and sought to become involved with the engineering curricula at one or more historically black colleges or universities (HBCU). In 1987,

---

1. Director, Division of Health Assessment and Consultation, Agency for Toxic Substances and Disease Registry, Atlanta, GA.

2. Chief, Environmental Engineering Branch, DHAC, ATSDR, Atlanta, GA.

representatives from ATSDR visited several engineering schools in search of qualified candidates for a summer internship program.

Cooperation of the schools proved to be vital to the success of the recruitment. Schools prepared for our visit by screening candidates, based on interest and appropriate background. This allowed us to interview only those who were suitable for the program. ATSDR was able to offer positions to two engineering students from Prairie View A & M University. Since then, the ATSDR summer internship program has expanded into areas including health science, toxicology, and epidemiology. Building on the knowledge and perspectives gained during the first summer internship program, we have been able to enhance the many different internships that have followed.

AGENCY PLANNING BEFORE THE STUDENTS ARRIVE

Several levels of planning must be done before the students arrive. Usually the students are new to the area and have no knowledge of housing, transportation, and other necessities.

Making the transition to this new environment is stressful for them. Realizing these stresses, the sponsoring agency or company must assist the students in meeting their needs. Obtaining appropriate living accommodations is usually the most difficult problem. Since the students will generally be located in an area for less than 3 months, apartment leases may be difficult to arrange. The sponsoring organization can assist the students by working with local universities, realtor associations, and other groups to obtain inexpensive but suitable housing. The sponsoring agency should keep in mind that these basic needs must be met to ensure a productive internship.

Whether the internship is with a government agency or private industry, a budget must be developed. This budget should take into consideration all travel, training, and incidentals and must be approved before the students arrive. If possible, the students should be allowed to manage some aspects of "their" budget in order to gain an additional "real world" experience. For example, if a certain amount is allotted for travel, the student could plan trips that do not exceed the budgeted amount. Regardless of the manner in which the budget is developed, the exact salary must be approved before the student arrives, and provisions must be made for the intern being paid at the end of the first pay period. Nothing detracts from a summer internship more than late arrival of the first few pay checks.

A major complaint of student interns has been that they did not know what to expect of the job or what the sponsor expected of them. By developing a workplan for the internship, by having both the student and management agree to elements of the plan, and then by adhering to the workplan throughout the internship, we have found that few, if any, of these problems arise. Workplans must be developed with the cooperation of all branches and departments that will participate in the program. To be successful the workplan must include these items:

1. Definite goals and objectives

2. A person identified as the intern's point of contact for the entire internship

3. A list of expected accomplishments

4. Flexibility

Goals and objectives for the program are important for the sponsor as well as the student. At the outset, the sponsor and the student must clearly understand that the program is not simply a source of cheap labor for the summer. Everyone involved in the program must adhere to the philosophy that the benefit must be first to the student, and second to the sponsor. At times the sponsor may not see a benefit in the time and expense of a particular phase of training. If, however, both the organization and the student are committed to the goals of the program, this will seldom occur. Furthermore, without goals and objectives, supervisors may tend to view the students as a resource to be used in reaching their own objectives. Goals and objectives also reinforce the student's active participation in the program. Without them, students become disenchanted or even demoralized, which quickly leads to failure of the internship.

Although they may have several assignments during the summer with different supervisors, the students must be able to relate to one person whom they can contact with any problems, whether work-related or not. This person who checks on the interns daily, ensures that such matters as time and attendance are properly handled, that individual supervisors are adhering to the initial training plan, and that interns are notified if new and unforeseen opportunities for training arise. This contact person will receive any products of the interns efforts and will use these products during final evaluation of the training program.

The list of expected accomplishments should include periodic reports from the interns on their travel, status reports of projects they are assigned to, or evaluations of their work on individual assignments. Such reports will allow the sponsor to evaluate an intern's progress and provide mid-course corrections that will maximize the value of the training. ATSDR uses these projected outputs as a means of treating the student as a professional. Any quality product that he or she prepares is endorsed and used by the Agency. Because the students actually become part of a multidisciplinary project team, this emphasis makes the internship a more realistic work experience.

Any employee workplan should be flexible, but this flexibility is particularly important for an internship. The student's workplan must allow for amendments made necessary by changes in work assignments, new initiatives by the sponsoring agency, or loss (through illness or reassignment) of the primary mentor. More importantly, such built-in flexibility will allow the student to take advantage of training opportunities that arise unexpectedly.

AFTER THE INTERNS ARRIVES

An initial orientation must be provided for the interns upon their arrival. A proper orientation is the key factor in enhancing the interns' understanding of both their role for the summer and the expectations the sponsor has for them. The overall training plan should be discussed with them, and they should be introduced to the supervisors they will work with during the summer. A list of expected accomplishments, projects, and assignments should be provided on the first day of the job. This will provide the interns with the "big picture" and help them understand how the various segments of their training are interrelated. The projected outputs should be directly related to their work assignments and should be due early enough in the training period to allow evaluation and discussion before the interns return to school.

Some more mundane expectations, such as attendance and the use of time should also be discussed with the interns so that no misunderstandings arise later. The sponsor should also provide some examples of day-to-day activities so that the interns understand typical office operations. In addition, the students should be introduced to the office staff they will work with during the internship. These introductions should reflect the respect and desire for positi interchange that will be the goal of the entire program.

ASSIGNMENTS

All assignments should agree with the workplan developed before the student arrives. Two very important issues must be kept in mind when deciding on intern assignments: (1) the work must be meaningful and worthwhile, and (2) the level of difficulty should not exceed the student's academic capability. For example, ATSDR's mandate includes the preparation of Health Assessments for hazardous waste sites. A student intern could be assigned a site and asked to prepare the outline for a Health Assessment in coordination with a knowledgeable preceptor. Depending on previous academic training, the student could be asked to evaluate environmental pathways, human exposure pathways, or both. These evaluations, completed with assistance from the preceptor, would become part of the final Health Assessment issued for that particular site.

Assignments given to summer interns should be designed so that they can be completed during the summer. Generally, one large project that requires attention for the entire summer will help the student remain focused. In addition, several smaller projects that take only 2-3 weeks each should be included. These assignments should be as close as possible to actual work assignments given to other members of the regular staff.

Obviously, student interns should be given every opportunity to expand their own expertise. There may, however, be sections, branches, or divisions within the sponsoring organization in which interns cannot work effectively without considerable additional training. In such cases, the interns can be exposed to the work performed by these groups through science seminars, regular staff meetings, or simply by meeting with group representatives who can explain their work. This provides a more complete frame of reference to help the interns put the importance of their assignments in context with the mission of the sponsoring organization.

The assignments, training, and internship as a whole have less meaning if the work performed is not evaluated. The preceptor should meet with the student every few weeks to evaluate progress. This provides feedback to the student at regular intervals and also ensures that the training is progressing satisfactorily. The last week of the training period should include a more extensive evaluation, preferably with a written performance critique and a discussion by the student of the value of the internship. This evaluates both the student and the internship program. Students should be interviewed orally in a relaxed environment, and their views should be used to modify any future internship programs. Each group that a student worked with should provide a critique of the intern's performance. This conveys a sense that the work was important and was not just busy work.

BENEFITS TO THE INTERNS

ATSDR has found that students appreciate and enjoy the "real world" work experience that an internship can provide. The closer the work is to the actual professional activities of the sponsoring organization, the greater benefit the student derives. Other obvious benefits include working in a professional environment (outside school), applying university training to actual problems, solving those problems using engineering logic, and seeing firsthand how government and private businesses operate on a day-to-day basis. Furthermore, if the workplan for the summer is developed properly, the interns can also see their work actually being used.

The purpose of the internship extends beyond the summer and impacts the students after they return to school. Often students will have learned things that will help them make decisions on what type of work they will choose after graduation. After seeing what the sponsoring organization has to offer, they may decide that additional education or other fields may better benefit their career. Providing the basis for such decisions is the essence of an internship. If a program is developed properly, planned properly, and operated by persons who understand its purposes and benefits, it will be a positive experience and can benefit the student for years to come.

BENEFITS TO THE SPONSOR

Several benefits can accrue to the sponsor of a summer intern. First, the sponsor can evaluate the work of a prospective employee for 10-12 weeks. Should this person later apply for a job, excellent reasons for hiring or not hiring are already on record. This is a rare opportunity because evaluating a potential employee who has had only academic training is very difficult.

Second, the summer internship program allows the sponsor to build a recruiting tool within the student's academic institution. Other students will learn about the sponsoring organization and seek employment opportunities there, thus lessening the need for recruitment advertising. All employers are in competition for the best talent. A reputation built through word of mouth by former interns may help overcome an employer's drawbacks, such as smaller starting salaries or fewer fringe benefits.

Finally, the sponsor can build a relationship with the academic institution itself. If a sponsor provides a positive learning experience for the interns, the staff of the educational institution is more likely to encourage students to look to that sponsor for a positive professional environment after graduation. The institution becomes a recruiter and the contact it has with the students can be more effective than a company's recruiting program.

CONCLUSION

Student internships can provide positive work experience for both undergraduate and graduate students. The sincerity of the sponsoring organization in its desire to assist students and the completeness of the planning effort that precedes arrival of the students are two very important aspects of this type of program. The work must be perceived by the interns as useful, and the evaluations at the end of a training period must be used to improve future training sessions. Summer internships can be useful to the students, to the training organization, and to the educational institution. Such programs should be made available for every university's engineering program.

PROFESSIONAL ENGINEERING EXAM
VERSUS
PRACTICE

James K. Cable[1]

ABSTRACT

The paper focuses on the concern of Midwest highway agencies over Professional Engineer examination passage rates. It examines some of the causes and ways the problem can be overcome. In the past many of the exam questions have been focused in a narrow are of Civil Engineering rather than a cross section of the profession. In a time when Civil Engineers are being asked to work in very specialized areas it is important that the examination recognize the background that each will bring to the test. It is the duty of the society and the academic community to develop test questions that evaluate the abilities of the person in the area of Civil Engineering in which they are practicing. This requires the continuing cooperation of those in education and practice.

INTRODUCTION

In 1987, highway agencies in the Midwest identified a problem in the passage rates on the Professional Engineering Exam in the Civil Engineering area. The Iowa State Board of Engineering and Land Surveying Examiners, The Iowa Department of Transportation and Iowa State University set out to investigate the causes and determine ways to improve the passage rate.

THE PROBLEM

In March of 1987, Iowa State University was contacted by the Iowa Department of Transportation to discuss the passage rate of Civil Engineers in the Professional Engineering Examination. In 1986, 63% of the Civils in Iowa failed the examination. This compared to some 83% failing on a national level. A study of the

----

[1]Professor, Civil and Construction Engineering Department, Iowa state University, Ames, Iowa 50011. Fellow and Chair (1989-1990) of Committee on Student Services (ASCE).

Civil Engineering passage rates from 1982-1986 shown in Table 1, illustrates some of the Iowa concern. The passage rates for Civil Engineers in 1982-1984 was 81% or greater and consistent with that of the other engineering professions. From October, 1984 through 1986, the passage rate fell dramatically to 32% for Civils versus 54% for all professions. A review of states bordering on Iowa in 1986 indicated Civil Engineering passage rates of 0 to 13%. These passage rates were a major factor in the decline in the overall professional passage rate from 91% to 54% in four years. It drew the attention of both public and private agencies in Iowa that produced the students and those who required licenses as a means of promotion and employment.

TABLE 1

	Civil			
	No. of Examinees	Passed	Passing Rate	All PE % Passed
April 1982	27	23	85%	88%
October 1982	21	20	95%	91%
April 1983	38	33	87%	87%
October 1983	18	16	89%	88%
April 1984	32	26	81%	76%
October 1984	11	8	73%	58%
April 1985	27	20	74%	75%
October 1985	15	10	67%	56%
April 1986	11	5	45%	63%
October 1986	19	6	32%	54%

University and Department of Transportation officials gathered with the Iowa State Board of Engineering and Land Surveying Examiners to determine the problem and seek possible solutions. They first considered the potential sources of the problem. These included items such as: 1. Lack of proper preparation in the university; 2. Improper or inappropriate test questions and/or grading policies; and 3. Improper preparation of the test participants for the examination.

CONSEQUENCES

The decline in passage rates came at a very inappropriate time. The excessive rate of retirements in the highway agencies, construction firms and private engineering firms was creating a large need for engineers with licenses. The license was necessary for one to be eligible for advancement. Many agencies were forced to leave engineering positions vacant, fill them with non-engineering administrators, or consolidate activities and delete the position. None of these alternatives serves to benefit the agency or the profession in the long run. Valuable expertise and professional competence is lost and cannot be easily recovered.

The lack of registered engineers to fill the vacancies in turn placed pressure on the young engineers and the agencies to concentrate efforts on the examination. Special time and effort was devoted to review sessions and the use of various techniques to prepare for the examination.

A second consequence involved the entry of non-engineering administrators into the vacant positions to provide supervision in many cases. This type of person is not often skilled in the engineering knowledge to meet the legal safety questions. Many felt that the public safety and trust in highway agencies began to suffer as a result.

Those who were unable to pass the test in several attempts began to become discouraged. The desire to become proficient and practice professionally began to decline. This was viewed as a threat to the profession and the public trust in transportation agencies.

REMEDIAL ACTIONS

Representatives of the Iowa Department of Transportation and the Civil and Construction Engineering Extension staff of Iowa State University, met with the Iowa State Board of Engineering and Land Surveying Examiners on April 9, 1987 to discuss the problem. The Board agreed with the analysis of the problem and agreed to assist in the examination of the potential causes of the low passage rates. A selected group of University and Transportation Department staff were allowed to review samples of the examinations given over the four period in question. It should be noted that Iowa is one of five states that limits the range of problems that the Civil Engineer can attempt. It is a way to direct the person to solve problems in the area only in which they "feel competent".

The test review revealed that the test did not contain an adequate number of Civil Engineering questions to address all areas of the practice equally. The majority of persons taking the exams are practicing in the areas of transportation, sanitation, public works, or consulting in Iowa. The questions on a given exam were focused heavily on only one of the practicing areas. Often the questions were reasonable in content, but limited in scope. The exams reviewed indicated a heavy reliance on the structures area in several successive tests. Since the Civils could not work problems in the areas set aside for other disciplines, the exam placed undue requirements on those who were not structural designers.

The review team also noted that most participants are unaware of the grading policies employed in the Professional Engineering Examination. The grading policies have been developed to test the ability of the participant to document the problem solving process. It is designed to measure ones' ability to develop the problem solution. It differs from the Engineer in Training Examination that is often viewed as a timed test that resembles the academic community examinations. In that case, the correct answer is equal to or more important than the method of solution. Each problem in the Professional Engineer Examination is worth ten points. Over half of the points may be obtained for the proper documentation of the assumptions made, and the development of the solution. Passing grades can be obtained for the proper development of an answer which does not have to be totally correct. There appears to be a lack of communication about the grading policy between those giving the examination and those taking it.

Several recommendations were made to the Board of Engineering Examiners. The number of Civil Engineering questions should be increased and an effort made to provide an equal number within each exam that applies to each area of the profession. The examination preparers were asked to review the scope of each question to assure the balance of areas covered in any one test. Agencies that trained personnel for the examination were urged to emphasize the method of how to analyze each question and properly document the process used to solve the problem. The use of University instruction versus specialized "in house" instruction using various types of support materials, delivery methods and amounts of review monitor time were considered. The use of sample problems, discussion sessions and various types of lectures or audio/visual presentations are adequate for most persons.

The study also pointed out something about the characteristics of those taking the examination in Iowa. Table 2 identifies the sources of those taking the examination. Note that the majority of those persons are involved in consulting, state or federal agencies. This is where most of the design work is done that requires the professional stamp of approval. The peak numbers in the state figures represent the results of changes in funding and public policy regarding staffing levels. The large numbers represent the rehiring of personnel after periods of staff reductions. The slow reduction of applicants in the county/city column can be attributed to declining budgets and reductions in the number of assistants retained. The overall number of applicants declined from a high of 50 in 1983 to a low of 18 in 1986 which can be attributed to a decline in the desire to obtain the license, the passage rate, and budget and hiring constraints.

TABLE 2
Types of Experience - Civil Engineering Exam Applicants

	PUBLIC		PRIVATE	
	City/Cnty	State/Fed	Consulting	Other
April 1982	5	4	8	6
October 1982	6	3	10	2
April 1983	3	9	17	5
October 1983	2	2	6	6
April 1984	3	10	9	8
October 1984	0	2	5	2
April 1985	0	7	8	8
October 1985	0	0	9	2
April 1986	2	2	3	0
October 1986	1	4	5	1
April 1987	3	11	6	4
TOTALS	25	54	86	44

The age, amount of experience and educational level of the applicants is illustrated in Table 3. Some 60% of the applicants were under 30 years of age and 35% were between 30 and 40 years of age. A total of 92% of the applicants had less than 10 years of experience and 6% had 10-15 years of experience prior to the examination. The figures pointed to the fact that some 96% of participants had engineering degrees. This data pointed to the fact that most of the persons were taking the examination as soon as they met the four years of practical experience requirement. The relatively constant number of persons in the 30-39 year age group represents those who waited until they had met other obligations such as the military service prior to taking the exam. Others in this group may represent those having problems with passage of the exam.

TABLE 3

	AGE			EXPERIENCE			Eng	No Eng
	25-29	30-39	40+	-10	10-15	15+	Deg.	Deg.
April 1982	13	10	0	21	2	0	22	1
October 1982	16	5	0	21	0	0	21	0
April 1983	28	5	1	23	0	0	34	0
October 1983	8	8	0	16	0	0	16	0
April 1984	19	11	0	29	1	0	29	1
October 1984	4	4	1	6	3	0	9	0
April 1985	14	8	1	22	1	0	22	1
October 1985	5	5	1	10	0	1	9	2
April 1986	3	3	1	6	1	0	7	0
October 1986	5	5	1	10	0	1	11	0
April 1987	11	10	3	18	5	1	7	4
TOTALS	126	74	9	193	13	3	200	9

Table 4 addresses the rate of passage and the number of attempts by participants. It notes the increasing number of failures on the first two or more examination attempts. It was one of the major alarm statistics that got Iowa agencies into the review process. The examination participants are usually young, have limited experience, are college trained, and failure to gain passage on the first attempt is compounded in the succeeding times. It is not a positive sign for the profession.

TABLE 4

				# Times Taken			
				Failures		Passes	
	# Exams	# Pass	% Pass	1	2+	1	2+
April 1982	27	23	85%	3	1	23	0
October 1982	22	21	95%	1	0	18	3
April 1983	38	33	87%	5	0	32	1
October 1983	18	16	89%	1	1	16	0
April 1984	32	26	81%	5	1	23	3
October 1984	11	8	73%	2	1	8	0
April 1985	27	20	74%	6	1	18	2
October 1985	14	9	64%	4	1	6	3
April 1986	11	5	45%	3	3	4	1
October 1986	19	6	32%	10	3	5	1
TOTALS	219	167	76%	40	12	153	14

CONCLUSIONS

If ASCE is committed to professional development, the membership must work to assure a steady flow of trained and qualified persons to become registered. Each member, whether they be part of the academic community or the practicing agencies, must contribute to both the training and the examination of the new engineers. The educational training should provide the person with the basic and broad knowledge of the various areas of Civil Engineering. It should also provide training in how to prepare to take an examination. Practical training must take the form of field experience under the guidance of a registered engineer. The hiring agency must also make allowance for time and assistance in the preparation of the participant for the Professional Engineer Examination. Both practitioners and the faculty must take an active interest in the submission of potential test questions and the review of the examinations to assure that they test for the types of work being performed and do not become narrowly focused. This type of approach applies to both the profession of Civil Engineering and Land Surveying. Cooperation between the parties is the key to success.

In a footnote, many of the review team recommendations have been implemented in Iowa and in the national examination. Passage rates for the Department of Transportation have risen from 40% in the Fall of 1987 to 87% in the Spring of 1989. Changes in the exam questions and methods of instruction are paying off.

THE FACULTY ADVISOR OF AN ASCE STUDENT CHAPTER

R. John Craig[1]
James K. Cable[2]

ABSTRACT
Recently, the position of the faculty advisor of the ASCE student chapter has been de-emphasized in the educational community. The role and incentives of the faculty advisor are discussed. Ideas from faculty and non-faculty members, given at an ASCE educational conference, are presented. Position and ideas of the Committee on Student Services (CSS) are expressed. Recommendations are given which have been formed by CSS from the CSS Committee and faculty advisors in ASCE. Hopefully, the ideas and comments will help the position of the faculty advisor with regard to helping the student who is just starting his/her professional life in Civil Engineering.

INTRODUCTION

With the emphasis of educational institutions at present on items such as research and scholarly work, CSS (Committee on Student Services) wishes to reflect on the importance of the faculty advisor as it pertains to the ASCE student chapter. There is an emphasis on teaching personnel to "shut their doors" and work on research and publications in journals. This is affecting about two-thirds of the faculty advisors at civil engineering schools. It is felt by CSS that the faculty advisors are not given proper incentives to be motivated to do a good job. This paper will discuss the faculty advisor of the ASCE student chapter with emphasis on role and incentives. A review of a discussion session at the ASCE National Civil Engineering Education Conference held in Columbus, Ohio (1985) will offer ideas and comments on this topic. Through this paper, CSS hopes to generate a possible

----

[1]Deceased Professor, Civil and Environmental Engineering Department, New Jersey Institute of Technology, Newark, New Jersey 07102. Member and Chair (1987-1988) of Committee on Student Services (ASCE).

[2]Professor, Civil and Construction Engineering Department, Iowa State University, Ames, Iowa 50011. Fellow and Chair (1989-1990) of Committee on Student Services (ASCE).

solution to giving proper recognition and importance to the faculty advisors of ASCE student chapters.

SOCIETY REQUIREMENTS

The ASCE Official Register - 1989 gives the objectives of the Student Chapter Program, as follows:

-The objective of the ASCE Student Chapter is to help students prepare themselves for entry into the Civil Engineering profession and society.

-Students must master the principles and techniques of engineering to earn their degrees. There are, in addition, important nontechnical subjects, generally called "professional matters," which are not usually covered in class work. Many aspects of these professional matters lend themselves to Student Chapter programs.

-Student Chapters of the ASCE also help civil engineering students begin those professional contacts and associations which, continued through life, are so valuable to the practicing engineer in serving mankind and the engineering profession more effectively.

-In addition, the Chapter is a good medium for exercising principles of personal and public relations. For example, Chapter members prepare, present and discuss papers, conduct Chapter activities, hold office, request and entertain guest speakers, visit engineering works under construction, prepare reports to the Society and read the publications of the Society. These and many other Chapter activities contribute to broad personal development by stimulating an early professional consciousness while the student is still preoccupied with techniques.

Student Chapter activities at any college or university should be designed specifically to complement the formal undergraduate education being provided. Chapter members, in conjunction with faculty members and practitioners, should be encouraged to examine the undergraduate education and to brainstorm a set of goals and activities for the Chapter to make this education more professionally complete. Thus, the role of the student chapter becomes that of a laboratory or course in professional life. Because of this role and its enormous potential for creative use, CSS believes that educational use and status of the Chapter in our colleges and universities must be elevated. The" life" of the Student Chapter is important.

FACULTY ADVISOR ROLE

Given the role of a student chapter as a laboratory or course of professional life, the effective performance of the faculty advisor is critical. The faculty advisor must have many skills and teaching aptitudes. It takes a highly motivated, dedicated, and talented individual to function effectively as a faculty advisor. This paper will bring out all of these ideas. CSS believes that faculty advisors are actually de-motivated by their circumstances at their schools. The paper looks at this also, and discusses the incentives for motivation. Recommendations are provided to assist changes in attitude in support of the important element in our educational system, the students--they are the ones who will benefit.

The ASCE Committee on Student Services (CSS) presented a session on April 12, 1985 at the ASCE National Civil Engineering Education Conference in Columbus, Ohio. The session centered on the role of the faculty advisor. There was a presentation of material on the faculty advisor's role by four speakers. The presentations were:

1. "The Role of A Faculty Advisor - Non Faculty Viewpoint" by D. McGinley.
2. "The Role of A Faculty Advisor - Faculty Viewpoint" by W.K. Wray.
3. "Why Not to Have Incentives for Faculty Advisors" by L. Lauskey.
4. "Incentives for Faculty Advisors" by R. Scranton.

NONFACULTY VIEWPOINT

Dan McGinley described the faculty advisor from a non-faculty viewpoint, as follows: "The faculty advisor was represented by a square with each side being one of the advisor roles. Those roles include information broker, facilitator, membership director, and consultant. At the center was the student. Unequal emphasis of any side will alter the program and proportionally reduce the effectiveness for the students. If any side is missing, the student, a future civil engineering professional, will be lost. It is important that the non-faculty person often expects these roles in the faculty advisor; and recognizes the faculty advisor's contributions to the profession in this area do not go unnoticed.

Information brokers are expected to know about; 1) deadlines and due dates; 2) chapter annual reports/records of the students to effect the most positive areas of activity and involvement.

Facilitators are expected to reduce the anxiety and encourage timely execution of projects by: 1) providing ongoing continuity; 2) gaining section support through Contact Members; 3) being knowledgeable of CE department interests and budgets; 4) maintaining communication with other faculty advisors, national officers and committee members, and headquarters staff; and 5) keeping abreast of local CE interest stories. They are expected to guide students to accomplish in one day what would have taken them seven.

Membership director duties include: 1) student participation motivation; 2) school procedures resource; 3) chapter regional conference motivation; 4) advisory personnel involvement solicitation; 5) available publications resource; and 6) code of ethics enforcement. This person must be a "walking" official register, student chapter handbook, regional conference guidelines, ASCE News, and Civil Engineering magazine.

Consultant duties require practical expertise when advising the students about: 1) activity planning to complement their formal education; 2) report and application preparation; 3) ASCE "visitor" protocol; and 4) school and chapter responsibilities and finances. The advisor must: 1) guide the student chapter direction; 2) communicate chapter activities to school administration; 3) motivate and administer national student membership, and develop student leaders.

The faculty advisor asks students to accept the challenges offered by professional society services which distinguish a professional civil engineer from persons similarly educated.

FACULTY VIEWPOINT

Warren Wray gave his viewpoint on the role of a faculty advisor as follows: This person should act as an ASCE professional role model, providing annual chapter continuity. They should only advise or suggest when necessary and adopt a "Hands off" attitude, unless action is needed. The faculty advisor must have the support of the department head, other faculty members, upper administration, practicing professionals, alumni, and family. The department head and faculty must be responsive to the needs of the ASCE group in the form of advice and assistance, commitment of department resources when required to meet chapter objectives, projects, or activities. They must supplement the faculty advisor expertise and assist in the completion of student projects. Upper level administration recognition of the advisor's efforts makes the job more rewarding. The local ASCE Branch and Section should recognize faculty advisor and contact member contributions to students activities. Alumni support rather than criticism of the faculty advisor's "leadership" toward receiving Certificates of Commendation or Letters of Honorable Mention is a must. Everyone should remember that a faculty advisor has a family and other interests besides teaching and research. Advisor teaching and committee assignment loads should be balanced with the chapter advising level of effort and time demands. This is critical to maintenance of advisor motivation and enthusiasm.

ADVISOR INCENTIVES

L. Lauskey and R. Scranton's incentive presentation centered on the following items: The student chapter is an unusual course. There is a faculty advisor, but not really an instructor. It has no syllabus, no grades, and no set meeting hours. Incentives should serve as encouragement for the faculty advisor to do a good job, and rewards are payment for a job well done. Without them, one has no right to expect performance. It is a valuable activity and should be rewarded by universities in the same manner as teaching, scholarly activity and service. It deserves incentives, rewards and, furthermore, it deserves expectation that the job will be done well. The advisor role is an opportunity in creative teaching. With no grade to give in the course, there is no threat to the students if they don't produce homework or on the exam. The faculty advisor must resort to his own wits, persuasive powers, and the help of others to make it a success. He needs support from administration and the practicing engineers in the form of contact members. It can be said about incentives, that if the job is done well, it will be rewarded and the promise of a reward is, in itself, an incentive.

The faculty advisor assists the chapter in defining its objectives, purpose and the path to accomplishing both. The student chapter should try to isolate and define deficiencies in the curriculum that either have been neglected or simply don't fit in. The student chapter should attempt to answer these needs with activities conducted in an informal atmosphere of mutual respect, where all students and all faculty are working toward the same goal. Establishment of objectives and activities which are going to benefit the student, is the goal. It requires the students to work, with the faculty kicking in and helping as well. All students should have some role, no matter what their commitments are. The students will make mistakes. The faculty advisor's job is to take their mistakes, which can be personally devastating to a student, and turn them into positive learning experiences for all of them. That's teaching! That's education! The student chapters are encouraged to participate in community service projects. These projects will typically require planning, design, financing, and construction. What's the faculty advisor doing? He is monitoring the efforts toward the goal to guard against failure, identifying the importance of the project for the students, and physically helping out on a project on a Saturday or a Sunday. The students appreciate this. The faculty advisor must act as a professional role model for these students if he is to expect and demand excellence.

With a clear picture of the faculty advisor, what are the incentives for the faculty advisor? ASCE could give free dues or free manuals to the faculty advisor. How about the anticipation of self-satisfaction? All the faculty advisors want that self-satisfaction, but that is not enough. There should be financial support for his travel, etc. and staff support for typing, mailing, etc. Faculty should encourage students to participate at all levels, offer suggestions and participate themselves. Consideration should be given in the merit review, tenure, and promotion process. Rewards and incentives should go hand-in-hand with expectations. The student chapter can be seen as a laboratory course in professional life. The objectives should be determined by the faculty and the students along with the syllabus, requirements, and excellence. Don't waste the opportunity simply because of failure to recognize a problem, monetary cost, reluctance to provide release time to some faculty member with the talent and the skills to handle the position.

DISCUSSION RESULTS

What incentives are given to the younger faculty advisors? Should research be more important than advising in promotion and tenure consideration? Why not disband the student chapter for better items such as research? It is not possible to substitute the position of faculty advisor for scholarly work. Actually, a teacher needs all three areas in education: teaching, research and scholarly activity, and service. The department chairman should make sure that balance is maintained. In service, though, doing a good job won't help, but doing a bad job will surely be detrimental. The faculty advisor must show an aptitude for research potential. Department chairmen are an important key as to how things are evaluated. What will happen at the institute level regarding the activities of the faculty advisor? It is research and scholarly work. The faculty advisor will not get any credit above the department level. Professors, including the faculty advisors, are role models for the young people, and they owe it to the ones who had helped them before. Perhaps a section person should be the faculty advisor because of the faculty load. How do we evaluate a good faculty advisor? Some ideas include: 1) lets the students do the work; 2) makes each year a better one than the year before, for the student chapter, and doesn't try to win the Ridgway Award; 3) sets and attains goals; 4) has a quality program; and 5) tries to interest students in participating in ASCE activities after graduation.

The student chapter and the role of faculty advisor are important. Research and scholarly activity are also important in survival in education as a profession. The faculty advisor should be encouraged to do a good job, but must always be aware of the "writing on the wall." The audience agreed with the presentations, but were realistic about the actual situation in which teachers find themselves in academics.

CONCLUSION

The faculty advisor position should be elevated in order to ensure that the full potential of the ASCE student chapter can exist. The faculty advisor should be encouraged with significant career enhancing incentives to be educationally creative and productive. The educational value and status of this activity as a leadership role needs to be elevated. Recommendations from a report on Incentives and Recognition of Student Chapter Faculty Advisors by Rick Scranton for CSS to the ASCE Board of Direction are cited, as follows:

-Encourage continued faculty advisor creativity.

-Convince university staff of the importance of the student chapter and encourage academic administrators and faculty to exploit its full potential.

-Attract faculty advisors possessing the proper talents.

-Provide a forum for the training of advisors and the exchange of Ideas between advisors.

Implementing the above objectives will take thoughtful and vigorous effort by many. Implementation steps include:

-ABET examination of the specific educational contribution of student chapter activity, and the recommendation of increased support activity when appropriate.

-Create an awareness in college administrator professions of the necessity of a creative educational student chapter activity. Identify the critical role of the faculty advisor. Proper individual selection and career-enhancing recognition for a job done well through direct communication between top level academic administrators and advisors. ASCE's representative knowledge and recognition of chapter activities in other successful chapters.

-Local Sections awareness of the educational value of student chapters and recognition of quality performance through academic administrators.

-Provision for interaction between new student chapter advisors and veteran faculty advisors, to provide for the exchange of ideas.

-Provide faculty advisor awards in conjunction with outstanding chapter activities from the section and include academic administrators in the ceremonies.

CSS hopes to help in working with the ASCE student chapter faculty advisors and ASCE section leaders to stimulate ideas and attitudes, to elevate the advisor role and rewards.

REFERENCES

1 ASCE Register, 1989

2.Scranton, R. (CSS), "Incentives and Recognition of Student Chapter Faculty Advisors," Report to ASCE Board of Direction, August 27, 1979.

CIVIL ENGINEERING EDUCATION: THE FUTURE IS NOW

Stewart W. Johnson,[1] Member, ASCE
John P. Wetzel[2]
M. Anis Johnson[3]

ABSTRACT

The United States of America as a nation faces a major challenge in improving its entire educational system, especially education in mathematics, science, and engineering. Our future in this technologically competitive world rests on young people now in our elementary, middle, and high schools and in our universities. Members of the American Society of Civil Engineers and related technical and professional societies have a responsibility to work for the improvement of our educational system. The future of civil engineering education must be considered in relation to the overall education and technology programs and their status in the United States.

INTRODUCTION

The authors of this paper have had valuable opportunities to interact constructively with students and faculty at the elementary, middle school, high school, and university levels. The Young Astronauts who enthusiastically met with astronaut Dr. James van Hoften, a civil engineer, at the Space 88 conference (Johnson and Wetzel, 1988) in Albuquerque, NM in August, 1988 were primarily elementary school students. They heard astronaut van Hoften tell about the space shuttle mission on which he repaired the Solar Max satellite. At the middle school level our efforts are in a company join-a-school program with a middle school in Albuquerque. The join-a-school program pairs businesses and schools and facilitates partnerships that help both businesses and schools improve their programs.

At the high school level we are developing a student paper session for the Space 90 conference (Johnson and Wetzel, 1990), as we did for Space 88, as well as an extensive exhibit of student art related to space exploration. Astronaut Dr. Mary Cleave, an environmental engineer, presented certificates to participating students at Space 88. We have also participated in tutoring programs and in technical presentations for high school students. The Space 88 student

---

[1]Principal Engineer, Advanced Basing Systems. BDM International, Inc., 1801 Randolph Road S.E., Albuquerque, NM 87106; and Adjunct Professor of Civil Engineering, University of New Mexico.

[2]Associate Staff Member, Space Systems. BDM International, Inc.

[3]Educational Consultant, Albuquerque, NM.

program was honored by the Albuquerque Public Schools Board of Education and the Superintendent at a Board meeting.

At the university level we continue to work with faculty and graduate and undergraduate students in several departments at the University of New Mexico and New Mexico State University on research efforts relating to extraterrestrial facilities engineering. A special emphasis in this effort is on the construction in the 21st Century of an astronomical observatory on the lunar surface (Johnson et al., 1989; Burns et al., 1990). The focus of these efforts is productive working relationships among industry, faculty and students in departments of civil engineering, geology, and astronomy. Continued support of education by the engineering and scientific community is important to the essential technological advance of our country.

IMPORTANCE OF TECHNOLOGY

We must not forget that the real basis of world leadership is science and technology applied to our civil engineering infrastructure, environment, medicine, energy, communications, transportation, agriculture, manufacturing, pharmaceuticals, space exploration, and defense (da Costa, 1989). Before Confucius, China was the world's leader in science and technology. Chinese astronomers, world explorers, chemists, metal workers, and medical people were the leaders of their time. When followers of the humanitarian philosopher Confucius sought the ideal social system for China they decided to expend energy on solving domestic problems and avoid wasting resources on exploration and science. The result was a decline of the power of the Chinese and eventual takeover by barbarian armies. This takeover occurred despite the fact that, early in the 15th Century, China led the world in gunpowder, printing, metallurgy, engineering, and medicine. China between 1405 and 1433 sent large fleets to open westward trade routes. An expedition of 62 ships reached the east coast of Africa by 1431. The Ming Emperors halted the voyages and, in an atmosphere of conservatism, Chinese science and technology faltered, trade initiatives ceased, and the Chinese became an exploited people (The Ming Dynasty, 1988).

We in the United States are at the crossroads. We are faced with difficult challenges in educating future engineers and in maintaining in our citizenry the basic understanding of engineering and science needed to make informed decisions on technology and technology applications. This literacy in science and technology is important. We believe that a robust space science and technology program in the United States is critical to our future as a world leader and that the civil engineering profession will have an important role in that program.

EDUCATIONAL TESTING RESULTS

The Educational Testing Service (ETS), with funding from the U.S. Department of Education and the National Science

Foundation, recently performed a study of 24,000 13-year-old students in the United States, the United Kingdom, Spain, Ireland, Korea, and four Canadian Provinces (Byrne, 1989). The tests given as a part of the study were translated from English into the native languages. Math achievement results were highest in Korea. The lowest scores in mathematics were by students from the United States and French-speaking students of Quebec.

In science the highest achievements were by students from British Columbia and Korea. At the bottom in science were students from the United States, Ireland, and French-speaking populations of Ontario and New Brunswick. According to Archie Lapointe of ETS, the results are reasonable indicators of what 23-year-olds of 1999 will be able to accomplish in science and mathematics. The American students were weakest in mathematics and science skill areas most often predictive of future careers in civil engineering and related fields.

U.S. EDUCATIONAL SYSTEM

The shortcomings of our United States educational system are widely recognized (Adams, 1989). Where these challenges are particularly significant for our future as a nation are in science and technology. In spite of efforts to help improve teaching and to upgrade courses, the National Science Foundation reports that "even our best mathematics students have the lowest achievement scores of any country for which comparable data are available". The National Assessment of Educational Progress (NAEP) has found that "only seven percent of U.S. 17-year-olds have the knowledge and skills to perform well in college level science courses". In mathematics, because of losses in the 1970's, recent gains in performance levels still leave us behind where we were twenty years ago. Blacks, Hispanics, and other minorities will constitute 35 percent of our population by 2020 (Tifft, 1989b). Blacks, Hispanics, and women continue to display lower performance levels in mathematics as they approach graduation from high schools.

Many students entering college have weak backgrounds in algebra, geometry, and trigonometry. Some calculus courses taught in high school give these students unwarranted confidence when they enroll in college calculus. The short supply of high school mathematics teachers exacerbates the problem. Computers and sophisticated hand-held calculators will force changes in the teaching of calculus (Cipra, 1988).

The teaching of physics has not been revolutionized in the way that the practice of physics has been revolutionized by the computer. Although research in physics is accustomed to rapid change, the rate of change in teaching is slow. Better access to computers, better software, and faculty members prepared to use computers in teaching are needed. The content of physics courses and physics textbooks needs revision. Consideration is being given to application of artificial intelligence techniques and expert systems, use of simulations as lecture demonstrations, and computer data acquisition, programming, and modeling. These applications in physics lectures and laboratories could lead to substantial improvements in the teaching of physics in coming years (Wilson and Redish, 1989).

Public comprehension of national scientific and technological challenges is important for our future. We have strong research departments, but responsible leaders in government and industry must be able to understand and implement research outputs in a climate that can be described as one of national science and engineering illiteracy. Our civilization rests on scientific and technological underpinnings. Can we sustain our leadership as the numbers of our scientists and engineers fall and the literacy of the general population in science and engineering remains low? Richard C. Atkinson, president of the American Association for the Advancement of Science, believes that shortly after the year 2000 our demand for Ph.D.'s in the natural sciences and engineering in the academic, governmental, and industrial sectors will be less than 60% filled by our graduation rates (Holden, 1989).

FOREIGN SCHOOLING

In the age of space we live in a competitive world which relates to many nations and technical disciplines. Figure 1 relates to the relative importance of engineering education in the USSR and Japan compared to the U.S. (Quayle, 1988). The Japanese soil produces few things that would sustain a nation so the country chooses to cultivate schooling. Tomonao Hayoshi, a Japanese space scientist, notes that the long-term emphasis has been in the primary education system with an emphasis on technology studies (Triplett, 1989). The view is that technology studies are of obvious practical value and can be adopted and digested very easily. The Japanese are increasingly aware of educational systems in other nations (Tifft, 1989a).

COUNTRY	NUMBER	% OF TOTAL GRADUATES
USA	78,000	8%
USSR	335,000	39%
Japan	72,000	20%

Figure 1. 1985 Engineering Graduates.

Japan's National Space Development Agency (NSDA) has become a major partner in the international space station. Their Japan Experimental Module (JEM) will be manned by three Japanese astronauts who are already in training. Plans also include microgravity experiments, a free-flyer, an unmanned shuttle, and eventually a space plane. Their H-II rocket is to have a lift capability close to that of the Ariane 4 of the European Space Agency (ESA). Their planning seems to focus on the mid-1990's although the H-II is scheduled to be ready in 1992 when they could begin offering international launch services (Triplett, 1989) because their licensing agreement with McDonnell Douglas expires. The Japanese continue to search for a program which has a consensus of popular support and will prove to be practical, profitable, and of benefit to everyone.

The Japanese construction industry has shown a keen interest in the United States and Japanese space programs. In March, 1988 the senior author was invited to lecture on lunar soil mechanics and construction and participate in discussions with leaders from Japanese construction companies on future construction opportunities on the Moon. The Proceedings of Space 88 (Johnson and Wetzel, 1988) have papers from the Japanese construction industry that explore these opportunities. This strong interest continues in the Space 90 program (Johnson and Wetzel, 1990).

CHALLENGES TO IMPROVE EDUCATION

Civil engineering education is a journey, not a destination. Education encompasses motivation and guidance of potential engineering and science students from elementary school onward and lifelong learning. There is a continuing need to attract more and better students to programs in science and engineering. Some bold initiatives are required (Massey, 1989):

1. Members of professional societies could ask their members to contribute several hours per week to improve science and mathematics education.
2. Professional societies' members should pledge involvement to drastically reduce attrition rates of students planning to major in science, engineering, and mathematics.
3. Professional societies' members could work to encourage cooperative programs for promising students hoping to transition from precollege institutions to institutions where they earn degrees in science and engineering.
4. Identify funds in research grants so that participation of undergraduates in research-related projects would be encouraged.

The individual initiatives suggested above can be facilitated by government support at the local, state, and federal levels. The nation's greatest challenges - being competitive internationally, being literate in engineering and science, having a capable work force, and meeting the needs of the infrastructure and environment as well as the space programs - all require science, engineering, and education.

In the report "Educating Scientists and Engineers, Grade School to Grad School" published by the Office of Technology Assessment in 1988 and in recent reports from the National Science Foundation, the challenges we face as a nation are clear (Holden, 1989). Estimates by the National Science Foundation suggest that the dip in the population of 18- to 24-year-olds to 24 million in the mid-1990's from 30 million in 1980 will lead to a substantial shortage of scientists and engineers by the year 2006. The numbers project a shortfall of 675,000 in the scientist/engineer population by 2006 and a shortfall in Ph.D.'s of 9000 by the year 2000. Studies show that white males now constitute 47% of the total workforce but by the year 2010 they will make up less than one-third of the college-age population.

There are, however, positive steps being taken to interest young people in new frontiers in engineering education. Some possibilities for future engineering education are presented by Ardis (1989). DeJarnette (1988) and Hartmann, et. al. (1988) present examples of university programs to which civil engineers can make significant contributions. Colorado State University and cooperating organizations (Sadeh, 1989) have recently developed curricula for civil engineering in space under a NASA space grant program.

ASCE AND NATIONAL SPACE POLICY

The ASCE National Space Policy Committee[1] and the Committee on Policy Review recommended, and the ASCE Board of Direction approved, that the American Society of Civil Engineers support U.S. National Space Policy. The key points in the ASCE policy statement include:

1. Support for space-related research and development including, for example, the NASA Project Pathfinder, which is a technology development program for the manned exploration of the solar system.
2. Support for U.S. near-term initiatives to achieve long-range goals in space
   - Earth-based research and development facilities
   - Space Station Freedom
   - Mission-to-Earth, which is the use of space technology to analyze and predict changes in the Earth's environment
   - Planning and technology development for activities on the moon and beyond.
3. Strong support for education at all levels and required technology development to support space exploration.
4. Continuation of ASCE activities and conferences that support national space policy, such as "Space 88" and "Space 90".
5. Sharing its intellectual resources with the Administration, Congress, and the space program to assist in making choices between competing national priorities.

In adopting this approach, the committee noted that ASCE and the U.S. space program have important common interests and objectives. A properly conceived space program promotes the progress of engineering, science, technology, national health, prosperity and welfare. Engineering research and education

[1]Space Policy Committee members are: W. David Carrier, III; Raymond S. Leonard; Stewart W. Johnson; Ted Johnson; James Naismith (chairperson); and Brenda Seymore Sanders (ASCE staff contact).

have important roles in advancing our society in a very competitive world where we lag in significant areas. Reestablishing strong support for space research and education will significantly contribute to providing the Nation innovative and creative civil engineers to serve the increasingly complex needs of society.

APPENDIX I. REFERENCES

Adams, Robert McC. (1989). "Smithsonian Horizons," Smithsonian Magazine, Vol. 20, No. 4, page 8.

Ardis, Colby V. (1989). "Teaching Civil Engineering in Space," Civil Engineering, Vol. 59, No. 9, page 6.

Burns, Jack O., Nebojsa Duric, G. Jeffrey Taylor, and Stewart W. Johnson (1990). "Astronomical Observatories on the Moon," Scientific American, to appear in March issue.

Byrne, Gregory (1989). "U.S. Students Flunk Math, Science," Science, Vol. 243, No. 4892, page 729.

Cipra, Barry A. (1988). "Calculus: Crisis Looms in Mathematics' Future," Science, Vol. 239, No. 4847, page 1491.

da Costa, Robert (1989). "The Currency of Leadership," Defense Science, Vol. 8, No. 4, May, 1989, page 5.

DeJarnette, Fred R. (1988). "Establishing a University-Based Mars Mission Research Center," Engineering Education, Vol. 78, No. 9, pages 23-25.

Hartmann, Melvin J., Frederick P. Povinelli, and Steven C. Hartman (1988). "NASA's University Space Engineering Research Centers," Engineering Education, Vol. 78, No. 9, pages 19-22.

Holden, Constance (1989). "Wanted: 675,000 Future Scientists and Engineers," Science, Vol. 244, No. 4912, pages 1536-1537.

Johnson, Stewart W. and John P. Wetzel (1988), editors. Engineering, Construction, and Operations in Space, Proceedings of Space 88, American Society of Civil Engineers, 1361 pages.

Johnson, Stewart W., et. al. (1989). "Lunar Astronomical Observatories: Design Studies at the University of New Mexico" in Engineering Excellence: People Make the Difference, Proceedings, 1989 Annual Conference in Lincoln, NE, American Society for Engineering Education, Washington, D.C., pages 12-14.

Johnson, Stewart W. and John P. Wetzel (1990), editors. Engineering, Construction, and Operations in Space, Proceedings of Space 90, Publication pending.

Massey, Walter E. (1989). "Science Education in the United States," Science, Vol. 245, No. 4921, pages 915-921.

Quayle, Dan (1988). "Can the U.S. Win the Other Space Race?: Space Science Education," Defense Science, Vol. 7, No. 9, pages 52-53.

Sadeh, Willy (1989). Personal communication.

"The Ming Dynasty," (1988). Time, Vol. 132, No. 16, pages 10-11.

Tifft, Susan (1989a). "Japan's Search for U.S. Colleges," Time, Vol. 134, No. 4, page 57.

Tifft, Susan (1989b). "The Search for Minorities," Time, Vol. 134, No. 8, pages 64-65.

Triplett, William (1989). "Japan is Knocking at Heaven's Door," Air and Space Smithsonian, Vol. 4, No. 3, pages 52-63.

Wilson, Jack M., and Edward F. Redish (1989). "Using Computers in Teaching Physics," Physics Today, Vol. 42, No. 1, pages 34-41.

CIVIL ENGINEERING AS A CAREER FOR WOMEN

Barbara Fox,[1] M. ASCE

Abstract

As a practicing civil engineer for over thirty years, the author's experiences and observations first as an engineering student and then in her engineering career provide the bases for suggested improvements in guidance and professional development efforts to attract and retain more women as civil engineering professionals. If we are to meet our need in the 90's and beyond for qualified, competent civil engineers in the United States, we must intensify our efforts today.

Introduction

Although white males dominate the engineering profession in the United States at the present time, the situation is changing, and in the future greater reliance will have to be placed on women and minorities to meet our civil engineering needs. Past efforts by civil engineering educators and practitioners to increase the number of women engineering students and assimilate women engineers fully into the profession have been inadequate. To meet our anticipated needs for qualified, competent civil engineers, we must learn from the past and adequately address this problem now and in the future by increased guidance efforts directed toward female students and increased sensitivity to women engineers' special problems and needs. This paper presents an insightful look into this problem by a woman civil engineer.

Preparation for College

By my junior year in high school, I knew I wanted to be a civil engineer. Math was my favorite subject

---

[1] Water Engineer, Bur. of Water Oper., Dept. of Water, City of Chgo., 1000 E. Ohio St., Chgo., IL 60611

in grammar school, and I had a very good foundation in it at my Catholic grammar school as well as excellent English classes. Our science instruction, however, was very limited. Because of my grammar school academic record, in high school I was placed in the college preparatory course which included four years of math, science and English.

We had no high school guidance counselors and our science teachers did nothing to encourage those who were taking four years of math and science to pursue scientific careers. However, I had a very good math teacher in freshman year and another nun who was an excellent math teacher and good friend my last three years. She was also my junior homeroom teacher, taught a slide rule class after school and gave us materials about career opportunities long before career guidance was "in". She encouraged my developing interest in civil engineering then, and she kept in touch over the years and applauded my accomplishments as my career progressed.

As a high school junior, I prepared a paper entitled "Civil Engineering as a Career for Women". It covered branches of civil engineering, educational/social qualifications necessary to be a civil engineer, advantages/disadvantages of civil engineering, types of employment open to civil engineers, women in civil engineering in World War II and in 1951 when the paper was written, and the outlook in the future for women engineers. With some updating, those topics would still be apropos for career guidance efforts today.

None of my relatives or family friends were engineers, but my family valued education. Also, careers for women were accepted in my family, and my mom provided the support I needed to pursue my goal. My dad died when I was in sixth grade, and my mom taught public grammar school for almost 40 years. At first she was somewhat skeptical about my wanting to be a civil engineer, but she supported me by not trying to dissuade me from my goal and by paying for my college expenses at Marquette University.

Studying Civil Engineering

I was accepted at Marquette and entered the engineering school in September, 1953. I was assigned a "big sister" to "show me the ropes", but by the time I arrived at Marquette, she had transferred to Liberal Arts. There were no other female engineering students in the freshman class of over 300, and a sophomore transfer student in mechanical engineering was the only other female student in the College of Engineering.

As the only two women engineering students, we had no role models to emulate, but we became good friends almost immediately. A theology class we shared in my sophomore year was my only class where there was another woman in my years at Marquette. In fact there were only 2 to 5 women in the engineering school of approximately 1,000 at any time during my years at Marquette. This was quite a change from my all girl's high school of over 1,000 students, and it took a while to adjust to the situation.

Soon I felt pretty well at home. The fellows would help me with classes with which I was having trouble such as engineering drawing, and I would help them with subjects I found easier such as algebra. Few of the male students or teachers resented a girl in their class, but I got the distinct impression the civil engineering department head looked unfavorably upon women studying civil engineering. As a female engineering student I received no special counseling, but school officials felt I should be more aggressive although I was doing well in class.

At Marquette, just as at so many other engineering schools throughout the country, the number and percentage of women entering and graduating in engineering has risen considerably since 1957. The freshman engineering class now has 220 males and 67 females, while the junior civil engineering class has 13 females out of 55 or 23.6%, and the senior class has 11 females out of a total of 55 (3). The engineering administrators at Marquette today actively encourage women to enroll in the College of Engineering as they do at so many other engineering schools.

There were no women teachers in the College of Engineering, and my only foreign teacher was an Austrian-American professor who was an excellent political science teacher. The large number of foreign graduate students and foreign faculty who are creating problems for women students, graduate students and faculty in some engineering schools today were nonexistent at Marquette in my day.

I attended college before coed dorms were established and curfews were abolished, so there was no one to study with late at night. Also, the engineering, medical and dental schools were on quarters, and the rest of the University was on semesters, so I had exams and school breaks at different times than my roommate and other dorm friends, and I took my required gym classes on the semester system with women students from liberal arts. Some consideration was extended for gym classes missed during engineering break periods, but the other problems I had to cope with by myself. A sense of humor, an independent nature, and the ability to adapt to a variety of situations were necessary attributes for woman engineering students and women engineers then and now. I was determined to stay in civil engineering in spite of the problems encountered, but some female engineering students who lack this determination may need special counseling to keep them from dropping out once they enter engineering school. The engineering administrators at Marquette recognized this determination by allowing me to participate in graduation ceremonies with my original class in June 1957 although I didn't finish my classes and graduate officially until December 1957, because I had viral pneumonia when I was a sophomore.

The lessons I learned from my educational experiences than and in engineering education related activities since then are many. The value of math, science and English education must be stressed as early as grammar school, and female students with an interest in and an aptitude for math and science must be encouraged to pursue further studies in these subjects in high school and college. Parental support and encouragement will be a big plus in these endeavors. ASCE must work with educators, school councils and parent groups, the Society of Women Engineers and minority coalitions and groups to make special efforts to target minorities and females and encourage them to prepare for and pursue education in civil engineering. All engineering colleges and universities must make special efforts to attract and retain female engineering students and provide special programs and counseling when necessary. Incentives must be provided and special programs must be undertaken to increase the number of female graduate students, teachers and administrators in engineering colleges and universities and female practitioner involvement in engineering education. Finally inequities regarding male and female graduate students and faculty members must be eliminated, and any problems created by foreign-educated faculty must be confronted (1,2,5).

As a sophomore I joined the ASCE student chapter and was elected secretary during my junior and senior years. The associations with my fellow students and the Wisconsin Section members of ASCE as a student chapter member were valued by me then, and I feel they were an important aspect of my professional development in my transition from engineering student to practicing engineer. The importance of active ASCE student chapters, well supported by their school administrations and local ASCE branches and sections, in the development of future active members of the profession and of ASCE cannot be emphasized too highly.

My Career and Professional Development Activities

The summer before receiving my BCE from Marquette, I began working for the City of Chicago, Department of Water as a Sanitary Engineer in the Water Quality Surveillance Section. At Marquette I took the Municipal Option (as opposed to the Structural Option) in the civil engineering curriculum and had already taken water related and environmental sanitation courses. I was told I could return there as a full-time employee after my December graduation. Because I liked my work and my co-workers, I was happy to return there after receiving my degree. I have continued to work for the Department of Water since then in various engineering and administrative positions in the Department's Water Quality Surveillance Section, Water Purification Division and Bureau of Water Operations. As Water Engineer I assist the Deputy Commissioner of Water Operations in engineering management duties for the Bureau.

I've enjoyed much of my engineering career, however, in some respects the "old boy syndrome" operated in the Department of Water for many years. There was a reluctance to give women engineers and laboratory personnel shift work and field experience which would help them advance to higher level positions, to hire women in the operating and trades positions, and to advance women in non-clerical positions beyond certain levels. Many of these barriers have been removed, but old attitudes die hard, and there are still some invisible, unstated barriers with which we women must deal.

The Society of Women Engineers (SWE), incorporated in 1952 as a non-profit educational organization dedicated to making known the need for women engineers and encouraging young women to consider engineering as a career, has been a very active, vocal force in toppling many of the barriers against women engineering students and women engineers throughout the country. SWE has always encouraged its members to belong to and actively support those societies representing their particular field of engineering, and most of the women who have become very active in professional organizations such as ASCE were first very active in SWE at the local and/or national level.

My own story is a good case in point. Although I was an active member of the Marquette Student Chapter, I didn't begin to get really involved in Illinois Section activities until 1970. This culminated with my service as the Illinois Section's first woman President in 1984-85. Nationally I served on my first ASCE committee in 1974, and in 1988 was the first women to serve on the ASCE Board of Direction.

Meanwhile I had joined SWE in 1958 and served as Chairperson of numerous Chicago Section Committees, in all Section offices and as Section Representative to National. Nationally I chaired five committees from 1962-67, was an Executive Committee Member, 1964-67 and served as Treasurer SWE, 1967-69.

In SWE we received the opportunity and encouragement that was often lacking elsewhere. Because of its size every member could become an officer or chairperson of one or more local committees, and most of us were involved in guidance activities such as science fairs, certificate presentations to outstanding female high school students, and professional and school presentations on women engineers. The organizational and leadership ability we developed then as well as our avowed support of educational and guidance activities was later put to good use in ASCE and other professional organizations. The percentage of women engineers involved with guidance related activities is much higher I'm sure than that of their male counterparts, and SWE deserves a great deal of credit for this.

The Outlook Today for Women in Engineering

SWE is still at the forefront of guidance activities directed toward increasing the number of women engineers, especially minority women engineers. ASCE and other professional organizations and educational institutions whose guidance efforts in those areas need improvement would do well to join forces with or emulate SWE. Implementation of all provisions in the ASCE policy statement entitled "Opportunities in Engineering for Women" and all suggestions made by women members of ASCE in response to a 1976 MADEX survey (4) would be a step in the right direction. Over the years SWE has had its detractors, including some members of the civil engineering community, who feel a separate society for women engineers is unnecessary. When the idea of women in engineering become natural to all segments of society, and women engineering students and women engineers no longer need a separate society sensitive to their special needs and problems such as sex discrimination, family issues, part time jobs, reentry problems, delayed and switched careers, and barriers to advancement to high level management positions, then SWE can disband. However, SWE which also serves as a source of role models and mentors for younger women engineers, women engineering students and grammar and high school students with an interest in math and science, is still needed today to champion the cause of women engineering students and women engineers.

Many of the stumbling blocks for women engineering students and women engineers have been eliminated or at least been brought down to a more manageable size in the past thirty years, however, much still remains to be done. Female engineers can play a leading role in these efforts, but because of our limited numbers and limited resources we need visible, vocal support and action from our male colleagues, professional organizations such as ASCE, engineering educational institutions, government agencies, the business sector and the public at large. With all of us working together, we can complete this task.

References

1. Braum, Eleanor, "Why So Few Women In Engineering", Engineering Education, Vol. 79, July/Aug., 1989, pp. 556-557.

2. "Briefings", Engineering Education, Vol. 79, July/Aug., 1989, pp. 530-531.

3. Faherty, Keith F., Personal Letter, Milwaukee WI, Sept. 21, 1989.

4. MADEX, "Women in Civil Engineering", Engineering Issues, ASCE, Vol. 104, July, 1978, pp. 213-228.

5. Vetter, Betty M., Finding and Keeping Technical Experts In the 1990s., 1989. Adapted from a presentation at the First International Conference on Human Resources Management for A/E Firms, San Francisco, CA, April 10-13, 1988.

QUALITY AND NEEDS OF CIVIL ENGINEERING EDUCATION

Paul Keim[1], Fellow A.S.C.E.
Ace Rasmussen[2], Member A.S.C.E.
Boris Korin[3], Member A.S.C.E.

ABSTRACT

As part of its concern for civil engineering education, the California State Council of ASCE has adopted a position paper regarding the quality and needs of civil engineering education. Eight points addressing the curricula of engineering schools, recruitment of teaching personnel, interaction between practicing engineers, students, and educators, and ASCE involvement in education are presented for consideration.

BACKGROUND

Civil engineers are concerned about the quality and content of engineering education because these relate directly to the practice of civil engineering and to the credibility of the profession with the public.

In general, education can be defined as the preparation for informed and intelligent living. Civil engineering education must prepare the student entering the profession to identify and solve problems, and to evaluate the social, as well as technical, implications of those solutions. To accomplish this end, the student must master the basics of chemistry, physics, and mathematics,

---

[1]Retired, c/o Los Angeles Section ASCE, 2550 Beverly Boulevard, Los Angeles, CA 90057

[2]Daniel, Mann, Johnson, Mendenhall, 3250 Wilshire Boulevard, Los Angeles, CA 90010

[3]LeRoy Crandall and Associates, 900 Grand Central Avenue, Glendale, California 91201

which are common to all engineering. In addition, the curriculum should include English, history, the arts, and the science of human behavior.

QUALITY

Curriculum There is growing concern that the distinction between education and training is blurring. Training prepares one to do something; education helps one understand the significance of it. Although universities both train and educate, the pressures of the business and industrial communities are forcing many universities to emphasize training. If this trend continues, the question is raised as to whether civil engineers can function as active, responsible participants in social decision-making. Without a basis in human sensitivity and social values, can civil engineers determine if their work achieves this goal?

The idea of preparing students for immediate "day one" production overlooks the concept that the engineering complex depends upon thinking, both in the short term and the long term. It would appear that it would be more productive over the long term if employers would seek out well-educated civil engineers with a firm basis in science, design, and economics; an understanding of the duties and functions of a practicing professional; an ability to express themselves clearly both verbally and in writing; and an ability to use their thought and observation faculties to evaluate the problems and situations they face.

The quality of education at the pre-high school and high school levels obviously impacts the college student. The cliche' "Johnny can't read or write" is not without foundation. High school teachers report that many students either can't or won't handle assignments. Standards are therefore adjusted to meet the demand for a passing grade. Students are being passed along whether or not they will do the work necessary for a passing grade. To cite a supporting fact: U.C.L.A. admits about 4,000 freshmen from the top 12% of high school graduates in California; approximately one-half fail the English proficiency test.

A common complaint of many engineering managers and senior engineers is that many young engineers cannot communicate. Oral presentations are poorly organized and lack preparation. Written reports violate the very basics, such as proper sentence structure and punctuation.

Faculty Qualified and dedicated faculty are at the root of quality engineering education. Larger salaries will help attract better staff. However, the number of persons of quality who are willing to commit to the demands of teaching is not great. One common complaint of many students is that teachers seem more concerned with their research and other affairs than with teaching. A faculty is often made up of people with PhD's but little or no practical experience. Often, professors do not belong to their professional societies, which does not impart to students any feeling of the importance of that association.

NEEDS

In addition to the basics of science and mathematics and the applications of these basics to design, civil engineering education should include and foster thought, communication, language and the liberal arts which are common to the disciplines that make for an educated person. The art of critical thinking and expression is the desired goal. Its development requires a thorough basis not only in the technical disciplines but also in the liberal arts and social sciences.

The universities must retain teaching personnel with an interest in teaching and an appropriate mix of practical and research experience. Instructors should be professionals who are supportive of their professional societies and participate in local Section activities.

The rapid changes in technology make it impossible for the universities to prepare students for all future technical developments. Practicing engineers must therefore continue their education to maintain their professional competence.

There is a need for better interface between the practitioner and the academic community. Professional societies need to take the lead in establishing minimum requirements for continuing education to maintain competence.

POSITION

The California State Council of ASCE recognizes the importance of quality civil engineering education, and takes the position that:

1. Civil engineers should be prepared to evaluate the social implications of their engineering solutions and to function as informed responsible participants in social decision-making.

2. ASCE, ABET and the academic community should collectively produce a definition of civil engineering education and develop specifications for procedure, content and expected results of that education.

3. ASCE should encourage a diverse curriculum providing a broad education in addition to specialty training for civil engineering students. The art of critical thinking and expression should be developed.

4. Curricula should help develop verbal and written communication skills.

5. ASCE should encourage recruitment of teaching personnel who are practitioners and active in their professional engineering societies, and should encourage practitioners to share their experience with students in an academic environment.

6. ASCE should encourage engineering employers to establish work programs for students so that the needs of civil engineering practice become part of the education process.

7. Programs should be developed to encourage interaction between practicing engineers, students, and educators.

8. ASCE should promote and sponsor workshops and seminars for continuing education. These workshops and seminars should not be limited to just technical subjects but should include sections on the social implications of the technologies being covered.

WOMEN IN THE CIVIL ENGINEERING WORK FORCE
TAPPING A NEW RESOURCE

Amy Spicer Smith, P.E.[1]

Abstract

Once the world leader in technology, the United States is rapidly losing its competitive edge. With the attrition of foreign born engineers to their homelands, a declining interest among students toward technical studies, and an overall decline in student population, the U.S. may be confronted with an alarming shortage of engineers by the year 2000. To circumvent this catastrophe, the U.S. must actively cultivate human resources for technical careers. One key group to be targeted for these careers is women. Historically, however, women have faced many barriers to pursuing careers or remaining in the engineering field. These barriers include lack of encouragement to enroll in the proper preparatory curriculum at the high school level, discouragement at the college level, and opportunity inequities at the business level. There are positive steps that educators, employers and professional societies can take to assist in destroying these obstacles, such as the support of educational programs and a dedicated effort on the part of business to offer equal challenges to their women engineers.

Introduction

With the growth of technology, advanced communication systems, and international corporations, the U.S. is becoming part of a complex global marketplace. To compete in this economy we will need thousands of innovative, highly trained individuals. But the alarming statistics reveal that our future resources, today's students, are failing miserably in comparison with the students of other countries. As reported by Time magazine, "In a survey of 17 countries published last year by the International Association for the Evaluation of Educational Achievement, U.S. ninth graders tied with Singapore and Thailand for 14th place in science".

---

1 Associate, Dannenbaum Engineering Corporation, 7950 Elmbrook Drive, Suite 250, Dallas, Texas 75247

Enrollment in engineering schools is declining as fewer students show an interest in technical careers. In addition, many foreign born engineers are returning to their homelands and many experienced engineers are nearing retirement. By the turn of the century, the U.S. will experience a critical shortage of qualified technical human resources unless action is taken to promote individuals to enter into technical fields.

To alleviate the shortage, women should be encouraged to enter these fields. However, women have generally shown less interest in pursuing engineering studies. Additionally, women often face gender-related obstacles throughout their academic and business careers, some which are self-imposed and some which are imposed by others. This paper examines the obstacles that have prevented many women from pursuing or remaining in the engineering field and presents ways in which educators, employers, and professional societies can remove these barriers and encourage women to become engineers.

Education

Women have traditionally not been encouraged by high school teachers and counselors to study the advanced math and sciences necessary to be fully prepared for a college engineering curriculum. An informal survey among female engineers conducted by this author reveals that most encouragement came from family members.

Many women drop out of engineering after the freshman year, often due to problems resulting from the portion of male faculty who feel uncomfortable with women in the engineering classroom. Foreign males, who often come from cultures that view women as subservient and not as professional colleagues seem to be especially hesitant to accept females into the engineering field. With only 2 percent of the faculty being female, undergraduate women have few role models and few persons to turn to resolve these problems.

Therefore, it is not surprising that although the number engineering degrees conferred to women increased substantially from 1975 to 1986, the numbers are on a decline. Unless a concentrated effort is made to recruit women, the percentage of female undergraduates in engineering schools is expected to remain at about 15 percent, and at only about 4 percent at the doctorate level.

Employment

Women working in a male dominated area are often confronted by prejudicial attitudes of some co-workers and clients. Women surveyed expressed the opinion that whereas a male engineer's knowledge of engineering is accepted unquestionably, a woman is under constant scrutiny until she proves her competency.

Other preconceived ideas about women's attitudes bar them from certain aspects of engineering, particularly field work. Men often assume that women will naturally dislike field work because of the dirtiness. On the contrary, many women prefer the casual dress and physical aspect of the construction site, and virtually all women engineers appreciate the valuable knowledge gained through construction management.

Additionally, women often feel alienated when, due to their gender, they are excluded from participation in certain business-related social functions. Discrimination is evident in corporate sponsorship of activities geared toward "men only" activities such as hunting and fishing trips. Company support of all male softball teams provides additional company benefits to a select group of employees, while not necessarily matching those benefits to the remainder of the staff.

Alienation continues with the lack of role models and female mentors. Not only does this shortage of senior women engineers create problems for young female engineers, but it presents little opportunity for male upper managers to interact and develop a professional respect for women engineers as peers.

Women In Management

Despite these obstacles, some women have gained the respect of their industry and have risen to management positions. As more women move into senior level positions, men will accept women as peers more easily.

Other women have opted to circumvent some of the barriers and have started their own companies. Between 1977 and 1983 the proportion of construction firms owned by women has more than doubled, from 1.9 percent to 4.1 percent. But being the owner of the firm does not automatically relieve a women from sexism, as many women report of having to constantly prove that their Women-Owned Business Enterprise (WBE) is not merely a front for a male-dominated firm. However, most women owners report favorably of WBE programs, since it gives the business community an incentive to hire women, and women the opportunity to prove their ability.

The civil engineering field does not require nearly as much brawn as it requires brains. Women have all the mental and physical capabilities to be engineers and managers, and with foreign competition eating away at U.S. technical strongholds, this nation cannot afford to waste anyone's talent.

Educational Reform

At the junior high and high school levels, teachers and counselors must emphasize the growing importance of math and science in the careers of the future. Girls in particular should be encouraged to pursue technical careers. In 1974 a very successful program called "Expanding Your Horizons" was started to introduce junior high and high school girls to women in math and science related professions during a one-day seminar. The morning session consists of speakers from technical fields. In the afternoon, the girls divide into groups and attend various workshops where they participate in experiments and activities intended to give them a hands-on experience with technology.

At the college level, action must be taken to emphasize to advisors, professors, and assistant professors that all students shall be treated equally, regardless of sex, race or nationality. Additionally, programs should be implemented to inform engineering faculty of the growing need to recruit women into the engineering field and of the special concerns of the female student.

Programs to assist female engineering students in overcoming obstacles associated with entering a male-dominated profession should be required of all freshman women. Purdue University has developed such a program for engineering women students. The required seminar helps them deal with special problems, develop self-confidence, and creates a support network among the students. This seems to be successful, with Purdue conferring 22 percent of it's engineering degrees to women versus the nationwide average of 15 percent.

In order to assure that women will be motivated to enroll as freshman engineering students, universities must actively pursue high school women. Several universities, Clarkson University, Purdue University and the New Jersey Institute of Technology to name a few, have instituted programs to invite high school women to their campuses and introduce them to engineering. Presentations at freshman orientation, women's dormitories, sorority houses and women's clubs offer additional areas for recruiting.

Another major step for universities is to increase the percentage of women faculty. As it may initially be difficult to find qualified women for these positions, women engineers from the business community should be pursued as part-time professors or as guest lecturers.

To have women with advanced degrees for future faculty positions, advisors must encourage their outstanding women students to consider prolonging their academic career in order to acquire masters and doctoral degrees, and should assist in securing funds and grants in order for women to pursue their studies. University regents and administrators should appeal to industry to sponsor scholarships for women; after all, it's an investment for them in future resources.

Reform in the Business Community

The public sector has been very conscientious in the hiring and promoting of women engineers. Most women surveyed who work for government agencies reported that they felt that their salaries and opportunities were equal to those of their male counterparts.

However, private industry, operating in a less regulated environment, is more likely to be guilty of discriminatory practices. Although most instances of discrimination may be subtle or unintentional, existing company policies and practices should be evaluated to determine their suitability to the changing gender mix in the business environment.

As more women enter the engineering profession, work attitudes must change. Women should not be "protected" from field work on the assumption that "they wouldn't like it" or "don't want to get dirty". Knowledge of construction procedures is an important aspect of every engineer's training.

Firms need to present their female engineers with the same challenges as those presented to the male engineers, allowing women the same opportunity for success or failure. It is through evaluation on these equal terms that the future company managers will appear, and management decisions will not have to be based on preconceptions and differing datums but on the competency of the individuals.

Additionally, the predominance of "men-only" activities should become a thing of the past and be replaced with activities that include both men and women, and encompass a wide range of differing entertainment tastes.

Corporate sponsorship of scholarships for women is not only a positive move towards the promotion of women engineers, but is also a wise investment by industry for future technical resources.

American Society of Civil Engineers

The American Society of Civil Engineers has many opportunities to make women more visible in the engineering world. The first and most important step is at the national level. ASCE should set a goal for the 1990's to promote women in the civil engineering profession. A committee to develop nationwide programs and work with local chapters to accomplish this goal should then be appointed. Some suggestions for programs follow.

Local chapters of ASCE should develop committees whose purpose is to visit area high schools and introduce girls to the civil engineering profession through guest speakers and seminars. Branch sponsorship of scholarships to local universities for women in engineering would be another means to attract women to the engineering field.

Branch chapters of ASCE should create a task force to work with local universities to develop a seminar for freshmen women entering engineering, using Purdue University's program as a model.

At all levels, national to local, women should be encouraged to become officers and committee chairmen. This would be extremely easy to the local level, where many official positions are readily available to those willing to take the responsibility. Many women would be flattered and enthusiastic to be asked to be an officer for their organization.

Program chairmen should make an extra effort to find women speakers for the monthly meetings, as well as to invite speakers to talk on topics related to the evolving gender mix in the work place.

ASCE Journal chairmen should actively solicate papers from women engineers and academians for publication. Suggestions for topics include "Choosing a WBE for your Project Team", and "The Changing Demographics in Civil Engineering".

Convention chairman should appeal to women to make presentations. Additionally, chairmen should take care that special convention activities, such as golf tournaments, are offered equally to both men and women attendees.

These recommendations will allow women professionals more visibility and will accelerate the acceptance of females to the engineering field. ASCE should implement the suggested programs immediately in order that, by the turn of the century, the goal to provide women with equal acceptance and opportunity in the civil engineering profession will be met.

Conclusion

The preceding ideas and recommendations are intended to increase the number of qualified civil engineers for our country's future. It is not enough to merely satisfy the quantitative need; as a nation competing in the global marketplace we must strive to have the best and brightest engineers working for us.

Appendix References

Rubin, Debra K., "Women Take On Construction", Engineering News Record, October 16, 1986, p. 24-29.

Smith, Emily T., LaPlante, A., Angiolillo, P., Cantrell, C., "The Women who are Scaling High Tech's Heights", Business Week, August 28, 1989, p. 86-88.

Tift, Susan, (1989) "A Crisis Looms in Science" Time, September 11, 1989, p. 68-70.

Vetter, Betty M., (1989) "Finding and Keeping Technical Experts in the 1990's", Journal of Management in Engineering, Vo. 5, No. 1, January 1989, p. 42-52.

SELLING CIVIL ENGINEERING TO HIGH SCHOOL STUDENTS

Karen C. Kabbes, Member ASCE[1]

Abstract

In the spring of 1989 the ASCE Committee on Career Guidance contracted for a market research firm to determine how top high school math and science students perceive civil engineering as a field of study and how those students obtain information and make decisions on what field of study to choose. This paper describes that research and the results.

Introduction

Declining civil engineering enrollments lead the American Society of Civil Engineers (ASCE) to create a Committee on Career Guidance (CCG). Created in 1987, CCG was charged with the responsibility of implementing, monitoring and evaluating specific activities to help assure an adequate number of highly qualified students to chose civil engineering as a field of study and to periodically review and recommend updates in ASCE's career guidance materials, presentations and programs (ASCE Official Register, 1989). One of the first functions of the CCG was to review the existing publications, pamphlets, movies and slide shows available through ASCE for career guidance. Many of the materials appeared dated and needed to be replaced. However, CCG was concerned that replacement guidance materials be designed to effectively reach the target audience, top high school math and science students.

Market Research Proposal

To learn about the target audience, CCG entered into a contract with Associated Research Group (ARG) of

1 Head, Northeastern Illinois Floodplain Management Unit
Illinois Department of Transportation
201 West Center Court, Schaumburg, Illinois 60196

Rochester, NY., a full-service market research firm. ARG was asked to answer several questions. These included:

1. How do high school students perceive the profession of civil engineering?

2. How do high school students perceive civil engineering as compared to other disciplines of engineering?

3. How do high school students perceive possible career choices in civil engineering?

4. Who influences students career choice?

Because of CCG's limited budget ($3,000 to $6,000), and the nature of the questions asked, ARG suggested the initial research consist of two parts.

One part would consist of a quantitative survey to measure students' beliefs, opinions and attitudes towards civil engineering and to understand student's perception of civil engineering as compared to other engineering disciplines. The second part would include two focus group interviews with students to explore their perceptions about civil engineering and obtain feedback on ASCE's existing career guidance brochure and video.

Survey and Focus Group Sessions

The survey and focus group sessions were held at the University of Illinois in Champaign-Urbana, March 22 and 23, 1989 and consisted of students attending the state final TEAMS (Tests of Engineering Aptitude, Mathematics and Science) competition held by JETS (Junior Engineering Technical Society). The competition spanned two days and included schools from all over the state. The students attending were generally in the upper 5% - 10% of their class, scholastically, and expressed an interest in math and science.

All students attending the competition were given a survey to complete at the end of the TEAMS exams. On each day 10 student volunteers were chosen to take place in a focus group session. So as to not predispose the students to specific answers about civil engineering, the students were told the purpose of the focus group was to discuss how students make career choices and to evaluate guidance materials prepared by a group whose job it was to make these materials.

Results of the Focus Group Sessions

As summarized in the final report prepared by ARG for CCG (ARG, May 1989), students chose a college major depending on a number of different factors, including:

1. The ability to use a subject learned in high school (ie. chemistry, biology, etc.) in which the student has a strong interest in, or aptitude for.

2. The potential a college major has to ensure employment with a high paying job upon graduation.

3. The challenge and personal satisfaction of studying and working in a difficult area.

Students indicated they received their information on possible careers from a variety of sources. ARG's report stated teachers, counselors, clubs and organizations, special school programs, and older friends and family members were noted by students as having the most influence on career decisions. Surprisingly, the students did not believe their parents were an important source of career information

Attitudes towards Civil Engineering

ARG's analysis of the focus sessions indicates students were not as knowledgeable or excited about civil engineering as compared to other engineering disciplines (ARG, May 1989). Students could talk about electrical engineering or mechanical engineering with a great deal of enthusiasm in the focus groups, but their knowledge of civil engineering was limited to construction related activities. In particular they thought of civil engineering as building things, dealing with "ground type stuff... highways, parks and dams", "basic drafting", "basic architecture". ARG concluded the students perceived the type of projects that civil engineers work on as "not particularly stimulating".

Students indicated that the potential income level for civil engineering and desirable job prospects were not as good as other engineering disciplines. ARG's report quotes one student as saying, "I heard it was low paying compared to other engineers and that civil is just building stuff." When pushed by the focus group moderator, students indicated that the primary advantage of a career in civil engineering was, "You can see your work and know it benefits the public". However, ARG concluded that students were not excited by this advantage and viewed civil engineers work as "finite", with solutions that are straight forward and logical. In other engineering disciplines solutions were viewed by students as being more innovative, requiring more creative thinking. Other engineering fields were viewed as more high tech, more on the leading edge of technology, than civil engineering.

Results of the Survey

In comparison to the focus group sessions that only dealt with a small number of students, completed surveys were received from 456 students (ARG, May 1989). The survey prepared by ARG asked students to distinguish by description between civil, mechanical, electrical and chemical engineering. Students were then given a list of professions consisting of computer programmer, mechanical engineer, civil engineer, architect, chemical engineer, technical sales representative and electrical engineer, and asked which profession was most likely and least likely to match a series of 18 descriptions. The last portion of the survey consisted of asking students how strongly they agreed or disagreed with six statements about civil engineering. Students were also asked their sex and grade level, if their school provided career information on engineering and if so, had they reviewed it, and if they knew a civil engineer and other type of engineer.

Students were generally able to accurately distinguish between the four areas of engineering described. ARG reported ninety-four percent correctly matched chemical engineer, to a description of what a chemical engineer might do. The profession least accurately matched with the description of what a civil engineer might do was civil engineering, correctly matched 85% of the time.

In choosing which type of engineers were most likely or least likely to match a series of descriptions, the students believed civil engineers were most likely to work outdoors, contribute to solving problems like hunger, housing and air pollution, be community leaders, and be managers. Among the types of engineers listed, the students believed civil engineers were least likely to work alone, least likely to be well paid, least likely to work with the latest technology and least likely to have high prestige among other engineers.

When asked to agree or disagree with the statement civil engineers are paid more than other engineers, ARG reported that 40% disagreed, 40% didn't know and 20% agreed. When asked if a civil engineer is likely to be well respected 51% agreed, 25% disagreed, and 24% didn't know. Asked if civil engineering degrees are tougher to get than other kinds of engineering degrees, 46% of the students disagreed, almost half of them strongly, and 41% didn't know. Forty-six percent of the students indicated they knew how to find out more about civil engineering as a profession, 19% disagreed and over one third of the students didn't know.

When asked if they were considering a career in civil engineering, ARG stated 68% of the students disagreed and most indicated strong disagreement (57%). Only 13% agreed to some extent, 5% strongly. One fifth of the students didn't know. Asked if civil engineering is an excellent career choice when compared to other engineering degrees, 41% disagreed, 40% did not know.

Conclusions

Because the students involved in this study all came from one state and were preselected by their involvement in JETS, the results of similar focus groups and surveys may differ if held in different parts of the country with different populations of students. However the results from this limited study do support the observation that a smaller number of top math and science students are choosing civil engineering as a field of study.

The results of this study suggest we need to do a better job at communicating the complexity of our profession and the creative challenges that face us. Building buildings and roads are activities that are clearly recognized as civil engineering, but it is only one part of civil engineering. Our message to students can no longer be centered on the role of the civil engineer as a builder. The students involved in this study did not find construction to be a very interesting profession. How many of those students, however, would be intrigued by the challenge and creative demands of developing a complex mathematical model of a region's transportation network to find a series of optimal intersection timing scenarios to reduce traffic congestion?

Success at attracting top math and science students may depend on offering them work that demands top math and science skills. The survey results indicated the students do not consider civil engineering a very demanding course of study and it has low prestige among other engineers. What applications in civil engineering require the use and understanding of mathematical modelling, calculus and differential equations, physics and chemistry? Those may be the areas of civil engineering we want to highlight in the future. However, if most civil engineers do not use these skills in his or her position, do we really need top math and science students to be civil engineers? Perhaps we need to examine our profession and determine if we are really using civil engineers in positions that may be more appropriately filled by civil engineering technologists.

Many of the students had done enough homework on engineering careers to learn civil engineers generally have the lowest starting salaries of engineers. Additional information about the range of salaries among people with civil engineering degrees, including individuals who had used their engineering degree to move into management of a large organization, would help clarify the potential earning power of a civil engineering degree.

The perception of low salaries and lack of creative challenges on the trailing edge of technology plague our image. If we are going to attract a share of the best and the brightest we are going to have to revise our sales message to give the most accurate picture of the salaries and the as yet unanswered technical challenges that face the engineer of today and tomorrow.

References

Associated Research Group, Final Report, May 1989.

Official Register, American Society of Civil Engineers, 1989, 170.

BENEFITS OF AN ACTIVE ASCE CHAPTER TO AN UNDERGRADUATE ENGINEERING COLLEGE

by

Kanti L. Shah, M.ASCE[1]

ABSTRACT: Inspite of the rural location, the ASCE Student Chapter of Ohio Northern University undergraduate engineering college has received top national honors because of its professional and public activities. Factors are discussed for the successful operation of the award-winning achievements. Benefits to the institution, faculty, and students have been listed. These are: professional development and better employment opportunities for students through increased contacts with area professional engineers, alumni pride and its role in donations and pledges to the institution, and excellent publicity for the institution's professional education.

INTRODUCTION: The purpose of this paper is to describe how an undergraduate college in a rural area has received national recognition because of its professional activities and public service and to list some of the benefits to the institution and students. Incoming students generally have little information, often inaccurate, which describes for them the role of an engineer in meeting the demands of society. One of the problems is the first two years of most engineering programs. These core programs consist of physical sciences, mathematics, mechanics, humanities, and social sciences. These are the basics for engineering courses, but they do not provide any real insight into the engineering profession. Students need to be shown relevancy in their education. This is especially true in undergraduate engineering colleges where opportunity for consulting and research work is limited. Many engineering educators evaluate the traditional approaches toward engineering education and develop innovative programs to serve changing needs of today's students. One organization that the T.J. Smull College of Engineering at Ohio Northern University (ONU) has used effectively to promote relevancy of the civil engineering education is the ASCE. Students are able to appreciate the many aspects of the civil engineering education because of the activities of the ASCE Student Chapter.

---

[1]Professor, Civil Engineering Department, Ohio Northern University, Ada, Ohio, 45810

ASCE Chapter at Ohio Northern University (ONU)

The Board of Direction of ASCE authorized the formation of student chapters at accredited schools in 1920 and the purpose was " ------ to help the student prepare himself for entry into the civil engineering profession and the Society."

The ASCE Chapter at ONU has strived to fulfill the goal set by the Board of Directors of ASCE. The civil engineering faculty, the dean, and the university administration support the chapter by attending monthly meetings. The Dean and the administration provide financial support for travel to the site of annual rigid regatta competition sponsored by the ASCE. Two professional engineers associated with the leading engineering firms in the area serve as contact members. They support the chapter's activities by attending the meetings and suggesting speakers and field trips.

The Chapter has flourished in every respect in spite of its location and size. Ohio Northern University is an undergraduate institution in a rural setting. In addition, the enrollment in the Civil Engineering Department has declined in recent years. However, the Chapter membership (both in quality and numbers) has gradually improved. Each member is encouraged to get involved. The Chapter has its own office and a phone. Many ideas for the Chapter's activities are generated in this office.

The primary objective of the Chapter is to supplement the classroom education and promote professionalism among the civil engineering students. This is done through public service and interaction with the practicing engineers and faculty from the neighboring engineering colleges. This in itself is not an easy task because of the location of this Chapter. The nearest university town is approximately seventy miles one way.

ACTIVITIES

Professional and Social

The Student Chapter holds monthly meetings with professional speakers and field trips. Occasionally, the speaker is a civil engineering alumnus who talks about his/her work. The Chapter also helps in the orientation of the freshmen civil engineering students. It sponsors social events such as picnics and a Christmas party. The Chapter members every year design and build a concrete canoe to enter the regional canoe race. In 1988, they also designed and built a steel bridge for the Second Annual Steel Bridge Competition sponsored by The American Institute of Steel Construction. The ONU team took first place in aesthetics and third place overall. The Chapter received an award of $100.00. Talks and papers related to civil engineering have been presented by the Chapter members. To provide an example of the Chapter's activities, the following table summarizes the 1988 activities.

PROGRAM:

Mr. L. Winkleman from S.E. Johnson spoke on asphalt pavement and Mr. Croft Merritt from ODOT addressed the seniors regarding a job.

Mr. Chris Brown, the Bucyrus City Engineer, spoke about business and civil engineering.

The ONU Student Chapter held its Annual Christmas Party and watched a videotape on hazardous waste disposal.

Mr. D. Girardot, Director of ONU Physical Plant, spoke about his involvement with the Army Corps of Engineers and his present duties as Physical Plant Director.

Mr. T. Allen from S.C. Johnson spoke about construction of the Huron Bypass

Joint Mtg with the Toledo Section of ASCE - Dr. David R. Warner, Professor of Law, spoke on "Law, Computers, Ethics, and Engineering"

Field trip to Churchill Asphalt Plant, Lima, Ohio

Prof. R.L. Swart spoke on accident reconstruction and showed a video tape by General Motors on the subject.

ASCE Chapter Picnic

Field trip to Envirosafe Landfill Facilities, Toledo

Five students discussed their summer job experiences relating to civil engineering

and build a concrete canoe to enter the regional canoe race. In 1988, they also designed and built a steel bridge for the Second Annual Steel Bridge Competition sponsored by The American Institute of Steel Construction. The ONU team took first place in aesthetics and third place overall. The Chapter received an award of $100.00. Talks and papers related to civil engineering have been presented by the Chapter members. To provide an example of the Chapter's activities, the following table summarizes the 1988 activities.

PROGRAM:

Mr. L. Winkleman from S.E. Johnson spoke on asphalt pavement and Mr. Croft Merritt from ODOT addressed the seniors regarding a job.

Mr. Chris Brown, the Bucyrus City Engineer, spoke about business and civil engineering.

The ONU Student Chapter held its Annual Christmas Party and watched a videotape on hazardous waste disposal.

Mr. D. Girardot, Director of ONU Physical Plant, spoke about his involvement with the Army Corps of Engineers and his present duties as Physical Plant Director.

Mr. T. Allen from S.C. Johnson spoke about construction of the Huron Bypass

Joint Mtg with the Toledo Section of ASCE - Dr. David R. Warner, Professor of Law, spoke on "Law, Computers, Ethics, and Engineering"

Field trip to Churchill Asphalt Plant, Lima, Ohio

Prof. R.L. Swart spoke on accident reconstruction and showed a video tape by General Motors on the subject.

ASCE Chapter Picnic

Field trip to Envirosafe Landfill Facilities, Toledo

Five students discussed their summer job experiences relating to civil engineering

ATTENDANCE AT ASCE OR OTHER PROFESSIONAL MEETINGS

Date	Number of Chapter Members Attending	Location	Sponsoring Organization
9/21/88	3	Tony Pacos, Toledo	ASCE Toledo Section
11/16/88	5	Holiday Inn Perrysburg, OH	University of Toledo ASCE Student Section
11/29/88	20	Faucet Hall OSU, Columbus, OH	ODOT

STUDENT TALKS OR PAPERS

Date	Name of Student	Subject	Occasion of Presentation
2/12/88	Richard Martin (C.E. Senior)	"Do People Really Care"	Submitted to Third National Hazardous Waste Essay Competition
5/13/88	Verna Babcock	"Design Curves for the Embedment Problem"	Engineering Advisory Board
11/14/88	Rob Miller Aaron Biedenbach	"Civil Engineering Studies"	Open House

COMMUNITY SERVICE PROJECTS

Fortunately, the Student Chapter offers much more than the usual activities, as valuable as they are. It has done well in the areas of community service projects. The Chapter's community services include design, construction, and the furnishing of playground equipment for Ada Elementary School and a bench for the Ada Park. Students also surveyed a proposed site for a nature center and an addition to a rural cemetery.

The Community of Ada has recognized the services of ASCE (ONU Chapter) as indicated by the news releases in the local papers and letters from the Ada School's principal (to the Chapter).

AWARDS

The Chapter's annual reports submitted to the national office of the ASCE have been rated very high for its activities. It has consistently received honors which include a letter of honorable mention and a letter of commendation. For the last two years the Chapter has been the finalist for the top Ridgeway Award. In 1988, the Chapter won the Vice-President's Zone II Award.

BENEFITS

There are several benefits of an active chapter to the institution and its students. The faculty of undergraduate engineering colleges, such as T.J. Smull College Engineering of ONU are not involved with research and consulting work to the extent that the faculty of the engineering colleges offering graduate work are involved. Students of these undergraduate colleges rarely derive the benefits from the research and/or consulting work of the faculty. An active ASCE chapter provides an opportunity for civil engineering students and faculty to hear and talk to professional engineers, research workers, managers, and civic leaders at its monthly professional meetings and social gatherings. This is very helpful in providing current view points regarding civil engineering and related matters, especially to students of an undergraduate engineering college in a rural area.

The joint technical and social meeting in March with the parent section and the university of Toledo and the frequent attendance of students at the monthly section meetings have initiated a significant interaction between students, professors, and practicing engineers. A ride in a university van or a faculty member's car provides a more relaxed and friendly environment than the usual classroom atmosphere. Such an environment, without doubt, is very conducive to the overall professional development of the students.

An active and award winning chapter brings recognition to its civil engineering department and university. It often becomes a source of good publicity for the institution's professional education. On many occasions, alumni have expressed their joy and pride in the ONU Chapter's achievements. It definitely plays a positive role in the pledges and donations by the alumni.

The Civil Engineering program at ONU is considered to be a very good one. An excellent curriculum and an active chapter are the main reasons for its good reputation. The graduates are readily employable.

CONCLUSIONS

Building an excellent student chapter program is an evolutionary process. The award winning tradition of the ASCE Chapter at ONU has created an environment which is very conducive to the excellent professional education. The public service by the chapter has created goodwill of the community toward the Chapter. As a result of the activities of the Chapter, the Civil engineering department and its students enjoy a good reputation among area professional engineers and other engineering colleges.

Effective Evaluation of Teaching Quality

C. W. Lovell[1], Fellow ASCE

Abstract

The situation which obtained at the author's university several years ago may have been typical of many departments of civil engineering. The evaluation of teaching quality was almost nonexistent, and the excuse was the perennial one: since this is so difficult to accomplish, we will assume that the teaching quality is uniformly high in all courses.

The paper describes how a departmental teaching committee was able to institute an effective system of evaluation of teaching quality by students and peers. Although the system is entirely voluntary, it is sufficiently dedicated to the principles of objectivity and confidentiality, to motivate nearly half of the civil engineering faculty to use it. Many of these faculty return to the evaluations each semester.

Introduction

Although the typical civil engineering faculty member spends well over half of his university time in the teaching function, effective evaluation of the quality of the communication effected in this effort is sparse or totally lacking. At least this was true in the author's department several years ago. Faculty objection to student and peer evaluation of teaching quality was so staunch that a required participation system was out of the question. Keys to a successful voluntary system proved to be: (1) confidentiality and (2) objectivity. Data from student evaluations were entered into a data

---

[1]Professor of Civil Engineering, School of Civil Engineering, Purdue University, West Lafayette, IN 47907

bank, and statistics were developed for the civil engineering group, as a whole. Evaluations were processed by a "disinterested" security person, selected by the department chairman. Only a single copy of the results of the evaluation was printed, and this was sent to the teacher under sealed cover.

Evaluation Groups

Groups from which evaluations are usually solicited are: (1) alumni, (2) peers, and (3) students.

Alumni were solicited via a questionnaire in a periodic newsletter. Response on teaching quality, as they recalled it, was interesting, but not very helpful. Only teachers who impressed them favorably were remembered, and only subject matter areas, not courses, were recalled. Teachers who taught them subject matter of great value to their practice were invariably teachers in their practicing specialty. For example structural practitioners received greatest benefit from teachers of structural subjects, environmental engineers from environmental teachers, etc. There were also the responses indicating that they liked certain teachers as persons, and remembered these contacts in a pleasant way.

The most obvious disadvantage of alumni evaluations is that they come well after the performance being evaluated. In the meanwhile the curriculum has changed, the emphasis in courses is different, and many of the faculty remembered have moved, retired, or become administrators.

The next time one of your colleagues categorically announces that the only valid evaluations are from mature engineers (alumni), remind him of the difficulties intrinsic to such evaluations. Overcoming these will be a formidable task.

Another mature group, which can rather more successfully participate in evaluations of teaching quality, is the teacher peers. Two subgroups have been successfully employed in the author's department. Peers who are specialists in instructional techniques can contribute valuable comments as a result of "sitting in" on a lecture. Technical peers (other civil engineers) contribute a different set of comments relating to logic of organization, technical comprehension, pace of presenting new information, and appropriate use of background material.

All such evaluations are voluntary, and are accomplished within a broad enough time period to preclude the teacher being able to specially prepare for

FIGURE 1

Offer of Services...Peer Review

I would like a peer review involving:

________ (A) a Civil Engineering faculty member

or

________ (B) a Center for Instructional Service representative

The following dates are acceptable for the critique (list at least four):

	Date	Time	Location
(1)			
(2)			
(3)			
(4)			
(5)			

the peer visit. See Figure 1 for the offer of services made by the departmental teaching committee. The system is confidential in that the peer reviewer gives his comments only to the teacher. Users of this type of evaluation have been universally enthusiastic about its value.

The potential for obtaining helpful teacher/course evaluations from students is enormous. The evaluation is timely; perceived faults can be adjusted for the next time the course is taught. The numbers are sufficient to produce meaningful statistics. Questions can be posed which are critically important to the process of communication, and which the students are qualified to answer. An example is: "Does the teacher appear to be knowledgeable about the subject matter?" Assuming that the teacher is knowledgeable, low ratings mean that the teacher needs to find ways to better emphasize his qualifications. If he fails to recognize this perception on the part of the students (albeit it in reality a misconception), and fails to act to remove it, the learning process is severely impeded.

To emphasize the point further, student opinion is important, even when it is erroneous! The teacher who is well prepared and qualified, but who is perceived by his students to be otherwise, must revise his teaching technique to remove the misconception.

FIGURE 2

Instructor/Course Evaluation
Question Responses

A = Excellent;B = Good;C = Fair;D = Poor;E = Very Poor

INSTRUCTOR

1. My instructor's preparation for the course was..
2. My instructor's knowledge of the material being covered was..
3. My instructor's acceptance of criticisms and responsiveness to suggestions was..................
4. My instructor's enthusiam for teaching was.........
5. My instructor's sensitivity to students' needs and interests was.......................................
6. My instructor's use of an appropriate variety of activities, approaches, methods, and materials was..
7. My instructor's organization in presenting material for this course was...............................
8. My instructor's office hours, appointments, and availability for help was.........................
9. My instructor's grading procedures were...........
10. When compared to other instructors I have had, this instructor was...................................

COURSE

11. As a learning experience, this course was..........
12. The textbook for this course was...................
13. The availability of resources for this course (i.e., TA's, readings, handouts, films, tapes, etc.) was..
14. As a requirement for my major, this course was..
15. When compared to other courses I have had, this course was...

The author has found the 15 questions of Figure 2 to be a valid set. Questions 10 and 15 might be objectionable to some teachers, but they seem to be common denominators for such questionnaires.

It is best that teachers and teaching assistants _not_ administer the evaluations. They can be proctored by members of the department teaching committee (or some similar group); sealed in front of the class as an evidence of confidentiality, and immediately delivered to the security person for processing.

Statistics generated for the teacher and the course include the mean for each question as well as the percentile values within the civil engineering population. The evaluation numbers are sent to only the teacher, usually after the course has ended. Thus the responses of the students are held in appropriate confidence.

Written comments are solicited on the back of the evaluation form, but these are screened for the occasional destructive comment, and are typed on the evaluation sheet which the teacher receives. (The security officer does the screening and transcribing.)

Interestingly enough, although there are critical student comments, including the occasional rude one, the majority of comments are favorable. This reinforcement of good teacher performance and effective courses is both helpful and appreciated.

Use of student evaluations implemented in the system described above has steadily increased with time. Elements of success seem to be the confidentiality of the transfer of potentially useful information.

What happens in the absence of written evaluations? The spoken word becomes the judgment, and this may be governed by the loudest, most opinionated voices, not the most knowledgeable and logical.

Teaching Assistants

The teaching committee which developed the evaluation program described above has also been instrumental in instituting a teaching assistant (TA) qualification program. This orientation occurs on three days immediately preceding the beginning of the fall semester. All new contact TA's must attend sessions which describe the academic procedures at the university and teaching techniques. Each new TA has the opportunity to make a 5-10 minute video tape which is subsequently critiqued by experienced faculty and TA's.

Foreign TA's receive special instruction on operating procedures, and, of course, must also meet all university language proficiency requirements.

Closure

Both peer and student evaluations of teachers and courses have been developed by a departmental teaching committee at the author's home university. With the evidence of both objectivity and confidentiality, the system has been used on a voluntary basis by more than one-half of the civil engineering faculty. Prior to the establishment of this system, use of such evaluations was essentially nil. An accompanying program of teaching assistant orientation and training is also believed to be highly successful in improving communication between teachers and students.

Gaining Experience in the Computer Age

Brian R. Brenner[1]

In recent years, advances in technology have had a tremendous effect upon the design profession. With the availability of powerful, inexpensive computers and effective software, Civil Engineers can now directly solve technical problems in a few hours, ten times as fast as the old manual solutions, which were less accurate. Much attention has been focussed on the great technical advances of computer use for Civil Engineering. Less has been said about another aspect of this subject that is having as great, if not greater, effect on the way we do our work. Computers are not only changing the technical quality of Civil Engineering work, but they are changing the structure and organization of the way we do engineering, especially with regards to the training of new engineers.

Traditionally, Civil Engineering education has been handled, in a broad sense, by a Master Engineer-Apprentice format. In most design offices, newly-graduated engineers needed several years of that magic, all important ingredient, "experience", to complete their education. The key elements of Civil design could not be learned through four or more years of formal education. Only through a sort of osmosis, only while assisting the experienced engineer, could the younger engineer gain the skills needed to make the jump from engineering graduate to design professional.

This arrangement has served the design industry well for many decades. In the past, all branches of Civil Engineering design have required weeks of manual calculations to solve design problems. Over the years, clever engineers developed short-cut methods,

[1]Senior Structural Engineer, Bechtel/Parsons Brinckerhoff, One South Station, Boston, MA 02110.

rules of thumb, and effective approximations to get at the required information. Still, the brunt of the work remained: days of tedious moment distribution, days of tedious Hardy-Cross analysis, days of tedious processing of geotechnical lab results. For the experienced engineer, this problem was actually an opportunity. Younger engineers would be assigned the task completing the manual calculations. In the process of trudging through this work, the new engineer indirectly gained the required knowledge that defines experience. By working through different analyses, the younger engineer learned not just how to do the analysis, but which method was appropriate to solve a particular design problem. By manually marking up drawings, the engineer not only helped complete the work, but also learned about appropriate details and how drawings were put together. For the senior engineer, it was a good arrangement: the cumbersome manual work provided a format to train inexperienced engineers, and the availability of the younger engineers to do this work freed the more senior designers to conceptualize the problem and guide it to solution.

Computers are changing the traditional process of completing design work, however. In minutes, computers can solve problems that used to require days to complete. For example, an inexpensive PC-based frame program can analyze the whole frame and even select the appropriate members. In this way, what was a week's worth of the young engineer's time is reduced to a few hours. Lengthy manual calculations are no longer necessary. On the other hand, no longer will the junior engineer gain that intangible, experience, by plowing through the tedious manual work.

Because, even with automation, it is still necessary for young Civil Engineers to make the leap from collegiate graduate to practising engineer, we must develop new methods and strategies for this to happen. As the structure of engineering design changes and we seek a new format to train inexperienced engineers, consider the following three points:

First of all, computers can be used to help train younger engineers. Junior designers can learn through iteration, because computers permit fast, efficiently prepared iterative designs. For example, in the same amount of time it takes to solve a complicated building frame by moment distribution, the young designer can run ten frames on the computer, varying assumptions and learning, by trial and error, how the structure behaves. A practised structural engineer would gain this type of knowledge over time via manual analysis methods, but by removing much of the dog work, computers can speed the learning curve.

Also, clever CE software writers can incorporate artificial intelligence methods into the programs so that, not just analysis techniques, but the decision-making process itself, can be learned. It is not enough, as an example, to change a column base from fixed to pinned in a computer model. The engineer must also understand the ramifications of this change: is a pinned detail available for this case, and does it really behave as a pinned end, or are there important secondary moments to consider? This type of rule-based knowledge can gradually be incorporated within increasingly sophisticated CE programs. While it is unlikely that computers will replace human beings as the decision makers in the design office, the machines, when creatively used, can speed the learning process, the gaining of experience.

For younger engineers to learn in this way, engineering managers must provide the time to do it. Previously, junior engineers involved in the preparation of manual calculations, however tedious, were considered to be doing "billable" work. Now, these tedious calculations are no longer required. However, the same accounting approach must be provided for periods of education.

Secondly, consultants and professional societies must develop detailed standards and guidelines for engineering computer use. Such standards can help form the framework not only for Civil Engineering computer use, but for the new, computerized education process for younger engineers. When prepared and applied properly, these standards will help fill the gap in the development of engineering expertise. They might take the form of recommended spreadsheet software, or lists of tried and true CE application software, or more standardized formats for CADD transfer. The standards should help engineers avoid repeating past mistakes. They must not be allowed to become too rigid. Because of the rapidly changing nature of engineering software and hardware, designers must be free to creatively adapt developing technologies to their needs. Computer standards must grow and adapt with the engineers, not stifle creative computer use.

Thirdly, junior engineers must adopt a more aggressive approach towards the gaining of experience. The days of leisurely, tedious manual calculations are ending, along with the framework in which new engineers learned their profession. In this period of change, with no firmly established process for gaining experience in place, young engineers must seek out engineering knowledge and not wait for it to fall into their heads. The new environment in consulting offices will provide less and less need for engineering dog work, and, thus, less and less need for lazy junior engineers. On the other hand, the situation provides great opportunity for industrious young engineers to gain professional insight and make substantial contributions to the design process sooner. Managers and senior supervisors must encourage this kind of contribution, and not fall back on the old, manual procedures as a way of maintaining control.

Out in the design office trenches, many veteran Civil Engineers have complained that today's recent graduates are weak in the fundamentals. Despite the great powers of automation, it is still necessary to understand how a bridge is built, for example, and how to reflect this knowledge in the structure's design and presentation in contract documents. However, some CE educators have reacted to the advent of computers by concluding that the fundamentals are outdated. So, they avoid teaching moment distribution and other manual methods, because nobody uses that today anyway, and drafting is no longer required (or even offered in some cases!) because engineers don't draw now. However, while moment distribution use may be going the route of slide rules because of computers, the underlying structural principles are still relevant. Young structural engineers must understand the concepts of structural behavior: bending theory, the Law of Statics, etc. To avoid these topics is the Civil Engineering equivalent of throwing out the baby with the bath water.

Civil Engineering educators can incoporate software use in the classroom in ways that emphasize understanding the fundamentals. Young engineers don't really need to be grilled in the mechanics of moment distribution. However, the approximate structural methods, such as the portal method, are still a valid tool for quickly learning structural behavior. Engineers probably won't be asked to draw with ink on mylar. However, understanding drawing presentation is an essential skill, and one that can be incorporated in the curriculum via AUTOCAD and other relatively inexpensive PC programs.

Consider the example of flow nets. In a Soil Mechanics class, professors would typically assign a flow net problem to present the subject of groundwater flow. In the past, flow nets were manually drawn, involving a great deal of time and tedious iteration. Today, flow nets can be prepared by computer. There is no intrinsic value to learning the tedious rules of manual flow net preparation. However, the proper response should not be to avoid this subject altogether because the computer does it all anyway. In a class I prepared, we developed several problems around a flow net software package. Students were required to solve many flow problems, varying parameters and geometry on the computer model and letting the machine do the dog work. Instead of spending hours learning how to draw and erase perpendicular flow lines, participants in this class gained experience with the behavior of different designs of permeable dams. So, the old manual methods for flow net preparation are not of great value to young Civil Engineers, but the underlying concepts of groundwater flow and the graphical representation by flow nets are essential knowledge.

Civil Engineering is the oldest branch of engineering. The Romans built aqueducts quite successfully without PC's. In a sense, we are quite set in our ways. In the modern CE design era, we have instituted and grown to rely on successful procedures to complete our work. Automation is rendering obsolete many of these procedures which performed so well in the past. Today, and increasingly so in the future, successful Civil Engineering practice will require learning how to harness the new technology not only to perform sophisticated technical analyses, but also to bring younger engineers up to design speed.

Telecommunications in Civil Engineering Education

C. Allen Wortley[1], F. ASCE

Abstract

Telecommunications (the electronic transmission of messages) is emerging as an important part of the University of Wisconsin's engineering outreach program. Technologies currently being used include videotapes, satellite and microwave transmissions, broadcast and cable television, telephony, computer networks, and communications audio bridges. The needs of undergraduate, graduate, and practicing civil engineers must be determined; and telecommunications programs, using appropriate technologies, implemented to meet their needs. As a profession, civil engineers are behind their other engineering counterparts in manufacturing and industry. Policies, procedures, and incentives which will attract students and faculty to telecommunications are required. Also required is the necessary telecommunications infrastructure.

Introduction

"Tele" is a combining form meaning "having to do with operating over long distances" and "communication" is a noun meaning "the act or fact of passing along, transfer." Nowadays, the noun "telecommunications" normally means "the electronic transmission of messages."

"Outreach" has become an important, identifiable and often referred to activity. Outreach is the College of Engineering's public service responsibility to instruct, inform, and transfer technology. The College should be able to simultaneously teach students on campus and off; link campus resources to all students and faculty, including industry; and receive and transmit information worldwide.

Technologies currently being used for telecommunications outreach include videotapes, satellite and microwave transmissions, broadcast and cable television, telephony, computer networks, and communications audio bridges. The College is also implementing a new, cost effective teleconferencing system, known as audiographic teleconferencing.

---

[1]Associate Dean for Outreach, College of Engineering, University of Wisconsin-Madison, 1513 University Avenue, Room 266, Madison, WI 53706.

Audiographic Teleconferencing

"Audiographic teleconferencing" is the combination of audio and computer graphics in a teleconferencing system. The current generation of personal computer-based audiographic conferencing equipment allows for the simultaneous transmission of voice and computer graphics over ordinary phone lines and for voice and graphic interaction among participants. The graphics may be drawn with a sketch pad or a mouse, may be scanned in, or may come from a video camera. Graphic images can be created on line, or they may be created ahead of time and stored like a slide show. If graphics are created ahead of time, a variety of painting or drawing utilities may be used, or files may be transferred in from a CAD program. When a conference is underway, all sites are live on the audio network, and several sites may use the graphics tablet at one time, creating a high degree of interaction.

Courses can be delivered anywhere there is a telephone line, and many sites can be served at once. Experts from any location can serve as course instructors.

Students can ask questions over the telephone and can supplement those questions with drawings or sketches on the graphics tablet. Interactive assignments may also be given. The system makes use of the computer by allowing the use of complex applications programs to develop homework or special projects. When these programs are used, files may be quickly transferred between students and the instructor for review.

Audiographic teleconferencing is an effective way to deliver continuing education; and is an alternative to the costly live, face-to-face, seminar or short course where engineers travel to the course location. It will also impact undergraduate and graduate education as students take courses from their campus libraries, dormitories, and homes; and as faculty teach from their offices, laboratories, and homes -- all away from the traditional classroom setting.

Identifying Learners' Needs

We all understand that technology moves rapidly, that obsolescence is easy to come by, and that to be competitive we must be up-to-date. Given these understandings, and given humanity's social instincts to discuss matters of concern, we have a market for live, and delayed, interactive professional training. Identifying and assessing the needs of adult distance learners, and hence the market for programming, is either done formally with survey methods or informally by intuition. We at the University of Wisconsin-Madison use both.

The College of Engineering has experimented with delivering graduate electrical engineering courses to local Madison industry via ITFS (instructional television fixed service, that is a narrowcast, microwave broadcast to line-of-sight receivers within 20 miles). Industry appreciated our efforts to provide them video and two-way, interactive audio in their on-site training rooms. The question now is, can the College launch an ITFS network in the Madison area, or throughout the State of Wisconsin by using land lines or satellite interconnects to local ITFS's.

This proposal is being formally evaluated by a market research firm conducting industry interviews and convening groups of industry people. They focus on and discuss the question of ITFS's viability to meet their employee's undergraduate, graduate and continuing education needs. This market research is addressing campus-wide ITFS instruction, not engineering alone.

The College of Engineering is now however surveying five thousand practicing engineers throughout the State about their total educational needs (and desires) and about their attitudes towards various delivery methods, including telecommunications. Previously the College telemarketed by phone key technical administrators in 64 state organizations, each with over 1000 employees. The administrators were asked about their perceptions of training needs, certification mechanisms, and delivery techniques. They favored direct instruction over the use of videotape and non-interactive, mediated instruction. Intuitively though, we believe this preference will change as telecommunications come of age and as total costs are fully considered.

We have launched a program of state-of-the-art live interactive, televised continuing education courses to industry throughout the nation via satellite. Intuition, rather than formal market research and survey methods, was used. Obviously, careful planning of the programs and observations of other's experiences have also been used. Large financial risks are involved. Enrollments in the first few programs have been good, and financial success achieved.

Evaluating Telecommunications Programming

Distinguishing which technologies are effective with specific student and adult learners, and the overall effectiveness of delivered programs, are essentials for sustaining an education program. Conducting such evaluations will strengthen future offerings, ensure a continued market, and produce ideas for new programs.

One measure of programming success for continuing engineering education is net program revenue. The College of Engineering functions at a very high level (usually 98 percent or more) of self-support generated through collection of program fees and tuitions. It is essential to conduct a program at a "breakeven" or a slight "profit".

Telecommunicated credit offerings are being evaluated through on-campus/off-campus student grade comparisons, student reactions and comments, numbers of enrollees in various courses, and in other ways. External standards of comparison are becoming available as more peer institutions participate in telecommunications.

Teaching Telecommunicated Courses

Faculty members in traditionally taught classes focus primarily on the presentation and delivery of lectures to groups of students meeting on campus. Telecourse faculty are more likely to focus on communicating with students in a more personal manner that is often tutorial in nature. Telecourse faculty generally depend to a much greater degree on phone and mail (now electronic-mail) contact with students than they do on class meetings.

One of our ITFS telecourse faculty members said,

> "I am thoroughly convinced that the presence of off-campus students enriches the classroom experience for the on-campus student. This occurs as the off-campus students raise questions and discussion relating to their actual work experience to the course material. Personally, I found the additional interaction very stimulating and rewarding."

Another university's faculty member observed,

> "Teaching television courses has broadened me as a faculty member because I do more individual in-depth work with television course students."

Motivating Instructional Faculty and Securing Their Participation

We as a College, building a telecommunications outreach program, which is under-funded or in most areas un-funded, are finding that it is a low priority with most teaching and research faculty. We are now investigating incentives and policies which are fair to all and which will attract the faculty and secure their continuing participation.

At present we have little to offer by way of effective policies now in place. We strive for policies that will encourage our faculty to become actively involved in teaching and course promotion; and that will give administrators control of production concerns (such as visual quality, course notes, and instructional methods) without unnecessary intrusion on the faculty's traditional instructional role.

Under general University policy there is no additional compensation for teaching credit courses as a principal instructional mission. This includes live televised courses or courses taped for subsequent viewing by students. University policy provides for, on a case-by-case basis, the ownership, use and control of instructional materials. In certain circumstances, funds may become available to aid the teaching mission via televised courses.

Sources of funds include: tuition remission to the College for credit students enrolled off-campus, contract fees and direct industry supports, unrestricted grants to faculty to assist their research programs (and hence instruction), per credit television surcharge fees, network subscription fees, and proceeds from subsequent videotape sales.

The following are some uses of funds to support the instructional mission of departments and faculty members: supports for employing research assistants, teaching assistants, graders, secretaries, and others; salary buyouts or summer supports for on-campus academic activities; base salary adjustments or overload compensation; purchase of equipment, texts, course materials, etc.; and discretionary funds for travel and support of other activities.

Telecommunications in Civil Engineering and Civil Engineering Technology Schools

In April 1989 I wrote to the department heads of 209 schools of civil engineering and 118 schools of civil engineering technology and asked if they would furnish me information on their current and proposed use of telecommunications in their undergraduate, graduate, and continuing education programs. I also asserted, "I have a feeling, but unsupported with data, that as civil engineering educators, we are not keeping pace with our counterparts in electrical and computer engineering, mechanical engineering and industrial engineering. Are we?".

As of October 1, 1989, representatives of 73 (35% of the 209) departments of civil engineering (or civil engineering combined with some other engineering discipline) and 18 representatives (15% of the 118) programs of civil engineering technology (or some other related technology program) responded. As to my assertion, that as civil engineering educators we are not keeping pace with our counterparts, 14 agreed, 2 disagreed, 74 did not choose to comment, and I felt that we shouldn't try to keep pace.

The use of telecommunications in civil engineering and civil engineering technology programs, both now and in the planned future, is pretty nominal. Some reasons for this include the college's resources, campus location, populace's needs, faculty interests, and newness of the technologies.

Many of the responding schools are doing a few telecommunication things, such as, electronic mail, computer networks, occasional live television, and videotaping. But the general characterization has to be that the activities of the responding schools is minimal. Also, one could probably correctly suppose that many schools did not bother to respond because they had nothing to report.

There are however some main exceptions, and those schools and programs are enumerated below. (Some schools not mentioned, do indeed have extensive telecommunications, but not in civil engineering, or did not respond.) They are:

University of Arizona, Microcampus Network
Colorado State University, SURGE Program
University of Florida, FEEDS Network
University of Idaho, Video Outreach Program
University of Iowa
Iowa State University
University of Missouri-Kansas City
Purdue University, IHETS Network
Stanford University, SITN Network
Southern Methodist University
University of Texas-Arlington

Some states are planning telecommunications networks, that include civil engineering. For example, Oregon and Wisconsin.

A complete analysis of the information received in this canvass about telecommunications is beyond my present scope. The information reported includes undergraduate, graduate and continuing education courses for credit and for no credit, and public service activities. Most of the reported activity, and activity of others such as the National Technological University, are in the fields of mechanical engineering, manufacturing systems engineering, electrical engineering, computer science and engineering, and materials science -- not civil engineering. The needs for telecommunications in civil engineering may indeed be less than these other disciplines, but not by the margins that presently exist.

Additionally, ASCE is not involved with telecommunication programming to the extent other societies are; for example, IEEE, ASME, and SME. This may be a reflection of the above.

Concluding Remarks

At this time, instructional television and telecommunications, within the College of Engineering, on the Madison campus, and throughout the University of Wisconsin System and the State of Wisconsin, are what I would characterize as being in their infancy, a bit confused administratively, under-funded and in most areas un-funded, a low priority with most teaching and research faculty, an issue un-addressed and not understood by many administrators and legislators; and, most importantly," ... a necessity if Wisconsin is to remain competitive in attracting and retaining world class scholars, in providing quality education to its citizens and in providing a communications infrastructure as important as physical highways and transport systems in the effort to retain existing jobs and attract new industry to Wisconsin." (1)

"The convergence of voice, data and video technologies presents profound challenges and opportunities for higher education. We must expand our thinking when we plan for the use of technology. We need to make it easier for individuals with experience, ideas and planning expertise to interact for the mutual benefit of all concerned. This collective task must involve not only those who will use the technology, but also those who must manage and pay for it. New strategies are central to the effective use of new technologies in instruction, research, administration, outreach, library and media services, and services to business and industry." (1)

I think what is good for Wisconsin, is good for other states; and what is good for civil engineers here, is good in other places. The College of Engineering plans to concentrate on the tremendous opportunities for continued and increased effectiveness in telecommunications. We will include civil engineers of course.

References

1. University of Wisconsin System Telecommunications Task Force. 1988. Telecommunications and the University: Planning for Change, the Report of the University of Wisconsin System Telecommunications Task Force, August 29, 1988, 84 pp.

Whole-life Career Planning

Professor James Armstrong FASCE*

History

The Institution of Civil Engineers in the United Kingdom was Chartered as a qualifying organisation in 1828, although its roots date back to 1771 when the Society of Civil Engineers was first established by Smeaton and others. The first address to the embryonic Institution in 1818 by Henry Palmer described the Engineer as, 'a mediator between the Philosopher and the Working Mechanic, and like an interpreter between two foreigners, he must understand the language of both.'

Most of us will recognise the significance of this remark, and will have sought to acquire both theoretical and practical knowledge during our careers. The creation and development of learned societies dedicated to the furtherance of the art and science of engineering has been based on this understanding. There is no more important task facing such organisations than the continuing sharing and updating of knowledge and skill.

The form of the delivery of continuing education and training varies with changes in the needs of society, and with changes in the nature of developments in engineering. In the U.K. one of the most recent developments has been an initiative by the Engineering Council, a coordinating and registering body for all engineering disciplines. This seeks to develop a greater degree of self-motivated training by the individual engineer.

*President, Institution of Structural Engineers, UK. Professor of Continued Professional Education, University of Leeds UK. 32, Langford Green, London SE5 8BX, UK.

Current provision of CPD

Professional Institutions in the UK have a variety of ways of testing competence, following the attainment of a first degree in engineering and three or more years of experience and training under supervision. Following this initial approval there are no further tests of competence, although advancement to higher grades of membership may require evidence of satisfactory mid-career training as well as achieving position of authority.

The form and amount of formal 'Continuing Professional Development' (CPD) taken up by individuals therefore varies considerably. Some engineers will continue their development by taking higher degrees in full time education, some in a part-time mode. Many will attend short courses of a few days or weeks duration, or will attend evening classes. These are likely to be sponsored in some way by their employers, but may be self-funded. In addition a great deal of private study takes place, as well as on-the-job training. Larger employing organisations will sometimes provide in-house training courses in their own areas of interest. The role of the 'Open University' and 'Open Tech' Institutions in the UK has not yet developed to be significant in technical training, but they are a valuable source of management training.

The Professional Institutions play an important part as enabling organisations in the provision and encouragement of such career development. This is done via specific programmes of educational activities, by their publications, and by collaboration with educational institutions. The volume and quality of such provision is impressive, but the take-up by members is very variable. Such training is also supplemented by Industrial or Trade organisations providing knowledge of specific new materials and techniques.

The key point to note in all these instances of CPD is the variability of take-up by the individual engineer.

The changing context

The rate of change of engineering technology is more rapid than it has ever been before. Developments in information technology, in increasing specialisation, in the scale of projects, in international collaboration, in an increased awareness of environmental concerns, all require the executive engineer to update his or her knowledge, but also make it increasingly difficult to find the time for mid-career training.

There is a need therefore for improved efficiency and organisation in such provision, and the individual needs to be both encouraged to maintain and improve his or her competence, and advised as to how best to do this.

Changing methods

In 1987\88 the Engineering Council in the UK studied the problem of the continuous updating of Engineers, leading to their report 'A Call for Action'. This was followed by the launching of a pilot scheme for Career Action Planning by individuals as a prime motivator of training efforts. The scheme has been promoted through several industrial and consulting organisations, with the cooperation of their staff. Subsequent discussions suggest that this may need to be given further impetus by some mandatory requirement, or peer group pressure, to achieve the level of career-long continuing education and training felt to be essential.

The pilot study is based on the use of a self appraisal booklet issued to all participants, which enables them to assess and record their position and needs against the following heads:

1.0 Identifying development needs for:-
- 1.1 present job performance
- 1.2 foreseeable future roles
- 1.3 personal and professional development

2.0 Creating a Career Action Plan

3.0 Recording achievement

Check lists of activities to be asssessed are given under three heads:

Technical\Scientific
Personal effectiveness\Communications
Commercial\Business

Individuals are asked to assess their current effectiveness against these items, and to identify their strengths and weaknesses for their present responsibilities. This indicates any needs for immediate further training or study.

A similar review is then undertaken of possible short and medium term prospects, and of the likely level and nature of skills required for these future roles, suggesting further development studies.

The third area of review of professional and personal aims seeks to highlight aspects of life which are of particular importance to the individual. This includes not only professional ambitions and work satisfaction, but family, social and sporting activities, preferred geographical location, etc.

It is proposed that a network of experienced Career Advisers be established to counsel engineers on appropriate lines of development and on available methods of obtaining the training they need. The analysis carried out will then be used to structure a Career Action Plan with a series of inter-related development actions drawing on available sources of instruction and study. Such sources would include:

> Job experience; reading; professional meetings; coaching; communication skills; in-house courses; external courses; open learning; additional academic qualifications; etc.

It is recognised that there are problems in the provision of CPD to all engineers. Some are in isolated postings, well away from centres of professional or educational activity, or operating as sole practitioners or as members of very small organisations, where it is difficult to take time-out during normal working hours. Developments of distance learning using recent developments in information technology and expert systems may prove helpful in these cases.

Whilst the formulated Career Action Plan needs to be flexible, it is felt that it gives structured guidance for the most effective use of resources and time by the individual engineer. The final plan includes realistic time scales by which the various goals might be achieved.

Finally the booklet contains space for a record of training undertaken, which it is intended would be of interest to future employers and perhaps to Professional organisations in considering applications for promotion to higher grades of qualification.

The individual's place in the Industry

The Career Action Plan seems to focus on the individual's own wishes and needs, and not to consider restrictions arising out of the needs of industry, of the professional organisations, and of society in general. Such individual developments can, and must, be considered in context. It is necessary to harmonise the aims and objects of each of the groups in which the Engineer is involved. In the table below an attempt has been made to set out the interests of all four parties under the heads of the Individual, the Professional Organisation, Industry, and Society. If programmes for the development of individuals take into account the interests and potentials of all sectors, then his or her development can be constructed to satisfy all the needs, and still lead to a fulfilling and rewarding life.

Aims, Objects, Methods and Facilities for Career development

	INDIVIDUAL	PROFESSION	INDUSTRY	SOCIETY
AIMS	Satisfaction	Competence	Performance	Contribution
OBJECTS	Skills	Up-date	Efficiency	Awareness
METHODS	Planning	Activities	Plans	Contexts
FACILITIES	'CAP'	Institutions	Associations	Gov't\Social

It is recognised that each sector affected by the engineers competence will have its own particular set of needs and methods of providing for those needs. This table attempts to show possible relationships between these, which need to be reconciled to optimise the satisfaction of all parties. Much collaboration already takes place, but this is not always deliberately planned. It is the Author's opinion that pro-active collaboration would be much more effective in achieving the desired aims of all parties.

An example of the possibility of integrating professional and commercial needs into the individual programmes could lie in the use of the ASCE document 'Quality in the Constructed Product.' In setting out the processes involved in project procurement from inception to final delivery and maintenance, clear indications are given of the level and type of competence required for differing processes and levels of responsibility.

Where next?

This conference is entitled 'Setting the Agenda for the 90's and Beyond'. In the UK we are very conscious of the challenges facing the Engineer. The rapid changes in social organisation and technology have already been mentioned. The contribution of the engineer to improving and maintaining the quality of life in communities is undoubted. If we are to make the effective contribution that we all want to make, then we must not only maintain our competence, but we must be seen to be doing so in order to generate the confidence of society. We may then be given the opportunities and responsibilities we require to direct and achieve the developments possible and essential to the future welfare of society.

As Henry Palmer said at the formation of the Institution of Civil Engineers in the UK in 1813 'the engineer stands as a mediator between the Philospher and the Working Mechanic and like an interpreter between two foreigners, he must understand the language of both.' Just as we integrate practising engineers in the initial formation process with undergraduates, then we must also realise that there is a continuing formation process, and that learning should be a life-time commitment for engineers. Continuing Professional Education and Training must become part of our underlying philosophy, an essential and automatic aspect of the engineering ethos.

Learning is living, and like living should only stop once in a life-time.

Technical Competence Among Civil Engineers

By John R. Henry, [1] M. ASCE

Abstract: Civil engineering is the broadest of all engineering disciplines, encompassing a nearly incomprehensible scope of knowledge. Consequently engineers must focus on a particular subject and develop expertise in that area. In the small, rural office, engineers sometimes tend to practice outside of their area of technical competence. Engineers should recognize that an important part of being a professional is to know ones own technical limitations, as well as when to seek advice from consultants, or obtain additional training. Recent state licensing trends are to limit civil engineers allowable scope of practice by statute law, and require specific licenses by examination in order to practice in areas such as land surveying, structural engineering, geotechnical engineering, etc. One reason for such a trend is that engineers don't use discretion in practicing within their own areas of expertise. Professional societies and academic institutions should strive to provide more continuing education programs that enhance, upgrade, and extend the knowledge of practicing engineers who wish to expand their scope of practice and keep abreast of changing technology. Lastly, ASCE should offer a specialty certification program that would serve to identify individuals who can demonstrate expertise in certain areas of engineering practice.

INTRODUCTION

In the paper, Professionalism and the Civil Engineer, by Harvey R. Gobas, the author points to several reasons for the decline in the stature of the civil engineering profession, and proposes ten commandments as a means of professional advancement. While the paper is entirely correct when it states, "...our potential demise is being caused from within - by civil engineers themselves," it overlooks one other major cause for a decline in our professionalism. The first priority of professional

---

1. Plan Check Engineer, Butte County Building Department, 7 County Center Drive, Oroville, California.

engineering should be technical competence. Without the necessary technical expertise to perform competent professional engineering, all other criteria are meaningless.

A significant number of engineers, perhaps the majority, are practicing in small offices in a rural setting. In this rural business climate, engineering offices cannot survive unless they offer to perform diverse engineering services, ranging from land surveying to structural design. Although projects tend to be small in scope and of limited complexity, it is within this climate that engineers tend to operate out of their area of professional competence and technical expertise. For example, engineers whose practice consists primarily of site design related to land development, such as boundary surveying and mapping, site planning, grading, and drainage, periodically may be confronted with a client requesting structural engineering design. Many do not have the experience, nor in some cases the academic training, to do structural engineering in a professional and competent way. Particularly in seismically active areas such as California, the requisite knowledge and experience to perform competent structural design is more extensive and demands more attention to detail.

Related to this is the fact that employment policies within small offices may not approach the standards established by ASCE Manual 55 in terms of basic salary compensation and benefits. Local economic conditions generally dictate compensation and benefit scales, and many owners of small offices are not familiar with or even aware of Manual 55. This fact tends to cause those engineers who have expertise in special areas of practice, such as structural engineering, to migrate to the cities which have larger firms specializing in these fields, and which offer more attractive compensation and benefit packages.

In addition, clients often select a consulting engineer based upon cost factors alone, with no regard for technical competence. The unsuspecting client assumes that all licensed engineers are of the same professional caliber. The following statement of prophetic wisdom from ASCE Manual 45, has countless times become all too real: "Selection based primarily on compensation factors, with less or no consideration as to competence, can result in unsatisfactory service to the client and in higher overall capital and life-cycle costs."

The primary objective herein is to remind engineers that

an important part of their professional expertise is knowing in what areas they are technically proficient and professionally competent. In addition, to practice out of ones area of competence is a violation of statute law in most states.

Secondly, professional engineers should know when to seek the advice of other professionals who are specialists in the fields in which they themselves lack proficiency.

Thirdly, it is incumbent upon the profession and the academic community to provide continuing education and training programs designed to upgrade and advance the knowledge of engineers in subjects which may be out of their usual area of practice.

THE BROADEST OF ENGINEERING PROFESSIONS

Webster's Dictionary defines a civil engineer as, "an engineer whose training or occupation is in the designing and construction of public works (as roads or harbors) and of various private works." Needless to say, this definition is so general as to imply that the design or construction of anything for use by or for the benefit of civilization falls within the province of civil engineering. This represents a tremendous scope of technical knowledge and discipline.

Presently, civil engineering encompasses land and geodetic surveying, mapping, site planning, highways, roads, bridges, buildings, dams, water supply facilities, wastewater disposal and treatment facilities, drainage facilities, hydrology and hydraulics, city and urban planning, earthwork and grading, environmental assessment, power facilities, foundation and geotechnical engineering, and most recently computer applications related to all other aspects of civil engineering.

Engineers cannot possibly be proficient in all of these areas, but must select only a few areas of interest and pursue them until they achieve a certain level of technical expertise. The most successful large firms are those who have staff that work in certain areas and confine their work to those fields in which they are competent. The firm principals know each individual's strongest subject, and ensure that they continually improve their special talents by staying within their respective areas of technical competence.

On the contrary, the small engineering office does

not have the necessary staff to have specialists for each discipline. With only two or three engineers, the small office is forced to be more diversified, and attempts to cover the areas of design as the demand for services dictates. It is in this competitive business environment that the small office finds itself taking design projects that may be out of their particular area of competence. When confronted with this situation, the small engineering office has three options. The most obvious is to inform the client that they do not do enough of that type of engineering to feel technically comfortable, then refer the client to another engineer within the community who is qualified. Another option would be to take the job, but consult with another qualified engineer, if economically feasibly and aggreeable to the consultant.

A third option available to the engineer is to seek additional training from professional societies, seminars, textbooks, or continuing education programs. Once the engineer feels technically comfortable, he or she could then accept design projects that are within his or her level of competence.

THE PEER REVIEW PROCESS

While occupying a position as an engineering plans examiner, the occurance of flagrant errors in the application of engineering principles and judgement has become much more common than anticipated. Of particular concern is when registered professional engineers lack an understanding of the principles of engineering, sometimes even at the fundamentals level. Another startling observation is that engineers with no academic training in a subject, would literally invent a methodology which in no way resembles acceptable engineering principles or practice. Thirdly, some engineers blindly follow design handbooks without any understanding or working knowledge of the terms in the formulas, and consequently produce calculations which are completely meaningless, sometimes even dimensionally incorrect.

Particularly diconcerting is the fact that vast regions of our state, in this case, California, have no technical peer review at all, and rely soley on the design professional and the fact that he or she is licensed by the state to practice. A license to practice civil engineering does not necessarily ensure that all engineers can produce proper designs in all areas for which they can legally practice.

A case in point is a local government which had virtually no staff to do structural plans examination. For years the jurisdiction didn't review the work done by design professionals where buildings were concerned. Recently the agency contracted to obtain an experienced professional engineer as plans examiner, only to find that many of the engineers who were providing design services in the area for years were not producing professional or structurally adequate designs or calculations. The response of the designers to what they perceived as "new requirements," was a sort of defacto precedent, that is, they had done it that way without objection for so long that it must be correct by prescription.

Engineers should be aware that state laws which govern licensing of professional engineers usually contain specific clauses relating to practicing within ones area of competence. For example, the California Administrative Code states:

"A professional engineer or land surveyor registered or licensed under the Code shall practice and perform engineering or land surveying work only in the field or fields in which s/he is by education and/or experience fully competent and proficient."

What engineers need to recognize is that practicing outside of their area of competence is a violation of statute law and could result in revokation of their license to practice.

STRUCTURAL ENGINEER, CIVIL ENGINEER, OR LAND SURVEYOR

In some states, such as California, there is an additional registered title of structural engineer. In California, one who has a license to practice civil engineering can do structural engineering for most structures with the exception of hospitals and public schools. This means that a civil engineer who opted in college to concentrate his studies on sanitary engineering, and did not take the full range of structural engineering coursework, could legally design any type of structure once licensed. That this actually occurs is evident when one examines engineering plans and calculations submitted by a wide range of engineering firms.

In order to become registered as a structural engineer, California requires an individual to be a registered civil engineer first. With additional experience requirements and comprehensive examinations, one can obtain a license to use the title of structural engineer, with an assigned registration number. The individual is then refered to as a "registered structural engineer." The requirements to become a structural engineer in California are so rigorous that those registered as such are generally recognized as being thouroughly competent to practice all phases of structural engineering.

There are, however, vast numbers of civil engineers who are entirely competent in all phases of structural engineering as well, but for one reason or another have not obtained structural registration. It is this group who stand to be hurt most by the reputation being developed by those civil engineers who are not fully competent.

Prior to 1982, California civil engineers were allowed to practice land surveying to its fullest extent, which means they could establish boundary. This assumed that they were completely knowledgeable of boundary law, legal principles, Public Land Surveys, water boundaries, legal descriptions, and state plane coordinate systems. In the academic sense, civil engineers generally have only a cursory exposure to surveying in college, which usually only covers the mathematical aspects of surveying. Much of the college coursework relates to civil engineering and construction aspects of surveying, which has very little to do with Cadastral or boundary surveying. Beginning in 1982, California law allows civil engineers to do only "engineering surveys" which is incidental to civil engineering projects. In order to establish boundaries, prepare maps, and set property corners, a civil engineer now has to obtain a land surveyor's license, which is obtained by passing a rigorous examination.

If civil engineers who are not technically capable continue to practice structural engineering, it may come to pass that the state will require all civil engineers to obtain structural engineering registration in order to do any structural design.

In essence, the state does not leave it to the discretion of the professionals to recognize within which areas they are competent to practice, but limits their range of practice by statute. The engineers, in effect, will have brought this upon themselves by not staying within their own respective areas of competence.

Most professional engineers recognize that obtaining an

engineering degree does not, in itself, make one an engineer. This is somewhat analogous to an auto mechanic. Simply buying a complete set of mechanic's tools does not make one a mechanic. An apprentice mechanic must learn to apply the tools, and to develop a mechanical thought process. This is done by working under the direction of a master mechanic who imparts to the apprentice the knowledge which has been gained from years of experience. It is the knowledge which is the difference between the master and the apprentice, not the tools.

CONCLUSION

Professional engineers should be able to recognize their own limitations with regard to the vast scope of knowledge encompassed by civil engineering, and practice within their areas of competence.

Our ability to recognize our own capabilities and to confine our practice to these areas should be sufficient, without the need for statute law to prevent us from exceeding these boundaries.

Professional societies and academic institutions should strive to provide continuing education programs that serve to enhance, upgrade, and extend the knowledge of practicing engineers who wish to expand their area of practice.

ASCE should offer a specialty certification program whereby members could obtain certification in specific subject areas of civil engineering. This specialty certification should be by examination, and should be comprehensive enough to demonstrate competence and proficiency in the particular subject.

ACKNOWLEDGMENTS

The support of Frank L. Tyukos is gratefully acknowledged for his continual dedication to maintaining professional standards among engineers.

APPENDIX. REFERENCES

Gobas, H. (1988). "Professionalism and the Civil Engineer.", Journal of Professional Issues in Engineering, ASCE, 114(2), 142-147

California Administrative Code (1987), Title 16, Chapter 5, Section 415.

21ST CENTURY LEARNING

Richard B. Lewis[1]

Abstract

Grades have been, traditionally, the results of homework, quizzes, perhaps a mid-term exam, and some form of final exam. Lectures, for the most part, contribute to the great majority of classroom activities, while the students are involved in very little of the learning process. Most of <u>their</u> learning occurs during the early morning hours when they attempt to read the text, match it with their notes (usually an impossible task) and complete the problems assigned.

Although the lecture is a vital tool in the dissemination of information, it does not take into account the many areas of teaching open to us that involve the student in both aural and visual learning.

This paper will address several approaches the teacher can pursue to involve the student in every course and in every class. These approaches involve the use of communications, computers, library research, humor, and a whole range of creative activities that can be adapted to suit any learning situation (not teaching situation!).

This paper will probe areas brought out in the ASCE Report "Civil Engineering in the 21st Century," November 1987, re "a visionary professional" with attributes that include "being innovative" and a team member (p. 7).

INTRODUCTION

The quest for excellence in teaching is a life-long process. For some, the label "Outstanding Teacher" may come in the early part of a career; for

---

[1]Assistant Professor of Civil Engineering and Coordinator of Construction Engineering Technology, Louisiana Tech University, Ruston, LA 71272.

others, much later. For most, never. Still others state they never think about it, or don't really care. But that's preposterous! Existing literature on how to become a better teacher is voluminous. From monthly publications to society proceedings, workshops to seminars, and Centers of Excellence to our own colleges of education, there is simply too _much_ information on the subject. The majority of professors I know are capable, innovative, creative and professional; some of them simply _cannot_ teach, and others struggle their whole careers--unsuccessfully--trying to be a better teacher.

What is the difference between a bad and good teacher; a good one and an outstanding one? Why do student grapevines invariably direct their friends to one professor, and leave the other one with only a handful of students?

THE PROBLEM

There are as many answers as there are professors! Yet, with some research and groupings, the trends are becoming clearer; with a great deal of scrutiny, it is possible to narrow the answers down to one statement: The emphasis has invariably been on the _teacher_ and not the _learner_. We need to turn that around! John Milton Gregory states it like this: "the way people learn dictates how you teach," (quoted in Hendricks, 1987) and calls this The Law of the Teaching Process. If, as we enter the classroom, we are absolutely certain of the objectives to be accomplished in the students' _learning_ activities, in lieu of the mindset, "what am I _lecturing_ about today?" the results will be revolutionary. But how is that accomplished? How can we turn the process around?

SOME SOLUTIONS

The better questions are: How do people tend to process information? and how can we provide students with enough breadth in _instructional_ styles to accommodate their _learning_ styles?

There has been an abundance of literature on these and related subjects written over the last 25 years, and especially in the last 15 years, that not only suggests, but compels, each teacher (at _any_ level of teaching) to read and heed. (Claxton and Murrell, 1987; Dixon, 1982; Grasha, 1984; Gregorc and Ward, 1977; Kolb, 1981; Pask, 1976; Schmeck, 1981) Engineering teachers, mostly, are not aware of this literature, and this unique ASCE presentation--a forum encouraging discussions rather than presentations--could be the vanguard in disseminating what some other disciplines already have in place. Furthermore, many of the participants at this forum will be involved, directly or indirectly, with continuing education. Although learning styles change throughout life, there is a leveling-out period during college-age adults that usually lasts into their 30s or early 40s. (Kolb, 1981)

In fact, this plateau-like period consists of changes in refinement within a learning style rather than a change of style itself. Thus, every engineer needs to know how he/she learns. This information will be of significant value when we attend our next seminar/workshop, and it most definitely will change our presentation style the next time we are called upon to stand before a group and persuade, challenge, inform, etc.

For college/university personnel (teachers, administrators, other staff), this knowledge will be an asset in student advising and counseling. To be aware of a student's learning style may suggest different approaches to counseling, aid orientation leaders to help students understand their own preferences and strengths in learning, and even guide teachers to provide stimuli for developing new ways of learning for their students.

Dr. David A. Kolb's Learning Style Inventory (LSI) (Kolb, 1985) has been tested, successfully, in a number of different teaching environments (including engineering), and pre-sented in many journals. One conclusion from his study is that each individual has his/her own learning style, and that style is set by the time most college students enter their junior year. From his studies, and the case histories of the many who have tested his theories, it is impossible to miss the red flag they are waving: The lecture method is the poorest method of teaching (on a sustained basis). Waldheim (Waldheim, 1987) further states that any blackboard work we do, using words, numbers, or letters, is an additional example of the <u>auditory</u> mode of perception (not visual, as one might suspect). McCart (McCart, et al., 1985) has reported extensive work with Kolb's LSI and then followed through with additional questions concerning best-learning-environments. The top preferences of each group studied were small group sessions, workshops, demonstrations, and self-study. Patricia Cross, in an absorbing study on ways to improve teaching, states "All too often, information flows from the notebook of the professor to the notebook of the students without going through the mind of either." (Cross, 1986)

We must expect great things from our students and not be satisfied with mediocrity. Hendricks writes, concerning the basic goals of teaching, that we should teach students to think, to learn, and to work. (Hendricks, 1987) Reinforcing the "learning" portion, Culver reports on "self-managed learning," in which the sequence of practice-to-theory-to-practice" be utilized; i.e., begin with a practical experience or demonstration, develop the theory, and give another, more difficult experience or demonstration that will be the basis for extension to other areas. (Culver, 1987) All engineering courses are fertile ground for innovative ways to teach without resorting to lecture. Consider the following data, from an article by James Stice: (Stice, 1987)

Learning Method	Retention by Learner
What they read	10%
What they hear	26
What they see	30
What they see and hear	50
What they say	70
What they say as they do something	90

The data clearly indicate that students need to learn effective communication, which involves four skills: listening, reading, speaking, and writing.

Listening is more than hearing, but less than judgmental evaluation. We must develop, in ourselves as well as our students, the capacity to absorb content, without attempting to speak so as to have the last word (either vocally or silently). Listening--without immediate evaluation or drifting off to personal thoughts--is an art that requires nurturing.

Reading allows the development of ideas, the increase of vocabulary, and the humbling experience of realizing how much we have yet to learn.

Speaking promotes intellectual development, the ability to convey thoughts accurately, and reinforcement of what we think we know.

Writing is a stimulating exercise and the culmination of all the above. It separates the knowledgeable from the intellectual, and opens new vistas that spark innovation, creativity, and just- plain-thinking.

Each and every engineering course can be "taught" in such a way that each and every student, no matter which "learning style" (or combination) he/she possesses, can perform to higher expectations than that of the lecture-homework-quiz syndrome, if nurtured along these paths of listening-reading-speaking-writing.

PERSONAL EXAMPLES

In the teaching of Statics and Strength/Mechanics of Materials, I require weekly journals (evidence of listening and reading), a term paper (evidence of writing), and a short 75-100 word essay on each quiz. Class participation and probing questions give evidence of speaking ability. The journals and term paper must be done using a word processor, and I give extra credit for a computer problem/solution they have created as a result of their present study.

I have taken Kolb's LSI; as a result, I continually update my teaching methods to reflect not only how I like to teach, but also some alternate methods that might reflect how students of opposing learning styles might like to learn (the assumption is that I have some of each style in a class of 30-50).

Why go to all this trouble? Frankly, it's a lot more fun! Teaching is always a challenge; but now, I can concentrate on the student in lieu of the material. The feedback I receive is that a great majority of the actual material I teach is never used again (following graduation); however, the principles and thought processes utilized in arriving at a decision are introduced, nurtured, tested, and honed in every class. When I probe, question, refute, argue, and ask "why" and "how", they are learning to do the same. True, there is a body of knowledge that must be put to memory. Aside from that, let's get them to the point where they can fend for themselves. Why do we put that off to graduate work, or a senior-level capstone design course? Thinking, reflecting, and making decisions in Statics can be exciting!

CONCLUSIONS

A teacher who is aware of his/her learning style and the probable disparity between his/her style and those of the students will find heightened interest and learning occurring in the classroom if some significant changes are made in the delivery of the material. The demand for "a visionary professional"--the 21st Century Civil Engineer--calls for imagination, critical and creative thinking, and leadership. The teaching profession must start now (we may already be too late) to prepare ourselves and the next generation of students to meet the new, unexplored and challenging tasks of the future. This may require some rethinking of our present academic postures, some retraining, and certainly some institutional adjustments to make it all happen. We must concentrate on the learner; we must!

REFERENCES

Claxton, Charles S., and Murrell, Patricia. 1987. "Learning Styles: Implications for Improving Educational Practices." ASHE-ERIC Higher Education Report No. 4, Washington, DC. Association for the Study of Higher Education, 1987.

Cross, K. Patricia. September 1986. "A Proposal to Improve Teaching." AAHE Bulletin 19:9-14.

Culver, Richard S. February 1987. "Who's In Charge Here? Stimulating Self-Managed Learning." Engineering Education 77:297-301.

Dixon, Nancy. July 1982. "Incorporating Learning Style into Training Design." Training and Development Journal 36:62-64.

Grasha, Anthony. Winter 1984. "Learning Styles: The Journey from Greenwich Observatory (1796) to the College Classroom (1984)." Improving College and University Teaching 22:46-53.

Gregorc, A. T., and Ward, H. G. 1977. "Implications for Learning and Teaching: A New Definition for Individuals." NASSP Bulletin 61:20-23.

Hendricks, Howard G. 1987. Teaching to Change Lives. Portland, Oregon: Multnomah Press.

Kolb, David A. 1981. "Experiential Learning Theory and the Learning Style Inventory: A Reply to Freedman and Stumpf." Academy of Management Review 6:289-96.

________. 1985. Learning Style Inventory. Boston: McBer & Co.

McCart, C. L., Toombs, W., Lindsay, C., and Crowe, M. B. 1985. "Learning Styles Among Established Professionals." Paper presented at the annual meeting of the American Educational Research Association, Chicago, IL, April.

Pask, Gordon. 1976. "Styles and Strategies of Learning," British Journal of Educational Psychology 46:128-48.

Schmeck, R. 1981. "Improving Learning by Improving Thinking." Educational Leadership 38:384-85.

Stice, James E. February 1987. "Using Kolb's Learning Cycle to Improve Student Learning." Engineering Education 77:291-96.

Waldheim, George P. February 1987. "Understanding How Students Understand." Engineering Education 77:306-08.

INNOVATIVE APPROACH TO CONTINUING EDUCATION

David G. Parker[1]M,ASCE and Sandra C. Parker[2]

Abstract

The Civil Engineering Department, in conjunction with the College of Engineering at the University of Arkansas, has developed a program for continuing education based on the use of video technology. This program serves practicing engineers throughout the state. The approach uses the 'candid classroom' concept in which selected classes on the Fayetteville campus are videotaped to allow offering the same classes at other locations in the state. The courses can be viewed at four Graduate Resident Center locations as well as at engineering firms in the state. Practicing engineers can take individual courses or can pursue a program leading to a Master of Science degree with an emphasis in Civil Engineering.

Introduction

The College of Engineering of the University of Arkansas is the only institution in Arkansas offering graduate engineering education, and the only Civil Engineering Department in the state is located at the University of Arkansas in Fayetteville. The College of Engineering is committed to providing continuing education opportunities to the engineers in the state. However, because the engineering population is widely dispersed throughout Arkansas, an innovative approach to the delivery of continuing education courses was needed.

---

[1]Professor, Civil Engineering, University of Arkansas, Fayetteville, AR 72701

[2]Associate Professor, Industrial Engineering, University of Arkansas, Fayetteville, AR 72701

Graduate Resident Center

In 1986 the Graduate Resident Center for Engineering (GRCE) was established in Little Rock by the University of Arkansas. The center now offers graduate engineering courses by video tape at four locations within the state (Camden, Fort Smith, Hope, and Little Rock) and at several industrial plant sites. Students taking courses through the GRCE take the same course and receive the same credit as students on the main campus at Fayetteville. Through this program, practicing Engineers can take individual courses or can pursue a program leading to a Master of Science degree in Engineering. One of the specialities that can be taken is Civil Engineering.

Video Instruction

Classes in the GRCE program are presented as videotaped lectures. The approach used is the 'candid classroom' concept in which classes are videotaped at the same time the class is offered on the Fayetteville campus. Tapes recorded on the Fayetteville campus as well as printed materials such as handouts and exams are sent weekly to the GRCE site locations. During each GRCE class, site coordinators at each location set up the tapes, hand out materials, and collect homework. The tapes may also be viewed at other times for the convenience of the GRCE students who may have to miss the evening session because of demands of work. Students may also choose to review the tapes a second time for some of the course material. Faculty visit the off-campus locations about four times per semester and are available by telephone during regularly scheduled office hours.

Video Classroom

The video production facilities used for GRCE courses are located at the College of Engineering on the Fayetteville campus. The facilities consist of a classroom studio and a control room. The classroom is equipped with three industrial quality video cameras, wireless microphones and a special monitoring system, all of which are controlled by a Production Technician located in the control room. The Production Technician can use a variety of on-line aids to record the classes, including slides and computer-generated graphic displays. One of the video cameras in the classroom is mounted overhead so that any material that the instructor places on a custom-built podium can be recorded and also simultaneously displayed during the lecture on two large monitors mounted at the front of the classroom. The podium contains a variety of equipment including controls and monitor for the overhead camera, a monitor showing what is actually being recorded, and a two-way setup for communicating with the Production Technician.

Student Statistical Data

The GRCE program was started in the Fall of 1986. Figure 1 shows the growth in student enrollment in the MSE program since 1986. Since that time, 75 students have been involved in the Civil Engineering option and thirteen Civil Engineering courses have been offered in the areas of environmental, structural, surveying and transportation engineering. Civil Engineers from state and federal agencies and private consulting and industrial firms have participated in the off-campus program. For the 75 students involved in the Civil Engineering option, the makeup of their education is as follows:

59% have BSCE degrees,
15% have MSCE degrees, and
6% have degrees other than CE.

For the entire GRCE program the degree seeking status is as follows:

54% are seeking a degree,
32% are taking courses for information only, and
14% are unknown.

For the entire GRCE program the tuition is paid as follows:

90% of the students have 100% of tuition paid by employer,

7% of the students have 50% of tuition paid by employer, and

3% of the students pay 100% of tuition.

Faculty Time

The amount of extra time required by the faculty involved in video classes varies depending on the individual and the nature of the material being presented. Most faculty report that video classes are much more time consuming and may require as much work as up to two regular classes. The extra time requirements are for visiting the off-campus classes, answering

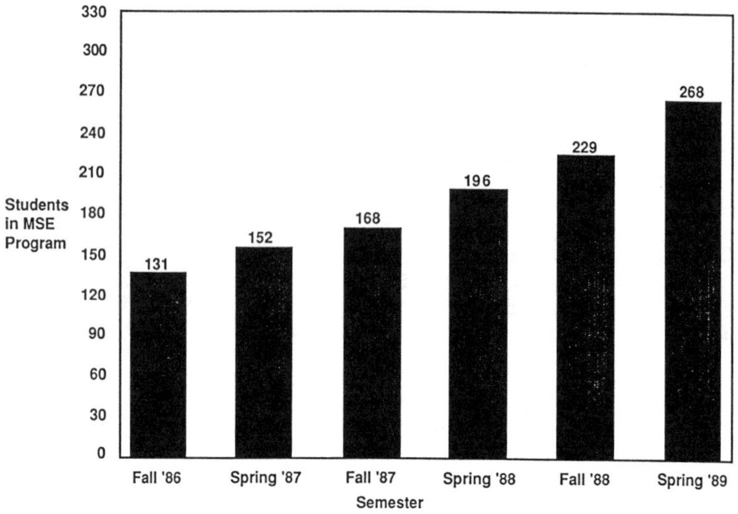

Figure 1. Students Enrolled in MSE Program

telephone questions, and preparing special course materials and handouts. The authors have found that the preparation of special course materials is the most time consuming part of preparing for the video classes. Class notes including text, figures, and calculations are prepared with special care before class so that they are easy to see on the monitors at the remote locations. In addition to projecting notes on the monitors all notes are copied and handed out to the off-campus students before each class.

Impressions of the Program

Most of the students at the remote locations seem to be favorably impressed with the quality of the program and are glad to have the opportunity to take the same classes that are taken by the Fayetteville students. However, some students have voiced complaints about the lack of interaction with the instructors and an occasional delay and/or mix up in sending materials between Fayetteville and the remote locations. The performance of the students at the remote locations tends to be more variable than performance of the Fayetteville students. It is not unusual for the remote students to have both the highest and the lowest grades in a class.

The following are seen by the authors as the advantages and disadvantages of the videotaped approach to providing continuing education to remote locations. Advantages:

(a) Off-campus courses are the same as on-campus courses.

(b) Tapes are available for viewing at times other than the regularly scheduled classes.

(c) Classes can be presented to many students at a variety of locations.

(d) Faculty commuting time to remote locations is reduced.

Disadvantages:

(a) A large investment in production equipment and staff is required.

(b) Interaction between faculty and students is reduced.

(c) An increase in faculty preparation time is required over conventional classes.

In general, the authors feel that the video program has been successful because it has allowed many practicing Engineers throughout the state to take classes without interrupting their careers. Future plans for the GRCE program include increasing the number of course offerings, increasing the number of off-campus locations, and possibly adding satellite transmission capabilities.

MANDATORY CONTINUING PROFESSIONAL DEVELOPMENT?

Bobby E. Price[1], P.E., F.ASCE

ABSTRACT

Mandating continuing professional development (CPD) for engineering registration renewal is a complex and perplexing issue. This paper addresses the purpose of CPD, its pros and cons, what the states are doing in this area, how the mandated program in Iowa is operated, what various professional organizations are doing and some questions that must be considered prior to initiating a mandated CPD program.

INTRODUCTION

Mandatory continuing professional development (CPD) for engineering registration renewal stirs the emotions of engineers in various ways. Some believe it is needed as a safeguard to the public welfare and the continued growth of the engineering profession. Others sense it to be an encroachment on the prerogatives of engineers to determine their own methods to maintain competence. Very few engineers take a "don't care" attitude.

No group of engineers are probably affected as much by the prospect of mandatory CPD as are civil engineers. Civil engineers are noted for their almost universal work on projects dealing with the public welfare and constitute one of the largest groups of registered professional engineers in the nation.

Several other professions/groups appear to be steadily moving toward mandatory CPD. The table below reflects changes from 1979 through 1989 in the number of states that have implemented mandatory continuing education for 16 selected professions/groups, including engineering.

---

[1]Professor of Civil Engineering, Louisiana Tech University, Ruston, Louisiana 71272.

1989 Mandatory Continuing Education Update
(After Phillips 1989)

Profession/Group	1979	1989	Profession/Group	1979	1989
Architects	1	1	Psychologists	7	19
CPAs	28	48	Pharmacists	21	42
Dentists	9	14	Phys.Therapists	3	10
Engineers (Prof.)	1	1	Physicians	20	22
Lawyers	8	33	Real Estate	11	33
Nurses	10	11	Social Workers	10	26
Nurs.Home Admin.	42	45	LPNs	8	12
Optometrists	45	47	Veterinarians	19	28

The purpose of CPD should be to increase the competency of engineers, to encourage engineers to maintain competency in a systematic manner and to let the general public realize that professional engineers are working toward improved performance. Competency is a combination of an individual's knowledge, experiences and skills. Performance is the outcome of how that competency is used. (Phillips 1987)

PROS AND CONS OF MANDATORY CPD

Some positive benefits of initiating mandatory CPD by the engineering profession are to a) expand the knowledge and performance of individual practitioners; b) increase the public's awareness and confidence in the engineering profession; c) improve recognition and personal awards received from maintaining competence and improving performance; d) follow the general trend of other professions in the nation to emphasize competency and currency of material; and e) preempt legislative mandates to initiate competency requirements which may be detrimental to the engineering profession. (Price, Mar. 1989)

Some of the negatives of establishing mandatory CPD are that a) the requirement may not be applicable to all engineers because of exemptions; b) mandatory CPD is difficult to enforce and considerable extra expenditures are needed to monitor and enforce the requirement; c) CPD is difficult to measure and evaluate; d) the significant costs of courses, seminars, travel, and lost personnel time to the registrant's company may be prohibitive; e) CPD may lead to future re-examination requirements to maintain registration; and f) CPD can be successful only when done in close cooperation with the various state professional engineering organizations and the state boards of registration. (Price, Mar. 1989)

WHAT THE STATES ARE DOING

A questionnaire was mailed to 52 of the state boards of registration and to 44 of the state engineering societies affiliated with the National Society of Professional Engineers (NSPE). There were 42 responses from the state boards and 33 responses from the state engineering societies. A vast majority of the responses from both groups to all the questions were either "No" or "Not Applicable". The questionnaire gave no indication that there is a ground-swell movement toward CPD. The state engineering societies appear to be taking the lead in discussing and reviewing the subject more so than the state boards of registration. (Price, Aug. 1989)

Only Iowa has a mandatory program in place and operating through its state board of registration. (See the next section on Iowa's program.) Michigan has had enacting legislation since 1980 to initiate a program of mandated CPD for reregistration. The Michigan Board of Professional Engineers has requested the Michigan Society of Professional Engineers to draft criteria for meeting these requirements. A final draft was to be submitted to the Michigan Board by September 1989. The Montana Board of Professional Engineers has a required reporting procedure for each registrant to submit to the Board verified statements that the registrant has maintained professional competency. The practice of engineering for a minimum of 520 hours during each of the two years preceding license renewal or a combined time of 520 hours per year of practice, formal course work, home study, and/or group study is acceptable.

The Alaska Legislature is considering a statutory change to require continuing professional development for engineers. The Kentucky State Board of Registration will seek legislation in 1990 to enable a mandated CPD program to be established. The New Jersey State Board of Professional Engineers is of the opinion "that there are no standards/testing utilized within continuing education units (CEU)", and that "it is simply a 'certificate of attendance' and therefore not necessarily beneficial to licensees". Wyoming has a provision in the state law to require CPD.

The Florida Engineering Society has a voluntary Certificate of Continued Professional Development (CCPD) Program offered to its members. Applicants document three years of CPD-type activities and receive a three-year certificate after approval. CPD activities are assigned varying point values. Applicants must earn 150 points over the three-year period. Points are allocated for experience, for professional/technical organization activities, for professional activities involving teaching or paper presentations/publications, and for educational/ development programs taken for academic credit, or for attending conferences, and self study programs. About 7.5% of the total membership participate in the CCPD program in Florida. (Cook 1989) The South Carolina Society of Professional Engineers has a procedure whereby members may submit a continuing education form to the state society to report CPD-type activities. Both the Georgia Society of Professional Engineers and the Louisiana Engineering Society are making feasibility studies on CPD. (See the section on LES Task Force on Continuing Professional Development for more details of the Louisiana study.) The Ohio Society of Professional Engineers is looking toward establishing a certificate type program.

IOWA'S MANDATED CPD PROGRAM FOR REGISTRATION RENEWAL

The Iowa mandated CPD program began with the 1987 state law that requires CPD for all licensing agencies in the state. The law applies to all professional engineers and land surveyors, but not to EITs. Some registrants like it while others state that a registrant will maintain competence without being forced to do so.

Registrants may attend technical or management type courses, sessions at national or state professional organization meetings, and even Monday morning staff meetings if there is a job-related program presented. The Iowa State Board of Engineering Examiners accepts a proper mix of technical and management programs. There is no limit as to the number of hours a registrant may take in either category; however, the Board may hold up the registration renewal if the registrant is too heavily involved in management courses. The Board accepts programs on how to manage an engineering firm, how to manage people, and on advanced computer techniques. Registrants must show on a registration renewal form satisfactory completion of 15 hours for each year for a total of 30 hours minimum over two years to be considered for renewal.

The Board's Executive Secretary, and sometimes the Chairman of the Board, reviews each listing for acceptance. The Board may audit the registrant's records. Reviewing the renewal forms takes considerable administrative time. The Board renews one-half of the approximately 5,400 registrants each year, at an estimated annual cost of about $12,000.

All registrants with the Board have to meet the same requirements, except exemptions are given to those in the military and those with temporary job assignments overseas. Permanent residents in other countries must meet the regular requirements unless an exemption is granted. Reciprocity with other states with mandated CPD programs is acceptable.

Iowa experienced a drop in the number of registrants after the requirement went into effect, but the decline has leveled off. Iowa lost some registrants who lived out-of-state and who felt that the CPD requirement was not worth maintaining their Iowa registration. Some of the inactive and/or retired engineers dropped their registration. (Peters 1988)

LES TASK FORCE ON CONTINUING PROFESSIONAL DEVELOPMENT

The author chairs the Louisiana Engineering Society Task Force which is studying the CPD issue. The questionnaire mentioned earlier, telephone calls and papers written on the subject were used to help gather information. The Task Force recognized early that CPD was broader in scope than just continuing education.

The Task Force has reviewed the results of the questionnaire, correspondence and literature received, and telephone calls made. A hearing has been set for the LES Annual Meeting in October 1989 for the Task Force to relate what it has learned to the LES membership and to solicit comments and opinions from the membership. The Task Force will eventually make a recommendation to the LES Board of Direction whether or not to pursue mandatory CPD, voluntary CPD or nothing at all. If CPD is recommended, the LES Board will then work with the Louisiana State Board of Registration to effect a proper course of action.(Price, Mar.1989, Aug.1989)

POSITIONS OF ASCE, NCEE AND NSPE

According to the Chairman of the American Society of Civil Engineers Committee on Continuing Professional Development, ASCE probably does not have a position on requiring CPD to maintain registration. (Walker 1989) The National Council of Engineering Examiners (NCEE) took a position in 1979 not to support mandatory CPD, but to encourage engineering societies to promote a strong and intensive research and public relations program on CPD. (NCEE 1979) There has been no change in the NCEE position paper since 1979. (Stricklin 1989) The National Society of Professional Engineers (NSPE) favors giving each engineer the responsibility for professional development, but does affirm the rights of each member state engineering society to support mandatory CPD and to revise the state engineering licensure law. (NSPE 1987)

CONCLUSIONS

Mandatory continuing professional development is a touchy issue and requires considerable study by professional organizations, state boards of registration and state legislatures. Questions which should be considered by those making the studies are:

1. What kind of continuing education courses or events will be acceptable? How many hours of course work/ activities will be required each year or every two years? What should be an acceptable standard for the courses to meet? Who will have the authority to approve and monitor acceptable courses?
2. What type of enforcement policy will need to be set to monitor compliance with a required CPD program?
3. Will exemptions be granted for temporary employment in another country and for military service?
4. Will requiring CPD cause registered engineers to drop their registration? Will a category be established for non-active registrants?
5. Will there be an adverse financial effect on the operation of the state boards of registration? The additional annual cost has been estimated to range from $25,000 upward.
6. Should CPD requirements be voluntary or mandatory?
7. Will specialty certification help or confound the issue even more?

A mandatory CPD program needs to be deliberated with an open mind to seek a proper position that will be beneficial to the profession and provide a safeguard to the public welfare. The American Society of Civil Engineers and the National Society of Professional Engineers should provide leadership in developing such a position, and encourage other engineering professional organizations to follow that leadership.

REFERENCES

Cook, C. E. (1989). "Florida Engineering Society Certificate of Continued Professional Development". Executive Director, Florida Engineering Society, Tallahassee, FL.

National Council of Engineering Examiners (1979). "Position of the National Council of Engineering Examiners on Continuing Professional Competence." Proceedings of 85th Annual Meeting, Clemson, SC.

National Society of Professional Engineers (1987). Policies--Professional and Administrative Handbook, PP No. 122-C, Alexandria, VA, 25.

Peters, P. (1988). Telephone Conversation in November 1988. Executive Secretary, Iowa Engineering and Land Surveying Examining Board, Ankeny, IA.

Phillips, L. (1987). "Thoughts on Improving Mandatory Continuing Education", Lifelong Learning Forum Newsletter, Vol. IV, No. 2, Univ. of Georgia, Athens, GA.

Phillips, L. (1989). "1989 Mandatory Continuing Education Update", Louis Phillips & Assoc. Newsletter, Summer Issue, Athens, GA.

Price, B. E. (Mar. 1989). Memorandum on Continuing Professional Development. Chairman, Task Force on Continuing Professional Development, Louisiana Engineering Society, Baton Rouge, LA.

Price, B. E. (Aug. 1989). Responses to the LES Questionnaire on Continuing Professional Development. Chairman, Task Force on Continuing Professional Development, Louisiana Engineering Society, Baton Rouge, LA.

Stricklin, R. (1989). Telephone Conversation in August. Executive Director, National Council of Engineering Examiners, Clemson, SC.

Walker, D. (1989). Telephone Conversation in January. Chairman, American Society of Civil Engineers Committee on Continuing Professional Development, New York, NY.

An Interactive Videodisc Training Program

James B. Martin, P.E.[1]

ABSTRACT

The purpose of this paper is to describe an interactive videodisc system and training program developed for the Federal Highway Administration by the North Carolina Institute for Transportation, Research and Education. The program includes instruction on various pavement structure repair techniques including: Pothole Patching, Crack Sealing, Maintenance of Gravel Roads, Ditch Maintenance, Shoulder Maintenance, Asphalt Chip Seals and Basic Traffic Control. The system hardware has been mounted in a heavy-duty shipping crate suitable for shipping. The system is designed for set-up and is directed toward highway maintenance engineers and workers.

INTRODUCTION

An interactive videodisc training program entitled "Pavement Structure Repair Techniques" has been developed by the North Carolina Institute for Transportation Research and Education (ITRE). This program was designed for highway maintenance engineers and workers in local highway agencies and is intended to instruct them in the proper way of performing various pavement maintenance activities. Many millions of dollars are spent annually on the maintenance of roads, streets and highways across the United States. Performing this maintenance in a proper and cost effective manner will prolong the life of these facilities and benefit the public. It is therefore important that the highway maintenance engineers and workers performing these activities be instructed in the proper techniques to be followed. In the past, highway maintenance training has not been a high priority item in many highway departments.

In January 1988, ITRE was awarded the Rural Technical Assistance Project (RTAP) 46A, Pavement Structure Repair Techniques by the Federal Highway Administration (FHWA). The original contract required the preparation of seven 15-20 minute

----

[1]Program Manager, Institute for Transportation Research and Education, 3900 Merton Drive, Suite 211, Raleigh, NC 27609.

professional-quality videotapes for instruction on seven different highway maintenance activities. These maintenance activities are:

1) Pothole Patching
2) Crack Sealing
3) Maintenance of Gravel Roads
4) Ditch Maintenance
5) Shoulder Maintenance
6) Asphalt Chip Seals, and
7) Basic Traffic Control

FHWA had previously completed an interactive videodisc training program on Work Zone Traffic Control. This training program has received favorable response from FHWA and local highway agencies. Therefore, FHWA was interested in developing other interactive videodisc training programs. It was decided that RTAP 46A was an excellent project to further utilize the interactive videodisc training technique. In addition to producing the seven videotapes, ITRE was responsible for preparing an interactive videodisc training program on the same subject areas. Ultimately, the intent is to test the effectiveness of the interactive videodisc training versus that of videotapes. All seven videotapes have been completed and are being widely distributed. The purpose of this paper is to describe the steps taken in developing the interactive videodisc program.

PROGRAM DESIGN

It was desirable to have all seven subject areas included on one side of a single videodisc. Compressed audio was a necessity to achieve this end product. ITRE subcontracted with United States Video Corporation in Vienna, Virginia to assist in the technical preparation of the videodisc. Using the EECO compressed audio technology it is possible to place up to 40 seconds of audio into one still-frame on the videodisc. This was a key factor in being able to present all seven subject areas on one side of a single videodisc. The final disc contains 864 slides and video stills, 674 character generated text screens, 100 segments of video totalling 21 minutes and 3 hours of linear and compressed audio. The program also utilizes the IBM Infowindows touch screen monitor. The touchscreen eliminates the use of a keyboard.

The program begins with a brief introduction on it's use. This is done to familiarize the student with the methodology used in the training program. They are given a brief explanation on what an interactive videodisc is, and are guided through some sample questions and responses.

Once a student has viewed the introduction he or she must identify themselves by their first, middle and last initials, and their birth month. This unique identifier allows the program to track student responses. Software is written to score each student's response to questions, and report these scores.

After the student has identified himself a main menu is presented. From this menu a student can select any one of the seven modules included on the videodisc. These modules are Pothole Patching, Crack Sealing, Maintenance of Gravel Roads, Ditch Maintenance, Shoulder Maintenance, Asphalt Chip Seals, and Basic traffic control. Upon selection of a module, the student encounters another menu. Each module is broken into segments. The student can then select the segment that he or she would like to view.

For example, in the Ditch Maintenance module the student would select from the following segments:

1) Introduction
2) Ditch Maintenance Using a Motorgrader
3) Ditch Maintenance With a Hydraulic Excavator
4) How Many Trucks?

The Introduction segment briefly describes the purpose and need for ditch maintenance. Segments 2 and 3 describe the details of two different methods of performing ditch maintenance. Segment 4, How Many Trucks?, describes an easy way to calculate the number of trucks required to perform the maintenance activity. The menu also contains a selection to exit the module and return to the main menu.

Each module presents instructional material in clearly defined segments. The instruction includes the use of still-frames and motion video to illustrate the maintenance procedures. Following the instruction material several questions are presented to the student. A total of 78 questions are included in the entire program. These questions are in the format of multiple choice or true/false. Many answers are listed on a text screen while others utilize split screen video images. The student must then respond to the question by touching the proper answer on the screen. If a correct response is chosen the student is congratulated and proceeds to the next question or segment of the module. If an incorrect response is chosen the student receives remediation and the correct answer is reinforced. After a review the student then moves to the next question or segment of the module.

When making any selection the student must touch the selection on the screen. Their selection is then highlighted by a box around their choice. The student then must touch an enter box that is generated by the computer graphics on the upper right hand portion of the screen. The student has the option of changing the answer any number of times until they press the enter box. This allows the student to change their mind or to make a correction if they hit the wrong answer by mistake.

In addition, an escape box is generated by computer graphics on the lower right hand portion of the screen. The escape box allows the student to interrupt the program at any point if needed. The student can then exit the system and return at a later time to complete that particular module. The response tracking program will save the student's responses to questions up to point they escaped the module. The software will monitor responses to the remaining questions once the student resumes the program.

EQUIPMENT

As previously stated, the Pavement Structure Repair Techniques interactive videodisc system utilizes the IBM Infowindow touchscreen monitor. An IBM PS/2 Model 30 microcomputer was selected because of its size. The EECO P-CAT 300 compressed audio decoder card is installed in the microcomputer to decode the compressed audio from the videodisc. A Pioneer LD-V6010A videodisc player is used. The IMSATT authoring system was used and the United States Video Corporation did the system programming.

The Federal Highway Administration intends to send this system to various locations throughout the United States. The system is to be tested at several of the RTAP Technology Transfer centers across the country. There are some 42 RTAP Technology Transfer centers in the nation. These centers are jointly funded by FHWA and state highway departments. ITRE is the RTAP Technology Transfer Center for the state of North Carolina. Most of these centers are affiliated with universities and are established to provide technical training and technical assistance to local highway maintenance agencies. Many of these agencies have limited training budgets and the Technology Transfer program provides quality training at a minimal cost.

For mobility the equipment is packaged in a heavy-duty shipping crate with dimensions of 27" x 45." The equipment was mounted inside the shipping crate by staff members in the Federal Highway Administration's electronics lab located in McLean, Virginia. The equipment is mounted on a frame within the crate that is held in position by heavy-duty shock absorbers. The monitor is mounted such that it is eye level when sitting. The crate and equipment weighs approximately 200 lbs. Heavy-duty wheels are mounted on the bottom for ease of movement.

All of the equipment is wired to a single power strip located in the rear of the shipping crate. The equipment is packaged for use with minimum set up. Once the equipment is received the front and back doors must be removed from the crate. A power cord from the power strip must be plugged in and the toggle switch flipped on. The system then will automatically boot up. The operator can then respond to the touchscreen monitor.

Initially, there was concern about the equipment being damaged in transit. However, the equipment has been shipped several times and no damage has occurred thus far.

TRAINING EFFECTIVENESS

The next step in the project is to measure the effectiveness of the interactive videodisc training versus training done with the videotapes. Test groups will be selected in North Carolina and in two other states. Half of the test group will view the videotapes in the more traditional training methodology. They will then be asked to respond to the same questions that are presented in the interactive videodisc program. The other half of the test group will work through the pavement structure repair techniques interactive videodisc program. Their responses will be tracked by the computer software and scored as to their percentage of correct answers. A direct comparison of the two groups will be analyzed. FHWA is interested in the effectiveness of this interactive videodisc training technique. They have recently purchased several systems for their in-house training use.

It is evident the Federal Highway Administration recognizes the advantages of interactive videodisc. The training is presented on an individual basis. The student proceeds at his or her own pace. The student is forced to necessarily participate in the training activity. When simply viewing videotapes the students may have a tendency to daydream and miss several important points in the material being presented. The interactive videodisc does not allow this to happen. Also, when a student makes an incorrect response to a question interactive videodisc presents an immediate review and remediation. It is believed that the student will retain more knowledge because of this.

FUTURE

The Pavement Structure Repair Techniques interactive videodisc program prepared by ITRE has been reviewed by many highway maintenance engineers. The response to the program has been positive. It must now be tested with the Highway Maintenance Workers. The Federal Highway Administration is hopeful that the interactive videodisc will prove effective in training these employees. In the long run proper and cost effective highway maintenance procedures will save money and provide a longer life span for the country's roads, streets, and highways.

MANAGEMENT EDUCATION FOR CONSTRUCTION PROFESSIONALS

by Donn E. Hancher[1], F.ASCE and George Stukhart[2], M. ASCE

ABSTRACT

Management education for practicing civil engineers has long been a topic of concern in civil engineering education. The objective of this paper will be to overview the various avenues available to provide civil engineers with management education, with major emphasis on professional development programs. Details will be given on two highly successful programs offered at Texas A&M University by the Center for Construction Education of the Civil Engineering Department. The Construction Executive Program is a four-week program for upper level managers, and the Construction Management Program is a two-week program designed for project managers.

INTRODUCTION

Civil engineers have always been heavily involved in management activities, both in private industry and in municipal and government agencies. Most will agree that the responsibilities of management are common for most practitioners during their careers; however, there has never been agreement as to the best approach to prepare them for these duties. This need for management skills involves all participants in the construction process, the designer, the owner and the constructor. A major finding of The Construction Industry Cost Effectiveness Project conducted by the Business Roundtable in the early 1980's was that many of the problems of declining productivity and quality in the United States Construction Industry can be attributed to ineffective management practices, by all the parties involved. There are other factors affecting this

1) Professor of Construction Engineering and Management, and 2) Associate Professor of Civil Engineering, Dept. of Civil Engineering, Texas A&M University, College Station TX 77843

decline, such as the "constant state of confrontation" and indifference to new technology, but the need for improved management is prevalent. (Bus. Roundtable, 1982)

Some civil engineering programs offer management topics in their courses, primarily construction courses; however, the vast majority offer little or none. Major topics of importance to engineering managers include: planning, accounting, finance, contract law, human resource management, negotiation, quality, marketing, risk analysis, decision-making and communications. They can be introduced at the undergraduate level, covered in depth in a graduate program, or can be handled in professional development programs. There is little agreement as to the "best way" to handle this education process. The authors feel that all methods have merit and should be utilized, especially the professional development approach.

UNDERGRADUATE EDUCATION

The four-year undergraduate engineering curriculum has been the subject of considerable discussion in the past decade. There are several conflicting objectives which most curriculum committees face:

- The need to keep subject matter current with a constantly expanding technology base.
- The need to meet accreditation requirements.
- The need for more liberal arts studies.
- The expressed needs of industry for graduates who have technical skills for immediate application.

Furthermore, many of the positions held by civil engineers in early years require management skills as well as technical. The National Research Council emphasizes the importance of interactive skills, because the engineer must be equipped to "represent ideas, needs, and objectives within the organization." (National Research Council 1985). Such interactive skills would include the ability to work in teams, to plan and organize tasks, and to prepare oral and written reports.

It is not possible to fully develop managerial skills and meet the objectives stated above within a four-year civil engineering program. For years, many have proposed a five-year undergraduate program to incorporate all of the additional desired subjects into the curriculum, but to no avail. Although some introduction to management can be included in traditional undergraduate programs, it remains infeasible in the foreseeable future to do more.

GRADUATE EDUCATION

Graduate programs have the potential to develop in-depth technical skills, increase interaction in the classroom, to develop interpersonal skills, and to introduce the topics of management. However, most engineering graduate programs neglect these latter attributes in favor of much greater concentration on technical proficiency. For the majority of civil engineers in industry, the technical graduate program enhances their immediate job performance, but still leaves a serious void in their management skills.

The MBA has long been regarded as a viable option for securing a well-balanced education in business to supplement the engineering degree. MBA programs are growing in enrollment and provide an opportunity for the transition from engineer to manager, especially for persons desiring to abandon their engineering career and pursue other vocations. Another major reason for the popularity of MBA programs is the relative geographical accessibility of such programs as compared to graduate engineering programs. Many of the engineers who pursue the MBA do so on a part-time basis, often at night, and will continue to be a popular option in the future.

An ideal program for civil engineers would be a graduate program that offers them the opportunity to strengthen their technical knowledge, but at the same time allows them to also strengthen their management knowledge. A logical solution would be an engineering management program; although few such programs exist, there are several construction engineering and management graduate programs in civil engineering departments which fulfill this need. Most of these programs offer a balance of technical, construction management, and pure business courses for their students.

It is commonly agreed that graduate education for engineers is more meaningful if they would undertake it after they have been in industry for five to ten years. Unfortunately, the vast majority of graduates who would be eligible for graduate school admission, but enter industry instead, do not return to school. There are several reasons for this, such as financial restraints and career interruption, but mostly it is a matter of overcoming inertia.

Another major problem for many practitioners pursuing graduate education is the problem of admission. Many schools will not admit persons unless they have strong undergraduate records, or can score high on graduate level aptitude tests, such as the graduate record exam (GRE). A viable option for these persons, or for those who do not feel it is feasible to return to graduate school, is the vast variety of continuing education opportunities for management education.

CONTINUING EDUCATION

Continuing education programs offer a feasible solution to the need for executive development and for maintaining currency in one's specialty field. The variety and purpose of such programs are numerous. Courses may last from one day or less to 22 weeks, the duration of the Harvard Business School's Advanced Management Program. Courses are offered by various sponsors: technical societies, consulting firms or individuals, company programs, or by universities and colleges. This paper emphasizes the value of continuing education programs which are several weeks long, where participants have some benefit of interaction. One or two day programs provide questionable lasting benefit:

"The disadvantage of short one or two day sessions is that, although the material is current and useful, it is often not retained by the students. The short one- and two-day programs do not have the same educational value as a more intensive several-week program in an academic atmosphere." (Business Roundtable 1982)

There are both advantages and disadvantages of utilizing continuing education programs for management education as compared to graduate school. Participants are usually more experienced and frequently are more highly motivated, programs are targeted to the particular audience, and require a much shorter period of time than graduate courses. On the downside, participants are not really required to attain academic skills and may not apply themselves. Also, the concentrated time period may frustrate some persons, and often places undue pressure on instructors to be super lecturers.

The Department of Civil Engineering at Texas A&M University, through its Center for Construction Education, offers two highly successful professional development programs for participants in the construction industry. Although the programs are primarily attended by engineers, other professionals also attend and blend in well with their classmates. The specifics of the two programs will be briefly outlined in the remainder of this paper.

CONSTRUCTION EXECUTIVE PROGRAM

The Construction Executive Program (CEP) is a four-week concentrated management course for senior executives in the construction industry. It is targeted at individuals who have fifteen years or more experience and who have attained or have the potential to attain senior executive positions in their companies. The primary objective of the CEP is to introduce the participants to state-of-the-art management concepts applicable to the construction industry. A secondary, but very important objective, is to promote a cooperative professional attitude between all participants in the construction process: owners, designers and constructors.

The CEP is conducted during January and early February in College Station, which is an opportune time for the participants to be absent from their jobs. The initial Texas A&M CEP was held in 1986 with 17 attendees and has grown steadily, with 22 in 1987, 31 in 1988, 32 in 1989, and 36 in 1990. The original limit for enrollment was set at 32 participants to provide for optimal interaction between participants, and between participants and instructors. The limit was extended to 36 in 1990 due to heavy demand with several applicants delayed until 1991. The CEP tuition is $9,900.00.

The program is designed to offer the participants a full range of educational formats, including classroom lectures, workshops, group work sessions, homework, and a comprehensive case study at the end of the program. Microcomputers are used extensively during the program. The instructors are selected for their expertise, teaching ability, and knowledge of the construction industry. Although some of the instructors are college professors, a majority are practitioners and consultants with extensive industrial experience. A detailed evaluation procedure is used to evaluate both the instructors and the course materials for each session of the program.

The key topics covered in the 1990 CEP are outlined below:

Strategic Planning: Managerial Finance, Construction Finance, Strategic Planning, Making Decisions Under Uncertainty, Organizational Design.

Human Behavior: Negotiation, Human Resource Management, Personality Profiles, Stress Management, Effective Communication.

Project Management: Integrated Planning and Control, Quality Management, Safety Management, Materials Management, Information Systems, Computer Uses in Management, Project Management.

Contracting and Liability: Perspectives on Construction Failures, Labor Relations, Construction Law and Contract Management.

There are several factors which contribute to the success of the Texas A&M Construction Executive Program. The Faculty Advisory Committee provides academic guidance and support to the program. The Industry Advisory Board approves program content and actively promotes the program. The Center for Construction Education Staff coordinates all of the arrangements for the program. Finally, the graduates of the CEP are its strongest supporters and provide tremendous assistance.

CONSTRUCTION MANAGEMENT PROGRAM

The Construction Management Program (CMP) is a two-week professional development program designed to enhance the management capabilities of persons involved in or entering into project managment of major construction projects. The program is also designed to benefit all participants in the construction of today's challenging projects, including owners, designers, contractors, suppliers and other professionals. It is anticipated that individuals who attend the program will have had some management experience with planning, design or construction activities in their organizations, but it is not mandatory. CMP tuition is $3,900.00

The CMP is operated on the same successful format as the CEP and utilizes a variety of instructional methods including lectures, case studies, and small group discussions. A balance of theory and practice is presented by the speakers who have been selected for their expertise in their respective subjects. Interaction amongst the participants, faculty, and staff is an essential element of the program. The CMP is offered twice each year, once in the spring and once in the fall. The number of participants has been about 20.

A special part of the CMP is the team presentations which are given during the final session and are made on a case study topic assigned to participant teams. The teams the past year have been divided into two problem categories, constructability and project safety. Some of the major topics covered in the CMP include:

Organizational Structure	Financial Issues
Managing Change & Conflict	Negotiation
Quality Management	Project Planning/Control
Contract Administration	Jobsite Labor Relations
Leadership and Motivation	Productivity Improvement
Contractor Prequalification	Human Resource Managment
Effective Communications	Microcomputer Applications

Although many of the topics appear to be the same as in the CEP, the emphasis is more directed toward project management than company management.

The success of the CMP is attributed to the same reasons as the CEP, a close-working partnership between academia and industry in identifying the program content, excellent speakers, and changes as needed to continually improve the program.

SUMMARY

The experiences of the Department of Civil Engineering at Texas A&M University in educating practitioners from the construction industry on management of their projects and organizations have been quite successful. Although the Department also offers construction management education on a formal basis for its undergraduate and graduate students, the impact on management improvement via the Construction Executive Program and the Construction Management Program is believed more effective. This is primarily due to the maturity of the participants in the two professional development programs and their previous industry experience. Other major factors contributing to the success of the programs is the financial commitment made by the school to initiate the programs, a strong relationship with industry members as program advisors and supporters, and a designated academic advisory committee and professional staff dedicated to the success of the programs.

APPENDIX I - REFERENCES

Business Roundtable (1982) "Management Education and Academic Relations, A Construction Industry Cost Effectiveness Project Report," Report A-5, The Business Roundtable, NY, NY, 10-11.

National Research Council (1985) "Engineering Education and Practice in the United States, Foundations of our Techno-Economic Future, National Research Council, Washington D.C., 117-120.

ARE GEOTECHNICAL SHORT COURSES OF VALUE?

John W. Williams[1]

ABSTRACT

In ever increasing numbers, geotechnical short courses are being advertised with the suggestion that participants can develop and upgrade critical professional skills. The reasons behind these increasing numbers of short courses range from genuine desires to increase the level of competency of the profession to purely profit for those organizations putting the programs together. Short courses have appeal because of their relatively narrow topic, practical orientation, compressed time schedule, and subject timeliness. The quality of offerings is highly variable with no adequate way for participants to record their approval or disapproval for future reference. A central clearing house (organization) is needed to maintain an up-to-date data base on the schedule of offerings, the course quality, and the track record of the organizations offering courses.

During the past five years, the number of fliers received by the typical engineer or scientist for geotechnically oriented short courses has grown tremendously. There is no indication that this trend will change in the immediate future. These observations stimulate some important questions including:

1. Why are all of these short courses being offered?
2. Are all of these short courses needed?
3. Are those being offering the ones most needed by the professional?
4. Is there any means of assuring quality control for these courses?

- - - - - - - - - - - -

[1]Professor and Chairman, Department of Geology, San Jose State University, San Jose, California, 95192-0102

A number of reasons may be offered to explain this rapid growth in the number of short courses during the past five years. The more notable influencing factors are recent and major changes in the geotechnical practice and the imposition of legal requirements as to how things are to be done. This has been stimulated by increasingly vocal public demands for accelerated clean up and protection of the environment accompanied by public funding for such projects. The economic and employment problems in the petroleum industry have created interest in geotechnical short courses.

Although concern for the quality of the environment has been expressed by a few individuals for decades, the interest of the general public exploded in the early 1980's. As more and more individuals became aware of the increasingly complex and toxic problems associated with waste disposal and water, the demands to "clean up this mess" grew. New and strengthened legislation was enacted to address many of the issues. Associated with this legislation has been the appropriation of billions of dollars of public money to aid in the clean ups. Typical of the use of this funding has been the designation of "Super Fund Sites", those sites considered most critical, requiring immediate and detailed attention. More than 20 billion dollars have been allocated for these projects, and it is estimated that the amount will grow during the next decade (Turner, in press).

With this renewed interest in the environment, groundwater, and hazardous wastes and with the availability of funding, many individuals and organizations have become very interested in working in these areas. Based on projections (Turner, in press), it is estimated that approximately five thousand geotechnically trained individuals will be needed to work in site clean up during the next ten years. Currently, adequately trained or experienced personnel in engineering geology, hydrogeology, geotechnical engineering, and related fields are not available to meet this demand. Indications are that demand will outpace the supply of adequately trained personnel for at least the next five years.

Since 1983, large numbers of geotechnically trained professionals have been laid off from the petroleum industry. Approximately 25 percent of that industry is either underemployed or unemployed. Although many of these individuals have not been able to secure new geotechnical jobs, large numbers retain their interest and desire to work in an allied area. Sectors of employment considered attractive are the waste and groundwater industries. These potential employment sectors are actively marketed by the petroleum industry as very viable possibilities for their former employees. Employees with a geotechnical background are required for many of these positions and because groundwater and petroleum involve working with the migration of fluids, the transition between the two fields of practice should be relatively easy. However, recent history records that the transition between the two areas of employment often is not easily made. Petroleum trained and experienced professionals generally need retraining to make a successful transition.

The geotechnical field has become much more complex during the past decade. Organic chemistry, transient groundwater flow, unsaturated contaminate flow, clay mineralogy, geochemistry, are subjects considered essential components of any comprehensive hazardous waste site investigation. These terms were used infrequently in the classical areas of geotechnical work such as foundation investigations. Statements are repeatedly made by representatives of the Regional Water Quality Control Boards that "if a consultant is using techniques which are several years old, then the study has a good chance of being out-of-date and inadequate."

From the examples mentioned, there are good indicators that large numbers of individuals are in need of and desirous of either initial or additional training in select geotechnical areas. Where do these people go to get this training?

Private firms prefer to have their employees participate in short course offerings rather than the traditional programs in universities and colleges (Turner, in press). The reasons for this are clear. Private companies often are unwilling or unable to allow employees the large amounts of time needed to attend traditional college courses. The schedules of the employees are such that attendance on any regular bases for extended periods of time (semester or academic year) can be difficult or impossible. Many practitioners consider course offerings at colleges and universities too theoretical and impractical. Some courses are simply not up-to-date with the current state of the practice or regulations governing the practice. Many, if not most, colleges and universities are having difficulty in getting qualified permanent faculty to teach the essential courses. Evaluations of the numbers of students currently getting PhD's in the appropriate disciplines indicate that this supply and demand crisis will become worse during the next decade (American Geological Institute, 1988). Some colleges and universities are reluctant to present practical, applied geotechnically oriented courses. This is particularly true of programs in traditional science or engineering departments where the offering of such courses often is viewed as "appropriate for a trade or technical school but not for a research oriented scientific or engineering department."

All of these factors combine to create a situation in which the return to the traditional college classroom is not the best and in some cases not even a workable solution to the problem of getting adequate training or retraining for the large numbers of people seeking it. In addition, the majority of private companies do not have the desire or the resources to provide extensive in-house training.

A logical solution to this dilemma is to seek the needed retraining via the short course offerings. In theory, the short course has many of the characteristics which are desired by the private sector; narrowly focused content, short-time commitment on the part of the participant (generally a few days to a week), up-to-date materials, and practical orientation to instruction.

Many organizations, including extension services of colleges and universities, private organizations, and professional societies, are presenting and sponsoring geotechnical short courses. It is prudent and very informative to analyze the reasons why any particular organization might offer short courses. Considerable time and effort are required to produce a good geotechnical short course. Several contributing factors might be suggested with the clear realization that a combination of factors influences the decision of any particular organization to offer a course. A partial list of reasons should include:

1. The genuine desire to provide quality continuing education for the members of the organization associated with the short course.
2. The profit motive encourages the development of short courses. Many courses are designed to be substantial money makers for the organizers.
3. The need to promote an organization - to increase its visibility, membership, and thereby its prestige, influence, and revenue base.

For many universities and colleges, revenues earned through the extension programs have desirable characteristics. In the case of state schools, the earned revenue is relatively free of the normal state regulations governing expenditures. Other advantages to colleges and universities include:

1. The offerings of short courses are a way in which regular teaching faculty at the university may supplement their base salaries.
2. The presentation of short courses are effective public relations activities for colleges and universities.
3. Many college and universities and associated professional societies presenting short courses provide the attenders with continuing education units (ceu). This enhances the image of the organization and strengthens the ties between the organization and the individuals taking the courses.

Many professional societies are facing increasing difficulties in meeting expenses from the revenues received from the payment of dues by their members. Costs for publishing journals, membership directories, travel for officers, etc., are growing faster than dues can be raised. Continually increasing dues can and often does lead to reduced membership, and, thereby, a failure to augment the organization's treasury. Every course offered cannot be guaranteed to be financial success, but the great majority are designed to be profitable for the group organizing the course. In some cases, specific percentages of profit are designed and set as goals.

The desire to improve the overall quality of the profession is an important reason for offering a short course. This is a factor and for some groups the most important element.

More geotechnical short courses are available. The basic question remains as to whether the individual is really "getting his or her money's worth." Several conclusions are reasonable including the idea that if one views the taking of a single or even several short courses as a sure path to career success, that person is going to be disappointed. Just as there is no magic diet pill that will instantly make a person thin, there is no magic short course that will instantly produce an expert in some new field.

Additional problems associated with geotechnical short courses include the timeliness of the offering in terms of topic and content. Some organizations, particularly large universities, conduct frequent and formal surveys of the profession to determine what is needed in course offerings. Most organizations are less organized or formal about deciding what will be presented. They rely on their best estimate (guess) as to what is needed or present what they have available with little regard for what is really needed by the profession.

The quality of the courses being presented varies greatly from those which are truly excellent, well scheduled, effectively taught, and up-to-date, to those in need of considerable improvement along with a few that should be considered an embarrassment. Most courses tend to be one of two general types; those which are repeated at different times during the year and/or in different locations of the country, and those which are given only once in a single location. Those courses which are presented only once do not have much of an opportunity to benefit from review and feedback. Those courses which are offered more frequently can benefit from thoughtful review comments assuming that an adequate means of providing participant feedback is provided. A common situation is that when a participant experiences a bad course, he or she simply writes it off to bad luck and hopes that the next one will be better.

Timeliness of course content can be a problem for courses which are repeated, particularly if repeated for several years. It is all too easy for the instructors to repeat the materials given the last time. Many college professors are accused of using notes that are becoming yellowed with age. Instructors of long-running short courses can be guilty of the same failing. In a discipline, such as geotechnical engineering which is undergoing so many rapid changes in the state-of-the-art, the presentation of dated materials is not appropriate.

Organizations exist which certify the appropriateness of short courses for continuing education units. Traditionally this has meant for each continuing education unit approximately 10 hours of instructional contact time is scheduled with some form of post-instruction examination or formal review. The qualifications of the instructors are reviewed and theoretically subject to approval. Beyond these minimal requirements, little is done. There exists only limited and ineffective opportunities for participant feedback and comment. Sanctions for producing poor short courses really do not exist. This element of the review process must be strengthened.

Simplistic as it may sound, what is needed is more cooperation and coordination and a realistic appraisal of why things are being done (the motivations for giving short courses). Cooperation and coordination is needed among organizations giving short courses to insure the following: the ones given are the ones needed; the offerings are not needlessly duplicated; and the resources are coordinated so that the offered courses are of the highest quality. Many will say this is desirable but unrealistic if one considers why organizations are interested in giving short courses.

As a start, a highly publicized, easily accessible central clearing house for short course offerings is needed. This organization would serve as a bulletin board for offerings for all organizations, a repository for participant feedback, and a collection of historical data on the "track record" of groups presenting short courses. It would be a "better business bureau" for geotechnical short course offerings.

Until we as members of professional organizations begin to do more of the things suggested, geotechnical short courses and short course offerings in general will continue to be a lottery for the participants. Some will be winners and some will be losers.

REFERENCES

American Geological Institute, 1988, Geoscientific employment and hiring survey: American Geological Institute, 54 p.

Task Force on Women, Minorities, and the Handicapped in Science and Technology, 1988, Changing America: The New Face of Science and Engineering, Interim Report: Washington, D.C., 62 p.

Turner, A. Keith, in press. Technical personnel requirements to support the national clean-up of hazardous waste sites, in Higgins, J. and Williams, J. W. (editors), Academic Preparation for Careers in Engineering Geology and Geological Engineering: Special Publication of the American Geological Institute.

Industry-Sponsored Continuing Education for Faculty

John P. Turner[1], A.M. ASCE and Scott Litke[2]

Abstract

This paper examines the concept of professional development for civil engineering faculty through industry-sponsored continuing education. The benefits to industry, faculty, students, and the civil engineering profession are discussed. An industry-sponsored workshop for geotechnical engineering faculty is described and proposed as a model for faculty continuing education. The factors that determine the effectiveness of industry-sponsored continuing education are identified and used to develop guidelines for successful programs.

Introduction

The diversity of technology in the field of civil engineering makes it difficult for most faculty members to develop and maintain the high level of expertise that is necessary to teach graduate and undergraduate students all of the most up-to-date information, even in the faculty member's subdiscipline of civil engineering. Industry-sponsored continuing education offers one approach to improving the teaching capabilities of engineering educators in areas outside of their own research interests, while also offering benefits to industry. Industry-sponsored short courses, seminars, and workshops can provide educators with technical knowledge and resources which otherwise may be available only in highly-specialized and disseminated research journals, or if the faculty member is involved in research on that topic. Industry benefits by becoming an active partner in training the

1 - Asst. Prof., Dept. of Civil Eng., Univ. of Wyoming, Box 3295 Univ. Station, Laramie, WY 82071-3295.
2 - Executive Director, ADSC, P.O. Box 280379, Dallas, TX 75228

future practicioners of the technology which supports that industry economically.

This paper describes a highly successful workshop sponsored by the foundation drilling industry for geotechnical engineering faculty. This workshop is used to illustrate how industry-sponsored continuing education can be an effective resource in improving all areas of civil engineering education as our universities and the construction industry move into the 21st century as partners in preparing civil engineers of the future.

Drilled Foundation Workshop

The Drilled Foundation Workshop for Engineering Faculty was conceived, funded, and organized by the Association of Drilled Shaft Contractors (ADSC), an international trade organization of foundation drilling contractors, drilling equipment manufacturers, and foundation design engineers. The workshop took place during July, 1987, at Colorado State University (CSU) in Fort Collins, CO and at the CSU Mountain Campus at Pingree Park.

Objectives. The overall objective of the workshop was to provide civil engineering educators with the knowledge and resources required to teach students modern methods of drilled shaft foundation design and construction. A survey conducted by the ADSC in 1985 indicated that of approximately 200 civil engineering programs in U.S. universities, only six offered significant coursework on drilled shaft foundations. It was noted further that most foundation engineering text books include much more coverage of other foundation types, such as shallow foundations and driven piles, than of drilled shafts. This lack of drilled shaft education at the university level was seen as limiting the advancement of drilled shafts as an important foundation type, even though drilled shafts have undergone tremendous increases in usage over the past 15 to 20 years.

To address the lack of drilled shaft education at U.S. universities, the ADSC began planning a workshop for select civil engineering faculty. The purposes of the workshop were to: 1) expose educators to the latest drilled shaft technology through an intensive week-long program of lectures and field demonstrations; and 2) provide engineering faculty with a complete educational package to be used for teaching drilled shaft design and construction to students.

Planning and organization. A workshop planning committee was formed in December of 1985, and consisted of members of the ADSC executive staff, ADSC contractor members, and four university professors involved in drilled shaft teaching and research. In early 1986, the first author was hired by the ADSC to be the workshop coordinator. A technical coordinator from the academic community was considered to be critical for a workshop aimed at the academic community. The coordinator was designated as the person immediately responsible for organizing and planning the workshop, which was accomplished between March of 1986 and July of 1987.

Planning and organization consisted of: 1) selecting the locations and dates, 2) selecting speakers and developing the workshop program, 3) developing a teaching package, 4) identifying and selecting the participants, 5) business arrangements, and 6) publicity. The criteria for site selection included adequate conference facilities, availability and convenience of transportation, cost, pleasant climate and surroundings, and a reasonable degree of seclusion. This last requirement was felt to be important to insure that participants could focus their full attention on the workshop without distractions, and was a primary factor in choosing the CSU mountain retreat at Pingree Park for the location of most of the lectures. The workshop dates were chosen to avoid conflicts with the nine-month academic schedule and with known scheduled conferences and other professional meetings of interest to geotechnical engineering academicians.

Important aspects of the program development included topics to be covered, selection of speakers, scheduling the topics in a logical sequence, and conducting field construction and testing demonstrations. In selecting topics, several factors were considered, including obtaining a good balance between theory and practice and between design and construction, the availability, reputation, and speaking skills of potential speakers, and the amount of material which could be covered reasonably within one week. After developing a list of topics, speakers with national and international reputations as experts in their field were recruited. Details of the program, including topics and speakers, are too lengthy to report in this paper, and are available in the workshop final report (Turner, 1987).

To identify and select participants, a workshop brochure/application was developed and mailed to 448 geotechnical engineering professors at 213 colleges and universities identified by a telephone survey conducted by the ADSC. 127 applications were received from 103 universities. Criteria for selection included the size and reputation of the university, the number of students enrolled in foundation engineering courses, quality of the applicant's resume, and geographical balance. The targeted optimum number of attendees was 60. Anticipating a 10 percent attrition rate, 66 professors representing 31 states and Canada were selected. The actual number attending the workshop was 58.

All business arrangements, including lodging, meals, transportation, production of printed materials, and field support, were handled through the executive office of the ADSC in Dallas. Each participant was responsible for their own travel arrangements to and from Stapleton Airport in Denver. Once in Denver, the ADSC covered all expenses. The workshop was promoted through several articles in Foundation Drilling magazine, the official publication of the ADSC. ASCE agreed to cosponsor the workshop and published announcements in Civil Engineering magazine and Civil Engineering News.

Teaching package. The teaching package was designed to provide professors with the resources required to teach drilled shaft engineering to students. Emphasis was placed on making the material easy to use while still being comprehensive and sufficiently technical. The primary teaching resource is a binder consisting of 675 pages of teaching notes and a set of 99 color slides accompanying the text of an introductory lecture entitled "Uses and Construction of Drilled Shaft Foundations". The teaching notes were authored by the workshop teaching faculty and include separate sets of notes for each topic and speaker. Most of the notes are in a format which permits them to be copied as overhead projections for teaching. Technical reports and papers recommended by the teaching faculty were provided as supplementary references.

Other teaching resources in the package included a textbook on drilled shaft foundations, a booklet on pumping concrete, drilled shaft construction and inspection specifications, and videotapes on drilled shaft construction and concrete placement. Following the workshop, this complete teaching package was made available free of charge to engineering faculty who were not able to attend the workshop and to the general engineering community for a nominal fee.

Workshop Sessions. The technical parts of the workshop consisted of classroom lectures, field demonstrations, microcomputer sessions, and panel discussions. The classroom sessions consisted of lectures by the teaching faculty and exposed the participants to the material in their teaching package.

A clear understanding of the behavior of drilled shafts requires engineers to be familiar with field construction and testing procedures. Observation of actual field operations is essential to the education of knowledgeable foundation engineers. The field demonstrations were designed to expose the workshop participants to this type of activity and to supplement information covered in the teaching notes. The field demonstrations consisted of a half day session on construction methods and equipment, including the actual construction of drilled shafts at a field site on the CSU campus, and a full-day session with demonstrations of nondestructive integrity testing methods and full-scale load tests on drilled shafts that had been constructed several weeks prior to the workshop.

A microcomputer laboratory at CSU was used to conduct a three hour session on the use of two computer programs for the analysis of drilled shafts under axial and lateral loads. The programs were installed on 28 microcomputers used by the participants, while the instructors used a computer equipped with a projector to demonstrate the programs on a large viewing screen.

Several contractor panel discussions were held during the workshop to discuss topics that were not addressed explicitly in the teaching notes, such as contracting and bidding procedures, field inspection, contractor-engineer relations, etc. Each session began with introductory remarks by experienced contractors, followed by a general discussion among all of the participants. The exchanges were lively and both academicians and contractors came away with a much greater understanding and appreciation of each other's viewpoints. Based upon the evaluations of the participants, as well as comments made by contractors and teaching faculty, the contractor panel discussions were one of the most valuable aspects of the workshop.

Discussion

The Drilled Foundation Workshop for Engineering Faculty provides an example of how industry can participate directly in engineering education by cooperating with universities to provide continuing education for faculty. The benefits to faculty, students, and industry are greater than can be achieved through more conventional approaches to faculty development. Faculty members benefit by being exposed to intensive instruction on technical topics important in their field, at minimal cost to the university. The technical content includes a large amount of material on "practical" aspects of the topic because the industry sponsoring the workshop is a major participant. In most areas of civil engineering, the practical topic of concern is construction, which often is perceived by practicioners as not being covered adequately in today's curriculum. Industry sponsorship also provides faculty members with contacts in industry who can be a valuable resource for cooperative activities in research, consulting, and teaching. Several workshop attendees have received research grants from the ADSC since the workshop.

It is estimated that the 58 workshop attendees will impact over 5,000 civil engineering students during the five years following the workshop. This is the most direct benefit to the ADSC and represents a very cost-effective method of promoting a technology that supports the foundation construction industry. Additional benefits to industry include the fact that several of the attendees are textbook authors who subsequently expanded the coverage of drilled shafts in their texts, and the ADSC was invited to develop a session on drilled foundations at the ASCE Foundation Engineering Congress at Northwestern University in June of 1989. Students have benefited from the increased teaching capabilities of their professors and also from $50,000 in scholarships awarded by the ADSC since the workshop.

Conclusions

The ADSC Drilled Foundation Workshop for Engineering Faculty was successful in meeting its objectives , which were to: 1) educate the educators on technology for the design and construction of drilled shaft foundations; and 2) provide educators with a complete package for teaching drilled shaft design and construction to civil engineering students. An additional benefit achieved by the workshop is the spirit of cooperation that was fostered between academics, practicing engineers, and contractors. We conclude that this workshop can serve as a model for cost-effective, industry-sponsored continuing education. To be successful, the following items are recommended:

1. The topic of continuing education must be clearly identified as being important to industry and of interest to engineering educators. This is necessary to justify the cost of the workshop.

2. Planning, organization, and conducting a workshop of this type must be a team effort to be successful. Hiring of a technical coordinator from the academic community is critical for a workshop aimed at professors.

3. Speakers should be selected carefully and should be the most outstanding individuals available. Speakers should be made aware of what they are expected to do, especially with regard to any published material they are to provide. Speakers should be compensated adequately for their time.

4. For civil engineers, the technical content should reflect a significant amount of input from practicioners in industry, providing a balance between practice and theory. Field construction and testing demonstrations are recommended for illustrating practical aspects of civil engineering.

References

1. Turner, J.P., "Drilled Foundation Workshop for Engineering Faculty", Final Report to the Association of Drilled Shaft Contrators, Dallas, TX, Jan, 1988, 52 p.

Consulting Reports As Refereed Publications

Thomas R. Currin, M.ASCE and Steven S. Schneiderman[1]

ABSTRACT

At the top of a university's promotional criteria are contributions to refereed journals. These propagate a body of knowledge under the auspices of renowned specialists. However, under some circumstance faculty are unable to generate scientific interpretations which form the basis of refereed publications. Engineering technology faculty are so disposed due to the dearth of basic research facilities available to them; due to the mission of engineering technology which is training technologists to apply science toward building, operating, and maintaining systems designed by engineers; and due to the traditional intensified teaching loads of engineering technology.

Still, engineering technology faculty often function as consultants to industry and governmental agencies. The generated consulting reports, proposals, designs, and data, fulfill many criteria established by journals to qualify as a refereed document. Also, engineering technology faculty's remuneration for such is no different than that from a grant for pursuit of basic research which yields summer salary to the principal investigator. In fact, consulting reports from

----

1. Thomas R. Currin, Ph.D., P.E. is Interim Head of Engineering Technology, Western Kentucky University, Bowling Green, KY 42101. Steven S. Schneiderman, Ph.D., P.E., is Chairman of the Engineering Technology Department at Murray State University, Murray, KY 42073.

engineering technology faculty are at least comparable to basic research. No speculation is allowed. The final document must be complete in regard to science and often with respect to finances. The engineering technology faculty consulting report must also be actionable toward the problem solved. Finally, consulting work occurs over a compressed time frame. Success is mandatory without the luxury of long term inquiry. Universities which consider refereed publication above all other contributions promote institutional disservice by not allowing equality for engineering technology faculty consulting reports.

Introduction

A university's list of acceptable promotion criteria is usually long and varied. There are many directions toward which a faculty member may contribute, but one stands peerless: the refereed publication. This development stems from the need for the propagation of a body of knowledge. It is assumed that old expertise guides new in a cycle which continually piles up information. At specific disciplines new concepts alter dogma and so, science marches forward.

The Setting

Institutional matriculation, married to such philosophy, has fostered a tiered structure from purely research institutions downward to the purely applied. In between are colleges and second tier universities which fulfill regional missions. Regionalism with few exceptions neither promotes the need for nor precludes basic research, so such institutions can be found determining their destiny via administrative and faculty will. While it may be said that each institution is governed by its stated mission, a case can be made that the mission whether developed internally as in private colleges or externally as in state institutions is interpreted to suit the needs and desires of those charged with accomplishing the mission.

When the will to expand beyond regionalism is manifest, promotion criteria often become aligned with those of higher tiered universities. The traditional approach has been to mimic the upper echelon until a similar publication and external grant rate define equality. A mind-set that expansion fosters academic excellence and better serves to accomplish the university's mission becomes epidemic.

Engineering Technology research and faculty promotion criteria represent a gray area where traditionalism should be, if not negated, then modified to facilitate need and reality. The correct modification is to accept faculty consulting efforts equally with refereed publications. Reasoning requires merely comparison of effort in light of university and publication policies.

The Refereed Publication

A refereed organ requires submission followed by expert evaluation followed by editing followed by resubmission, etc. Usually an editor determines whether the submission is publishable as received, requires some or substantial modification, or is inadequate [Lehr]. Referees are external specialists. These are normally not paid but review as guardians of science. Numbers vary from none, editor acceptance, to an average of under two [Juhasz, et al]. Publishable criteria include protection or expansion of a body of knowledge, innovative application of known science, simplification of complexity, or editorial comment. With the exception of editorial comment the time frame from authorship to publication occurs over months or years.

Refereeing policy is as varied as the sciences it protects. Some senior editors are solely responsible. Some publications call upon one recognized expert. Some call upon a committee. Some identify referees to the author and some do not [Juhasz, et al, Jackson]. Summarily, there is no consensus other than existence of the refereeing function.

The Consulting Report

If scientific application must be warranted and an open forum maintained, then a consulting report prepared by an engineering technology faculty is minimally equal to and at times, superior to a refereed article. A consulting effort at a specific discipline requires clear knowledge of one or more sciences, innovative application, protection of the body of knowledge, and in some cases its expansion. Additionally, a consulting report allows little room for speculation. It is generated to solve specific and general applied science needs. Results of scientific synergism must be known or predicted within reasonable engineering certainty. In short, no one hires a consultant expecting a report loaded with "maybe".

Additionally, engineering technology faculty consulting efforts stem from regulated professionals. The consultant must in most states be licensed as a professional engineer or surveyor. All states have licensing procedures for engineering graduates and forty allow BSET graduates from accredited programs an opportunity to sit for Engineer-in-Training and Professional Engineer examinations.

The forums of engineering technology are the implemented results of consulting efforts. These are constructed edifices, systems, equipment, educational programs, etc. Many occur in the public eye for adjudication by laymen and experts alike. And consulting efforts are equally important to the sponsoring university, especially a university in transition toward upper tierdom, as a basic science effort. In the transitional case consulting successes are the foundation upon which the reputation that a university can succeed at research is built.

Comparison of structure, publication vis a vis consulting report, reveals comparable effort. A refereed article flows from abstract to text to references. The consulting report includes an executive summary, text, appendices, and references. Differences are manifest in developmental time frames. Basic science inquiry must be independent of time, though granting agencies may beg to differ. Data collection rate is predictable but interpretations, the meat of refereed articles, cannot be elicited on demand. By contrast, the consulting agreement nearly always includes a deadline. Another difference stems from the economic analyses normally manifest within a consulting report. The cost of developing new or modified tenant is unpredictable.

Similarities occur at the proposal stage. Basic science proposals describe methodology required to look into some facet. These are based upon previous investigations and infer, but do not guarantee, that result will follow procedure. Basic science proposals include the cost of inquiry and milestones with the unwritten but universally embraced proviso that data analyses may dictate experimental, and so cost and time modification. Consultants submit similar proposals describing equally precise procedures but guarantee results within the allowable time frame at a specified cost.

Therefore, when a university decides to expand its horizon, the efforts of consulting efforts cannot be undervalued. Even in the few cases where policy does not promote expansion, a similar argument prevails.

The University Environment

From engineering/engineering technology perspective there are two physical situations which affect the promulgation of research. The first is at universities which support both engineering and engineering technology and the second is at institutions which support only engineering technology.

At universities which support both programs, engineering faculty are recruited mostly from their viability as researchers. Engineering technology faculty are recruited for their viability as educators. Where engineering faculty are welcomed after a direct collegial blitz, B.S. through Ph.D., engineering technology faculty are required to have worked in industry for a minimum of five years and achieved Professional Registration [A.B.E.T.]. Also, traditionally the terminal engineering technology degree has been the M.S.. At this type of institution engineering technology faculty have the opportunity to participate at basic and applied research. Analytical equipment and laboratories exist to support engineering faculty development. The engineering technology faculty can and do work with engineering faculty to develop interpretations which qualify for inclusion within refereed journals.

At universities which support only engineering technology, analytical and laboratory facilities are focused on applied science. The faculty is inhibited and often prohibited from obtaining external research funding because often the principal investigator has only the M.S. and research facilities are geared to the application of science. Faculty are further hampered by limited or non-exestent graduate programs. Lacking graduate programs the pool of highly motivated, enlightened graduate students available to conduct research is absent. However, such venues are well suited to accomplish consulting research.

Finally, it is not impossible for "refereeable" interpretations to be developed and published from engineering technology departments. However, the mission of engineering technology and the development of engineering technology students are more aligned with applied science. Consulting work promotes both. A university which can enhance its reputation and serve its region as a result is remiss by downgrading the value of such effort.

Appendix

Accreditation Board for Engineering and Technology, Inc, "Criteria for Accrediting Programs in Engineering Technology". New York: ABET, 1989.

Juhasz, S. et. al., "Acceptance and Rejection of Manuscripts", IEEE TRANSACTIONS ON PROFESSIONAL COMMUNICATIONS, April 1975. New Jersey: IEEE, 1975.

Lehr, J. H., "Conference Proceedings vs. the Refereed Journal", GROUND WATER Vol. 24, No. 2. New York: NWWA, 1986.

INCREASING & SUSTAINING DIVERSITY IN THE WORKFORCE

F. Peter Fairchild [1]
Billie Jean Potter [2]
Charles A. Parthum, P.E., Fellow, ASCE [3]

Abstract

Civil engineering as a profession is being influenced by a severe shortage of engineers, dramatic shifts in demographics, new economic realities, and changing social values. One of the most important changes that must be made in adapting to these forces is to improve the attraction and retention of women and minorities in the profession. This complex issue will take sustained efforts by universities, industry, and professional societies to build and preserve a more diverse engineering workforce. This paper outlines the issue of changing demographics and how the resultant changes will affect our industry and the professional development of current and future engineers. It also presents some examples of programs and initiatives that one firm is taking to combat the problem of a shrinking workforce. The paper concludes with some recommendations for the future.

Introduction

One of the major issues consulting engineering firms face is the need to attract and retain women and minority engineers in the environmental consulting business. In particular, we are not reaching black and Hispanic students and attracting them to the field of civil engineering in general, much less to the specialty area of environmental engineering. The mid to late 1980s saw increasing numbers of women beginning to study engineering and enter the field, but apparently they have become disillusioned. They are now leaving engineering and discouraging other future women engineers from entering the field. As a leading engineering consulting firm, Camp Dresser & McKee Inc. (CDM) is concerned with this shortage because our growth is limited not by lack of work, but by a lack of people to do the work. CDM is also concerned about this for the profession. There is an enormous challenge awaiting civil/environmental engineers: the restoration of our country's environment and infrastructure. To meet this challenge, this country will need many more engineers (as well as other specialists). The social and demographic facts indicate, however, that if more engineers are needed and if civil engineering wants to grow as a profession, there must be improvements in our ability to increase and sustain an expanded and more diverse engineering workforce.

The Nature of the Problem

Demographic changes are primary indicators of the problems that engineering firms will face by the end of this decade. In comparison to the past two decades, fewer workers will enter the labor force in the 1990s, and a much greater proportion of those entrants will be women and minorities.

---

[1] Vice President, Director, Corporate Human Resources, Camp Dresser & McKee Inc., One Cambridge Center, Cambridge, MA 02142.

[2] Senior Technical Editor/Writer, Camp Dresser & McKee Inc., One Cambridge Center, Cambridge, MA 02142.

[3] Senior Vice President, Administration and Professional Development, Camp Dresser & McKee Inc., Ten Cambridge Center, Cambridge, MA 02142.

The October 1988 issue of *ENR*, a leading "voice" in the engineering industry, focused on "Searching for the Best and Brightest." Citing *Workforce 2000*, a study done by the U.S. Department of Labor, and statistics from the National Science Foundation and the commission on Professionals in Science and Technology, author Debra Rubin describes the issues facing our industry, and the changes that will occur in the period from 1985 to 2000 as follows:

- The number of young workers aged 16 to 24 will drop by nearly 2 million (8%)
- As the number of white males, who had traditionally filled almost all roles in engineering and construction lessens, women, minorities, and immigrants will begin to replace them as the mainstay of the labor force
- Although women make up more than 40 percent of the workforce, in 1986, women made up only 4 percent of the engineering work force
- Today, blacks make up less than 2 percent of the pool of engineers, and Hispanics make up 5.4 percent

As can be seen from the demographics, consulting engineers face two interrelated, critical problems: sufficient numbers of young people—especially minorities and women— are not entering the engineering profession; and even if they do become employed engineers, they often do not remain. Although these trends are applicable across the board to all students, they strike particularly hard at those who will be constituting a greater percentage of the workforce — women and minorities. These demographic changes, therefore, have an immediate and critical effect on the backbone of the consulting engineering business—its people— and on the engineering profession as a whole.

The ASCE Task Committee on International Competitiveness appears to be taking the demographics quite seriously. The committee's final report, which assesses the new influences and pressures the global economy is exerting on the design and construction community, spells out its concern for these issues: *"The demographics are clear: unless we make rapid and significant progress in the engineering education of blacks, Hispanics, and women, we will not have a sufficient pool of engineers to compete in the international areas. . . . We must strengthen our educational system to produce superior students who are capable of completing science and engineering curricula through the doctoral level."*

Concurrently, consideration must be given to the situation a woman or minority engineer faces once she or he has made it through the educational process. Creating an atmosphere where employees can grow, and have opportunities for promotion to middle and upper management positions is necessary to ensure assimilation into the workforce. The February 1989 issue of *Graduating Engineer*, focusing on women's issues, contained an article on "The Best Companies for Female Engineers." In a telling paragraph about why some firms are less than the best, and a good explanation of why many women either never enter engineering, or leave it once they learn what their obstacles are, Lorraine Dusky writes, "Advancement is slow, sexist wisecracks are common, maternity leave policies are draconian. And by the time women hit middle management, salary disparity is the norm. It's still possible to find out that the guy at the next bench, who's not as capable as you, is making $500 a month more."

The problem of slow advancement is equally applicable to minorities. Notes a human resources manager from a leading engineering firm: "I often hear from minority employees that unless we reach a 'critical mass' wherein they feel more comfortable and less like outsiders, and also see others like themselves moving up, they are not apt to stay." Female minorities face double bias. Nehama Jacobs, author of *Success & Betrayal: The Crisis of Women in Corporate American*, comments on the situation in an article in the February 1989 issue of *Graduating Engineer:* "You can blame the promotion system we call 'BOGSAT', an acronym for 'Bunch of Guys Sitting Around a Table', for part of the role model problem. Senior managers throw out the names of promotable managers, and every time some brave soul ventures a woman's name, that woman has to be so perfect that management ends up promoting 'safe' women who fail to inspire the women below them." More women role models, and minority role models, i.e. "critical mass," will be helpful to all concerned.

In addition to the direct changes that these new demographics are creating, there are indirect challenges that management must face. They include the following:

Challenges to the status quo: The engineering industry is facing two inevitables: women and minorities are necessary to keep business going; and companies must reach a "comfort zone" in which their influx is accepted in order for work to proceed as usual. As Morrison, White, Velsor, and the Center for Creative Leadership indicate in *Breaking the Glass Ceiling,* "Corporations cannot manage attitudes, but they can manage behavior with accountability, rewards, and punishment, as in all other important areas of concern. What gets measured in business gets done; what is not measured is ignored." Eventually companies will have to allow women and minorities to climb the corporate ladder. Additionally, the atmosphere will have to be conducive for these employees to remain. Working mothers (and fathers) need assistance in child care issues—day care, flex time, better maternity benefits.

Changes in Our Clients and Customers: Just as the composition of the engineering workforce is changing, so is the composition of clients' organizations. Potential customers can no longer be assumed to be all white males - women and minorities are entering all fields, including corporations and municipalities, and especially, state and federal government. Exclusion of women and minorities within our firms is sure to be noticed by women and minority clients.

The Value of Diversity Goes Beyond Meeting Our Need for Numbers: Aside from the sheer numbers of women and minorities that consulting engineering firms will need in upcoming years, the issue worth looking at is whether maintaining a diverse workforce adds additional value to engineering firms. Minorities and women can make positive differences within the profession and within their respective firms. The strengths from these groups, such as differences of outlook, can directly contribute to business success. Womens' method of communication, which is often consensus-building, can greatly facilitate planning and integration of project work.

Attracting Women and Minorities to the Profession

Consulting engineering truly is a "people" business. The only product is the knowledge and ideas of the staff; consulting firms grow and prosper financially by adding more people. All the assets walk out the door and go home every night. When CDM started seeing shortages of environmental engineers several years ago, the firm became concerned and began to work at increasing the supply. Since women and minorities were apparently very much underrepresented in the workforce and at other engineering firms as well, the company began looking for additional ways to attract these particular groups to the profession. This section of the paper presents several examples of what CDM and other firms are doing. It is by no means an exhaustive list, since we all seem to be working independently of one another. There are undoubtedly many other viable ideas out there waiting to be publicized and shared, but presented here are some efforts that seem to be working.

1. Selecting a Civil/Environmental Engineering Major: CDM has become active in alumni activities with college and university engineering departments. Many schools sponsor alumni career events and almost student chapters of ASCE welcome recent graduates who want to speak about careers in engineering. Slide presentations showing what engineers really do are effective, as students today are concerned about challenging work that is also of value to society. Many of these events are open to freshmen and sophomore students, most of whom have not yet selected their engineering specialty.

Campus visits are also scheduled in the fall for student career fairs, in addition to our on-campus interviews in the spring. These fairs attract non-engineering students and give the opportunity to invite science and math-oriented undergraduates to explore engineering as a major. An article in the October 1989 issue of *Engineering Times* highlights the issue of perception in regard to choosing engineering as a profession: "A number of women currently in engineering believe that the lack of interest in the profession stems more from a lack of information about the field than any perceived notions that engineering is not a proper place for women." On-campus presentations can help address this perception problem. These programs also provide an exceptional opportunity to reach minority students who are determining their major, especially environmental engineering. We have found that frequently minority students do not use the services of college placement offices but do attend career fairs and open houses. In many cases, CDM is the only environmental firm present at these career events.

An observation made during these events is that minority engineering students think only of structural work when they think of civil engineering, yet their interest in working on solutions of urban environmental problems is striking. Students often return to the CDM career display with their friends so they can hear about environmental work. Students wishing careers in business are surprised to find how much "management and business" is involved in running a civil/environmental project.

This raises the issue of whether engineering students are being exposed to the full range of career opportunities available. At CDM employees are encouraged to maintain contacts with their college engineering departments and to be active information conduits for potential students.

2. *Engineering Faculty:* Another issue is the lack of women and minority faculty in our engineering schools and their continuing underrepresentation in new faculty. For example in 1988, this country graduated only 29 women civil engineering Ph.D's, 1 black Ph.D, and 4 Hispanic Ph.D's. The lack of role models among faculty and teaching assistants is a disincentive for women and minority undergraduates to pursue civil engineering. This problem may be further aggravated by the increasing proportion of foreign faculty teaching engineering. Difficulty with the language and feeling that there is a lack of empathy with students, especially women students, is an issue that should be studied further.

3. *Society Involvement:* Women and minority students are active in organizations that represent their professional and personal interests. Therefore CDM has increased participation and interaction with such groups, including the National Society of Black Engineers, the Society of Women Engineers, the National Action Council for Minority Engineers, Minority Engineering and Science Achievement, and the National Association for Minority Engineering Program Administrators.

4. *Exposure at the Junior High and High School Level*: Another approach is to work at bringing the message of civil and environmental engineering career opportunities to local junior high and high schools. The decision to major in engineering in college or even to take math and science courses in high school is made early in a student's career. If students and their high school guidance counselors do not know about civil/ environmental careers, it is very unlikely we will see students enter the field. Many employees have sons and daughters who are entering or are in high school; these employees are very willing to visit, once the need is called to their attention.

Recently *Engineering Times* reported on activities in New York City, such as after-school engineering clubs and engineering mentor programs. These activities bring seventh and eighth grade students into contact with practicing engineers who are succeeding in encouraging students to pursue engineering careers. CDM recently hosted a group of minority high school students who were participating in a Northeastern University/ASCE Summer Institute for minority students. For most of these students, it was their first exposure to environmental engineering; several expressed interest in looking at environmental engineering as a career. These students will be followed closely.

There are other examples of engineers reaching out to introduce high school students to the profession. These include programs such as Math Counts and JETS sponsored by the National Society of Professional Engineers (NSPE), special programs at schools during "Engineers Week," and "adopt-a-school" programs which link practicing engineers with students and teachers in a particular local school. Another possibility is a remedial program in math and sciences staffed by volunteer engineers. Statistics tell us that 80 percent of students are lost to sciences and engineering by 10th grade. If students are reached earlier, engineering as a career can be an option. All of these efforts can be particularly effective in reaching minorities because of the large minority population in urban schools. Twenty four of the largest urban school systems in the country are "minority majority" systems; that is, more than 50 percent of their students are minorities.

5. *Advertising:* CDM has significantly increased advertising efforts in the past few years to promote the company among women and minority college students. The Corporate Communications department has prepared full-color ads for publications such as *Graduating Engineer* (special issues for women engineers and minority engineers), *Black Collegian*, *National Society of Black Engineers (NSBE) Journal*, and the Society of Women Engineers convention program. These ads are aimed as much at publicizing what environmental engineers do as they are for recruiting students to CDM. They have been successful in making students more aware of the industry as well as increasing the number of women and minority job applicants.

Recently, ASCE has advertised in some national publications such as in-flight magazines on national airlines to promote the profession. This effort must be expanded. Perhaps ASCE could work with other professional societies that represent civil engineers to increase the funding available for this effort. It is particularly important to have some of the ads address career opportunities in the field, especially for women and minorities.

6.. *Role of Cooperative Education:* CDM has had a long history of participating in and supporting cooperative education programs. It is a primary source for evaluating and eventually hiring engineering graduates. Recently the firm has worked with the National Commission on Cooperative Education to encourage women and minority participation in Cooperative education (Co-op) programs. Co-op programs are an ideal way to introduce women and minority students into the engineering workforce in a way that allows both the students and the hiring managers to develop experience and comfort with increasing diversity in the workforce.

7. *Industry/University Interactions:* From a marketplace perspective, colleges and universities appear slow to react to changing job opportunities. Declining enrollments in civil engineering coupled with explosive growth in the need for talented civil and environmental engineers calls for a much higher level of active involvement and partnership between industry and universities. CDM has been taking advantage of opportunities to work with schools of engineering and civil engineering departments, providing guidance on curriculum and program changes. Many of the firm's senior officers participate on advisory committees and boards.

Retaining Women and Minorities in the Engineering Workforce

The solutions for attracting more people, particularly women and minorities, seem numerous and relatively straightforward, although not necessarily easy. In short, there needs to be a better effort of selling the challenges and opportunities of the profession, and of exposing potential engineers to the work of the profession. But attracting more engineers is only part of the solution to future workforce problems. Retention of this talent is essential; this appears to be a far more complex challenge. It touches on social, organizational, and personal issues.

This section presents some ideas and approaches that hold promise for success in the retention of women and minorities. These ideas also hold tremendous challenges to engineering organizations and to each professional.

1. *Opportunity and Visibility:* One of the most important actions for an organization to take to support workforce diversity is to enhance opportunities and visibility for women and minorities within the firm. Research and our own experience have demonstrated that there are no particular differences in capabilities between men, women, and minority engineers. What is most often missing for women and minorities is access to opportunities to progressively learn the skills they need to become experienced technical experts, and then to "show their stuff." Because of their smaller numbers and "newness" in our organizations, many women and minorities get overlooked when assignments are made for the kinds of challenging and highly visible situations that any professional needs to develop and advance.

One answer to this problem is job rotation. CDM is in the third year of a very successful Rotating Employee Program. New employees in the program rotate through three or four different assignments, which last six months each. Participants are assigned an advisor, and together they develop a plan that provides a variety of challenging work suited to the employee's interests. This program is an excellent vehicle to expose women and minorities to the variety of projects available. While they are not given preference in assignments, women and minority engineers are encouraged to participate, and the program advisors are asked to consider them when planning assignments. In addition, project managers receiving these rotating employees are more comfortable with the trial nature of the assignments—this has helped build a positive and receptive environment for minorities.

Another important action that CDM is taking is to recognize the visibility value of company task forces and committees and to encourage consideration of women and minorities in these efforts. This is aimed at trying to overcome the natural tendency to pick people who are well known and familiar for such assignments. When an effort is made to expand the list of candidates, new names get added and their visibility and commitment increases.

2. *Support and Reinforcement:* There is a major dilemma facing companies that are seriously attempting to retain women and minorities. On the one hand, it is recognized how difficult it is to be one (or a few) among many. Anyone who is in the minority can feel threatened, out of the mainstream and stuck in their career. They need reinforcement from the organization and support from their peers. On the other hand, women and minority engineers want to be accepted as competent professionals. They do not want special attention that somehow says they need help. In fact, they do not need help doing their jobs, any more or less than any other employee.

One approach is to seek their assistance in helping the organization adjust to changes in the workforce so other women and minorities have the chance for growth and advancement. Initially, we believe this is best done individually or in groups of two or three, rather than to create a women's or minority organization within the company. Another approach is to identify women and minorities already in positions of responsibility and power and seek their assistance as mentors and advisors for junior staff. It is vital that there be visible role models for women and minorities and that the company acknowledges this important developmental activity. If role models are not available, a strong mentoring program will be of great value.

3. *Organizational Commitment:* Business practices and personnel policies and programs reflect the culture and values of a company. Encouraging diversity in the workforce represents a change in culture for most engineers; therefore, it is important to review policies and programs to make sure they are consistent with this new goal. CDM has made numerous changes to be more responsive to the changing needs of our employees. These include:

- flex time
- alternate work arrangements including work at home and job sharing for certain positions
- establishing a "cafeteria-type " flexible benefits plan
- child care payment assistance through tax-free reimbursement accounts
- wellness programs such as pre-paid counseling through an employee assistance program, physical exam allowance for all employees, and expanded well-baby care payments
- cash incentives for employees who can be covered by a spouse's medical plan

Further, through employee surveys and meetings, the firm continually seeks better definition of employee needs and suggestions for improving human resources programs.

4. Leadership Commitment: While the suggestions above will be helpful in retaining women and minorities in the engineering workforce, they will not ensure success in the long run without one key condition: the commitment and personal involvement of the leaders in engineering firms and professional societies. An important responsibility for leaders is to articulate and reinforce a set of shared values for their organizations. The values for the engineering profession for the 1990s must include the respect for each person based on his or her own merits and the recognition of how important it is for the civil engineering workforce to become more diversified.

How well organizations adopt these values will be reflected in the personal example set by the leadership. For instance, they should personally encourage expanded visibility for women and minorities. And when the opportunity arises, they should be willing to promote women and minorities into positions of responsibility and influence. At CDM, it was shown that all the talk about a career ladder for technical specialists was meaningless until technical professionals were promoted to the top levels of the company based on their technical contributions, not management responsibilities. The same holds true if an organization wants to fully assimilate women and minorities.

Another important area for personal example is in dealing with inappropriate behavior within our organizations. Jokes, innuendos, racial and sexual stereotypes have no place in our profession, yet we recognize that they do exist. In fact, these types of behavior create a climate of obstacles—even overt hostility—that can un-do the best-planned efforts to increase diversity. For many years, the engineering workforce has consisted primarily of white males. As engineers now welcome more women and minorities into the workforce, it is very important that individuals' behavior does not create a hostile or unfavorable work environment. Leaders must practice what they preach in this area and be willing to deal personally and firmly with those who are not acting properly.

Finally, it will be most important for leaders in the profession to be visible and vocal in support of programs and initiatives that promote diversity. This means going beyond expressing concern for compliance with EEO laws. It means recognizing that the future strength and vitality of our profession depends on how well we attract and retain women and minority civil engineers.

Recommendations for Further Action

The following initial steps should be taken toward the long-term objective of attracting and retaining more minority and women engineers:

1. Include as one of the goals of the National Forum on Education and Continuing Professional Development for the Civil Engineer to develop a mechanism for continuing the dialog on this and other urgent issues facing the profession in the 1990s. If the forum is to be held only every five years, an annual roundtable or workshop to share ideas and successes is necessary.

2. ASCE should encourage the establishment of more formal links with other professional organizations so we can approach the problems of engineer shortages and workforce diversity on a coordinated, united front.

3. ASCE, NSPE, ACEC and other engineering societies should extend an invitation to the Society of Black Engineers, Society of Women Engineers, and others to work together to develop a blueprint for action in the areas of attracting and retaining minority and women engineers.

4. Participants in the forum should meet with the leaders of their organizations to explore the issues and discuss how best to address them within their particular organization.

Bibliography

ASCE Task Committee on International Competitiveness. (1989) "Final Report." 3-4.

Chapple, A. (1989) "Women engineers: Still few and striving for status." *Engineering Times*, October 1989.

Dusky, L. (1989) "The best companies for female engineers." *Graduating Engineer*, February 1989. 36-39.

Morrison, A.M., White, R.P., Van Velsor, E., and the Center for Creative Leadership. (1987) *Breaking the Glass Ceiling*. Addison-Wesley Publishing Company, Inc., Reading, Mass.

Rubin, D.K. (1988) "Construction moves to groom talent that will propel it to 21st century." *ENR*, October 20, 1988.

Technology Transfer in the 21st Century

E.M. Wilson* Member ASCE
and
J.A. Puckett** Associate Member ASCE

Abstract

The future needs of local managers and their level of expertise is addressed with respect to electronic networking and delivery systems. These systems will complement existing training techniques such as workshops, seminars, and short courses. These techniques will continue to be a primary training source for local road and street managers. Specifically, a Technology Transfer Network (T^2N) is described where database management, artificial intelligence (AI), and current research knowledge is effectively combined with manufacturer product information, and local users' (engineering) experience with the products, methods, and other state-of-practice information. This system is emphasized with respect to the information and personnel management activities required.

Background for Technology Network System

The transfer of technological information from researchers, product manufacturers, and end-users is presently critical for the effective design and maintenance of public works systems. As the infrastructure continues to age, more effective methods will be required in the 21st Century. A scenario for

---

* Professor and Director of the Wyoming Technology Transfer Center, University of Wyoming, Laramie, Wyoming 82071

** Associate Professor, Department of Civil Engineering, University of Wyoming, Laramie, Wyoming 82071

21st Century technology transfer for local road and street managers is presented. It is clear that the present communication technologies offer a plethora of high-tech solutions. But the critical aspect of system implementation will depend, as it always has, on the communication of the information. The envisioned T^2N is shown in Figure 1. [1,2]

As illustrated, a large rational database system will be developed to store near-raw information. Such information will include: manufacturer information, demonstration materials (both video and static information), state-of-the-art research, training courses, past and future development activities, and existing state-of-practice information. This system will contain information on almost every aspect of roadway design and maintenance, and hence will be of little use to the trained engineer or lay person not expert in database query methods.

It is anticipated that this database will be a combination of numerous databases supported and maintained by several entities. At this level of information management, timely posting and appropriate access is critical.

The key to the management of the system is twofold: the function of the filter and the role of Technology Transfer (T^2) Centers. The filter is a team of experts who organize the information in a rational manner to support users in many technical specialties and at many levels of expertise. For example, the county engineer should be able to gain information on the prevention of washboarding or potholes, the technician-level foreman should be able to receive training regarding the recommended practice, and the researcher should also be able to gain information on the same topics at yet a different perspective level. Clearly, the proposed system is much more complex than the traditional bibliographic databases.

Another critical feature of the filter is to gain local experience information and place it in the database. There is a wealth of information which the end-users have regarding the application of techniques and/or products. This information is perhaps of more value than the volumes of technical information generated annually by the research community. But how can this information be efficiently extracted? How can this information be validated and reformatted into a form suitable for use by others? The management structure

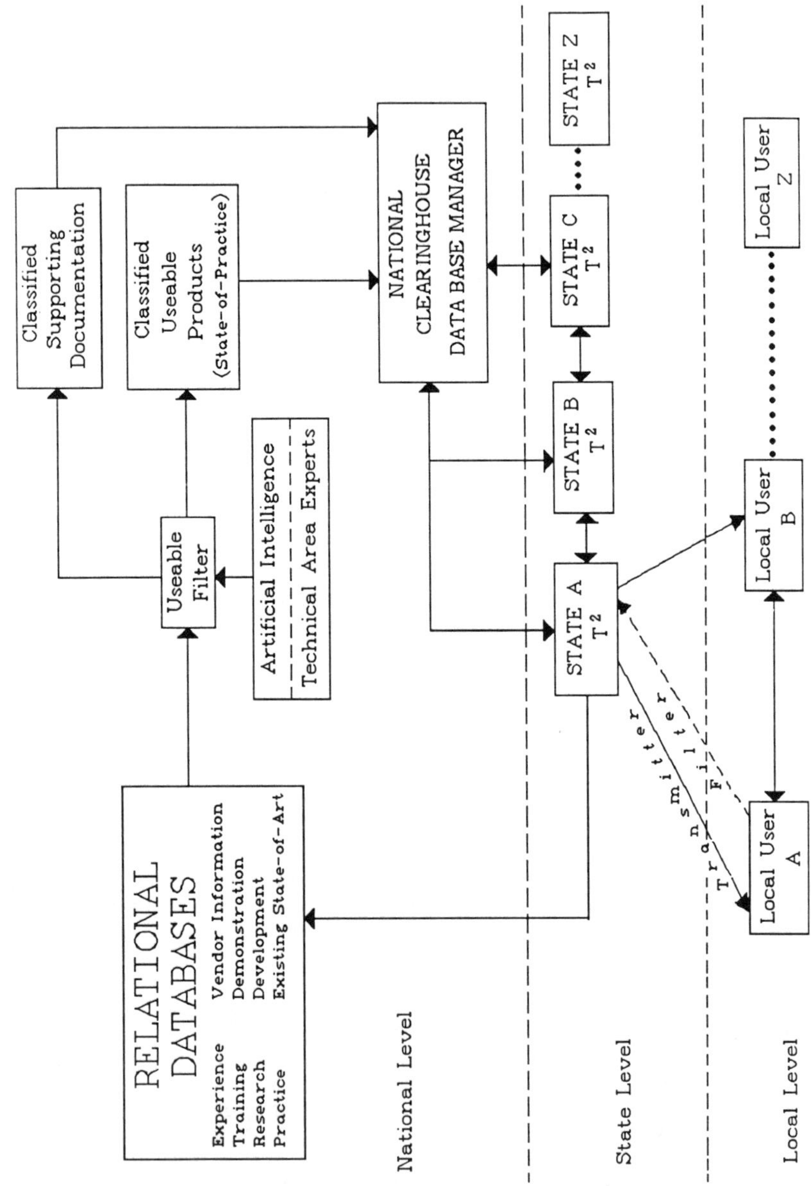
RELATIONAL DATABASES
Experience
Training
Research
Practice
Vendor Information
Demonstration
Development
Existing State-of-Art
National Level
State Level
Local Level
Useable Filter
Artificial Intelligence
Technical Area Experts
Classified Supporting Documentation
Classified Useable Products (State-of-Practice)
NATIONAL CLEARINGHOUSE DATA BASE MANAGER
STATE A T²
STATE B T²
STATE C T²
STATE Z T²
Local User A
Local User B
Local User Z
Transmitter
Filter

illustrated, in part, addresses these issues. The present T^2 Centers are well placed and are presently developing the end-user rapport required to provide this interface. The personnel are required to interact with the end-users; and therefore they must be well-accepted as local experts, have a significant track record in helping the local managers, and be able to screen and reformat local information for entry into the rational database. Finally, the T^2 must perform the traditional role of satisfying user inquiries.[3]

The T^2 Centers perform the role of application and resource experts. The Centers knowledge of the local environment includes techniques, successes, failures, climatic effects, and economic conditions. The Center must be able to screen, with the help of the database manager, information for the National system, and provide appropriate and relevant information to the local user. Further, the local experience must be qualified for local conditions prior to entry into the national system. Simply stated, what may work in one location may be totally inappropriate in another. Hence, any local experience entered into the rational information database at the national level must be categorized appropriately.

The T^2 filter provides high-level technical expertise in the areas of artificial intelligence and database management. National database management is required to organize and maintain the massive databases. Such databases will include traditional bibliographic information and will also include on-line video media, traditional catalog information and local experiences. These capabilities will be required for an expansion of today's hypermedia technologies.

The artificial intelligence expertise is required to develop and maintain the system necessary to support the T^2 Centers. The database must be queried for information based on end usage, local expertise, location, and also local experiences. Present day query methods are mostly inadequate.

The state technology transfer centers will be linked to local users and a national system which includes many areas of practice and research. The state centers are expected to perform a key function in the bi-directional information transfer. One function is to transmit information to the local users, and the second is to filter information regrding user experience. The former is a traditional role and the latter is a new and integral part of the T^2N. Here local user experience can be qualified (filtered or screened) and forwarded to the national system in the proper format with the proper level of significance. Once in this system, the user information, product information, research reports, etc. are categorized by topic, user expertise, function, and managed by an expert system whcih guides the local user inquiries.

To illustrate, the system must be able to meet a query such as:

- **Problem:** Gravel roadway rutting.
- **Location:** Mountainous.
- **Condition:** Extended dry periods.
- **Expertise:** Limited, non-engineering.
- **Construction Equipment:** Single blade.
- **Constraints:** Limited construction season, limited water, long hauls for materials.
- **Previous experience:** Bi-annual grading.

Such a request must be met by the T^2 personnel with the appropriate information for the situation at hand. Further, local experience involving similar situations should be made available (with appropriate contacts).

Educational Requirements

The operation and management of the system described will require new types of engineering professionals. At the filter level, it will require professionals who are experts in database management and AI. The filter will also require cross-disciplinary work in knowledge engineering, which is the process to gain and organize information and traditional fields. Presently, the AI discipline is in its infancy and is slowly evolving, producing the professional experts required.

The T^2 Center staff will also require technical expertise to support the development, operation, and maintenance of the national database. The staff must be able to efficiently access the required information and perhaps more importantly, add information to the database based on local experience. This aspect of the system management is critical and is one for which the present T^2 staffs are unprepared. Considering that the T^2 staff are preferably experienced engineers, the staff of the 21st Century are today's practicing engineers; and therefore have little expertise in the areas described.

Because of the critical role of the Centers, the development of this expertise will be crucial in the next decade.

In order for T^2 to fulfill this role, knowledge and credibility of T^2N are required. Everyone of the millions of potential users need not have technology training--only the T^2 staff at the state level and the national databases management personnel. This manpower training can be met concurrently with the development of the T^2N.

Summary

Solving these training issues will be critical to success in the 21st Century. Compounding the problem is the loss of expertise and the manpower shortage that will be faced in the profession. Both indicate the importance of having the T^2N in place. The T^2 credibility is dependent upon quick and accurate communication. The credibility of the T^2 Center is dependent upon continued successes. The statement "You are only as good as your last responses success", is true.

Without the T^2N, T^2 Centers will be in jeopardy. The major need is to provide the investment necessary to develop T^2N for the future local road and street manager. As each day passes, local experience is lost. Efficient technology transfer requires this effective communication network now. The projected needs of the 21st century are really here today due to these shortages. The need is now; and without recognition and financial support, the future is bleak.

Appendix

1. Wilson, Eugene M., "21st Century T^2 For Transportation Managers," Proceedings, Engineering 21st Century Highways, American Society of Civil Engineers, San Francisco, California, April 24-26, 1988, pp. 48-54.

2. Transportation Reseach Board, "Transportation Professionals: Future Needs and Opportunities," Special Report 207, National Research Council, Washington, DC, 1985.

3. Wilson, Eugene M., ""Take Advantage of T^2," American Transportation Builder, ARTBA, Second Quarter, 1987, pp. 12-13.

THE IMPACT OF GENERAL EDUCATION REFORMS ON THE CIVIL ENGINEERING CURRICULUM

Martin E. Lipinski, P.E., M. ASCE*

Abstract

The integration and acceptance of general education reforms in professional programs, specifically civil engineering, represent a challenge to those responsible for curriculum development. This paper provides a review of the general education reform movement, describes the development of the general education program at Memphis State University with emphasis on the integration of the program into the civil engineering curriculum, and summarizes the key lessons learned during the process.

Introduction

Reform of the general education component of the undergraduate curriculum continues to be a leading issue in higher education today. The controversial 1983 report, To Reclaim a Legacy: A Report on the Humanities in Higher Education (1), written by William J. Bennett, the then chairman of the National Endowment for the Humanities, received national attention by advocating a return to the "classic" undergraduate education focused on a "great books" curriculum. Integrity in the Classroom: a Report to the Academic Community (2) and Higher Education and the American Resurgence (3), were reports of significance which were issued within the following two years examining elements of the undergraduate curriculum. Best-sellers such as The Closing of the American Mind (4) by Allen Bloom and Cultural Illiteracy (5) by E.D. Hirsch supported the concept of a "liberal education" and the introduction of more humanities in the college curriculum.

On campuses throughout the country, students' basic competencies in mathematics and composition have been

----

*Professor of Civil Engineering, Memphis State University. Memphis, TN

evaluated and found to be deficient. Their awareness of the breadth of human knowledge and cultural achievement, and their ability to integrate learning experiences have been questioned. Subsequent debates resulted in the formation of university committees given the charge of developing general education reforms. At many institutions, revised curriculum which embody general education principles have been implemented.

One important challenge of general education reform has been its integration and acceptance into professional programs, specifically engineering. The Accreditation Board for Engineering and Technology (ABET) has long recognized the need for a broad education in its criteria. However, institutional general education reform, with its own set of objectives, may impose severe constraints on those charged with developing refinements in engineering programs. The key issue is not the merits of the liberal arts or the study of humanities, but how to incorporate the general education agenda into the engineering curriculum while providing a comprehensive technical education in a four-year period.

Focus of General Education

The term "general education" is subject to many interpretations. The task group of the Association of American Colleges has defined it as:

> "... the cultivation of the knowledge, skills, and attitudes that all of us use and live by during most of our lives - whether as parents, citizens, lovers, travelers, participants in the arts, leaders, volunteers, or good samaritans" (6).

This idealized definition has been modified by institutions into statements which are used to guide the development of specific programs. At Memphis State University, for example, the statement of philosophy of general education begins with the following:

> "The general education program is that part of the undergraduate curriculum that promotes a common background for all baccalaureate candidates" (7).

Agreeing that there is value to a common core of undergraduate experiences has been only the starting point for general education discussions. The major conflicts have been over what should be included in that common body of knowledge. One point which has received near universal agreement is that the core should insure that students achieve basic proficiencies in the areas of mathematics and written communications. Identifying courses that satisfy other goals, such as intellectual and cultural awareness, has generated heated discussion.

The most vocal proponents of general education have been those in the humanities and fine arts. Citing the benefits of a liberal education as the foundation for developing the values and attitudes needed to address complex societal problems, they advocate adding more liberal arts courses to the core. While their arguments have a solid philosophical basis, their actions are affected by other factors. Some see an opportunity in the general education program to increase their departmental credit hour production and attract students that have enrolled in increasing numbers in professional programs such as business and engineering. Others view general education as an avenue to introduce courses with a particular point-of-view, such as women's studies or cultural awareness.

Obviously, these deliberations affect engineering programs. While engineering departments are not providers of general education courses, engineering students will be subjected to university general education requirements. The engineering profession has long recognized the need for a broad education. For accreditation purposes, ABET requires not only basic mathematics and science but also specifies that "In the interests of making engineers fully aware of their social responsibilities and better able to consider related factors in the decision-making process, institutions must require coursework in the humanities and social sciences as an integral part of the engineering program" (8). Competency in written and oral communication is also specified as essential for the engineering graduate. Satisfying the objectives mentioned and simultaneously meeting other ABET requirements for laboratory experience, computer-based experience, and engineering design has, at many institutions, resulted in an increase in the number of hours in the undergraduate program. Institutional general education reform, with its own set of objectives, may add further to the hours required for graduation. The challenge confronting those involved in revising engineering curricula is not centered on the merits of the liberal arts experience for engineering students, but rather on how to incorporate the general education agenda into the engineering program when faced with other objectives and a limited number of hours to satisfy all of them.

The General Education Program at Memphis State University

General education reform was initiated at Memphis State with the appointment of a General Education Task Force in September, 1984. This group created a document setting forth the philosophy for establishing a program at the University. Subsequently, three additional committees were established before the general education curriculum was developed. The basic program was established in the spring, 1987. A permanent general education committee, with representation of all colleges, was created to approve courses which would satisfy the requirements. A general education director was appointed and the committee was given the charge to avoid the "cafeteria approach" and limit the number of courses which satisfy each requirement. The first general education courses were offered in the fall semester, 1989.

The curriculum accepted by the faculty and approved by the administration consists of both a lower and upper division component. The program requirements are:

- Basic College Skills
  - English Composition - 6 hours
  - Oral Communication and Rhetoric - 3 hours
  - Mathematics - 3-6 hours
- Human Studies
  - World Heritage
    - Literary Heritage - 3 hours
    - Fine Arts Heritage - 3 hours
    - Historical/Philosophical Heritage - 6 hours
  - American Heritage - 6 hours
  - Social Institutions/Social Sciences - 6 hours
- Natural Science - 8 hours
- Fitness and Wellness - 4 hours
- Writing Intensive - 3 hours
- Computational Intensive - 3 hours
- Integration - 3 hours

The program specifies that the last three requirements, writing intensive, computational intensive, and integration, should be fulfilled with upper division courses within the student's major.

Impact on the Civil Engineering Curriculum

The Civil Engineering curriculum at Memphis State University was revised and expanded to 141 hours prior to the establishment of general education requirements. A capstone design course and an upper division technical writing course were added in response to ABET criteria for upper division design experience and technical communication skills. Avoiding an increase in the number of hours required for graduation while integrating the desires of the faculty for the technical component of the civil engineering curriculum, ABET criteria, and the general education demands was the challenge faced by the department.

The majority of the general education requirements were already satisfied in the civil engineering program. All mathematics and science requirements were exceeded as expected. The basic English composition requirement was also met. Engineering students at all universities in the State of Tennessee were exempted from an American history requirement by law if they completed an American history course in high school. Therefore, the American Heritage requirement was waived.

The ABET criterion for one-half year of social science and humanities was part of the curriculum. Eighteen hours, which included 6 hours of English literature and 3 hours of economics, were required. However, the new general education requirements established categories which did not match with the existing program. Also, the existing civil engineering program followed the ABET requirement that a sequence be established in the social sciences/humanities with one course at the upper division level. The new general education curriculum did not include any upper division sequences. A problem was avoided because the development of the general education program contained a provision for course exceptions based on program accreditation requirements and consideration that general education would not increase the total number of hours required for graduation. A request was made to the general education director to allow engineering students a wider choice in courses which would satisfy the ABET criteria and meet the spirit of the general education requirements. After several meetings and discussions with departments offering appropriate courses, this exception was approved.

Technical communications was another area where an exception was requested. The civil engineering department was satisfied with an excellent upper division course offered by the English department which satisfied the ABET criteria. The department requested that this course be substituted for the oral communication and rhetoric general education requirement. This request was initially denied on the basis that the existing English course did not have an oral communications component. Again, after several meetings and discussions with the English department and the general education committee, a new course, technical communications, which included a speech component, was developed and accepted for substitution by engineering students.

The civil engineering department had no difficulty in satisfying the upper division general education requirements of a computational intensive course with any one of a number of its required technical courses.

The capstone design course was identified as satisfying both the integration and writing intensive general education requirements. This course requires students to integrate material from their previous undergraduate experiences, and students must prepare progress and final reports, write technical memorandum and make a final oral presentation.

While adjustments and compromises were necessary on the part of the civil engineering department and the general education committee, the general education requirements were absorbed into the civil engineering curriculum without increasing the number of hours required for graduation. Because the program is in its infancy, it is too early to evaluate any long term impacts. However, perceptions among the civil engineering faculty are that the communications skills of the students may be strengthened by requiring more writing throughout the undergraduate program. Also, by reducing the number of social science/humanities options and improving the courses available, students may be exposed to courses which satisfy the spirit of a liberal education.

Lessons Learned

Implementing the general education curriculum at Memphis State University has taken several years. Faculty had to be sold on the concept, and compromises among departments have been reached only after prolonged debate. The civil engineering department was able to integrate the new requirements into its program because it was involved in the deliberations from the beginning. It is imperative that engineering faculty actively participate in such discussions to share their perspectives on general education and the curriculum. Most faculty in other colleges do not understand accreditation requirements, such as those established by ABET, and are surprised by the breath of the engineering curriculum. They also are unaware of the demands made on engineering students. By serving on university-wide committees, engineering faculty can communicate their concerns and have a role in shaping the program to meet the needs of their students. This may be seen as self-serving, but it is necessary to avoid a situation where compromise is unworkable.

General education programs must be viewed as opportunities to improve our undergraduate civil engineering program, and not as an obstacle to overcome. The goals of general education are in concert with ABET requirements and with the demands of practitioners who hire students. As civil engineerings we are charged with developing solutions to complex problems which involve social and environmental issues. Strengthening the social science/humanities component of the program will improve students' abilities to analyze these problems.

Civil engineering faculty should lead the efforts to improve basic communications skills. To succeed, our graduates must have the ability to communicate the results of their technical analyses, both in writing and in oral presentations, if their recommendations are to be accepted.

Developing a general education program and integrating it into the civil engineering curriculum will involve frustration and compromise, but the end result will be beneficial. As stated recently in a report describing the integration of undergraduate liberal and professional study, " A student's whole education must be greater than the sum of its parts and is a joint responsibility of all faculty" (9).

REFERENCES

1. William J. Bennett, To Reclaim a Legacy (Washington, D.C.: National Endowment for the Humanities, 1984).
2. Association of American Colleges, Integrity in the College Curriculum: A Report to the Academic Community (Washington, D.C.: Association of American Colleges, 1984).
3. Frank Newman, Higher Education and the American Resurgence (Princeton, N.J.: The Carnegie Foundation for the Advancement of Teaching, 1985).
4. Allen Bloom, The Closing of the American Mind (New York, NY: Simon and Schuster, 1987).
5. E. D. Hirsch, Jr., Cultural Literacy (Boston, MA.: Houghton Mifflin, 1987).
6. American Association of Colleges, Task Force on General Education, A New Vitality in General Education (Washington, D.C.: Association of American Colleges, 1988).
7. Academic Senate Special Committee on General Education, Memphis State University, General Education at Memphis State University (Memphis, Tn.: Memphis State University, 1988).
8. Accreditation Board for Engineering and Technology, Criteria for Accrediting Programs in Engineering in the United States (New York, NY: 1988).
9. Professional Preparation Network, The University of Michigan, Strengthening the Ties That Bind (Ann Arbor, Michigan: The University of Michigan, 1988).

Is Environmental Engineering Separable from Civil?

Charles D. Turner[1], MB

Abstract

The design of water and wastewater conveyance and treatment systems has been part of Civil Engineering since its inception. During the last twenty five years, however, a tremendous increase in diversity, complexity, and sophistication of environmental systems has taken place. Environmental engineering is currently a separate undergraduate program at 12 institutions.

The purpose of this paper is to document the direction that environmental engineering has taken within civil engineering at the undergraduate level and compare this with undergraduate degree programs in environmental engineering. Engineering institutions are shown to have taken two paths: 1) environmental specialization in civil engineering graduate programs, and 2) the evolution of undergraduate environmental engineering degree programs. The differences between undergraduate civil and environmental programs are examined using degree program course material from undergraduate catalogs. The number of graduates from the various programs are summarized to provide a quantitative sense of program movement. Finally, the merits of program flexibility are reviewed in the light of Accreditation Board for Engineering and Technology (ABET) requirements.

Introduction

The future role of environmental engineering as part of civil engineering undergraduate programs is not clear. Should civil engineering departments modify their programs to include additional course work in environmental? Should existing environmental courses be shifted away from the traditional water and wastewater system design as the trend in many texts suggests? Should environmental engineering be separated from civil and those courses currently taught in civil engineering be deleted? Should the design/science core of courses be opened up to provide more technical electives to the student? Perhaps more importantly, what are the

---

[1] Prof. of Civil Engineering, University of North Dakota, Box 8115, University Station, Grand Forks, ND 58202

implications for the profession of civil engineering?

Since the turn of the century, world population has tripled, the consumption of fossil fuels has increased by a factor of 30, and industrial production has expanded 50 times; 80 percent of that expansion has occurred since 1950 (MacNeil 1989). Water quality and water supply problems have paralleled this growth. In order to address these challenges, the field of sanitary engineering underwent rapid expansion after World War II. This expansion is still ongoing as we attempt to provide a safe and healthy environment in which to live and work. In addition, the water and wastewater technical areas have been supplemented by solid wastes and air quality to include all forms of pollution. The expansion of knowledge and information in civil engineering has paralleled our economic growth.

The purpose of this paper is to document the direction that environmental engineering has taken within civil engineering at the undergraduate level and compare this with undergraduate degree programs in environmental engineering. To accomplish this, the growth of knowledge in the water and wastewater field is shown using the historic evolution of the seventeen editions of Standard Methods. In an attempt to deal with this knowledge explosion, engineering institutions are shown to have taken two paths: 1) environmental specialization in civil engineering graduate programs, and 2) the evolution of undergraduate Environmental Engineering degree programs. The differences between undergraduate civil and environmental degree programs are examined using degree program course material from undergraduate catalogs. The number of graduates from the various programs are summarized to provide a quantitative sense of program movement. Finally, the merits of program flexibility are reviewed in the light of Accreditation Board for Engineering and Technology (ABET) requirements.

Information Explosion

The explosion of information in the environmental area can be approximated by examining successive editions of Standard Methods (1905-1985). This lab text is a convenient reference for information growth because the fist edition was published in 1905 and it has been updated 16 times. The 17th edition was released in mid 1989 but a copy was not available at the time of this writing. Therefore, the 16th edition is used as the current reference mark. Figure 1 is a plot the number of pages per edition versus time. The curve 1 is an exponential curve illustrative of the tremendous increase in analytical knowledge that has occurred since 1905. Two organizational expansions occurred up through 1946, but the explosion in analytical methods occurred after World War II.

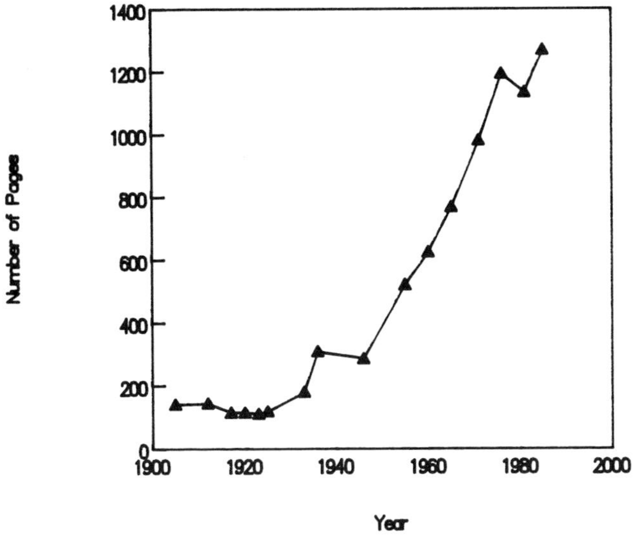

Figure 1. Page count of successive editions of Standard Methods.

The first five editions were published by the American Public Health Association (APHA). In 1925, the American Water Works Association (AWWA) joined the APHA for publication of the 6th edition. In 1936, the committee report from the Water Pollution Control Federation (WPCF) was included in the text and the volume jumped from 180 pages in 1933 to 309 pages. In 1947, the WPCF became a full partner with the APHA and the AWWA in publishing Standard Methods (1985).

Civil and Environmental Engineering Programs

The expansion of environmental engineering knowledge and the need for its application has spawned two different approaches to providing the necessary education. The first is expanded opportunities for graduate work in the environmental area. This option is a natural derivative that follows the traditional approach of obtaining a broad based undergraduate education and then specializing at the graduate level. The second approach is to provide a narrower or more focused undergraduate degree.

The 1988 Annual Report from ABET lists 200 schools that are accredited according to the undergraduate program criteria for civil engineering programs. The same report lists 12 schools that are accredited under the undergraduate environmental group with two of these having dual listing as combined civil and environmental engineering. These two schools are the University of Wisconsin at Madison and

Vanderbilt University. These two schools were not listed in Table 1 because their degree data could not be separated into environmental and civil areas. The two other schools listed by ABET but not included in Table 1 are Humbolt State University and California Polytechnic State University. Data for Humbolt was not available and California Poly listed its data as combined degree data similar to Vanderbilt and Wisconsin.

Table 1 provides data on ABET listed institutions awarding undergraduate degrees in environmental engineering. Data is also included on their master's programs and faculty numbers. Not included are 10 institutions that offer a Master of Science in Environmental Engineering (MSEE) degrees but do not offer a Bachelor of Science in Environmental Engineering (BSEE) degree. A total of 79 students received MSEE degrees from these institutions in 1987-88.

Table 1. Environmental Engineering Degree Data[a]

School	Faculty	Degrees Granted, 1987-88 Undergraduate	Graduate
Univ Central FL			
-Civil	9	30	5
-Envir	6	7	1
Univ of FL			
-Civil	24	83	50
-Envir Sci	17	26	22
Mich Tech Univ			
-Civil	16	91	8
-Envir	(16)[b]	9	-
MT Col Min Sci & Tech			
-Envir	6	14	2
Penn State Univ			
-Civil	22	110	28
-Envir	6	14	2
Rensselaer Poly Tech			
-Civil	12	24	7
-Envir	2	10	2
W M Rice Univ			
-Civil	5	10	5
-Envir Sci	5	2	2
Syracuse Univ			
-Civil	12	17	7
-Envir	(12)[b]	5	3
Total			
-Civil		365	110
-Envir		87	34

a. Data source: Engineering Education, March 1989.
b. Faculty for Envir program drawn from civil program.

A total of 87 BSEE degrees were awarded in 1987-88. The

same institutions awarded 365 BSCE degrees during the same time period. Only Montana College of Mineral Science and Technology offered an environmental undergraduate engineering degree and not a civil degree. Most of the BSEE degree programs have strong ties to civil and their curriculums reflect this. Faculty are often closely tied to both program areas. Michigan Tech and Syracuse draw their BSEE program faculty from their civil engineering faculty. Most BSEE degree programs are basic civil curriculums with structural components deleted and environmental course work substituted. The University of Florida Environmental Engineering Science program is the furthest removed from this. As its title suggests, the program is strongly science based and shares the fewest courses in common with civil engineering. The Florida program also has the largest number of faculty members and graduates, 17 and 22, respectively.

The overall picture is shown in Table 2. In 1987-88, 8798 BSCE and 3214 Master of Science in Civil Engineering (MSCE) degrees awarded. The number of undergraduate students taking environmental electives is not known. If the assumption is made that 20% of the students earning degrees had an emphasis in environmental, 1759 BSCEs had an interest in the environmental area. A review of university undergraduate catalogs indicated that most schools require two environmental engineering courses at the junior/senior level. Course work beyond this would come in the form of technical electives.

Table 2. Civil and Envir. Engineering Degrees - 1987/88[a]

	Civil	Civil/Envir.[b]	Envir.
B.S.	8796	1759	87
M.S.	3217	643	113

a. Data source: <u>Engineering Education</u>, March, 1989.
b. Values in this column are 20% of the civil column values.
c. Data in (a) above compares favorably with ASCE (1989).

Applying the same 20% assumption to MSCEs, 643 of the MSCEs were in the environmental area. Separate and not included in this total are 79 degrees from known programs that have listed a Master of Science in Environmental Engineering (MSEE) and another 34 MSEE degrees from programs with BSEE degrees. The total for separable MSEE degrees was 113.

ABET Requirements

ABET requirements provide flexibility for both civil and environmental engineering programs. The general curricular content criteria for all programs at the basic level is given in section IV.C.2.a. In accordance with section, course work must include at least one year of mathematics and basic sciences, one year of engineering sciences, one-half year of engineering design, and one-half year of humanities and social sciences.

ABET program criteria are submitted by the professional society associated with the degree. Program criteria submitted by the society(s) supplement the general ABET criteria listed in section IV.C.2. of the ABET annual report. The civil engineering program criteria are submitted by the American Society of Civil Engineers (ASCE). For civil engineering, the curriculum content must achieve a broad base of coverage with a minimum of four of the major civil engineering discipline areas included in each student's program. A major comprehensive design experience must also be included.

The ABET program criteria for environmental engineering programs are submitted by the American Academy of Environmental Engineers (AAEE) in cooperation with the American Institute of Chemical Engineers, the ASCE, the American Society of Heating, Refrigerating and Air-Conditioning Engineers, the American Society of Mechanical Engineers, and the Society of Automotive Engineers. The curriculum content must cover at least three areas from the following: air pollution control engineering, water quality engineering, solid wastes engineering, or environmental health engineering. As with civil engineering, these criteria supplement the general ABET criteria.

Membership in the technical society provides some insight into the historical base and identity of the program involved. ASCE has a membership 107,274 (Oct. 1988). AAEE has a membership of 2350 (Oct. 1988) (ABET 1988).

Conclusions

Undergraduate environmental engineering programs are limited to 12 institutions in the United States. Data from 8 institutions show that in 1987-88 a total of 87 BSEE were awarded while 8796 BSCE degrees were awarded. The BSCE degree has a far broader appeal than the BSEE degree.

The eight undergraduate environmental engineering curriculums varied greatly. Several of these curriculums vary only slightly from a BSCE degree while others, such as the University of Florida Environmental Engineering Science degree, do not contain the BSCE ABET required major discipline areas.

The MS degree is the most common vehicle used for obtaining additional expertise in the environmental engineering area.

The number of MSEE degrees awarded by schools granting BSEE degrees was 34. Another 79 MSEE degrees were awarded by institutions that do not offer BSEE degrees. A total of 3217 MSCE degrees were awarded. A significant percentage (20% is the author's estimate) of the MSCE degrees will be environmental in content. The MSCE degree remains the dominant degree because of its flexibility and recognition.

Environmental engineering can be separated from civil engineering at the undergraduate level. However, a trend in this direction is not likely to be established because of the strength of the civil engineering discipline. Therefore, the MS degree is likely to continue as the preferred method for obtaining expertise in the environmental engineering field.

References

Accreditation Board for Engineering and Technology, Inc. "1988 Annual Report." New York, NY. 66-71, 103-123.

American Society of Civil Engineers. "Civil Engineering Degree Data - 1988." memorandum dated Oct.20, 1989. New York, N.Y.

American Society for Engineering Education. (1989). "Engineering College Research and Graduate Study." Engineering Education, 79(2), 61-396.

MacNeil, Jim (1989). "Strategies for sustainable economic development." Scientific American, 261(3), 155-165.

Standard methods for the examination of water and wastewater. (1905-1985). 1st-16th eds., APHA-AWWA-WPCF, Washington, D.C.

AN ENGINEERING INTERNATIONAL RELATIONS MINOR

N. Bruce Hanes, Member ASCE[1]
Frederick C. Nelson[2]

ABSTRACT

In 1992 the twelve European Community countries will reduce, eliminate, or harmonize trade barriers between them. An agreement is now in the final stages that would create the "Euro-Engineer" who would be permitted to practice in any of the countries of the European Community. In addition, it is expected that our country will develop trade agreements in the Pacific and by 1995 our trade with Pacific rim countries will be 100 percent greater than with Europe. The American Engineering Education Community must respond to this trend toward globalization with appropriate programs. The International Relations Minor for students in the College of Engineering at Tufts University is a first step toward that response.

INTRODUCTION

Robert J. Samuelson noted in his syndicated column recently that "Globalization" is a fashionable word that people are now using even though they are not sure what it means. He went on to clarify its current meaning with numerous examples. One example, of particular interest to engineering educators and practitioners, is the world-wide competition of the multinational companies for "the scarcest resource of all--talent". One must ask, "What role does the American Society of Civil Engineers and the Education Activities Division have in the international search for talent?"

----

[1]Professor, Civil Engineering, Tufts University, Medford, Ma 02155
[2]Dean, College of Engineering, Tufts University, Medford, MA 02155

Engineers are people who build things. Increasingly the things they build are systems. Where engineers once built roads, they now build transportation systems, or even more broadly, integrated urban infrastructures. Successful system builders have to combine simultaneously technological matters with economic, social, and political matters. They must learn to engineer more than materials and equations.

At Tufts University we educate our students with this systems perspective in mind. We also provide a few managerial skills, at least the ones that can be taught in a classroom. This provision has been codified into the undergraduate Engineering Management Minor, far and away the most popular non-degree program in the College.

It is also increasingly clear that there is a global need for engineers who know another language, can live within a different culture, and can understand international economic, business, and trade practices. Such internationally-oriented engineers will be the leaders of increasingly numerous technological projects that involve many nations, many peoples, and many disciplines. This is an important, unfilled need that Tufts University, along with many others, can address. This proposal presents a plan to meet this emerging global need by introducing a second minor into the College, an International Relations Minor.

BACKGROUND

In 1992, the twelve countries of the European Economic Community will reduce, eliminate or harmonize trade barriers. This "United States of Europe" will have a larger population than the United States of America. The European engineering community is also in the process of eliminating barriers to engineering practice among the different countries. An agreement is in the final stages among the twelve European Economic Community countries as well as 10 other countries that will create the Euro-Engineer, an engineer who will be licensed to practice in any of these twenty-two countries.

In December of 1988, the Accreditation Board for Engineering and Technology (ABET) sponsored the first International Forum on Engineering Education. This conference provided a forum for the discussion of efforts to recognize engineering education on an international level. An important outcome of the Conference was an agreement among the nations of Australia, New Zealand, The United Kingdom, the USA, and the Republic of Ireland for mutual recognition of their respective accredited engineering degrees. The agreement has to be ratified and requires observation by ABET during accreditation visits in other countries. These visits are now under way. A similar agreement between the U.S. and Canada has been in place for a number of years. In January of 1989, ABET, along with the National Council of Engineering Examiners (NCEE) and the National Society of Professional Engineers (NSPE), signed a "Suggested Draft Agreement" with the European Federation of Engineering Associations which would recognize accredited engineering degrees between twenty-two European countries and the United States. The goal of the agreement is laudable and deserves the support of the engineering profession to solve the many problems involved in its ratification.

In a similar vein, UNESCO funded a symposium at Ohio State in July of 1989 to explore the ways in which engineering education and technological expertise can bring prosperity to developing countries. A major objective of this symposium was to compare engineering curricula in the economically developed and developing nations and to discuss to what extent these curricula should have similar educational content. Again, we see a major effort at globalizing engineering education, an effort that is concurrent with the globalization of technology-based industry.

In the arena of engineering practice, the Council for International Engineering Practices (CIEP) has been formed to provide for the exchange of qualified engineering services between the United States and other countries. The Council consists of six members, two each from NCEE, NSPE and ABET. The CIEP has commenced discussions with the Canadian Council of Professional Engineers (CCPE) relative to the professional engineering services section of the recently ratified Canada-USA Free Trade Agreement. This agreement provides that neither country will discriminate between providers of professional engineering services. Each provincial association and registration board is free to treat professional engineering services as it may choose as long as it does not discriminate between Canadians and Americans. A Memorandum of Understanding has been signed between CIEP and CCPE to insure regular communication and cooperation on all matters affecting the practice of engineering across international borders. It is anticipated that this agreement is the first of many that will be developed with different countries or groups of countries for the exchange of engineering services.

THE PROGRAM AT TUFTS

Engineering students minoring in International Relations will be required to learn at least one modern foreign language and to become acquainted with applicable areas of international relations: such as arts, literature, science, and technology. The Engineering International Relations Minor will be modeled on, but be less intensive than, the International Relations Program in the College of Liberal Arts at Tufts. In particular, it will require a minimum of six rather than twelve courses.

MINOR REQUIREMENTS

A minimum of six courses with:

(a) A core of five courses, one each from the five areas of economics, international history, international relations theory and systems, United States foreign policy, and foreign cultures. Each of these areas has multiple offerings listed on the Core Course List (Table I);

(b) Two of the core courses must be in one of the four participating departments for the international relations program--political science, history, anthropology/sociology, and economics. One course must be completed in two of the other three departments;

(c) Two of the core courses must be in one of the three regions listed below with one course in one of the other two regions;

Region

A	B	C
USA	Latin America	USSR
Canada	Caribbean	Eastern Europe
Western Europe	Africa	Peoples Republic of China
Japan	Middle East Asia, except Peoples Republic of China	

TABLE I

CORE COURSE LIST

Core List

Economics

Economics 1-2 Introductory Economics
Economics 5 Introductory Economics
Economics 113 Marxian Economics
Economics 132 Natural Resource and Environmental Economics
Economics 134 Economic Development
Economics 136 Topics in Economic Development
Economics 138 Economics of Food and Nutrition Policy
Economics 160 International Economics
Economics 161 International Trade
Economics 162 International Finance
Economics 163 Economics of the European Common Market
Economics 164 Comparative Economic Systems
Economics 166 The Soviet Economy
Economics 167 Communist International Economic Relations
Economics 169 Economics of Latin America

International History

History 7 Cultural Regions in the Medieval Period
History 8 Making of the Modern World
History 9 Europe to 1815
History 10 Europe since 1815
History 62 The Foundations of Asian Civilization
History 63 History of the Indian Subcontinent
History 64 Modern History of Southeast Asia
History 66 History of Japan and Korea
History 67 Chinese History I
History 68 Twentieth-Century China
History 72 Modern Africa
History 83 The Modern Middle East until World War I
History 84 The Modern Middle East from World War I
History 105 America in the Nuclear Age
History 120 Historical Marxism
History 129 European Diplomatic History:1870-1940
History 130 Contemporary Europe
History 134 Modern Southeastern Europe
History 136 Modern Russia to 1917
History 137 Soviet Russia
History 139 Modern Germany
History 140 European Fascism in Comparative Perspective
History 141 France, 1815-1914
History 142 Recent History of France, 1914 to Present
History 143 Modern Italy since 1870
History 144 London:World Metropolis
History 147 Revolution in Latin America:Mexico and Cuba
History 148 Revolution and Counterrevolution in Central and South America
History 150 Political Terrorism in Historical Perspective
History 155 Islamic History
History 165 History of the Chinese Communist Party and Revolution
History 166 The United States and Northeast Asia
History 176 Race, Class and Power in South Africa
Political Science 133 Non-Governmental Actors in International Relations
Political Science 136 Soviet Foreign Policy
Political Science 137 International Politics:The Middle East
Political Science 139 Chinese and Japanese Foreign Policy

International Relations Theory and Systems

History 129 European Diplomatic History:1870-1940
History 130 Contemporary Europe
History 134 Modern Southeastern Europe
History 149 War, Trade and Empire:The British Abroad from the American Revolution to the First World War
History 160 Colonial Asia and the West to 1950
History 161 Independent Asia and the West, 1950-Present
Political Science 21 Comparative Politics of Developing Countries
Political Science 22 Cross Cultural Political Analysis
Political Science 31 International Relations
Political Science 32 International Relations:Imperialism
Political Science 120 Political Economy of World Hunger
Political Science 122 Government and Politics of Soviet Union
Political 125 Politics in India
Political Science 126 Modern Asian Politics
Political Science 127 Latin American Politics
Political Science 128 Government and Politics of Eastern Europe
Political Science 129 African Politics
Political Science 131 International Relations: Theory and Practice
Political Science 132 International Global Human Rights
Political Science 134 International Organization
Political Science 138 International Law
Political Science 143 Marxist Leninist Political Thought
Political Science 168 Seminar in Comparative Communism
Political Science 172 Seminar on West European Integration
Political Science 174 Seminar on The North Atlantic Community

United States Foreign Policy

History 59 Cold War America
History 105 America in the Nuclear Age
History 108 The United States and Vietnam
History 110 History of American Foreign Policy in Theory and Practice:1750-Present
Political Science 101 The Presidency and the Executive Branch
Political Science 130 Force, Strategy and Arms Control
Political Science 135 American Foreign Policy
Sociology 120 Sociology of War and Peace
Sociology 140 The United States, Vietnam, and the War

(d) Any course may be counted three times in order to satisfy one core field, one departmental requirement, and one regional requirement;

(e) Students are encouraged to spend at least one semester abroad; the College of Engineering has student exchange programs with the University of Sussex, the Institut National des Sciences Appliquees de Lyon and the Korea Institute of Technology that may be used for this purpose. Other Tufts-approved programs may be arranged through the Office of the Director of International Programs in Ballou.

(f) The sixth course may be selected from any field, department or region. The Humanities and Social Studies requirement of the College of Engineering should be kept in mind when selecting this course.

(g) Students must demonstrate competence in a second language equivalent to four semesters of college courses. Secondary school transcripts are not accepted in evidence of language competence. Language competence may be determined on the basis of a placement examination taken at Tufts, CEEB scores, or a college transcript. Entering students who do not demonstrate the required level of competence must take college language courses until that level is reached.

Depending on the student's language proficiency, the election of the International Relations Minor may extend the degree program beyond thirty-eight courses.

SAMPLE MINOR PROGRAM

A number of different programs are possible. One acceptable program is presented in Table II.

Table II
Sample Minor Program

Course Work
- Economics 1 or 2 - Introductory Economics
- Political Science 31 - International Relations
- Political Science 131 - International Relations:Theory and Practice
- History 84 - The Modern Middle East from World War I
- History 110 - History of American Policy in Theory and Practice:1750-Present
- Chinese 116 - Classic Fiction
- Demonstrated Foreign Language Proficiency
- Optional Semester Abroad

CONCLUSION

The impact of advancing technology on the social-economic-political network of the globe is growing rapidly, yet its implications are far from adequately understood. For example, the increase in international technology transfer suggests that the totality of advances produced by the international fraternity of nations as a whole will far outstrip the new technologies generated by any single nation. Who are the engineers in the United States prepared to lead this nation's industries in exploring and exploiting this trend toward globalization? It is the premise of this paper that Tufts engineers with the International Relations Minor will be among the first in this emerging cadre of international engineering leaders.

The Global Classroom Project

Linfield C. Brown [1]
David M. Gute [2]

Abstract

This paper will present the didactic and experiential benefits gained from conducting an interdisciplinary academic program at Tufts University entitled the Global Classroom Project (GCP). The GCP emphasized bilateral environmental problems faced by both the US and the USSR and included two "Space Bridges" or live television transmissions linking the students and an expert panel in both countries. The Space Bridges allowed students the opportunity to have specific discussions within two topical areas; global climate change and water pollution control problems. In parallel to the Tufts course (N=216), a similar curriculum was offered at Moscow State University. The format for the GCP involved the coordination of five participating courses with a common weekly plenary session and individual section meetings. Core Tufts faculty were obtained from the Civil Engineering Department, the Environmental Studies Program, the Urban and Environmental Policy Department, the Center for Environmental Management, the Nuclear Age History and Humanities Center and the Fletcher School of Law and Diplomacy. Funding was provided by the US EPA, the United Nations Environmental Programme, the Sloan Foundation and the MacArthur Foundation.

Introduction

The offering of the GCP during the Spring 1989 semester at Tufts University coalesced from three enabling factors inside and outside the University. First and foremost the concept of the GCP was derived from a successful formula developed by Martin J. Sherwin, Professor and Director of the Tufts Nuclear Age History and Humanities Center for use in presenting nuclear age material. Dr. Sherwin's model was used in three separate semester long courses with active involvement with Soviet counterparts in each offering. The GCP, reported on here, represents the first application of this model to an environmental issue as well as the first participation of the Civil Engineering faculty in the activity.

[1] Professor and Chairman, Department of Civil Engineering, Tufts University, Anderson Hall, Medford, MA 02155

[2] Assistant Professor, Department of Civil Engineering, Tufts University, Anderson Hall, Medford, MA 02155

The second enabling factor is the rise in academic prominence of "the Environment" at Tufts. There are numerous activity centers within the University which contribute to this reality. For the purposes of this paper collaborative activities of the Civil Engineering Department will be highlighted but the reader should bear in mind the institutional priority placed on the study of environmental issues and policy options. This interest has culminated in the appointment of a University-wide Dean of Environmental Programs who will have as a primary goal the coordination of environmental programming. Within the Department of Civil Engineering the influence of the environment can be seen in the recent development of a one year Masters program in Hazardous Materials Management to complement existing Masters programs in Environmental Engineering, Environmental Health, Geotechnical, Water Resources and Structural Engineering.

The third enabling factor is found in the tremendous surge in interest in the environment nationally and indeed worldwide. Recent examples of such interest include the naming of the "Endangered Planet" as issue of the year by Time Magazine (Time Magazine,1989) the recent prominence of the environment as a subject of discussion at the recently concluded economic summit held in Europe, and the continuing litany of environmental stories which shape world headlines including: global climate change, the ozone "hole", and destruction of the tropical rain forest.

Rationale for Involvement of Civil Engineering

The tradition of strong involvement with the amelioration and control of environmental problems is consistent with the history of civil engineering, most particularly with the discipline previously known as sanitary engineering. The successes of the sanitary era have been extolled as the first public health "revolution" by many commentators (Terris, 1980). In fact, the advances achieved in the golden age of sanitary engineering, improved supplies of potable water, treatment of sewage, etc. have been credited with the lion's share of the improvement in the health status of Americans as contrasted with advances in medical treatment (McKinlay and McKinlay, 1977).

This primacy is currently in a re-birth given the many contributions that civil engineering can make to regional, national and increasingly international environmental health problems. This return is being seen in many different forums including three recent national workshops convened by the Bureau of Health Professions, US Public Health Service, to address the recruitment, training and development of environmental health professionals (Unpublished manuscript, 1989; Levine Associates, 1988; Pacific Environmental Services, 1987).

At these meetings the need for a greater supply of environmental health professionals, including civil engineers, has been consistently made. This call is played out against a background of declining enrollments for undergraduate civil engineering programs. Given this scenario the Civil Engineering Department has attempted to respond in creating academic programs which seek to instill in engineering students a familiarity and perhaps commitment to environmental issues and among the general student population a more heightened awareness of environmental problems and possible options for control and abatement. Participation in the GCP is consistent with this policy and has contributed to the goals stated above.

Description of the GCP

The faculty of the GCP was obtained from a variety of sources within the University. It had representation, as mentioned already, from the Department of Civil Engineering and the Nuclear Age Center History and Humanities Center, as well as the Center for Environmental Management, the Department of Urban and Environmental Policy, the Environmental Studies Program and the Fletcher School of Law and Diplomacy. The offering of the course during the Spring 1989 semester was preceded by a semester-long period of preparation which involved faculty and staff from all of the participating programs. This planning phase encompassed the setting of course objectives and the coordination with Soviet counterparts, most particularly with Evgeny Velikhov, Academician, Soviet Academy of Sciences.

The academic objectives for the GCP were as follows:

1. To introduce students to the arena of United States and Soviet national and international environmental policy.

2. To analyze the connections between environmental issues and international relations both with respect to the underlying causes of environmental problems and to efforts to manage them.

3. To examine the scientific and technological basis for environmental policy and the interplay between scientific knowledge and public policy decision making.

4. To examine selected global environmental problems and alternative approaches to addressing them.

The GCP consisted of five separate courses (one held jointly) around a common plenary session which met weekly for two hours (See Figure 1). In each of the four courses, which were geared to specific student populations, a weekly

FIGURE 1
STRUCTURAL FRAMEWORK FOR ENVIRONMENTAL "SPACEBRIDGE" COURSE
TUFTS UNIVERSITY - SPRING 1989

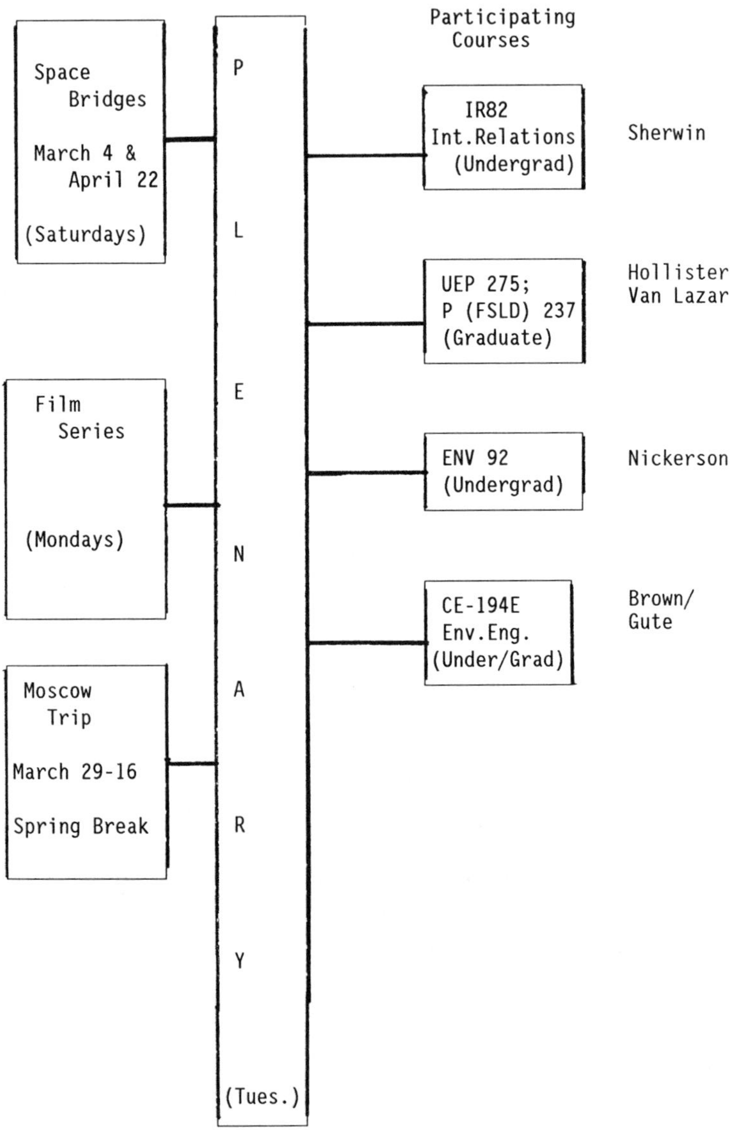

discussion session was also held. The plenary session was complimented by a weekly film series consisting of video material on a variety of environmental topics. At two junctures in the semester a "Space Bridge" or live television hook-up (via satellite) with the Soviet Union was achieved. The Soviet audience consisted largely of Moscow State University (MSU) students. These MSU students pursued a curriculum similar to the Tufts curriculum.

The usual format of the plenary session was to have a member of the core Tufts faculty, all six participated in lecturing before the full complement of students, or an invited lecturer speak for a period of approximately one hour; leaving the balance of the two hour class open for discussion.

Soviet participation was coordinated by the offices of Academician Velikhov, a member of the Presidium and chief science advisor to Mikhail Gorbachev. Additional working relationships were consummated with Gosteleradio which provided for national T.V. Soviet broadcasts of the Space Bridges. The Space Bridges in the United States were facilitated through personnel and equipment supplied by the Boston Public Television affiliate, WGBH.

The GCP was complimented by a visit to Moscow conducted during Spring break for a group of approximately 50 Tufts students, graduate assistants and faculty. The trip included visits to the MSU campus, meeting with Soviet students and professors and cultural and historical locations. The land portion of the trip was underwritten by the United Nations Environmental Programme. Airfare was the responsibility of the individual traveller.

Discussion

The GCP was conducted with a roster of 216 students drawn from the following disciplines;

Undergraduate	
Biology	120
Engineering	8
Political Science	49
Graduate	
Fletcher School	30
Environmental Policy	9

Individual courses were taught by the five following academic entities within the University; Civil Engineering, Environmental Studies, Fletcher School of Law and Diplomacy, International Relations, and Urban and Environmental Policy. The first two being primarily directed at an undergraduate student population, the last three being composed primarily of graduate students.

The Civil Engineering section provided a forum of discussion for topics encountered in the weekly plenary session. Students were evaluated via a written final examination, a 15 page paper selected in discussion with the instructors, and participation in discussion throughout the semester. Paper topics ranged from a comparison of water pollution problems in Lake Michigan and the Aral Sea to a review of the etiology and control measures associated with global climate change.

An evaluation of the GCP was conducted in the form of a questionnaire distributed to all student participants. In general, student comments were similar to those voiced by participating faculty in monitoring sessions held throughout and at the conclusion of the Spring semester. Strengths of the GCP were indicated as:

- Breadth of material-choice of the environment as a topic
- Ability to interact with Soviet colleagues via the Space Bridges and the Moscow trip
- Interdisciplinary nature of faculty and approach
- Participating in the Space Bridges
- A faculty benefit, in addition to those listed for the students, was the chance to work in a collegial fashion across disciplines on an environmental project.

Weaknesses of the course as elicited from the students included:

- Uneven level of instruction-sometimes too advanced, other times simplistic
- Lack of sufficient reservoir of scientific and technical materials from the Soviet Union
- Logistical problems surrounding the trip to the Soviet Union
- Too ambitious an agenda-perhaps more depth on fewer selected topics
- A faculty weakness was the lack of any mechanism to assure support of faculty time in relation to the GCP.

The course generated widespread regional interest with students attending the Space Bridge sessions from many area institutions including Harvard, MIT, Dartmouth, Boston University, Dartmouth, Williams College, and Brown University.

<u>Budget</u>

The overall budget for the entire GCP project totaled approximately $141,000. Each transmission associated with a Space Bridge cost approximately $45,000. The balance of the budget was used for honorariums for outside speakers, costs associated with the film series, support of faculty visits to the Soviet Union, salaries for the teaching assistants for the GCP, and the cost of duplicating teaching materials.

Conclusions

The GCP represented a successful attempt at fashioning an interdisciplinary course targeted at international environmental problems. Shortcomings of the course included an overly ambitious agenda and the lack of sufficient technical materials available from the Soviet Union. These drawbacks were more than outweighed by the positive reaction of the students and faculty involved to the material and concepts of the course. This held true for both the undergraduate civil engineering population as well as the general student body population. Future plans at Tufts call for the offering of two more Space Bridges with the Soviets on environmental topics in the Spring-1990 semester. The Space Bridges will be coordinated with existing courses as opposed to the developing of a specific course similar to the one reported on here.

In summary, the GCP was a ambitious, complicated, yet beneficial addition to the course of undergraduate and graduate education at Tufts. Participating undergraduate civil engineering students benefitted from the breadth of exposure to the material and were quite enthusiastic in their response to the GCP.

References

Levine Associates, Inc., Evaluating the Environmental Health Work Force. Prepared for: US Department of Health and Human Services. Public Health Service. Bureau of Health Professions. 1988.

McKinlay, J.B. and McKinlay, S.M. The questionable contribution of medical measures to the decline of mortality in the United States in the twentieth century. Millbank Memorial Fund Quarterly. 55:405-428. 1977

Pacific Environmental Services, Inc., Education and Training for Health Risk Assessment and Management of Hazardous Waste. Prepared for: US Department of Health and Human Services. Public Health Service. Bureau of Health Professions. 1987.

Terris, Milton. Epidemiology as a guide to health policy. Ann. Rev. Public Health. 1:323-44. 1980.

Time Magazine. January 2, 1989.

EDUCATING CIVIL ENGINEERS FOR GLOBAL PRACTICE

Russel C. Jones,[1] Fellow, ASCE

Abstract

Engineering education in the United States today must prepare its graduates for practice in an increasingly international era. This paper begins with a comprehensive description of the global environment in which civil engineers must practice, and the implications of this broader environment for the engineering profession and for engineering education. Patterns of engineering education practiced abroad, and ideas from those patterns which might be useful for engineering education in the United States, are also covered. Suggestions are given for ways in which engineering education must change in order to best shape the education of engineers well prepared for the current global environment.

The Global Environment

Engineering is clearly in an era of a global marketplace. Civil engineers engaged in the design and construction of major facilities are challenged by competent and aggressive firms from other parts of the world, in both American and international practice. Civil engineers in product design capacities must consider not only the markets in the country in which their products are designed and manufactured, but markets around the world as well. This global marketplace, and its associated economic concerns, has begun driving foreign policy decisions in the United States and in other countries. One response to such economic concerns has been the emergence of regional free trade zones, such as the European Economic Community, several common market agreements in Latin America, the Pacific Basin and other areas, as well as the recent free trade agreement between the United States and Canada.

It is generally felt that the United States is in decline as the leading technical and industrial pace setting country in the world. Major evidence to support that observation is the balance of trade deficit, the

---

[1]University Research Professor, University of Delaware, 205 McDowell Hall, Newark, DE 19716.

huge foreign debt balance, and the increasing amount of direct foreign investment in U.S. production facilities using foreign technology.

World wide communications, facilitated by satellites and by ease of international air travel, have clearly changed the world in which the engineer, and other citizens, live and work. Science and technology now operate in a global intellectual environment. Research and development results in one country are readily available to developers in other parts of the world. It is clear that immediate foreign translation of United States research and development results into products before we can react has disadvantaged our industrial manufacturing competitive position. Similar impacts are being felt in the design and construction sector. However it is also clear that many foreign countries now have strong research and development efforts themselves, so that technological innovation is also being done elsewhere. In fact, U.S. companies are now often buying technology from abroad, where selling it there was typical in the past.

The growth of international banking, and the trend toward multi-national business corporations, has accelerated the development of a global technological environment in which engineers must practice. There is concern that our public policy has not kept pace with these developments. For example, our federal science and technology policy is very heavily driven by national security issues, and global competitiveness concerns have not been adequately responded to by our national level decision makers.

The competition between developed countries in various parts of the world is further compounded by the fact that third world countries want also to be active in the global economy. Developing countries do not want to simply provide low cost labor for the manufacture of products designed and marketed in the more developed countries of the world, or carrying out the construction of facilities designed elsewhere, but want to be able to provide products and services of their own as interdependent players in the global economy.

Formal geographical boundaries which limit the spread of engineering practice are diminishing in several areas of the world. The previously mentioned free trade zones in Europe and North America are examples of areas in which former restrictions about engineering practice across national boundaries are decreasing. In these areas, accreditation and registration equivalency agreements are currently developing, to help bring about the concept of one single worldwide engineering practice modality which is based on equivalency of education.

Civil Engineering in the Global Era

Several observations point to the trend toward globalization of science, technology and engineering. More technical research and development is being conducted in other countries, such as Japan, the United Kingdom, and Germany. There is a significantly increased number of foreign graduate students at United States engineering schools, and a resulting increase in the number of faculty members at our engineering schools from non-U.S. backgrounds. Many components of complex products now come from around the world--for example components for aircraft, and for computers and other electronic products. Worldwide design teams are becoming more typical, with engineering on large or complex projects being done in several countries at one time. There are more international partnerships, and they are typified by partners having equal weight rather then being dominated by U.S. firms as might have been typical in the past.

Design and manufacturing firms in the United States have a major gap between the topnotch research and development typical in this country, and its implementation into practice and products. Our international competitors appear to do much better in this development cycle. We must develop mechanisms to connect research and development more closely with operating units in our technically oriented firms. In the manufacturing area, for example, many of our international competitors have found effective ways of integrating design and manufacturing to better control costs and quality. U.S. industry is now focusing on quality, but the necessary investments in automation and modern production facilities are still lacking. U.S. firms, both those engaged in manufacturing and those providing services such as design, need to move strongly in this direction also. The competitive challenge is to increase quality while decreasing costs.

Our current technological leadership grew up when the United States was the primary economic force in the world. Many of those leaders have difficulty recognizing current and future needs--based upon demographic changes, technological changes, and the development of global competition. We must strive to have those being prepared for future leadership positions in commerce, government, and education better prepared for this global era.

Ideas from Engineering Education Abroad

American engineering education should observe the shape of technical education in other countries, in order to identify possible directions in which our own engineering education might productively evolve.

In Japan, recent engineering graduates are sent for their early job assignments, to the industrial manufacturing floor, in order to get experience there prior to assignments in design or other parts of the engineering spectrum. It is felt that this experience is helpful in providing necessary practical background in such important areas as quality control.

Engineering schools in France now require two foreign languages of all students, one of which must be English. They also require six to twelve months abroad for each student, in a country which utilizes one of those foreign languages.

In Germany, engineering professors are an extension of the industrial system. They are closely tied to engineering practice professionally, and tend to be heavily involved in industry oriented research and development.

In Italy, the time period for the education of an engineer is two years more then is typical in the United States. One of those additional years is spent in secondary school, including a much heavier treatment of mathematics then is typical in the United States. The other additional year is spent at the college level.

In the United Kingdom, the formation of engineers involves a coordinated package of education, training, and practical experience. The formal education period is in fact one year shorter then is typical in the United States, but the training and practical experience portions are more highly structured.

In Latin American countries, collaborative programs across national boundaries have proved effective in advancing engineering education. Such organizations as the Pan American Federation of Engineering Societies, better known as UPADI, have been effective vehicles in organizing such collaborative programs. The length of the total academic program in engineering is the same as that of other professions such as medicine and law, allowing for a more comprehensive curriculum.

Countries in the Pacific Rim tend to have merit based entry systems for higher education, based on examinations on several levels. The freedom of choice of a professional education must be exercised early in the educational process, usually at the level equivalent to our ninth grade. Then the basics of science and mathematics can be strongly developed prior to college for those students destined to study engineering. These Pacific Rim countries tend to utilize their education systems to prepare graduates closely tailored for the needs of industry and commerce.

Japan has a popular second path to a doctoral degree in the engineering field. Graduate students complete an industrial project which is then written up and defended before a university faculty committee, instead of the traditional research oriented dissertation study.

Implications for Engineering Education

Future generations of American engineers will need to have a global perspective. They will need to be able to work with ambiguity, to understand and work within various cultures, and to practice their engineering profession against diverse sets of standards.

The engineering talent needed by the United States is in increasing short supply. A cumulative shortfall in natural sciences and engineering graduates is expected to reach approximately 430,000 by the year 2000. In approximately that same time frame, the doctoral shortfall in natural sciences and engineering is expected to reach 8,000. We must find ways of stimulating bright American students into this natural sciences and engineering pipeline.

The physical science and mathematics base for all engineers is universal, so engineers have little difficulty working across international borders in technical areas. Language and cultural knowledge, however, can often be a barrier to effective engineering across such borders. This knowledge base must be added to the engineering education process in order to develop graduate engineers who can work most effectively in the international arena. Technology knows no political boundaries, and new technological development moves very rapidly. Only those countries who have a science and engineering manpower base well equipped to move rapidly and adapt to new environments will succeed well in our highly competitive world.

Preparation of engineering graduates for this competitive world must be guided by substantial input from engineering practice, as it changes in response to international developments. Engineering graduates must continue to have a solid theoretical base, but increasingly need to be taught more about application of theories so that they are well prepared to utilize them in diverse situations. There also needs to be more emphasis on technology management, in that the majority of engineering graduates will be called upon to manage at least technical groups as they progress through their careers. Design education in engineering schools needs to be much more interdisciplinary, and involve teamwork--a better preparation for what the graduate will meet in practice than is currently typical.

More interaction between industry, universities, and government is also needed. One major area where such interaction needs to occur rapidly is to help improve primary and secondary education, so that the input to higher education is significantly better. Engineering schools can assist in this process by running summer programs for high school teachers, for example. This same broad spectrum of interests needs to work together to tap additional pools of people to meet the projected shortfalls in engineering talent anticipated over the years and decades ahead. Many women, minorities, and many of the brighter of the traditional flow of students are now opting for other fields. The engineering field needs to expand its activities in helping to run secondary school feeder programs, and in providing college level safety nets for non-traditional engineering students.

We need to keep the flow of engineering graduates into the workplace quantitatively strong, so that our technical workforce remains large enough to compete against major international competition. Among other needs, we need a dramatic increase in engineering faculty members who are women and minorities--both to serve as role models, and to meet the need for large numbers of new American faculty members in our engineering classrooms over the years ahead. It is particularly important to attract more American citizens into the engineering field for graduate education, to fill the needs of industry in research and development and to fill engineering school faculty positions. To make the engineering field more attractive to larger numbers of students in general, we also need to correct public image that engineers lack intellectual acumen and have no broad interests beyond the technical.

It is suggested that engineering schools develop masters level engineering programs which contain some business topics as well as technical topics. Graduates of such programs could then expect to get working assignments and rewards similar to those who come through MBA programs. Further, we need to prepare our top engineering graduates to move into chief executive officer positions in industry, government, and education. Our international competitors are often ahead of us in this respect. In Latin America, for example, a significant number of presidents and cabinet ministers come from engineering backgrounds.

Several specific directions are indicated for the preparation of engineering graduates for international engineering practice. It is strongly recommended that engineering students learn a foreign language, in order to provide receptivity to learning other languages later as career needs develop. It is also recommended that there needs to be in the curriculum some introduction into the unique and distinct conditions associated with international engineering practice. It is recommended that more study-abroad opportunities be developed for engineering students, and that more international exchange programs be developed for engineering faculty members. This direct involvement of our students and faculty members abroad could be the most effective way of preparing our engineering graduates for international engineering practice.

Summary and Conclusions

There is a new global environment within which our country, our engineering practice, and our engineering schools must now operate. The many complex directions of this new global environment are impacting the ways in which engineering practice is conducted. Such changes in engineering practice require revisions in our engineering education system, in order to assure that future engineering graduates are well prepared to work and contribute to the betterment of humanity in the international arena.

Patterns of engineering education in other countries can provide models or directions for the fruitful evolution of engineering education in the United States. From whatever sources fresh ideas emerge, international competitiveness requires that engineering education be increasingly responsive to the needs of American industry and engineering practice.

Acknowledgements

The perspectives in this paper derived primarily from the "International Forum on Engineering Education (IFEE)", convened by the Accreditation Board for Engineering and Technology (ABET) in Washington, DC in November 1988. A more comprehensive overview of IFEE, based on a summary presentation made by the author at its conclusion, may be obtained from ABET (345 East 47th Street, New York, NY 10017).

Design Content in the Undergraduate Curriculum

Vernon B. Watwood Jr.[1], Fellow ASCE

Abstract

Because of various pressures on civil engineering undergraduate curricula over the past thirty years, the proportionate share of the curriculum devoted to design has dropped at most universities and is a subject of concern. A so called "capstone" or major comprehensive design experience has been proposed by some educators to strengthen the design content. The writer contends that some aspects of this type of course make it of less benefit than more conventional design intensive courses. The typical curriculum is very tight which means that some other design course would have to be removed to introduce a new one. Therefore, the capstone design experience is not a solution to the problem.

Introduction

The proper role of design in the civil engineering undergraduate curriculum has been the object of much discussion in recent years. This can be easily verified by a quick perusal of the recent engineering literature. For example two of the four papers in the spring 1989 edition of Civil Engineering Education are on the topic of the teaching of design. An examination of the proceedings of any recent ASEE conference will show the presence of many design oriented papers.

Part of the reason for the present discussion of this issue stems from changes to the typical civil engineering curriculum which occurred during the late fifties and the sixties. These changes were in response to many forces, not the least of which was a perceived weakness in the basic and engineering science abilities

---

[1]Prof. and Chmn., Dept. of Civ. Engrg., Michigan Technological Univ., Houghton, Mich. 49931.

of engineers practicing in the United States. A second influence was the heightening of social consciousness that took place in the sixties. A third factor was an increasing emphasis on research at most universities which was brought about at least partly by the increased research funding made available in response to the Russian launch of the first Sputnik.

The net result was a strengthening of the mathematics, chemistry, physics, and engineering science courses such as thermodynamics, mechanics (both solid and fluid), and basic material behavior. Additional course requirements were added in the humanities and social sciences areas. Such changes were not uniform across all universities and programs but most were affected to some extent. The course credits necessary to effect this change had to come from somewhere rather than simply added on and as a general rule design courses suffered.

Sometime in the late seventies or early eighties many educators began to realize that the pendulum had swung too far and began to take corrective measures to balance out the emphasis on design versus theory. In recent years courses have been proposed which have been labeled "Capstone" or "Senior Design Experience". In some cases these courses have been formulated in a way to make them very useful. Unfortunately in many other situations they have been proposed in a format that makes them less useful than other alternatives.

Crowded Curriculum

If a new course is to be added to most civil engineering curricula, something must come out. Engineering faculty generally understand this simple statement well. Many engineers outside of academe do not understand it at all. Even many faculty outside of the school of engineering do not understand it. Most disciplines other than engineering have large elective components so that a small addition to the required component can easily be achieved.

The reason for the tightness in engineering programs is quite simple. It stems from the requirements set forth for ABET accreditation and the need to provide a minimal level of education for an engineer to begin his or her career. There is no room here for an in depth analysis of the ABET requirements. Suffice it to say that three fourths (three years) of the curriculum is prescribed. That leaves one year to squeeze in other requirements such as computer literacy, development of oral and written communication skills and to acquire an understanding of ethical, social, economic, and safety considerations in engineering practice. In addition to the ABET requirements, university general education requirements must also fit into the unspecified portion and the credits required are often not trivial.

In addition to the above, in many universities there are enormous pressures to provide more free electives for the student. Some engineering schools have gone to great lengths to create a curriculum that appears to have great flexibility. However, a close inspection will reveal in most cases that the rules are such as to allow little room for any truly free electives.

The net result of this situation is that much thought must be devoted to any modifications to the typical civil engineering curriculum and in particular one must give careful consideration to any addition since something will have to be removed to make room for the new course.

The Capstone Design Course

A certain mystique has evolved regarding what constitutes a "capstone design experience". The phrase is not clearly defined by ABET even for the programs that explicitly require a capstone course. However, there are several descriptor phrases that are used in the literature when capstone courses are described. These are (Wetzel, 1986, and others): team approach, major comprehensive design experience, engineer-client relationship, realism, ethics, plans and specifications, practitioner involvement, multidisciplinary, creativity, communication skills, interaction with society, and the need to introduce little or no new engineering material, i.e. the course is to be purely synthesis of previously learned information.

Many of these concepts are being considered for adoption in the ABET program criteria for civil engineering. For example, the proposed program criteria for civil engineering and similarly named engineering programs for 1990-91 uses the phrase "...must culminate in a major comprehensive design experience." No amplifying explanation is given to better define what the phrase means. However, since the statement is in the program criteria it implies that it is an amplification and a more specific definition of the regular ABET design criteria which already specifies that the curriculum must contain at least one upper-division course which is primarily design and which draws upon previous course work in the relevant discipline. In particular, a strong impression is given that what is really being referred to is a "capstone" design experience and this raises serious questions regarding what will actually be expected by an ABET reviewer.

There are about seven programs that explicitly require a capstone design course in the present ABET program criteria. Most, if not all, of them are much more narrowly focused programs than civil engineering. In some of these programs the capstone design experience is a senior project that many civil engineers would classify as research rather than design.

A Complete Engineering Education

In the concluding sentence of a paper about the capstone course, Ring (1987) states "The (soon-to-be-a-professional) senior civil engineering student who has not experienced a complex 'design' project, either in the university or in on-the-job summer or COOP experience, is not fully prepared to enter the marketplace." This is the type of thinking that is driving the concept of the capstone experience and which, in the writer's opinion, is mistaken.

First of all, virtually no student is assigned the task of performing a complex design in a summer job or COOP experience. In most situations the student employee performs the duties of an inspector, or serves in a survey party, or performs routine tasks at a drafting board or computer. Such experience is extremely valuable to the student to help him or her better understand the overall role of engineering. However, it is not in any way a complex design experience in which one is responsible to make significant decisions regarding the synthesis of a project.

If the statement "not fully prepared to enter the marketplace" is translated to mean not fully prepared to practice all aspects of professional engineering then one must certainly agree but the reasons go much beyond the lack of a single course. It is not practical to think in terms of providing fully trained civil engineers who are ready to take responsible charge of complex design projects after only four years of education past the present high school degree. To provide such an education would require, as a minimum, a seven year curriculum which would culminate in a one year internship. This would require each school of engineering to set up a consulting engineering firm with the faculty as project engineers in much the same manner as medical schools have one or more hospitals associated with their programs. This setup would produce engineers who would be capable of being fully licensed upon gradation and who would be fully capable to design reasonably complex projects in their area of specialization.

If on the other hand the statement "not fully prepared to enter the marketplace" is translated to mean not fully prepared to begin work as a junior engineer (or engineer in training) then this writer must disagree and is not alone in that opinion. One practitioner (Lindsey, 1986) states, "...an in-depth understanding of the theory is the most fundamental thing a graduate must have." He goes further to say "Only introduce the student to the design process so the consultant can teach him how to design." A few years ago the national ASEE meeting had as its theme "Academe and Industry - Partners in Education". With the present four year system that partnership is absolutely essential and most of the content of the capstone design experience can be taught most effectively by industry.

However, there is no question that the engineering graduate must have received some training in the design aspects of engineering. Even though a thorough grounding in the fundamental theory will go a long way toward providing the necessary preparation for an engineering career, there is a definite need as well as an expectation on the part of most engineering employers that newly graduated engineers have some knowledge of the design process. In particular, most structural engineering employers would be very disappointed if a newly hired civil engineer were completely unaware of the major structural design codes and had no idea of how such codes are used in design.

<u>What is to be Eliminated?</u>

Eventually the problem of adding a capstone course reduces to the necessity of eliminating some other course to make room. Only one half year of the curriculum is available for the study of design related material. Decisions must be made regarding how those credits will be utilized to meet the design requirement.

At the writer's institution, the present ABET requirements are met by designating several senior courses as design courses. These meet the existing ABET criteria of being primarily design and drawing upon previous course work. Each student is required to take at least one of these courses.

Some of these courses would probably meet the new ABET Civil engineering program criteria of being a major comprehensive design experience but there are others which probably would not. One course (among several) which in its present form would probably not meet the criteria is Timber Design. This is a three credit course (quarter system) which for all practical purposes is entirely design except for a few lectures on wood properties. No single large design project is required. Instead the students are given a series of smaller design projects (one per week) which could best be described as component design.

With the proposed set of rules this course would probably be lost to a large segment of our student population. If the course is not redesigned it would be pushed out of the curriculum by a requirement for the capstone type course. If it is redesigned to fit the capstone mold, much of the material regarding timber design, per se, will have to be removed to make room for the integration of the major aspects of a capstone course. In fact one of the distinguishing features of a capstone design course is that little new material should be brought in. The emphasis in a capstone design course is supposed to be the integration of previously learned material.

Hence it should be clear that the imposition of a capstone course will extract a price. The civil engineering community should consider very carefully whether the capstone course is worth more than the course it replaces.

Optimization

Once it becomes clear that the education of an engineer must, under the present set of constraints, extend beyond academe, then it is necessary to ask - what should be taught where? In the writers opinion the "total picture" as it is sometimes called must be brought together after the young engineer has left the university.

In the on the job setting there is no problem creating realism. The need for problem definition becomes obvious. The constraints to the design problem will be set by reality, not by a professor trying to size the project appropriately for a three or four credit course. The team approach required is settled by the reality of the size of the employing firm and the project and can be from one to twenty or more in size.

The engineer-client relationship would be real not contrived. The need for good plans and specifications becomes obvious to the learning engineer and he or she has assistance from technicians in preparing them.

Interaction with society will take place as needed. This concept was forced upon designers by a public that was angry with the way designs were being imposed upon them, not as a result of actions by academe. Hence practicing engineers are at the forefront in this arena.

If engineering educators can really teach creativity, they will be unique in the academic world. Creativity can only be encouraged and such encouragement can easily happen outside of a capstone design course.

Summary

Design, as well as a solid preparation in engineering fundamentals, definitely belongs in a civil engineering curriculum. If the proper circumstances exist, then a design course with some of the capstone type attributes has a place in the design portion of the curriculum. However, an unqualified "capstone" design course should not be required. This writer believes such a requirement will be counter productive. The primary ingredients of the capstone design experience can best be learned after the student graduates.

If the presently proposed ABET program requirements are adopted, they should be spelled out with no ambiguity so that an ABET visitor will not assume that all of the usual capstone type course requirements are intended.

References

1. Lindsey, S. D., "Engineering Education - A Consultant's Viewpoint", 1986 ASEE Annual Conference Proceedings, pp. 442-443.

2. Ring, Stanley, "A Highway Design 'Capstone' Course For Senior Engineering Students", 1987 ASEE Annual Conference Proceedings, pp. 1598-1600.

3. Wetzel, R. and McCullough, E. S., "Capstone design Courses in Civil Engineering", 1986 ASEE Annual Conference Proceedings, pp.436-441.

Undergraduate Civil Engineering
Hydrology Course Content

James R. Groves, M.ASCE, and
Edward B. Jernigan, AF.ASCE[1]

Abstract

A survey of civil engineering educators revealed that there was general consistency in the topics covered in introductory courses in hydrology, which was also reflected in the typical text book contents.

Introduction

In response to an interest expressed by members of the American Water Research Association and the Universities Council on Water Resources, instructors at colleges and universities across the U.S. were contacted and asked to submit course outlines of their undergraduate, graduate, and continuing education courses in surface or ground water hydrology. The survey included questions on text and reference books used, course prerequisites, topical coverage with numbers of classes, and computer usage. Announcements calling for responses were placed in the journals of the two sponsoring associations, survey notices were mailed directly to members of the associations, and individuals were contacted by telephone and requested to respond.

Because of the sampling procedures used, it is not possible to specify precisely how many inquiries were distributed, nor to comment on whether the sample of responses was representative of the current state of education in hydrology. However, a total of 39 comprehensive responses was received, representing colleges and universities in 21 states, from New

[1]Associate Professors of Civil Engineering, Virginia Military Institute, Lexington, VA 24450

Hampshire to Hawaii. Of the total, twenty responses were from non-civil engineering curricula, such as forestry, geology, geography, rangeland resources, earth sciences, agricultural engineering, and some graduate programs.

This course content summary for undergraduate hydrology courses in civil engineering curricula does not include responses from non-civil engineering curricula or graduate courses. The results of the overall survey, including similar course outline comparisons in graduate and continuing education courses will be available in 1990 from the Publications Office of the American Water Research Association.

Course Content

The results of the survey are summarized in Table 1, which shows the total number of class hours normally spent on each of eight major categories of topics in the undergraduate hydrology courses, as well as hours spent on review and testing and on miscellaneous topics. The typical 3 credit hour course meets for 50 minutes, approximately 43 times in a semester. Courses of 3 credit hours conducted on the quarter system or meeting twice a week typically meet approximately 30 times.

In general, the hydrologic topics listed in Table 1 follow typical text book content, as summarized below:

Hydrologic cycle: abstractions
Meteorology: precipitation measurement, distribution
Statistics: probability, frequency analysis, regression
Rainfall-runoff relationships: rational, SCS, unit hydrographs, streamflow
Routing: hydrographs, open channel flow
Hydrologic design: maximum precipitation and flood events, spillways, reservoir yield
Watershed models: simulation, basin models
Groundwater: infiltration, well flows, models
Snow: measurement, runoff
Other: legal aspects, ethical considerations, water quality, collection systems, special projects, weather modification

The hours shown in Table 1 were generally taken from course summary sheets such as are typically prepared for accreditation review visits. Inconsistency of reporting was evident in such categories as hours for review and testing, and in total course hours. In some cases it was difficult to distinguish between hours studying the hydrologic cycle and meteorology; and

TABLE 1. Class Periods Devoted to Hydrology Topics

School	Hydrologic cycle	Meteorology	Statistics	Rainfall-Runoff	Routing	Design	Watershed models	Ground water	Snow hydrology	Other	Exams & review	Total periods
A	4	3	5	8	3	7	7	3	1	-	4	44
B	3	4	4	13	7	8	2	*	-	1	3	45
C	2	3	7	6	6	3	2	1	2	2	$	34
D	5	4	6	17	3	2	-	*	-	4	4	45
E	3	5	3	6	3	3	3	7	-	8	2	43
F	9	3	4	10	3	-	3	5	-	1	6	44
G	2	3	8	5	-	8	-	-	-	-	2	28
H	2	6	2	10	3	7	2	3	-	4	4	43
I	3	4	4	14	4	-	4	-	4	-	4	41
J	5	6	6	12	5	5	-	6	-	-	$	45
K	3	3	5	7	4	2	-	4	-	-	$	28
L	1	4	4	6	3	7	-	2	-	-	3	30
M	5	6	4	9	3	3	-	3	3	3	$	39
N	6	6	6	6	6	5	-	6	-	-	$	41
O	5	-	-	5	5	4	-	4	-	2	2	27
P	3	3	3	12	6	12	-	3	-	-	3	45
Q	9	3	2	13	-	21	-	4	-	-	2	54#
R	3	3	3	9	3	3	-	3	-	3	$	30
S	1	2	5	8	4	3	-	3	-	-	4	30

* School offers separate ground water hydrology course.
$ Information not provided.
\# Four credit-hour course.

between time studying rainfall-runoff relationships and watershed models. This may account for some of the significant variation in time attributed to these topics. It was also noted that at least two of the four colleges which allocated no periods to ground water hydrology offer separate courses in this subject. Other schools may have a similar situation, but did not indicate it on the survey form.

It was interesting to note that geographic location had little effect on the teaching of snow hydrology. Some schools located in the deep south included it, while schools located in the north did not.

The most common topics related to the application of hydrology, listed under the "other" category, included unspecified special projects, and the hydraulics and design of storm water collection systems.

Other Survey Results

One prerequisite listed for the undergraduate hydrology courses was consistent, with 14 of the 19 respondents requiring fluid mechanics or hydraulics. Four of the schools required statistics, three respondents did not indicate prerequisites, one listed calculus only, and one listed concurrent registration in a planning/decisions course. Also, three of the schools had computer literacy listed as a specific prerequisite.

Computer use was prominent feature in only three of the courses, with lists of specific software to be used, programs to be written, and/or problems to be solved. Eleven of the remaining respondents simply indicated computer usage was included in the course, one indicated it was "encouraged", and one that it was "not required".

The text book competition was wide open, with five of the nineteen schools using Viessman (1989), four using Lindsey (1982), three using Chow (1988), three using Bedient (1988) and the remaining three using McCuen (1989), Hjelmfelt (1975), and bound course notes. One school did not indicate what text was used.

Few of the survey respondents indicated whether or not the hydrology course was required, so a reliable percentage could not be determined. However, a survey of civil engineering education reported by Wadlin (1985) indicated that 49.2 percent of the respondents required a course in hydrology for their batchelor's degree, which was an increase from 44.3 percent that reported this requirement in 1978.

Conclusions

There were few surprises discovered in the review of undergraduate civil engineering courses in hydrology. If anything, the consistency could be considered remarkable, without major variation that might have reflected the background and interests of the faculty, or the local or regional hydrology. There appears to be a consensus on what topics should be included in an introductory undergraduate course.

Appendix - References

Bedient, P.B., and W.C. Huber (1988). Hydrology and Flood Plain Analysis. Addison-Wesley, New York.

Chow, V.T., D.R. Maidment, and L.W. Mays (1985). Applied Hydrology. McGraw-Hill, New York.

Hjelmfelt, A.T.Jr., and J.J. Cassidy (1975). Hydrology for Engineers and Planners. Iowa State University Press, Ames, Iowa.

Lindsey, R.K., M.A. Kohler, and L.H. Paulhus (1982). Hydrology for Engineers. Third Edition, McGraw-Hill, New York.

McCuen, R.H. (1989). Hydrologic Analysis and Design. Prentice Hall, Englewood Cliffs, New Jersey.

Viessman, W.Jr., G.L. Lewis, and J.W. Knapp (1989). Introduction to Hydrology. Third Edition, Harper and Rowe, New York.

Wadlin, George K.(1985). "Survey of Civil Engineering Education - 1984" in Challenges to Civil Engineering Educators and Practitioners - Where Should We Be Going?, Proceedings of the Conference, Ohio State University, April 11-13, 1985. American Society of Civil Engineers, New York.

Civil Engineering Education - Where Do We Put What?

Roger E. Smith,[1] Member, ASCE
Charles H. Samson,[2] Fellow, ASCE

Abstract

Calls for change in the civil engineering curriculum are coming from all corners, and these generally call for more to be taught each year. The profession must identify and support an approach that fits its needs and has a realistic chance of being adopted. Application oriented graduate degrees are available and provide a realistic option.

Calls for Change

There appears to be general feeling among civil engineers and educators that there is a need to review, and probably change, our approaches to educating civil engineers. A recent item in the News Briefs of Civil Engineering reported that "wrenching changes" would be required in the education of civil engineers if they are to compete in "the global economy of the future" (ASCE 1989a). Nearly every edition of Engineering Education contains an article on changes needed in engineering education. Van Valkenberg provided a series of articles on curriculum change (Van Valkenberg 1989a & 1989b). An American Society for Engineering Education task force report identified "The Overburdened Curriculum" as the first element (David 1988).

---

[1]Assoc. Prof., Dept. of Civ. Engrg., Texas A&M University, College Station, TX 77843.
[2]Prof., Dept. of Civ. Engrg., Texas A&M University, College Station, TX 77843.

The Need for Change

Since institutions began educating civil engineers in the 19th century, the profession has grown not only in size, but also in specialization and complexity. The knowledge in many areas is growing at an exponential rate. Many of the new techniques which allow us to be more accurate in our analysis and design also require additional education. Computers and calculators have not only replaced the slide rule, they have opened up a whole new area of education that the civil engineering student must study. It is true that students no longer spend time studying the use of the slide rule; however, the computational techniques replacing slide-rule calculations require even more educational effort. Computer usage is being integrated into nearly all phases of civil engineering education.

Interdisciplinary education is also becoming more important (Luberhoff 1989). Many projects which address the complex problems, including maintenance and rehabilitation of the infrastructure and construction in space, will require specialists from many disciplines working together to solve the problems. Communication among team members, as well as the need for someone to integrate the efforts of the team members, will require more interdisciplinary education for civil engineers, if they are to be participants, let alone leaders, in these teams.

Civil engineering is becoming more international with each year (ASCE 1989b). Foreign firms are not only competing with American firms abroad, but they are also competing for work in this country. Civil engineers need education in the languages and cultures of foreign competitors and foreign countries where they may work.

ASCE's president has stated that one of the needs in civil engineering is to develop the citizen-engineer (Focht 1989). Others have argued that the changing world and the changing attitudes toward engineering require that we provide a broader education to develop engineers more sensitive to society's needs (Wenk 1988). This requires more education in citizenship, management, environmental studies, and communication.

At some institutions, university-wide core curricula have been created to provide a measure of "breadth" to the education of all students. While this serves one aspect of the needs of engineering students, it also reduces the hours available within the four-year program for technical course work. There are also calls for new subjects to be added to the technical course work, the latest of these being design and construction in space (Ardis 1989).

All of the above leads to the simple conclusion that the amount of knowledge that needs to be taught to the civil engineer has increased tremendously over the last few years compared to that required, or available, a few years ago. Accomplishing this within a four-year program may not be possible.

The Texas A&M Approach to Change

At Texas A&M University, a committee has been formed to study the civil engineering curriculum to insure our graduates are provided the education needed. Key undergraduate curriculum concerns include the amount of specialization that should be allowed, the amount of time that should be spent on science versus application-oriented education, and the amount of time which should be spent on non-civil engineering subjects such as communication, management, foreign language, and citizenship.

Specialization allows in-depth education in one, or a few, areas of civil engineering. As the amount of knowledge increases and other areas of education take more time, the specialization can become more extreme. A broad education provides a background in all fields of civil engineering upon which the graduate can build after graduation and mobility within the civil engineering field. Some engineers will be generalists rather than specialists, and must be knowledgeable in the "systems approach" to integrate the work of specialists into complete systems.

The concern over how much science and theory versus how much applied engineering should be included in the curriculum appears to be as old as formal engineering education (Florman 1989). We now require that the engineer have a strong background in basic science believing that engineers apply basic laws of science, and that this knowledge of science and basic engineering fundamentals separates engineers from technicians. On the one hand, teaching how to use available techniques to solve today's problems to the detriment of the theory and concepts behind them may result in engineers who cannot adapt to tomorrow's problems. On the other hand, those who leave the education system with only the theory and concepts may not be able to make the immediate contribution that employers need.

The Texas A&M civil engineering curriculum study committee intends to work with the faculty to complete an analysis of the education needs of our students. An effort will be made to provide an efficient curriculum which meets the goals of the students, the profession, and the faculty. While an attempt will be made to fit the material into the existing baccalaureate degree, the committee is anticipating that their analysis will show a need to teach more than can be packed into a four-year program, no matter how efficient the curriculum.

How to Handle What Doesn't Fit

A major question facing the profession is how to handle teaching the material which does not fit into the four year program but is needed to be an effective practicing engineer. The call for a five year bachelor of science program is again being heard (Wenk 1988). This was tried several years ago but failed. Major disadvantages include the fact that students are less likely to enter an area of study which requires five years for a bachelor's degree while all other fields require only four and that administrators don't support such programs.

A number of writers have proposed approaches that would retain the four-year baccalaureate degree but require additional work toward an advanced degree which would be regarded the first professional degree (Pletta 1987, Samson 1989a, 1989b). This approach falls within the existing education pattern, and changes to the professional registration requirements are decisions the profession must make. A possible approach would be to change the requirements for professional registration to an ABET recognized baccalaureate degree and an ABET recognized master's degree with three or four years of experience.

Educational institutions have always provided additional engineering education through the master of science and doctor of philosophy in civil engineering. However, these two degrees have been seen by many practicing civil engineers as oriented too much toward research. Some institutions have modified the master of science degree to allow it to be a more practice oriented degree. Others, such as Texas A&M University, have developed practice oriented master of engineering and doctor of engineering degrees. The master of engineering requires more course work than the master of science degree, including course work outside the department; instead of writing a thesis, the student prepares an engineering report. For the doctor of engineering degree, the student is required to complete work in design, professionalism, and business and then serve an internship in industry. At the completion of the internship, the student must prepare an internship report.

There will always be a place for the traditional master of science and doctor of philosophy degrees. A shortage of engineering faculty is developing, and the doctor of philosophy is needed for the faculty positions (Brighton 1989). Not all institutions need offer every kind of graduate degree; diversity of approaches among institutions may be healthy (Troxler 1989).

There is a current need for continuing education throughout an engineer's career which will increase in the future (Willenbrook 1988). Continuing education programs must be considered an essential element in the total educational process for a civil engineer, especially in relationship to specific analysis and design techniques. New analysis and design techniques will continue to proliferate as industry continues to take advantage of computer developments. By working with the academic institutions, practice oriented continuing education courses can be developed which build on the foundations provided and keep the faculty in touch with the practicing engineer. Continuing education could also be used to bridge the period of change to a master's degree as a requirement for registration by requiring six or eight years experience with a set number of continuing education hours from ABET-approved institutions for registration.

Practicing engineers have a unique opportunity, even an obligation, to help define the direction of civil engineering education for the future. It is apparent that changes are needed, and many civil engineering programs are searching for ways to modify their curricula to keep pace with the changing needs.

Need for Industry Support and Direction

There are several questions that practicing engineers who hire civil engineering graduates must consider, including:

1. Would employers prefer practice-oriented education prior to hiring the graduate, or would employers prefer to provide their own practice oriented training?

2. How much continuing education are employers willing to support?

3. Would employers pay a higher salary to those holding practice-oriented graduate degrees?

Whatever the general attitude of practicing engineers, they must support it to educators and administrators. Currently, many administrators give less support to practice-oriented degrees. Tenure and promotion criteria generally either ignore, or give reduced credit to, work with students involved in practice-oriented graduate degrees. This tends to discourage civil engineering educators from promoting the practice-oriented graduate degrees in favor of the more traditional research-oriented degrees.

Summary

Although the faculty may make valiant attempts, the knowledge that practicing civil engineers of the future will need to be competitive in the world market will not fit into a four year baccalaureate program. Five year baccalaureate programs did not work in the past, and there is no reason to believe they will work any better now. Practice oriented master of engineering and doctor of engineering programs have been developed to compliment the more traditional master of science and doctor of philosophy degrees.

The profession should identify an approach that supports its needs and promote it with faculty and administrators, as well as registration boards. The application-oriented graduate degrees are often given less support by the administration and faculty, and professional support will be needed to enhance their use.

Appendix. References

Ardis, C. V. (1989). "Teaching civil engineering in space," Civil Engineering, ASCE, 59(9), 6.

ASCE (1989a). "Report urges changes in engineering education," Civil Engineering, ASCE, 59(8), 11.

ASCE (1989b). "Task group offers advice on regaining U.S. Competitive edge in world design/construction market," ASCE News, ASCE, 14(10), 1.

Brighton, J. A. (1989). "Engineering faculty for the 1990s and beyond," Engineering Education, ASEE, 79(5), 538-539.

David, E. E. and Willenbrook, F. K. (1988). "One year later: the ASEE action agenda," Engineering Education, ASEE, 78(11), 151-155.

Focht, J. (1989). Presentation made to civil engineering faculty and students at Texas A&M University, 26 September.

Florman, S. C. (1989). The civilized engineer, St. Martin's Press, New York, NY.

Luberhoff, B. J. (1989). "It's an interdisciplinary world," Engineering Education, ASEE, 79(4), 463.

Pletta, D. H. (1987). "Uninvolved professionals and technical disasters," Journal of Professional Issues in Engineering, ASCE, 113(1), 23-31.

Samson, C. H. (1989a). "Engineering education - focus on the future," The Bent of Tau Beta Pi, 80(2), 31-33.

Samson, C. H. (1989b). "Engineering education - where do we go from here?" Proceedings, ASEE Southeastern Section Meeting, Tuscaloosa, AL, Apr. 2-4, 98-102.

Troxler, G. W. (1989). "Engineering in the '90s: strength in diversity," Engineering Education, ASEE, 79(1), 25-30.

Van Valkenberg, M. E. (1989a). "Unjamming our curriculum," Engineering Education, ASEE, 79(3), 456.

Van Valkenberg, M. E. (1989b). "Are we ready for top-down curricula?" Engineering Education, ASEE, 79(4), 524.

Wenk, E., Jr. (1988). "Social, economic, and political change: portents for reform in engineering curriculum," Engineering Education, ASEE, 78(10), 99-102.

Willenbrook, F. K. (1988). "The need for career-long education," Engineering Education, ASEE, 78(10), 90-91.

REEXAMINATION OF CE CURRICULUM FOR THE 21ST CENTURY

Clinton E. Parker,[1] M. ASCE, Guy E. Jester,[2] M. ASCE, Chandra S. Desai,[3] F. ASCE, and James T. P. Yao,[4] F. ASCE

Abstract: The need for the civil engineering profession to restructure the curriculum and make it more relevant to the modern technological world is addressed. Criteria and qualifications required of modern civil engineers are delineated and elements essential to the reorganization and restructuring of the curriculum to assure that civil engineering graduates are prepared to perform challenging tasks within the complexity of a changing modern society in the 21st Century are presented.

Introduction

The civil engineering profession is changing rapidly. Consequently, the civil engineering curriculum, the way course material is organized and taught must be closely examined and carefully reorganized to provide modern civil engineers with the qualifications required to meet challenges of the 21st century.

Opportunities and problems of increasing complexity will offer challenges requiring a different engineering education than offered today. Therefore, it is both important and timely to examine restructuring of the current curriculum and to plan for future educational programs in the civil engineering profession. Restructuring cannot be simply reordering or making cosmetic changes in the contents of existing courses, and adding or deleting of a few courses. In order to address the problems of today and tomorrow in a realistic manner,

---

[1]Prof. and Chairman, Dept. of Civil Engineering, Univ. of Texas at Arlington, Arlington, TX 76019
[2]Vice President, J. S. Alberici Construction Co., Inc., St. Louis, MO 63121
[3]Regents' Prof. and Head, Dept. of Civil Engineering, Univ. of Arizona, Tucson, AZ 85721
[4]Prof. and Head, Dept. of Civil Engineering, Texas A & M Univ., College Station, TX 77843

the restructuring must be substantive, fundamental, and in some aspects radical.

Civil Engineers of the future should be able to specialize in their areas of interest for professional considerations. However, their overall education must prepare them to think and understand basic theories, unified analysis and design and applications as "whole" or "total" problems. Thus, they must be taught to treat the problems in a "holistic" manner, rather than their independent components.

In November 1987 the American Society of Civil Engineers (ASCE), with support from the National Science Foundation, held a workshop, Civil Engineering In the 21st Century, to examine the profession. Hall et al. (1989) reported on the issues examined at the ASCE workshop. In their report they envisioned that "The very nature of civil engineering practice and construction will be totally changed through the application of information technologies" and that "The future civil engineer, in many cases, will be a systems professional" because "Full vertical integration of projects will combine conception, design, production, maintenance and use into a single integrated function".

Grayson (1989), President of the American Society for Engineering Education, emphasized the need for rethinking how engineers are educated. The Western-Southwestern Civil Engineering Department Heads, at their 1989 annual meeting, submitted to ASCE a resolution expressing the importance and urgency of reexamining our curriculum. At the 1989 International Forum on Engineering Education, the President of the National Society of Professional Engineers (Chapple 1989) expressed the need for "--- a new breed of engineer to meet the challenges of a true worldwide economy in the 21st century".

The objective of this paper is to identify curriculum criteria and qualifications required of modern civil engineers. Philosophical ideas and concepts essential to the reorganization and the revitalization of the curriculum are presented. In addition, we suggest how the overall curriculum may be reorganized and reshaped to incorporate needed material so that civil engineering graduates are prepared to deal with new realities and new challenges of the future as outlined by recent views and reports.

Examination of CE Curriculum

Jester (1989) concluded from a review of civil engineering curricula in 1968, 1978, and 1988 that they were virtually the same and current graduates are not as well prepared as they should be because "--the curriculum has not changed as rapidly as society and the profession at work--". He observed that engineering courses are taught as a "collection of boxes of subject matter". An independent study by Birdwell (1988) supports Jester's findings. He compared the current curriculum at Texas A & M University with the curriculum of the late forties and found them to be quite similar.

Hall et al. (1989) expressed a concern for the amount of material that is attempted to be covered in a four-year curriculum. They reported that "-- there is an expanding interest on the part of educators, students and practitioners to include newer technologies--in the civil engineering curriculum". Also, a concern was expressed about the lack of preparation of graduates for dealing with public policy. Grayson (1989) stated his concern for leaving areas of knowledge outside the technical arena "--to be learned in a loose series of humanities and social science courses" and students' lack of understanding cultural and ethical values.

Although current criteria for accreditation requires an integrated educational experience, most curricula are being offered as a block of courses in topical areas. Faculty members have become more specialized and are busy with their research interests. Consequently, few if any faculty members are motivated to look at the "whole" or willing to take the necessary time to reexamine the best way to integrate new technologies into the curriculum. Professionals outside of the education community see deficiencies in graduates and advocate additional courses or subject matter but do not see how anything can be removed from the program. The general reaction to external pressure to changes in the curriculum (e.g., the state legislature or university or college administration may mandate additional courses or a reduction in credit hours for graduation) is often one of compromise. Reorganization of the curriculum often becomes a "bargaining" process resulting in some "fair" distribution of the total reduction required. As a result of this process, we have lost insight as to what the total educational experience should provide.

The civil engineering curriculum has evolved into what it is today because it is usually easier to delete or add a course than to identify a common thread with which it can be integrated into a logical series of courses. In many cases we have become oblivious to common fundamental science and engineering principles that underlay and cut across areas of specialization and other engineering disciplines. Because of our specialization and our lack of communication with others both within civil engineering and other fields of engineering we may not know the detailed course content of courses we ourselves do not teach. For example, the instructor of a structural design course may not be up-to-date with the contents of courses on pavement design, dam design, foundation design or wastewater treatment plant design. As a result the student (or even the professors) may not see the common threads that link the same subject matter as taught in different areas of specialization. The same is true for engineering science and analyses courses. For example, fundamental principles applicable to structural analyses or fluid mechanics are not seen by the student to be applicable in electrical science. The net results are that (a) similar fundamentals are not reinforced, (b) differences are not explained and (c) the curriculum becomes unnecessarily cluttered with repetition of the same material in different forms.

Required Criteria and Qualifications

Close examination of recent views and reports focusing on engineering education reveals that a consensus on required educational criteria and qualifications for future engineers has developed. Based on the vision of the future shared by both practitioners and educators within the profession not only must the curriculum be redesigned but also the way we deliver the material and the context of the delivery require a fundamental change.

The report of the National Science Foundation (1989) on the 1988 NSF workshop on undergraduate engineering education delineated goals for the educational process: (1) integrative capability; (2) analysis capability; (3) innovation and synthesis capability; and (4) contextual understanding capability. It points out that the accomplishment of these goals will contribute to the "ability to learn on a career-long basis". The report states, "Present and future demands of engineering practice and societal participation require engineers to have well developed learning skills and to be adaptable to rapid technological change as well as change in the global marketplace" and recommends "sweeping changes in engineering curricula". Among the recommendations are "--programs that lead to a better understanding of foreign cultures and languages and the practice of engineering on a global scale" and "--focusing on the integrative nature of engineering to serve the -- purpose of starting an engineering student's education with the core methodology of the engineering process--".

Jester(1988) suggests that the entire curriculum be redesigned starting from scratch. He calls for "--a fundamental change in the way we conceptualize the engineering sciences, as well as engineering analysis and design" and "--courses based on teaching concepts rather than on a narrow subject matter and data bases". He emphasizes the need to present themes as a unified body of knowledge and the necessity for more breadth of coverage in engineering science courses and the understanding of interdisciplinary concepts. He states, "Creative design, systems perspective, management and communication principles should permeate the entire engineering curriculum".

In the ASCE workshop report, Hall et al. (1989) point out that "Fundamentally different approaches are necessary to be successful and competitive in the future". An example of the technological change that is emerging is the creation of new materials that must be "--designed and produced as specific solutions for individual applications". The authors call for an engineer to be knowledgeable about a diversity of technologies and institutions, with a worldwide orientation and capable of functioning in a global community, and with a perspective beyond traditional technical boundaries. They note that these professionals must be prepared for lifelong learning and capable of participating and directing an integrated team approach to obtaining engineering solutions.

Based on the article by Chapple (1989) covering the ABET sponsored International Forum on Engineering Education, the participants were able to identify many of the causes of this country's competitiveness problem and, to a large degree, placed the blame on all levels of education. With regard to engineering education, a technically competent "new breed" of engineers educated in behavioral science, management, economics, and international societies and cultures was a focus of the discussion. Also, the need for enough breadth in engineering education so that engineers can function in political, social, economic, fiscal, technical and family environments and can continue to enhance their education through lifelong learning was advocated.

Thus it is of paramount importance that the curriculum criteria upon which the profession must focus its attention be delineated. The curriculum criteria must address more than the topical matter of course content. It is absolutely essential to bring into the curriculum continuously, in a dynamic manner, all possible and relevant scientific and technological developments, such as information and computer technology, innovative utilization of existing and new materials, uncertainty and risk analysis and management technologies, so that the student is prepared to solve the problems of today and tomorrow.

The student should be taught, at an early stage, how to conceptualize CE problems. We should emphasize the need for human thinking, judgment, and creativity by using innovative methods in presenting material. As most problems involve interrelationships of a wide range of disciplines, the civil engineer should be educated to address the whole problem as a system. The conception, planning, analysis, design, and operation of engineering systems may require the use of the principles of physics, mechanics, chemistry, mathematics, computers and materials science as well as engineering. However, any amount of theoretical knowledge by itself will not be able to provide the fuller comprehension of complex problems that civil engineers will face in the 21st Century, nor will it alone lead to their solutions.

Experience and feel for the response of CE systems and materials composing them are vital. Hence, the student must be provided with sufficient hands-on experience in laboratory and experimental techniques. The lack of these experiences appear to be one of the major drawbacks of the education of the U. S. students. Yet, it is feasible to integrate experimentation and demonstration in much of the basic and engineering fundamentals course work with little effort. Course work should be presented with a system prospective, minimizing lecture and emphasizing analysis of the total problem, integrating experimentation and design.

Civil engineers must interact increasingly with modern society. This interaction involves changing social, economical and political factors, as well as special attention to cultural, ethical and legal considerations. Modern technology and its products straddle intricately the boundaries of various countries. Because of this, the student must be introduced to the culture, languages, and socioeconomic conditions of the world. It is desirable to guide engineering students to read, discuss, and learn more about literature, arts, philosophy and metaphysics so that they can deal with people and society as intelligent well rounded individuals.

Reorganization of Curriculum

To meet the required criteria, we must challenge the underlying assumptions and purposes of the present civil engineering curriculum and develop a reasonable goal that recognizes the needs of the present world in which we live as well as the requirements of the 21st Century. In addition to the development of the ability to apply pertinent knowledge to the practice of engineering, engineering programs will have to involve the broad educational objectives expected in modern post secondary education.

Although the present curriculum can be used as a starting point from which to build a completely new curriculum based on a unified system approach, there must be a fundamental change in our conceptualization of the engineering curriculum -- a complete transformation that will require the definition of new disciplines and completely redesigned courses as the building blocks (Jester, 1988). We are not proposing here to "load up" the curriculum with additional courses. We are proposing a fresh look at the topical subject areas included in the curriculum, a change in the way material in the subject areas is taught and bringing to the curriculum material that is essential for the practice of engineering in a changing technological, multiculture world environment. We strongly believe that vertical integration of the curriculum will eliminate unnecessary duplication of material, provide the student an understanding as to how many fundamental principles cut across traditional disciplines and enable graduates to analyze engineering problems and propose solutions based on the "total" environment.

The fundamental objective of an effective curriculum should be to integrate the backgrounds in basic subjects and socioeconomics and humanities into concepts and analysis pertinent to the design of solutions for civil engineering problems. Basic subjects should involve a thorough study of mathematics, natural and engineering sciences, interdisciplinary areas such as topics from computational methods, probability and stochastic methods, systems engineering, and concepts for problem solving. Topics from innovative and emerging technological developments and experimental techniques should also be included. Socioeconomics and humanities should involve appropriate study of philosophy, literature, art, the social sciences, management, technology and society, cultural, ethical and legal issues, and global interrelationships.

Knowledge of basic theories, concepts and models, and socioeconomics and humanities should be integrated so as to obtain successful solutions of various CE problems. Basic theories of unified analysis and design must be taught followed by their application to specific problems. Concepts thus learned should culminate in a final stage of unification in which problems treated as systems involving specific components are solved. Frontiers of civil engineering such as new materials, influence of global changes, environmental concerns, new construction in space, oceans and polar regions, and infrastructure problems constitute a important set of subjects that should be addressed.

The following addresses specific areas within the curriculum and are offered as starting points for reorganization and restructuring.

Computer Usage: The use of computers must be integrated in all applicable courses as an information and analysis tool not an adjunct or added attraction. Their usage must become second nature to the student not just in their engineering and scientific courses but also in their liberal arts and business courses. Hands on use of computers must start from day one and be part of the entire educational experience, especially in laboratory, analysis and design courses.

Humanities and Social Sciences: The humanities and social sciences should challenge the student and be relevant to society as a whole. They should be fully integrated into the curriculum and offer substance to the total understanding of man's environment and relationship with others. Human history and human behavior are important elements. Human history should include history of technology (technological advances through the ages, including the role of technology in a modern society); and dilemmas of a modern technological society, i.e., depersonalization, abstraction and dehumanizing effects. Human behavior should encompass ethical conduct, professional judgment and integrity; values of people from different callings in a global society; general and specific cultural values; and values typical of various professions.

Communications: Although effective communication must be emphasized throughout the curriculum the student should also be provided with a comprehensive program that includes: effective writing; analytical reading; critical listening; persuasive speaking; function of languages (attitudes and values); languages of other cultures; and awareness of verbal and visual scientific statements. Speaking, writing, listening and reading skills must be imparted very early in the curriculum and required as graded exercises in all courses throughout the educational experience.

Engineering Sciences: The traditional approach to engineering science does not provide the student an understanding of the common fundamental science and engineering principles that cut across engineering disciplines. In fact, because of the way engineering sciences are currently taught, student are lead to believe at an early critical stage in the educational process that the various disciplines share very little in common. This must be changed if we are to produce graduates that are capable of working beyond traditional technical boundaries and if we are to provide a basis for lifelong learning.

The engineering sciences should be restructured so that fundamental principles common to traditional disciplines will constitute a unified body of knowledge showing commonalty. This may require the rewriting of text books. It will definitely require the engineering faculty to reexamine the way they teach and to view engineering as a whole and not as fragmented elements that support a narrow interest. Three distinct actions are required in the restructuring of these courses. First, a matrix of the fundamental principles in all engineering science courses should be prepared to identify those principles in currently offered courses that are common and are repeated. Second, we need to examine the content of the subject matter that uses common principles and develop courses to present this content in a context that cuts across all the traditional fields of engineering. Third, we should introduce a core methodology of the engineering process and integrate into it the conceptualization of problems and solutions that use the principles taught in engineering science.

<u>Analysis and Design</u>: Courses can no longer be separated into "this is analysis and this is design". Nor can civil engineering courses that rely upon the same principles of analysis and design continue to be taught by area of specialization. The basic principles of analysis and design must be presented in a unified manner. Courses we now teach identified as an area of specialization should be examined and reorganized using essentially the same approach as suggested under engineering science. New courses should be developed that utilize common analysis and design principles and examples should cut across all areas of specialization. Material design should be integrated into the courses and become a part of a unified systems design approach.

<u>Conception, Synthesis, Planning and Design</u>: Although in all engineering courses the student should be taught to look at the "whole", it is of course impossible to bring together all aspects into a fully integrated project until toward the end of the terminal semester of the degree program. Here concepts learned can be

brought together in the final stage of unification in which problems treated as systems involving specific components are solved. Problems should be analyzed and solutions generated in a total system context. Political, economic, social and cultural, as well as technical, aspects of specific problems should be recognized as both part of the problem and part of the solution. They must be dealt with as part of the system.

It will not be an easy task to put all the subject areas together so that the basic blocks build on each other while integrating the needed laboratory and computer experiences, team building, communication skills, systems analysis and design necessary for treating problems in a "holistic" manner. It will take the full commitment of the profession and strong dynamic leadership to initiate changes needed in both course material and the way it is presented.

Summary

The basic Civil Engineering Curriculum should provide the graduate with an understanding of themselves and of the world in which they live (Grayson 1989), in addition to providing the framework for a lifetime of learning from experience, study and additional education. More than being technically competent is required. The graduate must also be able to operate in a highly technical world with a global economy in which most decisions are based on political and/or economic considerations. This requires continuous interaction with personnel of other professions and skills, people of other cultures and sociological outlooks as well as politicians, corporate decision makers, architects, and engineers with other technological specialties. As one author wrote, "New hires must be technically competent, they must be behavioral scientists, and they must be pragmatic managers all rolled into one package" (Chapple 1989).

For the most part, introduction of topics or subjects presented here are not new to Civil Engineering and, as a result, can be easily agreed upon as worthy of inclusion in the curriculum. The challenge is to integrate course material across subject areas and design new courses and new approaches of instruction so that the student can develop the fundamental tools and disciplined thinking early enough to conceptualize engineering problems and develop logical solutions while, at the same time, develop the leadership skill necessary to fully integrate project analysis and design in a complex multiculture world.

Is it possible to meet all of these requirements? Yes, with completely redesigned courses and a restructured curriculum we should be able to approach these ideals. Because of already an overcrowded curriculum, this will indeed be a difficult task. At issue will most certainly be the question of a four- or five-year program and whether the bachelor of science should be the first professional degree. This issue is not addressed here, however, it is one that needs to be closely examined in our pursuit of a curriculum that can serve the profession well into the 21st Century. Regardless of the length of the program(s), restructuring of both courses and curricula is a necessity. Prior to full implementation of an entirely new program some experimentation will be required. It is hopeful that we are not so rigid that exemplary curricula or courses can be developed and experimented with in a number of universities without jeopardizing accreditation.

Appendix. References

Birdwell, J. R. (1988). private communication with J. T. P. Yao. Texas A & M University, College Station, TX.

Chapple, A. (1989). "ABET Forum: Engineering Education Gets Well-known Diagnosis, But Cure Elusive." Engineering Times. National Soc. Prof. Engineers, 11(1), 1-3.

Grayson, L. P. (1989). "Officers Report: It is Time to Rethink Engineering Education." Engineering Education News. Am. Soc. Engineering Educ., 15(11), 3.

Hall, W. J., et at. (1989). "Civil Engineering in the 21st Century: A Vision and a Challenge for the Profession. Am. Soc. Civil Engineers, New York, N.Y.

Jester, G. E. (1988). "The Curriculum for Civil Engineers of the Future: A Practitioner's Viewpoint." Presented at the Engineering Education Breakfast at the Annual ASCE Convention and Exposition, St. Lewis, MO.

"Report of the 1988 NSF Workshop on Undergraduate Engineering Education." in Report on the National Science Foundation Disciplinary Workshops on Undergraduate Education (1989). National Science Foundation, Washington, D. C.

"Resolution on Curriculum" (1989). Western-Southwestern Region Civil Engineering Department Heads Meeting, Tucson, AZ, April 8-10.

Essential Legal Study for the Civil Engineer

Mark S. Nadler, P.E.[1]
James E. Lake, P.E.[2]

I. Abstract

There is a need for an introductory course of legal study in the civil engineering curriculum. In this paper, the authors, both of whom are licensed professional civil engineers who practiced in the design and construction management fields prior to attending law school, assert the need for the course and suggest its topical outline.

II. Introduction

Engineers, as applied scientists, occupy a unique and valuable position in our society. They bridge the gap between theory and reality by solving technical problems through the application of scientific principles. Engineers prepare for this unique and important position by extensively studying science and technology. To better prepare the engineer for his or her professional career, it is essential that the engineer have at least a basic understanding of the legal framework within which he or she will operate.

One would not think of obtaining a driver's license without learning the traffic laws. Yet, engineers routinely obtain licenses to practice engineering armed with an impressive array of scientific knowledge, but little knowledge of the laws that

---

[1]Mark S. Nadler, P.E., Member Washington State Bar, Associate ASCE, Bogle & Gates, Two Union Square, 601 Union Street, Seattle, WA 98101.

[2]James E. Lake, P.E., Associate ASCE, James E. Lake, P.E. Consulting Civil Engineer, 6824 19th Street West, #108, Tacoma, WA 98466.

profoundly affect their professional life. In practice, engineers routinely encounter the law in many forms. They deal with contracts of employment, manage and perform contracts for professional services, manage and perform construction contracts and deal with administrative agencies.

In order to provide the engineer a basic grasp of the areas of the law which have such a major impact on almost every aspect of the engineer's practice, the engineering student should be offered the opportunity to take an elective course that provides a survey of the fundamental areas of law that the practicing professional is likely to encounter. These fundamental areas include contract, business, tort, administrative and property law. The following discussion suggests such a curriculum.

III. Contract

A. Contracts are ubiquitous. In a sense, contracts are to the business world what molecules are to matter. Every business is nothing more than a series of contracts built on other contracts. Every person that practices engineering makes and performs contracts. In fact, many engineer's entire practice consists of writing contracts, advising clients regarding the effect of the contracts they write, and supervising the work performed under their contracts. The design, construction engineering and inspection of almost all civil engineering projects involves these contractual elements. Consequently, it is of obvious importance for civil engineers to have a basic understanding of what a contract is and how it works.

B. The essence of a contract. A contract is an offer to do something, which someone has accepted, accompanied by a promise to do something in return, where the promises are among those which society has decided to enforce. The life of a contract can be conceptualized as having two phases: formation and performance.

A law course for the engineer should begin with the formation of a contract, which is simply the acceptance of an offer. This topic should deal with the aspects of contract formation as they usually arise in the engineering context including the competitive bidding process, professional services contract negotiations and employment contracts.

Next, the course should address contract performance including modifications (change orders).

contract performance is the phase in which the parties act to fulfill their respective promises. Unfortunately, many contractual relations do not survive the performance stage without difficulty. When a party fails to perform its obligations under a contract, it may be considered to have breached, thereby giving rise to certain remedies for the performing or non-breaching party.

In many instances the rights of the non-breaching party depend upon whether the breach was material or non-material. Frequently, engineers acting as project managers, engage in negotiations with the contractor either prior to or concurrent with the occurrence of the breach. Consequently, the course should give the engineer a sense of what may constitute a material breach, so that the engineer will be in a position to conduct informed negotiations.

The course should also provide a brief introduction to the various types of remedies available. Remedies include damages, rescission and specific performance. Damages vary by jurisdiction, ranging on one hand from awarding the non-breaching party a sum of money intended to put that party in the position it would have been in if the contract had been performed, to the other hand where the remedy is restoring the non-breaching party to the position it would have been in if it had never entered the contract at all. In some jurisdictions, punitive damages may also be available. Rescission is the annulment of the contract, and specific performance requires the breaching party to fulfill its promised performance.

C. Miscellany. Along these same lines, the engineer ought to be introduced to other subjects such as substantial completion, mistake, the parole evidence rule, the statute of frauds, the Miller Act, and subcontracting.

IV. Business Law

This topic should build on the basic concepts of contract by presenting business structures and sales transactions. Under this heading the engineering student should learn about the basic business organizational structures that the engineer is likely to encounter in practice, including the sole proprietorship, partnership, the corporation and the professional services corporation. The concept of agency which is applicable to every form of business structure should also be introduced.

A. Organizational form. Many engineers in private practice, practice alone or with only a few others. In this setting they frequently utilize either the sole proprietorship, partnership, or professional services corporation organizational form. Although these may provide tax advantages to the owners of the business, they also provide, in many cases, a means of making the various owners liable for the malpractice or other tortious acts of any of the other owners and employees. This potential liability places not only the engineer's investment in his or her firm at risk, but also all of the engineer's personal assets. Understanding the limitations and advantages of the various organizational forms will enhance the engineer's ability to choose between them.

Many engineers will inevitably wind up in management. A basic understanding of the corporate form would enhance the engineer's ability to meet this responsibility, especially for those working in large manufacturing and industrial companies.

B. Agency. As business managers, engineers must understand the basic concepts of agency. An agent acting on behalf of a principal can make the principal liable to third parties. It is important that the engineer as a potential principal, realize that although agents frequently have <u>true</u> authority, an agent without <u>true</u> authority can bind a principal using <u>apparent</u> authority without the principal's knowledge or permission. This may be especially troublesome in the instance of engineers in private practice, because every owner or employee is an agent of the firm and the other owners may be personally bound by the agent's acts. The same is true of engineers acting as the owner's representative in construction management.

C. Sales. The area of sales contracts is somewhat specialized and would likely be of interest primarily to engineers intending to work in a commercial or industrial setting where they have responsibilities for procurement or sales. The blue print for this area of law is the Uniform Commercial Code.

V. Tort

A. Torts for Engineers. Tort makes you responsible, and in some instances liable for what you do and for what others do when acting in your behalf. A cursory understanding of the underlying concepts of tort is essential for those who employ others, those who provide professional services to others (and are therefore potentially subject to malpractice liability), and those who design products which are sold to the consuming public (products liability).

B. Fault-based tort. In its simplest sense, all fault-based tort, including negligence, consists of an actor's duty to another party, which the actor breached, thereby proximately causing damages to the other party.

This is the blueprint for malpractice, which is merely a form of negligence. The engineer has a duty to his or her clients to practice in a non-negligent fashion. Thus, the engineer should be given an idea of what negligent practice is if he or she is to avoid practicing negligently. The course should explain that negligence is only a label that may be placed on the engineer's work after it has been judged. The engineer's work will be judged according to the standards of the practitioners in the field and geographic area in which the alleged malpractice took place, even if it is a discipline other than that of the engineer (for example, a civil engineer designing power distribution grids may be judged using the same standards as an electrical engineer doing the same work).

C. Product Liability. It seems only reasonable that designers and manufacturers may be found liable for damages caused by their negligently designed or manufactured products. However, it may come as a surprise to find they can also be found liable <u>without fault</u> for damages their products cause, in those instances where the courts decide that social policy requires it. The engineer is entitled to at least an introduction to this notion of strict liability.

D. Miscellany. The engineer should be introduced to other areas of tort including the concepts of *respondeat superior*, trespass, nuisance and the so-called toxic torts.

Respondeat superior is the legal doctrine which makes employers liable for the torts of their employees. Engineers should understand that in many instances they could be liable for the negligence and other torts committed by their employees. These may include not only their employee's malpractice, but other torts such as negligence, trespass or, even assault.

Trespass and nuisance are of special interest to engineers involved in construction management, because they frequently assist in obtaining easements for construction work, and they subsequently supervise the work within those easements. These subjects are also important in the context of hazardous waste sites.

VI. Administrative Law

An administrative agency is any department, bureau, administration or other entity operating under the executive branch of either the state or the federal government. For example, the Board of Registration for Engineers and Land Surveyors, and the Environmental Protection Agency are administrative agencies. Administrative law deals with the procedural requirements imposed on these agencies, and the underlying substantive regulations they administer. Every engineer should be introduced to the concepts of administrative law because every engineer will deal with an administrative agency at some point in his or her career.

Although their structures may vary, all agencies are created by a legislature through the passage of the agency's organic statute which sets forth the mission and structure of the agency, and empowers the agency to act. The manner in which agencies may act is set forth in the Administrative Procedures Act (APA) and its variations. The APA sets forth two headings for the agencies' activities: "rule making" and "adjudicative procedures."

It is in their rule making mode that agencies promulgate regulations (which carry the force of law). Adjudicative procedure, on the other hand, includes all other agency decision making, such as decisions to grant or deny a license or permit. Most of these decision making activities involve public input including significant input from the engineering community, especially where the agency is one which deals with technical subjects such as the EPA. It is important that engineers understand that they have a right and a responsibility to work with the agencies, and in so doing, they have the opportunity to tangibly shape the regulations that profoundly affect much of our lives.

VII. Property.

Property law may be divided into two basic categories, real property and personal property. Real property deals with land and interests in land. Personal property includes all other forms of property including intellectual property (patents and copyrights), certain intangible rights (such as allocations of available air shed), and all tangible items other than land.

Property law, especially real property descended from the English feudal system and seems almost anachronistic because it contains aspects that are steeped in antiquity and may appear to make little sense today. Consequently, much of it is difficult to understand unless one realizes that its various aspects arose in response to problems that may no longer exist. Once these legal solutions are adopted, it is difficult to change them because property law must remain stable in order for society to function (as one can imagine, a lack of certainty in property law would make it impossible to invest the huge sums of money necessary for development). Consequently, this area of law is slow to change.

There is considerable variation in real property laws from state to state, but certain underlying concepts emerge. The engineering student should be introduced to these fundamental concepts because civil engineers routinely become involved in making improvements to real property. The introduction should include the concepts of eminent domain, adverse possession, easements, covenants and licenses, each of which carries a specific legal meaning.

Although perhaps not as important, at least a passing understanding of the basic concepts of personal property would also be beneficial.

VIII. Conclusion.

The authors of this paper do not advocate turning the engineering curriculum into a curriculum in legal studies. We do not even assert that the curriculum we have outlined above should be a required course. We merely point out what we, as practicing professional engineers and attorneys, perceive as a gap in the engineering curriculum that we studied. Our own experiences prove that our performance as engineers would have been enhanced if we had had a stronger legal background. Consequently, we suggest that the civil engineering curriculum would be improved if engineering students were allowed the option of being introduced to the rules they will be expected to follow as practicing professionals.

BACCALAUREATE CET PROGRAMS: WHERE & WHY ?

David E. Hornbeck, Member, ASCE [1]

Abstract

Baccalaureate Civil Engineering Technology (BSCET) programs were developed during the early 1970s and have continued to exist since that time. Employers have utilized BSCET graduates successfully, as reported in broad-based studies (Keating, 1984). The investigation reported in this paper attempted to determine if a clear philosophy, unique to BSCET and distinct from civil engineering (BSCE) and associate-degree civil engineering technology (ASCET), has been accepted by engineering practitioners and graduates.

Introduction

Four aspects of BSCET education were examined:

1. co-existence with BSCE curricula,
2. influence of PE registration requirements,
3. geographic distribution of BSCET programs, and
4. use of graduates by employers.

Procedure

Co-existence with BSCE curricula was investigated using data from the Engineering Manpower Commission (1988, 1989) and population estimates (U.S. Bureau of the Census, 1989) to determine a per-capita distribution of graduates. For calculation purposes, the mean annual production of graduates for two years, 1987-88, was used. In the analysis, state-by-state data were sorted by number of BSCE graduates per capita, by number of BSCET graduates per capita, and by the combined BSCE+BSCET graduates per capita. For geographic distri-

---

[1]Professor of Civil Engineering Technology, Southern College of Technology, 1100 South Marietta Parkway, Marietta, GA 30060-2986

butions, population estimates and regional divisions of the U.S. Bureau of the Census (1989) were used. Since it has engineering schools and PE registration, the District of Columbia was treated as a state.

State engineering registration requirements were obtained from the National Council of Engineering Examiners (1988) and from either questionnaires or telephone calls to the state boards of registration.

The use of BSCET graduates by employers was investigated through questionnaires to employers of BSCET graduates of the Southern College of Technology since June, 1983; some respondents also employed graduates from earlier years. This information was supplemented by questionnaires to graduates in 1986 and 1989.

Results

Summaries of results of the surveys, including job profile information, are shown in Tables 1 and 2.

Table 3 shows distribution of graduates per capita in each of the 50 states, plus District of Columbia, which are ranked in order of production of BSCE graduates. Table 3 also indicates availability of PE registration for BSCET graduates. Table 4 shows regional per-capita production of graduates and status of PE registration. The ten regions in Table 4 were derived from nine regions defined by the U.S. Bureau of the Census (1989).

Graduate Surveys Mailed	168
Responses Received	52
PE Required for Advancement	33%
Currently have EIT status	23%
On a scale of 1-10, how closely does your job relate to your education ?	8.2

Areas in which you frequently work:

Project Planning	63%	Site Development	46%
Construction	60%	Highways	40%
Estimating	56%	Constr. Materials	33%
Design	54%	Water/Wastewater	33%
Surveying	50%	Drafting	29%
Management	50%	Structures	25%

Table 1: SUMMARY OF SURVEY RESPONSES, BSCET GRADUATES, 1986-1989

Employer Surveys Mailed --------	119		
Responses ---------------------	41		
Grads. Employed by Respondents--	177		
PE Required for Advancement ----	15%		
PE Desirable for Advancement ---	68%		
On a scale of 1-10, how closely did BSCET education relate to jobs in your organization ? --------------------------			8.1
Level of responsibility assigned BSCET compared to BSCE working for the same company:		Higher ------------	6%
		Same --------------	78%
		Lower -------------	9%
		Different Areas ---	7%
Fraction of engineering work in your organization performed by BSCET graduates --------------------------			81%
Areas in which BSCET Graduates were utilized:			
Design -------------	63%	Highways -----------	37%
Project Planning ---	56%	Structures ---------	27%
Estimating ---------	51%	Water/Wastewater ---	24%
Construction -------	44%	Management ---------	22%
Surveying ----------	46%	Repair & Maint. ----	17%
Site Development ---	39%	Geotechnical -------	15%

Table 2: SUMMARY OF SURVEY RESPONSES, EMPLOYERS OF BSCET GRADUATES SINCE 1983

Observations

There were 35 BSCET programs reported in 22 states in 1988; eight states had two or more programs. Five states (PA,GA,MA,CO,OR) accounted for 49% of the national total of BSCET degrees.

No BSCET programs were reported in 29 states, including the ten states having the highest per-capita production of BSCEs and including five of the ten states having the lowest per-capita production of BSCEs.

National production of BSCET graduates is 2.4 grads/yr/M.pop., representing 7% of the total national production of BSCET+BSCE graduates.

BSCET graduates raised the per-capita total of BSCE+BSCET graduates significantly in states shown in Table 5. The largest proportion of BSCET graduates was in Georgia, where 42.8% of the BSCE+BSCET degrees were in Civil Engineering Technology.

STATE	Populat. (x 1000)	BSCE/ M. pop.	BSCET/ M. pop.	CE+CET/ M. pop.	PE Avail.
DISTR/COLUMBIA	622	143.9	0.0	143.9	YES
NORTH DAKOTA	672	120.5	0.0	120.5	YES
SOUTH DAKOTA	709	105.8	0.0	105.8	YES
DELAWARE	644	93.9	0.0	93.9	YES
ALASKA	525	90.5	0.0	90.5	YES
WYOMING	490	84.7	0.0	84.7	YES
UTAH	1,680	69.0	0.0	69.0	YES
VERMONT	548	64.8	0.0	64.8	YES
MONTANA	809	64.3	0.0	64.3	YES
NEW HAMPSHIRE	1,057	57.7	0.0	57.7	YES
MASSACHUSETTS	5,855	55.5	10.6	66.1	YES
NEBRASKA	1,594	52.1	0.0	52.1	YES
COLORADO	3,296	51.4	11.1	62.5	YES
NEW MEXICO	1,500	50.7	5.0	55.7	YES
OREGON	2,724	48.5	16.0	64.4	YES
LOUISIANA	4,461	47.5	0.0	47.5	YES
WASHINGTON	4,538	46.9	0.0	46.9	YES
CONNECTICUT	3,211	45.6	0.0	45.6	YES
RHODE ISLAND	986	43.6	14.2	57.8	YES
HAWAII	1,083	43.4	0.0	43.4	YES
WISCONSIN	4,807	41.7	0.0	41.7	YES
TENNESSEE	4,855	41.7	2.4	44.1	NO
VIRGINIA	5,904	40.7	3.3	44.0	YES
IOWA	2,834	36.5	0.0	36.5	YES
INDIANA	5,531	36.1	0.3	36.3	YES
WEST VIRGINIA	1,897	35.8	10.8	46.7	YES
MAINE	1,187	34.1	0.0	34.1	YES
SOUTH CAROLINA	3,425	33.3	5.1	38.4	NO
PENNSYLVANIA	11,936	33.0	7.1	40.1	YES
MISSOURI	5,103	32.9	0.4	33.3	NO
ALABAMA	4,083	32.8	3.3	36.1	YES
TEXAS	16,789	32.2	0.5	32.7	YES
KANSAS	2,476	31.9	0.0	31.9	YES
NORTH CAROLINA	6,413	31.4	3.7	35.1	YES
ARIZONA	3,386	30.9	3.2	34.1	YES
MICHIGAN	9,200	30.1	0.0	30.1	NO
IDAHO	998	29.6	0.0	29.6	YES
OHIO	10,784	29.4	1.7	31.1	YES
CALIFORNIA	27,663	28.9	0.2	29.1	YES
NEVADA	1,007	27.3	0.0	27.3	YES
NEW YORK	17,825	25.5	1.8	27.3	YES
NEW JERSEY	7,672	23.7	2.5	26.3	YES
FLORIDA	12,023	23.0	0.8	23.9	NO
MISSISSIPPI	2,625	21.5	0.0	21.5	YES
OKLAHOMA	3,272	18.9	0.0	18.9	YES
ILLINOIS	11,582	17.9	2.3	20.2	NO
MARYLAND	4,535	17.2	0.0	17.2	YES
GEORGIA	6,222	14.7	11.0	25.7	YES
KENTUCKY	3,727	14.4	3.4	17.7	NO
ARKANSAS	2,388	14.2	0.0	14.2	YES
MINNESOTA	4,246	13.9	0.0	13.9	NO
Nat'l Totals:	243,399	32.5	2.4	34.8	

TABLE 3: STATE-BY-STATE PER-CAPITA ANNUAL BSCE & BSCET GRADUATES, 1986-88

REGION	Popul. (x 1000)	BSCE/ M. Pop.	BSCET/ M. Pop.	CE+CET/ M. Pop.	States w/o PE
New England (CT,ME,MA, NH,RI,VT)	12,844	50.7	5.9	56.6	
Middle Atlantic (NJ,NY,PA)	37,433	27.5	3.7	31.2	
E. North Central (IL,IN,MI,OH,WI)	41,904	28.6	1.1	29.8	2
W. North Central (IA,KS,MN,MO, NE,ND,SD)	17,634	36.8	0.1	36.9	2
Upper S. Atlant. (DE,DC,MD,VA,WV)	13,602	39.4	2.9	42.3	
Lower S. Atlant. (FL,GA,NC,SC)	28,083	24.4	4.3	28.6	2
E. South Central (AL,KY,MS,TN)	15,290	29.2	2.5	31.7	2
W. South Central (AR,LA,OK,TX)	26,910	31.5	0.3	31.9	
Mountain (AZ,CO,ID,MT, NV,NM,UT,WY)	13,166	46.8	4.2	51.0	
Pacific (AK,CA,HI,OR,WA)	36,533	33.9	1.4	35.3	
Nat'l Totals	243,399	32.5	2.4	34.8	8

TABLE 4: REGIONAL PER-CAPITA ANNUAL BSCE & BSCET GRADUATES, 1986-88

STATE	BSCE+BSCET/yr/ M. Populat.	BSCET/yr/ M. Populat.	% OF TOTAL from BSCET
Georgia	25.7	11.0	42.8%
Oregon	64.4	16.0	24.8%
Rhode Island	57.8	14.2	24.6%
West Virginia	46.7	10.8	23.1%
Kentucky	17.7	3.4	19.2%
Colorado	62.5	11.1	17.8%
Pennsylvania	40.1	7.1	17.7%
Massachusetts	66.1	10.6	16.0%

TABLE 5: STATES WITH HIGHEST PROPORTION OF BSCE+BSCET DEGREES COMING FROM ENGINEERING TECHNOLOGY

Regionally, as shown in Table 4, the Lower South Atlantic area has the lowest per-capita generation of BSCE+BSCET graduates at 28.6 grads/yr/M.pop. New England has the highest production at 56.6 grads/yr/M.pop.

With the exception of one state (GA), the nine states having the highest per-capita BSCET production all have a per-capita rate of generation of BSCE graduates above the national mean.

Thirty-one states produce above the national mean of 32.5 BSCE grads/yr/M.pop.; 29 of those states permit PE registration for BSCET graduates. The 21 states having the highest per-capita production of BSCE graduatees all permit PE registration for CET. Four of the nine states having the lowest per-capita production of BSCE graduates (MN,KY,FL,IL) do not permit professional engineering registration for BSCET graduates.

Forty-two states and the District of Columbia permit PE registration for BSCET graduates; this is an increase by two over that reported by Frank and Ziegler (1984).

Collectively, the eight states who do not permit PE registration for BSCET graduates produce 25.1 BSCE grads/yr/M.pop., which is 23% below the national mean. The same eight states produce 1.5 BSCET/yr/M.pop., which is 38% lower than the national mean.; the rate of production of BSCE+BSCET is 24% below the national mean.

Employers indicated that 84% of the time BSCET graduates were assigned engineering responsibilities equal to or greater than that assigned BSCE graduates. Overall, employers felt that 81% of their engineering work could be successfully handled by BSCET graduates. Both employers and graduates felt the BSCET education was closely related to job responsibilities.

Approximately a third of the graduates felt PE registration was required for satisfactory career advancement, whereas 15% of employers indicated it was "required" and 68% indicated it was "desirable." Some employers provided a dual-track system so BSCET graduates could advance along a PE track or a track for technologists/designers.

Conclusions

BSCET graduates are used successfully by employers in most of their engineering work.

BSCET graduates supply a significant supplement to BSCE graduates in several states.

BSCET programs are most popular in New England, the Lower South Atlantic, and the Rocky Mountain states.

BSCET production per-capita is lower than the national average in states where the PE is not available.

Per-capita production of BSCET graduates appears

to be higher in states having BSCE production higher than the national average.

Based on the information from this study, it appears there is not a clear distinction by employers between BSCE and BSCET graduates.

REFERENCES

1. Engineering Manpower Commission, ENGINEERING AND TECHNOLOGY DEGREES 1987, PART III, 1988, American Association of Engineering Societies

2. Engineering Manpower Commission, ENGINEERING AND TECHNOLOGY DEGREES 1988, PART III, 1989, American Association of Engineering Societies

3. Engineering Manpower Commission, ENGINEERING AND TECHNOLOGY ENROLLMENTS, FALL 1987, PART I-ENGINEERING, 1988, American Association of Engineering Societies

4. Engineering Manpower Commission, ENGINEERING AND TECHNOLOGY ENROLLMENTS, FALL 1987, PART II-TECHNOLOGY, 1988, American Association of Engineering Societies

5. Engineering Manpower Commission, ENGINEERING AND TECHNOLOGY ENROLLMENTS, FALL 1988, PART I-ENGINEERING, 1989, American Association of Engineering Societies

6. Engineering Manpower Commission, ENGINEERING AND TECHNOLOGY ENROLLMENTS, FALL 1988, PART II-TECHNOLOGY, 1989, American Association of Engineering Societies

7. Keating, J.L., "CE and CET Graduates," Proceedings of the April, 1985 ASCE National Education Conference, Columbus, Ohio, pp. 30-37.

8. National Society of Professional Engineers, "State by State Summary of the Requirements for Engineering Registration," Fourth Edition, 1988

9. United States Bureau of the Census, Statistical Abstract of the United States: 1989 (109th Ed.), Washington, D.C., 1989

10. Frank, R. and Ziegler, T.,"Registration of Engineering Technology Graduates," Journal of Engineering Technology, Vol. I, No. 2, 1984, American Society for Engineering Education

Common Engineering and Technology Courses

By Ajit S. Gill[1], Fellow, ASCE

Abstract

This paper discusses the merits of modifying some technology degree courses to bring these to a level equivalent to that of engineering degree courses. The suggestion, if adopted, would lead not only to strengthening the technical base of technology programs but also enable more technology graduates to enter engineering degree programs thus promising a more rewarding career to such aspirants and producing more civil engineers to meet the increasing needs of the nation.

Introduction

The career of an engineering technologist is one of the few which promises a degree holder little advancement in level of responsibility from the starting stage of a person's professional life until possibly his retirement. The role of a technologist requiring him to work under the direction of an engineer severely limits the prospects of his advancement. It appears desirable to modify the curricula of engineering technology programs so as to enable a capable technologist to earn an engineering degree by entering a program of studies suitably tailored to accept a technologist at an entry level higher than that of an associate degree or diploma holder who only fulfilled basic requirements of general education at college in addition to some low level technical courses. This does not appear to be an insurmountable problem considering the fact that a graduate of civil engineering technology program approved by the Technology Accreditation Commission of Accreditation

[1] Professor, Engineering Technology,
Florida A&M University, Tallahassee, FL 32307

Board of Engineering (TAC/ABET) would have completed a four year degree course of nationally recognized technical standard. The intention in making this suggestion is not to lower the standard of engineering degree requirements, but to raise, if necessary, standards of civil engineering technology instruction in order to affect reasonably smooth transition into a higher level engineering degree program.

Technology and Engineering Courses

As at present, most engineering degree programs do not recognize any technical courses of engineering technology programs as equivalent to theirs, thus necessitating for a person with a technology degree to start from ground zero in respect of technical courses in order to work up to an engineering degree. The argument that is most commonly put forward by engineering educators against accepting for transfer credits of technology courses is that a technologist's training is inadequate in mathematics and in general sciences. This is a valid criticism to some extent; however, it does not apply to all courses covered in an engineering curriculum. Consider the following comments of some authors of standard textbooks of engineering subjects:

- "The material presented in the text (on statics) and most of the problems require no mathematical knowledge beyond algebra, trigonometry and elementary calculus" (Beer and Johnston 1987)

- "It is expected that students using this text (on strength of materials) will have completed a course in statics," (Beer and Johnson) 1981. It is noteworthy that no other requirement is mentioned to pursue this course.

- "This text may be used by students of all engineering disciplines for the first course in fluid mechanics. Any student who has had basic engineering mechanics should be able to read and understand it" (Robertson/Crowe 1980).

- "This book (on highway engineering) can, in the main, be used by those who have very little background in technical engineering. However, some knowledge of route surveying, dynamics and fluid mechanics is necessary for a complete understanding of some topics" (Oglesby and Hicks 1982).

Similar comments would apply in support of background knowledge required for courses in engineering drawing, engineering economy and first level courses in surveying and soil mechanics. The laboratory work in engineering drawing, first level courses in surveying, soil mechanics, engineering materials, hydraulics and strength of materials requiring either practice in use of instruments(in surveying) or determining physical properties (in soil mechanics and engineering materials) or verifying laws of applied sciences is essentially similar in engineering and engineering technology curricula. Thus it appears that, with suitable modifications in theoretical contents, a number of courses for engineering and engineering technology could have common content making it a feasible proposition for a technology graduate to secure some transfer credits into an engineering program. This would require a bold approach by both engineering and technology educationists to meet the situation head on: willingness on the part of the former to accept course transfers where objectivity so dictates and action on the part of the latter to upgrade the contents, where necessary, of such courses which could qualify for acceptance for transfer. In order to determine whether any improvement in technology level courses can realistically be affected, let us examine the causes of deficiency of technology programs in comparison to engineering programs.

i. In general, the academic skills of technology students are lower than that of engineering students. This objection would probably not apply to a technologist attempting to enter an engineering program since he would most likely belong to the group of higher than average academic skills. As an additional requirement, a minimum grade point achievement level could be set by various institutions for accepting course transfers from technology programs.

ii. A technologist's training is inadequate in mathematics and physcial sciences. The usual requirement for mathematics in an engineering degree program includes an advanced course in calculus covering differential equations and one course in applied engineering mathematics. Most civil engineering technology programs do not require these courses. This criticism is valid in not so much that these courses ought to be pre-requisites for all common engineering and technology courses but that a strong foundation in mathematics is a great asset to anyone preparing for a career in engineering. A university degree in a mathematically-oriented field like that of engineering technology ought to require greater competence in mathematics than that expected of an associate degree holder. Applied mathematics and differential equations are essential tools to enable one to understand the principles of natural sciences and applications of these principles in the practice of the profession. The tremendous explosion in knowledge during the last quarter century has been largely in field of pure and applied sciences in each of which mathematics is a vital ingredient. Not to acquaint a technologist with applied mathematics and differential equations is a disservice to the engineering profession. Thus a justification exists for including these two courses in technology curricula.

Once the standard of mathematics requirements for technology program is raised, necessary adjustments in technical content of some technology courses can be affected to raise their standard equivalent to that of engineering degree curricula. A comparative evaluation of technology and engineering course contents of <u>some</u> technical subjects would reveal that it should be relatively easy to accomplish this objective.

iii. In some cases, the level of competence of instructors of technical courses in technology programs is open to question. TAC/ABET could conceivably examine the matter critically in order to tighten further the minimum academic requirements of instructors. Similar to the requirments in force for professional registration for teaching some engineering courses,an identical requirement could be laid down for teaching of some specified technology courses.

iv. A typical civil engineering program requires in general sciences two courses each in physics with calculus and in chemistry. The author agrees with the need for identical requirements in physics for a technology program but has some reservations about the necessity of two courses in chemistry since this would be of benefit in study of subject like materials science and environmental science. These subjects would be of a higher level or of a more specialized nature than expected for technology programs. As such, a second course in chemistry would not be of critical importance in the context of this paper.

3. Some Possible Common Courses

In light of the arguments presented, it appears that, with suitable modifications, a number of courses of engineering technology programs may be brought up to engineering degree course standards without burdening the instructional aims of either discipline. A list of such common courses would probably comprise first level courses (where there is more than one sequential course) in engineering drawing, engineering economy, statics, dynamics, engineering materials, strength of materials, fluid machanics, (or hydraulics), surveying, soil mechanics and highway engineering. Not all of these courses from any single technology program would probably be of transferable standing; each institution would no doubt make decisions best suited to its particular mission. Conceivably, from a good technology program about 20 to 27 semester hour credits could be of transferable standard. Thus, a full time attendance of two to three semesters in an engineering institution would enable a capable technologist to earn a degree in civil engineering.

The emphasis on "hands on" training of a technologist requiring considerable exposure to laboratory and field techniques would include testing of engineering materials, field and laboratory testing of soils and would probably be more intensive than that compared to engineering students. Such exposure would be of some advantage to a technologist undertaking an engineering degree course.

It is difficult for any institution to embark upon such a program without a general consensus amongst educators and support of ASCE that this proposal would result in advancing the aim of the profession by producing more civil engineers from the ranks without in anyway sacrificing the quality of their education.

4. Conclusion

Attempt has been made in this paper to remedy the present situation which prevents engineering technology degree holders to secure some transfer credits to civil engineering degree programs. This is detrimental to the ambitions of capable technologists who attempt to earn engineering degrees. It has been suggested that educationists and ASCE consider the feasibility of accepting in principle transfer of equivalent level engineering technology courses to those in engineering. Correspondingly, this would require educationists in engineering technology to raise the standards of such courses to a level equivalent to those of engineering.

APPENDIX. REFERENCES

1. Beer, F.P., and Johnston, E.R., Jr. (1987). "Mechanics for Engineers Statics." McGraw-Hill Book Company, New York

2. Beer, F.P., and Johnston, E.R. Jr. (1981) "Mechanics of Materials." NcGraw-Hill Book Company, New York

3. Oglesby, C.H. and Hicks, R.G. (1982). "Highway Engineering." John Wiley and Sons, New York

4. Robertson, J.A. and Growe, C.T. (1980) "Engineering Fluid Mechanics." Houghton Mifflin Company Boston

Geological Education of Undergraduate Civil Engineers

Christopher C. Mathewson, MASCE[1]

Abstract

The geologic environment forms the basic foundation upon which the Civil Engineer designs, creates and maintains the modern built environment. Geological education of undergraduate Civil Engineering students is all too often taught by unqualified faculty and limited to the memorization of facts unrelated to the practice of Civil Engineering. The failure of geological educators to present geologic courses for engineers that are relevant to their practice has resulted in a significant deficiencies in the Civil Engineer's geologic education. This lack of a geologic awareness by the young engineer results in over design, construction claims, and a decrease in the quality of work performed by the Civil Engineer. ASCE must take a stand and demand geological education designed to meet the needs of the young engineer. Properly designed and taught geology courses for engineering students can be developed that will enhance the quality and academic preparation of future Civil Engineers.

Introduction

The foundation upon which the Civil Engineer designs, constructs, operates and maintains the built environment is provided by the geological environment. The history of education and practice in Civil Engineering indicates that as engineering technology increased the engineer's geologic understanding decreased. Karl Terzaghi, considered by many Civil Engineers to be the father of modern Soil Mechanics, is also recognized by many geologists for his contributions to the understanding of landslide processes. Arthur Casagrande, who developed the Atterburg Limits is also recognized for his geologic understanding of the geochemistry of soil-water reactions. Today, however, the "new" approach to design is based on probabilistic methods which often assume that the geologic system is a random system.

---

[1]Past President, Association of Engineering Geologists; Professor of Geology and Director, Center for Engineering Geosciences, Department of Geology, Texas A&M University, College Station, TX 77843-3115

This change from "earth based" engineering to a more analytically and statistically based engineering philosophy reflects our modern advances in computer technology and the growing litigious society. Many of us still remember the Log-Log Double Duplex as the primary symbol of the engineer, today it is the calculator or even the portable computer. "Slide rule engineers" could not afford the time to run numerous computations to attempt to obtain the "right" answer. We learned to do the thinking and then provide the "best" answer that logically fit the problem. Engineering was still very much an "art" with engineering judgement playing an important role in the practice. As a result, engineers were required to be aware of the idiosyncrasies in the relationships between the natural and engineered systems. Thus, "slide rule engineers" were forced to understand, or at least better appreciate, both the natural and built environments.

The advent of computers made it possible for the "computer engineer" to develop analytical techniques that can solve complex problems by mathematical iterations. An ability to obtain a quantitative answer to an indeterminate problem gives the engineer a greater feeling of confidence. Technology to generate quantitative answers, public demands for a risk-free environment and the increasingly litigious society forces the engineer to remove or severely restrict the use of engineering judgement.

Historical Development of Geology and Engineering

The decline in geological education and training for undergraduate Civil Engineering student can be related to the increase in the education and training of Engineering Geologists. As engineered works became more complex and as the most accessible land was developed, a greater need for geological investigations and interpretations increased. Complex sites combined with the litigious society required the engineer to include engineering geologists as part of the engineering team. A review of the development of Engineering Geology shows the field's close tie to the practice of Civil Engineering.

Engineering Geology was formalized when Penning (1880) published his text in England. Professor W. O. Crosby initiated the first training course in applied geology for engineers in the United States at the Massachusetts Institute of Technology in 1893. A similar course was offered at Cornell University by Professor R. S. Tarr shortly thereafter. Reis and Watson (1914) published the first American text in engineering geology. These texts and early academic programs established engineering geology as "the application of geology in engineering." They filled the need for the Civil Engineer's education and training in geology so that the engineer could interpret the natural system.

The catastrophic failure of the St. Francis Dam in California on March 12, 1928 was a key event that led to the recognition of the value and need for engineering geologists as part of the construction engineering team. Berkey (1929) published a paper on the responsibilities of the geologist at an American Institute of Mining and Metallurgical Engineers symposium on geology and engineering for dams and reservoirs. The character of engineering geology for the next thirty years was thus, shaped by an understanding that the failure was geologically predictable. Academic training and research carried out during this period concentrated on the applications of geology in construction engineering.

Kiersch (1958) described engineering geology as an applied practice of geology associated with the constructive efforts of engineers. He further defined the other applied fields of geology, mining, petroleum and ground-water geology, as applied practice associated with the destructive efforts of engineers.

Another California disaster, the floods and landslides following the winter storms of 1951-1952 in the Los Angeles area, further expanded the scope of engineering geology. Los Angeles Ordinance No. 100,347 required a geologic report of hillside conditions prior to land development (Slosson and Krohn, 1977). This ordinance added the responsibility for the protection of the health, safety and welfare of the general public to the practice of engineering geology. These events introduced the concept of geology in urban planning and led to numerous urban geological mapping projects. McGill coined the term "Urban Geology" in 1963 and stated that Urban Geology "has been with us for many years, yet possesses a timeliness and an urgency that is regrettably little appreciated" (McGill, 1964). The practice of engineering geology continued as applied geology, however, it now incorporated the practice of Urban Geology. Today many cities have both a city geologist as well as a city engineer; thus city planning, development and growth is evaluated by a team.

The January 28, 1969 blowout of an oil well in the Santa Barbara Channel, California brought concern for the environment to national attention. The nation's eyes were opened to the environmental crisis and the need for Environmental Geology (Muskie, 1972). President Nixon's 1970 address to Congress made the environment a national issue which has continued to the present. The field of engineering geology was further expanded to include environmental problems, such as hazardous wastes and ground-water protection. Environmental Geology, as a specific practice of geology, was formalized with the publication of the first textbook in Environmental Geology by Flawn (1970). Today's environmental problems are approached through a team composed of geologists, engineers and others.

As the practice of engineering geology expanded from the application of Geology in Engineering, to include Construction Geology, Urban Geology and Environmental Geology, the field became increasingly more interdisciplinary and the socioeconomic responsibilities of the practitioners increased significantly. This has resulted in the introduction of numerous related "subfields" such as Engineering Geomorphology (Fookes and Vaughan, 1986) and Engineering Geochemistry (Prokopovich, 1988). Kiersch (1964) suggested that engineering geology could be termed "Physico-geology" to reflect a physical geology base of the science.

The disastrous landslides and debris flows of the 1970's and 1980's in California, Virginia, Colorado and Utah, however, introduced a new emphasis to Engineering Geology: the scientific investigation of the phenomena of earth processes which affect the public health, safety and welfare, engineered works and the environment. Only recently has the field of engineering geology begun to develop as a "science" rather than as "applied" geology. The modern undergraduate engineering geology student is being educated as a complete geologist with additional engineering courses, such as engineering mechanics, strength of materials, and soil, rock and fluid mechanics. The intent of their education is that they **are not engineers**; they are **geologists** who understand the philosophy of engineering.

Undergraduate Teaching of Geology for Engineers

The recent development of Engineering Geology as a specialty "science" within Geology rather than the simple application of geology to engineering has resulted in a severe shortage of academicians who are qualified engineering geologists. As a result, the classical "Geology for Engineers" course is all too often taught by someone other than an engineering geologist and limited to the memorization of geological facts that the undergraduate engineering student soon forgets. Very few Civil Engineers will ever be concerned about the chemical formula for the mineral Feldspar but they may face the problem of an excavation in a decomposed granite, rich in Feldspar. The ability to recognize a sandstone or shale without understanding the three dimensional distribution of the rock in space provides little information for the design engineer.

With an increasing use of analytical and probabilistic techniques, the Civil Engineer must understand the natural variability of the earth system. A "geological judgement" must be incorporated into the interpretation of the results. Such questions as "What are the most favorable conditions to be expected at this site?" and "What are the most unfavorable conditions?" must be asked before the design is finalized. The Civil Engineer who lacks a basic understanding of the geological environment is unable to ask or interpret these questions.

The increasing interdependence between the geologist and engineer in the design, construction and maintenance of an engineering project requires that both professionals be able to communicate. Just as the Engineering Geology degree program is designed to produce geologists that understand engineers, the Civil Engineering degree program must be designed to produce engineers that understand engineering geologists. This is even more important with the increasing number of states that register professional geologists. True, the liability is now shared! Geological liability is the responsibility of the geologist and no longer the Civil Engineers' in states that register geologists, but failure to communicate may in itself cause the failure that affects both professionals, be it over design, excess cost, or structural failure.

The common engineering joke "give me a one-handed geologist, so I can get an answer without the other hand" should be revised. The engineer should always ask "where's the other hand?" After all, you need two hands if you are going to define the range of values or the degree of variability. Without this interpretation of the variability of a site, how can the engineer apply proper engineering judgement?

Conclusions

If the future civil engineer is to interpret and evaluate a geological report or discuss it with an engineering geologist, the engineer must understand it. To understand the report the engineer needs more than a list of memorized facts! They need to understand the philosophy of geology; "Why does the geologist need both hands?". Once the engineering student realizes that geology is the interpretation and not the analysis of the earth system, they can relate the natural and built environments. A properly designed and taught "Geology for Civil Engineers" course can provide this critical education.

The American Society of Civil Engineers and the Association of Engineering Geologists need to work together to develop a basic Geology for Civil Engineers course that meets the needs of the Civil Engineering profession. This course should concentrate on the physical and mechanical properties of earth materials and the phenomena of active surficial earth processes. The course instructor must also understand the fundamental differences between engineering and science.

References

Berkey, C. P., 1929, Responsibilities of Geologists in Engineering Practice: *American Institute of Mining and Metallurgical Engineers*, Technical Publication 215.

Flawn, Peter T., 1970, *Environmental Geology*: Harper and Row, New York, 313 p.

Fookes, P. G. and Vaughan, P. R., Editors, 1986, *A Handbook of Engineering Geomorphology*: Chapman and Hall, New York, 343 p.

Kiersch, George A., 1958, Quantitative Trends and Research Needs in Engineering Geology: *Bulletin of the Geological Society of America*, Vol. 69, p. 1597 (Abstract).

Kiersch, George A., 1964, Academic Trends and Engineering Geology: *Journal of Geological Education*, Vol. 12, No. 1, March, pp. 14-21.

McGill, John T., 1964, *Growing Importance of Urban Geology*: U.S. Geological Survey Circular 487, 4 p.

Muskie, Edmund S., 1972, Forword, in Tank, Ronald W., Editor, *Focus on Environmental Geology*: Oxford Press, New York, 2nd Edition.

Penning, W. Henry, 1880, *Engineering Geology*: Bailliere, Tindall and Cox, London, England, 164 p.

Prokopovich, Nikola P., 1988, Geochemistry in Engineering Geology: *Bulletin of the Association of Engineering Geologists*, Vol. XXV, No. 2, pp.269-276.

Reis, Henrich and Watson, T. L., 1914, *Engineering Geology*: John Wiley and Sons, New York, 565 p.

Slosson, James E. and Krohn, James P., 1977. Effective Building Codes: *California Geology*, June, pp. 136-139.

CIVIL ENGINEERING EDUCATION: WEAK OR MISSING LINKS

by Charles H. Samson,[1] Fellow, ASCE

Abstract

Civil engineering education is undergoing critical examination concerning the content needed to prepare a person for civil engineering practice. Consideration is given to the essential element links in the educational process. It is concluded that the strengthening or adding of some elements is needed.

Introduction

As members of a learned profession, civil engineers are expected to have formal engineering education needed to serve the public in the practice of engineering. Obligations of the civil engineer include possessing and applying technical competence, practicing the art and science of engineering in a professional and ethical manner, and providing leadership to society in the effective use of technology toward an enhanced quality of life (Samson 1989a). Much debate has occurred in the engineering profession concerning the elements essential to fulfilling these obligations. Opinions are often based upon individual engineering experiences - what was needed or not needed in meeting the responsibilities of that individual. Opinions also vary as to what knowledge and skills should be acquired from college, and what from actual work experience. The whole problem is further exacerbated by the continued reduction in available time for engineering technical study and the continuing increase in knowledge that occurs in the engineering field.

---

[1]Prof. and Assoc. Head, Dept. of Civil Engineering, Texas A&M Univ., College Station, TX 77843

Element Definitions

For purposes of the present discussion, elements of the engineering educational process are categorized as follows:

Mathematics and Basic Sciences
Engineering Sciences
Basic Engineering Analysis
Basic Engineering Design
Project Design
Communication - Oral, Written, and Graphical
Humanities and Social Sciences
Professionalism and Ethics
Leadership
Quality Management
Systems Approach and Problem Solving
Computer Programming and Applications
Probability and Statistics
Engineering Economy
Systems Engineering

While most of the above elements are used here in the sense as defined by ABET, it will be helpful to provide some elaboration:

Basic Engineering Analysis and Basic Engineering Design. - The modifier "basic" is used in each case to indicate emphasis on analysis or design of components of systems as distinguished from systems themselves.

Project Design. - "Project design" refers here to the design of a complete project or system.

Leadership. - "Leadership" is distinguished from "management," although leadership is generally regarded as an important attribute of an effective manager. The effects of technology - both beneficial and detrimental - have become increasingly significant to the future of society. As never before, engineers must play an important role in providing leadership in critical decision making - in arriving at the compromises that best serve society as a whole.

Quality Management. - Various terms are used to refer to the approach to management espoused by W. Edwards Deming, Joseph M. Juran, Armand V. Feigenbaum, and others. In-depth discussions are provided by Deming (1982), Juran (1989), and Feigenbaum (1983). Quality management is based upon statistical concepts, and is an elaboration of the management principles fostered by Deming and Juran in this country and in Japan in the early 1950s, but adopted most extensively by Japan. In recent years quality management has been attracting rapidly increasing attention in the United States to the point that many - perhaps most - large companies are endeavoring to adopt this approach to some degree. Notwithstanding the incorporation of the term "management," quality management is intimately related to systems engineering in many manufacturing and service industries.

Systems Approach and Problem Solving. - Much of civil engineering deals with "systems" and much involves problem solving. These are treated here as a single element in that the systems approach is an orderly, rational approach to engineering problem solving.

Systems Engineering. - Hall (1969) represents systems engineering by means of a three-dimensional "morphological box." He defines seven phases of the time axis as (1) program planning, (2) project planning (and preliminary design), (3) system development, (4) production (or construction), (5) distribution (and phase in), (6) operations (or consumption), and (7) retirement (and phase out). He defines seven problem-solving steps of the logic axis as (1) problem definition, (2) value system, (3) systems synthesis, (4) systems analysis, (5) optimization, (6) decision making, and (7) planning for action. A third dimension represents professions that might be involved in the systems engineering process. Systems engineering embraces not only different phases, such as project design, but also various systems-analysis and decision-making methods. Mar and Palmer (1989) discuss systems engineering and its importance to the practice of civil engineering. Samson (1985) discusses the need for educating engineers in the systems approach.

The Weak or Missing Links

Figure 1 is a model (similar to that provided by Jewell 1986) of the educational process that incorporates the element links listed earlier. The term "element" does not necessarily refer to a single course, but rather a category of knowledge, competence, or skill essential (in the writer's view) to the formal educational preparation of a civil engineer.

The following comments provide amplification:

1. The diagram indicates the kind of content that should be regarded as essential, but not the quantity. It is recognized that giving proper attention to all these elements may require more than a four-year program. The diagram recognizes

only some of the primary interactions between elements. Many other interactions exist.

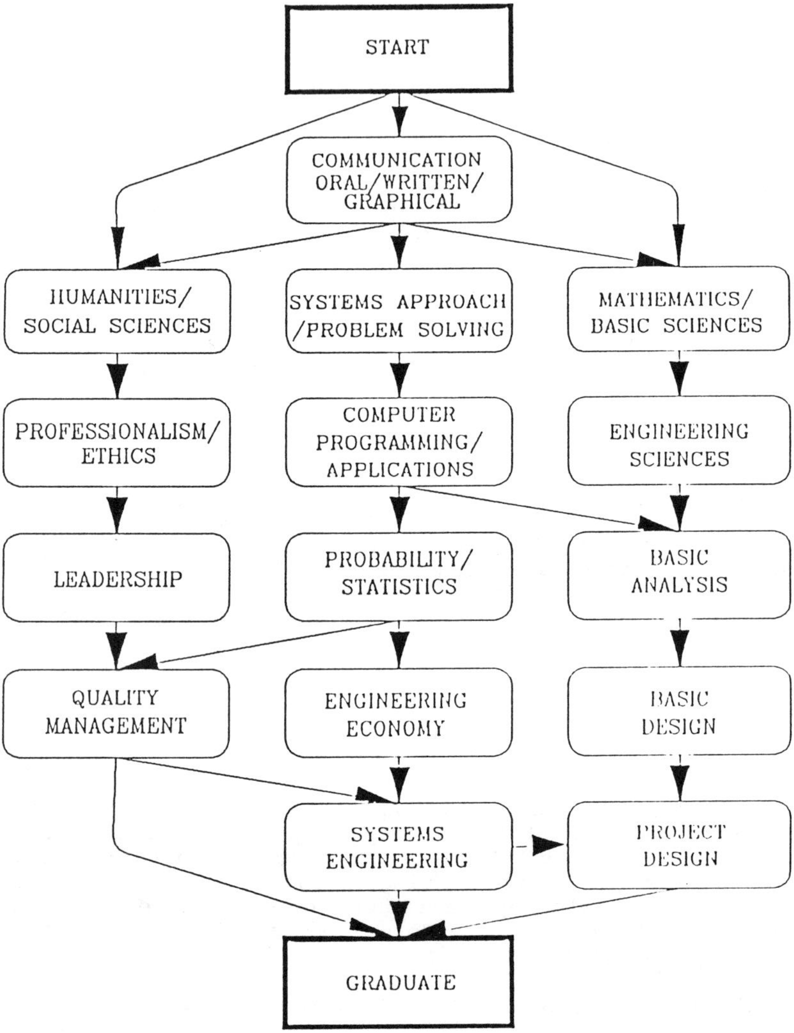

Figure 1. Element Linkages in Civil Engineering Education

2. ABET requires that "in order to promote breadth, the curriculum must include at least one engineering science course outside the major discipline area." This sometimes requires a choice between certain basic topics such as thermodynamics and electrical and electronic circuits. In view of the fact that students are not ordinarily in position to make a judicious choice, it would seem logical, in the writer's opinion, to require at least some work in all the engineering sciences.

3. It appears that as further time constraints are imposed on the civil engineering curriculum, undesirable compromises are often made in the coverage of basic engineering design.

4. The design of complete projects is often not adequately addressed, if considered at all. While requiring work on a wide variety of projects is not feasible from a time-constraint standpoint, it is important that civil engineering students have at least one design experience related to a total civil engineering system.

5. A continual complaint made by engineering employers against civil engineering graduates, and engineering graduates in general, is the lack of communication skills. The need for stronger preparation in this area seems evident.

6. Although some engineering curricula offer elective courses in professionalism and ethics, most do not require such work.

7. There appears to be no focus on leadership in engineering curricula.

8. Civil engineering curricula have not recognized quality management as an essential element in the civil engineering educational process. As previously noted, quality management and systems engineering have a close relationship; this should be recognized in civil engineering curricula.

9. Civil engineering curricula typically do not provide students an introduction to thinking in terms of civil engineering systems before the students encounter "pieces" of the civil engineering process. Jewell (1986) diagrams a suggested integration of systems concepts into the civil engineering curriculum.

10. ABET criteria specify: "Additional work is encouraged in one or more of the subjects of probability and statistics, linear algebra, numerical analysis, and advanced calculus" (ABET 1988). Requirement of some work in all these areas would certainly strengthen the mathematics content. Because of the critical importance of the probabilistic and statistical aspects of many civil engineering functions, this topic is regarded as an essential element.

11. Civil engineers may be involved in any or all phases of the systems engineering process. Indeed, an argument can be made that every civil engineer ought to be concerned about the total life cycle of every civil engineering system, and the implications of activities in every phase on the activities of every other phase. While some civil engineering curricula provide work in systems engineering tools, many do not stress the overall attention to total life cycle considerations.

Conclusions

1. The elements identified in Fig. 1 should be incorporated as links into the civil engineering educational process. The question of time and effort devoted to each of these elements, and the ABET accreditation implications, should be addressed separately. Adequately incorporating all elements may require more than the nominal four years. (See, for example, Samson 1989b.)

2. The culmination of the basic formal education should focus on the merging of project design and quality management with the systems engineering process as illustrated in Fig. 1.

3. An introduction to the systems approach and problem solving should occur early in the curriculum.

4. The other essential elements of Fig. 1 should bridge the gap between an early appreciation of engineering systems and the final understanding and application of the civil engineering systems process.

Appendix. References

ABET (1988). "Criteria for accrediting programs in engineering in the United States." 1988 Annual Report, ABET, New York, NY.

Deming, W.E. (1982). Out of the crisis. MIT Press, Cambridge, MA.

Feigenbaum, A. V. (1983). Total quality control. 3rd ed., McGraw-Hill, Inc., New York, NY.

Hall, A. D. (1969). "Three-dimensional morphology of systems engineering." IEEE Trans. Syst. Sci. Cybern., SSC-5, 155-160.

Jewell, T. K. (1986). A systems engineering approach to civil engineering planning and design. Harper & Row, Inc., New York, NY, 5.

Juran, J. M. (1989), Juran on leadership for quality. The Free Press, New York, NY.

Mar, B. W., and Palmer, R. N. (1989). "Does civil engineering need system engineering?" J. Prof. Issues in Engrg., ASCE, 115(1), 45-52.

Samson, C. H. (1985). "Engineering and the systems approach," Proc., International Conf. on Gen. Sys. Res., Los Angeles, CA, May 27-31, 1985, 720-722.

Samson, C. H. (1989a). "Engineering education - focus on the future." The Bent of Tau Beta Pi, 80(2), 31-33.

Samson, C. H. (1989b). "Engineering education - where do we go from here?" Proc., ASEE Southwestern Section Meeting, Tuscaloosa, AL, Apr. 2-4, 1989, 98-102.

Integrating Engineering and Architecture Education

Ernst W. Kiesling[1], F. ASCE,
Jimmy H. Smith[2], F. ASCE, and
James R. McDonald[3], M. ASCE

Abstract

The dual-degree (ARCE) program at Texas Tech University offers an integrated program leading to accredited baccalaureate degrees in both Civil Engineering and Architecture (Structures Option). The curriculum requires simultaneous progress in both engineering and architecture courses. The full design content in both fields is required. The ARCE program provides graduates with a broad in-depth perspective in the building sciences field. By becoming registered in both engineering and architecture, professionals enjoy leadership advantages and career flexibility available to few practitioners.

Introduction

For the past decade, Texas Tech University has been producing graduates with accredited baccalaureate degrees in both Civil Engineering and Architecture (Structures Option). Upon the program's inception in 1978, a separate major (ARCE) was established at Texas Tech. The integrated curriculum requires simultaneous progress in both engineering and architecture courses;

---

[1]Professor of Civil Engineering; Associate Dean for Applied Research. Formerly Chairman, Civil Engineering Department, 1969-1988, Texas Tech University.

[2]Professor of Civil Engineering; Director, Murdough Center for Engineering Professionalism, Texas Tech University.

[3]Professor and Interim Chairman, Civil Engineering Department, Texas Tech University.

the full content for accreditation of degree programs of both fields is required.

The program has proven attractive to students despite requirements in excess of 190 semester hours. This paper presents features of the program and summarizes career paths chosen by graduates.

Background

Since the department's inception in 1927, the Civil Engineering Department at Texas Tech has had a strong emphasis in structural engineering. In recent years, undergraduate students have been given more choices of emphasis, but 8 hours of structural analysis and 4 hours of structural design are required of all civil engineering graduates. Those students choosing the structures emphasis typically have a total of 8 hours of structural design plus 8 hours of design in other areas. Currently the civil engineering curriculum comprises 140 semester credit hours, exclusive of physical education, band, or R.O.T.C.

The College of Architecture continues to offer a Bachelor's degree in Architecture with two specialties--architectural design or architectural structures. The architectural structures specialization curriculum includes the same mathematics courses required of civil engineering majors (through differential equations and Laplace transforms) and civil engineering structural analysis and design. Eight hours of physics plus three hours of basic science electives are required.

Even before the dual-degree ARCE program came into being, a few students earning a bachelor's degree in architecture (structures option) earned a second degree in civil engineering. Some graduates earning only the architecture (structures option) degree have, in the past, become registered in both engineering and architecture.

ARCE Program Overview

In 1978 an integrated program was inaugurated which allowed students to simultaneously earn a B.S. degree in Civil Engineering with a structures emphasis and a Bachelor of Architecture (Structures Option) degree. Primarily for administrative purposes, a separate major designated ARCE was established for students enrolled in the dual-degree program. The integrated curriculum has students enrolled in both architecture and civil engineering courses each semester, simultaneously acquiring presentation skills and computer skills while taking basic science, mathematics, and humanities courses. While continuing to take courses in both architecture and engineering, emphasis shifts from architecture to engineering in the fourth and fifth years.

The dual-degree program is shown in the 1989-90 catalog as 5 academic years plus 3 summers, the equivalent of 6 academic years; it consists of 195 semester hours. Elective choices are essentially eliminated.

Dual-Degree ARCE Curriculum

The ARCE curriculum from the 1989-90 undergraduate catalog is shown in Table 1. Careful analysis reveals the sequencing of subject matter to permit a continuous building upon knowledge and skills acquired in both architecture and engineering. Curriculum changes effective in the next issue of the catalog are intended to strengthen and extend yet further this integrated, progressive sequence.

Curriculum revisions effective in 1990-91 expand the design requirements in architecture, including a 6 semester hour design thesis. The semester hour requirement for the dual-degree program rose from 195 to 207. It is too early to assess the long-term impact of the curriculum changes.

Career Paths of Graduates

Since 1982 when graduation from the organized ARCE program began at Texas Tech, 62 people have simultaneously earned accredited degrees in both architecture and civil engineering. Five earned both degrees before 1982. Of the 67 graduates, 56 are men, and 11 are women. Ten (10) graduates, including one woman, are foreign-born.

Career paths pursued by graduates will likely change with time, but employment after graduation gives some prediction of careers ARCE graduates will choose. Figure 1 shows the distribution of employment ARCE graduates have chosen among selected major categories of employment. This employment data is based on a sample of approximately half the graduates.

In view of the number of graduates who are working predominately in engineering fields, would it not be advantageous to concentrate academic preparation in one

TABLE I

Dual-Degree Curriculum, Bachelor of Architecture (Structures Specialization) and Bachelor of Science in Civil Engineering.

	FIRST YEAR		
Fall		*Spring*	
ARCH 1331, Intro. to Arch.	3	ARCH 1432, Intro. to Arch. Des.	4
ARCH 1351, Arch. Graphics	3	C E 1305, Engr. Anal. I	3
ARCH 2312, Hist. Ancient-Med.	3	POLS 2302, Amer. Pub. Pol.	3
MATH 1350, Anal. Geom.	3	MATH 1351, Calc. I	3
POLS 1301, Amer. Govt., Org.	3	ARCH 1341, Arch. Freehand I	3
P.E., Marching Band, or ROTC		P.E., Marching Band, or ROTC	
	15*		16*
	SUMMER SESSION		
First Term		*Second Term*	
ENGL 1301, Ess. Coll. Rhet.	3	ENGL 1302, Adv. Coll. Rhetoric	3
MATH 1352, Calc. II	3	MATH 2350, Calc. III	3
	6		6
	SECOND YEAR		
Fall		*Spring*	
ARCH 2313, Hist. Renaissance	3	ARCH 2432, Arch. Des. II	4
ARCH 2431, Arch. Des. II	4	ARCH 2252, Bldg. Syst. Integr.	2
ARCH 2251, Bldg. Syst. Perform.	2	PHYS 2301, Prin. Phys. II	3
ARCH 1342, Arch. Freehand II	3	PHYS 1106, Prin. of Phys. II (Lab.)	1
PHYS 1308, Prin. Phys. I	3	C E 2301, Statics	3
PHYS 1105, Prin. of Phys. I (Lab.)	1	ARCH 2341, Arch. Freehand Draw. III	3
	16		16
	SUMMER SESSION		
First Term		*Second Term*	
U.S. History	3	U.S. History	3
	THIRD YEAR		
Fall		*Spring*	
ARCH 3431, Arch. Des. III	4	ARCH 3432, Arch. Des. III	4
ARCH 3382, Intro. City Plan.	3	C E 3303, Mech. of Solids	3
ARCH 2311, Hist. of 19th & 20th Cent.	3	C E 3305, Mech. of Fluids	3
C E 2210, Proc. in Prob. Anal. I	2	C E 3210, Proc. in Prob. Anal. II	2
C E 3201, Const. Materials	2	Elective (Approved Basic Sci.)	4
MATH 3350, Math. for Engrs. I	3		16
	17		
	SUMMER SESSION		
First Term		*Second Term*	
CHEM 1307, Prin. of Chem. I	3	M E 3321, Thermodynamics	3
CHEM 1103, Prin. of Chem. I (Lab.)	1	CTEC 2301, Surveying	3
C E 3302, Dynamics	3		6
	7		
	FOURTH YEAR		
Fall		*Spring*	
ARCH 4331, Arch. Des. IV	3	ARCH 4383, Urban Des. I	3
ARCH 3351, Const. Drawings	3	ARCH 3352, Const. Draw. & Docs.	3
C E 3440, Struct. Anal. I	4	C E 4340, Struct. Anal. II	3
C E 3371, Environ. Engr. I	3	C E 3341, Prin. Struct. Des.	3
C E 3306, Stress Analysis	3	C E 3321, Intro. to Geotech. Engr.	3
	16	C E 3121, Geotech. Engr. Lab.	1
		C E 3130, Civil Engr. Seminar	1
			17
	FIFTH YEAR		
Fall		*Spring*	
ARCH 4395, Design Program	3	ARCH 4333, Design Prob.	3
ARCH 4363, Contemp. Arch. Issues	3	ARCH 4391, Arch. Prof. & Process.	3
C E 4343, Des. Concr. Struct.	3	C E 3272, Water Syst. Anal.	2
C E 4342, Des. Steel Struct.	3	C E 4321 or I E 3322	3
C E 3354, Intro. to Hydrology	3	E E 2303, Elec. Sys. Anal.	3
C E 4361, Transp. Engr.	3	C E 4330, Des. Engr. Systems	3
	18		17

Admission and progression requirements must meet exact requirements of both Architecture and Engineering.
*Minimum hours required, exclusive of P.E., Marching Band, or ROTC—195.

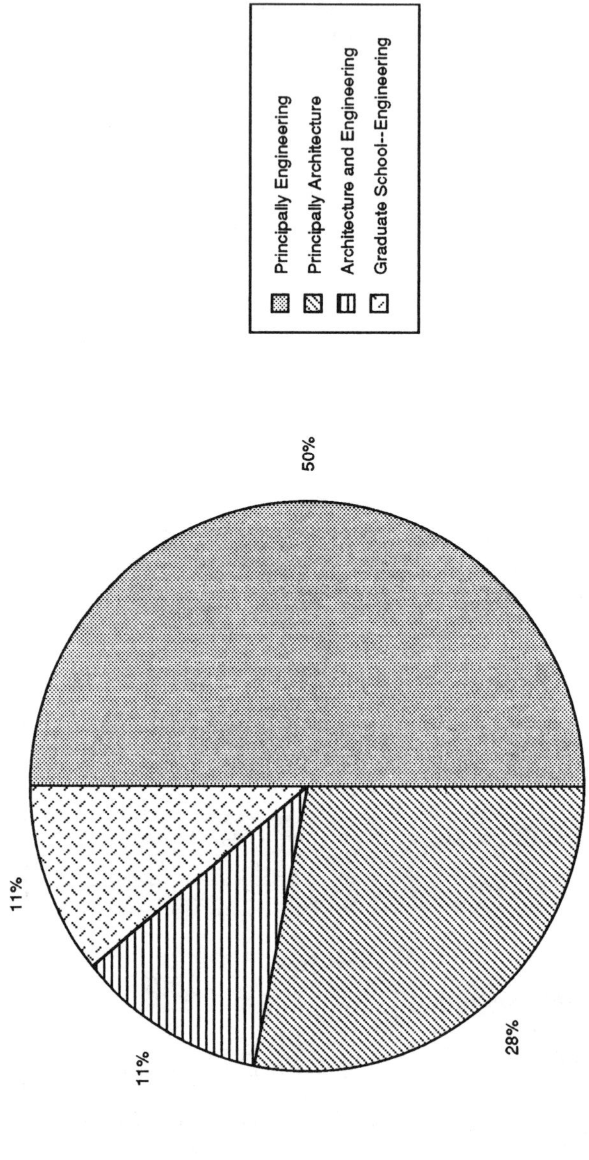

Figure 1. Distribution of Employment of ARCE Graduates

field, complementing undergraduate education with a graduate degree in the same or related field? Obviously the answer to that question lies with each individual and is influenced by opportunities and circumstances. But the fact remains that for one interested in professional practice in the broad field of building sciences, the dual-degree ARCE program at Texas Tech University provides broad-based, in-depth education with few, if any, parallels.

Professional Advantages Offered

Advantages of the dual-degree ARCE program have become evident to faculty even during the last 2 years of students' undergraduate experience. Compared with engineering education alone, architectural training gives ARCE students superior presentation skills, judgement in aesthetics, project planning and organizational skills, and superior ability to synthesize the many elements of design for total performance. Engineering education, on the other hand, produces a broad knowledge of basic and engineering sciences, and specialized structural design knowledge, affording the student greater depth of understanding and confidence in design than most architecture students possess. Surely the added maturity of students in the fifth year and beyond helps distinguish dual-degree ARCE majors from single-degree majors in either field. The added confidence which comes from broader education and greater maturity gives dual-degree majors advantages not only in scholastic endeavors, but also in leadership roles. These advantages continue into the professional careers of graduates of the ARCE program.

We have observed through the years that only a small percentage of graduates follow the career path they predicted for themselves as undergraduates. Not only do personal interests and circumstances influence choices but changes in the national economy and other factors influence career opportunities. Because the dual-degree ARCE program provides thorough education leading to accredited degrees in related, interactive--yet very different--disciplines, it affords rare career flexibility in practicing engineering, architecture, or both. For some, this flexibility might be the most important benefit of the dual-degree program.

Effective Teaching Using Semi-Notes

M. U. Hosain[1], M.ASCE

Abstract

This paper provides an introduction to an instructional technique which has proved to be very effective in teaching engineering courses. The technique involves the use of semi-notes, i.e. prepared lecture material in which blank spaces are left in intentionally. The students are only required to fill in the blank spaces while the lecture is being presented. This half-finished format is a compromise between providing students with no notes and providing them with a complete set of notes. The semi-note technique tends to increase the level of attention in the class room and improves the teaching effectiveness of the instructor. The method also allows an instructor to reduce the time required to cover course material, without sacrificing the quality of instruction. It provides a suitable start for a novice professor and is ideal for an instructor who is not that proficient in the language of instruction.

Introduction

Over the past decade in most universities in United States and Canada the undergraduate civil engineering curriculum has been subjected to some sweeping changes. Besides having a firm foundation in engineering sciences, an engineering graduate must be well trained in computer technology, be an effective communicator and a capable manager. In addition, the curriculum must include many nontechnical courses so that the students can acquire an understanding of environmental, social, economic and political issues. With all these new areas of study, the time available for teaching technical subjects has been considerably reduced within a four - year program of study. The

[1]Professor, Civil Engineering Dept., University of Saskatchewan, Saskatoon, Canada, S7N 0W0.

method of instruction must be improved so that students can acquire a clear understanding of the essential elements within the reduced time limits. After 20 years of teaching, the author is convinced that the desired efficiency can be achieved by utilizing "semi-notes".

Since the development and welfare of society depends to a large extent on engineering, the demand for engineers is steadily increasing. According to Brighton(1989), the number of engineers employed in the United States increased by 7% annually from 1972 to 1986 and that this figure is expected to reach 9% in the near future. Brighton(1989) also reported that as many as 25% percent of the present engineering faculty in the United States would reach retirement age within the next 6 years. These vacancies as well as new faculty positions, resulting from an anticipated increase in engineering enrollment, will be filled by young engineers with postgraduate degrees. Because of a reluctance on the part of U.S. students to enter graduate schools, most of these positions will likely be filled by engineers who received their undergraduate training in countries where English is not the common language. Unless there is a sudden change in the attitude of U.S. students, this situation may not be avoided. Most of the novice professors will plunge into teaching without any formal training in the art of teaching. Lack of experience, coupled with the fact that most of them may not be proficient in the English language, will probably produce poor results. Adoption of the "semi-note" technique by these novice professors is likely to improve the situation to some extent.

The objective of this paper is to provide an introduction to an instructional technique involving the use of "semi-notes" and to discuss it advantages and limitations.

Semi-Notes

As the name suggests, semi-notes consist of prepared lecture material in which blank spaces are left in intentionally. Fig. 1 shows a typical page from the author's structural analysis class notes. The distinguishing feature is the half-finished format. Students are only required to fill in the blank spaces while the lecture is being presented. This half-finished format is a compromise between the practice of providing the students with a fully prepared set of notes and the presentation of blackboard oriented lectures without the use of any prepared lecture material.

The author originally adopted this technique mainly due to

two reasons. First, he was convinced that a totally blackboard oriented lecture is not suitable for engineering courses even if the lecturer is a good communicator. This is due to the fact that engineering lecture material invariably contains complicated diagrams, tables etc. Copying

2.16

Example 2.8

GIVEN: A simple truss with two load cases. Assume AE = constant.

REQUIRED: (a) The horizontal deflection at point 3 for load case 1
(b) The vertical deflection at point 2 for load case 2
(c) Explain why the two values are the same
(d) Verify the results using the computer program PTRUSS
(Assume A = 2 000 mm^2, E = 200 000 MPa and L = 3 000 mm)

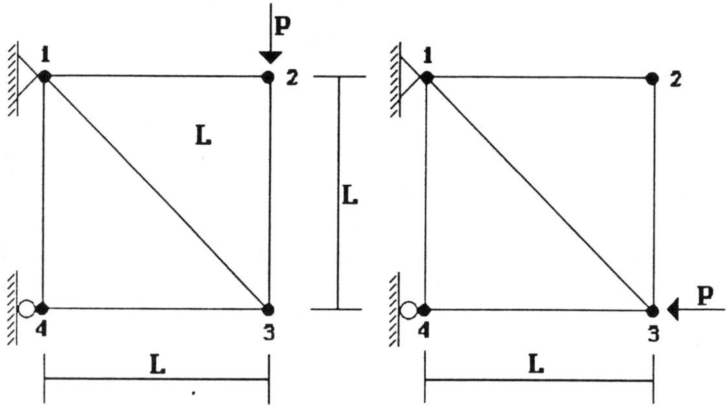

Load Case 1 **Load Case 2**

SOLUTION: Use Dummy Unit Load Method $\delta = \Sigma S\mu L/AE$

STEP 1 Case 1:

(continued)

Fig. 1 An Example of Semi-Note: Half-finished Format

material off the blackboard becomes the principal task of the student leaving no time to follow and understand the lecture material. This style of lecturing depends too much on the lecturer's eloquence and vocabulary, and therefore should not be the choice of a person who has difficulty in communicating in the language of instruction. For the author, who speaks English with an East Indian accent, this type of lecturing technique was ineffective. Moreover, a totally blackboard. oriented lecture is very time consuming.

Secondly, the alternative popular method of providing the students with a complete set of notes and going over the notes using overhead transparencies did not work out for the author either. This method encouraged absenteeism and somehow failed to generate an enthusiasm among those present. Lack of eye contact is partly to blame. Utilization of semi-notes provided a compromise between these two techniques. It permitted rapid progress while allowing eye contact between the instructor and the students.

Advantages of Using Semi-Notes

The most important advantage is a noticeable improvement in teaching effectiveness. This is mainly due to the fact that the use of semi-notes is synonymous with "organization" which is the key to a succesful lecture. In survey after survey, students at the University of Saskatchewan have identified "organization" as the most important common trait amongst superior lecturers. The use of semi-notes compels even the most disorganized professor to become systematic. Careful preparation of semi-notes ensures that important items such a diagrams, equations and anything that may cause ambiguity are already clearly defined (Fig. 2).

Another major advantage of using the semi-note technique is an efficient utilization of the allotted lecture time. As indicated earlier, changes in engineering curricula have placed new demands on the present generation of students and instructors. It has become necessary to instill in our students essential knowledge in a shorter period of time. Major emphasis must still be placed on the teaching of engineering fundamentals (Hosain 1985). Utilization of semi-notes allows an instructor to reduce the time required to cover course material, without sacrificing the quality of instruction.

The two advantages discussed above are very important. However, the author feels that the "essence" of semi-note technique lies in its ability to increase the "attention level" of the students. Education psychologists (Hilgard and Bower 1975) have found that the level of attention usually ranges from:

(a) Euphoric excitement
(b) Alert Attentiveness
(c) Relaxed Wakefulness
(d) Drowsiness
(e) Deep Sleep

3.32

Example 3.7

GIVEN:

3/4 inch diameter A-325 HSBs designed in bearing, threads excluded from the shear plane. Plate A is 6 mm thick and the column flange(plate B) is 13 mm thick. The steel is G40.21M-300W type.

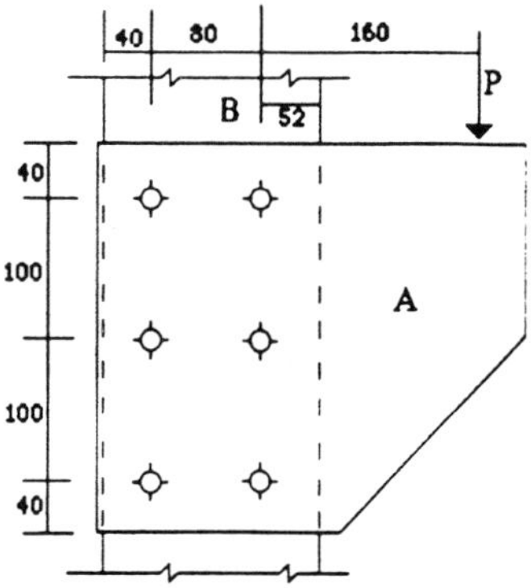

REQUIRED: The maximum factored load P on the basis of the fasteners
(a) Use the instantaneous centre approach(Handbook Tables)
(b) Use the traditional approach
(c) Explain why there is a difference in the answers

SOLUTION: Part(a): Instantaneous Centre Approach

<u>STEP 1</u>

(continued)

Fig. 2 A Typical Page from Author's Steel Design Course

The first state normally lies in the domain of rock concerts and some religious gatherings. Also, some politicians have the ability to bring the audience into a state of frenzy. In an average class-room, most students are, depending on the time of the day, either in a state of "alert attentiveness" or "relaxed wakefulness". However, a monotonous lecturer can quickly alter the situation. In fact, some professors are known to put students in a state of "deep sleep" on a regular basis. After 20 years of teaching, the author has arrived at the conclusion that a certain amount of copying off the blackboard, as required in the semi-notes method of instruction helps to keep the students attentive. This is perhaps also the reason why simply going over prepared lecture material using an overhead projector does not produce good results.

The author prefers to use the blackboard for routine lectures although a companion set of the lecture notes is available on transparencies (Hosain and. Hauer 1984). The transparencies are used only for reviewing material covered in the previous lecture before a new lecture begins. Adoption of this concept has improved teaching effectiveness considerably as evidenced by the results of course evaluations conducted by the department head over a number of years.

Other Advantages

1. Temporary absence of the instructor due to other obligations causes little disruption. Provided with the filled-in notes, a graduate student can perform quite adequately. On many occasions, the author received positive responses from the students concerning the performance of graduate students. Comments such as " with the type of notes we have, how can he go wrong" are not uncommon.

2. The author had no difficulty finding replacements when he went on sabbatical leave. Three of his former students, working with local consulting engineering firms, promptly agreed to teach the required courses on a part-time basis provided the filled-in notes are given to them.

3. Once the lecture notes have been carefully prepared, very little time is required to prepare day-to-day lectures. Thus, more time can be devoted to research and other scholarly activities.

Disadvantages

1. Lecture notes may not be published "as is" as a book.
2. Certain amount of wastage of paper and extra xeroxing costs are integral part of semi-notes.
3. Takes considerable time and effort to develop a good set of semi-notes.

Conclusions

1. The semi-note technique increases the level of attention in the class room and improves the teaching effectiveness of the instructor.

2. Not being wholly dependent on verbal communication skill, the semi-note technique is ideal for professors who are not that proficient in the language of instruction. On the other hand, utilization of semi-notes makes it easier for foreign students to follow lecture material and do well in the course.

3. The method would provide a suitable start f or a novice professor At the beginning, the notes may simply contain a fews headings and diagrams. Gradually these pages can be converted into a decent set of semi-notes.

4. It is ideal for professors who are heavily involved with research or who must be away a lot. Graduate students will be able to fill in for them quite adequately. Occasionally, xeroxed copies of the filled-in. notes may be provided if a substitute is not available.

References

Brighton, John A. 1989. Engineering Faculty for the 1990s and Beyond. Engineering Education, ASEE, Vol 79, No. 5, July/August, pp. 538-539.

Hilgard, Ernest R., and Bower, Gordon H. 1975. Theories of Learning, Prentice-Hall, Fourth Edition.

Hosain, M.U. 1985. A Simple and Effective Computer Aided Instruction System. Proceedings of the Second International Conference on Civil and Structural Computing, CIVIL-COMP 85, London, England,December, pp. 479-481.

Hosain, M.U., and Hauer, Pamela 1984. Utilization of Microcomputers in Teaching a Course in Structural Analysis. 1984 ASEE Conference Proceedings, Salt Lake City, June, pp. 850-854.

THE CIVIL ENGINEER: SCIENTIST OR TECHNICIAN?

J. M. Roesset,[1] F.ASCE, and J. T. P. Yao,[2] F.ASCE

Abstract

Civil engineers ". . . plan, design, and supervise the construction of facilities essential to modern life in both the public and private sectors . . ." (from "Is Civil Engineering for You?", ASCE, 1988-89). As such, we need to have (a) a good understanding of basic sciences; (b) a broad-based coverage of interdisciplinary concepts; (c) an efficient and effective way of written, oral, graphical, and computer communications; and (d) the capacity to become leaders of project teams (Guy Jester, 25 October 1988). In this paper, an attempt will be made to delineate civil engineering education from those of pure sciences on the one hand, and low-level technology on the other. The main purpose of this paper is to stimulate discussion on the development of civil engineering curricula with which we may educate engineers who not only can solve societal problems of today but also face the many challenges of tomorrow.

Introduction

By definition, civil engineering deals with the application of scientific knowledge to the planning, design, construction, and operation of physical systems and facilities. Considerable differences of opinion exist among engineering educators as well as between academicians and practitioners on (a) the relative importance and proportion of basic courses versus practical design and construction offerings, (b) the amount of liberal arts courses that should be required, (c) the relative emphasis on the "why" versus the "how to" of engineering curricula, and (d) the distinction between education and training.

---

[1]Department of Civil Engineering, The University of Texas, Austin, TX 78712-1076.

[2]Department of Civil Engineering, Texas A&M University, College Station, TX 77843-3136.

Our mental image of the ideal engineer is often a function of our own experience and needs in research, education, or practice. Yet, the tremendous diversity of endeavors and the ever-increasing breadth of the civil engineering profession dictates that there is a continuous spectrum between the applied scientist and the technician. In fact, our society does need many different kinds of civil engineers.

The civil engineer, who would be in control of a complete project from the planning stages and the conceptual design to the detailed analyses and dimensioning, the construction, and the eventual operation and management of the facility, needs the assistance and cooperation of many experts in the different phases of the project. As an example, the critical infrastructure problems facing the nation today call for educational programs in high-level inspection and monitoring, system identification, reanalysis and redesign, rehabilitation, private and public financing, and management, in addition to the more traditional ones in planning, analysis, design, and construction. It would be unrealistic to expect that any civil engineer can become an expert in all these different disciplines.

The main purpose of this paper is to stimulate discussion on the development of future civil engineering curricula by attempting to delineate the essential qualities for all civil engineers and to identify those factors that affect engineering education. We must remember that civil engineers are concerned primarily with the basic needs of mankind: shelter, water supply, waste disposal, transportation, and preservation of the environment. As Harold T. Larsen pointed out in his article on engineering in the <u>Encyclopedia Americana</u>, "Modern engineering is different from ancient engineering mostly because the pattern of living is different, and engineering in the future again will be different in the same degree, proportionate to the utilitarian needs of a changing world." Societal needs have a strong influence on the role and education of civil engineers, and the values and goals of a society equally influence the ability to recruit good students for an engineering career and the type of education these students are looking for. It is thus important to look at the history, evolution, and future of civil engineering in the context of the social environment.

Scientists Versus Technicians

Some engineers in both academia and practice feel that anyone who knows how to evaluate an integral is a scientist. Others believe that anyone who is not conversant with Hilbert spaces is merely a technician. These two extreme viewpoints are rather comical, albeit real, distortions which are not worthy of further discussion herein.

A more common complaint among practicing engineers is that graduates of some of our best schools are of little use in practice because they require one or two years of practical training on the job before they can become productive. The implication is that these students have a strong background in applied science but very little exposure to the practical aspects of engineering. The usual response is that any student who would be of immediate use in a design or construction firm, after only four years of study, would be little more than a technician, and that given the limitation in the amount of knowledge that can be transmitted in an undergraduate curriculum, it is more appropriate and natural for universities to concentrate on theory and for industry to provide the practical experience. These two viewpoints deserve more careful consideration because there is a substantial amount of truth in the arguments presented by both sides.

Most undergraduate curricula have all too often a very limited practical component and this is even more so in programs where a large fraction of the students are expected to go on to graduate school. Some graduate programs do emphasize design, construction, or management with a substantial practical content. Others are intended to provide the students with more in-depth knowledge of a scientific discipline or of new important and emerging methodologies. If we associate the concept of an engineering scientist with an orientation towards research, we must admit that many students are indeed better prepared to become high-level specialists than practitioners. On the other hand, if we associate the concept of technician with a person trained to be an expert in a rather narrow area with more emphasis on his ability to do things according to present standards and codes than on the understanding of the underlying reasons and principles, we must also admit that students trained to satisfy the immediate needs of industry at a particular time, whether in routine reinforced concrete and steel design, in construction supervision, or in computer programming, are basically technicians. So are some engineers with graduate degrees who specialize in the use of a particular computer program or in the use of specific methodologies. This in itself is not bad since our society needs all these different types of persons, and all of them can make significant contributions. Traditionally, civil engineers provide the link between the scientists and the technicians and assume the leadership role in the planning, construction, or operation of large civil engineering works. The question is how much of the education of these engineers can be provided by a university or acquired through practical experience.

Quoting again Harold T. Larsen from his article in the Encyclopedia Americana, "Although the distinguished feature of modern engineering is its dependence on scientific fact, an engineer starts down the road to failure if he considers his professional equipment complete when he has passed a required number of courses at college. The hallmarks of ancient engineering, native wit, ingenuity, and practical experience, still are necessary ingredients and without them engineering falls short of being a profession." This statement points out again the fact that the education of an engineer cannot be complete after four, five, or even eight years of college education. It also indicates that the desirable qualities of an engineer, in addition to some scientific knowledge, are native wit, ingenuity, creativity, and experience. One could add to this list the ability to think, synthesize, and lead. These are desirable qualities in all forms of human endeavor. The remaining questions are then whether these are qualities which are innate or which can be taught. More importantly, whether our present educational system tends to enhance or rather to inhibit these desirable qualities.

Education Versus Training

Ideally, the goal of a university should be to provide an education, which involves developing in the students the ability to (a) think; (b) perform critical analysis in order to identify the reasons and controlling factors of various situations; (c) predict the consequences of different courses of action; (d) synthesize the acquired knowledge and the results of the analyses; (e) develop self-confidence, tempered by a realization of one's own limitations so as not to become arrogant; (f) appreciate and enjoy the cultural legacy of mankind; and (g) make decisions and assume responsibility for one's own actions. Education can thus be viewed as the enhancement of the human mind and spirit, learning how to deal with a variety of situations one may encounter in life, and how to behave in the social environment. Universities could fulfill this role very well when only a small percentage, the elite, of the population had access to higher education.

The role and mode of operation of universities underwent drastic changes with the access of large segments of our population to higher education. On one hand, education was no longer perceived as a goal in itself but as a means to secure a job in a competitive market. The implied need to introduce and emphasize more practically-oriented courses in developing specific skills makes the distinction between education and training more difficult. On the other hand, the large enrollments in the most basic courses have made it extremely difficult for faculty members to provide students with the desirable personal contact, supervision, and tutoring. The teacher assumes sometimes the role of an examiner, whose duty is to weed out a certain fraction of the class, rather than an educator. This represents a clear deterioration of the quality of education, and it is most regrettable that it affects primarily the basic and most essential courses.

A final significant factor has been the changing emphasis of many universities from educational to research institutions and the perception of research as a source of needed revenues, rather than a scholarly activity. Research is a vital and necessary component of graduate education but generally it has little effect on undergraduates at present. The best researchers, who are also often the best teachers, may do little teaching since their time is considered more valuable in other activities. The mounting pressures on young faculty members to generate research funds quickly and to produce in a few years a large number of publications provide them with little incentive to devote a major effort to course preparation or interaction with undergraduate students. New faculty members are normally hired directly out of graduate school with minimal exposure to engineering practice. Moreover, they are often selected on the basis of their expertise and specialization in a new methodology or emerging technology, which is considered important to an ongoing research effort or to the generation of new research funding. The result is that the number of engineering educators who have had any involvement with practical issues, beyond consulting activities in highly specialized topics, is rapidly decreasing.

All these changes have resulted in a deterioration of the quality of general undergraduate education, while improving perhaps our ability to produce researchers and high-level specialists.

Society's Influence

A number of programs have been created at the national and local levels to attract good students from secondary schools to engineering by informing them and their teachers about the jobs that engineers do and how they serve society. While all these programs are worthwhile, their impact is likely to be limited in a society where success and respect are measured in terms of financial status. Starting salaries in traditional civil engineering jobs are low in comparison to those of other professions. Thus, if we want to attract the brightest students to civil engineering, it will be necessary not only to make them aware of the accomplishments of the profession but also to improve the financial rewards and to make clear the possibilities of advancement. In societies where the basic needs of housing, transportation, water supply, and energy are still unfulfilled, civil engineers have a higher social status, and, correspondingly, a much larger proportion of the brightest students are interested in pursuing this career, as illustrated by the large number of talented foreign students in many civil engineering programs.

Students enrolled in civil engineering tend to select courses leading to specialties where they perceive there are more abundant or better paying job opportunities. In many cases these are not the more traditional offerings. Thus, if there is an increase in the number of high-level specialists on one hand, and technicians on the other, it is not only the result of the changes in the orientation of universities but also a consequence of society's demands and opportunities.

Civil engineers must be able to communicate effectively. The actual extent of the communication skills and general knowledge of a nontechnical nature, which may be required, is a function of the social environment. The engineer who wants to play a leadership role in the planning, construction, and operation of important facilities and in the society must be able to understand political and financial trends and to be aware of legal constraints. Nevertheless, it would be shortsighted to consider the liberal arts component of an engineering curriculum only in terms of its utilitarian value in the practice of a profession, ignoring their formative value and the enrichment of the student as a human being.

A society which is becoming increasingly prone to litigation also becomes naturally risk averse. There is a reluctance to assume responsibility and to make decisions based on one's judgment, relying instead on numerical values whether they make sense or not. Creativity and originality become dangerous since the tried and true routine solutions offer the smallest risk. This type of environment affects seriously the very qualities we are looking for in a leader.

We must also realize that the curricula of universities must reflect the degree of preparation of the students coming out of secondary schools. The quality of secondary education is again a function of society's values.

University-Industry Collaboration

The ultimate responsibility for the planning of engineering curricula must rest on the shoulders of the faculty members. Meanwhile, it is desirable and necessary to seek the advice and viewpoint of the profession whenever we plan significant changes in engineering curricula.

If we accept the premises that there is an insufficient amount of real life, practical exposure in undergraduate curricula and that the education of a civil engineer cannot be complete after taking a required number of courses at a university, we must conclude that there is a need for a stronger collaboration between universities and industry. A number of such collaborative efforts are already in place. It would be interesting to evaluate the success of these programs and the degree of satisfaction on both sides and to determine whether they should be expanded.

Many universities make use of practicing engineers to teach some engineering courses as adjunct professors. A possible solution is to have an arrangement where practicing engineers return for a semester or a full year to a university to help in team teaching of design synthesis courses with a faculty member, bringing in real-life cases and the practitioner's viewpoint, and taking in return courses on new developments or advanced methodologies. The practicing engineer could also participate in ongoing research projects which might be of interest to the firm. Such an arrangement can be highly beneficial for both sides if carefully planned and implemented.

Faculty members could also spend their sabbatical leaves working in engineering firms and getting the exposure to current practice. A point of concern is whether spending this time getting practical training might be considered a worthwhile activity by the university.

Conclusion

In his book From the Earth to the Moon, Jules Verne wrote "The Yankees, the first mechanicians in the world, are engineers - just as the Italians are musicians and the Germans metaphysicians - by right of birth." We feel that this sentence would not have been written today because the quality of American engineering is falling behind that of several other countries. The practice of civil engineering is very different today from what it was in the past and will be very different again in the future. Change for the sake of change is a waste of time and effort. However, we must adapt to changes and attempt to influence it in the right direction. If we do not, many initiatives and challenging tasks facing the profession will be lost to engineers in other disciplines by default. It is thus timely and important for us to begin to reevaluate our goals and objectives and to review in detail our civil engineering educational programs (undergraduate, graduate, and professional development). It appears in particular that we should:

1. Review carefully the quality of the education in the basic sciences and liberal arts, looking not only at their utilitarian purpose but also at their formative value. Because these courses are not taught by civil engineers, all too often, we tend to disregard them or to consider them as unpleasant requirements.

2. Reevaluate the common core of courses that all civil engineers must take and the specialization electives in light of the changes in civil engineering practice.

3. Review in detail the content of all courses and their mutual interactions.

4. Introduce as much as possible a practical component in all courses. It is important, for instance, in courses that deal primarily with methodology to select examples and homework which illustrate real-life applications.

5. Reintroduce in civil engineering curricula synthesis courses where students get exposure to all the phases involved in the planning, analysis, design, and construction of a given type of facility.

6. Decide, in collaboration with industry, how much of the education of a civil engineer can be accomplished in a university environment and how much must be completed in engineering practice.

7. Establish better programs of collaboration between the universities and industry on undergraduate education, reeducation of practicing professionals, and joint research efforts.

8. Decide whether four years are sufficient to provide civil engineers with the academic education which is required. In case of need, all universities should act in a decisive and concerted way to define new undergraduate programs.

9. Review graduate programs from the viewpoint of their coherence and their relation to the research effort.

10. Review continuing engineering education programs.

Many of the reviews recommended above are being conducted periodically at various universities. Due to time constraints, they are often performed routinely without the in-depth and far reaching treatment. It should also be noted that the ideal curricula or modes of operation are not necessarily the same for all universities. Private institutions with high selectivity in their admission procedures face a very different situation from some state universities with other admission requirements and much larger enrollments in accomplishing the noble and worthy goal of mass education.

A Required Capstone Design Course-Is It Worth It?

John W. Smith, MASCE[1]

Abstract

The faculty of the Civil Engineering Department chose to implement a senior level design course to allow the student the opportunity to experience the practice of engineering prior to his/her graduation. This summary or capstone course was implemented at some cost to the curriculum and considerable effort by the faculty to develop a meaningful course. Beginning with the ABET definition of design and drawing on the practical experience of the faculty, a course was molded which continues to be modified five years later. The course centers around the concept that to learn design one must do design. Verbal and written expression is a major thrust of the course as indicated by its inclusion in the writing intensive portion of the University's General Education Requirements. In addition to the traditional learning mediums of lecture and homework, the students participate in buzz sessions, team design projects, and discussion sessions with practicing engineers. The products of the course include engineering reports, plans & specifications, team design participation, and the experience of being a "design engineer". Only time will tell if the course is truly worth the curriculum revisions and the work by the faculty. The product,i.e., the young engineer, has found ready acceptance in the job market. The appreciation of the student for the opportunity to "put it all together" may be the best measurement of whether it is worth it or not.

---

[1] Professor, Department of Civil Engineering, Memphis State University, Memphis, TN 38152

Introduction

The Civil Engineering program at Memphis State University began in 1969 as a traditional program of its time and has been fully accredited by ABET ever since. The program included the traditional design courses ("steel design, reinforced concrete design, wastewater system design") coupled with design emphasized in many courses.

The faculty recognized the need for the opportunity for the student to "put it all together" in 1983. The impetus for this action was the belief of the faculty that the students would benefit from the opportunity to bring together the analysis competencies and design fundamentals obtained during their undergraduate career. Given the "traditional" 132 semester hour program, an obvious conflict existed over what do you give up to incorporate such a course. Do you delete a required course? Do you expand the curriculum and depart from the norm of your peer departments? The faculty chose to increase the requirements for a degree to include not only the capstone course but also a communication course with the total required course credit rising to 140 semester credit hours.

Looking back over five years, was the decision worth it? Has the course accomplished what was envisioned? Has the cost of the increased number of required credit hours impacted enrollment? To answer these and other questions, one must look at the course development, the course content, the course expectations, and finally the course products.

Course Development

The course as it is taught today is both similar and different than the original course offered in the Fall 1984 term. The basic definition of what is meant by "design" remains the same. Using the ABET concept as a guide, design is taught as including the process of defining a problem, generating alternatives, analyzing alternatives based on pre-set criteria, defending the selected alternative, and performing the detailed design of the recommended alternative including plans and specifications. The class room approach and use of reference materials have changed markedly in the past five years. In addition, the course has recently been designated as a writing intensive course for the University general education program which places greater emphasis on the written portion of the course.

> The ABET accreditation state:
> ". . . the engineering design component of a curriculum must include at least some of the following features: development creativity, use of open-ended problems, development and use of design methodology, formulation of design problem statements and specifications, consideration of alternative solutions, feasibility considerations, and detailed system descriptions. Further, it is desirable to include a variety of realistic constraints such as economic factors, safety, reliability, aesthetics, ethics, and social impact.
> and
> Courses that contain engineering design normally are taught at the upper-division level of the engineering program. Some portion of this requirement must be satisfied by at least one course which is primarily design, preferably at the senior level, and predicated on the accumulated background of the curricular components."

Due to University requirements of implementing new courses, the capstone course was offered as an elective for the first three years before it became required for all Civil Engineering students.

From the beginning, the course has centered around an actual design. The selection of a text book was difficult in 1984 and is difficult today. The text written by Jewel (A Systems Approach to Civil Engineering Planning and Design) has been used most often. Recently, the ASCE publication, A Guide to the Construction of A Quality Project, has been used for portions of the course. To allow the students the maximum opportunity for guidance and interaction with other students, the course was originally developed as two fifty-five minute lecture periods per week coupled with one three hour laboratory period. The laboratory time is used for computer access, outside speakers, team design meetings, and discussion sessions.

The latest major development in the history of the capstone design course has been the designation by the University of the courses as a writing intensive course for the General Education Requirements. In essence for a course to qualify for this purpose, it must require the student to write frequently, to write voluminously, and to write varied types of assignments.

Course Content

The major thrust of the course has to be the design experience. To complement this experience, as shown on the following condensed syllabus, the lecture periods are used to present material on systems approach to design, design/decision criteria and where it comes from, alternative generation, quality in a design, design team selection and responsibilities, risk and uncertainty and statistics, and specifications from the design engineer's view. In addition, students are required to sit for a fundamentals of engineering exam to reinforce their understanding of engineering basics. This component of the capstone course was originally intended to force the student to study for the EIT exam; however, it has served as a useful vehicle to reinforce the fundamentals as they apply to the final design product and their importance.

CONDENSED SYLLABUS
Civil 4199

Subject	Lecture Periods	Lab Periods
systems approach	2	1
design team concept	3	
alternative analysis	6	2
data analysis	4	2
project management	4	2
quality in design	6	2
design review	5	2
exams		2
presentations		2

Students are required to complete and submit bi-weekly time sheets accounting for their time in the capstone class and their effort on the two designs for which they are responsible. This small element of reality of time accountability is used to move the student toward the "real world". In addition, each student is required to complete a "design journal" which is essentially class notes and notes gathered while the student does background work related to their design. This journal is periodically collected and reviewed by the instructor.

The lectures and exams are almost secondary to the major thrust of the course - two design projects. Each student is required to complete two designs during the term, one an individual design and one a team design. The student can propose an individual design or one is assigned to him/her. This project is carried to completion by the student with assistance as needed by the instructor. The student is responsible for all data collection, design criteria determination, decision criteria, etc. Periodic project reports and a final design report are submitted. and graded for not only content but also presentation. In addition, each student presents his/her design to the class and an invited jury of practicing professionals and faculty. Typical of the individual projects is a simple retaining wall, a landfill, sludge composting process selection, hydraulic and structural design of a culvert, column design for a two story office building, and the traffic control system for a street intersection to name a few. Each design is defined by the student with assistance from the instructor.

The second design is a team design in which the students are separated into discipline related teams (i.e., transportation, structures, foundations, etc.) to attack a design as a class. Students must depend on the other teams for data and system compatibility. One person or team subset is sometimes designated as project manager and assumes the responsibility of coordinating all the other teams. The team design is also presented in both written and oral formats by the students. Team designs are defined by the instructor as the owner who includes any constraints for the design team, i.e., economics, time, space, etc. Typical of projects undertaken by the class are bridge designs, a parking garage, a research facility, a landfill, and a wastewater treatment facility.

In both design settings, the student is required to carry the design from conception to plans and specifications. The importance of a good engineering report to support the plans and specifications is also a part of the design experience of the capstone course.

Course Expectations

Students successfully completing this course will be able to perform analyses of alternatives, prepare a detailed design of the recommended alternative, describe the design procedure and detailed design both orally and in writing, and demonstrate a mastery of selected engineering fundamentals. Students will be expected to submit all non-problem assignments in typewritten form, preferably using a word processor on a P.C. The use of a P.C. for alternative analyses and detailed design is encouraged.

Because this course is considered as a "writing intensive" course under the General Education requirements at Memphis State University, all assignments are graded for style, grammar, spelling, etc., in addition to engineering content. All work is corrected and returned to the student and (if necessary) required to be resubmitted. Students are expected to satisfactorily write a letter of transmittal, a letter of inquiry, progress reports (at least 2), final design reports (2), and a set of specifications.

Products

The products of the course are both tangible and intangible. Each student will leave the course with an engineering report he/she has prepared for his/her individual project. The goal of the course is for this report to be of such quality that it can be used as a marketing tool for the student to use in seeking employment. The student will prepare specifications for both design products. Typical plans are prepared so as not to turn the course into a drafting experience.

The intangible experience of "doing a design" is also a product of the course. Allowing the student to work through a real life problem under controlled conditions increases the confidence level of the graduate. The intangible experience of working as part of a design team is also a product of the course. Experience gained in dependence on others in the team cannot be measured, but is important once the student leaves the University.

Is It Worth It?

This is the question asked initially. In all probability, only time will answer this question. The perceived cost of the expanded curriculum did not materialize. Other peer programs have come to the same realization as MSU and have expanded their number of hours for a degree. The anticipated work load to develop the course was underestimated; however, this is part of academia.

The value of the course is best seen in the appreciation the students have developed in their ability to tackle a real live problem and bring it to completion. The realization of what they have not learned from their dispersed "analysis & design" courses is also of value. The acceptance of the product by the engineering community has been excellent. The students who have participated in the design experience have proven to be productive to their employers in very short time periods.

Practitioner Participation in Design Education

Fred Videon, M.ASCE[1]
Ken Munski, M.ASCE[2]
Gary Simonich, M.ASCE[3]

Abstract

Civil engineering design at Montana State University is taught in a one-year course using projects provided by engineering practitioners. Students must design a multifaceted project which involves defining the scope of work and managing the project from its inception to the completion of design documents. Design teams of three or four students must prepare a proposal for engineering services giving the detailed approach to the design. Then they must complete the preliminary and final design following the schedule outlined in their proposal. Written and oral reports are required at several stages during the design. Students must describe their work to non engineers in some of the reports. Ethical and professional issues frequently must be addressed during the performance of the design.

Introduction

During the past several years, increased emphasis has been placed on having open-ended design included in engineering curricula. The ABET General and Program Criteria regarding design at the senior level in engineering programs have provided the impetus for engineering faculty to develop innovative approaches for design education. These criteria recommend design projects that have many of the elements of design creativity found in engineering practice. The ABET recommendations have been implemented in the senior design course at Montana State University which was developed by a team of practitioners and faculty. The purpose of this paper is to describe the development and implementation of this course.

---

1. Professor Emeritus, Montana State University, Bozeman, MT

2. Manager, Chen Northern, Great Falls, Montana

3. Director, Structural Division, HKM Associates, Billings, MT

Background

The senior design course at Montana State University developed from a senior project course in structural engineering that had been taught for several years. Originally the structural design course was a one-quarter course that dealt with the final design of a small structure. However, since this course only lasted one-quarter, it was found that students were not exposed to many aspects of planning and conceptual design that are so important in the actual design process. Therefore in 1984 the course was modified to be a full academic year. In 1986 a further modification was done to change the course to general civil engineering design. During the initial planning for this final modification, it was concluded that practitioners should work with the faculty to develop a course that would represent to the students all facets of the design process. Thus, the Montana Section of ASCE organized a committee of practicing engineers and faculty to develop such a course. The present senior design course was planned by this committee.

The ASCE Practitioner/Faculty Committee concluded that the senior design course should include elements that are normally encountered during an actual design project. These elements are:

1. Planning the scope of work for the design of a project,

2. Data acquisition,

3. Conceptual design,

4. Preliminary design,

5. Final design of the recommended alternative, and

6. Preparation of a final design report.

Further, it was concluded that students should learn to deal with engineer/client and engineer/public relationships during the design. Also, they should learn to work as a member of a design team and serve as a project manager for a portion of the project design. Finally, they should learn to communicate the results of the design at key stages during the design process.

The recommendations of the ASCE Practitioner/Faculty Committee were used to develop the design course at Montana State University. The course has been taught for three years as an elective course and will become a required course in 1990.

Description of Senior Design

The course, Senior Design, is taken for a full academic year. Students normally register for two credits per quarter. Occasionally one or two students will register for only two consecutive quarters and, in such cases, the project design teams must scope their work accordingly. Projects for the course are furnished by practicing engineers. Students work in teams of three or four on these projects under the direction of a faculty adviser and a practitioner adviser. The course is preceded by a two-day workshop on professional engineering practice. After completion of the workshop, the course format generally includes the following items: 1) formation of design teams, 2) development of project proposals, 3) conceptual design and preliminary engineering and 4) final design.

Professional Practice Workshop

Most students have had little or no experience working with an actual engineering design project. Therefore, students learn how a design project is organized in a professional practice workshop that is taken before they begin the design projects. The workshop is taught by two practitioners and one member of the faculty. Usually one of the practitioners is from a consulting firm and the other from a government agency. The workshop is a two day session that begins the day before classes begin. This schedule is very efficient in terms of the time commitment of the practitioners who teach the workshop and the students who take it.

The professional practice workshop emphasizes planning for the engineering design of a project. A case study of an actual project is used by the instructors to demonstrate how engineering tasks are identified and how the scope of work is determined. The primary objective of the case study is to develop a project proposal which includes a detailed plan of work for accomplishing the design tasks. Following the presentation of the case study, the students are assigned a similar project and asked to prepare a "Proposal for Engineering Services". Students work in teams of three or four to prepare their proposals. During the final four hours of the workshop, the teams give oral presentations of their proposals to a review committee composed of the course instructors. These presentations are followed by a critique and general discussion of the proposals. Students use the comments made during the critique to refine their written proposals which they have about ten days to complete and submit to the faculty instructor.

Formation of Design Teams

Design teams of three or four students are organized during the first two weeks of the course based on student's interests and qualifications. Students are assigned to projects based on their responses to "Solicitations for Engineering Services" announcements for the projects that have been furnished by consulting engineering firms. A typical announcement for a project is given below.

> "The Board of Directors of the Highland Subdivision Corporation is soliciting interest and qualifications from engineering firms to provide engineering services for the design of a subdivision of 30 to 35 units. The subdivision is to be on a 9.6 acre tract bounded on the east by Highland Boulevard, on the north by the Deaconess Foundation property, on the west by Josephine Park and on the south by Hilgard Subdivision. Services required include subdivision layout including location of streets and utilities and design of streets, sewer, water and storm drainage."

Each student will select a first and second choice project and prepare a statement of interest and qualifications for their selections. Experience has been that students have always been assigned to the project of their first or second choice.

Development of Project Proposals

After the design teams are organized, each team receives a "Request for Proposals" (RFP) for their project. The RFP's are detailed problem statements which outline the engineering services required by the client. The RFP's give information such as project location, required items of work, sources of information, prior studies, client contact and total project budget. The design teams use the RFP's to prepare their "Proposals for Engineering Services" for the projects. These proposals are similar to, but more detailed than, the proposals that the students prepared during the Professional Practice Workshop.

The design teams spend four to six weeks developing proposals. During this time each team will visit the project site and meet with the practicing engineer adviser. The proposals are a vital part of the project since the finished proposals contain the work plan and schedule for completing various phases of the design. Consequently, the proposals receive careful review by the faculty and practitioner advisers to insure that each team has the necessary resources to accomplish the work they have proposed.

Design teams present their proposed scope of work to the entire class which plays the role of client's representative. Teams normally have a full class period to present the detailed scope of work and to respond to questions. The written proposals are used by the design teams and the advisers as the official planning document for the design tasks.

Conceptual Design and Preliminary Engineering

Conceptual design and preliminary engineering begins after the "Proposals for Engineering Services" have been approved. This phase of the design normally requires a full academic quarter. During the conceptual design phase, student design teams gather data and begin the analysis of the project to identify various design concepts and alternatives. Preliminary design includes evaluation of the design alternatives with preliminary cost estimates. A "Preliminary Design Report" is written and presented to the client. This report contains a discussion of alternatives and gives recommendations for the final design.

Design teams give two or three oral reports during the conceptual design phase. These oral reports are supplemented with written progress reports and the "Preliminary Design Report". Scheduling of the reports is easy since the deadlines for various phases of the design are contained in the "Proposals for Engineering Services". The progress reports are very important since they provide incentive for the design teams to meet their deadlines and they are a forum for review of the progress and the technical content of the engineering that the teams have accomplished. Upon approval of the "Preliminary Design Report", the design teams begin the final design.

Final Design

Final design involves working out the details of the project. Requirements of this phase of the project include a final design report and a set of plans. Normally, complete specifications are not required although specification are discussed and the students are required to become familiar with example specifications for similar projects. The final design report and the plans must be submitted at least four weeks before the end of the course. By the time the design teams have reached the final design phase, the project has been reduced to a series of text-book design problems where the "givens" and "to finds" are clearly defined. Therefore, the students have considerable experience in the type of work then required.

The final design report and the plans are submitted to the project advisers for review and comment at least four weeks before the end of the course. The final design report contains a detailed description of the technical approach that the teams have used to accomplish their designs. A consulting firm would prepare this type of report for technical review by the client and the quality assurance review team. The report discusses technical details relating to the project including design assumptions, approaches to the work, technical references used and any other information that provides specifics about the detailed design. The project plans are submitted with the final design report. Since the plans are the end product of the engineer's work, detailed plans for the project must be submitted by the design teams. The method used for drafting the plans is left to the design teams. Many teams use CAD which is available to all engineering students. Teams are expected to submit plans are in sufficient detail for construction of the project. Review comments from the project advisers are submitted to the design teams so that they can make revisions and corrections to the design report and plans prior to resubmitting them for final approval.

As with other phases of the course, the student design teams must give oral presentations and submit written progress reports as they progress through the final design. Normally, each team reports twice during the design. The reports contain more technical details than the preliminary reports since the purpose of the reporting is for quality assurance review.

Discussion

In order to accomplish the objectives of the senior project design, it has been found that the course must extend over at least two consecutive quarters and it is preferable for the course to be taken for a full academic year. This gives the design teams time to plan, review and revise their design work much as would be done on the design of the actual project. Approximately 200 hours of effort are required of each student. Therefore, it is important when evaluating the proposals at the beginning of the course, to insure that the scope of proposed work can be completed by the three or four person design team with an effort requiring 600 to 800 man hours of work. The projects should be selected with care by the faculty and practitioners to insure that they meet these time constraints.

During the three years the course has been taught, design projects have been provided by ten different consulting engineering firms. The projects are drawn from actual projects that the consulting firms are doing. Design may have been completed by the consultant but in no case will the construction be done. The consulting firms are contacted by a general mailing each year and asked to submit projects. So far, all of the projects needed have been provided without any additional solicitation. Practitioners who have furnished projects and advised student design teams have been enthusiastic about the course and have asked to be included in the future.

Senior project design at Montana State University provides students with an excellent introduction to the planning and design of a real project. It is one of the few courses in the B.S. program in civil engineering that requires students to define the problem, to develop a plan for its solution, to obtain data needed to investigate possible solutions and to submit a completed design. The course uses student's abilities in the engineering sciences and in basic design along with economics and written and oral communications and provides a realistic exercise in tying together many aspects of engineering education into engineering practice.

The Professional Education of Civil Engineers

by Tung Au,[1] F ASCE and Chris Hendrickson,[2] M ASCE

Abstract

This paper discusses the objective and content of a set of undergraduate courses to further the professional development of civil engineers. These courses are directed toward the development of the ability to apply pertinent knowledge to the formulation and solution of practical problems within the discipline. While some are classified as design courses in the broad sense, all of them promote the philosophy of the ABET for professional growth.

Introduction

The primary objective of engineering curricula is to develop a solid foundation for professional practice. As stated in the latest criteria issued by the Accreditation Board for Engineering and Technology (ABET 1988): "A significant measure of an engineering education is the degree to which it has prepared the graduate to pursue a productive engineering career that is characterized by continued professional growth." The underpinning focus of an integrated educational experience is directed toward the development of the ability to apply pertinent knowledge to the formulation and solution of practical problems within the discipline.

At the basic level, ABET expects the curricular content of a program to include at least (1) one year of an appropriate combination of mathematics and basic sciences, (2) one and one-half years of an appropriate combination of engineering sciences and engineering design, and (3) one-half year of humanities and social sciences studies. The determination of engineering science content is based on the extent to which there is an extension of knowledge toward creative application. Engineering design is viewed as the

---

[1]Professor of Civil Engineering and Public Policy, Carnegie Mellon University, Pittsburgh, PA 15213

[2]Professor of Civil Engineering, Carnegie Mellon University, Pittsburgh, PA 15213

vehicle by which students are expected to develop creativity, solve open-ended problems, consider feasible solutions and devise detailed system descriptions, subject to a variety of realistic constraints such as economic factors, safety, reliability, aesthetics, ethics and social impact.

Additional criteria developed by ASCE and approved by ABET for civil engineering programs further specify that "a minimum of one-half year is required in civil engineering courses with a broad base of coverage, rather than one or two specialty areas" (ABET 1988). It is not stated explicitly whether this requirement may be satisfied by targeting specific courses within the categories of engineering sciences and engineering design, or should be satisfied by an additional set of civil engineering courses. In any case, it is important to examine what constitutes civil engineering courses with a broad base of coverage which would provide the underpinning focus required by ABET.

Paradoxically, if more civil engineering courses with a broad base of coverage can be designed to satisfy the categories of engineering sciences and engineering design, more time slots will be left for civil engineering courses in specialty areas. Each institution must decide for itself what is the best approach to meet its objectives in providing such coverage.

In a perceptive article on engineering design, Liebman(1989) has highlighted the problems of teaching design and observed that the development of additional theory and methodology is within the province of engineering research. He also expounds a broader view of professional practice as "problem solving" in these words:

> ....It is argued that since design is what engineers do, design is what they should be educated for, at least in their professional as opposed to general education. Unless one accepts a very broad definition of design (certainly broader than that in most of these discussions), this view far too narrowly bounds the province of the engineer. A brief look at what engineers actually spend their time doing, from determining causes of malfunction or failure, to advising presidents and legislators on national technology policy, shows that many of the most important functions of engineers involve technological analysis without the objective of creating an artifact (which is the common image when one uses the term "design".)

The authors concur with this broader view on professional development of engineers and recommend a set of broad-based civil engineering courses to fulfill this objective.

More importantly, a set of courses with a broad base of coverage should also serve the goal of producing graduates with a general education that will serve them well in their professional, social and personal roles. A liberal professional education should indeed provide a carefully crafted program which integrates general education with technical specialties. Boyer and Hechinger (1981) have eloquently reiterated the goals of higher education with an awesome challenge:

> ...The aim is not only to prepare the young for productive careers, but to enable them to live lives of dignity and purpose; not only to generate new knowledge, but to channel that knowledge to human ends; not merely to increase participation at the polls, but to help shape a citizenry that can weigh decisions wisely and more effectively promote the public good.

These goals have in effect long been advocated by the civil engineering profession which at its best embodies the art and science of directing the resources of nature for the use and welfare of mankind.

Broad-Based Civil Engineering Courses at Carnegie Mellon

The professional development of civil engineers depends not only on academic course work but also on extra-curriculum activities such as participating in professional society activities. Nevertheless, course work is the focus of ABET attention for accreditation. This paper will discuss a set of courses which would satisfy the additional criteria proposed by ASCE and ABET to foster professional development. At Carnegie Mellon University, the following broad-based courses are offered by the civil engineering department:

- Creative Engineering Design and Innovation (elective course for freshmen)
- Engineering Economics (required course to be taken in any semester from freshman through junior years)
- Computer Applications in Civil Engineering (required course for sophomores)
- Engineering Synthesis and Design (required course for juniors)
- Project Management for Construction (elective course for seniors)
- Computer-Aided Tools for Civil Engineers (elective course for seniors)

In addition to these formal courses, a two semester project is required of all students during the Junior and Senior years.

As long as proper student advising is provided to ensure that ABET criteria are met by all students, some of the requirements may be met by elective courses. For example, in addition to the three required courses listed above, any two of the three elective courses may be included to fulfill the minimum requirement of one-half year of broad-based civil engineering courses. The contents of this set of courses will be explained in detail below.

Creative Engineering Design and Innovation

This introductory course encourages the development of personal qualities that psychologists have identified as traits that are found in creative professionals. While above average intelligence and extensive professional training are necessary conditions, they are not sufficient. Other important qualities include openness to experience, ability to manipulate concepts, independent judgment, aesthetic sensibility, and perseverance.

The emphasis of this course is placed upon problem definitions and the creative development of alternative solutions. The applications of basic principles are introduced in a variety of open-ended, realistic engineering problems to give the students an understanding of the use and role of analysis courses that follow in the curriculum. Some people may argue that creativity cannot be taught, at least not in the current state of knowledge. No one should underestimate the difficulty of teaching creativity, but in the hand of a skillful instructor, it can be done in a modest way (Romualdi 1979). Creative engineering is stressed by the application of solution steps and class discussions about creative process and the mechanics of overcoming roadblocks to creativity.

Engineering Economics

For civil engineering students who aspire to plan, design and/or construct facilities, concepts of costs and benefits should be studied in depth, particularly with respect to the tradeoff between initial construction cost and subsequent operation and maintenance costs. An introductory course in engineering economic analysis not only can serve this purpose, but also address some issues pertaining to risks, ethics and social impacts of technological development (Au and Au 1983).

This course emphasizes the understanding of economic decision criteria in the selection of alternative engineering facilities as well as the mechanics of discounted cash flows. It highlights the differences in taxes, financing costs, required rate of return and other factors in public versus private ownership of facilities. Concepts of risk and uncertainty are introduced so that the effects of various parameters may be evaluated by sensitivity analysis. The selection of a social discount rate and the use of grant subsidies for public projects raise the ethical issues of intergenerational transfers of costs and benefits and redistribution of incomes among social groups. These discussions broaden the horizon of students.

Computer Applications in Civil Engineering

This course is intended to introduce a set of problem-formulation methods and computer-based problem-solving tools that students can use to enrich their learning experience in subsequent courses (Fenves et al. 1988). The problem solving tools consist of generic tools including text editors, spreadsheets, equation solvers and computer graphics. These tools share a common characteristic: problem formulation and initial solution time is comparable to the time it takes to produce a solution manually, while subsequent solutions with different data values require essentially no further effort. As a result, assignments can include sensitivity analysis and parameter selection.

This approach of emphasizing problem formulation and using computer-based tools forces students to generalize the assigned problems rather than just grind out a number for a specific case. Hence, the traditional distinction of "given" and "required" items corresponding to input and output data respectively would largely disappear. The course meets ABET criteria of a computer science course (other than computer programming skills) that can be classified as engineering sciences when it "carries knowledge further toward creative application."

Engineering Synthesis and Design

In this course, students are exposed to the design process of formulation, synthesis and evaluation through short design projects. While students may make use of the computer-based problem solving tools familiar to them, standard design or analysis packages are not usually available. The projects change from year to year, and each project lasts three to four weeks. For example, one project called for designing a small pump-storage energy generation facility, while another involved design of retaining walls. A more open-ended project involved redesign of traffic flows to alleviate congestion near campus, focusing on safe and efficient movement of vehicles through a parking area. Students often relied on generic tools for each project, and they used spreadsheets to perform relevant calculations and to prepare cost estimates. With these aids, students could readily select design parameters and investigate alternative configurations within the specified time constraint.

More recently, some formalism in design methodology resulting from research (Maher 1989) has been added to the course. Some features of non-routine design methods such as mutation, analogy and inversion are also included. Such an approach to engineering synthesis and design will enable students to work individually or in small groups where deemed appropriate. It also emphasizes student creativity, use of open-ended problems, development of design methodology, and consideration of alternative solutions.

Project Management for Construction

The improvement of construction efficiency and productivity is one of the most important factors in enhancing the competitiveness of U.S. industries since construction is a particularly seminal industry. Hence, an integrated process for design and construction should be instilled in all civil engineers whether or not they will be engaged in construction eventually. A broad-based approach with emphasis on the basic philosophy and implementation of this integrated process will nurture the understanding of realistic constraints such as economic factors, safety, reliability, aesthetics, ethics and social impact of engineering construction.

This course covers project management for construction from the owners' perspective so that the life cycle costs and benefits from initial planning through operation of a facility are relevant to decision making (Hendrickson and Au 1989). It is through the understanding of the entire process of project management that planners, architects, engineers, constructors, material suppliers, financial analysts and others can respond more effectively to the owners' objectives and improve the productivity and quality of their work. This approach will enhance the breadth of civil engineering students who may choose to pursue any specialty area.

Computer-Aided Tools for Civil Engineers

The objective of this course is to introduce students to representative, professional quality computer-aided engineering tools in civil engineering (Fenves et al. 1988). The course is organized around a sequence of engineering projects in which students are given real-world problems in specialty areas such as structures, geotechnical engineering, water resources engineering and construction management, and they are asked to develop solutions using available domain-specific professional quality software. Assignments take the form of written project reports.

Projects are open-ended and tend to be loosely defined. The assignment format enables students to work in an unstructured problem solving environment as they might encounter in practice. Students are asked to describe how they will model and solve a given problem, independently of the computer tools. Next they will convert their idealized model into data for the specific computer program, perform the necessary runs, and obtain and analyze the results. Part of the course is also devoted to a general discussion of computer use and computer literature, which gives students an opportunity to compare experiences and critique systems in a computer based environment. The approach used in this course can better prepare students to make appropriate judgments and have realistic expectations of the capabilities of computer programs to solve practical problems.

Conclusions

The broad objective of ABET criteria for curriculum development call for nothing less than the overall advancement of professional practice. One must take a broader view beyond the narrow interpretation of design to describe the activities of true professionals. Educators in the civil engineering profession must look forward beyond traditional courses in engineering sciences and engineering design in promoting the philosophy of the ABET requirements. It is believed the set of civil engineering courses described in this paper will serve the purpose of advancing the professional education of civil engineers.

Acknowledgment

The authors wish to acknowledge their colleagues in the Department of Civil Engineering, Carnegie Mellon University, for their contributions to the development of the courses described in this paper.

References

1. ABET, Criteria for Accrediting Programs in Engineering in the United States, Accreditation Board for Engineering and Technology, (Approved as of November 13, 1987), Washington, DC, 1988.

2. Au, T., and Au, T.P., Engineering Economics for Capital Investment Analysis, Allyn and Bacon Inc., Newton, MA, 1983.

3. Boyer, E. and Hechinger, F., Higher Learning in the Nation's Service, Carnegie Foundation for the Advancement of Teaching, Princeton, NJ, 1981.

4. Fenves, S.J., Hendrickson, C.T., Maher, M.L., Rehak, D.R., and Thewalt, C., "Two Undergraduate Courses in Computer-Aided Engineering" Engineering Education, November 1988, pp. 122-126.

5. Hendrickson, C., and Au, T., Project Management for Construction, Prentice Hall Inc., Englewood Cliffs, NJ, 1989.

6. Liebman, J.C., "Designing the Design Engineer", Journal of Professional Issues in Engineering, ASCE, V. 115, No. 3, 1989, pp. 261-270.

7. Maher, M.L., "Synthesis and Evaluation of Preliminary Designs", in Artificial Intelligence in Design (J.S. Gero, Editor), Computation Mechanics Publications, Boston, and Springer-Verlag, Berlin, 1989, pp. 3-14.

8. Romualdi, J.P., "Teaching Creativity in Undergraduate Engineering Education", Proceedings of ASCE Conference on Civil Engineering Education, April 1979.

Preparing the Engineer to Deal with Aesthetics

Frederick M. Law, F.ASCE[1]

Abstract

The civil engineer has a responsibility to formally consider aesthetics as part of every design. To help prepare the civil engineering student to fulfill this responsibility, Southeastern Massachusetts University offers a course dealing exclusively with aesthetic considerations.

Introduction

The civil engineer, more than any other design professional, has the power to shape man's environment. This power carries with it, however, a responsibility to shape man's environment in a way which is aesthetically pleasing.

Recognizing the importance of aesthetic considerations in civil engineering, the Department of Civil Engineering at Southeastern Massachusetts University has, for the past nineteen years, offered a three credit undergraduate elective course dealing with aesthetic considerations in civil engineering.

Course Outline - CEN 493

I. Definitions of Aesthetics
(Philosophy, Psychology, Science)

II. Elements of Visual Design
(Line, Form, Tone, Texture, Color)

III. Basic Principles of Visual Design
(Balance, Rhythm, Emphasis, Proportion, etc.)

IV. Theories of Design and of Visual Design

V. Practice in Visual Design
(Simple Geometric Designs)

VI. Social Implications and Humanistic Meanings of Engineering Design

---

[1]Professor and Chairman of Civil Engineering, Southeastern Massachusetts University, No. Dartmouth, MA.

VII. Specific Designer's Descriptions and/or Design Criticisms of Engineering Works

VIII. Practice in Visual Design
(Civil Engineering Designs)

Course Description

The course begins by making the student aware of the broad range of definitions of aesthetics; from a branch of philosophy, to a branch of psychology, to the science of the beautiful. This is followed by familiarizing the student with the specific elements of visual design: line, form, tone, texture, and color.

Next, the major principals of visual design of rhythm, proportion, balance, and emphasis are defined and discussed as are the minor principles of repetition, alternation, sequence, radiation, parallelism, transition, symmetry, and contrast. Finally, various theories of design are explained and discussed. These theories range from very general theories of design (Vitruvius 1914 or Aesthetics Committee 1978), to specific theories of visual design (Faulker, Ziegfeld and Hill 1963), to very pragmatic approaches to specific designs (Leonhardt 1968). See Table 1 for an outline of these theories.

Table 1. Examples of Design Theories

..."aspects of design"...
1. Usefulness
2. Strength
3. Beauty

(Vitruvius 1914)

..."three criteria ... (for) form"...
1. Efficiency (in materials)
2. Economy (in construction)
3. Elegance (in form)

(Aesthetics Committee 1978)

..."three basic design principles"...
1. Balance
2. Continuity or Rhythm
3. Emphasis

(Faulker, Ziegfeld and Hill 1963)

..."aesthetics of bridge design"...
1. Good Order
2. Good Proportions
3. Simplicity
4. Good Service
5. Easy Fabrication

6. Economical
7. Present Technology
8. Creativity and Intuition

(Leonhardt 1968)

With this background, the student then undertakes several simple, two-dimensional geometric design problems. For example, each student may be given a set of paper geometric shapes and asked to compose them in some visually pleasing arrangement as shown in Figure 1. Each student's arrangement is then discussed in terms of the specific principles of visual design employed and critiqued in terms of the various theories of design previously explained.

As part of at least one of these simple geometric designs, the student is asked to convey a specific meaning through his design (e.g., to convey a "sense of motion"). This assignment is intended to dramatize the fact that "meaning" can be conveyed through even a purely geometric design. The idea that all designs, and in particular civil engineering designs, may have "humanistic meanings" as well as "social implications" is further reinforced through a set of assigned readings.

This phase of the course ends with the students undertaking several three-dimensional geometric design problems. Three solutions to one such problem employing styrofoam blocks, are shown in Figure 2.

Next, the students are given the opportunity to observe a number of specific civil engineering creations and the designer's description of the work and/or a design critique of the work. For example, the most recent American Institute of Steel Construction's Prize Bridges may be studied, or specific examples of bridges from "The Tower and the Bridge" (Billington 1983) or from "Bridges" (Leonhardt 1984) may be selected for study.

Finally, each student is asked to prepare a preliminary design of one or more civil engineering works, such as a water storage tank (Law 1972), a small bridge, or a site plan for a large project. An example of one of the student designs for a footbridge completed in this part of the course is shown in Figure 3.

To critique these civil engineering designs a special jury is drawn from the members of the faculty of the College of Engineering, as well as from the College of Visual and Performing Arts. The critiques are conducted in front of the class as a whole so that all students benefit from each critique.

Compose these shapes in a visually pleasing arrangement:

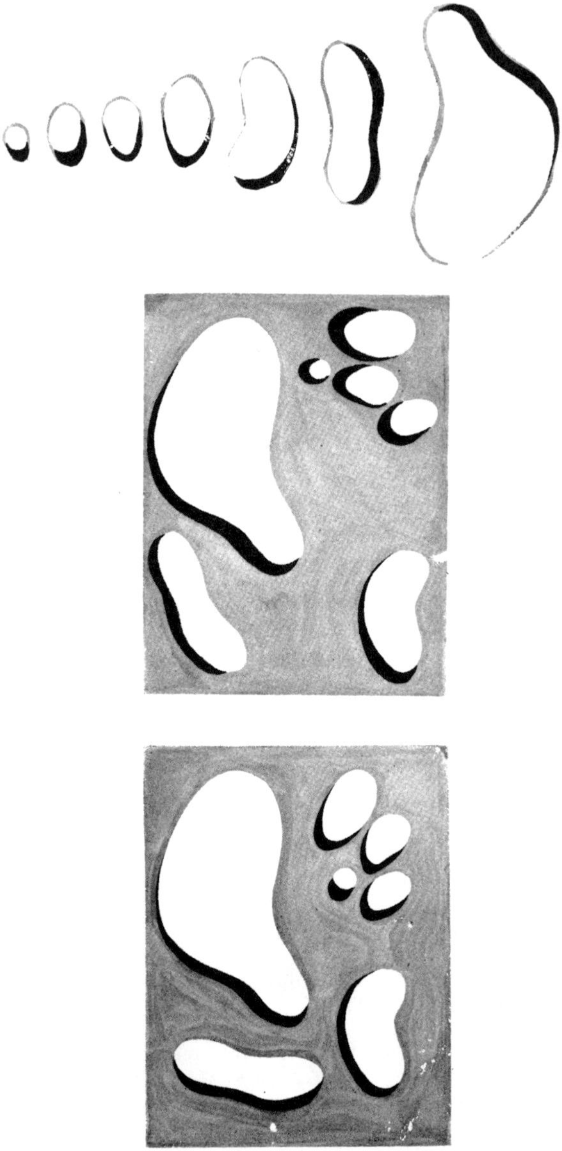

Fig. 1. Simple Geometric Design Problem and Solutions

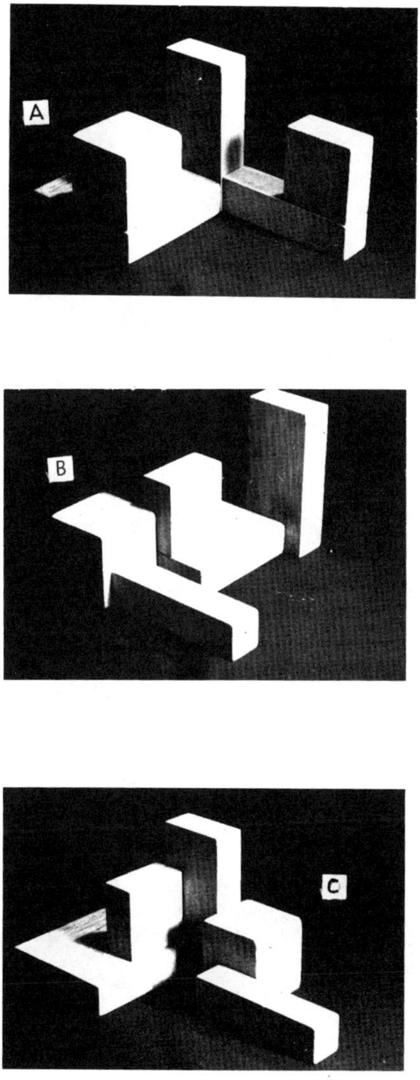

Fig. 2. Three Solutions to a Three Dimensional Design Problem

Fig. 3. Student Design of a Footbridge

Emphasis throughout the course is placed on building a firm foundation upon which an aesthetic sensitivity can be developed by each student as his professional career develops.

Conclusions

Although the success of such a course is difficult to measure, it has been found that the aesthetic perception of the students, as measured by the Meier Aesthetic Perception Test (University of Iowa 1963), has been raised each time the course has been given. Perhaps of greater significance, students claim that after completing the course, they are unable to pass a structure or other civil engineering work without trying to analyze it aesthetically.

Finally, unless a formal attempt, such as CEN 493, is made to introduce our civil engineering students to aesthetic considerations, few can be expected to give serious attention to aesthetics in their future professional work.

Appendix 1. References

Aesthetics Committee, Structural Division, American Society of Civil Engineers (1978). Unpublished committee report by David Billington, Princeton University, N.J.

Billington, D. (1983). The Tower and the Bridge, Basic Books, Inc., New York, N.Y.

Faulker, Ziegfeld, and Hill. (1963). Art Today, Holt, Rinehart and Winston, New York, N.Y.

The Meier Art Tests-II, Aesthetic Perception (1963). Bureau of Educational Research, University of Iowa, Iowa City, Iowa.

Law, F. (1972). "Learning to Create Attractive Water Storage Facilities", Public Works, Sept., pp. 100-101.

Leonhardt, F. (1968). Aesthetics of Bridge Design", Prestressed Concrete Institute Journal, Feb., pp. 14-31.

Vitruvius. (1960). The Ten Books on Architecture, trans. M.H. Morgan, Harvard University Press, Cambridge, MA, 1914; reprinted by Dover, New York, N.Y.

SENIOR DESIGN COURSE: 30-YEAR EXPERIENCE

Robert D. Miles, Fellow ASCE[1]

Abstract

The Civil Engineering Design Project was made a requirement for graduation in the Fall semester 1959. The 5-person design teams divide design responsibilities into environmental, geotechnical, hydraulic, transportation, and structural systems.

Approximately four weeks are concerned with planning, and a report is required discussing the site plan. Ten weeks are devoted to component design. A final report is submitted one week before the end of classes. Each team must form a conclusion about engineering, environmental, and economic feasibility. The last week is used for one-hour design presentations. Four staff members hear the reports and ask questions. The reports are graded by eight to ten staff persons. Seventy percent of the grade is individual effort and 30 percent is team effort.

The issues related to the course are staff effort, student procrastination, incomplete data base, and student concepts of design.

Introduction

The faculty of the School of Civil Engineering, Purdue University, made a major revision of the undergraduate curriculum in 1955. The basic concept that civil engineering is a design profession required a capstone course in design. The bulletin of the Schools of Engineering gives the following course description.

> CE 498 Civil Engineering Design Project Sem. 1 and 2, Lab. 9, cr. 3 (8 CE). Prerequisite: CE 393 (CE Systems Analysis).
> Planning, design, and analysis of a civil project; integrated and realistic group project involves as much as possible all major aspects of the civil engineering profession.

---

[1]Professor and Research Engineer, School of Civil Engineering, Purdue University, West Lafayette, IN 47907

The students apply the knowledge of their program of study to the planning and preliminary design of a project of the type presented to a consulting firm.

The Design Projects

The Head, School of Civil Engineering charges the Senior Design Committee to select two design projects each year. They recommend staffing assignments for the professional staff and the teaching assistants. The committee consists of six faculty, two students, a graduate student, and the course coordinator (ex officio). They examine the projects for the planning and design challenges. They determine whether the project requires a fixed site or that teams select the site. The committee determines the number of staff persons (0.3 FTE) and the number of teaching assistants (0.25 FTE) to recommend to the Head. The course coordinator has a 0.5 FTE assignment. The School Head notifies the Area Heads of the staff they are to provide.

The project is presented to the student in a "Request for Proposals". Titles of 20 projects of the 60 completed are listed in Table 1.

TABLE 1
SENIOR DESIGN PROJECTS 1959-1989

1.	Locate and design candy factory	Fall 1959
2.	Plan and design reservoir	Spring 1961
3.	Plan and design belt-line railroad	Fall 1963
4.	Locate and design tri-county airport	Spring 1963
5.	Plan and design light rail system	Fall 1967
6.	Urban Redevelopment-Lafayette	Fall 1968
7.	Southwest Lafayette Expressway	Fall 1969
8.	Safety Rest Area, Interstate	Fall 1970
9.	Wabash River Canalization	Fall 1974
10.	Hog Abattoir	Spring 1976
11.	Shopping Center-Cumberland Ave.	Fall 1976
12.	Caterpillar Industrial Park	Spring 1978
13.	Tippecanoe State Park	Fall 1982
14.	Calvert-Baker Animal Science Center	Fall 1983
15.	Wabash River Dam and Parkway	Fall 1986
16.	Regional Transfer Station	Spring 1987
17.	Undersea Habitat	Fall 1987
18.	Resource Recovery-4-county	Fall 1988
19.	Tippecanoe Truck Emporium	Spring 1989
20.	Heliport, National Guard	Fall 1989

The Class Sessions

The class meets for laboratory sessions from 8:30 to 11:20 a.m. each Tuesday, Thursday, and Saturday. Two Saturdays are mandatory when lectures on professional registration and ethics are conducted.

The Tuesday and Thursday sessions start at 8:30 a.m. in a lecture room. Attendance is taken. Outside speakers discuss the project under consideration. The staff present a discussion of design concepts. All formal talks are listed on the class schedule. In the laboratory each team is assigned six drafting tables. Attendance is taken in the laboratory and students are required to sign before 9:45 a.m.

The Student Teams

The design teams are organized as five-persons. On occasion, because of class enrollment, four-person and six-person teams need to be formed. The members are stratified by the course coordinator by areas of course work and by graduation indices.

The curriculum is over weighted with students in the structural area (70-80 percent) and geotechnical area (50-60 percent). It is underweighted in transportation (10-20 percent), environmental (5-15 percent), and hydraulics (5-10 percent). Some students are required to work in a design area (hydraulic, environmental or transportation) outside their major or minor.

Staff

The ideal staffing requirements are a generalist/specialist in each area - environmental, geotechnical, hydraulic, transportation, and structures with a teaching assistant in each area. The course coordinator has requested that, whenever possible, professional engineers be assigned to staff and TA positions. This helps to establish role models for the students.

The Written Reports

Planning. Each team is required to prepare a conceptual planning report that is submitted as a team effort during the fourth week of the semester. This is a formal report with an abstract, a discussion of the project supported by a conceptual and functional site plan. The planning report includes a Gantt chart listing the team planning tasks and design tasks by areas with a double-bar time frame for anticipated time and for actual time as the project develops. Each member of the team also submits a resume for staff critique.

Progress. Two progress reports are required during the 7th and 11th week. The reports are two-page summaries of design tasks, percentage of task completion, and future tasks. The progress report contains an individual Gantt double-bar chart listing planning and design tasks. It also contains a table of cost of the various efforts of the engineer. The hours of effort are recorded by the designer. A specific cost per hour is assigned for each effort ($60 per hour for engineering, $5 per hour for typing) and a factor for overhead and profit are selected to determine the total cost. The first report covers the first seven weeks and the second covers week seven through the tenth week.

The progress reports help the student allocate time and comprehend the costs involved in engineering effort. They are asked to review ASCE Manual 45 to obtain some idea of how a consultant charges for services.

Final. Each team submits (14th week) a preliminary design report. This report contains an executive summary. The report is a discussion of alternatives with recommendation of preliminary designs supported by figures, tables, plates and appendices. The reports are usually organized into an introduction and five chapters with a summary. Each chapter contains an abstract, introduction, purpose, scope and a discussion of the alternatives considered with analysis of designs recommended. The report is supported by figures and plates to show ability to draft and communicate designs. Each designer prepares a cost estimate. The team, in the summary, develops the total cost of the project to include final design costs.

The Oral Reports

Progress. The first class after the submission of the progress report the team presents a 30-minute report. Each person has three minutes for show and tell. Five teams are able to report in a 170 minute class session. If more than five teams exist, the staff is divided to conduct two sessions.

Final. Each team member is allocated 12 minutes to make an oral report during the fifteenth week of classes. A five-person team has 60 minutes for presentation and 20 minutes to reply to questions. Two teams report during the 170 minute session. The staff is divided into two or more groups (4 or 5) to hear simultaneous oral reports. If a design area is not represented, the area heads assign additional persons to hear the reports. The minimum number of staff to hear a report is three. The assigned staff (9 to 10) is able to conduct three rooms with hearings simultaneously. In three sessions with four rooms, twenty-four teams can present reports.

The recent small classes required two rooms each day for reports. This allows video recording of presentations for review by the student in consultation with a staff person. This staff person works with the students on the preparation of written and oral reports starting their sophomore year. This effort has added significantly to developing the written and oral communication skills of the student.

The Grading

The grades for the senior design course are based on the items in Table 2.

TABLE 2
SENIOR DESIGN GRADE SUMMARY

Progress Reports (Average of 5 grades)	10%
1. (Staff)(1 or 2 grades)	
2. (Staff)(1 or 2 grades)	
3. (English Review)(1 grade)	
4. Team Oral (4 to 6 grades)	
5. Team Oral (4 to 6 grades)	
Planning Report (Average of 8 to 11 grades)	10%
Oral Report (Final)(Average 3 to 5 grades)	20%
Written Report (Team Average)(8 to 10 grades)	20%
Individual Report (Chapter)(1 to 2 grades)	30%
Attendance	5%
Peer Evaluation (Special Form)(4 evaluations)	5%
	100%

A grade form is prepared for each student by the course coordinator. The written reports are evaluated during the 15th week by the assigned staff. Each staff person is asked to read the executive summary, the introduction, the purpose, the scope, the abstract of each chapter, and the cost summary for each team report and give a team grade. The staff person is asked to read in detail and evaluate carefully the chapter in their design area and give a grade to the individual who prepared the chapter, the appendix, and the plates in support of the chapter. The staff is asked to allocate about two hours for evaluation of each team report. The planning report grade consists of evaluations by the assigned staff. In addition, the special staff in charge of developing the written and oral communication skills of the students evaluates the report and assigns a grade. Each progress report is evaluated by the staff in the individual area of design of the student. For example, the two staff persons in structural engineering evaluate the progress report of the team member doing the structural design. In addition, one of the two progress reports is evaluated by the special staff in communications.

Feedback to the students is important for the planning report and the two progress reports. The students discuss the grading with the technical staff and meet individually with the communications staff person.

The Issues

It is obvious that the staff effort required for this course conducted in this manner is extensive. At full complement of 10 persons the effort is 3.2 FTE (Staff 1.7, TA 1.5). Experienced design staff with broad generalist background and extensive consulting exposure with an A&E firm might be able to reduce this to 1.8 FTE consisting of two professors (0.8 FTE) and three TA's (1.0 FTE).

The problem of students doing preliminary design in an unfamiliar area is of concern. The student has 30 credit hours of elective subjects.

"The elective course program shall be consistent with career objectives. For instance, one may elect to concentrate on a major in a specialized area with an integrated sequence of courses, or may choose a general program in civil engineering by taking courses in several areas. At least 21 credit hours must be CE designated courses which must include two integrated sequences with a minimum of six credit hours in each." (Schools of Engineering Catalog 1989-1990).

Students are advised to take four areas in civil engineering but the majority take three. About 70 to 80 percent of the Seniors are strong in structural engineering with usually additional courses in the geotechnical area and construction area. About 15 to 25 percent of the class have four or more courses in construction and management and are very weak in design knowledge. Five to 10 percent have design courses in environmental engineering, hydraulic engineering, or transportation engineering. Sometimes one or two members of a team are required to work outside their areas of study. The staff works closely with these persons and may advise them of a specific design procedure. Many students learn more about themselves with this type experience and accept the challenge.

The incomplete data base is a problem. Topographic maps, soil profiles, land use maps, aerial photographs, etc. are not always available for design work. The students work hard to interview the local sources for information and collect data for the project. The course coordinator is frequently asked by state, city, and county employees to collect the required data and supply it to the teams. Seventy-five students from 15 teams appearing in local offices asking for data is burdensome.

Conclusion

Most students look with favor on the course particularly when at a placement interview the interviewer explores the student's design responsibility. After graduation and with experience, the former students report that the design experience put them ahead of those graduates without such a course. Some hate the course and particularly those who procrastinate the most who have to work 48 to 54 hours to complete the project. The procrastinators let their team members down as well as themselves. The course usually ends with an "allnighter" and in 30 years no solution to this problem exists.

The civil engineering design project course is a significant design experience for the student. It simulates civil engineering as a design profession. It is a valuable course and should be a part of the last semester of each student's undergraduate program of study.

The only way to make the course more meaningful is to require a five year course of study. Each student then would have a minimum of four design courses in the various areas of specialization in the profession of civil engineering.

AWARENESS OF QUALITY CONTROL

BRIJ B. GOYAL* Ph.D.; F. ASCE

Quality control is the key to success in a design firm, or truly speaking, in any profession. Quality control does not mean catching errors, but is a process to assure that all phases, from planning to final construction, have been properly managed to satisfy the project requirements and client's needs. Quality is readily recognizable in retrospect, but is difficult to articulate in advance. To achieve quality results requires effective management of people and their skills; it also requires proper management of methods and time.

In recent years, we have become more aware of quality because of a number of dramatic failures that have lead to liability problems and increasing insurance costs. Currently, claims made and pending against consulting engineers are higher in number and dollar value than they have ever been[1]. This is an alarming situation. Small and large consultants are taking on the high risk of bankruptcy by not carrying insurance or by having large deductibles due to the increased cost of insurance. Thus, nearly all consultants have intensified their quality control programs.

The quality control is achieved through the efforts of professionals using their technical skills and judgement based on experience and effective management. It should be woven into the fabric of a firm's operation through emphasis on education and training programs. This requires a reorganization of priorities. The process of quality control education begins in the engineering schools. The efforts of instructors devoted to teaching must be supported by constant input from the design and construction

---

*Senior Professional Associate, Parsons Brinckerhoff Quade & Douglas, Inc., Chicago. IL

industry. Teaching, design and construction cannot be self-contained independent cells. Rather, constant exchange of information is necessary.

ENGINEERING EDUCATION

"An old definition of an engineer is this: he is one who applies science for man's use in an economic manner"[2]. In order to satisfy the above requirement, he should acquire the knowledge of material science and human behavior--a broad spectrum of knowledge. Engineering education today consists of courses in humanities, sciences and engineering. It has been recognized that mathematics and physical sciences form the basis of engineering, and that the liberal arts (humanities) courses provide the student with an opportunity to expand his knowledge of communicative skills--a large load for a four-year degree program[3]. This leaves limited time to devote to technical specialization in an era of accelerating technical knowledge. The Final Report of the Goal Committee[4] recommended that engineering programs be expanded to five years so that adequate time could be devoted to technical specialization. If we compare the engineering profession with medicine, dentistry, law or business, it is very clear that engineers spend less time developing their knowledge compared to other professionals (See Table 1). The trend toward achieving a higher education in engineering has been growing steadily in the last 2-3 decades but is far behind other professions. At present, approximately half of the staff of a large consulting firm has an advanced degree.This number drops to roughly 20 percent for smaller consultants (per a telephone survey by the author).

There is another aspect of engineering education which requires reconsideration. This pertains to faculty. The quality of engineering education depends upon the technical competence and practical training of the educators. Most members of the faculty are technically competent, but their experience in practical design or construction is limited. Programs should be developed to ensure continued upgrading of knowledge through the exchange of technical information between the educators and the practicing engineers. In an old survey, "In response to a question regarding the proportion of engineering faculty which ought to have industrial experience, 21 percent of the study committees responding felt that all faculty members should have some industrial experience, 74 percent felt that at least three-fourths of the faculty should have industrial experience, and 87 percent felt that at least one-half should have industrial experience"[4]. The need for industrial experience today is felt far greater due to the advancements of technical knowledge. The faculty doors should be opened for qualified and experienced engineers from industry to teach at the universities.

TABLE 1

REQUIRED YEARS IN EDUCATION AND TRAINING AFTER HIGH SCHOOL

Profession	Years in School After High School	Years After Graduation in Formal Training	Remarks
Medicine	8	3	More training for specialization
Dentistry	8	1	More training for specialization
Law	7	1	
Business	6*	-	Many schools recommend training before graduate studies
Engineering	4	-	No requirement except 3-4 years experience for state registration

* with graduate degree

Presently, the faculty is placing greater importance on the use of computers and readily available software to meet the needs of the future. The use of computers is excellent for engineers; but, it is taking away from their understanding of practical considerations required for design and construction. The results thus obtained, may not be directly applicable for use in a practical situation.

In a recent debate[5] at the University of Illinois, it was pointed out that the University has been criticized for emphasizing research at the expense of teaching. The faculty dedicate most of their time to research and personal development thus leaving the teaching to assistants. With due respect to teaching assistants, the professors pass on their responsibility of teaching (prime function) to young graduates who neither have advanced technical knowledge nor practical experience. Thus, the undergraduate engineering students may not necessarily, get the opportunity of learning from experienced professors before they reach the job market. The teaching in the fields of medicine, dentistry and law is conducted by full time professors. All these professors not only have practical experience but have licenses to practice.

PRACTICAL TRAINING

Engineering education does not stop after obtaining a degree; instead, it must continue while working within the industry. The best form of industrial education is practical training within the firm. The training should be taken seriously by the entire industry otherwise the severity of claims against the consultants could continue to grow. Consultants have always argued that they cannot provide educational training in their own offices due to the following reasons:

- o Inadequate budgets - the clients do not support training programs. The competition has led to bare bone fees.

- o Increase in overheads - loosing competitive edge in negotiations.

- o Busy schedule of senior consultants. They cannot spare time from routine administrative duties.

- o Inadequatly trained professionals in practical fields in the universities.

The above reasons are not entirely valid. If desired, alternative methods can always be developed. Consultants can organize cost-effective training by arranging luncheon seminars (brown bag) or instruction after normal working hours. This will involve some sacrifice by everyone, but the time spent will be well rewarded in the long run. In addition, state and federal government projects, and other major private projects should have some fee set aside to help consultants train young graduates.

In addition to the practical training provided within the firm, the professional engineering societies should become involved in setting industry-wide standard. The American Society of Civil Engineers, in association with the American Society of Engineering Education and the National Society of Professional Engineers, should provide guidance and binding regulations in regard to training. This idea is similar to that supported by the American Medical Association and the American Bar Association. No engineer should be allowed to take the licensing examination unless he/she has undergone a formal training program monitored by the state. Unlicensed persons should not be allowed to teach in any engineering college/university, nor should they be given the title, "engineer". This will help educational institutions in obtaining registered instructors having some field experience.

PROJECT MANAGEMENT

The art of project management is one of the best forms of quality control. However, the key[6] to producing quality products is to develop project managers from within the ranks of the company, as opposed to hiring administrators inexperienced in the field of engineering. A project manager should not only be a good administrator, but should also be technically competent in the fields related to the types of projects he/she manages. The project manager has the responsibility to set standards of performance and to assemble an appropriately skilled team to complete the work within the assigned time and budget constraints.

Another aspect of project management is to make sure that all aspects of quality control have been implemented -- quality assurance. The project manager should ensure that technical experts with long standing have reviewed the design concept memorandum and the final design package for completeness.

CONCLUSION

The engineering education which began in the schools should not end after a brief training at the start of one's career. Instead, engineers must be kept constantly informed of the latest developments, at all stages, to produce worthy products for their clients. Mr. Richard G. Folsom[7], President of Rensselaer Polytechnical Institute, in a special panel discussion stated that he constantly reminds his students "---if you're in this material, in this subject matter, you are in school for life. You have got to study continuously in order to be effective in your field." After listening to this statement, Mr. Lew Adler, of CBS Radio, questioned whether engineers realize that they must keep up with the latest techniques; "--- Is this why they fall behind?" Continuing education, and more so, continuous training in the practical field with the experts is probably the answer to quality control.

Without quality, quantity is meaningless. Awareness is the key step to quality control. Emphasis on engineering skills with appropriate training is more important than project management alone. Both must be encouraged to fully develop effective engineers to ensure quality products for the clients.

ACKNOWLEDGEMENTS

The author is grateful to the management for providing encouragement for his professional development. He is also thankful to Chuck Kohler and Alka Goyal who have given help and ideas in the preparation of this paper.

REFERENCES

1. American Consulting Engineers Council, "Liability Report", October, 1988.

2. Vannevar Bush, "Listen to Leaders in Engineering", edited by Albert Love and James Saxon Childers, Tupper and Lover Atlanta.

3. Jon Van, "Engineer Numbers Not There", Chicago Tribune, February 19, 1989.

4. American Society for Engineering Education, "Goals of Engineering Education, Final Report of the Goals committee", Jan, 1968.

5. David Hackell, "University Debates Best Diploma Route", News Gazette, a daily newspaper, May, 1989.

6. Myron Tribus, "The Quality Imperative", The Bent of Tan Beta P: Spring, 1987.

7. Lew Adler; W. A. Douglas; et al, "Engineering: A Profession in Trouble", Proceedings College Industry Education Conference, January 14-16, 1976, American Society for Engineering Education.

QUALITY IN EDUCATION AND PROFESSION

Ravi D. Kanvinde[1], M. ASCE

ABSTRACT

This paper presents a comparative study of the quality process at educational institutions and professional workplaces. In recent years, increasing attention has been given to quality management in design and construction industry. However, the majority of engineering graduates entering the job market have not received formal quality related training due to unavailability of courses at the universities. This gap between education and practice has created some interesting problems and challenging opportunities for all the participants in this field. This paper focuses on the need for quality related courses as part of the engineering curriculum. The discussion also addresses continuing education in quality as a necessary aspect of professional development.

INTRODUCTION

Quality Management is emerging as a basic philosophy of numerous design and construction organizations. Many companies have given top priority to quality process including separate departments created to administer quality practices. Investments in quality programs are considered essential to successful project operations.

Several questions arise from this scenario:
.. Why is there growing interest in quality?
.. How is the work environment affected by this current philosophy?

----
[1]Principal Civil Engineer, Design Engineering Dept, The M.W.Kellogg Company, Three Greenway Plaza, Houston, TX 77046

.. What do the companies expect from their employees?
.. How can an engineer entering the job market get prepared for the quality oriented environment?

This paper attempts to resolve some of these questions and formulate an agenda for education of quality. The author has had the opportunity to coordinate the quality process for Civil/ Structural technology at The M.W.Kellogg Company, a leading international engineering and construction firm. The observations, evaluations and recommendations presented here are based on survey of engineers, both entry level and experienced.

QUALITY - DEFINITION AND DIMENSIONS

What is quality? It is often stated that people may not be able to define quality, but they know it when they see it. This is because quality is a comprehensive term, a process rather than an entity. Its interpretation becomes important if a successful project is the ultimate goal. In the practical sense, it is the educated choice among various alternatives. From the economic standpoint, it is the optimum use of resources. In management terms, quality can be defined as conformance to agreed upon requirements. The goal of quality is not perfection, but a facility which will perform safely and efficiently, and meet the client's business objectives. Other essential elements of quality include, planning, communications, team effort, compliance with regulations and an obligation to society.

Quality in the design and construction industry is characterized by skillful applications of performance tested materials such as reinforcing steel and varying technical factors such as soil conditions. This unique combination of established criteria and untried data demands team effort by competent professionals committed to quality practice. The first step to achieve a successful project is to educate the team members in understanding quality process and organize them towards a common goal. It would be useful to describe quality in definitive terms before proceeding with the educational aspects.

Consideration of quality should begin at project's inception. Proper planning during the initial phase of a project has the greatest positive impact on quality. These planned activities necessary to provide a product conforming to established requirements, are termed "Quality Assurance" (QA). The specific procedures which include inspection, testing and other necessary actions, to verify conformance are called "Quality Control" (QC). The overall term, "Quality Management" (QM), concerns QA, QC, and all other activities involved in producing a quality facility.

As observed above, a successful program begins with education. It is worth noting that the quality management principles which are needed to plan and manage projects effectively should be learned during ones educational development in order to facilitate transition to the professional work environment. Before discussing how education can contribute to quality process, it is first necessary to address the issues that affect quality in design and construction industry.

QUALITY PRACTICE IN INDUSTRY

Traditionally, QA/QC practices have been part of all industry operations. Checking and inspection functions always existed as follow-up activities of design and construction phases. In the past, these programs were established by individual companies for their internal use with very little interactions and communications in terms of overall common philosophy.

The quality management movement that is sweeping the industry today is a relatively recent development. This surge in quality awareness is the result of realization during the past decade of American products' inability to compete in the world markets. In the period following World War II, American technology dominated the world. This technological leadership continued through innovation. However, prominence in marketing and commercializing products based on new technology has shifted to other nations, most notably Japan. It is worth noting that the foreign products have enlarged their market share here despite the narrowing cost difference. This change is most evident in the automobile industry. This phenomenon is also passing to other industries, including design and construction, to a point where it is necessary to analyze the significance of this change. The primary reason for advancement of some overseas products is customer satisfaction, which is a direct result of their quality. The American automobile industry was probably the first to recognize the quality issue and initiate a formal quality management policy. The design and construction industry is gradually riding the new wave of quality awareness.

Quality in design and construction industry is defined as conformance to the predetermined requirements. These requirements are translated into specifications and standards. Procedures are developed to reach the defined goals of a project. Teamwork and communications among project members are considered absolute requisites to reach the goals. Quality experts from within and outside the organization are training the employees. Companies are going extra lengths to institute quality programs including helping customers and suppliers to accept the quality management concept. The ability to proactively managing quality has become an important tool in the operations of a design and construction company.

EDUCATION OF ENGINEERS

As described in the previous section, quality is becoming an integral part of engineering design and construction. Project team members receive the necessary training to work in a quality oriented environment. They routinely follow quality management procedures to achieve a common goal of a successful job. In view of these developments, one obvious but often ignored fact surfaces, "Where do the new engineering graduates fit within this picture?" In general, their first encounter with the quality process is through the procedures manuals and brief instructions from their functional supervisor. This initial exposure is followed by formal training which commonly occurs after working on a project for some time. Additional training is mostly voluntary and may lead to study and understanding of quality. This reverse sequence of learning, with education as an option, often results in lost interest for probably the most important aspect of engineering field.

What can be done to improve this situation, where a definite need exists for education about the goals and procedures of quality? The understanding of the quality process must be introduced by means of education. Universities are the most effective places to achieve this first objective of quality literacy. It must be noted that the purpose of this education is not to seek mastery of the subject. The important thing is that engineering students should learn enough about quality as to be able to treat it as a necessary tool in their professional careers.

It is proposed here that quality related education should be required at the university level. There are several possible ways to institute quality instructions within the engineering curriculum. An extensive application may call for separate courses in quality. A limited exposure could be offered as part of a related course such as engineering management. Likewise, it may be beneficial to set aside time in almost every engineering course to establish quality principles.

Two major questions arise from this proposal, "Who should teach these courses and is there interest among faculty and students?" The answers depend on the type of courses offered. A seminar type study would be most useful, at least on a starting basis, for such an evolving subject. A teaching staff member, interested and familiar with the subject, would act as a moderator. Outside quality experts should be invited on occasions in order to enhance the ideas. As the quality field advances to the stage of reference and text book availability, the course should be converted into a regular subject. The other major question is the interest among students. This area will require some efforts on the school's part to publicize the importance, advantages and practical applications of the quality management process. ASCE could offer some assistance in promoting the idea and advertising for the schools. This proposed scheme will result in some challenging opportunities.

CONTINUING EDUCATION OF PROFESSIONAL STAFF

The design and construction industry is different from other manufacturing type operations, such as automobile plants, in the sense that the projects are each a first-time full-scale prototype. There is no opportunity to test the prototype. The facility must function in the intended manner when it is completed. This unique character gets magnified with the diverse international locations and varying regulations for each project. In view of these conditions, the quality program becomes an important requirement.

Engineers generally agree about the nature of quality itself, however, they often disagree about how to achieve quality. For this reason, organization and planning of a project are essential to achieve quality. A continuing education and training program is necessary to establish the guidelines for a successful project, remain informed in technology advancements and regain the edge on foreign competition.

AGENDA FOR EDUCATION IN QUALITY

The preceding sections present concerns for lack of quality related education and offer suggestions to improve the status. The quality education mission can be defined as development of engineers who are educated to think conceptually about quality prior to being trained to do and manage quality work. State of the art engineering and construction constitute quality management. It is therefore, imperative that education should support the industry requirements for their quality processes. It is the universities' responsibility to install the ideas of quality not as an addition to engineering, but as a vital part of the overall engineering function.

The following list of actions are based on principal themes that ASCE should initiate and pursue:

.. Define the expectations and objectives of university level education in quality.
.. Recommend the scope of coursework.
.. Encourage discussion among faculty and students to recognize the need for the quality courses at the university level.
.. Support the concept of education and continued training in quality practice.
.. Establish communications with the design and construction industry to evaluate the idea of education and research in quality.
.. Explore possible industry support.
.. Publicize quality education and available schools in ASCE journals.

It is anticipated that once interest is generated in the quality education, the idea will be supported by both students and the universities. Industry support would follow due to the benefits derived from quality educated staff.

CONCLUSIONS

The primary goal of university level education in quality management is to build adequate familiarity with basic concepts. This quality awareness is necessary for new graduates entering design and construction industry governed by quality practice. This method of formally educating engineers to think in terms of quality is considered vital to the overall success of the quality process.

REFERENCES

1. "Quality in the Constructed Project", Manual of Professional Practice, Vol. 1, American Society of Civil Engineers, New York, NY, May 1988.
2. "A New Agenda for Science", Sigma Xi, The Scientific Research Society, New Haven, CT, 1987.
3. "Quality Assurance Procedure", Civil Design Practices,
The M. W. Kellogg Company, Houston, TX, May 1989.

EDUCATION FOR DESIGN/SPECIFICATION INTEGRATION

C. Wayne Dillard, P.E., M.ASCE[1]

Abstract:

This paper is a commentary on the need to improve both initial and continuing education to better integrate specification preparation into the engineering design process. It discusses deficiencies which the author considers to exist in both the initial education of student engineers and the continuing development of practicing engineers for integrating the Project Manual (Bidding Requirements, Conditions of the Contract, and Technical Specifications) preparation in the design process. The author's experience as a design professional in preparation and review of contract documents and administration of construction contracts is the basis for commentary.

Introduction:

Preparation of specifications for construction projects is frequently approached as an afterthought, rather than fully integrated into the overall project design process. This approach frequently leads to conflicts, duplications, or omissions in the contract document package used by the contractor for bidding and constructing the project; and by the owner, engineer, or construction manager for monitoring and accepting the contractor's work.

This paper will present some concerns regarding the specification process and discuss some of the potential means for addressing those concerns to improve the overall specification production process among practicing engineers. It is not the intent of this paper to list all problem areas which may be encountered in contract documents nor to be all inclusive in suggested improvements in the educational process. The primary purpose is to stimulate thought and discussion and bring the overall problem into better focus.

[1] Senior Project Manager - Environmental, HOWARD NEEDLES TAMMEN & BERGENDOFF, 5900 Lake Ellenor Drive, Suite 600, Orlando, Florida 32809

How do we change the longstanding pattern of treating specification writing as an afterthought and bring about its complete integration into the design process? We educate both the young engineering student and the practicing engineer to the importance of specifications. We show each how the preparation of specifications can be accomplished concurrently with plan preparation and coordinated with the plans as they are developed. We demonstrate how problems that occur during construction because of conflict, duplication, and omission can be reduced to a minimum by proper integration of the specifications with the overall design.

Problem Areas:

In order to bring about improvements, major problem areas must be identified and a consistent approach to correcting them established. Representative problems which the author has encountered in using and reviewing contract documents prepared by various individuals and organizations include the following:

1. Construction details frequently are prepared and shown on the drawings without coordination with or reference to the specification sections describing the materials and the workmanship standards required for an acceptable installation. The absence of coordination between the drawings and the specifications may create a direct conflict between the two or completely omit an important component.

2. Engineers frequently use "front end documents" without consideration of the impact they may have on technical sections produced for the specific project. The results can be disastrous when an obscure provision of the "front end documents" takes precedence over the technical sections and allows construction of components with materials or workmanship standards of a lesser quality than the engineer intended.

3. Reference standards, usually publications of recognized associations or technical groups, are frequently incorporated into Contract Documents by reference without review of the standards by the engineer. This may result in references to obsolete or inappropriate standards with an undesired result on the construction of the project. Even when a copy of a reference standard is reviewed in the specification writing process, it may be an out of date version which is radically different from the current edition.

4. Manufacturer's data is often accepted and incorporated into contract documents without adequate review and evaluation by the engineer. Products may not stand up to the conditions imposed by the project, claims asserted in manufacturer's standard data notwithstanding. This situation most frequently occurs when the engineer, through either limited education

or experience, lacks confidence in his ability to develop an adequate technical specification for the needed item.

5. Use of manufacturer's standard equipment or material specifications in the Contract Documents may unintentionally exclude a competitive product from consideration by including a requirement that is manufacturer specific but does not affect the product's actual suitability for the intended application.

6. When multiple disciplines are involved in the design of a project, several individuals are required to produce the specifications covering the complete design. Often no one is assigned to coordinate the efforts of the different parties involved with a resulting lack of coordination between disciplines. Conflict or omission can result.

7. Engineers often do not make a distinction between "Master" and "Standard" specifications. As a result, inappropriate items are retained when "Master" sections are edited for the project or needed items are missing when "Standard" sections are used without being modified to meet project requirements.

Educating the Engineering Student:

The engineering student at a typical university or technical institution is taught the basics necessary for designing the component parts of a project, whether the project is primarily structural, transportation, environmental, electrical, or mechanical in nature. Rarely is the same student provided more than a brief introduction to the fact that the plans and details alone will not suffice as the complete package of information needed to implement the design intent.

A similar situation occurs when the young engineering graduate starts work on the first design project with his or her first employer. Few, if any, practicing engineers relish the idea of writing the specifications needed to implement their design. An approach often taken is simply to "cut and paste" the most recent set of documents used on a similar project. The new graduate quickly accepts this approach as the normal way of writing engineering specifications.

At least one introductory course in the basics of specification writing should be included in the undergraduate civil engineering curriculum. The course should introduce the young engineering student to the general types of contract documents in common use in the practice of civil engineering. The introductory course should provide a basic familiarity with common standard specifications which the student may expect to encounter when he or she enters the world of engineering practice.

The student can be introduced to the concept of standard specifications by reviewing ones which are frequently used by state departments of transportation, larger private or municipal utility providers, and the many agencies of the United States government involved in construction of civil engineering works. The student needs to appreciate that a different approach to preparing the project contract documents will be necessary when working with standard documents than will be the case when producing project specifications for a client who does not have such documents.

For structural engineering students, who may spend many of their years in design working as subconsultants to architects, an introduction to the standard format used by the American Institute of Architects may be necessary.

All students, regardless of their area of intended specialization in civil engineering, need to be introduced to samples of both good and bad specifications. The samples, whenever possible, should be taken from real projects where the impacts of good and bad techniques can be demonstrated from actual results. Practicing engineers should assist by making such documents available for use in the educational program.

Most important of all, engineering students need to be taught that a complete design includes more than the lines drawn on the plans and the calculations listed on the computer printouts or tabulated on calculation pads by hand. The student needs at least a brief introduction to thinking of a project in terms of a complete and coordinated set of documents which begin with the basic layout and calculations forming the design and are presented as drawings and specifications for implementing the design.

Engineering educators, while generally very competent in their fields, often have limited practical experience in engineering design and construction administration. The shortfalls in the education of engineering students in the preparation and use of specifications is not the sole responsibility of the educational institution. Practicing engineers need to make themselves, or other qualified members of their organization, available to serve as visiting instructors at engineering schools in their geographical areas of practice.

Continuing Education Opportunities:

The Construction Specifications Institute (CSI) offers two separate programs of education and training related to specifications production, both of which are available to members and nonmembers of CSI. One program consists of training seminars on the sixteen division CSI Format for specifications to practicing professionals. The seminars provide general introductions to the format and content of each of the divisions, but offer little in the way of "hands on" assistance in the development of specific technical sections typical of civil engineering public works projects. The emphasis of the seminars, and apparent major successes of CSI in developing individual specification writer use of the format, is directed more toward architectural applications.

The second CSI program is a certification program to identify individuals who have demonstrated competence in preparation of contract documents using the CSI format. Certification is obtained by passing an examination which is offered at numerous locations throughout the United States each year. A self study course may be taken to prepare for the examination. Local chapters of CSI frequently offer classes to their members covering the same material as the self study course. An individual who passes the examination is entitled to designation as a "Certified Construction Specifier" and may use the letters "CCS" after his name in the same manner as engineers and architects use P.E. and A.I.A. Recertification at specified intervals is required to retain the CCS designation.

Both CSI programs, when combined with their regular application in the creation of contract documents as part of project design activities, can be of significant value to an engineer interested in improving his specification writing techniques.

The Engineers Joint Contract Documents Committee (EJCDC), a joint endeavor of ASCE, NSPE, and CSI, offers standard documents and guides to preparation of other documents which may be used with technical sections to produce a complete set of contract documents. The documents produced by this joint committee have been endorsed by the Associated General Contractors and are familiar to many of the general contractors engaged in construction of public works projects. Short seminars to familiarize interested parties with the contents and concepts of the standard documents are regularly offered at different locations throughout the United States.

Summary:

There are many continuing education programs available to the practicing engineer interested in maintaining or upgrading skills in specification preparation. A few of them have been discussed in this section. But perhaps the best means of improving the process involves on the job training as we go about our daily engineering practice. Begin to think of specifications not as a necessary evil to be postponed as long as possible and then grudgingly prepared by cutting and pasting the last set of similar documents, but as an opportunity to express your engineering skills by making certain that your design is implemented as you really intended it to be.

Spend some time reviewing the last set of specifications which your design office produced. Have a set of the construction drawings at your disposal and compare what the drawings show with what the specifications say. Are they consistent? Do they properly complement each other? Have they been properly coordinated throughout? If the job has been constructed, what problems surfaced during construction that were the result of inadequate drawing/specification coordination?

Review your office's specification development processes and procedures. Are the people who are assigned the responsibility for producing the specifications an integral part of the overall design team? Are they as well trained and experienced as the people making the basic design decisions and detailing the construction? Do they know why a particular piece of equipment or a specific construction material is needed for this job? Have they been involved with the project from the very beginning and been a part of the decision making process? If you have answered no to any of these questions, your specification process may be questionable.

The author has attempted to highlight some of the potential problems that can develop when preparation of specifications for construction contracts is not made an integral part of the design process. Some suggestions have been made regarding changes in the educational process, both at the undergraduate level in our universities and technical institutes, and in continuing education for practicing professionals. This paper is intended to stimulate discussion and thought among the panel members and other interested parties; students, educators, and practicing engineers alike.

Appendix I. References

MASTERFORMAT, Master List of Titles and Numbers for the Construction Industry (1988), The Construction Specifications Institute, Alexandria, Virginia.

Standard Forms of Agreement, Engineers Joint Contract Documents Committee.

Back to the Future - Give Us a Profession, or Give Us a Trade

C. Gary Kellogg, Member ASCE[1]

Abstract

The professional school of civil engineering is recommended as an undergraduate and a practitioner-oriented, graduate teaching model. Literature is surveyed which outlines the need for development of design judgment in the education of civil engineers. In the profesisonal school concept, design and professional instruction is promoted as the domain of qualified practitioner-teachers. An operational framework is offered for the professional school, and variety in educational choices is encouraged.

Introduction

Amid the broad societal pressures of the 1980's, introspection gripped the minds of American educators (Hanson,1986). Whether this inwardness has produced significant outward educational improvement is yet to be judged (Kellogg,1989). Engineering educators, too, have been examining various proposals for change which are driven by the needs of a dynamic culture. These needs can be fulfilled using educational models from other fields. So common in law, medicine, and architecture, the professional school could be profitably borrowed by engineers as a system of education.

Engineering education involves inculcating certain attitudes, knowledge, concerns, and ethics into students who desire to practice the various fields of engineering. Even more definitive, civil engineering education deals with those engineering practices which most often bear upon the public at large. The professional civil

----

[1]Senior Structural Engineer, Spiegel and Zamecnik Inc., 918 16th Street NW, Washington, D.C. 20006.

engineer, more than any other, faces the immense responsibility of providing society with works of such great scale that there can be no or little opportunity for testing or prototypes. Civil engineers work in extrapolations. Consequently, of all engineers, the civil engineer should face the stiffest training and responsibility, the highest standards of a profession. "The fate of any profession is determined by the manner in which its members practice, and the societal environment in which it exists" (Pletta and Gray,1971).

For years civil engineers have gone unnoticed in the public press at best and have been chastised at worst. Instead of being the answer to environmental problems, engineers, that is civil engineers, have been portrayed as those who spawned these problems. In some minds civil engineers have caused everything from cancer to structural disasters. Obviously in such a relational background, the profession cannot flourish (Pletta and Gray,1971). Since economic worth is a matter of perceived value, this degradation of image has had serious consequences in the attraction of talented minds and adequate compensation to the civil engineering profession. It is no secret that civil engineers do not receive the highest salaries among engineers, much less among managers in general. Yet civil engineers are responsible for projects which have some of the highest failure consequences of economic and life loss. This responsibility has not been compensated in the marketplace.

At least in part, this malaise has its roots in the education of the civil engineer. Since Sputnik, the emphasis in engineering education has been more and more on the scientific approach of research (Schrader,1971; NSPE,1988b). Although practicing engineers had historically defined the educational agenda, the distinction between engineering and science was blurred in the engineering schools of the land. Research was touted as the answer to the technical problems of society. And, of course, this is partially true. Most recently, the American Society of Civil Engineers (ASCE) initiated the Civil Engineering Research Foundation (CERF). The first president of CERF, Harvey Bernstein, stated, "We [civil engineers] no longer wish to be viewed as the invisible engineer. We must re-establish civil engineers as the primary link between construction related technology and society" (ASCE,1989).

Mr. Bernstein is right. But research is not necessarily the means to advance civil engineering. Research is a process of analysis with a predilection to look at knowledge as parts, not a whole (Kellogg,1987; Naumann,1986). Civil engineering involves the integration of components. It is the putting together of the pieces into a whole, a synthesis, that is most often lacking in modern civil engineering education. This is engineering design and judgment.

Research Teachers

The Accreditation Board of Engineering and Technology (ABET) recognizes the importance of design content in engineering curricula and now requires a design capstone course which integrates the knowledge obtained in previous engineering coursework (ABET,1985). These directives are beneficial to students, and yet can only be as good as the execution in the classroom (Snell,1987).

For the most part, research professors do not have the time, enthusiasm, or experience needed to implement the ABET directive (ASCE,1986). There are exceptions, but it is difficult for a professor who makes a living in research to become an expert in design overnight. The current university environment militates against rewarding design teaching (Schrader,1970). The pressures are the other way or in many other ways in the modern, state-supported university. It is nigh impossible to teach design in an environment which is fragmented and driven by political pressures to produce other results (NSPE,1987b).

The research professor has most likely never experienced the marketplace pressures of the design professional and lacks judgment to quickly formulate design problems and solutions. Again, there are exceptions. But the current career track at the university is not conducive to meaningful design experiences for professors. They cannot teach something they do not know, nor can they invoke a passion for professionalism in their students (NSPE,1988b). A trade mindset of irresponsibility can prevail.

Design Teachers

Go back to the future; the time is in the past of practitioner-teachers and yet in the near future. The need is for qualified teachers of professional design (Guy,1986; Guy,1987; Kerr,1986). Such teachers are competent in their chosen fields of civil engineering design, successful in public or private practice, and experienced in the daily business of civil engineering. They are people who, from experience, understand the complex interaction and responsibility of civil engineers in today's world and are expert in judgment or the ability to formulate problems and solutions. Such teachers have at least 10 years of practical experience, equal to a full professor in the academic community. Their competence is not judged by the university community but by their completed engineering works and recommendations.

Moreover, a premium is placed on the character of these design teachers. Since ethical standards and professional attitudes can never be completely wrought by law, these individuals demonstrate sound professional ethics through reference and interview.

These design teachers are communicators. They are people who see civil engineering design as an avocation. Far from boring students, these teachers motivate and inspire students because they live in the design environment.

Professional Schools

The profession benefits by separating civil engineering programs from the university system and establishing professional schools of civil engineering (Pletta and Gray,1971; Guy,1987; Kellogg,1987). Professors with Ph.D.s are welcome in these schools, but this is the abode of the practitioner-teacher. The professional schools are autonomous and not subject to the systematic pressures of the university.

Most design courses are taught by the practitioner-teacher, and students are advised into a co-op experience (Al-Nassri and McBean,1986). Preferably the practitioner-teacher provides the co-op work experience for some of the students, a most effective combination. In particular, the design teacher influences undergraduate students and those graduate students desirous of practical training (NSPE,1988a). The highest level of teaching by the Design Professor conceivably occurs at the Doctor of Engineering level.

For those civil engineers interested in research, institutes are formed for that singular purpose. Ph.D.s are required to work at these institutes, and research is the main mission of the institutes. Such institutes are funded by either public or private sources. In this manner, researchers do what they do best, and practitioner-teachers teach design and professional studies. Frustration levels are reduced.

At the same time civil engineering is enhanced as a profession because students enter the profession with responsible attitudes. They have a vision to promote better relations with society at large, truly professional work. As relations and image improve, so does compensation.

The potential low overhead costs of design-oriented schools of civil engineering generate new markets for the education of civil engineers. Equipment costs have always been a roadblock in engineering education, especially in procuring the latest equipment. Through agreements with local governmental or private engineering laboratories, students are exposed to the latest in techniques and equipment without resorting to the purchase of expensive equipment which gets little use at the undergraduate level. Classroom instruction is implemented cheaply at almost any location. The increase in educational choices benefits those served - the students. The professional school exists for one purpose, to educate civil engineers.

The inertia of self-serving interests may be against the professional school concept (NSPE,1987a). There may be outcries for "quality", "competence", "rigor", and the like. But the detractors must answer for the competence of the products of today's engineering curricula. Technical tradesmen are not competent professional engineers.

> We must rededicate our university system to the notion that the purpose of science is not merely to unearth facts but also to show the relatedness of facts, and we must also recognize the degree to which the engineer can and must contribute to the unity, illumination and ultimately the harmony of our world (Schrader,1970). Together we must all practice engineering in the broadest sense or perish (Pletta and Gray,1971).

These words were written almost twenty years ago, and little has happened in this direction. Civil engineering or any profession cannot continue, much less grow, without the daily infusion of the thoughts and ideas of those who practice the profession. Professional schools of civil engineering are needed now.

Conclusions

ASCE and ABET must foster and encourage the professional civil engineering school concept. This format could flourish in current universities and/or altogether new educational systems. Narrow foresight is not providing civil engineering students with the options of education which they deserve. Given the choice, both potential design teachers and students will provide the engineering community with enhanced educational products -- the professionally educated student.

Professional schools of civil engineering, staffed by qualified practitioner-teachers, will supply the United States with a new body of engineers ready for today's competitive world market. Since this education will concentrate on design judgment, these future professionals will be posed to step into the civil engineering profession of planning, design, construction, and national leadership. Although education is a lifelong endeavor, formative instruction sets the tone for a career. Specialist researchers do not initially make good practitioners. Society is demanding change as expressed in the unending onslaught of litigation. Civil engineers have let this happen, failing to understand the broad impact of their work. Practicing civil engineers taught in professional schools will better influence society than those taught by researchers. Unless civil engineers act now, as leaders, the profession could languish into a trade-status death.

Appendix I. References

ABET (1985). Criteria for Accrediting Programs in Engineering in the United States. ABET, p. 7.

Al-Nassri, S., and McBean, E. (1986). "Work-Term Effectiveness in Co-operative Civil Engineering Education." Journal of Professional Issues in Engineering, ASCE, 112(4), pp. 296-305.

ASCE (1986). "So You Can Come Here to Teach?" Civil Engineering, ASCE, 56(8).

ASCE (1989). "CERF's Up and Running at ASCE's Washington Summit." ASCE News, ASCE, 14(7).

Guy, L.L., Jr. (1986). "I Accuse U.S. Practicing Engineers, Myself Included, of Shameful Neglect of Education!" Engineering Times, NSPE, 8(9).

Guy, L.L., Jr. (1987). "Educating Engineers for the 21st Century." Engineering Times, NSPE, 9(1).

Hanson, P.O., Jr. (1986). "Education Crises." The Bent, Tau Beta Pi, 57(4).

Kellogg, C.G. (1987). "Analysis vs. Synthesis." Engineering Times, NSPE, 9(1).

Kellogg, C.G. (1989). "Home-Grown Learning." Engineering Times, NSPE, 11(3).

Kerr, A.D. (1986). "A Question of Balance." Mechanical Engineering, ASME, 108(7).

Naumann, E.G. (1986). "Letters and Comment." Mechanical Engineering, ASME, 108(10).

NSPE (1987a). "First-Ever National Congress Convenes to Consider Engineer Ed." Engineering Times, NSPE, 9(1).

NSPE (1987b). "Narrowly Defined Course Work Squeezing Life From Undergraduate Studies?" Engineering Times, NSPE, 9(3).

NSPE (1988a). "Young Engineer of the Year Urges Peers To Become Mentors." Professional Engineering News, NSPE, 5(4).

NSPE (1988b). "Design: The Skill Slips in Training U.S. Engineers." Engineering Times, NSPE, 10(2).

Pletta, D.H., and Gray, G.A. (1971). "To Practice or Perish." ASCE National Structural Engineering Meeting, April 19-23, 1971, Baltimore, MD.

Schrader, C.R. (1970). "Engineering Education and a Professional Ethic." Stanford University ASCE Student Chapter Conference on Professionalism, May, 1970, Stanford, CA.

Snell, L.M. (1987). "Innovative Approaches in Concrete Education." Concrete International: Design and Construction, ACI, 9(6).

Uncertainty, Risk, Variability: Design Concerns

Ramey O. Rogness, Member ASCE [1]

Abstract

One of the major difficulties that is facing both the education and continuing development of the Civil Engineer in terms of design is the related problems of understanding and incorporating risk and uncertainty in the analysis and design process. This is especially true in terms of design optimization. Many students exposed to deterministic modelling interpret the design process as giving one "best" solution. Variability, risk and uncertainty are not well developed as to how they affect the data and design parameters.

Introduction

With the presence of variability and uncertainty not only in the design parameters but also in the future behavior and operational performance characteristics of the design product, the student and the design professional need to be made aware of the effect of variability and uncertainty on the design "solution." This is especially true with the use of the newer optimization procedures that have arisen with widespread use of microcomputer based programs and analysis packages. Rarely does a student or a design professional recognize that an optimization program or procedure that makes a solution the best (typically least cost or most effective) also reduces the inherent factor of safety that historically has been included in the design process.

Factor of Safety Considerations

This inherent factor of safety is in addition to the assigned factor of safety and arises from the design professional using rounded values that with respect to the design process permits redundancy and compensation with respect to variability. With the use of optimiza-

---

1 Assistant Professor, Department of Civil Engineering, The Ohio State University, Columbus, OH 43210.

tion procedures, such as structural optimization, what happens is that the optimized procedure reduces any "unneeded" or "extra" parameter value to that required (with the assigned factor of safety). What results is a true best design but with a historically reduced factor of safety that is closer to the assigned factor of safety. For those who do not recognize the phenomena, shortcomings in the safety or operational performance of the design product could result.

With more complicated design tools and design applications, the difficulty of addressing and overcoming these deficiencies is a major issue facing the future development of the Civil Engineer. In order to take advantage of advancing technology, new or modified procedures and techniques must be developed. This also means that understanding the inherent variability and uncertainty within the design process and the design parameters is necessary. These factors must be taken into account. The means to provide coverage and understanding is incumbent for the design professional.

Difficulties

A more major problem is attempting to provide understanding to a student or design professional with respect to uncertainty and risk that arise from variability. Unfortunately, in most of the civil engineering design applications used for students, the analysis and design procedures use deterministic parameter values. In using applications with uncertainty and variability, the student has difficulties not only in the means to analyze the data but conceptualizing that there is not one "best" solution. This is complicated by needing to provide the understanding that the design variables are not always exact which is also true for the behavior of design parameters during project life.

The best means to address this shortcoming is through the use of examples that exhibit those characteristics that a student or design professional can relate to. Examples of these include those related to water resources because variability and risk can be explained for the uncertainty design aspects. Two specific situations are the risk of dam overtopping and reservoir management (flood and drought versus supply).

Other examples include the selection of design parameters and analysis of what segments of the potential population range might be excluded for a design and the associated ramifications. A third example would be the use of simulation models to illustrate stochastic behavior of various models, design values and failure.

While several of these techniques can be used, the need to intervene early in the student design component courses and how to relate the topical area to the experienced design professional is complicated by the traditional educational design process. Advanced technology means present a major risk of future and present design professionals being less capable of designing safely and effectively. The difficulty is not the efficiency of the design process but the results of the design process. Can a student or a design professional using a new more complex procedure or technique determine the assumptions, limitations and internal manifestations of the procedure as it gives design results.

This is an ironic possibility considering the improved analysis approaches that is not recognized or appreciated in the civil engineering community and specialty areas. Applications arise in the majority of the specialty areas. There are inherent risks.

Errors

The "risk" is that the design professional will not understand the ramifications of the design process and inappropriately apply the technique, therefore resulting in under or over design. Because of the newness and differential behavior of the these techniques, the designs produced will not always match a designer's expectations or intent. This is especially true for newly adopted procedures and guidelines for practice that utilize unfamiliar or more complicated techniques. The chance for error or mistake is present and the identification of a mistake or error is not as obvious as with previous techniques that the professional has had experience with. Not much less of a problem is the identification of means to determine errors that may exist. This is similarly related to risk analysis.

Risk

The need to include risk analysis and uncertainty assessment in the design process is necessarily an important future direction for civil engineering education. With the enhancements from advanced technology, the ease of analyzing the subtle complexity of the design relationships presents the opportunity to greatly enhance the analysis procedures using these techniques. The difficulty arises in the proper means of the incorporation of uncertainty and risk into the educational design process and how to prepare the design professional for its usage. Implementation presents difficulties that may not be apparent to those unfamiliar with the concepts of risk analysis, uncertainty assessment and failure. The consideration of variability in the design process is presently not well understood or well used.

The Role of Simplifications

All too often the actual variability of data and performance characteristics are ignored in the design process. While these simplifications have been a proper approach in the past to permit analysis and proper selection of design parameters, it is meaningful to recognize that it is less likely to give proper results with new techniques. Part of the problem is the optimization techniques themselves are based on variability assessments and factors. The other, more significant, part is that the variability inherent for the design parameters which in the older design procedures were taken into account by redundancy and factors of safety, is not taken into account externally but more internally with the newer optimization techniques. The design professional must be made to recognize the difference, not only in the procedure and its approach, but in the design assumptions, and their ramifications. Unfortunately, this is not the case at the present and presents major risks for the future.

Infrastructure Considerations

One of the best illustrations of the difficulties faced is that of the deteriorating infrastructure. Two illustrations can be made to demonstrate the problem faced and the complications arising in trying to overcome the inherent shortcomings in the existing design and analysis process. Solutions to the problems rest on the availability of accurate historical data. Unfortunately, today the historical data do not exist.

The first situation is how to predict the actual performance of the facility, component or element. This can be approximated by direct or indirect means. How can its deterioration over time be known, much less modelled appropriately. What is the shape and smoothness of the deterioration relationship. What are the important factors that caused the deterioration and what are (or were) their values and relationships over time. Are there (complex) interactions that may not be apparent or observed. What are the model terms and significance.

What are the range of design parameters and variables that are typical. What can be extrapolated accurately and when are assumptions grossly out of range. Can design professionals determine or assess these answers. Are their interpretations subject to question. How can these issues be addressed in terms of educational needs.

How is uncertainty, variability, and risk included and incorporated into the process. Lets look at the design and analysis process for existing facilities. A major task faced is the challenge of making proper repair/rehabilitation/reconstruction choices for an in-service facility. What is always the desire is to have an (cost) effective and adequate safety design.

Lets look at the initial design. Every design decision which is made involves some element of uncertainty or risk. Some of this risk arises from the design parameter variability, simplifications in the modelling, and assumed performance and behavior of the facility during its design life. Uncertainty also arises because the final state of the decision depends on the subsequent occurrence of combinations of parameters which accumulate over time. Yesterday, a deterministic set of parameters was established and used for the design. Uncertainty also arises from the subsequent occurrence of parameter values being different than estimated and their cumulative effect.

To illustrate the difficulties arising, lets look at a situation with certainity and then uncertainty. With certainity, the designer can decide in advance the optimal design for all rehabilitation stages (choices) of a design. With certainity, the designer does not have to subsequently change the choice or decision point in terms of rehabilitation. Obviously, the design behavior and performance is not with full certainity. Under uncertainty, the designer should not commit to a specific rehabilitation choice or when it should be done. The decision actually could be postponed if the designer had more accurate information on the future design performance, than he has at the initial design point. Uncertainty is the norm.

One important fact that is not well developed is that uncertainty, variability and risk are not a single value. It should be obvious, but is not, that variability values and risk factors are not the same for each component of a design. Some data is more exact, while other data we have less confidence in. In fact, the uncertainity for various variables probably are independent. So how can a single assumed value be used. Only if the variability is the same for each, which is unlikely, or that the variability is attributable equally to all the design parameters. When is this correct? Probably never. Only if it is a composite (weighted) value might be true. This is not even being addressed.

To provide an example, the condition of a structure can be exactly designed today based on design load and performance assumptions. What will be the exact condition of the structure ten years hence. With historic data average conditions for similar structures can be predicted. A random result is the variability that occurs. The designer assumes it will be a specific deterministic value from a (postulated deterioration model) but it can not assured that this will be true.

Rehabilitation Choice

Now lets look at the situation for an existing structure or element. What is known? Its existing condition? With what confidence and risk? What are the design values to be used? To illustrate the difficulty lets look at what is know. At this time, the existing structure has not failed. From its known condition (as measurable), parameter values can be estimated. With what accuracy? What is the design period to use? The real problem is what is its design life for the future.

The existing structure has been in service, what is its remaining design life. Has it already exceeded its initial assumed design life. If it has, what is its remaining design life without doing anything. What is its design life if a rehabilitation or repair alternative is selected. What is the risk inherent in the design process. It is not the same as for a new or original design. Lets look at the situation and possible outcomes. A simple example illustrates the point.

If the repair or rehabilitation is not done, it would be likely that the structure would fail some time in the future. With repair, it is less likely to fail in the same period of time (under the same conditions). What are these time periods and the associated risks? What is the proper rehabilitation choice?

Failure and Remaining Life Concept

What constitutes failure and what are its criteria? Undesirably, for most students and many practitioners, the concept of failure is not well understood. Inherent is a desire to avoid catastrophic failure. Unfortunately this desire has not been reached in totality. In actuality, the practitioner indirectly deals with risk but in an nonobjective manner. Design life, design loads and performance limits are selected (by code, practice, choice or experience) along with tradeoffs. What is needed is that risk assessment and risk analysis be done more objectively as part of the design and analysis process to determine design parameters needed and to more accurately perform a design and recognize its tradeoffs. This is not the case now.

Thus an important factor that needs to be more closely addressed is the concept of remaining life, especially for the situations where original design life has been exceeded. This is clearly not understood by the profession. The selection of the "failure" criteria directly affects the remaining life determination. A better appreciation of "failure" and how it is addressed in the design process is also needed. These are difficult to accomplish but the needs are too obvious. With the maturity of the infrastructure, the profession is now directly facing these issues. More effective, efficient and appropriate design tools incorporating the concepts of risk and failure in the design and analysis process need to becoming forward now. The challenge is already upon the profession. How to handle it and the possible ramifications and shortcomings from improper development and implementation of new procedures is what is addressed in this paper. Extension to the problems with quality assurance concepts are obvious and direct.

The Need

The need is to incorporate risk and uncertainty concepts along with variability measures in the design and analysis process. The tools to accomplish incorporation are already available. The risk is two sided. One risk is the inability to implement correctly the needed concepts. The second risk is the possible implementation of improperly utilized techniques and procedures. Both are challenges that face and must be addressed by the profession. They are already present. The irony is that there has to be an uncertainty factor if the civil engineering profession can successfully address the need. The real risk is failure not to. The issue is how to address the need successfully.

UTILIZING PRACTITIONERS IN A SENIOR DESIGN COURSE

Raymond J. Kipp[1] and Luther W. Graef[2]

Abstract

The Civil Engineering Department of Marquette University has enlisted the assistance of professional engineers in presenting a functional and practical Senior level Design Course as required by the ABET General Criteria. This paper will describe the methods used as they have been refined over the past several years to provide a meaningful "true to life" design experience to civil engineering students.

Course Objectives

For the past three years, the Civil Engineering Department has required all students to complete a four credit semester course entitled "Civil Engineering Design" prior to graduation. The objectives of the course are:

1. To allow the student to apply engineering skills developed in basic and advanced engineering course work to obtain the design of a civil engineering facility;
2. To develop the student's initiative, responsibility and resourcefulness;
3. To allow the student to strengthen his/her ability to work as a member of a design team;
4. To improve the student's ability to write concise and accurate reports; and
5. To discuss ethics, professionalism, and those topics important to the development of the graduate engineer as he/she enters the engineering profession.

---

[1]Professor of Civil Engineering, Marquette University, Milwaukee, WI 53233. Fellow, ASCE.

[2]Principal, Graef Anhalt Schloemer and Associates, Inc., Milwaukee, WI 53233. Fellow, ASCE.

Course Description

The course meets for three lecture periods each week for fifty minutes and one three-hour laboratory period each week for three hours. The credit for the course is four semester hours.

Each student is required to participate as a design team member in producing a final design project as assigned. In addition, each student is required to attend lectures and prepare a written report on a number of the outside speaker's presentations. Each student is also required to complete a semester paper of approximately 5-6 pages on a subject related to professionalism or ethics.

The laboratory sessions were scheduled to afford an opportunity for design teams to meet with their professional engineer-client and to monitor the progress of each team. Oral reports were required during these lab sessions so that each student in a design team was required to present the progress of his/her team.

The smaller laboratory sections provided an opportunity for a discussion of projects and other questions related to the objectives of the course. Laboratory sessions during the final week of the semester were used for the presentation of each design team's final design. Each team was allowed 30 minutes with an additional 15 minutes for questions. These final designs are presented orally to the three faculty involved in the class, to other faculty from the department, to the other students in the class, to one or two invited professional engineers and to the practitioner-clients who wish to attend. Rating forms were completed after each design team's presentation. The grade for each design team was determined by the three faculty after reviewing these rating evaluations and the final design written report.

Project Design Teams

During the previous fall semester, the students who will enroll in the design course in spring were invited to form three-person project teams, with team embers whose optional courses during their studies have covered structural, environmental and transportation engineering. The students were asked to preregister as a team and thus form a design team during the spring semester.

Early in the semester, each of the professional engineers described for the entire class the design project he/she will be involved is as the practitioner-client. After all of the projects have been described, each of the student teams was asked to rate the projects the team wants to participate in on a 1-5 scale. The faculty awarded the design projects to the design teams as fairly as possible, attempting to award the projects selected if possible. (Our last experience with 54 students and 16 projects was that 14 design teams were assigned to one of their top three choices.)

The small number of students who fail to pre-register as a member of a design team were assigned to a team by the faculty.

Each design project was either a recently completed project, one which was currently in construction, or one that was being designed at the same time in the consultant's office. Titles of the sixteen design projects that were assigned last spring were:

- Sewage Lift Station, Wetwell, Forcemain
- Four-bay Fire Station
- Apartment complex on 33 Acre Site
- Reconstruction of Existing Main Thoroughfare Over An Existing River Crossing
- Renewal of Aging Downtown Area in City of 100,000
- Facility Planning and Design of Municipal Wastewater Treatment Facility
- Design of Auto Dealership
- Design of New 25,000 sq. ft. Office Building
- Water Treatment and Supply (Wyoming, MI)
- Water Storage and Distribution (Wauwatosa, WI)
- County Trunk Highway "Q" (Kenosha County, WI)
- General Mitchell International Airport - Master Plan Update
- State Highway 164 (STH 1990 - STH 74)
- County Trunk Highway "T" Interchange at I-94
- Second Creek Stormwater Project
- Landfill Development Plan

Faculty Involvement

The course is administered by a senior professor, assisted by two junior faculty. Each of the faculty is responsible for one laboratory session per week. All faculty involved attend all of the lectures. Each laboratory instructor is responsible for five or six design teams and grades all written reports, the semester paper and oral presentation by design teams in his/her laboratory section. Grading of the preliminary progress report is based on the average of the individual evaluations by the three faculty members involved. The final grades assigned to each student are also the responsibility of the three faculty.

The lecture periods are used for the following specific topics, many of which are presented by experts from several professions:

- Professionalism, Professional Societies
- Registration as a P.E., Engineer-in-Training
- Overall Design Process
- Engineer-Client Relationships
- Written Agreements, Contracts, Consulting Fees
- Technical Writing, Proposals, Reports
- Oral Communications
- Ethics in Engineering
- Safety - Risk Analyses
- Risk Assessment - Risk Management
- Cost Estimating - Sources of Accurate Cost Figures
- Municipal (State) (Federal) Codes, Zoning, Permits
- Department of Natural Resources, EPA Regulations
- Environmental Impact Assessment and Statements
- Flood Plains - Historical Sites - Restrictions
- Professional Liability - Liability Insurance
- ASCE Manuals - Quality in the Constructed Project
- Patents - Copyrights - Trade Agreements

Several lecture periods are used for the description of the design projects by the practitioners involved. Each design team also presents a preliminary progress report to the entire class during lecture periods. This report occurs approximately six weeks prior to the end of the semester.

Each student is required to summarize the presentation on a specific topic by an expert. These written reports are carefully reviewed and graded by the faculty.

Practitioner-Client Involvement

The involvement of the professional engineers and firms begins during the Fall semester when the senior professor contacts the firms and requests that two of the firm's design engineers submit at least one possible design project that one of the senior design teams will complete during the Spring semester. After the projects have been submitted and reviewed by the faculty involved in the course, the professionals are invited to a meeting, usually a breakfast, where a description and the objectives of the course are reviewed. A time schedule is explained so that each firm and professional is aware of the commitment expected of them. The professional is asked to act as a client to the design team assigned to his/her project and to meet with the team as often as necessary. A meeting should be held at least once every two weeks, although most of the practitioners have met every week with their team. We encourage having these meetings at the professional's office so that the students see the professional engineers in their workplace. The students are provided with the practitioner's phone number so that they can call if questions arise. The professional is also asked to be present at the lecture when the team presents its preliminary design report. At the final design presentation, the professional is asked to grade the oral presentation and to review the final report. One copy of the final design is given to the professional by each design team.

After the semester is completed, a second meeting is held with the professionals and faculty involved. This meeting is a luncheon meeting where the professionals are asked for their input on the course with criticisms and suggestions for improvements. The firms who have been involved are also asked if they would volunteer for this involvement in the future.

The involvement of each professional varies, but we estimate that each engineer contributes between 15-30 hours of their time assisting in this design course. This involvement of practitioners and design firms began several years ago with two firms and common projects for several teams. The response of practitioners has been such that the program has been expanded to numerous firms and a practitioner/client with a separate project for each design team. The support of the college staff has created a professional relationship which seems to invite and encourage the practitioners to want to be a part of this exciting education/partnership relationship.

The practitioners can see several distinct advantages to being a part of the program. First, and probably the most important, is the opportunity to return something to the profession by helping to train future engineers and to make the transition from an academic life to a full-time working atmosphere easier for the student. An excellent working relationship and rapport is established between the professors involved and the practitioner. The practitioner has the opportunity to meet, relate to, and communicate with the entire senior class as the projects develop and the reports are presented. Networking works both ways as the students and practitioners develop relationships before graduation rather than after. This also seems to better introduce students to ASCE and the profession and the ethical and working concerns of the practitioner, i.e. codes, public review bodies, environmental impacts, existing infrastructure, social concerns, etc.

Last, but certainly important is the opportunity for the practitioner to review in an objective way the design project from a new perspective. The students, with this opportunity, reproduce the design from a new vantage point, offering new solutions and ideas and questioning some traditional solutions. When completed, the project offers new insights to all who have participated - students, faculty, and practitioners. Everybody wins.

Grading Procedures

The final grade submitted for each student is determined by individual assessment (50%) and the design team assessment (50%). The breakdown used during the past year was:

Individual	
• One page written reports on invited lecturers	10%
• Semester Paper on selected topic	20%
• Oral Progress Reports	10%
• Participation and Professionalism	10%
Design Team	
• Written Progress Reports	10%
• Final Design - Report and Presentation	40%

Results

A comprehensive student evaluation was prepared to solicit student opinion and reaction. Surprisingly, a large majority of the most recent class had positive feelings and believed the course would be beneficial to their future employment. The opportunity to work closely with a design professional along with the opportunity to meet with the client in the office was appreciated. Practitioners offered suggestions, but primarily allowed the design teams to produce their own designs. Each design team was forced to consider alternatives and to determine preliminary estimates of the cost of their design.

The majority of the class believed that their writing skills had improved. Faculty and practitioners noted the vast improvement in both written and oral communications by almost every student. Students became quite proficient in the use of the computer's word-processing function in their writing.

Working with a team was another positive benefit, although there were some complaints that certain members did more work than others. Most of the projects involved not only structural engineering, but also environmental engineering. Several of the projects were primarily transportation and planning. The students realized that design involves many different specialties and learned the need for the skills of others. Students felt that more of the grade should be awarded for individual performance and less for the group effort.

Several students had completed four semesters of our cooperative program, employed with a civil engineering firm. They were more comfortable in working with a student design group and responded in the evaluation that this design course was very close to the real world experience they had encountered in their cooperative job.

The involvement and interest of sixteen professionals from seven consulting engineering firms and one state divisional transportation department along with an additional twelve experts who lectured in their particular specialty had a very positive influence on the class. The contribution of these professional engineers exemplified many of the attributes that the faculty endeavored to instill in the students regarding professionalism.

Improvements

Based on our experience over three years, we are planning several changes to improve the design course.

1. Identify the professionals and firms that are to be involved at least two months before the beginning of the semester. Have each professional submit a brief description of two possible projects. This would assist the faculty to select a variety of projects.
2. Assign the design projects to the student design teams as early in the semester as possible.
3. Assign the title of the semester paper within the first two weeks of the semester. Require each student to select a topic for his/her paper from a recommended list of topics related to professionalism and ethics in engineering.
4. Emphasize the use of visual aids in making a professional presentation of the final design product.

Acknowledgements

The Department of Civil Engineering acknowledges the following firms and agency and the professional engineers from these groups for their participation and contribution to this design course:

- CH2M-Hill
- Donohue Engineers and Architects
- Graef Anhalt Schloemer and Associates, Inc.
- Howard, Needles, Tammen and Bergendoff
- Opus North Corporation
- Ruekert Mielke, Inc.
- J. C. Zimmerman Corporation
- Wisconsin Dept. of Transportation
  District #2 Division Office

Creativity: A Neglected Fundamental

Theodore B. Feldsher[1]

Abstract

A fundamental connection exists between civil engineering and creative, imaginative thought--a connection that deserves serious consideration by planners of undergraduate engineering curriculi. Current engineering education programs do not recognize that civil engineers need and use creative, "right-brained" thinking skills as much or more than technical skills. A direct and effective way to address this need--to improve our non-technical, non-analytical skills and abilities--is to increase our exposure to liberal arts and humanities education, in parallel with engineering education. The potential benefits of such a change include not only increased effectiveness and personal fulfillment for individual engineers, but also an improvement in the stature of the civil engineering profession as a whole.

Introduction

My dictionary defines a creative person as one who has or shows "imagination and artistic or intellectual inventiveness." The stereotypical civil engineer, in contrast, is often perceived to be pragmatic, conservative, and narrow-minded--but hardly imaginative or artistic. Most civil engineers know that such stereotypes are false, but far fewer seem to be cognizant of the deep and fundamental connection between civil engineering and creative, imaginative thought. I have a vivid memory of one of my professors telling us that, out in the real world, 85 percent of a civil engineer's time is spent on non-technical activities--the kind that he or she was unlikely to have learned about in engineering school. After several years at a large consulting firm,

[1] Staff Engineer, Woodward-Clyde Consultants, 500 12th Street, Suite 100, Oakland, CA 94607

it has become clear to me that one of the most important of such non-technical activities is communication, in its wide variety of forms, without which even the most sophisticated analyses and the most brilliant ideas are of little value. Virtually all communication requires synthesis of ideas and information, a creative process. The report that communicates the results of a complex engineering study and the drawing that communicates the engineer's design intent are examples of the connection between civil engineering and creativity, because they both require the engineer to use "imagination and . . . inventiveness" in their preparation.

Creativity and Civil Engineering

The most interesting and stimulating projects for many civil engineers are those that present new and unique challenges, for which no "cookbook" solution is available. Such projects test our creative talents as well as our technical expertise. These are the projects for which we sometimes have to develop our own method and approach, for which we have to imagine potential solutions and alternatives, and then lay out and follow the paths to reach them. The parts of our work that require us to think creatively and to use our imaginations are most interesting for precisely that reason--they make us think with our entire mind, not just with our logical, analytical side. On a deep and fundamental level, the civil engineering impulse is to imagine, to design, and to create; to solve a problem and fulfill a need; and to thereby make a difference in the quality of life in this world. On that fundamental level, creativity is at the core of the civil engineering experience.

Engineering judgement is another of our non-technical activities that is firmly rooted in synthesis, and thus creativity and imagination. Judgement is a skill we develop by becoming students of our own experiences, a skill that allows our mind to bypass the usual logical, rational, and analytical pathways in the process of coming to a decision. Creative thinking is a basic component of and is closely related to judgement. Both take place in the same part of the mind, the part that performs synthesis rather than analysis. The judgement process is complex, intuitive, and nonanalytical, because it relies on the imaginative and creative abilities of our mind to locate relevant information in our experience "database" and to sort and combine it in a way that helps bypass the logical decision process. Without such synthesis of information, our judgement would be much weaker, because then we would be much less able to extrapolate beyond our experience, an essentially imaginative process. In the real world, no two projects or situations are completely the same, and so, to use our judgement most effectively, we must undertake a creative, intuitive, and imaginative thought process--very similar to that of a writer searching for a metaphor to illuminate an unfamiliar scene.

Samuel Florman has bluntly characterized much of today's engineering practice as a continuing struggle to avoid mistakes due to "human error, lack of imagination, and blind ignorance" (1976, p. 33). As discussed, imagination is a crucial component of judgement, which is itself a cornerstone of virtually all civil engineering practice. The mistakes and failures that might be attributed to lack of imagination have their counterpart in the cases where bold, far-sighted, creative engineers have achieved precedent-setting successes. But these successes are not common enough that we should divert our attention from the struggle--and the mistakes and failures demand that we examine it more closely. How is the civil engineering profession working to improve the effectiveness of the engineers within it? The American Society of Civil Engineers has recently published a Manual for Quality in the Constructed Project (ASCE, 1988), which is clearly a positive step. However, what Florman called the "lack of imagination" is not being addressed by today's civil engineering curriculi.

Civil Engineering Education

A relatively small number of engineering schools in the United States do offer students the opportunity to spend more time studying liberal arts and humanities before, or in some cases after, concentrating on engineering. This type of program typically consists of a 5-year schedule for the bachelor's degree, for example at Dartmouth's Thayer School of Engineering (Florman, 1987). Some other engineering schools have begun to realize the importance of developing communication skills during the undergraduate education, and have thus instituted heavier English and writing requirements, such as at the University of California, Berkeley. However, the main focus of the engineering educations offered by these schools has remained the same, and the additional liberal arts and humanities exposure appears principally intended as a broadening influence, to produce more well-rounded engineers. Even the progressive schools do not recognize the fundamental connection between creativity and engineering; and so they continue to institute programs that are out of balance and that produce engineers who are less creative and thus less effective than they have potential to be.

It is clear that one purpose of undergraduate civil engineering education is to provide young engineers with the tools needed to pass their professional licensing examinations. Such exams are written to test a broad base of knowledge, but not to great depth, and they concentrate almost exclusively on technical problem-solving abilities. In-depth knowledge of narrow topics and non-technical abilities gained from practical work experience are generally not a focus of P.E. exams. For this reason, it is much more appropriate and efficient to learn the civil engineering necessary to pass the P.E. exam in school than on the job.

Some might argue that the reverse is true for non-technical skills and abilities; that it is more appropriate and efficient to learn and refine them after one has entered the profession and is better able to appreciate their relevance and importance. By this line of reasoning, valuable undergraduate time should not be wasted attempting to foster creativity and imagination that will be developed anyway if they are truly needed. I submit, however, that such an approach--which is basically what we have now--does not work, because the creative skills are needed, but working engineers often lack the time and energy necessary to develop them. The unfortunate reality is that there is little immediate financial motivation for the significant task of expanding one's creative and imaginative horizons. The most appropriate and efficient time to improve these crucial non-technical skills in young civil engineers thus seems to be while they are still in school, and still intensely focused on learning.

Education in Creativity

Learning is a lifelong process for everyone, a process particularly important to those of us in the constantly changing and growing field of civil engineering. Engineering educators recognize this when they plan undergraduate programs that teach us not only engineering, but also how to think and how to continue learning. In essence, our undergraduate work is the beginning of a career-long education in engineering, a "launching phase" (ASEE, 1989) that can give our careers both energy and direction. Because it is so important, I believe that imaginative and creative learning also belongs in this "launching phase" of our education.

Creativity and imagination are basic human abilities which we all possess and which we use, to some extent, every day of our lives--whether we are consciously aware of it or not. But like any other of our abilities, they can be poorly or well developed; of little use or very powerful and important tools. Unlike experience and judgement, however, our other, more "artistic" and intuitive skills are not as likely to be improved during our civil engineering careers. Experience and judgement are gained by most observant engineers who enter professional practice, as they are exposed to new projects and problems to be solved. Creativity and imagination, on the other hand, tend to fade if neglected, especially if they are weak from the outset. Pressures for efficiency and conservatism, which inevitably come hand-in-hand with professional responsibility, do not usually leave much room for imaginative thinking.

Paradoxically, it seems that the need for creative thought increases over the course of our careers, as we gradually deal with more complex problems, or spend more of our time on non-technical activities, such as management and commmunication. Following the "launching-phase" concept, one could argue that creativity training thus belongs later in our engineering careers. It would no doubt be valuable for senior engineers to refresh their imaginative skills, but these skills are important throughout the development of our careers, and so should not be neglected until we reach advanced levels. If these important skills are not healthy and strong, we run the risk of reducing both our effectiveness and our fulfillment as engineers.

Right-Brained Engineering

If, as argued, creative and imaginative abilities are in fact important components of our everyday performance as civil engineers, and improving those abilities before or during our undergraduate education is not only possible but desirable, then the question that follows is how should current engineering curriculi be modified to address this need?

Several authors have suggested that broadening and liberalizing the scope of undergraduate engineering education could result in more well-rounded engineers, better equipped to fill leadership roles in society (Hutchinson and Muller, 1988; Florman, 1976; etc.). I maintain, however, that the primary benefit of a liberalized curriculum is that it can make us better civil engineers, better able to carry out the non-technical tasks that dominate our day-to-day work. The premise that education in the liberal arts and humanities has the potential to develop and enhance our creative and imaginative skills is central to this argument. Such an education, in subjects that require reading, writing, and discussion about art, literature, history, or human values, exercises and strengthens a different kind of thinking than is required, for example, by calculus, dynamics, or structural design. It is well-documented that two regions, or sides, of the brain govern two very different and separate modes of thinking (Edwards, 1989). A balanced education should stimulate and engage both sides of the brain and both modes of thinking.

According to the two-sided model of the brain, the "left" side governs rational, logical, analytical thinking--such as that required to solve a calculus problem. Most engineers are heavily "left-brained," for obvious reasons. The "right" side, which governs intuitive, nonrational, creative thinking, is in comparison sorely neglected by most engineering educations. The practice of civil engineering--aptly described in the title of a U.C. Berkeley course as "The Art and Science of Civil Engineering"--calls on both sides of the brain, and I believe that it would be beneficial to both the profession and society as a whole to expand the scope of civil engineering education to more directly address our important creative, imaginative, and intuitive side.

Perhaps the most significant objection that has been raised to the liberalizing of engineering education is that there simply isn't enough time to produce quality engineers and to expose them to humanities and the liberal arts, all in four years. This argument presupposes that time spent on non-technical subjects, beyond a nominal requirement, is detrimental to the production of quality engineers. As this paper has attempted to show, however, quite the opposite is likely to be true. Humanities and liberal arts education can make us better engineers, by improving our ability to use and apply all of our skills--creative and analytical, intuitive and rational--since our work really does include both the "art" and the "science" of civil engineering.

Conclusion

I think that many of the young people who develop an interest in the profession do so partly out of their desire to envision and create great works like the ones they see around them--works that both inspire and serve mankind--even though they may not understand how such works were created. The connection between the creative impulse and civil engineering is probably as clear in these young people as it ever will be. The San Francisco Bay Area is filled with monuments to the vision and boldness of civil engineers: the catenary curves and slender towers of the Golden Gate Bridge, the graceful, tapered Transamerica Pyramid, and the Caldecott Tunnels that dive beneath the Berkeley hills are just a few. The engineers who conceived and designed these works had to think both creatively and analytically, and to use both new ideas and engineering experience. They had to supplement and reach beyond their very likely unbalanced educations to successfully meet the challenges they faced. More liberal arts and humanities coursework in the civil engineering curriculum, even at the price of an additional year in school, would be well worth the extra investment to us as individuals, to the profession, and to society, if it facilitated the creation of visionary, successful projects like these. I believe, however, that the most important payoff of such an educational investment would be a greater degree of personal fulfillment and satisfaction for civil engineers which would have a direct and positive impact on the quality of our work.

References

American Society of Civil Engineers, Manual of Professional Practice: Quality in the Constructed Project--A Guideline for Owners, Designers, and Constructors. New York: American Society of Civil Engineers, vol. 1, Preliminary Edition for Trial Use and Comment, 1988.

American Society for Engineering Education, A National Action Agenda for Engineering Education. Washington, D.C.: American Society for Engineering Education, 1988.

Edwards, Betty, Drawing on the Right Side of the Brain. Los Angeles: Jeremy P. Tarcher, Inc., Revised Edition, 1989.

Florman, Samuel C., The Existential Pleasures of Engineering. New York: St. Martin's Press, 1976.

______________, The Civilized Engineer. New York: St. Martin's Press, 1987.

Hutchinson, Charles E., and Carol B. Muller, "Educating Engineers: In Praise of Diversity," Issues in Science and Technology, vol. IV, no. 4, Summer 1988, p. 71-74.

Methods for Practice-Oriented Engineering Education

Edward A. Beimborn[1]

Introduction

In recent years, the pace of change in all aspects of technology and society has become rapid. Changes are occurring in technology, economic conditions, resource availability, life styles, institutions and other areas, all of which will have profound effects on the future practice of the engineering profession. Engineers of the future will need to be a highly adaptable group that can quickly respond to change and have the capability of dealing with a broad spectrum of problems and issues. Yet while these changes are occurring, our education of engineers has drifted away from a broad based, real world orientation towards a highly abstract view of the world. Relevancy in engineering education seems to have declined and it appears more and more difficult for universities to remain in touch with the people who use their research or hire their graduates.

Substantial criticism can be made of our educational programs and their loss of relevancy. A high emphasis has been placed at universities upon extramural funding of research and scholarly publication. Nearly all new faculty enter engineering colleges directly from Ph.D. programs with a background of basic research and little if any real world experience. Few, if any, mechanisms exist for bringing in people with extensive professional experience to participate in the education of young engineers. Some would argue that universities have lost touch with reality and live in a simplified world where they have become out of touch with the users of their products. Basic research is done by pretending that it is applied. Applied research is done without ever talking to the people who will apply it. Students are taught to do things their instructors have never done themselves. Reports are written that no one can read or understand, and time is divided into so many little pieces that nothing can ever

---

[1]Professor, Department of Civil Engineering and Mechanics, and Chief Muckety-Muck, Center for Urban Transportation Studies, University of Wisconsin--Milwaukee, EMS E387, 3200 N. Cramer Street, Milwaukee, WI 53211.

be done completely or well. Relevancy to real world problems and has been lost at a time when it could not be more important.

This paper will explore the issue of relevancy and practice-oriented civil engineering education. This will include a discussion of student attitudes towards relevancy and practice-oriented civil engineering education and a description of efforts that have been underway at the Department of Civil Engineering and Mechanics of the University of Wisconsin--Milwaukee (UWM) to meet these needs. These efforts include the introduction of a senior design course into the curriculum, the use of improved telecommunications technology for course delivery, design contests, EIT requirements, and experiential assignments. The paper will describe these efforts as well as problems encountered and their level of success.

Student Attitudes

For several semesters a survey has been distributed to senior civil engineering students which asks them their perceptions of the relative importance of fifteen different course attributes. These attributes include course relevancy, instructor capabilities, comfort factors and course administration (exams, textbook, etc.). Students were asked to rate the level of importance on a scale of 1 to 5 with 5 representing the term "very important", 3 representing "neutral" and 1 representing "very unimportant". The results of these surveys are indicated in table 1.

Table 1
Student assessment of importance of course attributes
(1 = very unimportant, 5 = very important)

How important are the following to you when you take a course?	Spring, 1987 N = 25	Spring, 1988 N = 26	Spring, 1989 N = 33	3 year Average
Use of practical examples	4.51	4.76	4.70	4.66
Material is relevant to future career	4.43	4.73	4.70	4.62
Instructor has practical experience	4.17	4.35	4.64	4.39
A good textbook	4.17	4.31	4.67	4.38
Homework is corrected	4.35	4.20	4.45	4.33
Convenient schedule	3.79	4.38	4.45	4.21
Use of computers	3.90	3.76	4.03	3.90
Use of projects	3.64	3.31	3.61	3.52
Comfortable room	3.44	3.27	3.75	3.49
Use of guest speakers	3.13	3.60	3.70	3.48
Small class size	3.72	3.27	3.35	3.45
Frequent exams	3.64	3.04	3.27	3.32
Having lab experience	3.03	3.30	3.53	3.29
Use of slides and films	2.85	3.33	3.34	3.17
Use of field trips	2.85	2.88	3.18	2.97
Average	3.71	3.77	3.96	3.81

The two factors rated as most important by students in all cases were "use of practical examples" and "material is relevant to my future career". Both of these were rated near the top ("very important") portion of the scale. Other factors, "having an instructor with practical experience", "a good textbook", "homework is corrected" and "convenient schedule" were rated between "important" and "very important". The remaining factors such as class size, room comfort, frequency of exams, field trips and use of slides and films which primarily relate to how courses are administered were rated from "neutral" to "important". The results were fairly consistent over the three years in which the survey was used.

These results seem to clearly indicate that students place a high degree of importance to the relevancy and practicality of their education, i.e. its professional content, rather than how they receive the material. Thus efforts to increase the relevancy and practicality of a course would have more payoff in the eyes of the students than efforts to provide better administration. Discussions with practicing civil engineers and the people who hire our graduates have shown agreement with the feelings of our students. Experience in engineering practice is highly valued by potential employers even if only in a classroom situation. Teamwork, communications skills, the ability to deal with people, environmental and global awareness all are characteristics that are highly sought after. There appears to be a need to develop mechanisms to bring practice-oriented education into our curricula without a major sacrifice in technical content. A variety of approaches have been tried to do this at the University of Wisconsin--Milwaukee. These are described below.

Senior Design Project:

CE 395--Senior Design Project is a 3 credit, one semester course required for graduation for all civil engineering students. The course is designed to provide a comprehensive design experience for CE students just prior to their graduation. The course meets on a regular schedule three hours a week with approximately one-third of the time spent in lectures and two-thirds of the time for project activity. The major focal point of the course is the design project. The project is a semester long effort either done individually or by a group. The project is used to integrate student knowledge from other courses into a comprehensive effort that closely simulates actual engineering practice. Students are expected to spend about 120 to 150 hours on the project. A project proposal, weekly project reports, an interim report, a final project report and an oral presentation of the project are all required. There are no exams. Student projects have dealt with a variety of topics covering all aspects of civil engineering planning and design. Some examples include: Design of a camp for Afghan refugees, site planning for a university soccer stadium and practice area, neighborhood traffic control plans, structural design of warehouses, and reconnaissance studies for the rehabilitation of a dam and hydroelectric plant.

The lecture portion of the course is an important component which can be used to fill in voids in the curriculum. Topics that are discussed include project management, budgeting and resource allocation; engineering communications; professionalism and ethics; innovation and overcoming resistance to change; design procedures, and decision making. The classes are small and taught by senior faculty and provide an opportunity to help students to learn how to "think like engineers" and to understand the world in which they will function following graduation.

Student feedback from the course has been very good. Often students will spend additional time on the projects beyond the minimum required. The work produced has been of good quality and students feel a strong sense of pride in the final result. In addition, several projects have led directly into jobs and/or resulted in higher starting salaries for students who used the projects during their job search. More importantly, however, the opportunity for a comprehensive project experience appears to provide the students increased confidence in their ability and a better understanding of the real world of engineering professional practice.

Telelectures

Under an NSF sponsored project entitled "University/Industry/Government Partnership for Quality Engineering Personnel," UWM worked with Harza Engineering Company of Chicago to utilize advanced telecommunications technology to deliver lectures to the university from remote sites. A series of ten lectures were broadcast from Harza offices in Chicago to UWM and elsewhere via satellite with one-way video and two-way audio. These lectures were given by senior staff at Harza and covered a variety of topics including: planning and design procedures, field investigations, applied hydraulics and applied hydrology. Broadcasts were received by approximately 15 universities in North America.

The purpose of the project was to increase the practical orientation of CE students through a greater awareness of real world problems and procedures. The project was also used to test the feasibility of the use of improve communications technology to instill a better understanding of the design process and its relation to operation and maintenance of engineering systems. The programs were coordinated with the senior design course as described above and also included input from practicing engineers into student projects.

The experiment was evaluated through student surveys before, during and after the project, by the instructors involved and by a panel of recent graduates and other practicing engineers. Results of the project indicate that it is technically feasible to provide lectures to universities in such a manner and that there was an overall positive impact on increasing the students' real world knowledge of engineering practice. A telelecture system has a high potential in that it can provide access to new topics, provide a breadth of coverage on topics where faculty expertise is not available and give students a better understanding of the real world of engineering practice.

Design Contests

Students at UWM have eagerly participated in competitions such as the concrete canoe races and bridge design contests. These competitions require students to design and build something and to compete against other schools for prizes. Enthusiasm for the contests is generally very high and students will spend long hours, often with little sleep, to construct their entry. The project experience provides them with an opportunity to learn teamwork as they work with other students as well as to work under deadline pressures and to have to accomplish a goal with very limited resources. While exhausting at times, the competitions provide an enjoyable, real world experience that stays with the students well into their professional careers.

EIT Requirements

Students in Civil Engineering at UWM are now required to register for the EIT examination prior to graduation. This was done to help students to take the proper steps towards professional registration at an appropriate time. It has had the side benefit, however, of creating a high awareness of the role of the P.E. in civil engineering practice. Many students spend considerable time thinking about professional registration and the process involved and discussing it with their fellow students and faculty. Reaction to this requirement by the practicing engineering community has been also highly positive.

Experiential Assignments

Experiential assignments could be defined as ones that require students to take on a different role in order to better understand a particular problem or function. For example, civil engineering students were recently required to make a trip across campus by wheelchair in order to better understand the concept of barrier free design for the handicapped. By actually encountering doorways, elevators, uneven sidewalks, ramps, classrooms, etc. students gained an understanding of barriers to mobility far beyond that which would occur otherwise. Another example involved students taking on the role of a transit management team faced with a severe financial crisis and the need to present recommendations to a citizens' board in a short time frame. Experiential assignments of this type have been used in business schools for years as a major educational tool, but have had little application in engineering. They help to build problem recognition and definition skills, important aspects of professional practice, which are seldom discussed in engineering curricula.

Conclusions

The mechanisms discussed in this paper--senior design course, telelectures, design contests, EIT requirements, and experiential assignments--

all serve to help bring about a more practice-oriented education for civil engineering students. They provide an awareness of real world problems and a broadening of capabilities for students into topics of problem definition, professionalism, and decision making. Reaction by both students and practicing engineers have been good.

Unfortunately an emphasis on practice-oriented education tends to be an exception rather than the normal way of doing things. It runs counter to the mentality and pervasive forces which push engineering education in other directions. The pressure for external funding and for scholarly publication, especially on young faculty, make it very difficult to introduce such relatively simple innovations into our curriculum. What is needed is a pressure in other directions to create a balance of activities (and people), some of which are scientifically oriented and some of which are oriented to professional practice. Pressure is needed from inside our universities to make these changes. But even more important is outside pressure from the profession. Practicing engineers need to make their voices heard to deans, department chairs and faculty directly if they feel that the system is out of balance. Professional organizations such as ASCE, ITE, etc. need to place pressure on accreditation boards and upon funding agencies for university research to provide a more practice-oriented education. Creative efforts by students should be encouraged with aggressive efforts to help universities to develop realistic projects, assignments and competitions for their students. These efforts can all help to bring about a more practice-oriented education and to educate people to have a positive impact on the world of the next century.

References

Beimborn, E. A. "Things We Never Teach: The Decline of Relevancy in Engineering Education," **Proceedings**, ASEE North-Midwest Section Annual Meeting, October, 1988.

Beimborn, E. A., and G. M. Karadi. "Civil Engineering Education Through Satellite Television Linkages," **Proceedings**, 1980 ASEE Annual Conference, June, 1988.

Curriculum Wide Integrated Design Project

Everett E. McEwen[1]

Member, ASCE

ABSTRACT: A comprehensive design project is begun in the junior year and a portion of the design carried out in each of the subsequent civil engineering courses. The design project is completed in a capstone design course where student design teams write technical and final project reports. Practicing engineers participate in the program and also judge student presentations at the completion of the project. A reservoir-dam project and the design of a new civil engineering building are currently in progress.

INTRODUCTION

The civil engineering program at URI has always had a substantial design content; however, the design courses have been, until the initiation of this program, the more traditional courses in reinforced concrete design, foundation engineering, highway engineering, etc. where a portion of a civil engineering system was designed and a minimum of effort devoted to report writing, oral presentations, and cost estimates. In some cases projects were limited to one aspect of civil engineering.

A different approach to the teaching of design was initiated at the University of Rhode Island in the fall of 1988. Instead of requiring a comprehensive design project in a single capstone design course, the design project was integrated into all of the required and most of the elective civil engineering courses. The design project is introduced to the student in an early course (sophomore or junior). A portion of each of the succeeding courses is devoted to the design project. In the final semester of the senior year all students are required to take a capstone design course. In this final course design teams complete the project, writing reports on each of the major technical aspects of the project and a final project report including cost estimates and environmental impact. Each team will also make an oral presentation. The assignment of the design project is made near the beginning of the program to focus the students' thinking on the relation of each course to the final design. It also gives the

[1] Prof., Civil Eng., Dept. URI, Kingston, RI 02881

student a better understanding of analysis and design and the interaction of courses covering seemingly unrelated material.

In addition to civil engineering, both the mechanical engineering and chemical engineering departments have initiated similar programs. Their projects involve the design of a light-weight airplane and the design of the chemical process plant to produce plastic.

This paper describes the first two projects which are now underway and the progress to date. This program is a first attempt to integrate design throughout the curriculum and modifications are expected as we gain more experience. The need to evaluate this effort is very critical and means to do so are discussed below.

ORGANIZATION OF DESIGN PROJECT WITHIN CIVIL ENGINEERING CURRICULUM

Figure 1 shows the organization and course sequence. At the present time all of the required courses and most of the elective courses in the junior and senior years are included. At a later date sophomore courses will be added.

COURSE	*FUNCTION*
INTRODUCTION TO CVE DESIGN	*PLANNING AND MOTIVATION*
CVE REQUIRED COURSES	*PRELIMINARY DESIGN OF ALL SUBPROJECTS*
SOILS STRUC TRANS ENVIRON *CVE ELECTIVES*	*DETAILED DESIGN OF ALL SUBPROJECTS*
COMPREHENSIVE DESIGN PROJECT	*FINAL DESIGN, REPORTS, PRESENTATION*

FIGURE 1

The first course in the sequence is CVE 304 Introduction to Civil Engineering Practice. This course consists primarily of field trips and outside speakers. Its purpose is to introduce the overall design program, the specific design project for that class and to carry out the initial project planning. This course is also intended to motivate the student and to generate increased interest in the program.

Methods of analysis and introductory design techniques are presented in each of the required civil engineering courses followed by final design of each of the subprojects in the elective courses. During the summer of 1988 the faculty wrote course modules which pertain to the design project for each of their courses. The material in these modules are modifications of normal course content oriented toward the design project. For example, in the steel design class the instructor may lecture on bridge design and assign as a course project the bridge required for the project. The course modules vary in length from one to five weeks with most courses devoting about two weeks to the project.

The culmination of the program is CVE 498, Comprehensive Design Project. In this course the class will be divided into design teams of five to six students. One student will serve as project manager and have responsibility for the overall completion of the design, an initial planning report, a final written report, and a team oral presentation. Each of the students will serve in turn as project engineer for one technical aspect of the design and be responsible for a technical report including plans. All students will work on all phases of the design.

A design reference center has been set up in the civil engineering student lounge. Copies of final plans and reports have been obtained from local design firms and, along with design manuals, will be available as references. Drafting tables will be available.

Increased interaction with industry is one of the long range goals and discussions with local industry have indicated strong support for this program. The final oral presentations will be judged by a group of practicing engineers and faculty and prizes awarded.

KENT-WASHINGTON COUNTY RESERVOIR PROJECT

Since the practice of civil engineering is so broad, a major consideration in project selection was a project which involved equally several major technical areas. The project selected for the class of 1990 was the design of a reservoir, earth dam, spillway, water treatment plant, bridge over the spillway, and related road work. To make the project as realistic as possible a real site located in a rural area about ten miles from campus was selected, where environmental and site data could be obtained by the students.

The project statement which was presented to the junior class in the fall of 1988 in the course, Introduction to Professional Practice, is shown in Figure 2.

PROJECT STATEMENT

REQUEST FOR A PRELIMINARY ENGINEERING DESIGN FOR THE KENT-WASHINGTON COUNTY RESERVOIR AND DAM PROJECT

The South County Water Authority, 0001 Upper College Road, Kingston, Rhode Island, 02881, hereinafter referred to as the AUTHORITY requests the development of preliminary site and engineering plans for the design of the Kent-Washington County Reservoir and Dam from CVE Design Teams, Inc., 211 Bliss Hall, Kingston, R.I., 02881, hereinafter referred to as the ENGINEER. The ENGINEER is asked to develop a comprehensive plan and preliminary design drawings for a drinking water reservoir, dam, and treatment facility on the Fisherville Brook in Exeter and West Greenwich, R.I.

A recent study by the AUTHORITY has indicated the need for a reservoir to serve the need of a population of 5000 in the Town of Exeter by the year 2000. The water source is surface runoff from the Fisherville Brook watershed with an area of about 5 square miles. To contain sufficient water it is anticipated that an earth dam approximately 40' high and 2000' long will be necessary. Fill material for the dam is to be obtained from local materials on the watershed.

The dam design should include a concrete spillway capable of passing a 100-year rainfall event without overtopping.

The water treatment plant should have a design capacity of at least 1 MGD and be designed for the removal of suspended solids, tastes, odors, and colors. Provision for emergency short-term treatment in the event of reservoir contamination must be included. Redundancy in individual unit processes and operations to handle routine maintenance and emergency situations is a requirement.

The AUTHORITY wishes the ENGINEER to evaluate the feasibility of a low-head hydro electric plant at the site based on a cost-benefit analysis considering the reservoir stage. If feasible, the AUTHORITY will instruct the ENGINEER to proceed with the design.

It is expected that Pardon Joslin Road will be relocated. The ENGINEER should give consideration to locating the roadway on the top of the dam and of crossing the spillway with a bridge. Additional roads, parking areas, and bridge structures for project are to be included.

The ENGINEER will prepare an environmental impact analysis of the effect of the project on human, wildlife, and plant features at and adjacent to the site.

The ENGINEER will provide the AUTHORITY with written reports on each technical phase of the project as well as a final project report including cost estimates. Preliminary design drawings should be included with the report. In addition the ENGINEER will present an oral presentation for review by the AUTHORITY.

Figure 2

In Introduction to Professional Practice the class visited similar projects, some of which were under construction, and attended lectures by practicing engineers on the planning of these projects. The students also visited a large consulting firm and were able to see a comprehensive project under design. The course met once a week and the schedule is shown below.

September 8 Introduction to Design Project
September 15 Site Planning
September 22 Kent-Washington County Reservoir Project
September 29 Field Trip to Scituate Reservoir
October 13 Reservoir and Dam Planning (Mr. J. Miller, Fenton Keyes Assoc.)
October 20 Field Trip to Highway Construction Project
October 27 Highway and Bridge Planning (Mr. J. Giordano, G. Archibald, Inc.)
November 3 Field Trip to Site of CVE Project
November 10 Field Trip to Civil Design Firm (Maguire Group)
December 1 Earth Dam Construction (Mr. F. Pierce, Maguire Group)
December 8 Course Critique

During the 1988-89 and 1989-90 academic years design modules have been introduced into the following courses:

Civil Engineering Lab
Surveying
Environmental Engineering
Geotechnical Engineering
Structural Analysis I
Structural Analysis II
Highway Engineering
Hydraulic Engineering
Traffic Engineering
Design of Metal Structures
Design of Concrete Structures
Water and Wastewater Transport Systems
Water Quality Sampling and Analysis
Water in the Environment
Foundation Engineering

The students are just now entering their senior year, where the course modules will address the design of the reservoir, dam, water treatment facility, spillway, bridge and roadways. The tentative outline for the capstone design course which will be taught in 1990 for the first time is shown below

Class	Topic
1 - 5	*Project Planning and Site Planning*
Site Planning Report Due	
6 - 10	*Reservoir and Spillway Design*
Reservoir and Spillway Report Due	
11 - 14	*Dam and Foundation Design*
Dam and Foundation Report Due	
5 - 18	*Water Treatment Facility Design*
Water Treatment Report Due	

19 - 22	*Highway Design*
Highway Design Report Due	
23 - 26	*Bridge Design*
Bridge Design Report Due	
27	*Final Report Preparation*
28 - 29	*Team Oral Reports*
Final Project Report Due	

NEW CIVIL ENGINEERING BUILDING PROJECT

The project selected for the class of 1991 is the design of a new Civil Engineering building located adjacent to the existing engineering buildings. It will house all of the activities of the Civil Engineering department. The facility will include an auditorium, laboratories for teaching and research, classrooms and offices. A parking facility for department faculty, staff and visitors will also be designed. The added demand placed on university utilities will be studied and recommendations made for improvements.

The first step in this project is the functional planning and the development of floor plans, structural framing systems and building elevations. This planning phase was carried out in CVE 304 Introduction to Professional Practice. The initial planning lectures will be given by the principal architect and structural engineer of a local consulting firm. The class was divided into teams of three students and will make both written and oral reports at the completion of the course. A schedule for CVE 304 is shown below. The class met once a week for three hours and much of the work was carried out during class.

Week	*Topic*
1	*Introduction to project, client-engineer relationship, existing data*
2	*Review Uses of Existing Building*
3	*Lecture--Planning--Chief Architect*
4	*Lecture--Planning--Chief Architect*
5	*Lecture--Planning--Chief Structural Engineer*
6-7	*Interviews with clients to discuss laboratory requirements*
8-9	*Work Sessions*
10	*Lecture on Report Writing-work session*
11	*Lecture on Oral Presentation-work session*
12	*Work Session*
13-14	*Oral Presentations*

During the summer of 1989 course modules for the previous project were modified and new modules written for this project. This fall students began work on the project in the following junior level courses.

CVE 322 CVE laboratory--design of concrete mixes for foundations and structural testing lab.

CVE 315 Surveying--topographical survey of building site

CVE 352 Structural Analysis I--approximate analysis of building frames, computer software for structural analysis

CVE 374 Environmental Engineering--water supply distribution systems

CVE 381 Geotechnical Engineering--soil exploration, properties of soils, laboratory tests (a local drilling company will set up a boring rig on the site)

SUMMARY

Two comprehensive design projects have been integrated into the civil engineering program. With modifications these projects will be repeated at least once with future classes and a third project developed.

It is too early to make firm conclusions as to the advantages of this approach to teaching design. The comprehensive capstone design course needs to be taught at least once before deciding any modifications in the program. However, there are several positive aspects of the program which are apparent even at this early date.

1) The students probably are participating in as comprehensive a design project as is possible under the normal restrictions of a four year program.

2) The design team approach in the capstone course closely approximates the situation in industry.

3) Since all students are involved in all design aspects through the use of course modules, they develop a better appreciation of the interaction of various technical areas.

4) The student has a meaningful product at the completion of his degree and has the opportunity to write a comprehensive report and present his ideas to a critical audience.

5) The curriculum wide integration of the project has brought together faculty with quite varying technical interests.

ACKNOWLEDGEMENT

The development of this program was partially funded under an Incentive Fund For Excellence from the Rhode Island Board of Governors for Higher Education, and is under the general direction of Hermann Viets, Dean of Engineering. Professors W.D. Kovacs, D. Karamanlidis, K.W. Lee, L.T. Thiem, G. Tsiatas, D. Urish, G. Veyera, and R. Wright were responsible for writing the course modules.

The Murdough Center for Engineering Professionalism

Ernst W. Kiesling,[1] F. ASCE and
Jimmy H. Smith,[2] F. ASCE

Abstract

The Murdough Center for Engineering Professionalism at Texas Tech University, Lubbock, Texas, is designed to actuate and focus greater emphasis on ethics and professionalism in engineering on a national scale. Programs being developed include symposiums; mock trials; inventory of existing instructional materials and development of new ones; and assisting in various ways with ethics instruction in the classroom, professional society meetings, and the engineering workplace.

Introduction

The Murdough Center for Engineering Professionalism was organized at Texas Tech University in 1986 to actuate and focus a greater emphasis on ethics and professionalism in engineering. Program developments are aimed at broadening the sphere of influence in order to have national impacts.

This paper presents our definition of professionalism and a brief history of the Center followed by descriptions of programs and future goals.

---

[1]Professor of Civil Engineering and Associate Dean for Applied Research. Formerly Chairman, Civil Engineering Department, 1969-1988, Texas Tech University.

[2]Professor of Civil Engineering; Director of the Murdough Center, Texas Tech University.

Professionalism Defined

We define professionalism as encompassing but transcending legal, technical, and ethical considerations. As we will use the term, professionalism portrays an individual's character--the set of values or the "plumb line" which guides one's thoughts and actions. The hierarchy of professional concerns might be distinguished by how a given action is judged. For example, the public, through the court system, decides whether an action is legal or illegal. Higher on the professionalism scale than legal considerations are technical and ethical ones: professional peers must be called upon to determine whether work is technically correct or whether an action ethical. But professionalism, as used here, includes not just the actions of a person but it includes also the intent, motivation, and expected rewards for one's actions. Only the individual can establish and assess levels of professionalism, although actions will flow out of, and will reflect, one's attitudes and values, i.e., one's professionalism.

Setting

Many considerations combined to make the mid-1980's the appropriate time to launch a center to promote engineering professionalism at Texas Tech University. Fluctuations in the economy were producing numerous business failures with accompanying pressures. International competition and influence were changing the very culture of commercial, industrial, educational, and professional entities; these changes brought with them new perspectives on ethics and values. Then, too, some highly visible events in state, national, and international government focused attention on the importance of ethical and moral values in our leadership. Responses of engineering colleges to meet increased demands placed on graduates were many, but design content and computer skills, among other things, dominated curriculum revisions rather than ethics and professionalism. The need among engineering students and practitioners for increased emphasis on ethics and professionalism seemed obvious; the time was right.

The Civil Engineering faculty at Texas Tech had demonstrated their supportive attitude toward instruction in ethics and professionalism in several ways. For several years the undergraduate curriculum included four required courses, totalling eight semester hours, bearing directly on law, ethics, professionalism, and the relationship of technology to society. Faculty professional involvements and their own conduct clearly demonstrated their professionalism. Thus we had the human resources and the environment to make a program take root and flourish.

It was in this setting that the Murdough Center for Engineering Professionalism was born. J.H. Murdough, the chairman of the Civil Engineering Department from 1927 to 1962 and a legendary figure in the minds of all graduates during the Murdough era, exemplified the ideals which we envisioned for the professionalism program. Thus Murdough became the eponym for the Center.

While numerous articles and books are available on subjects of ethics and professional conduct, there are few films, tapes, or other instructional materials to stimulate thought or discussion in classes or group meetings. Development of such materials became a motivation for creating the Murdough Center for Engineering Professionalism.

Purpose, Goals

The major objective of the Murdough Center program was to make a positive impact on the ethics and professionalism in engineering practice. The strategy was to begin with civil engineering students at Texas Tech, to spread the influence to other engineering disciplines at Texas Tech, and then to impact engineering practice in the state and nation. Objectives included development of resource materials for instruction in ethics and professionalism and the support of studies and activities in the subject matter among engineering faculty and students. By establishing awards, professorships, and a chair in the name of J.H. Murdough, we hoped to honor and perpetuate his memory and, at the same time, to facilitate program endeavors.

Birth of the Murdough Center

The concept of the Murdough Center and its objectives were reviewed with the faculty and with members of the Civil Engineering Advisory Council in 1985. The Center was launched in the spring of 1986 with a banquet and public statement of objectives. A portion of the building housing the Civil Engineering Department, under renovation at the time, was designated for the Murdough Center. A bronze bust of Mr. Murdough, unveiled at the banquet, was positioned in the designated building area known as the Murdough Center.

Unrestricted contributions for support of Center activities total approximately $45,000 to date, two-thirds of which remains in endowment. Additionally, about $60,000 has been contributed by alumni for endowments to fund undergraduate scholarships and faculty development.

Program Development

A symposium which focuses on national engineering issues and which features widely known speakers is held annually on the Texas Tech campus. The symposium is scheduled to coincide with events such as the Distinguished Engineer awards ceremony, an event which attracts many "prime movers" among alumni to the campus. Departmental and College of Engineering Advisory Councils typically are on campus at that time. Symposium proceedings are videotaped for a variety of possible future uses.

Other programs include mock trials involving law students and engineering students, review sessions and study helps to those taking the Fundamentals in Engineering examination, and a faculty/practitioner discussion group involving both engineers and architects aimed at defining needs and opportunities for the Center. Also, a Standing Committee on Professionalism and Ethics has been established; membership includes faculty and students.

Very recently the Texas State Board of Registration for Professional Engineers has begun sponsoring a multi-phase Professional Development Program. Phase I of this Professional Development Program will use the facilities and personnel of the Texas Tech University Murdough Center for Engineering Professionalism. An advisory council consisting of representatives from each engineering college in Texas, industry representatives, and engineers in private practice has recently been established; the first meeting was held in Austin and was attended by 20 members. The Murdough Center serves as overall coordinator of this Advisory Council on Professional Development.

As part of this phase of the program, a survey will be conducted of all engineering colleges in Texas to determine current status and future program needs in professional development. The survey will include a written instrument as well as personal visits to selected engineering colleges, industries, and engineers in private practice. Resource materials in professionalism/ethics will be assembled and cataloged.

The Center will develop a "trial" correspondence course in engineering professionalism and ethics tailored to the needs in Texas. The correspondence course will be used on a trial basis in Texas and possibly a few other states.

To assist engineering faculty in their presentation of ethics-related studies in technical courses, the Center will develop short "packaged" course materials and/or case studies in selected engineering disciplines. Several engineering schools outside the state have agreed to participate in evaluating some of these packages.

In cooperation with the Texas State Board of Registration for Professional Engineers and the referenced Advisory Council, plans will be initiated for a state or national meeting dedicated to Professionalism to be held during Spring or Fall 1991.

Resources that can be used as educational programs in professional development for professional society meetings and/or seminars will be developed. Periodic newsletters and publications are planned for effective communication of developments and results of the Center's efforts.

Future Plans

The scope of activities and the sphere of influence for the Murdough Center for Engineering Professionalism will be expanded as resources and opportunities permit. Emphasis will be placed on developing instructional materials and in creating a vehicle to offer assistance to practicing professionals in their professional development and in dealing with concerns over legal or ethical issues.

Pursuit of the goal of becoming a national center for engineering ethics and professionalism will require widespread participation by practicing professionals in resource acquisition and program development. Those currently affiliated or associated with the Murdough Center feel that they can make a significant contribution to engineering professionalism in the future.

Implementation of the New ASCE QA/QC Manual in University Curriculums

Wayne S. Smith, Ph.D.[1]

Abstract

The new ASCE manual, Quality in the Constructed Project was incorporated into an existing course in the civil engineering curriculum at the University of the Pacific (UOP). The course format developed was based on the participation of practicing designers, owners and constructors as guest lecturers using a combination of individual presentations and panel discussions. The support and participation of practicing designers, owners and constructors was overwhelmingly positive. A similar level of support was also available from ASCE. All course participants derived substantial value and the benefits extended well beyond the participants themselves. The benefits obtained indicate that more universities should endeavor to incorporate the content of the ASCE manual in their curriculum and that they should strongly support the publication of the revised final version of the ASCE manual as an excellent educational tool.

Introduction

In May, 1988 the Preliminary Edition for Trial Use and Comment of the ASCE Manual of Professional Practice entitled Quality in the Constructed Project - A Guideline for Owners, Designers and Constructors, Volume 1 was published. The manual is the product of the efforts of over 1,000 individuals and organizations over a four year period. The trial and comment period for this edition ends in December 1989 and a revised final edition incorporating the comments received will be considered for adoption in 1990. While the manual and its contents have not been welcomed with open arms by all parties, primarily due to potential liability implications, there appears to be general agreement that the manual content will be effective in achieving a higher level of quality in constructed projects.

The trial use and comment period has spurred much discussion between members of the civil engineering profession as well as among other design professionals, owners, and constructors. This discussion has been healthy for all involved through its initiation of dialogue and achievement

---

1 Associate Professor, Department of Civil Engineering, University of the Pacific, Stockton, CA 95211

of consensus. Coincidentally, these are also some of the main goals of the manual in terms of constructed projects.

The manual provides guidelines and recommendations for owners, design professionals and constructors on how to provide quality in constructed projects. The roles, responsibilities and limits of authority are clarified and defined for all participants in the constructed project. The topics covered include contracts and agreements, methods of accomplishing communication between members of the construction team, the importance of team work, project management, the bidding process, selection of project personnel, operations and maintenance. Thus, all major topics involved in constructed projects are addressed. It should be noted that Volume 2 will address recommended practices for specific disciplines and has not yet been published.

Incorporation into the Curriculum

As part of the manual implementation efforts during trial use and comment period, faculty advisors of ASCE student chapters were encouraged to introduce the manual to civil engineering students and were provided a complimentary copy of the manual. It was through this means that I first became aware of the manual and initiated efforts to reveal its contents to students. Although ASCE suggested that the student chapters develop an instructional program utilizing the manual, I found that this task was too onerous for our student chapter. I suspect this would also be true for all but the largest ASCE student chapters. Since I felt that the manual contents were important to include in the education of civil engineering students, I examined the civil engineering curriculum at the University of the Pacific (UOP) to find a place to incorporate the manual content in a required course. At UOP this could best be initially accomplished in a civil engineering course entitled "Engineering Administration".

The engineering administration course at UOP traditionally covered engineering economics and construction contracting. The ten week four unit course used seven weeks for engineering economy and three weeks for construction contracting. The course content was revised to replace the construction contracting text with the new ASCE manual. The manual covered much of the same material included in the construction contracting text and also provided many new dimensions, perspectives and a wider range of topics. Although the manual was much less specific than the previous text in many areas, from an undergraduate engineering student viewpoint, the broader perspective and additional topics covered offered a more useful overview of the components of, participants in, and relationship between members of the construction team.

It should be noted that a thorough evaluation of the course content change was also planned and implemented as part of the course revision. This was accomplished by extensive feedback questionnaires completed by all course participants. Due to length restrictions, the results obtained from the course feedback questionnaires will be reported in a separate publication.

Although curriculum at all universities are not identical, they have many similarities. Thus, even though there may not be a course exactly like the Engineering Administration course at UOP elsewhere, there are almost always similar courses. It is suggested that the most appropriate place to insert the material contained in the ASCE manual in the curriculum is in an existing required course which deals with construction considerations. These may be called construction management or quality assurance/control at other universities.

A second place to include the ASCE manual content in the curriculum would be in a capstone design course offered by all ABET accredited universities. UOP is considering presenting the manual content in both courses. Repetition of material in different courses has been found to be good means to reinforce learning of the material as well as presenting different perspectives on the same material.

Course Format

The topics covered in the ASCE manual are of such diversity that only a few university professors could be expected to have substantial knowledge and/or personal experience in all areas covered. Furthermore, the usual lecture format of presenting a summary of the manual content would be of little value and/or interest to students. Therefore, an avenue to incorporate experienced individuals in the course presentation and a means to have the manual content come alive was thought to be essential. The avenue selected to accomplish this was to present the manual content in a series of panel discussions by invited guest lecturers from practice.

The manual content was subdivided into eight topics and associated with specific individual manual chapters. Practicing engineers, owners and constructors were then invited to participate in the course as guest lecturers. Each of the eight topics were to be presented and discussed by the guest lecturers not based on covering the manual content per say, but rather by discussing their personal experience on actual projects which related to the manual content. They were also asked to comment on the relevance and applicability of the manual content in their practice.

Each of the eight topics covered had at least two guest lecturers from the three different disciplines; designer, owner and constructor. In addition, for some topics, representatives of the insurance industry and legal profession were also invited as guest lecturers. Each guest lecturer presented a fifteen to twenty minute individual presentation which was followed by a panel discussion question and answer period. It might have been more interesting to have all three disciplines represented at each topic, but the time limitation of each course meeting (65 minutes) prevented this.

Participation and Support

The course format selected required the participation of at least sixteen guest lecturers from outside the university. Because of the diversity of the topics covered and the need to keep the burden on guest lecturers reasonable, it was decided to invite different individuals for each topic. A major question was "Would the practicing community respond adequately to allow the planned guest lecture format to work?". After all, each guest lecturer would require time to read the relevant manual chapters, prepare their presentation and participate in the actual class session. This would require at least five hours of their valuable time without compensation (except for a free lunch compliments of UOP School of Engineering!).

I am happy to report that the response of practicing engineers, project owners and project constructors was overwhelmingly positive! Not one individual contacted turned down the offer to participate in the course as a guest lecturer (except those that had a prior commitment on the scheduled class presentation day). In fact, some individual requests to participate had to be turned down or delayed until the next time the course was offered!

In addition to the community support from guest lecturers, ASCE provided course material and advice that was invaluable. This support was available through Margie Schaheed and Carl Hirsch of the Professional Services Department of ASCE Headquarters in New York. In addition, members of an ASCE committee entitled "Task Committee to Implement the Trial Use of the Manual of Professional Practice" (IMPL) also provided timely and appropriate support. In particular, Don Kline was extremely supportive and thorough in his assistance. In addition, ASCE provided complimentary copies of the manual to guest lecture participants as a token of their appreciation.

Value of the ASCE Manual and Course Format

The value of the incorporation of the ASCE manual into the university curriculum and utilization of the guest lecture format can be considered from several different perspectives; student, faculty, guest lecturers, university, civil engineering profession and society.

The student received value in that they were introduced to topics and information normally acquired after graduation and over a five to ten year period. This gave them a jump start on contributing to achieving quality in the constructed project and associated career advancement potential.

The civil engineering faculty in general and the course instructor in particular also received great value. Although the guest lecture format used was much more demanding of the instructor than the usual lecture format, the effort involved was well worth it. It gave the instructor and other faculty attending the course sessions the opportunity to learn about topics outside their normal area of expertise. It also revealed the different perspectives of each member of the construction team on each topic. In addition, it gave faculty the opportunity to meet and interact with practicing engineers, owners, and constructors. These contacts can often lead to unanticipated side benefits for both faculty and guest lecture participants.

The guest lecturers benefited from their participation through their interaction with both students and faculty. It gave them the opportunity to share their personal experiences with interested parties and some level of recognition of the value of their experience and knowledge. For those guest lecturers that were university alumni, it also gave them the opportunity to observe first hand the changes in the university, renew acquaintances with their former professors and discuss topics of mutual interest.

The university itself benefited from both the interaction with the community and alumni. The university was perceived as being interested in and involved with the community and alumni. University faculty could use this opportunity to explain the goals and needs of the school to the guest lecturers as well as to tune in to the educational needs affecting current practice. This helped identify current needs in terms of research topics and/or modification of curriculum.

The civil engineering profession benefited through the interaction of designers, owners and constructors by their joint participation in the class sessions. A heightened level of communication and understanding between construction team member disciplines was achieved. This can only lead to a higher level of quality in the constructed project and the ensuing benefits to the civil engineering profession.

Benefits to society will also be achieved through the increase in the level of quality in constructed projects. This will lead to greater societal value derived from the constructed projects. Society will enjoy the benefits of a well constructed project for many years after completion of construction.

Conclusions

Based on the information presented in this paper, the following conclusions can be derived.

1.) Written documentation of the means and methods to achieve quality in the constructed project presented in the ASCE manual provided engineering students with a textbook to learn the material presented while obtaining their degree as opposed to its acquisition after many years of practice.

2.) Since current students provide the future for the practice of civil engineering, it is highly desirable that they learn the content of the ASCE manual as soon as possible and preferably while obtaining their degrees.

3.) The content of the ASCE manual can be incorporated into the current civil engineering curriculum without adding new courses or deleting other significant material.

4.) A format for presenting the content of the ASCE manual utilizing guest lecturers from practice was found to be highly satisfactory for all participants; students, faculty and guest lecturers.

5.) Practicing designers, owners and constructors were very willing to participate in the course presentations and did an excellent job of bringing realism to the content of the ASCE manual.

6.) Excellent support for faculty choosing to introduce the content of ASCE manual in the curriculum was available from ASCE.

7.) All participants in the course; students, faculty, and guest lecturers received significant value from their participation.

8.) The benefits of incorporating the content of the ASCE manual in the curriculum using a guest lecture format extended beyond the course participants themselves to the university and will extend to the civil engineering profession and society in the future.

9.) The UOP experience indicates that more universities should incorporate the content of the ASCE manual in their curriculum.

10.) Finally, based on the above conclusions, it is evident that the development of the trial edition of the ASCE manual has been an extremely worthwhile effort in terms of an educational tool and the subsequent publication of the revised final version should be strongly supported by the university community.

WIN-WIN ATTITUDES AND PROCEDURES IN THE BUILT ENVIRONMENT

by: John P. Gnaedinger (1)

Abstract:

Civil engineers, during the educational process, as well as during professional practice and ASCE activities, have always focused on science, and on engineering theory, analysis and design. In fact, civil engineers rate a grade of "A" for research and technical accomplishments, with no end in site.

However, at civil engineering schools, undergraduate and graduate alike, there is a noticeable failure at almost all engineering schools to even include a single course for students on problems of construction and professional practice, which necessitates input from lawyers, contractors, insurance experts, mediators, code officials, marketing advisors and other participants in the construction of the Built Environment.

Even beyond the process itself, however, are the attitudes and human relationships that determine specific actions and results of the process in the real work of construction, with the frequent deficiencies in quality, the increasing numbers of lawsuits, and the obvious need for change, reactions to such notwithstanding.

Introduction

The Model of the Industry

I. Human Needs Identification [e.g. Housing, Food, Energy, Health Care, Education, Hotels, etc. (The Free Market Place)]

[1]Chairman, STS Consultants, Ltd., 111 Pfingsten Road, Northbrook, IL 60062

II. <u>Systems for Serving Needs - the Owners</u> [e.g. Home Owners, Hotel Owners, Apartment Developers, Manufacturing Companies, Retailers, Wholesalers, Transportation Companies, School Districts, Hospitals, Etc.]

III. <u>Project Development</u> [Architects, Planners, Developers, School Boards, Governmental Units, Consulting Engineers, Financial Institutions, Lawyers, Etc.]

VI. <u>Project Design</u> [Architects, Consulting Engineers, Structural, Mechanical, Electrical, Civil, Landscape Engineers, Code Writers, Building Officials, Etc.]

V. <u>Project Construction</u> [Estimators, Construction Managers, Contractors, Subcontractors, Equipment Suppliers, Material and Component Suppliers, Engineering Observation and Testing, Insurance Companies, Etc.]

VI. <u>Project Utilization</u> [Utilization of facilities by organizations serving human needs, such as power companies, or humans directly in terms of housing, transportation, communication, health care, hospitality, etc. plus Operations and Maintenance.]

The above concepts of the Model of the Built Environment Industry are, of course, brief and sketchy, and are not all inclusive. Nor are they intended to be. However, it is important to recognize that we civil engineers are not the whole show at all. We are only a small part of the "team" whose goal is, or should be, to serve human needs. The Built Environment Team consists of all the organizations listed above, but only one of most of the types of organization, are actually "working together" on each specific project.

It would be nice if "paper" organizations could be robotized for "working together". But, of course, it is only <u>people</u> who do things, decide things, react to each other, communicate, agree on prices, file suits, make mistakes, cover up mistakes, conceal information, fail to pay bills. Furthermore, the impacts of these attitudinal factors on the cost and the performance of the Built Environment in serving human needs, is far greater than the impact of a possible deficiency in well established analytical and design procedures for civil engineers.

Peer Review

Continuing the thought of the previous sentence, it is a frequent occurrence that even famous engineers, including professors, can and do make mistakes in calculations, though a review of such calculations by another in a design office usually will correct such an error. However, errors in judgment are not as routinely identified in an internal peer review at an engineering design office. This is particularly true where such design assumptions are made by the "boss" in an hierarchically-organized firm, in contrast to one functioning under "sapiential authority", where the one who knows the most about a design matter is involved in all related decisions. Even recent graduates can be the ones with sapiential authority, such as with regard to computer-based analyses, which are unfamiliar to the chief engineer with 35 years experience.

Anyone whose goal is to best serve the client, who in turn is in business to serve human needs, is a win-win person, if he or she takes practical steps to maximize the quality of the design. Unfortunately, there are many civil engineers, of mall ages, whose goal is to make a maximum profit rather than serve the clients need, who will cut corners in services performed. Others, including some architects and other professionals, are so arrogant that they don't want anyone to review their work. In other cases, this resistance to review is based on fear of being "found out" as not having the proper knowledge or judgment necessary, or simply fear of having mistakes found out.

While some of these win-lose attitudes can be attributed to "too much self-confidence:, there is proof of a win-lose attitude when that person attempts to, or actually does cover up the mistake, even where there are great risks involved in doing so. Such efforts to cover up mistakes are not limited to civil engineers, or even just design professional. Owners can conceal information, or change their minds and attempt to deny a change in instructions to designers. Contractors can conceal information from subs, but blame subs for the consequences. Concrete test data is known to have been faked, on a few jobs. Soil boring logs have been presented where the borings were never even made. Caissons have been approved where bells were known to have been omitted.

Some civil engineers are opposed to the ASCE Quality Manual because they think it is nobody else's business on their jobs. While there may be many changes necessary in the Manual, there should be no one who does not consider quality to be a worthwhile goal. Those who do are prime candidates for a win-lose label, disdaining concern for the client and for serving human needs.

Every civil engineering school should have a required course regarding professional practice, including peer review. Case histories of actual problems on jobs should be infused into the analysis and the design courses. Reliance on practicing engineers, discussing problems with students, should become the norm, even if only travel and hotel expenses are paid to the practicing engineer as evidence of his or her sense of professional responsibility.

Perhaps these elements might be termed "Ethics" by some. However, they might at best be considered "Situational Ethics, " requiring a case by case approach, without the simplistic "moral" approach of many ethical prescriptions.

Practicing engineers would be role models for the students, in addition to those professors who also serve that need, in addition to just furnishing information and testing memories.

Communications and Teamwork

We are faced with an increasing number of failures and lawsuits in today's construction processes. Many of these result from failures to communicate and the absence of team work. Present trends appear to be exacerbating the problems, by use of contract documents which busily define each parties mutually exclusive responsibility. Communications are all but eliminated.

Contingency-fee lawyers are permitted to sue every party to a construction project, including those obviously having no knowledge or involvement in a given issue or occurrence. Contracts may be written so that this cannot happen. There are existing laws of Barratry and Maintenance, which are never enforced by the courts against the contingency fee lawyers perse (Maintenance), or their shot-gun suits (Barratry). Contracts relieving one party from responsibility may be helpful during trial, but by that time $100,000 may have been spent on legal fees, not counting hundreds of hours of principals time in depositions, reviews, correspondence and study.

What can be done? In civil engineering school courses in human relations, including communications and team building (not hierarchical systems) should be developed and taught to all civil engineers whose goal is practice. Role models should be used as a standard practice in all schools.

The Japanese are credited with widespread use of teamwork theory in running their companies, if not their construction projects. Peter Drucker, among others, is the person often given credit for team theory of management.

Mediation-Arbitration (2)

Mediation-arbitration, if used to resolve all disputes on a construction project, would remove the risk of a shot-gun suit, and permit project structural engineers to openly strive to improve job safety, rather than avoiding any involvement with safety so that a contingency-fee lawyer can't include them in a personal injury suit, along with the subcontractor who erected faulty scaffolding. The effect of these Scaffolding Act suits, at least in Illinois, is not to reduce failures and injuries, but to deprive a project of the best judgment of the bright structural engineers. The only beneficiaries are the contingency-fee lawyers, who only take cases with big claim potential, and pocket 40% of the money collected.

Win-Win Jobs

How does one decide which projects to get involved with, to avoid the win-lose projects? The first decision is with regard to the win-win attitude of the owner. Without that, one is at risk. While some public owners claim they must get the cheapest price for engineering services, this is not true in many states. Private owners who select the architects, engineers and other professionals based on the lowest fee, rather than based on professional competence and experience, and on availability of engineers necessary to manage and perform services required, are win-lose. The goal of both parties should be selection of a fair fee; not the least fee from the owners viewpoint, nor the maximum fee for the consulting engineer.

In fact, the owner has the option of determining a fair fee in advance, and selecting the consultant most competent to perform the services, at the stated fee.

These are win-win selection processes. Selection based on cheapest fee is win-lose, and likely to be a predictor of further win-lose actions on the part of the owner. Such might include demands for additional services beyond origin requirements, for no additional fees; or delays in payments; or failure to pay for services rendered, or even a lawsuit.

There is no sure way of selecting a win-win client, but it is a worthwhile goal for all civil engineers. Where a win-lose architect or construction manager must be dealt with, even if the owner is win-win, it is equally risky business. Whenever one is asked to agree to indemnify an owner, or architect, or construction manager against their own errors or omissions, one is well advised to either ask them to delete the win-lose section, or look for another client.

Reputation can be a big help in identifying win-lose firms or individuals to be avoided. Let them work with your win-lose competitors, whose goal is to maximize profits or even sales, not to serve the needs of their clients.

A win-win job, however, is not assured by win-win owners and professionals, where win-lose contractors are selected, usually as low bidders. Such low bidders may count on reduced quality, quantity or lawsuits to come out whole or to make a profit. Private win-win owners will select contractors and subs based on past performance and reputation as quality performers with win-win character. Some general contractors bid low, get the job, and make their profit by chiseling their subs, or by non-payment of contract amounts to subs. The "culture" of the construction industry generally results in many win-win teams on private work, with a concentration of problems on jobs with developers, government or institutional owners, except where project managers resolve problems promptly and fairly.

Material suppliers are not without a percentage of win-lose firms, conspiring to fix prices, leaving some cement out of concrete, "adjusting" delivery tickets for fresh concrete to suit the wishes of the contractor, not the word of the spec., adding too much water, etc.

Equipment manufacturers are likely to have the highest standards of performance, but are prime targets as deeper pockets to be seized, fault not required.

Win-Win Attitudes and Performance

Standing above all of these problems with contractors, subs and material suppliers, frequently are arrogant professionals, including many civil engineers, who consider that they have "status" far above contractors working with their hands in actual construction. To find a win-lose job developing on a job with such arrogance obvious is no surprise. This is something everyone, including civil engineers, can do something about, in their own behavior.

How?

Believe, and act upon that belief, that the purpose of one's economic activity is to serve the needs of one's clients, on a win-win basis. Providing the client is served, and done so efficiently and with quality, one will earn a profit. Profit should not be the goal, but the reward, for a client well served.

The commitment to the client can only be fulfilled when it extends throughout the organization. This, in turn, can only be accomplished with a win-win management attitude toward all levels in organization, not just the management and/or the registered engineers. It must be a win-win-win team! An opportunity for all members of the staff to purchase shares in the firm is evidence of a team commitment.

Fulfilling the above may not be the only prerequisites to earning a profit. One must also watch costs and the performance of the entire organization, including one's own performance.

A commitment to our society is also important to civil engineers; "The only truly happy people are those serving others." USA has a free market system that is the best in the world, though not without many faults, such as gambling, currency trading, and other zero-sum, win-lose activities. It also recognizes freedom of the individual, more than any other nation. However, this freedom, usually unstated, has a level of responsibility attached, to the nation, to individuals, and to society. Win-win character implies a concern for society as a whole. This, perhaps, is best expressed by volunteerism and charity, in behalf of our citizens, our country, and the world.

Finally, each civil engineer should have a win-win attitude toward his or her profession. This should not be to the exclusion of the commitment to the society. Concern should be to maximize the service of the profession in serving human needs, directly or through clients with a similar commitment. Favoring the profession to the exclusion of the client' or public interest is win-lose unionism, not professionalism.

Conclusion

This is a conference on Civil Engineering Education. The typical narrow curriculum on science and technology should be expanded to introduce concepts of win-win professional activities, and recognition of the broader commitment to serving human needs in the Built Environment.

Civil engineers should also become activist, starting in college; (1) introducing project peer review to improve designs; (2) introducing team concepts and communications in project design and construction; and (3) in utilization of the win-win conflict resolution processes of mediation-arbitration, to be included in original contact documents with the name of the med-arbs; and mediation, to be used immediately in lieu of med-arb, when a dispute on a project arises during or after construction.

Finally, (4) in selecting win-win employers and win-win clients, win-win staff. First of all, civil engineers should behave in a win-win manner themselves, at all times.

The Role of Systems Concepts in Learning Design

Thomas K. Jewell, Member ASCE[1]

Abstract

Systems analysis and the systems approach are tools that assist in organizing and quantifying design. A knowledge of systems concepts helps students understand and appreciate the design process. Yet, a recent survey found that many civil engineering programs do not require course work in systems concepts. Prime reasons for not requiring systems concepts training were the lack of room in the curriculum and the lack of a specific requirement to do so by the Accreditation Board for Engineering and Technology. Responses to the survey are analyzed, and recommendations made for further action.

Introduction

It is my firm belief that systems concepts, systems analysis, and the systems approach should be required parts of any civil engineering curriculum. Systems concepts include both the systems approach and systems analysis. The steps of the systems approach are problem definition, data gathering, formulation of alternatives, evaluation and choice of alternatives, and implementation. These steps are necessary components of any effective design process. Systems analysis provides tools that help the engineer chose the best design solution using quantitative, rather than qualitative, criteria. Systems analysis tools include economic analysis, optimization, data analysis, project management, and decision making under risk or uncertainty. A knowledge of systems analysis and the systems approach helps undergraduates appreciate the concepts of design. It also increases their understanding of how to deal with problems that have multiple possible solutions.

Most civil engineering programs in Canada require the teaching of systems concepts as part of the design process. In the United States, civil engineering departments do not emphasize the systems approach and systems analysis to the same extent. I conducted a survey of civil engineering departments in 1981. Only about a third of the respondents thought that they were adequately integrating systems analysis and the systems approach into their undergraduate design experience. At the same time the Accreditation Board for Engineering and Technology (ABET) started putting

[1] Associate Professor, Department of Civil Engineering, Union College, Schenectady, New York 12308

increased emphasis on design. I believed then that an increased awareness of systems concepts on the part of faculty, and the emphasis on design by ABET, would lead to much wider inclusion of systems concepts in the curriculum. However, this does not seem to be the case. Discussions with my colleagues, both at Union College and at other undergraduate institutions, have led me to conclude that systems concepts are still viewed by many faculty members to be of secondary importance. To find out why this might be the case, I recently sent a questionnaire to 178 civil engineering departments. Eighty questionnaires were returned. This paper will analyze the results of the questionnaire and make recommendations for future actions.

Survey Results: Adequacy of Systems Education

The questionnaire sent to the civil engineering departments had eight general questions on the adequacy of systems education. It also contained a table in which respondents could indicate if and where specific systems concepts are covered in their curricula. This section contains the results of the general questions, and the next section will present the results of the tabular responses. In general questions one through six, all 80 responders answered each of the questions. Questions seven and eight requested answers from those responders who felt that systems concepts were not adequately covered in their curricula. Multiple responses were possible for questions seven and eight.

QUESTION 1: Does your curriculum include a formal course in systems analysis? Fifty-four percent of the curricula had a systems analysis course. Twenty-eight percent of the curricula had a required systems analysis course. The other 26 percent had an elective course.

QUESTION 2: Does your curriculum include formal instruction on the systems approach? Sixty-three percent of the curricula included formal instruction in the systems approach.

QUESTION 3: The systems approach and systems analysis should be integrated together in the same course. Twenty-one percent of the responders strongly agreed with this statement, and 44 percent agreed with it. Twenty-five percent were neutral. Ten percent disagreed, while no responders strongly disagreed.

QUESTION 4: The systems approach is adequately covered by your curriculum. Six percent strongly agreed with this statement, and 35 percent agreed. Twenty-one percent were neutral. Thirty-five percent disagreed, while three percent strongly disagreed.

QUESTION 5: Systems analysis is adequately covered by your curriculum. Ten percent of the responders strongly agreed with this statement, while 32 percent agreed. Fifteen percent were neutral. Thirty-nine percent disagreed, and four percent strongly disagreed.

QUESTION 6: The systems approach and systems analysis are adequately integrated together in your curriculum. Nine percent of the responders agreed strongly with this statement, and 21 percent agreed. Twenty-four percent were neutral. Forty percent disagreed, while six percent strongly disagreed.

QUESTION 7: If the systems approach is not adequately covered in your curriculum, please indicate all the responses below which indicate why you believe it is not adequately covered. Some respondents gave multiple reasons. A total of 75 reasons were given. Fifty-three percent of the responses indicated there is no room in the curriculum for teaching the systems approach. Another twenty-eight percent indicated that training in the systems approach is not specifically required by ABET. Four percent thought that systems approach concepts are too advanced for the undergraduate curriculum. Another four percent said that the concepts are not important to the undergraduate education. Eleven percent said that the concepts could be learned on an "as needed" basis.

QUESTION 8: If systems analysis is not adequately covered in your curriculum, please indicate all of the responses below which indicate why you believe it is not adequately covered. Fifty-seven percent indicated there was no room in the curriculum for systems analysis concepts. Another 31 percent indicated systems analysis concepts were not specifically required by ABET. Three percent thought the concepts were too advanced. Five percent did not think they were important. Three percent thought the concepts should be learned on an "as needed" basis.

Survey Results: Systems Concepts Covered in the Curriculum

The responders were asked to indicate how the individual concepts associated with the systems approach and systems analysis were covered in their curricula. Multiple responses were possible. Table 1 shows the number of responses for each concept and category, as well as the total number of responses for a particular concept.

Survey Results: Comments from Respondents

There were not many additional comments on the questionnaires. However, some comments were interesting. I believe they are indicative of the feelings of other respondents who did not take the time to write their thoughts down.

One respondent indicated that his department was planning to make room in the curriculum for systems analysis. Another reported that his department had just stopped requiring systems analysis.

Two respondents felt that civil engineering faculty are not educated in systems concepts, thus they do not truly appreciate the need for them.

The following comments are paraphrased from four different respondents.

We have an underfunded university. In the race for research dollars, curriculum and teaching innovations are low priorities.

Whenever possible, systems analysis and especially the systems approach should be integrated with specific subjects such as transportation and structures. Many subject courses do not incorporate the approach because the instructor does not take the time to step back to look at the "forest" rather than teaching the "trees."

In the "best of all possible worlds" we would have adequate time to expose students to the systems approach. How can this be done when ABET places more and more time demands on the curriculum?

I would like to see ABET require a course in systems for undergraduates.

TABLE 1: Coverage of Systems Concepts in Curriculum

	Required Course	Module in Required Course	Elective Course	Module in Elective Course	Not Avail. to Undergrad.	Should be Required	Should be Elective	Total Response
A. Systems Approach	13	17	12	10	8	4	5	69
DATA ANALYSIS								
B1. Statistics	30	29	12	10	1	8	0	90
B2. Hypothesis Testing	16	28	7	12	6	7	3	79
B3. Regression	17	35	9	11	3	6	0	81
C. Mathematical Modeling	11	28	12	15	7	3	3	79
OPTIMIZATION								
D1. Linear Programming	8	19	18	20	12	5	4	86
D2. Nonlinear Programming	4	9	15	13	23	5	4	73
D3. Integer Programming	2	10	13	10	24	4	5	68
D4. Dynamic Programming	3	11	14	14	22	5	4	73
D5. Post Optimality Analysis	3	10	13	9	25	3	5	68
D6. Multiobjective Analysis	5	8	13	14	23	3	5	71
ECONOMIC ANALYSIS								
E1. Feasibility	24	33	8	13	2	1	1	82
E2. Comparison of Alternatives	23	31	8	16	1	1	1	81
E3. Benefit/Cost	24	30	10	15	1	2	1	83
E4. Taxes	19	21	10	12	8	3	1	74
E5. Inflation	19	25	8	9	7	2	0	70
E6. Break-even Analysis	17	27	9	8	7	4	0	72
E7. Utility Theory	12	16	8	13	15	3	0	67
E8. Depreciation	22	29	8	12	4	3	0	78
PROJECT MANAGEMENT								
F1. PERT and CPM	13	22	15	22	7	4	3	86
F2. Resource Management	8	14	11	24	11	4	3	75
F3. Time-Cost Optimization	6	15	10	23	12	4	3	73
DECISION MAKING UNDER RISK AND UNCERTAINTY								
G1. Decision Analysis	6	15	14	19	18	4	6	82
G2. Uncertainty in Project Management	6	11	11	22	16	5	5	76
G3. Simulation	5	13	11	18	20	3	8	78

Interpretation of Results

Question 1 shows that only 28 percent of the curricula require a systems analysis course. Forty-six percent of the curricula do not even offer a systems analysis course. Forty-three percent of the respondents felt that systems analysis is not adequately covered by their curricula.

Thirty-seven percent of the curricula did not have any formal training in the systems approach. Thirty-eight percent of the responders felt that the systems approach is not adequately covered in their curricula.

Sixty-five percent of the respondents either agreed or strongly agreed that the systems approach and systems analysis should be integrated together in the same course. Forty-six percent of the respondents felt that the concepts are not adequately integrated together in their curricula.

The lack of room was the most often cited reason for not including systems analysis and the systems approach in the undergraduate curriculum. The second most cited reason is that the concepts are not specifically required by ABET. The other reasons lagged far behind in number of responses.

Data and economic analysis were the systems concepts most often required in the undergraduate curriculum. However, a majority of the occurrences of these concepts were as modules in required courses rather than as separate courses. Elective courses were evenly distributed among the major concepts. The concept most often listed as not being available to undergraduates was optimization theory.

Conclusions and Recommendations

Thirty-eight percent of the respondents reported that they do not feel the systems approach is adequately covered in their curricula. This indicates that they doubt the concepts are being adequately covered in the traditional design courses. Their students get bits and pieces of the systems approach in different design courses. They may also go through most of the systems approach in a senior-level comprehensive design project. The systems approach may be implicit in traditional design courses. However, the students must realize they are learning a process that is applicable to any design or planning problem. Students may not step back and look at the general process without being instructed to do so. A formal block of instruction in the systems approach will teach students the concepts. Once they learn these concepts, students will recognize the elements of the systems approach when they see them in their other design courses.

About a quarter of the curricula surveyed require a course in systems analysis. Another quarter offer it as an elective. However, only 43 percent of the respondents felt that systems analysis is adequately covered in their curricula.

Nearly two-thirds of the respondents felt that the systems approach and systems analysis should be integrated together in a systems course. Only 30 percent of the responders felt that the concepts are being adequately integrated in their curricula. This is the same percentage of respondents whose curricula require formal instruction in the systems approach.

The most often cited reason for not adequately covering both the systems approach and systems analysis in the curriculum was the lack of room. Very few of the respondents answered that the concepts were not important. However, indicating there is not room in the curriculum implies that the groups making curriculum decisions feel that other concepts are more important. Over the past twenty years many colleges and universities have reduced the number of courses or credit hours needed to acquire an engineering degree. Thus, there is a continual debate as to which concepts and courses are important enough to be included in a shrinking curriculum.

A surprising number of respondents said they do not include systems concepts in the curriculum because they are not specifically required by ABET. This reveals a perception of the accreditation process shared by many civil engineering educators. Their perception is that program evaluators are looking for specific types of courses. However, the ABET criteria are purposely vague concerning the specific methods that programs may use to meet the accreditation standards. Very few courses are specifically required. General ABET requirements do take up three years of the four-year curriculum. They take up even more if you include such mandated topics as computer training. If we want to give our students any latitude in the form of free electives or additional liberal arts courses, we have to be careful about how we structure the rest of our curriculum. To make it easier to show program evaluators that a program meets ABET criteria, departments tend towards a standardized set of courses. The thought is that these courses are accepted by most evaluators. Courses that do not fit the standard mold, although they may be perfectly valid and highly innovative, are avoided. Once a department establishes the standard set of courses, there is little room left in the curriculum. I have observed this process from the perspectives of an educator, a department chair, and an ABET program evaluator. On the whole, I sympathize with the feelings of the people in the trenches.

It is unlikely that accreditation criteria will ever require training in specific systems concepts. However, many of the components of the systems approach and systems analysis are mentioned in the criteria. The criteria state that students should develop and evaluate alternatives in their design training. Students should also be asked to evaluate the economic feasibility of design projects.

Systems concepts help students organize and quantify the design process. However, program evaluators may not recognize them as design when they are checking curricula. It is much easier to substantiate traditional design to evaluators who are not familiar with systems concepts. This is despite the reality that open-ended projects undertaken in systems courses may more closely meet the spirit of the ABET criteria for design than the more traditional courses.

The results given here reflect the state of systems education in civil engineering curricula. They do not necessarily reflect what should be. I have been teaching a systems approach and analysis course for twelve years. Any number of students have come back after graduation and told me how valuable the concepts and philosophy they learned in that course have been to them since graduation. Unfortunately, in a recent curriculum review, I had to yield to the wishes of my colleagues. We switched our systems course from a required course to an elective. Two reasons were presented. One was that some of the concepts in the systems course, particularly optimization and decision analysis, were less critical to the undergraduate curriculum than other concepts. The second reason was that we had to increase the design content of our curriculum to meet ABET requirements. Other courses were easier to substantiate as design.

It is my conviction that systems concepts should be a required part of every undergraduate's education. However, I do not want to increase the already stringent requirements set by ABET. Without a mandated requirement, it is up to those of us who are familiar with systems concepts to educate our colleagues as to their importance. In so doing, we will also convince our colleagues who are ABET program evaluators of the value of systems concepts. This would significantly increase the ranking of systems courses in the curriculum priority hierarchy.

I welcome the comments of any of my colleagues on their interpretation of the survey results. I hope that this paper will generate a dialogue on how we can increase the perception of systems concepts as an important part of the undergraduate design experience.

Acknowledgements

I wish to thank all of my colleagues who took the time to respond to my questionnaire.

Bringing the Practice of Engineering Into the Civil Engineering Classroom

Joseph S. Ward, P.E.[1]

Synopsis

There is a need today and into the 21st Century to bring practical concepts of design and construction into the civil engineering curriculum. This curriculum should be practice-oriented and can be accomplished by educator/practitioners presenting courses in the design sequence, by the universities employing adjunct faculty who are in current civil engineering practice, and by motivating the students to the practical aspects of civil engineering through participation in ASCE Student Chapter and Club programs.

Introduction

There is a current tendency in a number of institutions offering civil engineering curricula to present courses in the design sequence from primarily a theoretical basis without the application to practical design and construction.

In many instances, the civil engineering graduate emerges from the classroom with a good grasp of engineering fundamentals and a theoretical approach to design but with little or no understanding of the application of these concepts to the "real world" of civil engineering.

In a large part this is due to the relative inexperience in the practice of engineering by those on the civil engineering faculty. Many of these educators have completed their undergraduate and graduate studies, have conducted research for their doctoral dissertations and then, armed with a newly acquired Ph.D., they have entered the classroom with the responsibility of educating future civil engineers.

----

[1] Consulting Engineer, Stuart, Florida. Past President, American Society of Civil Engineers and National Academy of Forensic Engineers

It is acknowledged that there are a number of civil engineering professors who have a background in design and construction but these are generally few in number. The majority of our civil engineering educators today are recycled from student to teacher. This is particularly true among the younger faculty members whereas the "old timers" who have had exposure to design and construction are facing retirement or have decided to leave the academic environment to resume their engineering practice.

This situation must change. Otherwise, we will be producing civil engineers who may have a good theoretical background but who are ignorant in the practical application of the courses they received in their undergraduate and graduate programs.

Implementation

Educator/Practitioner - The most favorable approach to bringing the practice of engineering into the classroom is for the faculty member to have been exposed to the engineering application of theoretical concepts. These individuals may have been in practice prior to entering the educational arena or they may have been a part of the real world of engineering through summer employment and through engineering consultation and design during their tenure on the academic faculty.

These educator/practitioners can present to the students examples of the application of theory to design and construction. They can cite case histories that are directly applicable to the course material being presented. In design courses, they can demonstrate to the students important elements that must be considered during the design process, and the subsequent field applications for constructing the engineered facility.

I am a former educator/practitioner. In teaching undergraduate and graduate students I know that I became a better educator when I could directly relate my experience on specific projects that were applicable to the theoretical concepts described in the textbooks and in the equations I derived on the blackboard. I have known this to be true of many other educator/ practitioners who have successfully taught undergraduate and graduate civil engineering courses and have produced graduates who have been highly successful in their engineering careers.

In the past, these individuals were more prevalent than they are today but there is no reason why all civil engineering educators cannot be exposed to a practical orientation.

Adjunct Staff - To either augment the courses offered by full-time civil engineering faculty or to offer a substitute for the educator who is lacking the application to engineering practice, the universities should employ adjunct professors and lecturers who would be brought into various portions of a civil engineering course to explain how the theory can be applied to engineering design and construction. In many cases, these adjunct faculty members could be local engineers currently in active practice or could be those who have retired from practice, either of whom would be willing to donate their time as a contribution to the education of the civil engineer.

Adjunct faculty, as well as full-time faculty members, can demonstate to the students some of the pitfalls that can be experienced in the real world of design and construction. The concept of failures, of whatever nature, can be brought into the course sequence as an illustration of errors that may have developed during the design phase of a project as well as in its implementation during the construction activity. I am sure there are a number of forensic engineers who would be willing to participate in such an educational process for a further determination of aspects to be considered in the practice of engineering.

Student Chapters and Student Clubs - There is a great opportunity of initiating both undergraduate and graduate students in the practice of engineering through presentations made by engineers and constructors at periodic meetings of Student Chapters and Clubs sponsored by the American Society of Civil Engineers. Not only can practical illustrations be brought to the students' attention but field trips could be arranged by practitioners for direct exposure of the students to witness the construction of an engineered design. ASCE has been extremely helpful in this regard through their program of Student Services and their encouragement of Civil Engineering departments and Sections and Branches to appoint contact members and associate contact members who are non-educators and who can arrange for Student Chapter and Club programs.

Summary

There currently, and in the future, is a desperate need for bringing into the civil engineering curriculum, practical concepts in design and construction that follow the theory presented in the classroom. If such a need if not fulfilled, the civil engineering profession will be receiving entry individuals who do not have a knowledge of the practical ramifications of design and construction based on the classroom theory.

We must embark today on fulfilling this need or the civil engineering profession will suffer in the future.

Field, Laboratory and Classroom Design Projects

Joseph C. Cataldo[1]

Abstract

Ongoing design and field surveys have been used as classroom and laboratory projects, ultimately leading to masters thesis and ASCE publications. The modeling of stratified flow in combined sewer overflow(CSO) was used as a design and laboratory project by a group of civil engineering students. Both field and laboratory model studies were used to determine the CSO storage facility. Similar stratified studies were conducted on Lake Ontario by students for a civil engineering design course.

Introduction

The concept of design and physical modeling has been the concern of the hydraulic/environmental facility at Cooper Union for over a decade. How to teach modeling and incorporated this subject matter into design courses is not always an easy task. Especially, when the ultimate goal is to prepare the civil engineering graduate with an understanding of how and when one uses physical models to simulate prototype behavior that will culminate in a final design.Unlike fluid mechanics which can be thought out of a text book, design and physical modeling must be studied by hands on experience. The concept of physical models and how to use them in design has become an important subject to civil engineers in the last decade. Two physical models have been built at Cooper Union and used to model stratified flow. These models were the direct result of field studies on Lake Ontario and Jamaica Bay. The models have been used and are being used in undergraduate and graduate courses to study fluid/environmental phenomena and are the subject of this paper.

----

1 Professor, The Cooper Union for the Advancement of Science and Art. New York, New York

Modeling Design Experience

It has become increasingly important for a civil engineer to understand how physical models simulate prototype behavior and how to use the model experiments in the design of the prototypes being studied. Since many phenomena are difficult to estimate by mathematical models, a physical model, may be constructed and tested. The ideal experience is to have the student study the prototype in the field. Many practicing civil engineers have difficulty in planning field studies, interpreting data, and using this data in design. The Cooper Union has a series of required design courses that the civil engineering student takes in his/her senior year. The title of these courses are "Civil Engineering Design 1 and 2" and "Civil Engineering Experimental Projects." The objective of these courses is for the student to conduct an engineering project from the conception, through the planning, design and reporting phasis (both oral and written presentations). The students work in the field, laboratory, and office environment. Frequently this entails interacting with engineers at the government level and in private practice. What makes these courses unique is that the students have had the opportunity to work on prototype field studies, construct models and use the field and model data to determine a prototype design or redesign. Two models constructed in the hydraulics/environmental laboratory were used by a number of civil engineering student groups in their senior year (thermal plume studies on Lake Ontario and Stratified Flow in combined sewer overflow on Jamaica Bay). The details of these modeling projects are presented in the next two sections.

Thermal Plume Studies

A Cooper Union research team, consisting of three senior civil engineering students and the author, measured water temperatures and velocities from two boats at a number of fixed locations in the Ginna thermal plume. The Cooper Union research team was assisted by Rochester Gas and Electric personnel. The Ginna Nuclear Power Plant is situated on the southern shoreline of Lake Ontario. At full power, 490 MWe (net), the plant removes water from Lake Ontario at a rate of 25.2 cms and heats it to a temperature approximately 11°C above ambient lake temperature before returning it to the lake as a shoreline surface discharge. The discharge channel is 6m wide at the base with side slopes of 1:1 at the lake entry. The lake is shallow at the discharge, generally less than 2m deep, and gradually slopes downward away from the shoreline. The heated water discharge at the point of exit to the lake is perpendicular to the shoreline. The data from these surveys was used to define the dynamic behavior of the plume, thermal fronts, and to design a physical model of the discharge and lake. Thermal fronts are rapid temperature oscillations superimposed on a normally steady thermal plume. They are generally observed in the near field of a thermal plume for calm receiving water conditions and extend over the depth of the thermal plume.

The students used the field data to redesign the Ginna discharge to obtain a more efficient mixing pattern of heated and cooler lake water.

The 9.4 m long by 4.9 m wide main tank contained the water and equipment which comprised the model and receiving water body. Into this tank was discharged a continuous flow of heated water which formed the model thermal plume. The heated water was introduced into the main tank by means of a trapezoidal flume geometrically scaled to the dimensions and shape of the Ginna surface discharge structure at its point of entrance into Lake Ontario.

The main components of the electronic data acquisition system developed for the model studies were an array of 64 Yellow Spring Instrument Series 427 thermistors used to sense water temperatures, a Disa 55M constant temperature anemometer system (14M bridge and a temperature compensating probe) used to sense water velocities, and a microprocessor unit programmed to sample and record on magnetic tape the output signals of all the temperature and velocity probes every 1/4 of a sec. The physical model was calibrated by matching the Ginna surface isotherms obtained from infrared measurements and thermal profiles. The 1-100 scale model could be run forover 12 min. before the model boundaries began to distort the outer edge of the plume.

For the prototype studies the discharge Froude numbers, F, and discharge Reynolds number, R, ranged from 6.9-12.5 and 2×10^6-2.3×10^6 respectively. For the model studies the values of F and R range between 2.2-44.1 and 1,800-5,400. These model discharge flows are within the turbulent regime.

This model has been used by five graduate students for masters thesis topics and is presently being used by all of the civil engineering undergraduate students in a required hydraulic laboratory course.

The result of the thermal front studies may be found in reference 1. The important points of this study are summarized below.

Temperature oscillations were observed with amplitudes of over 3°C measured from the mean with periods ranging from approximately 100-700 sec. in the Ginna Nuclear Power Plant thermal plume. This physical model geometric scale ratios of 1-50 and 1-100 were used to model the plume. Temperature versus time for several positions inside the model thermal plume exhibit a clear oscillatory pattern with amplitudes varying from approximately 1-5°C (average differences between the discharge and ambient were 11°C). The discharge and ambient temperature traces revealed no significant temperature fluctuations. Predominant periods ranged from 10-70 sec in the model and approximately 100-700 sec in the prototype. The power of the dominant oscillations tended to increase with depth at the same position inside the plume as did the magnitude of the temperature variations. The mechanism responsible for the thermal fronts is believed to be due to a Kelvin-Helmholtz instability. This instability results in vortex formation at a shear layer due to the movement of the heated surface plume across the cooler ambient lake water. The Richardson number (J) computed from the measured temperature and velocity profiles, for locations in the near field plume with temperature oscillations, fell well within the unstable range when compared to J for the neutral curve developed for the Kelvin-Helmholz instability.

Dynamic Behavior of Stratified Flow

A survey team consisting of ten Cooper Union students and the author conducted field studies of the combined storm overflow (CSO) into Fresh Creek, a tributary of Jamaica Bay, New York. The Cooper Union research team was assisted by New York City personnel.

Fresh Creek Basin is located in southeast Brooklyn, New York. It is a closed-end estuary of Jamaica Bay, which is a well-mixed estuary of the Atlantic Ocean. Nearly 2,130 m in length, Fresh Creek varies in width from 25 to 210 m. The mean tidal range in Fresh Creek, as in Jamaica Bay, is 1.6 m. This large tidal range, coupled with the relatively shallow depth of 1.5 to 4.6 m, inducing mixing. Field surveys conducted during a period of no CSO discharge verified that Fresh Creek Basin is well-mixed. There is virtually no freshwater flow into Fresh Creek.

During periods, of heavy rain, combined sewage is discharged to the upstream end of Fresh Creek through two sets of tide gates. The CSO flows through two parallel 31m long rectangular channels 4.8 m wide and 3 m high before discharging into Fresh Creek Basin.

The data from this field study was used by several groups of civil engineering students for term projects. The field data was also used to design a physical model of the Fresh Creek discharge flow and mixing patterns were studied in the physical model to evaluate the potential of new method of off-line CSO storage. This method known as the flow Balancing Method, relies on complete isolation of the CSO in curtainwall compartments within the receiving water. This method has been successfully implemented in fresh-water lakes but has yet to be tested in an estuarine environment.

The physical model is a 1:36 scale model of the upstream section of Fresh Creek CSO discharge. Models of the rectangular discharge channels and the curtained area of the flow Balancing facility were fabricated of plexiglass and placed in a two foot wide channel in the laboratory. The blackish water of Fresh Creek was modeled by a solution of calcium chloride. Fresh water was used to simulate the CSO. Similarity was ensured by maintaining a densimetric Fronde number F (ratio of inertial to buoyant forces) equal to F calculated on the basis of field data. This model has been used by a number of undergraduate students for term projects.

The results of the stratified flow studies may be found in reference 2 and is summarized below.

Data collected during storm events of various intensities indicate that stratification usually occurs during CSO discharge into Fresh Creek Basin, a tributary of Jamaica Bay, New York. During all storm events, including intense storms that may cause plug flow, salinity intrusion occurs rapidly with a rising tide. It was found that plug flow can be induced in a long channel by placing baffles (lateral obstructions forcing flow beneath or over them) in the flow field in a laboratory model. A series of under baffles was found to be more effective than a single baffle; alternating under and over baffles were found to promote stratification. An existing two-dimensional, unsteady laterally averaged mathematical model was applied as a predictive tool.

Conclusions

A unique educational experience has been used over the past decade at Cooper Union by having the civil engineering students work on joint field and laboratory projects. These projects culminated in a design or redesign of a prototype structure. All of the projects that the students worked on were existing or proposed structures. Two projects that consisted of field, laboratory and design were considered in this paper;

1. Ginna Nuclear Power plant discharge into Lake Ontario, and

2. Fresh Creek basin CSO discharge into Jamaica Bay, New York.

These projects gave the students an understanding of how field surveys are conducted and how field data is used to plan, design and build a physical model to test prototype conditions. All of the civil engineering undergraduate students presently conduct experiments on the Lake Ontario physical model. The students' laboratory experience is being extended into ground water modeling. A group of three civil engineering seniors and the author have conducted groundwater studies in central Massachusetts to determine aquifer properties and groundwater pollution. Physical laboratory groundwater model studies are presently being conducted in the Cooper Union laboratory.

Although, the Cooper Union projects were the result of research grants and or private consulting studies, the author has found that private consulting engineering companies and government organizations (Army Corp Engineers, NYCDEP, etc.) have been very willing to work with undergraduate students by suggesting and supplying term projects.

References

1. Cataldo, J.C., "Dynamic Behavior of Thermal Plume", Journal of the Hydraulic Engineering, ASCE, Vol. 110, No. 1, Jan 1984, pp 18-36.

2. Cataldo, J.C., Ronan, A, Thatcher, M.L., and Ahmad J., "Modeling Stratified Flow in Combined Sewer Overflow", ASCE Vol. 113, No. 2, Feb. 1987, pp 207-224.

Shouldn't We Have Professional Schools For Professional Engineers?

Christian Meyer [1]

Abstract

After defining the terms *engineer*, *professional*, and *professional engineer*, it is found that only a small fraction of practicing engineers, who are in sufficiently independent positions, can qualify for the label "professional". Our existing educational system does not appear to be appropriate to cope with the tasks of supplying engineers for the next century. Arguments are brought forth to educate two kinds of engineers. Engineering technologists and technicians can be trained within the traditional curricula of four years or less to acquire the skills for most routine engineering tasks that will not be automated within the foreseeable future. The professional engineers should be educated in six-year programs, consisting of a two-year liberal arts and pre-engineering education and a four-year technical education in selective and truly professional schools comparable to those in law and medicine, followed by a one-year internship of on-the-job training, possibly under academic guidance.

Are Engineers Really Professionals?

Before answering such questions, it is prudent to define an *engineer* and a *professional*, and maybe just as well a *Professional Engineer*.

What is an *engineer*? Let me subscribe to the following definition (1): "a person who with judgment applies knowledge of the mathematical and natural sciences gained by study, experience and practice, to develop ways to utilize, economically, the materials and forces of nature for the benefit of mankind."

This definition excludes a lot of people who call themselves engineers, such as the fellow who used to shovel coal into the furnace of a steam locomotive or the operator of a modern diesel engine, and clearly also the fellow with the screwdriver sent by the front desk to fix your hotel room air conditioner. Also the retired carpenter, who inspects single family homes on behalf of a realtor or prospective buyer, does not fit this definition, no matter how worthy or capable an individual and how

[1] Professor, Department of Civil Engineering and Engineering Mechanics, Columbia University, New York, NY 10027

knowledgeable in his field he may be. This definition even excludes the Romans, in spite of their awe-inspiring proofs to their genius they left behind, because theirs, like that of the master builders of the Gothic cathedrals, was an empirical art, developed over time by experience and passed from one generation to another. Only since the emergence of modern science can we talk of truly engineering disciplines. For example, structural engineering dates back only to the 18th century, when the laws of mechanics and strength of materials were first used to rationally predict the behavior of structures and to design them to satisfy specified criteria, such as factors of safety against failure.

What is a *professional* or a member of a *profession*? Flexner (2) has listed essential prerequisites, namely that "the activities of a professional, 1) be intellectual; 2) that these activities, because they are based on knowledge, be learned; 3) that a professional be practical, as opposed to academic or theoretical; 4) that it have teachable techniques which are the work of professional education; 5) that it have a strong internal organization; 6) that altruism be a motivating force for professional work, the practitioner seeing himself as providing for the good of society". Let us add that, 7) the professional needs a certain degree of autonomy in order to make decisions based on his best professional judgment, without undue pressure exerted, for example, by superiors.

This definition clearly focusses on a relatively small number of groups. It clearly does not apply to just any profession, meant as synonym for occupation, such as bank teller or toll collector. The English language is very permissive in the use of this word. I remember when I picked up my car in a body shop and could not see what part of it had been painted, whereupon the painter's breast swelled with pride: "Sir, I am a professional." Also members of the "oldest profession" are lacking several of the above prerequisites to qualify under our definition. Actually, very few professions would. The medical and legal professions are the first ones that come to mind, sometimes the teaching profession and, well, even the engineering profession.

When we talk about *professional engineering*, there is fortunately less confusion and likelihood that groups of unqualified people usurp this label, because now, at least in a number of states, we are dealing with a legally protected title. Yet, protected or not, we should hesitate applying it in certain situations. For example, is the Thiokol engineer, who was told to exchange his engineering hat for his management hat and, against his better engineering judgment, approve the launching of the Challenger shuttle, a *professional engineer*? He violated the demand to use his engineering knowledge for the common good by serving the interest of his employer first. But more significant was the very fact that his position of dependency deprived him of the autonomy necessary to make an "independent" engineering decision.

By our strict definition there are indeed preciously few true professionals. Even the other traditional professions show signs of "softening up". Doctors increasingly give up part of their autonomy by working for large organizations, thereby submitting themselves to extraneous pressures such as cost-cutting. These are often exerted by non-doctors like accountants and business managers, for whom the bottom line and quarterly profits are more important than quality medical care. Also among the legal profession, it is not uncommon that large organizations exert influences that are foreign to strictly legal service matters but can affect individual professional decision making (3).

How does the engineering profession satisfy the seven prerequisites listed above in order to qualify for this label? 1. Intellectual - no doubt, our activities are of an intellectual nature. 2. Learned - yes, our knowledge is acquired through learning, at least to a large extent, augmented through practical experience. 3. Practical - surely so. 4. Teachable - yes, almost 300 colleges and universities with engineering schools are sufficient proof. 5. Strong internal organization - here the problems start. There is no single organization like an AMA. In fact, there is a multitude of different engineering organizations. Even within the civil engineering specialty, there are scores of organizations with often diametrically opposed goals and objectives. We cannot even agree on a common definition of "engineer", and consequently millions of people go by that name. What single organization could possibly speak on behalf of all of us, including construction managers, designers, draftsmen and university professors! 6. Altruism and autonomy - this prerequisite is met by maybe one out of five engineers, because the other four are in dependent positions, working for an organization or other engineer, thus lacking the autonomy necessary to make judgments primarily on the basis of professional knowledge and to do this, if need be, for the good of society, without worrying about the person who signs his paycheck. This dilemma arises whenever the interests of an employer and of the public at large appear to clash, regardless whether the employee is a licensed professional engineer or not. He simply lacks the hallmark of a true professional, namely the freedom to exercize his professional judgment free from non-professional factors, such as the bottom line. Some such constraints are not unknown to members of other professions either, for example, medical doctors, who, against their professional judgment, order unnecessary tests simply to protect themselves against potential malpractice suits. In this case it is the litigious segment of our society which challenges the professional's autonomy and claims part of it (3).

In summary, then, are engineers really professionals? Only a small number of them are, namely those in sufficiently independent positions who do not have to fear reprisals of any kind if they were to make "altruistic" decisions based on their best professional judgment.

What Kinds of Engineers Should We Educate?

Both producers of graduates (that is us, the educators) and the consumers (the employers) agree that the present education system needs improvement, but there are differences of opinion about the desirable objectives and the ways to achieve them. Many practitioners base their calls for educational reforms on the present needs as they perceive them. They respond to the alarmist projection of an impending engineering shortage (4), by calling for a large increase in output of our engineering schools. This requires two comments.

1. In a free-market economy, supply and demand are extremely efficient regulators. If not enough young Americans can be motivated to enter engineering, then there is but one explanation for this: the job market is not lucrative enough. Engineering employers demand all kinds of skills, knowledge, and experience from job applicants, but pay them less than what most workers get who build their designs. It is a wonder that the supply side has been as good as it had for so long, under such conditions. Immigration has helped. Without it, many of our schools would have no graduate programs to speak of, and there would be no faculty to teach the next generations of engineers. Industry simply does not offer sufficient com-

pensation to entice American-born students to pursue graduate degrees, because for less effort they can get substantially larger rewards elsewhere, for example, in other professions, which offer more autonomy and respect beside pay. Of course, our society at large is responsible for this problem, since it is not willing to pay more for engineering services, presumably because there is an *oversupply* of people offering such services.

2. If the supply versus demand of engineering manpower is at present unsure, the future has some surprises in store for us. Just as the industrial revolution has eliminated to a large extent the need for muscle power, the current computer revolution is now reducing the need for brain power. Already gone are the large halls with scores of engineers at their drafting boards. Gone is the need for dexterity in operating a slide rule. The skill of quickly finding the right pages in the AISC manual will be next. Computers are driving a wedge right through our profession, separating the professionals from the technicians. Yet, just as unskilled labor is still needed in spite of available machine power, there always will be a need for technicians of limited educational background in spite of computer-aided design capabilities, to perform routine tasks that do not require a deeper understanding of the underlying principles. The activities of professionals do require this. They will be the ones on top of the development and possibly even driving it, for example, by developing automated design systems and applying them in actual practice. This division of our profession has been taking place for some time now, but the computer revolution will accelerate it. (Less than 15 years ago engineers laughed at the prediction that soon each employee would have his own computer at his disposal. Today those engineers are laughed at whose employees do not each have a computer on their desks.) It is irrelevant to discuss the merits and demerits of such a "two-class" system, because it will develop anyway and is not unknown in other professions. For example, in the medical profession, a small number of doctors is supported by a large number of nurses and technicians. These are not in the least considered "second-class professionals", just because they are not doctors.

To Produce Professional Engineers We Need Professional Schools

Our engineering profession is being challenged by our competitors abroad. Both in Japan and Continental Europe, engineers enjoy a markedly higher prestige and income than their counterparts in the US, enabling them to attract the best and the brightest. If we want to compete successfully with them, first in international markets, but already increasingly on home turf, we will have to have an infrastructure in place to produce both a small number of highly educated professional engineers and an appropriate number of skilled technicians and engineering technologists for support. Many of the necessary skills of the latter ones can be acquired in two-year engineering technology curricula and trade schools, complemented with on-the-job training and practical experience.

Professional engineers cannot be educated within the framework of our existing schools. We will need exclusive professional schools to prepare them for the tasks of the 21st century (5). These are some of their required features:

1. At least two years of general liberal arts education prior to entering the professional school to prepare engineers for the enhanced role they will have to play in society and the professional responsibilities they will be facing. This should include communication skills, law, economics, social science, public policy, etc. This

requirement is necessary to avoid the problems arising in a society that depends increasingly on technology, yet excludes scientists and engineers from any decision-making, because they are perceived as technocrats who are unable to look beyond their limited world. We have to broaden our cultural background so that we can reach out to other segments of society. This is taken for granted in countries like Japan and France, but not in the United States, where each engineer who is elected to public office makes big news.

2. Four years of concentrated technical education, with emphasis on the basic sciences and principles of engineering practice to teach students to use their heads and not just learn how to use design handbooks or run computer programs. The key focus should be on knowledge with long half-life and to avoid overspecialized technical information that would for example be taught in courses on highway paving materials or on post-tensioning systems. The half-life of such material is very short, therefore it should always be acquired on the job. The essential courses of instruction can be incorporated into a four-year program with relative ease, something that is outright impossible within our present system.

3. A one-year internship of on-the-job training, preferably under academic guidance.

4. Admission to the PE exam following the internship.

5. The faculties of such professional schools should also engage in research to ensure that students are exposed to and actively involved with material on the cutting edge of technology. Research funding will have to be increased in any case. It is symptomatic of the shortsightedness of our construction industry, the largest single industry in the US economy, that it invests only .39% of sales in research and development (6,7). The entire industry seems to have been corrupted by the business mentality that idolizes quarterly profits at the expense of long-range investments.

The professional schools should be selective and there should be considerably fewer of them than the number of today's engineering schools to assure a supply of highly qualified and motivated students, with guaranteed job opportunities comparable to those in law and medicine.

In recent years we have witnessed the decline of the US from first to second rate in many important industries. It is debatable whether this applies already to civil engineering. But there is no question that unless we undertake very soon a major overhaul of our profession along the lines described herein, Japanese or German engineers will be designing our buildings and bridges, and Americans will be just good enough to supply the labor. We have already lost our competitive edge in so many disciplines. Unless we wake up and do something now, the construction industry will be next.

References

1. Kersten, R.D., "Professional Schools: The Debate Continues (Pro)", ASCE Nat. Conv., St. Louis, Oct. 1981, Preprint No. 81-518.
2. Flexner, A., "Universities", Oxford University Press, New York, 1980.

3. Kanigel, R., "Angry At Our Gods", Columbia University Magazine, October 1988.
4. Engineering News Record, Oct. 20, 1988.
5. Guy, L.L., "Educating Engineers for the 21st Century", Engineering Times, January 1987.
6. Wiggins, J.H., "Construction's Critical Condition", Civil Engineering, ASCE, October 1988.
7. Engineering News Record, 9/28/89.

Moral Values and the Civil Engineer

P. Aarne Vesilind *

Abstract

Although engineers strive to place public good paramount in their professional practice, they often find themselves in conflict with the public. This friction may in part be because engineers tend to be positivist/utilitarian, while the public at large is more likely idealist/egalitarian. Thus disagreements can be attributed not so much to technical misunderstandings between engineers and the public as to the clash of values. In education, it is necessary to teach ethics from both perspectives and to make the idealist/egalitarian principles useful ethical tools for future engineers.

Introduction

Engineering has always been grouped with the physical sciences, since engineering is supposed to deal with <u>things</u>, not people. Chemistry, for example, deals with atoms and molecules -- things, while geology deals with soils and rocks -- things. Engineering is generally considered to be involved with the application of these sciences, and therefore it is accepted that engineering is also all about <u>things</u>.

But chemistry would exist without people. Water would still be two atoms of hydrogen and one of oxygen, whether people knew or cared about it. The rocks and soils would also be there without people,

----

* M. ASCE, Professor and Chairman, Department of Civil and Environmental Engineering, Duke University, Durham, NC 27706

and geology would still be a science. A chemist and geologist labor in the laboratory not for the benefit of humanity, but rather to discover new knowledge.

But engineering without people cannot exist. Engineering is the application of the sciences for the benefit of people, otherwise there would be no logical reason for engineering. Engineering always involves people, because the point of it all has been to create a higher standard of health, comfort and living.

If we agree that engineering is inextricably tied to people, then I would like to argue that engineering cannot be classified only as an applied physical science, but should be considered to be an applied social science as well.

Consider the possibility that one reason engineers often have difficulties in their interactions with the public may be precisely because they misinterpret their own profession. They think that they and their work can be safely hidden from public concern, happily working with their machines, buildings, and refineries. But engineers cannot always be thus insulated, since their entire raison d'etre is to serve the public. It's no wonder the public will not leave engineers in peace.

In this paper, I argue that there seems to be a fundamental difference in what engineers think of themselves and what the public expects of engineers. This difference is most notable in the basic philosophical approaches engineers use in their profession and the philosophical outlook of the lay public.

Utilitarianism vs. Egalitarianism

One potential point of conflict in the engineer/public interaction is the differing value systems held by engineers in their professional work, and the public at large. Engineers tend to be utilitarian in the performance of their professional duties, seeking the greatest good for the greatest number. (Nelson and Peterson 1982) To utilitarians it is acceptable, for example, to place a value on a random human life in order to make policy decisions. This allows the EPA, for example, to calculate the risk associated with various pollutants and to place an "acceptable" risk at a probability level of 1 in 1×10^6. Benefit/cost analysis is a utilitarian concept, in that both costs and benefits can be distributed through society.

Utilitarianism is a form of teleological ethics, in which the result of one's actions are what matters, and not the action itself. Utilitarians believe that people are rational and that their decisions are both predictable and reasonable, and that in all decisions it is the outcome that matters and not the individual act. The good in various outcomes can be calculated by summing the advantages to the individuals and then choosing the best alternative. Jeremy Bentham, one of the founders of utilitarianism, called this process the "hedonistic calculus".

Utilitarianism appeals to engineers for numerous reasons, including its attempt to quantify benefit and cost. Unfortunately, such calculations often include the benefits and costs to persons who have little say in the final decisions. Thus when engineers build a dam, for example, it is the engineers who calculate the benefits and costs, not the persons involved. Although such a system is highly paternalistic, and places engineers into a welfare role, they are smug in the knowledge that they are doing what's best for society.

But why do engineers then often have such disagreements interacting with the public? Why do engineers so often seem to be the villains in our society? One reason for such friction is most people do not use utilitarian thinking in making everyday decisions. Most likely, people tend to be egalitarians, believing that everyone should be treated equally and that no single individual should be asked to bear the burden for society.

Egalitarianism is grounded in formal ethical theory in John Rawls' Theory of Justice (1971) which seeks to provide equal opportunity to all. Rawls' egalitarian ethical theory is deontological in that it is not the result of an action that matters, but the action itself. The act in itself is wrong or right, just or unjust, not so much the outcome of the act. Egalitarians may agree that a certain action may result in only one person in a million being harmed, but they hold that for that one person, the harm is unfair. As Rawls argues, "..in a just society the rights secured by justice are not subject to political bargaining or to the calculus of social interests."

Benefit/cost analysis is an unethical process to egalitarians, since the costs and benefits can (and most likely will) be unequally distributed. An egalitarian will also consider it unethical to place a value on a human life, or to even discuss the one in a million death for a given environmental contaminant.(Mingle 1989) Egalitarians wish to be fair to everyone, and not take advantage of a single individual for the benefit of many.

Egalitarianism differs sharply from utilitarianism, since the latter seeks to maximize benefits for society as a whole, and does not distinguish the individual persons within that society. In a public hearing, therefore, the engineer may in all honesty say that the net detrimental effect of the emission from a proposed combustion plant is to increase cancer deaths by one in a million, and consider this to be quite acceptable. But many members of the public, not appreciating the utilitarian calculus, will see this as grossly unethical and unfeeling.

Such a disagreement has often been attributed by engineers to "technical illiteracy" on the part of those who disagree with them. But this explanation is most likely incorrect. It is not technical illiteracy but a different ethical viewpoint that is in question.

Positivism vs. Idealism

The second problem engineers have in interacting with society is that engineers are without exception positivists.

Positivism is an attempt to apply the empiricist traditions of Newton, Locke and others to all elements of life. (Smith 1983) Positivists believe that all sciences are going through an inevitable evolution from the theological to the positivist position, where all knowledge is based on the scientific method. Mathematics is clearly the most developed of the sciences, and engineering, which is firmly grounded in mathematics, is perhaps second.

Positivists believe that objects of study in the social sciences should be treated in the same manner as objects of study are in the physical sciences, that is, the observer is able to remain apart and separate from the object being studied. Further, social investigations can be conducted in a value free environment, and social scientists should eliminate all bias in the experiments and not be emotionally involved in the studies. Positivists hold that if there is disagreement between any two scientists, all one needs is more data and better experiments in order to resolve the conflict.

Positivists believe that "The physical sciences provide the paradigm of objective knowledge." (Baum 1974) Science is value-neutral, and ethical concerns are only expressions of emotions. Positivists believe that since true science is objective, and since ethics cannot be "done" using the scientific method, ethical considerations have no bearing on science.

Taking inspiration from Albert Comte, many social scientists challenge the attempt to apply positivism to the social sciences, arguing that social sciences are not concerned with things, but rather the human mind, with its notorious subjectivity. In their view, the observer must become an integral part of the study since his/her presence influences the outcome. This so called "idealist" school does not accept the possibility of developing a set of universal laws which can explain and predict human conduct, since the very act of obtaining the data is dependent on who the data gatherer is. In this context, any observation is a subject-subject event, not a subject-object encounter as argued by the positivists. Social science, by this argument, must be descriptive instead of explanatory or predictive. Idealists hold that when we wish to make sense of the actions of others we must recognize that its meaning lies in what was intended.

In idealist theory, disagreement can only be resolved by appealing for understanding, since no new data could possibly be useful. Data are useful only within the context of how it was obtained. A certain situation can never be recreated. As Taylor (1977) observes, "The superiority of one position over another will thus consist of this, that from the more adequate position one can understand one's own stand and that of one's opponents, but not the other way around".

In the idealist school, it is impossible to separate facts from values, and every scientific experiment is value laden. First the scientist decides what experiment of all those which could be performed he/she actually chooses to undertake. A decision has to be made on how the experiment is to be conducted and what information is to be obtained. Scientists value truth, in that they must trust the results of previous studies. Finally, they choose what results to publish and what conclusions to draw based on their work.

The idealist points out that since the axioms underpinning positivism cannot be independently proven, they must be accepted on faith. How does the positivist position then differ from the message of religious missionaries who, being absolutely sure of their own faith, populate our television screens or travel to undeveloped parts of the world to indoctrinate the masses? Is the public being indoctrinated by the positivists in the same manner?

Engineering as Applied Social Science

Martin and Schinzinger (1983) suggest that the profession of engineering can best be explained as social experimentation. Engineers do not have all the answers at hand when a problem is to be solved, and cannot often perform full scale experiments in order to obtain such answers. For example, if a suspension bridge is to be constructed, the engineer cannot construct a full-scale model of the bridge in order to test out the design. He/she uses the best available knowledge, and designs according to proven principles or extrapolates using sound judgement. Sometimes the extrapolation is faulty, and the experiment fails. Engineers learn from such failures, and reconstruct the design process for subsequent projects.(Petroski 1985) In engineering, the objects of experimentation are in the public realm.

But engineering is more than just bricks and mortar. Engineering experiments by their very nature affect humans in many ways. A road is built, but this might be at the cost of community disruption, or the onset of traffic noise, or the cost may mean that other perhaps more pressing needs cannot then be met. A reservoir is constructed, but at the cost to recreational whitewater rafting, or despoilation of a bottomland virgin ecosystem, or the destruction of an ancestral homestead. The engineer must anticipate these effects, and somehow deal with these concerns.

The modern engineer must realize the difficulty his/her own upbringing and indoctrination has on his/her effectiveness as an engineer. Engineers have been carefully taught to be pragmatic, logical, rational, sensible, and systematic in their approach to problems. While these are useful attributes for solving technical problems, they also tend to alienate the engineer from society.

Many engineering decisions are clearly value laden, and positivist thinking helps very little in resolving such dilemmas. Since engineers have been taught to uphold positivism as a truth, they are frustrated when their logical approach does not result in an acceptable resolution to moral problems. The positivist cannot assist in responding to the question "All things considered, what <u>ought</u> I to do?"

Conclusion

In summary, engineers are utilitarians and positivists, and have been functioning as applied physical scientists, where decisions are made on an object-subject basis and where values do not enter the decision making machinery. As Nelson and Peterson have pointed out (1982B), one of the reasons that engineers are utilitarians might be precisely "because they cannot be held morally accountable afterwards". They have, in their view, functioned as trustees for the greater public good.

But the public is not utilitarian, and positivism may not be the appropriate theoretical system for all engineering decisions. The public at large tends to be idealists, and leans toward egalitarianism as its ethical philosophy. It is this clash of philosophies which has made it so difficult for engineers to understand their own function in society, and for the public to appreciate the role of the engineer.

One solution to this problem would be, of course, to educate the public to be more positivist and utilitarian in its outlook. In the 1920's and 1930's, the technocracy movement had many believers, and we even elected our only engineering president. But the best efforts of our colleges and universities, not to mention the charitable foundations and the NSF, have not seemed to advance the cause of positivism among the general public. Could it be that the public simply does not want to be converted? Perhaps it does not <u>want</u> to believe in the positivist religion.

If this is the case, engineers ought to recognize and appreciate the differences in the two philosophical outlooks. It could well be that engineers should at times practice their profession as applied <u>social</u> scientists instead of applied <u>physical</u> scientists. By so doing, they would be operating in the paradigm which is most familiar to the general public and thus would enhance their value and worth to society.

Opening up the mind to understand the idealist/ egalitarian philosophy is not a simple matter for engineers. But, quoting Pascal, (in Camus 1961) --

"A man does not show his greatness by being at one extremity, but rather by touching both at once."

References

Baum, R. (1974) "A Philosophical/Historical Perspective on Contemporary Concerns and Trend in the Area of Science and Values" Science, Technology and Human Values

Camus, Albert (1961) Resistance, Rebellion and Death Alfred A. Knopf

Martin, M.W. and R. Schinzinger (1983) Ethics in Engineering McGraw Hill

Mingle, J.O.(1989) "Inchemacy: An Ethical Challenge for Chemical Engineers" Chemical Engineering Progress August

Nelson, C. and S.R. Peterson (1982A) "If You're an Engineer, You're Probably a Utilitarian" Issues in Engineering, ASCE v.108 n.EI1

Nelson, C. and S.R. Peterson (1982B) "A Moral Appraisal of Cost-Benefit Analysis" Issues in Engineering, ASCE v.108 n.EI1

Petroski, H.(1985) To Engineer is Human St. Martin's Press

Rawls, John (1971) A Theory of Justice Harvard University Press

Smith, J.K.(1983) "Quantitative Versus Qualitative Research: An Attempt to Clarify the Issue" Educational Researcher March

Taylor, C. (1977) "Interpretation in the Science of Man" quoted in Social Theory, Wilson, J., Prentice Hall, 1983.

EDUCATING ENGINEERS FOR LOCAL GOVERNMENT

Dennis A. Randolph, P.E.[1]

ABSTRACT

Civil Engineers have traditionally filled senior administrative positions such as City Manager, Public Works Director, and Chief Engineer in city and county governments. However, there is a trend to fill these positions with non-engineers. Three reasons why this change may be occurring are: Civil engineers may lack some of the essential skills needed to administer and guide municipal public works and engineering functions; exceptional graduates from civil engineering programs may not see local government employment as an attractive longterm career, resulting in a shortage of qualified engineering applicants; there may have been failures on the part of previous position holders to recognize or accomodate changes in the relationships between engineers, the public, and elected officials. These failures may leave officials hesitant to hire people with similar backgrounds. All of these deficiencies can be attributed in-part to a failure of the engineering education system to provide broadly educated, humanistic engineers who are willing to respond to the changing needs of society. In the coming years, local government agencies will find competition for engineering staff increasingly keen. While the number of engineering positions will increase, the number of engineers coming into the work place will be less than needed. With a smaller pool to select from and a more informed and demanding citizenry to contend with, the content and form of civil engineering education must be reconsidered.

INTRODUCTION

Over the next few years a large number of working Civil

[1] Member, ASCE; Chief, Highway Operations, Fairfax County Office of Transportation, 4050 Legato Road, Fairfax, Virginia 22033

Engineers will be retiring. As a result, an opportunity for significant change in the civil engineering profession will occur. Change is difficult but the past few years have shown that some change in the way in which civil engineers look at and perform their work must take place. An informative example is the Interstate Highway Program experience.

Completion of the Interstate Highway System will be a significant event. In the rush to build, however, full consideration has not always been given to the impacts of design and construction. In response to this lack of consideration, a backlash occurred that was carried along and intensified by the almost revolutionary ideas and emotions of the 1960's and early 1970's.

This backlash has constrained the engineer and forced thorough consideration by the engineer of the impact of his work in social, economic, and environmental terms. The massive regulatory network that must now be traversed to build or repair nearly every infrastructure component is the product of this backlash. But in addition to these regulatory constraints there have been other more subtle impacts that directly affect the engineering profession.

THE IMAGE OF THE CIVIL ENGINEER

The battles that have accompanied the large infrastructure construction programs such as the Interstate Program have tarnished the image of the civil engineer in a way that is profoundly affecting the profession. Many citizens see civil engineers as rigid, almost mechanical, individuals more concerned with building a finished product then with developing people oriented facilities. Government organizations in particular have reacted to the negative image that civil engineers have acquired. In many instances managerial positions previously filled almost exclusively with civil engineers (for example: City Manager, Director of Public Works, Director of Transportation) are being filled more and more by non-engineers. This role shift is pushing the civil engineer into a more technician-like role consisting primarily of the rote application of standards.

But even though the image of the civil engineer has been tarnished, the engineering profession is still quite well regarded. This is true especially at the local government level where the civil engineer continues the long line of individuals, stretching back to ancient times, who have built and maintained the facilities needed by the public (1,4). However, the general bashing of the government employee that has been engaged in since the start of the Reagan Presidency has placed local government engineers in a very tenuous position, They must constantly fight and win the battle of shrinking resouces and deteriorating infrastructure, and at the same time present to the public a level of consciousness that is higher than ever demanded before. If they fail in either regard, the image of the engineer as the rigid, mindless bureaucrat immediately comes to the surface.

Added to the problem of image is a second, more internalized problem. Unlike engineers in private practice who must balance multiple major bottom-line objectives (profit, client satisfaction, quality of work), engineers in local government have only one major bottom-line objective ... serving the public good. Having only one major objective is in some respects a disadvantage because it can lead to a narrow viewpoint that harms the decision making process. As a result, there is the danger that the local government engineer will acquire the attitude that they are the only ones capable of knowing what is good for the public.

THE CHALLENGE FOR ENGINEERING EDUCATION

The problem that must be faced by civil engineering educators is how to produce civil engineers who have the skills to construct and maintain public facilities, and know how to determine just what is in the public interest. This problem is really one of emphasis and represents a conflict over the amount of time allocated between learning technical and non-technical skills.

In most cases, the civil engineer in local government is in a position which requires strong technical skills. These skills are needed because engineers must be able to provide traditional civil engineering services for the local population they are serving. In addition, these engineers serve as reviewers and regulators for the many civil engineers in private practice who are either providing services that impact the general public through private sector development or acting as consultants to the public sector.

Civil engineers in local government also need to be familiar with a wide variety of non-engineering skills. Most important among these skills are those used to communicate and interact with the citizens whom they work for and serve. Unlike the engineer in private practice, state and federal government, or academia, the engineer in local government makes contact daily with citizens seeking a solution to the everyday problems of life. For many citizens, their only interaction with civil engineers may be with those members employeed by local governments. So to a great extent, the opinion of the general citizen regarding civil engineers is dependent upon how well local government engineers can communicate.

EDUCATIONAL EMPHASIS

Producing a civil engineer with a balance of both technical and non-technical skills is a problem of resource allocation. There are simply too many things to learn and too little time available. This is of special concern because most civil engineering educational programs are not balanced, but weighted heavily in terms of technical skills and training. This weighting becomes even more unbalanced at the graduate level.

The emphasis on technical skills instead of non-technical skills is especially troublesome because many times the new engineering graduate is expected to meet with citizens on a daily basis and work out their problems with them. Also the nature of the work performed, whether design, regulation, or review is influenced heavily by continual and close contact with citizens, the single outstanding characteristic of a local government engineer's job. Even with the tremendous advances in the technical aspects of civil engineering, this emphasis will not change.

THE SHAPING OF A LOCAL GOVERNMENT ENGINEER

Because of the people emphasis of the local government engineer's work, the civil engineer interested in a career in local government service must be trained to be people oriented. Students can be guided in civil engineering training programs so that their studies reflect such an orientation. Such guidance can be a central theme throughout all of their training.

In addition to a people orientation, civil engineers in local government also need certain other abilities to be successful. Following are ten characteristics that will help make graduates effective in local government service from their first day. Even if not learned in school, the development of these characteristics early in a career is the key to a good local government engineer (2,102):

The ability to use information - The engineer in local government needs to be able to use what information is available and to find missing information needed to answer a question. In many instances it is simply a matter of good detective work, and skills such as inductive and deductive reasoning can be useful tools. Courses emphasizing investigative experience and logic would be useful.

Interpersonal skills - Can the engineer relate to all types of people under many types of situations? The effective engineer in local government must have the ability to quickly relate to all varieties of people for two reasons - to learn what they want, and to convince them that they are going to be helped. A basic understanding of sociology and psychology would be helpful in this endeavor.

Marketing skills - Engineering at the local government level is a continuous job of selling ... to citizens, councils,boards, and other engineers. An engineer needs to be able to sell ideas in order to get the job done. Marketing and advertising skills, such as those taught to sales persons, would be used on a daily basis.

The ability to manage change - Both engineering and society are rapidly changing and the engineer must be able to recognize these changes and adapt to them. The skills honed by the politician would be, useful, as well as those used by the historian to analyze change over the course of time. In addition, strong leadership training would be invaluable.

Time management - The success of the engineer in local government is dependent upon the ability to spread their time among many groups. General business and sales skills would be useful.

Team player - Local governments are not carried by a single person, and a successful engineer knows when to work with a team. Political science and public administration training, especially the aspects relating to coalition building, would directly serve this need.

Knowing when to act and when to respond - Many times engineers at the local government level must take vigorous action, while other times just listen. Success means being able to tell which time is which. Much of this skill relates to leadership, but in addition a background in the physical and mental nature of the human being would be useful.

Knowing how to become involved - Engineers in local government must know when to become part of the community, rather than remain the emotionless by-stander. Philosophy and theology could help make the engineer more understanding and aware.

Computer literacy - Engineers need to be able to recognize the computer for the tool it is. More than anything, this skill involves a desire to apply any and every tool to solve a problem.

A sense of personal accountability - Above all, the engineer must be ethical and have a strong sense of integrity. Courses in ethics and philosophy could at least provide some awareness of these factors.

A review of these ten traits suggests that they are not unique to the engineering profession, but actually provide a measure for any well rounded generalist.

This maybe the real key to developing engineers who can succeed at the local government level. Such engineers would have a broad based educational background which has an emphasis on the human being and the way that humans think and live. Achieving such an education may mean reorienting the engineering curriculum to be more like that of the attorney or journalist. In those professions, the emphasis is toward developing skills that assist in the understanding of a very broad range of topics and the ability to ask questions that quickly uncover the real problem or concern.

The Liberal Arts emphasis described above may also mean that the budding local government engineer may be more able to succeed if the first part of their undergraduate studies were in a traditional liberal arts program. Technical training could be limited to the last half of the undergraduate program and to graduate programs that again would be split evenly between technical and non-technical subjects.

CLOSING

Civil engineers in local government will always have to be technically competent. However, the skills emphasis for engineers in local government must be in non-technical areas because of the close and continuing inter-action that they must maintain with citizens. Engineering education programs have traditionally emphasized technical skills with the result that civil engineers in local government are being increasingly pigeon-holed into technical roles outside of the main-stream decision-making process. If this trend is to reverse, engineering programs must be changed to produce a new breed of people-oriented engineers, technically astute, yet capable of generalizing. Above all these engineers must be able to clearly and effectively communicate with the citizens whom they are serving.

REFERENCES

1. Management of Local Public Works; International City Management Association; Washington, D.C.; 1986; Page 4-11

2. Personnel Journal; Volume 66, Number 10; October 1987; Page 102

PUT THE $ SIGN IN THE EQUATION

Alfred C. Ackenheil, Fellow ASCE

Background

In 1940, there were 100 of us sitting in a meeting room at the Boston Society of Civil Engineers. We were engineering students from the Harvard Graduate School of Engineering, Massachusetts Institute of Technology, Northeastern, and other schools. Hardy Cross was speaking about "The Moment Distribution Method." He was an excellent speaker, but the only thing I can remember he said was, "Don't forget to put the $ Sign in your equation."

Course Notes and the $ Sign in the Equation

After the meeting, I got out my old class notes from my B.S.C.E (Structures) at the University of Pittsburgh ('39). I was curious to find out if any of my homework or project assignments had the $ Sign in any of the equations. None of my Civil Engineering class notes did. There were specific courses in accounting and finance which did, but the textbook examples and the class problems were not related to Civil Engineering projects.

Student to Teacher to Practice to Owner

After completion of the Master's Degree at Harvard (Soils), I began work in the summer of 1940 as a structural designer. Within six months I was invited to a full-time Civil Engineering faculty appointment at

---

Ackenheil & Associates Inc., Consulting Engineers, 1000 Banksville Road, Pittsburgh, PA 15216.

Pitt to teach Structural courses. The war was on and I combined my teaching with part-time structural design work for a local consulting firm.

After two years in the U.S. Navy Civil Engineer Corps I returned to Pitt in 1946 and set up the undergraduate and graduate programs in Soil Mechanics and Foundations together with the laboratories.

Industry needed geotechnical assistance and for the next 10 years I combined full-time teaching with part-time consulting. The consulting paid about three times my teaching salary. I wove my consulting projects into my lectures. By 1959 I couldn't continue to perform my part-time consulting work alone. I decided to incorporate a small firm and hire some employees. At this point I became an employer and an owner at the same time. It meant a modest investment of capital in office and laboratory equipment and putting some dollars at risk. It also involved performance of all the administrative functions of personnel, finance, public relations and physical plant. For the first time in my life the $ Sign became an important part of my equations. If you don't make a profit, you're out of business. By 1966 the proportions in my total workload had to shift from full-time teaching and part-time practice to full-time practice and part-time teaching.

For years the Civil Engineers at Pitt had been taking a required elective offered by the Geology Department. It was a good course for beginning Geology students, but it lacked real Civil Engineering applications. The C.E. Dept. Chairman was willing to drop the required elective given by the Geology Department and establish in its place a required course in Engineering Geology taught by the C.E. Dept. This was my part-time teaching assignment.

I planned the course lectures to follow the steps that would be followed in a design office to perform a project. The students played the role of project engineers in a geotechnical consulting firm. They were taught (1) handling the initial call, (2) office preparation and initial site visit, (3) establishing purpose, scope, proposal and contract, (4) subsurface exploration and testing, (5) analysis and design, (6) report preparation, and (7) field follow-up. At the same time, mineralogy, petrology, stratigraphy and similar geologic concepts were fit into the matrix. A 10-minute slide presentation on actual projects exemplified the subject for the day. It was fun to teach the course. The students enjoyed it. Nobody went to sleep.

$500,000 in the Equation

In 1966, I was quietly working on a project at my consulting firm when a police officer walked into my office to serve me with a legal notice. I looked at it and discovered that my firm was being sued for $500,000 because of a landslide on a football playing field project we had worked on five years before. This was a startling event. One that I had never experienced as a faculty member or designer.

I hired a lawyer to defend the firm and five years later at the State Supreme Court level the judge decided that we should not have been included in the case. The defense was costly, but judgment against us could have meant bankruptcy.

My insurance agent told me that no one, not even Lloyd's of London, would provide professional liability insurance for geotechnical engineers. The hazards and uncertainties of our work were too great. In 1968, I met with a small group of other Geotechnical Consultants in Chicago to discuss how we might protect ourselves. We began with an informal arrangement to insure each other. By 1971 this led to our founding the Terra Insurance Company.

But insurance was only the last line of defense. We formed the Association of Foundation and Soil Engineers (ASFE) to study methods of Loss Prevention. Other firms joined this group which today represents about 90% of the $ sales in the U.S. of Geotechnical Consulting. Loss Prevention involved all aspects of practice and ownership/management. Many publications were developed for the use by ASFE members.

After 1966, liability, litigation, and loss prevention began to be part of my lectures to the students. These items could be large numbers in the $ Sign equation.

From Student to Practicing Engineer: Some $ Problems

When students graduate and begin to practice as engineers they want to do a good job. What's a "good job" has all sorts of definitions and cost. Excellency is a goal that has a high $ cost. Normal practice could cost less. A business decision has to be made to select the quality and quantity of engineering work to be done to fit a client's budget. Students tend to do a "gold plate" job when the budget is only "silver plate."

Most students enter practice unaware of dangerous words they are using. Categorical and superlative words in reports can be embarassing in court if they can't be backed up. "No settlement" is different than "may be some settlement." "Assure" is an insurance company word, not an engineering word. $ costs of all the phases of disputes and litigation can be very high.

Risk evaluation needs to be understood by students. Normally engineers in practice are trying to reduce risk, not eliminate it. If the client wants to totally eliminate all risk of a landslide the construction $ costs can be enormous.

Good communication is essential to good engineering practice. Students take notes as they listen to lectures. These notes can be anything from very complete to none. In practice it is important to take notes on both sides of conversations on a telephone, to keep a well-organized correspondence file, and to be carefully recording who said what at meetings. Verbal communications with clients, the A-E team and contractors require care and caution. What is said usually has $ cost implications in liability or additional work requests. They must be documented.

Some Suggestions for the Decade Beginning 1990

There needs to be a reasonable balance between theory and practice in engineering education to prepare students for practice in the the real world. That doesn't mean teaching 50% theory and 50% practice in the undergraduate Civil Engineering curriculum. Reasonable balance might be 95% theory and only 5% practice. But the % of practice instruction must increase beyond what it has been to date.

Here are the suggestions:

1. Think $ Costs: Include in the design courses problems or projects which require the use of the $ Sign in the equation. Ask the students to make a preliminary estimate of Cost of Engineering Services and Cost of Construction.

2. Practice Manuals: Use at least one lecture or Student Chapter meeting to introduce the students to ASCE Manuals of Practice No. 45 and 55, and the ASFE "Introduction to Professional Practice."

3. Faculty Balance: Hire faculty with some practice experience. If research and publications are part of tenure requirements, why not have practice (say a minimum of 100 hours a year) as a requirement too.

4. Textbook Balance: Select books written by persons with a reasonable amount of experience in practice.

5. Quality Work, On Time, Within Budget: Encourage students to perform quality work on time. For budget, get them to set a $ value on their own time, benefits and overhead. Include a profit.

6. Library and Practice: Provide practice oriented books and publications in the engineering library.

7. Student/Practice Interchange: Have students visit consulting engineering offices and local construction projects and write a report including $ values of items observed.

8. Faculty/Practice Interchange: Establish a two-way interchange between faculty and local practicing engineers. Have faculty visit and/or work part-time for local firms. Have practicing engineers visit and/or give lectures at the school.

Conclusion

It has been an interesting challenge and very rewarding experience to sequentially become a student, faculty member, practicing engineer and owner of a consulting firm. Even with the three academic degrees, a lot of education came through the school of hard knocks and bitter experience in the real world. Let's work together in the 90's to help prepare the engineering students to have a better $ perspective as they enter the real world of practice.

Engineering Pathology: Case Studies in the Classroom

Kenneth L. Carper,[1] M.ASCE

Abstract

Case histories from the field of forensic engineering, integrated into the civil engineering curriculum, can provide motivation for learning technical concepts. Failure case histories can also be used to encourage classroom discussion of important non-technical issues, including ethics and other professional practice topics.

Introduction

Documentation and dissemination of the results of forensic investigations can lead to improvements in design and construction practices. Forensic engineering contributes to improved engineering design in the same way that medical pathology contributes to improved medical practices. In both professions, medicine and engineering, public welfare is enhanced by the willingness of practitioners to share information and to learn from failures.

Forensic engineering case histories, integrated into the civil engineering curriculum, can provide motivation for learning technical concepts. Case histories can also serve as a vehicle for discussing ethics and professional practice issues. Methods for integrating case histories into the curriculum are reviewed in this paper, along with a discussion of existing and developing resources.

---

1 Associate Professor, School of Architecture, College of Engineering and Architecture, Washington State University, Pullman, WA 99164-2220.

Why Study Failures?

The discussion of failure case histories need not be disillusioning to students, nor need it diminish their creativity and enthusiasm for the profession. In my experience, such discussions have stimulated greater enthusiasm and commitment to professionalism. Others have also experienced success with the introduction of failure case histories into the classroom (Chesson 1979). Students are enlightened regarding the consequences of decision deficiencies and are motivated to understand the technical concepts for preventing recurrence of the failure.

Specialized knowledge is an essential component of professionalism. An equally important component, sometimes overlooked in professional education, is skill in managing interpersonal relationships. Discussion of a forensic investigation, especially when litigation or liability issues are involved, provides the opportunity to introduce these professional topics that transcend the specific technical causes of failure.

As professionals struggle for survival in a litigious society, it becomes increasingly important to include these non-technical issues in the professional curriculum. One can no longer avoid litigation through intellectual brilliance and technical competence alone. An insurer of design professionals notes that, "Failures in interpersonal relationships, more than any other consideration, are at the root of conflicts" (Howell 1980). Failure case presentations stimulate the development of interpersonal skills, along with respect for technical competence.

Failures: Technical Causes, Procedural Reasons

Failures result from both technical and procedural deficiencies (Khachaturian 1985). Technical explanations can always be given, but the reasons failures occur often involve procedural problems. Even at the leading edge of technology where the frontiers of knowledge are advancing most rapidly, human performance deficiencies can be as important in contributing to a failure as are newly-discovered technical problems.

For example, the 1986 space shuttle accident uncovered a connection detail problem in the booster rocket (the technical explanation). But important questions were also raised regarding contributions of quality control deficiencies to the failure (procedural reasons). The monitoring of human performance is as critical as the monitoring of material properties or design calculations.

In the construction industry there are numerous opportunities for human performance failures. A building or bridge is a unique project, unlike a mass-produced product. In its design and construction, a great number of specialized consultants, contractors, subcontractors and suppliers are brought together. Constant pressures exist to reduce construction time and costs. Escalating litigation encourages confrontational rather than cooperative relationships among the various parties, most of whom are trying to turn the lowest bid into a profit. The fact that the system works at all is surprising. The fact that it works most of the time is a testament to the perseverence of dedicated professionals working in the industry. Students entering the field should have an appreciation for the possibility of failure and an understanding of the professional engineer's role in mitigating risk.

Forensic case histories involving procedural issues usually illustrate the need for interdisciplinary coordination. Use of these case histories in the classroom can expand the professional curriculum into disciplines outside the student's particular specialty. This exposure encourages the student to explore how his or her specialized field interfaces successfully with the work of other specialists.

The Crowded Civil Engineering Curriculum

The curriculum is frustratingly, but necessarily, crowded. New courses may not be introduced without removing other essential topics. Yet it is increasingly important that the student be prepared for the world of practice (Ward 1987). Forensic case histories can be used efficiently to address professional practice issues such as ethics, risk allocation, environmental impact, liabilities and professional values. Where case histories are integrated into existing courses, the goal should be to facilitate a discussion of these topics; failure mitigation should be emphasized rather than failure investigation.

Several schools offer courses on forensic engineering, usually at the graduate level. These courses serve as excellent electives, but they may not be available to all students. While these specialized courses deserve continued support, it is my judgment that all civil engineering students should be exposed to the topics addressed in this paper. Given the crowded state of the existing undergraduate curriculum, this can only be done by integrating case histories into existing required courses.

In my structural engineering courses, I have used case histories to illustrate fundamental theoretical principles. While the mechanics of failure investigation may require advanced study, the technical causes of failure often reflect basic principles easily understood by beginning students. For example, the connection failure responsible for the 1981 Kansas City Hyatt Regency Hotel walkway collapse is appreciated by students taking their first course in statics. This case can be used to illustrate the necessity of following a force through a connection as it is transferred from one member to another. Furthermore, the case provides an opportunity to discuss the construction project delivery system: shop drawings, change orders and professional attitudes (Rubin and Banick 1987).

The Hartford Civic Center Coliseum failure is useful in illustrating stability concepts: buckling of unbraced slender compression members (LZA 1978). This case also illustrates the risk inherent in non-traditional approaches to project delivery. The Station Square Shopping Center roof collapse in Burnaby, British Columbia is another excellent example involving instability: lateral buckling of the compression flange of a steel beam (Closkey 1988).

Some case histories involve many interrelated contributing causes. A parking garage collapse, for example, may result from a combination of design errors, construction deficiencies, materials problems and operational or maintenance errors. Presenting such a case history stimulates discussion of the many factors contributing to risk analysis and reliability.

Integration of forensic case histories into existing required courses is an efficient way to include the material and an effective means to reinforce concepts already in the curriculum.

Available Resources

A large inventory of failure case histories currently exists and efforts are underway to expand and increase accessibility to these resources (Carper 1987). The new Education Committee of the ASCE Technical Council on Forensic Engineering (TCFE) is compiling a bibliography of case histories and will be working toward development of resources for the classroom.

The Association of Soil and Foundation Engineers (ASFE) has published more than 40 case histories involving site selection errors and geotechnical failures (Gnaedinger 1987). The National Institute of Standards and Technology (NIST) has prepared investigative reports on several structural failures and several videotapes suitable for classroom presentation (Scribner and Culver 1988). Case histories are accessible through the Architecture and Engineering Performance Information Center (AEPIC) at the University of Maryland.

Engineering News Record is an outstanding source for current construction failure information. The local newspaper is also a good source. Events that happen close to home are of special interest to students. Reference to these "instant" news sources must be qualified with the understanding that the investigation is not yet complete. Care must always be exercised to protect the professional reputations of innocent parties.

Several journals regularly publish forensic case histories. These include the Journal of the National Academy of Forensic Engineers, published by NAFE, and Forensic Engineering, a Pergamon Press publication.

One journal is exclusively devoted to case histories from the construction industry: the Journal of Performance of Constructed Facilities. Now in its fourth year of publication, this quarterly refereed technical journal is published by ASCE. An interdisciplinary editorial board, working under the direction of the ASCE Technical Council on Forensic Engineering, includes representatives from throughout the design and construction community. Besides case studies, papers on litigation, dispute resolution, insurance and other related professional issues are included.

An extensive list of references and a number of case histories from a variety of engineering disciplines are included in the book, Forensic Engineering (Carper 1989).

Conclusions

Professionals have a responsibility to learn from failures as well as success stories, using the lessons of experience to advance their professions (Carper 1986). Discussion of forensic case histories in undergraduate education can reinforce technical concepts throughout the curriculum. Failure case studies also provide opportunities for interpersonal and professional skill development. These opportunities, and the attitudes they encourage, may prove to be as important to professional survival as the technical skills.

Appendix: References

Carper, K. L., ed. 1986, Forensic Engineering: Learning from Failures, ASCE, New York, NY.

Carper, K. L. 1987, "Failure Information: Dissemination Strategies," Journal of Performance of Constructed Facilities, Vol. 1 No. 1, February, pp. 1-10, ASCE, New York, NY.

Carper, K. L., ed. 1989, Forensic Engineering, Elsevier Science Publishers, Inc., New York, NY.

Chesson, E. 1979, "How Not to Do It," Proceedings of the National Conference on Engineering Case Studies, American Society for Engineering Education (ASEE), Washington, D.C.

Closkey, D. J. 1988, Report of the Commissioner Inquiry: Station Square Development, Burnaby, British Columbia, Queen's Printer for British Columbia, Victoria, B.C.

Gnaedinger, J. 1987, "Case Histories: Learning from Our Mistakes," Journal of Performance of Constructed Facilities, Vol. 1 No. 1, February, pp. 35-47, ASCE, New York, NY.

Howell, E. 1980, Untangling the Web of Professional Liability, Design Professionals Insurance Company, Monterey, CA.

Khachaturian, N., ed. 1985, Reducing Failures of Engineered Facilities, ASCE, New York, NY.

Lev Zetlin Associates, Inc. 1978, Report of the Engineering Investigation Concerning the Causes of the Collapse of the Hartford Coliseum, City of Hartford, Connecticut.

Rubin, R. and L. Banick. 1987, "The Hyatt Regency Decision: One View," Journal of Performance of Constructed Facilities, Vol. 1 No. 3, August, pp. 161-167, ASCE, New York, NY.

Scribner, C. F. and C. G. Culver. 1988, "Investigation of the Collapse of L'Ambiance Plaza," Journal of Performance of Constructed Facilities, Vol. 2 No. 2, May, pp. 58-79, ASCE, New York, NY.

Ward, J. S. 1987, "Non-technical Aspects of Engineering Practice in Civil Engineering Education," Proceedings of 1987 ASEE Conference, ASEE, Washington,D.C.

The Making of a Profession

Dennis Polhill, P.E.[1]
James C. Pequette, Ph.D.[2]

Abstract

Engineering through history has gone through many stages of metamorphosis. Significant advances in the profession occurred in the mid 1800's when a rebellious new nation, unrestricted by tradition, invented engineering as a specialized field of study within the university structure. A hundred years later another major step to improve the profession evolved with the advent of state licensing requirements.

The stage has been appropriately set. Before us lies the opportunity during the 21st century to evolve engineering into a true profession. The achievements of the profession have been phenomenal, but we must re-commit to excellence and look forward into the future. It is time that we re-evaluate and re-think the charge of the profession, the needs of society and of our practicing professionals, and the structure of the registration and educational process. By fine tuning our approach at creating professionals we can evolve engineering into a true profession and adequately respond to the demands that the future will

---

[1]Consulting Engineer, Pavement Management Systems, Inc., 245 S. Benton St., Suite 230, Denver, CO 80226. Member of APWA's National Committee on Graduate Public Works Education and President of the Graduate Public Works Alumni Association at the University of Pittsburgh.

[2]Regional Manager, Pavement Management Systems, Inc., 245 S. Benton St., Suite 230, Denver, CO 80226. President-Elect of the Colorado APWA Chapter; Chairman of the Colorado Chapter Professional Development Committee; and former University Dean.

impose upon engineers.

Introduction

The growth and development of the United States was guided by some very wise men who achieved enormously powerful long term goals through the exercise of their profound wisdom. National policies of the 1800's that helped America evolve to be the World power that it is today include: (1) westward expansion to the Pacific Ocean, (2) an open door immigration policy unprecedented in World history that yielded multiple compoundings of our population, (3) the Homestead Act that motivates the work ethic and innovation of individuals to new heights, (4) the enormous federal subsidy to railroad development, (5) expediting the initiation of a large army of technical experts by breaking with the traditional concept of liberal arts in university education and creating for the first time university level "engineering" education at land grant universities.

All five goals were exceedingly successful. Number 5, the creation of university level "engineering" education was the first step in the evolution of the engineering profession.

It took the better part of a century for the engineering programs to evolve and to come firmly institutionalized. Step No. 1 in the development of the engineering profession was complete.

Step No. 2 was the development of professional licensing. It was recognized that the health, safety and welfare of the general public deserved protection from those who would practice engineering without regard to their level of technical competence. Each State developed its own approach to licensing and over time the variations in licensing requirements from State to State diminished. Some work remains, but step No. 2 is effectively in place.

It is time to review our achievements and make minor modifications in both engineering education and licensing, where appropriate, to position ourselves to evolve into a full fledged "profession" during the 21st Century.

Definition

Engineering has been defined by the U.S. Engineers Council for Professional Development as "the creative application of scientific principles to design or develop structures, machines, apparatus, or manufacturing processes, or works utilizing them singly or in combination, or to construct or operate the same with full cognizance of their design; or to forecast their behavior under specific operating conditions; all as respects an intended function, economics of operation and safety to life and property."

Public Perception

More important than the Engineers Council for Professional Development definition is the general perception of the people of America. Engineers are aware of this perception. We regularly complain about the perceived status of engineers in relation to other groups such as attorneys and physicians. We know too well how society values our contributions. The parameters of salary, career options, political representation and input are a sad and unmistakable tribute to the status of engineering as a profession.

Trend

Engineers are generally not perceived as professionals and the trend suggests that our status is continuing to diminish. Salaries since the early 1970's have not kept pace with inflation. Engineering enrollments at universities have dropped. The number of engineers in leadership positions continues to drop both in the public sector and the private sector. In Japan, corporate leaders tend to have backgrounds in engineering. In the U.S., corporate leaders tend to be marketers and/or financial experts. Between 1970 and 1980 there was an 11% drop in the number of engineers holding municipal public works director positions. At the beginning of the century most city managers were engineers. Today it is rare to find an engineer-city manager.

Infrastructure Crisis

America's infrastructure crisis is a premier example of our engineering community's failure to provide the technological leadership needed. In a report published by the National Academy of Public Administration, The Next Generation in the Management of Public Works: Getting Some of it Together, the perception of public works managers is openly stated: "The current system of public works management has been likened to sending a drunk down an alley. He somehow reaches the other end, but not in a very efficient or graceful manner. The nation's infrastructure system has progressed on an uncertain course. Management doctrine has been developed in an ad hoc fashion, but is now increasingly coming unstuck from its traditional base. It is still a little light headed. Reorienting public works managers, especially, to think about management ahead of construction will involve a considerable cultural shift. Engineers and public works directors think of themselves as builders, not maintainers and managers. They live capital-intensive fantasy lives. Replacing the 'edifice complex' with a passion for management will require major changes in the education and acculturation of those who lead public works organizations and those who educate them."

Technological Implementor vs. Technological Advocate

The mainstream practitioner engineer of today is a technological implementor. Someone decides that a bridge is to be built or some other technical problem is to be solved. Then the technological implementor accepts the "givens" provided and solves the problem to the best of his ability within the defined limits. Probably about 90% of practicing engineers are technological implementors.

About 10% of practicing engineers are or have the capacity to become technological advocates. America's economy is in desperate need of technological advocates. As reflected by salaries (supply and demand) the need for technological implementors and general salary levels has been on the decline since the early 1970's. There has been an oversupply of implementors and an under supply of advocators. Salary trends adjusted for inflation are in direct proportion to demand.

During the next 20 years, all indications are that implementors will be in undersupply. Thus, demand and, in turn, salaries can be expected to reverse. However, this does not solve (and may in fact augment) our national shortage of technological advocates because most individuals will be attracted to traditional engineering curricula. Engineering programs must change to offset this probable trend.

The upside potential for engineers who carry management, leadership and communication skills is exciting. The universities that decide to address this need are destined to a future of prestige and national recognition.

Action Plan

For the engineering community to become the technological leaders and advocates of our society, we need to make immediate but gradual changes in our approach to engineering education and registration. In addition to the current emphasis on technical skills, equal emphasis is needed in the areas of leadership, communications, creativity, risk taking and entrepreneurship. Some would argue that some of these are personality traits and are not teachable. If so, they must be incorporated in our student recruiting and screening efforts. It is the authors' collective opinion that the educational process can be effective at enhancing and encouraging skills in these areas.

Action is required on 3 fronts:

- continuing education
- engineering specialty registration
- basic engineering education

Continuing Education

Continuing education is an excellent example of the lack of leadership in the engineering community in America. We acknowledge that our university training is obsolete within 5 years, but do little to stay current with changing technology. In fact, by our actions that resist continuing education we are contributing to the problem. Continuing education is a necessary condition for professional and technical competence and, therefore, must be required.

Engineering Specialty Registration

In systems analysis courses it is taught that "to recognize the broadest scope of alternatives the system problem must be defined in its most basic terms."

The issue of specialty engineering registration

is such an issue. In one context it seems appropriate. In another context it does not seem appropriate. When the universe of professional engineers is divided into two groups (technology implementors, and technology advocators) the issue is less clouded.

Among technological implementors, the benefits to the profession, to society and to the individual of specialty registration are clear. Engineers can focus on the technical specialty of their choice and learn it more completely. There will be more incentive and it will be easier for engineers to maintain a knowledge base that is up-to-date and current. Those engineers within a specialty who create the most value will be more widely known, sought after, and compensated.

For technological advocators, specialty registration is not appropriate. To an increasing extent technological advocators are not even engineers. The trend is away from engineers. If engineers are to assume these roles in the future, we must prove ourselves capable. The most basic characteristic of a technological advocate is a quantitative thinker who will aggressively pursue the implementation of technology.

Basic Engineering Education

John Naisbitt, author of best seller Megatrends said, "Education in America must be reinvented rather than simply improved." Naisbitt is advising us to be absolutely clear about what we are trying to achieve.

The extent and intensity of engineering curricula has been debated. It is time to decide to act. To become a profession, basic engineering education must be changed to a two step process. Step one is a four year (120 sem. hr.) general engineering degree. The curriculum would provide basic engineering education in all of the sciences, computers, mechanics, etc., along with a broader scope of liberal arts type courses. The students' exposure to various technical specialties would be limited to introductory courses that introduce the various engineering technical specialties and provide enough information to assist the student with a choice of engineering specialties. Some have suggested that a modified liberal arts degree would better serve the purpose.

Step two is a two year professional degree in a technical specialty selected by the student, i.e., structures, soils, transportation, surveying, management, public works, etc.

The result of this program would be that probably twice as many people would go through the general engineering program. About one quarter of the graduates would go on to various technical specialty masters programs. Probably one quarter would stop after the B.S. degree and become engineering technicians. About one quarter would pursue various types of management training, e.g., MBS, MPA, MPW, etc., and the final one quarter would complete their B.S. degree, and go into unrelated fields, e.g., banking, real estate, business. Obviously, some modification of licensing is needed. All-in-all, a world with more engineers would have to be a better world.

More importantly, engineering will evolve to be a true profession and engineers will benefit.

Conclusion

The challenge is to decide. Once we decide to become a profession the problem reduces itself to one of implementation. Implementation is something that engineers do very well; deciding is something that we do not do very well. A wise mentor once gave advice on goal setting. He said that a goal to have items of materialistic value was superficial and doomed to ultimate failure; but a goal to be something that contributes value to others will bring with it materialistic rewards.

Engineers desire to be more highly compensated and consulted. Similarly, America seriously needs technological leadership. I submit that we, the engineering community, should commit ourselves to be the technological leaders and agents of technological advocacy of the next century.

It is our challenge to decide and to take action immediately. We must implement the action plan presented to become a "true profession."

Engineering Ethics and Professionalism Course

Robert D. Miles, P.E.[1]

Abstract

The National Endowment for the Humanities, in 1978-80, supported the National Project on Philosophy and Engineering Ethics at Rensselaer Polytechnic Institute. This project brought together engineer/philosopher teams to discuss ethical issues in engineering. As a result of the experience, a three credit course on Engineering Ethics and Professionalism was developed. The course content is discussed in this paper.

The course meets for two 75-minute sessions per week for 15 weeks. It is divided into five subject areas: (1) engineering as a profession, (2) philosophical ethics, (3) engineering ethics, (4) environmental ethics, and (5) tests and reports. The purpose is to consider and examine the profession of engineering under the law and professional practice where ethical questions and questions of responsibility to the individual, the profession, the employer, and the public interest are involved. Enrollment started with eight students in the 1979 Fall semester and has increased to about 150 students per semester. The course is offered to the ten professional schools and departments in the Schools of Engineering. The largest enrollment comes from Electrical Engineering and Mechanical Engineering. Students have suggested that it be a required subject for all engineering students.

Introduction

The author participated with a philosopher friend (James Otten, Ph.D., MBA) in the summer of 1978 at Rensselaer Polytechnic Institute in the National Project on Philosophy and Engineering Ethics supported by the National Endowment for the Humanities. The 1970-80 period showed public concern about business, governmental, professional, and private ethics and questioned the impact of the engineering profession on the quality of life. This project brought

---

[1]Professor, School of Civil Engineering, Purdue University, West Lafayette, IN 47907

together teams of engineers and philosophers to discuss issues. For many years, the author had discussed professional licensing and the code of ethics in a senior design course in the School of Civil Engineering. In the early seventies, a formal course on professionalism for all engineering students was proposed but was not developed. The proposed project of the engineer/philosopher team from Purdue University was the development of a formal course on Engineering Ethics and Professionalism as an experimental course for 1979-80.

The first course was announced as to availability as an elective course to undergraduate students in the School of Engineering for the 1979 Fall semester, and eight students enrolled. For the 1980 Spring semester the course was co-listed in the schedule of classes by CE, EE, and ME, and 53 enrolled. Interest in the course increased each semester, and current enrollment is about 150 students each semester. Almost 1400 engineering students have participated in the ten year history.

The course is scheduled for 75-minute periods at 11:30 a.m. to 12:45 p.m. each Tuesday and Thursday during the 15-week semester. The length provides more opportunity for discussion than the normal 50-minute session three times a week for the three hours of credit.

The subjects for discussion are divided essentially into five groupings as indicated in Table 1.

TABLE 1: ENGINEERING ETHICS AND PROFESSIONALISM SUBJECTS FOR DISCUSSION

75-minute periods; 30 Sessions

Part A:	The Engineering Profession	
Five Sessions:		What is a Profession
		Professional Registration
		Process of Registration
		Technician Certification
		Specialty Certification
Part B:	Philosophical Ethics - Normative Concepts	
Three Sessions:		Ethical Egoism
		Ethical Conventionalism
		Utilitarianism - Act/Rule
		Kantianism - Duty
		Rights Ethics
		Virtue Ethics

Part C:Engineering Ethics - Content Uniqueness
Twelve Sessions: Review Codes, Rules of Conduct, Corporate Policy
Canons and Guidelines
"Where Do You Draw The Line" Game
"Privacy" Game
Professional Liability, NSPE, TV
Product Liability
Copyrights & Intellectual Property
Computer Ethics
"60 Minutes Our Reply" Illinois Power, TV
Corporate Ethics, TV excerpts
TRUESTEEL AFFAIR, TV

Part D: Environmental Ethics
Four Sessions: Environmental Issues
PLAGUE ON OUR CHILDREN, PBS/NOVA
GILBANE GOLD, TV

Part E: Reports and Tests
Six Sessions: Technical Society Report
3-Person Case Study (2 sessions)
Interview Report-Interview (2 sessions)
Two Tests plus Final Examination or
EIT as substitute

A. The Engineering Profession

The concept of engineering as a learned profession and the public interest in the application of engineering knowledge in their interest is discussed. Engineering, the only learned profession with a knowledge base obtained in a four-year undergraduate curriculum, is discussed as to its impact on society and the quality of life where change results from the artifacts of engineering effort. Comparisons are made with the medical and legal professions. The concept of a 5-year program is discussed.

A copy of the Indiana registration act is provided each student. The Delaware registration act is discussed by a Purdue graduate who has served on the Council of the Delaware Association of Professional Engineers that administers the act. The uniqueness of the Delaware law is emphasized. Quotations from the California, Florida, Illinois, Oklahoma, and Texas acts are compared to the Indiana law.

The process of registration is explained with emphasis on the Fundamentals of Engineering examination and the need for the individual to take the initiative relative to the Professional Practice examination after a minimum of 4-years of practice. The responsibility of the National Council for Engineering Examiners to assure quality in the uniform examinations and the debate on a national exam is explained.

The program of the National Institute for the Certification of Technicians and Technologists is discussed. Skill certification for engineers is not appropriate under state law; therefore, the debate on specialty certification by the technical engineering societies is discussed. This movement has grown significantly in the 10-year history of the course. The input of the graduate engineer and the professional engineer to achieve specialty certification after licensing under the law, as in environmental engineering and forensic engineering, is important for the future of professional engineering and public safety.

B. Philosophical Ethics

Three class discussions are used to review normative concepts of ethics as a basis for further discussion of engineering codes of ethics and rational reflection upon professional behavior. The subjects summarized from the writings of the philosophers are egoism, conventionalism, utilitarianism, Kantianism, rights, and virtue.

The textbook used is "Ethics in Engineering" 1988, second edition, by Mike W. Martin and Roland Schinzinger. They were a philosopher/engineer team associated with the NIH National Project. Suggested additional readings are indicated in William W. Lowrence, 1985, "Modern Science and Human Values"; James H. Schaub and Karl Pavlovic, 1983, "Engineering Professionalism and Ethics"; and Stephen Unger, 1982, "Controlling Technology".

C. Engineering Ethics

Twelve sessions are devoted to discussions of the content and uniqueness of professional ethics. The ASCE, IEEE, NSPE Codes of Ethics and the Indiana Rules of Professional Conduct are examined. A few examples of corporate policy statements on conduct are perused. Points of possible conflict of a code of ethics, rules of professional conduct, and corporate policy statements are discussed. Outside speakers lead discussions on product liability (a lawyer), copyrights and intellectual property (corporate representative), and computer ethics (computer scientist).

Four video programs are used to include: (1) the National Society Professional Engineers program on PROFESSIONAL LIABILITY, (2) the Association of Professional Engineers of Ontario, TRUESTEEL AFFAIR, (3) the Illinois Power Co./PBS program, 60 MINUTES OUR REPLY, and (4) excerpts on corporate ethics from Ethics in Business presented by PBS National Narrowcast Service in association with Ethics Resource Center, Inc. and WTVS-TV/Detroit. These are very valuable classroom sessions that illustrate ethical issues as every day affairs of concern.

Two games are played with the class organized as six-person discussion groups. These games were given to the Schools of Engineering by AT&T who had Simile II (DelMar, CA) develop the material. One game is entitled "Where Do You Draw The Line" and has five scenarios for four characters. For example: one character has five ways of taking ten dollars varying from the Robinhood syndrome to thievery. The second game is concerned with privacy in the work place and also provides an opportunity to discuss harassment.

D. Environmental Ethics

Undergraduate students are interested in environmental issues. It is important to have a discussion of engineering knowledge and development of change produced by such application of knowledge on the total environment - ecological, economical, political. A lecture is used to show the development of environmental awareness from the book "Silent Spring" by Rachel Carson to current issues of ozone depletion and electro/magnetic effects. Two video programs are used to assist in developing discussion particularly engineering ethical dilemmas working with and for various stakeholders. The two-hour PBS/NOVA video program "PLAGUE ON OUR CHILDREN" is somewhat outdated and controversial but it elicits discussion. The NSPE's National Institute for Engineering Ethics (NIEE) video entitled "GILBANE GOLD" is concerned with environmental ethics and engineering ethics and the interplay of private, business, and professional ethics which ought to be identical.

E. Reports and Tests

Three reports are required. The first report requires each student to examine the history, goals, and membership requirements of a professional or a technical society. A significant percent are not members of their student technical society and this report generates interest.

The second report is a three-person case study of ethical issues and dilemmas of concern in developing artifacts of engineering technology. Some of the projects selected for analysis over the years are listed in Table 2. Six to 10 subjects are used each semester.

TABLE 2: THREE-PERSON CASE STUDY TITLES

1. Political Contribution	6. Fuzzbuster
2. Automatic Brake System	7. Air Bag Safety
3. Fusion Technology	8. Hazardous Waste
4. Nuclear Waste Transport	9. Crash Avoidance
5. Cellular Phone Charge	10. Air Pollution

The third report is the summary of an interview with a graduate engineer and preferably a Professional Engineer concerning an ethical problem encountered in their career.

Three or four class sessions are used for oral reports of some of the three-person case studies and the interview reports. Usually eight case studies are presented in two oral sessions by teams selected on the basis of ethical analysis and depth of issues discussed. Ten or twelve interview reports can be presented in each oral session (one or two) at the end of the semester. Ethical reasoning is emphasized in these reports and oral presentations.

The preparation and grading of reports and tests for about 150 students per semester is the most difficult task associated with the course. Since essay tests are too time consuming, the first two of three tests consist of 50 true/false questions and 50 multiple choice questions. One or two graders have been hired to assist in grading for the last two semesters.

The final examination raises for class discussion ethical questions of fairness and equality as well as bribery versus a gift. If students are graduating and take the eight hour FE examination, they receive 100 points and are exempt from the final examination. If not graduating, the students must answer six or seven essay questions. The class debate about fairness and equality usually ends with the essay test being designated a take-home provided the last day of class. The gift is 100 points if they answer all ethical questions in reasonable form.

Grading for the course is based upon 900 points consisting of 300 points for three reports, 300 points for three tests, and 300 points for attendance. Attendance is taken each class session (30) by handing out a question on either professionalism or on ethics for the student to answer in longhand, date, and sign. If they are picked up from a stack the temptation to sign for another is too great and 82 forms appear with a body count of 76 in a class of 87. To overcome this problem, the daily is handed out by a grading assistant about 10 minutes after the class starts. Good control is exercised.

The course has a problem with the availability of files on tests and on reports. Previous tests are made available for reproduction at a central office on campus but the students complain about not being provided the answers. The files on reports were not thought to be too much of a problem but a few identical copies of old reports (interviews) surfaced this last semester. The students (senior 8) lost two grades and agreed to the penalty without complaint. To avoid this problem the tests must be revised each time and the report subjects must be changed each semester.

Feedback from students and former students are favorable relative to the course content and method of presentation. The part on philosophical ethics and normative concepts receive the most negative comments and requests for more case studies. A large number of students have commented over the 10 years that all engineering students should be required to take this course in the general format here presented. It is an easy course compared to the technical subjects, but it relates to their area of interest more so than an ethics course in the School of Humanities and Social Science. It is these factors that have caused its growth in interest these 10 years.

ENGINEERING ETHICS IN EDUCATION

Gennaro Gerald Marino, Ph.D., P.E.[1]

The Importance of Ethics in Engineering

The purpose of this paper is to demonstrate the importance of acquiring a higher level of ethical awareness from education, and its practice by engineers. The necessity for mandatory education in ethics, especially for engineers, is exemplified by examples and overviews of the range of existing world conditions.

Engineers have a tremendous effect on the advancement and development of society and therefore, if they like it or not, have inherited an important role. As shown in Figure 1 the engineer has a role from the conception to the final stage of the project. He has the technical awareness from which policymakers count on to assess the viability of project ideas. In engineering research and practice, project developments have purposes aligned to an individual, a group(s), and/or mankind. A consciousness of ethics, therefore, in engineering can have great economic and sociologic benefit. Many times, however, evaluating and weighing the outcome of these factors does not result in an analytical decision by the engineer.

Engineers tend to divorce themselves from ethics decisions and maintain a passive philosophy using numerous justifications. But, to the degree that an engineer does not make conscious decisions in how he practices he becomes - to that magnitude - a puppet and a victim of living for someone elses or for some other groups' interest. One can easily see how relevant a sense of ethics is to an engineer's perspective and their decision making process. Nevertheless, courses in ethics

---

[1]Geotechnical Consulting Engineer, 1601 Parkhaven, Champaign, IL, 61820.

are not required in educational institutes at any level, yet essential in engineering.

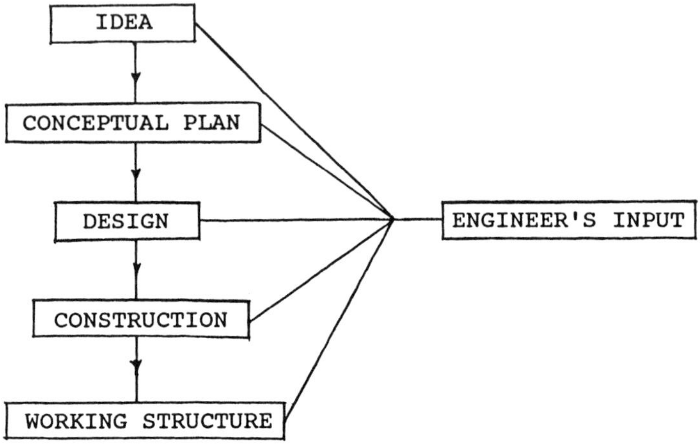

FIGURE 1 AN ENGINEER HAS CONSIDERABLE INPUT AND RESPONSIBILITY FROM THE BEGINNING TO END OF THE PROJECT

Ethics Definition for Engineering

Ethics can be defined in engineering as a system of thought the individual acts upon, in research or practice, for mainly the betterment of himself, his family, the group(s), and mankind.

Figure 2 illustrates the cause-effect relationship of the engineers' decision and the major areas of the project and life that it would affect. As an example, decisions 1 or 2 could be made to handle the same project situation are plotted in the figure. Decision 1 shows immediate remuneration for self and the family unit and actually has a negative affect on the user of the project (e.g. organization that owns the poorly engineered building); user recipients (e.g. leasees) and mankind (e.g. overall effect on the economy). For Decision 2 the immediate remuneration to self might not be as high as Decision 1 but positive and higher effects are caused across the remaining compartments (or Dynamics).

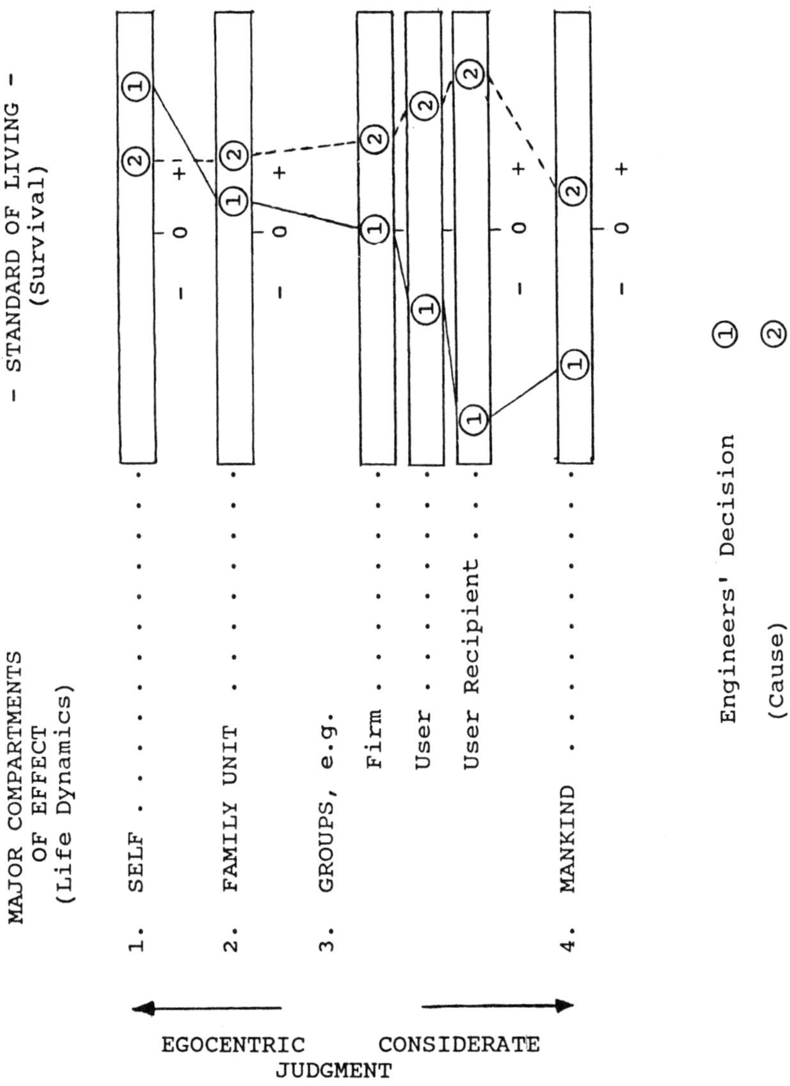

FIGURE 2 CAUSE-EFFECT RELATIONSHIP OF AN ENGINEERS' DECISION

Status Quo

In engineering projects, short-term solutions have been implemented for profit on the First Dynamic (see Figure 2) that many times contain risks of having long-term problems (e.g. chemical waste dumps and discharging untreated wastes in bodies of water, designing faulty systems, falsifying or suppressing data bias to the analyses, engineering systems that are workable but are far from optimum, etc.). An engineer not being true to himself and not taking full responsibility for the project are major ethical errors. As a result of this, the project becomes liable to questionable/under-engineering. A common mode of operation which occurs here is one is persuaded by his environment to compromise his own reality (and becomes "reasonable" about things) in the presence of pressure to cut costs, or for profit, or for some other aberrated reason(s). In civil engineering, knowingly underestimating construction costs to assure economic viability; basing a foundation design on an inadequate subsurface investigation; designing substandard structures; approving insufficient quality control during design or construction; and short-sided environmental and loading factors used in structural design are all examples of underengineered projects. Effects from these types of ethical errors are major construction delays; lawsuits; economic losses incurred on the project as well as the ramifications which result in areas where the needed additional project funds were removed; remedial repair of the structure; loss or diminution of life; etc. all solely at the expense of irresponsible engineering.

Currently the complexity of the present technology tends to cause engineers to concentrate or specialize without having a broad perspective. A common position taken among engineers and technicians is "I'm just a specialist. The purpose, operation, and application of these systems I'm involved in is beyond my scope of responsibility." An extreme example of this condition would be a project engineer presenting to a suppressive government the economic advantages of his plane in battle over his competitor's. This person, if he's less than a psychopath, is not conscious of that governments intentions and considers that this 'defense' system is never to be employed or has carelessly not thought about his actions.

This "specialist" viewpoint may be justified by the engineer through taking up an isolationist philosophy. We could say to himself "it's enough just to take care of myself." However, conditions do not stay the same, and when one does nothing to confront his environment and positively cause it to change he will be leaving change to chance or to an act of God. This is a dangerous situation. The isolationist has not seen or been exposed to the gratification and sense of joy in providing help to others. It is as if attempting to change his environment for the better, and being responsible for his work, would only be a detriment to his own survival.

Application of Ethics in Engineering Practice

Prosperity, high morale, and cost effective products are all indications of a highly ethical engineering condition. Thus applying ethics in engineering practice will have benefit to humanity, to the group(s), to his family, and to the individual. A practical education of ethics aids in spotting troubled areas in production, in pointing out potentially unwanted conditions, and in improving conditions. In the field of production, personal and group ethics relates to upgrading products or services and increasing the flow of them. This would involve using statistics in monitoring all aspects of the production line, rewarding groups or individuals with improved statistics and remedying downward statistics, setting effective policy, etc. In a construction project, for example, the costs can be minimized and a well built structure will be realized when: policy is set and followed, the structure is properly designed and checked, statistics are applied to monitor production of all construction elements, quality control is kept in, and schedules are enforced with suppressive persons or groups to production determined and handled.

As noted in Figure 1, an engineers' input is required from the conception to the completion of the project. Thus he has a responsibility to communicate to all relevant lay persons all the vital ramifications perceived to exist or to possibly exist. Furthermore the communication is not achieved until understanding is accomplished to the necessary level of authority. It is irresponsible to merely tell an associate of his concern and to use this as a justification for having communicated. Even engineers simply refusing to work on jobs with inhumane purposes will obviously cause a digression from these activities. But before ethics can be significantly practiced by engineers, the confidence and motivation must be instilled in the engineer as an individual. These qualities are vital since he serves society to such a large degree, and can be instilled if a firm grasp of engineering ethics is obtained from his education.

Education of Ethics in Engineering

In order to increase the consciousness or awareness of our engineers in society, engineering ethics course-work must be a requirement during their college training. Comments such as "how would you fit this into an already crammed course curriculum" seem irrelevant when one considers we are graduating engineers with the 'tools and no instructions'. It is suggested, therefore, that an introductory and applied course in ethics be a requisite. The introductory course would optimally fit after the sophomore year and at the beginning of their specialty while the applied course taken in their last semester.

Taken at the beginning of their engineering curriculum, the first course will provide a basic moral or ethical perspective to the engineering students while taking courses in their field. This introductory course should include:

- What is ethics,
- What are right and wrong acts and how to identify them,
- How does reason relate to ethics,
- Is man basically good or evil and does this apply to ethics,
- How does ones religion and philosophy fit in,
- Ethics in science and technology.

The purpose of the senior level course would be to establish and motivate a sense of ethics in the student engineer just before entering society. At this level the course-work should involve:

- A review of ethics.
- Establishing a code of ethical behavior in engineering.
- Related engineering case histories where ethical errors existed and where ethics was applied correctly.
- Practicals on current ethical issues related to the student's field of

engineering. (In addition to education in the area of application these practicals will demonstrate to the student how he can have an effect upon his environment.)

Summary

An increased sense of ethics in our engineers in society will have great socio-economic benefits. Yet today, training of ethics in engineering is not required in our higher educational institutions. The current structure of our educational system, in fact, gives an impression that a sense of ethics is not a vital role of the engineer. Presently, the only possibility that engineering students will be exposed to ethics is through selection of their 'humanity' electives which rarely even touch on the question of ethics in engineering.

The purpose of this paper is to recommend that training in engineering ethics become a mandatory part of the engineering curriculum. Items which should be covered in the course-work on ethics within the respective engineering departments are presented in the paper.

Acknowledgements

The ethics data presented in this paper was taken from works written by L. Ron Hubbard.

SIX MUNZ AGO I CUDN'T SPEL INJUNEER NOW I ARE ONE!

by

Dr. Francis E. Griggs, Jr., F.ASCE
Professor of Civil Engineering
Vice President for Academic Affairs
Merrimack College
North Andover, MA

How many times have you seen the title of this paper as a caption under a caricature of a civil engineer and along with everyone else broke into laughter, or at least into a grin? As in most caricatures, it works because the viewer believes that there is some truth to the image and expression. How many times have you heard it said by others and even other civil engineers that engineers can't write, can't speak, etc. Doesn't it bother you as an educator, practitioner and civil engineer that we all seem to be aware of the problem but do little about it? Doesn't it bother you that this image of civil engineers is widely held by many? This paper will describe the state of affairs existing in our colleges and profession today and indicate how we at Merrimack College are attempting, and with some success, to prepare our students to enter a profession which in the 21st century will not tolerate engineers who do not have the ability to communicate effectively.

WHERE WE ARE:

Recently a series of national reports on the state of higher education in the United States received the attention of the media and critics of higher education and unfortunately of too few colleges or faculty members around the country. Those reports, "To Reclaim a Legacy: A Report on the Humanities in Higher Education" (1984)[1], "Involvement in Learning", (1985)[2] and "Integrity in the College Curriculum: A Report to the Academic Community", (1985)[3] all stated that students, in general, are not leaving our colleges with an ability to communicate effectively. "Integrity in the College Curriculum" states:

> *Evidence of decline and devaluation is everywhere. The business community complains of difficulty in recruiting literate college graduates. Remedial programs, designed to compensate for lack of skill in using the English language, abound in the colleges and in the corporate world. Writing as an undergraduate experience, as an exploration of both communication and style, is widely neglected*"

"Reclaim a Legacy" states:

> *Impressionistic or anecdotal evidence for the decline of the humanities surfaces every time I talk with college professors, academic officers and students. Such evidence is familiar: students who graduate from college*

unable to write lucidly or reason clearly and rigorously---

"Involvement in Learning" states:

> *Students' performance of 11 of 15 major Subject Area Tests of the Graduate Record Examinations declined between 1964 and 1982. The sharpest declines occurred in subjects requiring high verbal skills.*

It should also be obvious that these indictments are aimed at many institutions which are mainly liberal arts colleges where most students take a large portion of their curriculum in courses which oriented more towards words and to a lesser degree towards math and science. It is my belief, and I think that of many firms which employ our graduates, that the situation is even more serious among graduates of engineering programs. Many surveys of employers have been made over the years asking employers which skills they felt were most important for graduates to possess and which skills they felt the recent graduates were most deficient in. In almost every survey, communication skills were at the top of both lists, e.g. they were the skills most needed and also the skills in which they were the weakest. My former colleague Dr.Thomas Jewell from Union College and I presented papers[4,5] at the 1983 and 1985 Engineering Education and Civil Engineering Education Conferences in which we quoted two (2) articles written by Kimbel et al[6,7] in 1979 which quantified this concern of the profession. More recently the IEEE, through its excellent journal, Professional Communication and national conferences, on the topic of professional communications, has been very active in describing the need for enhanced communications programs and in describing actions being taken by a few colleges and universities to address the perceived shortcomings of electrical engineering graduates. ASCE might want to consider the development of a professional communications journal. It is my belief that such a journal would be read widely and could have a major impact on the communications skills not only of our graduates but of all practitioners as well.

The Accreditation Board for Engineering and Technology (ABET) previously ECPD[8] began to realize that the communication skills of graduates from accredited programs in engineering were not up to the standards required by the profession and first inserted specific communications criteria into the standards for accreditation in 1979 under the Humanities-Social Sciences category. In 1980 a separate communications criterion stated:

> *k. Competency in oral and written communication in the English Language is considered by ECPD to be essential for the engineering graduate. Although the requirement of specific coursework may serve as a foundation for such competency, the development of communications skills should be demonstrated by student work in engineering courses.*

In 1983 the criteria seemed to take on a greater emphasis as instead of a small "k" designation it changed to:

4. Communications, Humanities, and Social Sciences

a. Communications

Good oral and written communications are considered by ABET to be a necessary achievement of a college graduate. Technically trained individuals should not be considered educated regardless of the depth of their technical capability if they cannot communicate, both orally and in writing, their technical findings, thoughts, and philosophy to others around them. Since it is by practice that the real importance of a specific aspect of educational endeavor is demonstrated to the student, a good technical educator will insist that reports be neat, grammatically correct, and lucid. Thus the visiting team will be looking for evidence that both oral and written communications have been taken into account in the review and evaluation of student work.

Unfortunately, in my opinion, the criteria was de-emphasized once again in 1988 as follows:

h. Competency in ***written communication*** *in the English Language is essential for the engineering graduate. Although specific coursework requirements serve as a foundation for such competency, the development and enhancement of writing skills must be demonstrated through student work in engineering courses as well as other studies.* ***Oral communication*** *skill in the English language must also be demonstrated within the curriculum of each engineering student.* (1988)

There is a pinprick of light coming from ABET on the issue as a proposed change (effective 1990 visits) to the guidelines under **3. Student Body** section c. states:

These items, which include the competence of students in both subject matter areas and ***communication skills must*** *(was should) be made available to the visiting team.*

Having been on the receiving end of many accreditation visits over the years I have never, I repeat never, had a team discuss and surely not critically evaluate the actions a college has taken to ensure that their graduates have good communication skills. This is in spite of the criteria and the comments the profession has made about these deficiencies. It seems as if the teams are counting on the Humanities and Social Sciences courses to provide all of the exposure to effective communication that our students need. In my opinion, this is an unjustified assumption as the amount of oral and written communications required in these courses is frequently minimal.

In summarizing, I am convinced that our graduates, in general, do not have as good communication skills as they had in the past. I am also of the opinion that it is now a necessity that they be fluent in both writing and speaking and in addition that they be exceptionally good listeners. We as engineering educators are partly to blame as we have not acted to address this deficiency by either working with our colleagues in the humanities and social sciences or by modifying our curricula to train our graduates in the art of effective communication. Given this dismal state of affairs is there anything we can do to ensure that our engineering graduates are literate and able to **"communicate, both orally and in writing, their technical findings, thoughts, and philosophy to others around them"?**

THE MERRIMACK PLAN- ONE SMALL STEP FOR THE COLLEGE AND A "POSSIBLE" GIANT LEAP FOR ENGINEERING EDUCATION

While a bit overstated (maybe a great deal) I am making the point that if schools of engineering commit themselves to take small steps towards emphasizing communications throughout their curriculum the net effect will be a giant leap for engineering education. In 1985 the Civil Engineering Department at Merrimack reviewed its curriculum in preparation for an ABET visit. As a follow up to recommendations made by the visiting team, we completed this review in the spring of 1986. We determined that our students needed a stronger exposure to the fundamentals of effective communications and a "little" more design. The department at that time required a full course in graphics and a full course in fortran programming for all freshmen. After some debate it was decided to institute a two (2) course sequence in Communications in place of the graphics and fortran courses. The rationale for this change rested on the premise that a full course in graphics, in which we taught the students how to be draftsmen was inappropriate for the 1980's with the development of CAD software and hardware. In addition, with the development of the Personal Computer and a growing availability of software for the PC we did not believe that an entire course in a higher level language was necessary, and that much of the computing to be done by Civil Engineers in the future would be on PC's.

Based upon the above we developed a two (2) course sequence in which we treat all of the encoding and decoding skills needed for effective communication:

- Written communications
- Oral communications
- Graphical communications (sketching and CAD)
- Graphical communications (graphs-data presentation)
- Listening
- Man Machine communications (programming and use of PC's)

The courses are scheduled twice a week for 1 hour and 15 minutes for a period of 13.5 weeks. The class size has varied from 15 to 25 students with an ideal class size being 20 students. The course syllabi are prepared for the entire year and passed out to the students on the first day of class so that they can see the rationale for the sequence and how we keep coming back to the various modes of communication. The courses are taught by senior members of the department usually by myself or Prof. Westerling, the Department Chairman. This commitment by the Department is intended to illustrate to the student the importance which we attach to the sequence.

In order to describe the courses I will indicate below how we treat each mode of communication throughout the sequence.

WRITTEN COMMUNICATIONS - The Freshmen are taking two (2) required

Rhetoric courses concurrently with the communications courses so they are getting a double dose of writing assignments. We have found, however, that the assignments are significantly different and as such reinforce one another without being duplications. Their first assignment is aimed at emphasizing the importance of the audience to whom they are writing. We ask the student to review an article that was written for *Science* and rewrite it so that a second year liberal arts student could understand it. We find in this first assignment that students have been trained to **assemble** information, but they have not been trained to **assimilate** it before they start to **write.** In other words they write without knowing where they are going. This paper is returned for rewrite several times before the student starts to understand the significance of assimilation and organization. The next assignment deals with the writing of definitions. This is a classroom exercise where I give the student the name of an object and instruct them to write a definition of that object. After a 10 minute period they have to read their definition to the class who in turn "guess" what object is being described. If the class is not able to determine the name of the object we, as a class, indicate to the definer where his definition is unclear. The next assignment is a technical description of an object or a process. In this exercise the use of illustrations is treated as a means of effective communications. We then have an exercise on writing a set of instructions. We talk about the importance of good instructions and everyone has their own favorite story of a situation where they have received a poor set of instructions. In a classroom exercise I ask groups of 2-3 students to write the instructions on how to tie a knot of their choosing. They then swap instructions and send along one of their group to watch the other group try--the word is try--to tie their knot. This exercise, with its built-in feedback loop, impresses on the students how easy it is to be misunderstood and hopefully forces them to follow, in the future, all of the rules for the effective writing of instructions. In order to give the student an appreciation of their heritage as Civil Engineers they are asked to use the resources of the Library to research the career of a famous civil engineer and submit a report as well as to give an oral presentation on that engineer.

As a part of their preparation of a spreadsheet template to perform all of the calculations required to balance a traverse they prepare a complete documentation report which could be used by anyone to balance a traverse of any number of sides. This documentation is given to another student, not in the class, to determine if the instructions are clear and easy to follow. The last writing assignment they prepare is a report to accompany the final design project. All of these writing assignments must be prepared on a word processor and are returned for rewrite when necessary.

ORAL COMMUNICATIONS - The students make a series of oral presentations throughout the course, the first coming on the second day of class. The topic varies, but it usually deals with a summer experience, why they came to Merrimack, etc. This presentation which lasts two minutes is videotaped on a cassette which the student brings to class every time they make a presentation. They will have, therefore, a history of their presentations throughout not only these courses but throughout their college career. The next presentation is associated with a mini-design problem given about one month into the first course. In this presentation the students in two person teams make use of models, visual aides, and other "props" necessary to effectively communicate their design. A five minute presentation on an engineering project currently in the news is assigned as the last oral presentation of the first course. Since this is a longer presentation the students must **assemble** more information on their project and **assimilate** it so they can make their point in a limited amount of time. As can be seen the topics range from one the student knows more about than anyone in the class (his experience) to one in which he has to become "expert" in a new area. In the second course the students make a presentation on a famous engineer (along with a written report), participate in a mini-debate with a classmate on a controversial topic in the engineering field, and finally make a presentation of their capstone design project.

GRAPHICAL COMMUNICATION (sketching and CAD) - The students are given instructions in freehand sketching and the orthographic, isometric, oblique and perspective projections. It is emphasized that engineers frequently communicate with sketches and that final drawings are now prepared using CAD technology. The student moves from sketching what they can see in front of them, to what they have seen and stored in their mind to what they create in their own mind (their designs). We introduce the students to MacDraft^{tm} in the first course as it is a fast learn and excellent introduction to CAD. We have the students initially reproduce some of their sketches on the Mac and then move to other civil engineering drawings such as floor plans, site plans, sections, and schedules on our DesignPro software using four (4) stand-alone CAD workstations. All of their designs after the first mini-design are prepared on the CAD systems. This year we will be training the students on a Mac II using AutoCAD. As a spinoff of this training, we have found students using CAD for lab reports and presentations in other courses to such a degree that professors are now requiring that drawing be done on the CAD systems.

GRAPHICAL COMMUNICATIONS (GRAPHS-DATA PRESENTATION) - We utilize Excel software to train the students in using graphs as effective ways of communicating numerical data. We begin by having them become aware of the gallery of charts available and when to use each type of chart. The greatest value of this portion of the course comes when we present them with tables of data which are not simply plottable as x-y plots. In these exercises the student must assimilate the messages which lie within the table and determine the best way to communicate this information in a graphical format. Students are slow to learn this material but when the projects are returned several times they, with time, start to really analyze the data and come up with the most effective way to present it graphically.

LISTENING - Perhaps the most neglected of the communications skills, listening is covered in the second course by a series of exercises in which we help the students become aware of the obstacles to effective listening and train them in ways in which they can maximize their retention of orally transmitted information. To help us in this portion of the course we use materials and tapes developed by TelStar Productions. These Effective Listening tapes come with tests which can be given immediately after listening to the tapes so that the students become aware of their listening deficiencies and begin to analyze why they are not picking up the information which is being transmitted to them. As in most skills building courses, the knowledge that they are not as effective as they thought in decoding oral information is the first step in the development of better listening habits and skills.

MAN MACHINE COMMUNICATIONS - The use of the MacIntosh computer along with the MacWrite II, Excel, MacDraft, and Eureka software introduces the student to word processing, spreadsheets, graphing (charting), database management, declarative (equation solving) languages and

CAD.[9] By the end of this portion of the course the students are fully trained in these packages which are all purpose utility software packages that can be used in many courses. We chose not to train the students on packages which have a unique purpose as these would be covered in later courses. Our goal was to have the student become aware of the various computational tools available to solve different types of problems, e.g., the calculator, the PC or the main frame computer and to be able to determine which tool is most effective in solving specific problems. I am happy to report that this goal is being met as we find students making the right computational choices in subsequent coursework.

As we set the agenda for the future we must place the development of effective communication skills near the top of our list of things to do. I would suggest that ASCE, ABET, and Civil Engineering faculty consider the following steps:

(1) that Civil Engineering professors take it upon themselves to emphasize the importance of effective communications in their course work;
(2) that ABET instruct its visiting teams to evaluate, critically, the extent to which our Civil Engineering programs foster effective communications skills;
(3) that ASCE, like its sister organization the IEEE, begin a Journal on Technical Communications;
(4) that practitioners provide feedback, on a regular basis, to colleges on the level of communication skills exhibited by recent graduates.

If we want our graduates to be appreciated for both their technical expertise and their ability to communicate we must, as engineering educators, take a portion of the responsibility on our own shoulders and develop courses such as described above. If we do our job our graduates will surely be able to: **"communicate, both orally and in writing, their technical findings, thoughts, and philosophy to others around them"!**

References:

(1) Bennett, W.J., "To Reclaim a Legacy - A Report on the Humanities in Higher Education," Nov. 1984.
(2) "Involvement in Learning: Realizing the Potential of American Higher Education, National Inst. of Education," U.S. Dept. of Education, Oct. 1984.
(3) "Integrity in the College Curriculum - A Report to the Academic Community," Association of American Colleges (1985).
(4) Griggs, F.E., Jewell, T.K., "Communications Courses - A Nicety or a Necessity?" Proceedings Annual Conference, ASEE (1983), pp. 401-407.
(5) Griggs, F.E., Jewell, T.K., "A Program to Enhance the Communications Skills of Civil Engineering Students," Challenge to Civil Engineering Educators and Practitioners-- Where Should We Be Going?" Proceeding (1981) pp. 154-163.
(6) Kimbel, W.R., and Monses, M.E., Hunze, J. "Are Today's Civil Engineering Graduates Qualified?", Civil Engineering Education, Vol. 1, 1979, pp. 68-75.
(7) Kimbel, W.R., and Monses, M.E., "Engineering Graduates: How Good Are They?" Engineering Education, Vol. 70, No. 2, Nov. 1979., pp. 210-212.
(8) Annual Reports of ABET (ECPD) 1982, 1983, 1988, 1990
(9) Griggs, F.E., Westerling, D., "Communications in Engineering Education," Proc. Annual Conference ASEE 1989, pp. 609-612.

PROFESSIONAL LIABILITY OF THE CIVIL ENGINEER

George T. Taoka[1], Member, ASCE

Abstract

Undergraduate civil engineering students should be educated in the subject of professional liability of engineering practice. The volume of lawsuits have increased exponentially in the past decade. Most practicing engineers are generally ignorant of and unprepared for the sequence of events that they are subjected to when they become involved in litigation. If future engineers are given the opportunity to gain an understanding of how the law affects their profession in undergraduate school, they will be better informed about their exposure to liability and may be able to avoid becoming involved in costly lawsuits in their private practice. When they do find themselves being sued, their knowledge of this subject will be useful in helping their attorneys to present strong defenses against the claims brought against them. This paper discusses a course on this subject that is presently being taught to senior engineering students.

Introduction

In the past decade, spectacular failures such as the collapse of the Kansas City Hyatt Regency walkways (Pfrang and Marshall 1982) and the construction lift slab collapse of the L'Ambience Plaza apartment building (Wonder 1988) have focused public attention on the problem of professional engineering liability. Lawsuits filed against engineers have been increasing at an exponential rate, and C. Roy Vince of Professional

[1]Professor of Civil Engineering, University of Hawaii at Manoa, 2540 Dole Street, Honolulu, HI 96822

Liability Brokers and Consultants (Zeitlin and Hinckley 1985) estimated that the chance of a practicing engineer having a claim made against him in 1985 was 43 percent. It is, therefore, imperative that students should receive training in the topic of professional liability before they begin their engineering careers.

Engineers are generally uncomfortable, and sometimes even hostile and suspicious, when they become involved with attorneys in legal proceedings. This somewhat unfavorable opinion is frequently the result of an ignorance that most engineers have about how the law affects their profession, and what attorneys do within the legal system. Only through education can a practicing engineer gain a working knowledge of engineering law and help him overcome his negative attitude toward the legal profession.

Course Content, Texts, and References

A course on professional engineering liability is presently being offered at the senior-graduate level at the University of Hawaii. The primary emphasis of the course is on tort law as applied to malpractice suits filed against design professionals. Fundamentals of contract law governing the duties of owners, design professionals, and contractors are also discussed. Since most engineering lawsuits result from design and/or construction failures, the topics of failure analysis and engineering probability are also included in this course. The role that expert witnesses play in lawsuits is also explained, and the need for credibility and impartiality of engineering experts is emphasized. Finally, the sequence of events associated with different stages of a lawsuit from the initial filing of a claim up to and including the trial, are carefully explained, so that students who complete this course will gain a useful overall understanding of basic engineering law.

The basic topic headings included in this course are listed below.

1. Responsibilities of Professional Registration
2. Contractural Duties of Design Professional
3. Probability, Safety Factors, and Failure Analysis
4. Codes, Standards, and Construction Practices
5. Tort Law Involving Professional Liability

6. Standard of Care and Engineering Negligence
7. Professional Liability Insurance
8. Filing of Claims and Discovery of Evidence
9. Role of Expert Witnesses in Lawsuits
10. Arbitration of Disputes
11. Pretrial Conferences and Settlements
12. Trial Procedure
13. Liability and Computation of Damages
14. The Appeals Process Following a Trial Verdict

The required text for this course is "Professional Liability of Architects and Engineers" (Streeter 1988). Other texts used as references are "Legal Aspects of Architecture, Engineering, and the Construction Process" (Sweet 1985), and "Legal Aspects of Engineering" (Vaughn 1977). These references explain the basic theories underlying engineering law, and contain numerous case histories illustrating how the law has been applied to a variety of real life situations. The study of case histories serves the same function in this course that numerical problem solving serves in traditional engineering courses. Case histories show how legal reasoning is applied to specific cases in the same manner that problem solving shows how engineering theory is applied to actual problems. Two other useful references are "Construction Disasters: Design Failures, Causes, and Prevention" (Ross 1984) and "Structural and Foundation Failures" (LePatner and Johnson 1982). Both contain numerous case histories of engineering failures.

Course Material Supplements Legal Advice

The knowledge gained from taking a course in professional engineering liability is not meant to replace the advice and services of a competent attorney. Rather, its purpose should be to supplement and to increase the usefulness of the services that an attorney can provide. As has been pointed out by both Vaughn (1977) and Streeter (1988), an important value of this course is to suggest to the practicing engineer those occasions when it is especially important and critical to seek legal services. The design professional will be aware of how the law affects the engineering profession, and will recognize the difficulties and pitfalls in his work which can eventually lead to litigation. He can seek legal advice as a preventive measure before claims are filed, rather than as a necessity to defend a lawsuit that has been filed against his firm. He is also in a better position to prepare a strong defense against any claim filed for engineering malpractice should a lawsuit be filed.

Need for an Adjustment in Thinking by Engineering Students

A fundamental difference in the reasoning required in a legal course as opposed to a traditional engineering course should be noted. Engineering students traditionally are taught to solve problems by a systematic form of deductive logic, where an "optimum" or "best" design to a given set of constraints is obtained. Engineering principles do not vary from problem to problem, so students expect a form of regularity in analyzing technical problems.

The reasoning applied to legal work is generally of a different type of complexity than that used in engineering work. As Bockrath (1986) has pointed out, in law the logic of the process is secondary to the fact of the result. This fact is embodied in a famous quote attributed to Justice Oliver Wendall Holmes, "The life of the law has not been logic; it has been experience" (Corley, Holmes and Robert 1983). Judicial reasoning is frequently based on analogy using previous decisions as precedent. This form of analogy is closer to inductive reasoning than deductive reasoning, and emphasizes the empirical and pragmatic aspects of the law. There seldom exists a well defined set of steps which will lead to a legal conclusion. Rules of law are full of exceptions, and the way a particular law is interpreted will usually vary with the facts of each individual case. The laws themselves will vary within different jurisdictions, so the regularity of analysis which students have used in other courses does not exist in a course on professional liability. They must adjust to the ambiguities and refinements in interpreting the law depending on the particular circumstances of each case. They should also learn to view legal issues from several points of view, to understand the legal strategies being employed by opposing consuls in a given malpractice lawsuit. A certain amount of unpredictability and uncertainty is always present in the law.

Engineering Probability and Failure Analysis

The topics of engineering probability and failure analysis are discussed in this course. Students are required to compute both the nominal as well as the actual factors of safety in engineering design. They are also introduced to different levels of failure, from a minor nonconformance of design requirement to the complete collapse of a civil project. The author has found the text "Uncertainty Analysis, Loads, and Safety in Structural Engineering" (Hart 1982) to be an especially useful reference on these topics. It is important for students to recognize that most design codes are changing or have changed their design criteria from deterministic to random values to be evaluated by probabilistic analysis. Designing for random loadings will, therefore, result in more realistic factors of safety than designing for deterministic loadings on structures.

Professional Liability Insurance

The types of liability insurance available to design professionals is discussed in this course. The cost, availability, and dollar coverage of various policies are reviewed, and the professional activities covered by these policies are explicitly staged. The difference between "occurrence" and "claims made" policies are clearly defined. Specific deduction and exclusion clauses are also carefully explained. The importance of carrying this type of insurance is stressed in this course.

Engineering Expert Witnesses

The role that expert witnesses play in professional malpractice suits is also discussed in this course. Since malpractice lawsuits of this type frequently require relatively technical testimony on a complex engineering level, opposing counsels will usually employ the services of expert witnesses to explain to the court the technical details involved in the lawsuit. These experts are also frequently asked to give their opinions on whether certain parties to the suit may have been negligent in the performance of their duties as professionals. It is important to stress to students that expert witnesses should be well qualified, impartial, and well prepared when giving testimony in court. They should conduct themselves with dignity and confidence, and answer all questions clearly and truthfully in all legal proceedings.

Resolution of Claims

The greatest portion of a course on professional liability should be devoted to the resolution of the claims which is the basis of malpractice lawsuits. The three ways a lawsuit is resolved is by compromise settlement, by arbitration, and by going to trial, also called litigation. Most lawsuits are resolved by compromise settlement, where the defendant or defendants agree to compensate the plaintiff or plaintiffs an amount that is less than the amount desired when the lawsuit was originally filed. This agreement is usually reached after the stage involving discovery of evidence is completed. Both sides of the lawsuit are aware of the strengths and weaknesses of their respective positions and are, therefore, willing to reach a compromise settlement of the original claims.

Arbitration is the submission of a dispute to a disinterested party who has special training and can act as an impartial judge in reaching a decision on the case. Its advantages over litigation is that it can resolve the claim speedily and inexpensively because its proceedings are less formal than those of a trial. The decision of an arbitrator may or may not be binding on all parties, however, depending on the jurisdiction of the courts. If the results of an arbitration are accepted by both sides of a lawsuit, both time and money is usually saved by all parties to the lawsuit. In some states, construction suits must be submitted to arbitration before it can be litigated.

The final topic that is discussed is that of litigation. Trial procedure is formalized, starting with a pretrial conference with the judge and the attorneys involved in the law suit. Then comes jury selection, perhaps the most critical part of the trail for everyone involved. Then each attorney gives an opening statement summarizing his theory of the case. This is followed by the presentation of evidence through witnesses. After all witnesses have testified, each attorney may present a closing argument. The judge then instructs the jury on points of law, and the jury is allowed to deliberate. When the jury deliberation is finished, the verdict is read to the court. In jury-waived trials, the judge renders the verdict. Frequently, the losing side of a trial will appeal the verdict to a higher court.

Conclusion

A course on professional engineering liability which is presently being offered has been described in detail. The need to include this type of course in the undergraduate civil engineering curriculum has been discussed. The course content and the text and references used have been described. The basic differences between this course and other traditional engineering courses is noted. Students are required to analyze case histories to see how legal principles are applied to individual failures.

References

Bockrath, J.T. (1986), "Contracts, Specifications and Law for Engineers," McGraw-Hill, New York, N.Y.

Corley, R.N., Holmes, E.N., and Robert, W.J. (1983), "Principles of Business Law, 12th ed., Prentice Hall, Inc., Englewood Cliffs, N.J.

Hart, G.C. (1982), "Uncertainty Analysis, Loads, and Safety in Structural Engineering," Prentice-Hall, Englewood Cliffs, N.J.

LePatner, B.B., and Johnson, S.W. (1982, "Structural and Foundation Failures," McGraw-Hill, New York, N.Y.

Pfrang, E.O. and Marshall, R. (1982), "Collapse of the Kansas City Hyatt Regency Walkways," Civil Engineering, 65-68, July, 1982.

Ross, S.S. (1982), "Construction Disasters: Design Failures, Causes, and Prevention," McGraw-Hill, New York, N.Y.

Streeter, H. (1988), "Professional Liability of Architects and Engineers," John Wiley & Sons, New York, N.Y.

Sweet, J. (1985), "Legal Aspects of Architecture, Engineering, and the Construction Process," West Publishing, St. Paul, MN.

Vaughn, R.C. (1977), "Legal Aspects of Engineering," Kendall Publishing, Dubuque, IA.

Wonder, D.R. (1988), "What Happened at L'Ambience Plaza,"Civil Engineering, 58(10), 68-71, October, 1988.

Zeitlin, L. and Hinckley, J.M. (1985), "Learning from Failure," Consulting Engineer, 64(6), 52-57, June, 1985.

Addressing Nontechnical Issues in C.E. Education

J. Richard Cheeks, P.E., M. ASCE[1]
Vincent P. Drnevich, P.E., F. ASCE[2]

Abstract

Presentation of professional, nontechnical issues to civil engineering students at the University of Kentucky is done with an introductory course, coverage as appropriate in technical courses, and with an exit course. Some topics are covered in all three which gives reinforcement and builds confidence.

Introduction

In today's litigious society, it is no longer adequate for an engineer's formal education to focus exclusively on technical issues. Rather, this technical training must be enhanced with an introduction to the nontechnical issues which impact the practice. Research over the last twenty years demonstrates that the vast majority of claims filed against engineers occur from failings in nontechnical areas such as communications; i.e., not saying what is meant, not meaning what is said, or not documenting conversations which three years later are the focal point of dispute. Misunderstandings about common law, and the impact of a single word in a contract also are common sources of claims.

---

1. President, Stokley-Cheeks and Associates, Inc., 132 MacArthur Ct., Nicholasville, KY 40356 and Adjunct Assistant Professor of Civil Engineering, University of Kentucky, Lexington, KY 40506-0046.
2. Professor of Civil Engineering and Acting Dean of Engineering, University of Kentucky, Lexington, KY 40506-0046.

In the early 1980's, the University of Kentucky's CE program recognized these needs and made the changes in the curriculum. The philosophy driving these changes was that the nontechnical issues should be addressed in more than one course, integrated into technical courses, and include practicing engineers in the classroom. The effectiveness of this effort has been enhanced by the use of field trips, role playing, and case history study.

Nontechnical Issues in C.E. Education

A list of nontechnical issues that need to be addressed during an undergraduate civil engineering education are given in Table 1. For each item in this table, an "X" in the adjoining columns indicates where in the curriculum at the University of Kentucky these topics are given attention. Some of the topics, such as written communication, are covered in general university courses explicitly designed for those topics. Their coverage in the courses listed in Table 1 then is focused on practice, application and reinforcing their coverage elsewhere.

Table 1. Matrix of Issues and Where Addressed in a Civil Engineering Program

Issues	Intro. Course	Tech. Courses	Exit Course
What is a Profession	X		
Professional Engineering	X		X
Oral Communication Skills	X	X	X
Written Communication Skills	X	X	X
Documentation of Activity	X	X	X
Codes of Conduct	X		X
Time Management	X		X
Goal Setting	X		X
Career Planning	X		X
Job Seeking			X
Professional Ethics	X	X	X
Management Skills	X		X
Interpersonal Relationships	X		X
Conflict Resolution		X	X
Professional Liability			X
Loss Prevention		X	X
Failures	X	X	X
Expert Witness Services			X
Continued Competency		X	X

The Introductory Course in Civil Engineering

At the University of Kentucky, this course is called *Introduction to Civil Engineering Systems*, is three credit hours, and is taught to entering freshmen. The stated objectives of this course are to:

- o Introduce the student to the profession of civil engineering;
- o Provide information on civil engineering through contact with other faculty, practicing engineers, professional and technical organizations;
- o Describe the curriculum and its purpose;
- o Begin developing skills with some of the methods, systems, and equipment used by the civil engineer; and
- o Provide opportunities to improve oral and written communication skills.

Significant use is made of personal computers for word processing and for calculations by use of spreadsheets. A personal computer with a projection device is available in the classroom. By the end of the semester, the students feel very comfortable in using the computer and tend to use it automatically for other courses where it may not be required.

Guest lectures by faculty and practitioners on the various disciplines are followed by homework assignments introducing a typical problem or design procedure of that discipline. At the following session of class, individual or groups of students present to the class their approach, successes, and difficulties in doing the assignment. The instructor and other students assist the presenter(s) with overcoming difficulties and the assignment then becomes due at the following meeting of the class.

Ethics is introduced by use of a questionnaire patterned after the one presented by Onsrud (1987) which places a young engineer on a construction site as an inspector and who is offered various items the by contractor which range from a soft drink to a new automobile.

Then the case studies introduced by Ende and Visilind (1985) are presented without comment. Additional real and hypothetical cases are presented, at least one of which focuses on improper student conduct (typically, cheating in some form). A link is forged between ethics in the classroom, ethics in engineering practice, and personal ethics using the material presented by Ende and Visilind (1985). Students are encouraged to discuss these ethical situations among themselves and with their friends and families in order to work out possible courses of action. Subsequent class sessions are devoted to evaluating courses of action in light of codes of conduct.

This section of the course concludes by reviewing the questionnaire presented at the start and examining what the ethically correct answers should have been. It is shown to students that subscribing to a system of ethics that has been codified is the essence of being a professional.

There is significant variety within this fast paced course. Much of the course requires the active participation of the students, occurring with a non-threating tone ("If you get stuck, the rest of us will help get you back on track.") Some sessions involve more formal presentations and others utilize role playing. The course is intensive and time consuming for both the instructor and students but is well received because of the learning, confidence, and skills it provides.

Use of Technical Courses

The technical courses in civil engineering at the University of Kentucky are similar to those found in most civil engineering departments. However, faculty are encouraged to utilized opportunities in these courses to enhance and reinforce nontechnical topics as they naturally occur. An example of this would be the discussion of failures in civil engineered facilities. Besides the technical components, it is natural to discuss the legal and professional aspects as well to provide the case in its whole context.

Faculty also are encouraged to have students enhance their oral and written communication skills. In several of the laboratory courses, oral presentation of results and their use for engineering design are required.

The Exit Course in Civil Engineering

The exit course is called *Seminar* and is one credit hour, meets once a week for two hours, and is taught to seniors during their last semester. It is built primarily around the text Introduction to Professional Practice developed by the ASFE Institute of Professional Practice (1988) and covers the following major issues:

- o Professionalism and Ethics
- o Professional Liability and Loss Prevention
- o Litigation and Alternate Dispute Resolution Procedures.

The format of the class includes group discussions, role playing, case history analysis, and the use of audio-visual teaching aids. The class also addresses many other related issues and concerns. The direction of these discussions are guided by the instructor and the interests of the students.

Professionalism and ethics discussions initially focus on the elements which distinguish a profession. Discussions are held on the duty of care owed by members of the profession to clients, other professionals, and the public. Codes of ethics, the guide for professional actions and interactions, are emphasized with three video tape presentations, a slide presentation on the Kansas City Hyatt disaster, and an essay assignment.

Professional liability and loss prevention techniques are presented with a series of six ASFE audio tape programs. These programs address the following subjects: Qualification Based Selection (QBS); Duty of Care; Liability Insurance; Construction Monitoring; Documentation; and Free Advice. Each audio program uses real case history examples to illustrate the topics. In addition, the students are given suggested exercises and quizzes developed specifically for each program.

Litigation and alternative dispute resolution (ADR) techniques are presented. The students learn about the litigation process, including the time and cost associated with discovery, various pleadings, trials, and appeals. They also are exposed to a variety of ADR techniques that are available, with a discussion of the benefits and shortcomings of each. This portion of the course culminates in two major events. The first is a team exercise where students engage in role playing to resolve a construction dispute. Teams of architects, geotechnical engineers, and contractors, on advise of their lawyers (also played by students) must resolve the matter, or face certain litigation where none of the groups can win financially.

The second event consists of showing the class a series of mock trials on video and for each, the class is required to serve as judge or jury before hearing the outcome. This portion of the course concludes by examining the role of the expert witness in the dispute resolution process.

Summary and Conclusions

The education of a professional engineer must include a number of topics which are essential to the engineer's performance as a professional. In this paper, the authors share their thoughts and experiences in delivering nontechnical, profession related topics to the students at the University of Kentucky. The approach is to utilize an introductory course and an exit course for the direct coverage of the topics with reinforcement of the topics as appropriate in the technical courses of the curriculum.

The approach used has been successful. Civil engineering graduates begin their professional careers with an understanding that their success is affected by nontechnical elements of practice. This knowledge is an important complement to the technical training and is necessary if students are to be successful practitioners in the litigious society in which they will work.

Acknowledgments

The authors wish to acknowledge their colleagues and students who over the years have helped improve the coverage of these topics by their active participation and by sharing their experiences.

References

ASFE Institute of Professional Practice (1988), Introduction to Professional Practice, Draft No. 2, January, 770 pages.

Ende, E.M., and Vesilind, P.A. (1985), "Ethics in the Field," Civil Engineering, Vol. 55, No. 12, December, pp. 64-66.

Onsrud, H.J. (1987), "Approaches in Teaching Engineering Ethics," Civil Engineering Education, ASEE, Vol. IX, No. 2, Fall, pp. 13-27.

Preparing Undergraduates for Professional Practice

Michael J. Moreau,[1] Member, ASCE

Abstract

The paper documents what a sample of design professionals consider are challenges to current professional practice. The objective of the paper is to document and demonstrate the need to familiarize students with positive responses to "real world" professional practice situations.

I. Purpose and Research Methods

The purpose of my research project was threefold:

* To determine what architects and engineers believe are the most significant challenges to professional practice.
* To determine what these professionals believe they can do about these challenges.
* To determine what these professionals are presently doing to meet these challenges.

Data for the research was collected through 28 personal interviews. Design firm principals were targeted for the research. The author sampled a cross section of the consulting industry in the area (southern Maine) by interviewing electrical, mechanical, and civil (geotechnical, structural, environmental, transportation) engineers and architects. Their combined average age and years of experience are 43 years old and 19 years experience, respectively.

II. Discussion of Most Significant Challenges

The three prominent challenges to professional practice in order of frequency of mention are:

* Maintaining Quality
* Liability Issues
* Policing the Profession

----

[1]Associate, Geotechnical Engr., Robert G. Gerber, Inc., 17 West Street, Freeport, ME. 04032

Maintaining Quality

Many factors influence quality in todays design and construction environment. Fiscal and scheduling restraints are the mechanisms most often mentioned. Fiscal and scheduling restraints limit the time design professionals spend designing, creating, and innovating. In response to these constraints, design professionals may perform little or no coordinated plan/specification reviews. They will have a tendency to "over-design" to eliminate costly analysis. They often will not explore alternate designs.

Other factors affecting quality are competitive bidding and work performed by unregistered service groups. Competitive bidding does a disservice to both the owner and the professional. The owner believes that all consultants are the same and that each respective design product is of equal quality. This is not so because each situation forces the design professional to do one or more of the items I have described in the preceding paragraph.

Unregistered individuals or service groups degrade design standards when they perform work normally requiring engineering talent of a registered professional. One interviewee cited a situation where a manufacturers' representative provided a detailed electrical device design for a nominal design fee. The actual cost of design was probably buried in the cost of the device. Other side effects are reduced quality of the design/construction product and increased liability exposure.

Finally, today's designs, particularly building designs, are very complex. Design review pressured by scheduling restraints may often be less than desirable. The situation is exacerbated by field changes or conditions and their necessary, but lagging, review.

Liability Issues

The second-most often cited challenge to professional practice are liability issues. Professionals confront liability issues on two fronts. First, architects and engineers must deal with the present condition of our litigious society. For example, all design team members may become parties in a frivolous and/or joint and several liability suit when the air conditioner fails.

Second, our present legal system does not assign responsibility for damages equitably. Even though 90 percent of cases don't go to court, the threat of litigation forces professionals to settle or spend time and money in their defense.

Our present day litigious society encourages large suits. This intimidates design professionals. They may often be unwilling to execute an innovative design or incorporate a new construction technique for fear of performing outside the standard of practice. This leads to more expensive, conservative design.

Designers must spend more of their time dealing with liability issues. Architects and engineers must be extremely careful to qualify what they say or design. They must evaluate projects to find anything with the slightest chance of going wrong. They must prepare and review detailed 30 page contracts vs. two page outline scopes used in the past. This leaves less time for design work, creativity, innovation and increases the cost of providing a service. Time spent preparing for court cases is of course, non-productive.

The sad result is that professionals may not be innovative and may not use new tools that become available to the profession. Innovative designs or the use of new techniques involves greater risk because the design has not been tried before. The end result is uncertain. Some clients may not understand or accept these greater risks. Thus, the design professional will tend to lean toward proven methods. There will always be experts who will define the standard of practice and testify that the new design is outside that standard.

Policing the Profession

The third-most prominent reported challenge to professional practice is policing the profession. Present market conditions offer opportunities for design/build contractors, unregistered service groups, firms, and vendors. They provide work products or services, sometimes illegally, which normally require engineering talent from registered professionals.

Enforcement of existing registration laws is lacking. Inadequately funded and staffed registration boards cannot consistently bring the strength of the law to bear. The design profession is as much or more at fault as is the legislature or the general public for the present state of registration boards. Attorneys General must deal with what they perceive as more serious crimes, not recognizing professional registration infractions as a hazard to public safety.

III. Responding to the "Real World" Challenges

The following text summarizes the design professional's actual and suggested responses to these "real world" challenges. Herein lies the body of knowledge which may be presented in an undergraduate course to prepare students for professional practice.

Maintaining Quality

Public safety relies on professional design. True professionals realize this and, therefore, attempt to maintain quality of the design/construction product in several ways. Architects and engineers carefully select competitive bid projects. Some A/E's avoid market sectors where competitive bidding is the process used. Professionals often told me that they refuse to respond to competitive bid projects. Some individuals use competitive bid solicitations as an opportunity to educate the owner/client about the inappropriateness of the process.

Design professionals who are concerned about maintaining quality are continually educating clients, regulators and, to a lesser degree, the general public. They explain the benefits of qualifications based selection of A/E services and the hazards and limitations of fee based A/E services selection. Some firms make formal commitments to inform lawyers, regulators, public works departments, etc., by sponsoring seminars. These actions will foster more responsible A/E services selection in the future.

A design professional's input is vital during all phases of work for which they are either responsible or in control. Design professionals try to inform potential clients about this by indicating what a project really entails and the benefits of a professional product/service. In some instances, they walk the client through the process to show them the necessary steps.

Greater efforts by design professionals to lobby for, and pass, Brooks Law-style legislation would benefit professionals who work on public projects. Design professionals working in the private sector would benefit from increased use of mutual work scope development.

Other ideas recommended by interview candidates to improve quality in the designed/constructed product are summarized below:

* tax incentives to build value into a building
* increase use of CADD- they produce exactness in a short time
* peer pressure- professionals should use peer review to help maintain the standard

* restrict supply of A/E's- prequalify individuals by education and experience
* seek a quality client base
* market the firm's reputation

Liability Issues

Architects and engineers are dealing with our litigious society through educating clients and, to a limited extent, by supporting tort reform. A/E's attempt to educate potential clients about liability exposure, contract law and review, as well as technical aspects of a project. These design professionals carefully word contracts with owners/clients to eliminate unreasonable liability and to limit reasonable liability. They often have these contracts reviewed by legal counsel. One firm has gone so far as to hire permanent legal counsel staff.

Professionals draft well defined scopes of work to spell out what they will do and what they won't do. They then record scope changes after the project begins. They educate their staff to understand what you do and don't do at a construction site. Unfortunately, some clients expect a design professional's work to be perfect. In these instances, A/E's attempt to inform the owner/client that construction is an imperfect science. Clients also often do not understand that every situation/site is different.

Many professionals felt that tort reform is best handled by professional organizations. These individuals possibly felt that professional organizations had more time or resources to address the situation. Only one interview candidate actually contacted legislative officials to sponsor tort reform. One firm is committed to tort reform through correspondence with, and participation in, professional organization activities.

Design professionals must increase their tort reform efforts. More professionals must lobby their federal and local legislative representatives. Tort reform is the only vehicle to establishing equality of responsibility by eliminating unreasonable liability and limiting reasonable liability.

New legislation to eliminate frivolous and joint and several liability suits is necessary. Legislators must control or limit allowable lawyer contingency fees. Limiting amounts paid in certain kinds of suits is another method of establishing equality of responsibility. Brooks Law-style legislation is necessary to eliminate competitive bidding as a viable A/E selection method by government entities. A more aggressive approach to enforcement of professional registration laws is necessary.

Other ideas suggested by A/E's to combat our present litigious climate are:

* avoid market sectors, e.g. avoid design/build projects where strict liability laws apply
* participate in peer review programs and stress in-house review to tighten quality control

Policing the Profession

Professional registration boards are reportedly ineffective in policing the profession. One interview candidate cited a situation where he reported obvious registration law infractions to the state board. The state district attorney reportedly did not prosecute because of the need to deal with "more serious crimes." Law enforcement officials presently do not see professional registration infractions as a threat to public safety as A/E's do.

Enforcement of existing registration laws and/or defining professional responsibilities through legislation would improve the situation a great deal. Convincing Attorneys General of the need to prosecute professional registration infractions would also help.

The design community must work more closely with the boards. Design professionals should use their registration boards as a resource. For example, one state requires collection of quite a bit of information when filing a complaint. Increasing professional registration board funding and staffing would improve follow-up of reported professional standards infractions and require less time on the part of the complainant. Another state publishes a quarterly newsletter. The newsletter contains names of individuals whose licenses have been revoked or re-instated and other professional registration news. This maintains awareness in the design community.

Active enforcement of registration laws and/or drafting effective new laws would restrict unregistered service groups or individuals from performing services which normally require registered professionals. Degradation of design standards, price competition, diluted quality, and increased liability exposure often result from the work performed by these groups.

One design professional suggested registration of subdisciplines. This would better define one's area of expertise. Some professionals attempt to educate clients and regulators to understand that qualified, registered professionals should perform consulting work. Some A/E's indicate the types of work which registered professionals should perform. Another individual suggested putting building inspectors in the line of responsibility for building problems.

In summary, design professionals do very little to address this challenge. Most individuals rely only on their own sense of ideals or code of ethics and standards.

IV. Conclusion

The most crucial challenge to professional practice is the continued education of clients, regulatory agencies, and the general public regarding the importance of professional A/E services. The public health and safety is at risk. The profession must show these people that procuring A/E services is more than providing a product for a price. Qualifications based A/E selection imparts a professional background and credentials which overcome the hazards and limitations of fee based A/E selection.

Design professionals must increase their tort reform efforts. The profession must help nurture legislation to eliminate frivolous and joint and several liability suits. Lawyer contingency fees and settlement limits for certain types of suits should be controlled through responsible legislation. Enactment of Brooks Law-style legislation will eliminate competitive bidding by government entities. Aggressive enforcement of professional registration laws is needed.

Finally, we as design professionals have allowed the evolution of the profession to its present state. We must strive to maintain the ideals of the profession. If we are to improve the character of the profession, we must recognize our responsibilities. We must continue to educate the general public about the importance of professional A/E services. We must increase our tort reform efforts. We must adhere to the standards of the profession and be willing to report violations of those standards. The public's image of the design profession directly emulates our will to address these responsibilities. If we become unconcerned about, or unwilling to administer our professional responsibilities, our profession will, in time, diminish to a trade.

ENGINEERS AND NEW TECHNOLOGY: LEGAL CONSIDERATIONS

E. Mabry Rogers,*
J. David Pugh,** M.ASCE

ABSTRACT

Society benefits when a new technology enables us to produce better, stronger, or safer structures in less time, or for less cost. Research developments, the use of high-speed computers, and advances in electronic communication have resulted in rapid advances in technology, some of them in the design and construction industry. Society's expectations regarding this progress have created a dilemma for design professionals who must recognize the potential liability risks that arise (a) when a design advance, building method or construction material that is the product of a new or rapidly evolving technology is specified and, alternatively, (b) when such new technologies are not utilized. Not only must design professionals recognize the potential liability risks, they must be aware of methods of alleviating these risks by considering: (1) the standard of care expected of the design professional; (2) the potential liability risks facing the design professional; and, (3) means of insulating the design professional from such risks. It is important that today's engineering students be made aware of this dilemma so that they may avoid it when they become tomorrow's practitioners.

INTRODUCTION--THE APPEAL OF "STATE OF THE ART"

The old adage "build a better mousetrap and the world will beat a path to your door" is no less

*Attorney and Partner in the law firm of Bradley, Arant, Rose & White, 1400 Park Place Tower, Birmingham, Alabama 35203

**Attorney and Associate in the law firm of Bradley, Arant, Rose & White, 1400 Park Place Tower, Birmingham, Alabama 35203.

applicable to a design engineer than it is to a manufacturer. Design engineers often find it more difficult to distinguish themselves from the competition than the manufacturer or builder who offers a tangible product for sale to his customers. Design firms must constantly strive to offer their clients state-of-the-art services for competitive prices to avoid losing clients to the competition. Competitive pressures often create a race among design firms to see who will be the first to complete a design which incorporates the newest, most innovative construction material or building method, particularly if the new technology translates into cost savings. Winning such a race can result in new business and publicity in trade journals and other publications. Unfortunately, the use of innovative technology can be as risky as it is rewarding. While potentially offering great rewards, caution must be used to avoid the risks inherent in a pioneering use of state-of-the-art technology.

Any designer intends that his designed system or structure will function as designed. Problems arise for the designer when anticipated or promised results are not realized. For example: Is the designer liable to the owner for damages if the HVAC system can not be operated cost-effectively because new, lightweight, synthetic fiber insulation specified by the HVAC engineer fails to perform as promised by its manufacturer? What if the contractor's expedited slab-pouring schedule is dependent on the slabs attaining the design 28-day strength much earlier through the use of an untried, exotic chemical admixture which the structural engineer specified? What if the designer specifies a new building method that results in a structural failure causing injury or death? Before questions like these can be answered, one must understand how the engineer's duty to perform arises and the standard by which his performance will be evaluated in the event of a claim.

<u>THE ENGINEER'S DUTY AND STANDARD OF CARE</u>

An engineer's duty to perform is defined by a written contract executed by the engineer and the individual or entity who is paying for the engineer's services, usually the "Owner" of the designed system or structure. The contract expresses what is to be accomplished by the engineer. Failure to perform work according to the contract is a breach of the contract resulting, potentially, in a claim by the Owner against the engineer. The issue is simply: Did the contract require the performance of "X" and did the engineer do "X"?

An engineer also has the duty -- frequently unwritten -- to perform his services without negligence. A claim against an engineer alleging negligence is usually based "in tort" and the issue is: Did the engineer perform his design services consistent with the standard of care imposed upon an engineer? Whereas the engineer's duties as written in the contract usually extend only to the other party to the contract, the engineer may owe a duty to perform without negligence to third parties such as other contractors or members of the public.

The engineer's conduct during performance of his contract must reach or exceed the applicable standard of care, or he may be found either in breach or negligent. Most jurisdictions apply the same standard to engineers (often grouped with architects and referred to generically as "design professionals") as is applied to other "professionals," or those who offer highly-skilled services for a fee, such as accountants, attorneys, or physicians. In a recent opinion of the Supreme Court of Alaska, the court described the standard of care for engineers as follows:

> In performing professional services for a client, an engineer has the duty to have that degree of learning and skill ordinarily possessed by reputable engineers, practicing in the same or a similar locality and under similar circumstances.
>
> It is his further duty to use the care and skill ordinarily used in like cases by reputable members of his profession . . .

<u>Clark v. City of Seward</u>, 659 P.2d 1227, 1230 (Alaska 1983).

There are three separate components of this standard of care. First, the engineer is not expected to be perfect, he is merely expected to exercise the level of skill ordinarily possessed by his peers. Second, the engineer is only expected to exercise that level of skill common in his specific locality. The traditional rule has been that an engineer's conduct will be compared against other engineers in the same geographic locality. As engineering education has become more nationally uniform, as communication capabilities have been increased, and as more uniform design standards are published, the "locality" limitation has been expanded beyond the confines of an engineer's particular community. Often, a "national" standard is applied by courts, particularly if the project is fairly conventional or if the project is for the federal government. The expansion of the scope of the relevant locality makes it imperative that design professionals stay abreast of the latest developments in their respective specialties on a national, and even international, basis.

The third and final component of the standard of care is the "similar circumstances" limitation. This component limits the scope of the standard to other projects of a similar nature. Thus, the level of care expected for the design of a nuclear reactor containment building may differ from that for the design of a parking lot.

THE STANDARD OF CARE APPLIED - CODES AND MORE

Obviously, if liability will be determined after the fact by comparing the designer's conduct against a certain standard of care, then the designer can minimize risks by conforming his conduct to that standard. The engineer is expected to "possess that degree of learning and skill ordinarily possessed by reputable engineers." The designer must realize that if he is specifying a new material, others in the locality will naturally possess very little, if any, learning and skill as to how to use the product effectively and safely. Thus, the designer will likely be expected to perform testing or research sufficient to determine the safe uses of the product. The designer should research carefully all the available data relevant to the particular technology concerned. Depending on the potential dangers of using a particular technology, the search should be expanded well beyond the limits of the community, state or even the region in which the engineer practices. Consider, for example, a new chemical concrete additive or a new welding alloy. There may be no other engineer in the state who knows anything about the use of the new product, but, it is likely that reports exist that describe the product's development. Such reports should be obtained and reviewed if at all possible. The reports may reveal that the welding alloy is virtually useless for welding certain metals or that the admixture's performance is dependent on maintaining the temperature of the concrete during placement well below temperatures commonly suggested by the American Concrete Institute code.

Code compliance is a critical issue. The engineer has a duty to comply with applicable design and building codes. St. Joseph Hospital v. Corbetta Construction Co., Inc., 316 N.E.2d 51, 63 (Ill. App. 1974). Failure to comply with building codes usually constitutes negligence per se, which enables a plaintiff to create a rebuttable presumption that the defendant was negligent by showing that (1) a code or statute was violated, (2) the code was promulgated to protect a class of people including the plaintiff, (3) the injury or defect was of a type contemplated by the code and (4) the code violation proximately caused the injury. See Burran v. Dambold, 422 F.2d 133 (10th Cir. 1970). This doctrine greatly reduces the plaintiff's burden in that he does not have to prove the existence of a duty and a subsequent breach of that duty. The engineer's negligence is summarily established, and the lawsuit turns to the remaining issue -- whether the engineer's negligence caused the specific injury or defect.

With new technology, the question may be: How does the code apply if the new technology being considered for use by the designer would make the building code irrelevant? For example, consider the 311 Wacker Drive Tower in Chicago, Illinois, which is the world's tallest all-reinforced concrete structure. During the design phase, the structural engineers for the project, Brockette/Davis/Drake, Inc., determined that applicable codes did not address the heavy stresses which certain concrete structural members in the Tower would develop, because existing codes did not anticipate the use of 12,000 psi concrete mixes made possible by the use of chemical admixtures. As a result, the designers virtually had to develop their own codes and seek approval from the relevant regulatory entities after extensive wind-tunnel testing on scale models of the Tower and surrounding structures and testing of various concrete mixes in the lab and in the field. "All concrete design fits bill for 946-ft. tower," ENR, Cover Story, pps. 42-47, Feb. 4th, 1988.

On the other hand, code compliance alone may not be enough to satisfy the designer's duty because of the lag-time between new developments in the profession and their subsequent incorporation into standardized building and design codes. The sufficiency of a design will be evaluated in accordance with the standards of the profession at the time the design was completed. Plant v. R.L. Reid, Inc., 365 So. 2d 305 (Ala. 1978). If the codes do not reflect the current standards of the profession, then more than mere code compliance will be required. Consequently, engineers have a duty to stay abreast of the current advances in the engineering profession, including those advances which are not reflected in codes, if they wish to avoid liability for not taking technological advances into account. J.T. Banner v. Town of Dayton, 474 P.2d 300 (Wyo. 1970). This rule is not a new one; in a leading case, a ship owner was held liable for not purchasing ship radios at a time when it was proved that the shipping industry did not generally require the installation of such receiving sets. The T. J. Hooper, 60 F.2d 737 (2nd Cir. 1932). See also, Helling v. Carey, 519 P.2d 981 (Wash. 1974) (ophthalmologist held negligent for not giving glaucoma test to patients under 40 even though test not considered necessary according to then prevailing standards).

The advent of "canned" design software presents the dilemma of new technology in a clear light. The use of such software can decrease the chance of making design errors because it can minimize the "human error" element. Because the software is so accurate and because its use is so prevalent in the profession today, the failure to use such software could be deemed negligent in an ensuing lawsuit. On the other hand, one of the dangers of using "canned" software is failing to understand the basic theories that serve as premises for the computer program. If the engineer does not understand the program, there is a great danger that the program will be misapplied. Consequently, the engineer could face liability for not using the new technology and, simultaneously, could face liability if he adopts the new technology without fully understanding its use and limitations.

THE CONTRACT AND THE STANDARD OF CARE

Perhaps the most important single way to insulate against the risks of using new technology is fair and candid disclosure, first to the Owner and then, if adopted, to the contract team, of (a) the fact that a new technology is being used and (b) the limits of the deisgner's knowledge about the technology. With adequate disclosure, it is only fair that the Owner bears the risk of using the new technology, because he will reap the economic rewards from the commercial use of the finished product. Disclosure can be enhanced with carefully drafted contract language which describes the use and limitations of the new technology.

Additionally, an engineer should learn to recognize and avoid signing contracts containing language that attempts to establish a higher standard of care than that which is ordinarily imposed. Avoid contracts with language such as ". . .will use his best efforts. . ." or ". . .will use the utmost care. . ." Remember, the negligence standard requires one to exercise no more than that ordinary level of care exhibited by the average professional in the community under similar circumstances. While one may strive for more, one should not contractually agree to do more. Arkansas Rice Growers v. Alchemy Industries, Inc., 797 F.2d 565 (8th Cir. 1986).

Another method of minimizing risks is seeking indemnity, that is, contractually requiring the other party to the contract to defend and pay damages in the event of a lawsuit. Some jurisdictions will even enforce contract provisions that indemnify the engineer for his sole negligence, but courts are reluctant to enforce such provisions. Therefore, to be fully effective, the indemnification provision should be carefully drafted. If part of the design work is subcontracted, or if, for instance in a design-build contract, the engineer also procures materials and equipment directly, the engineer should attempt to negotiate an indemnification agreement with his subcontractors and suppliers. Unfortunately, in some jurisdictions there are cases holding that the engineer's duty to independently exercise his professional skill and judgment is a non-delegable duty, regardless of contrary contractual provisions. Mayor and City Council of Columbus, Mississippi v. Clark-Dietz, Inc., 550 F. Supp 610 (N.D. Miss. 1982).

Another potential risk minimizer is the inclusion of a limitation of liability in the contract. Such a clause can be used to place a cap on total damages of any kind or simply to limit the kinds of liability or circumstances under which a party will be liable. A contractual liability limitation is only enforceable against the other parties to the contract, however.

CONCLUSION

New technology, like fire, can provide a great deal of comfort and security or it can burn you. The designer must stay abreast of recent developments in order to determine when the use of a new material or method is premature and when it may be appropriate. The keys to proper use are adequate understanding of the new technology, adequate research and testing efforts to determine performance limitations and code compliance, complete and fair disclosure of the intended use of the new technology, and the use of risk-limiting language in the contract.

Professionalism, Ethics & Image of Civil Engineers

By Enno Koehn,* F. ASCE and
Ricky J. Bourque**, M. ASCE

Abstract

Engineers, as reported by numerous surveys, are considered by the citizenry, to be honest, ethical and to exhibit a great deal of integrity. However, since the 1970's there appears to be increasing concern involving the public's image of the engineer as a professional.

This paper summarizes observations, over a period of years, by the authors concerning some of the problems which may contribute to the perceived "negative public image of engineers." Among the areas of concern are public communications, the attitude of the engineer toward the services he provides and the profession, and the education of engineering students.

In addition to noting the problems, the paper places a perspective on the public image issue. Questions raised include: What is the image problem? Does the image need to be improved? And what can be done and what benefits might be gained through raising the professional image of the engineer?

Introduction

Engineers, in general, and civil engineers in particular, because of their association with construction and/or the public sector, have long been the target of angry citizens expressing their frustration regarding changes in the environment. Tending to perceive themselves as the builder and maintainer of

*Professor and Chair, Department of Civil Engineering, Lamar University, P. O. Box 10024, Beaumont, TX 77710.
**Department Manager, Southwestern Laboratories, 1850 IH10, Beaumont, TX 77707.

housing, highways, and other components of modern life, civil engineers have naturally become upset when denounced as being indifferent to the wishes of the general population (Conner 1976). Nevertheless, the public often views civil engineers as being committed to building as a way of life and as being insensitive to other aspects of development. For example, citizen information programs generally tend to be regarded as justification, rather than educational in nature, and low attendance and suspicion often is the net result (Willeke 1974).

It has been observed that engineers due, in part, to their apparent narrow background often lack a broad view and generally experience difficulty dealing with the public (Godfrey 1979). The American Society of Civil Engineers (ASCE) and numerous professionals have recognized the communication problem between the engineer and the general citizenry. In fact, it is often suggested that the engineer must become better aware of society's needs in addition to understanding that their purpose, as a professional, is to serve society. In general, the professional image of the engineer is portrayed in the attitude of the engineer toward his work. This image must begin to be nurtured and developed during his years of formal education.

The paper explores, in particular, the areas of communication, work attitude, and education as they relate to the enhancement of the professional status, image, and prestige of civil engineers.

Communication

Public relations is vital in civil engineering. Today, engineers can no longer assume that the public will automatically accept and admire the results of their projects. Due to the anti-technological attitude of many individuals, the profession needs to become more involved in the communication process (Klotz 1978). It has been recommended that, in order to reduce conflicts, engineers must learn to speak in a language the people can understand and to which they will respond. They must also begin to establish a working relationship with the press and learn how people think and feel. Both the civil and construction engineer should accept the primary role in the information flow process and must not be satisfied until a true exchange of ideas, comments, and viewpoints between the professional and the general populus takes place (Conner 1976).

Good communications between engineers and the general

population can help bridge the traditional gap between technically oriented and non-technically oriented individuals. Reducing the possibility of conflict between the two groups may tend to enhance the status of the civil engineering profession. It is noteworthy, however, that consultants generally perceive that the communication skills of newly hired graduates is deteriorating. In fact, at a recent Education Forum held in conjunction with a Texas ASCE Section Meeting, improved communications (oral, written and graphical) was considered to be of utmost importance.

It is unfortunate that speakers at engineering banquets such as those held during Engineer's Week or professional conventions are often not engineers. Is there a scarcity of engineers that have the background and broad view of the profession as well as the necessary communications skills to be able to present an interesting and entertaining talk? This may indeed be the case and represents an example of the communications limitations of the profession. It has been written, however, (Hughes 1985) that since 1945 inventors and invention have been pushed by society into the shadows. "The individual inventor (engineer) is often patronized as eccentric, even comical." This image may explain, in part, the apparent lack of interest in using engineers as banquet and/or keynote speakers.

Engineering Attitude

In an article published in Civil Engineering, it was reported that some individuals perceive that the engineer is just a technical specialist, not a professional (Godfrey 1979). Nevertheless, as an engineer, it is not enough to "know how to do the right thing" but also to determine "the right thing to do." Merely number crunching with little or no opportunity to correct design errors is not engineering. In fact, this is an activity in which a computer can excel. The essence of engineering, therefore, is not only analysis and design but the application of ethics and professional judgement.

A civil engineer, contractor, author, and concerned professional, Samuel Florman, has written that engineers in industry often act not as university graduates but as if they were high school graduates that have received some additional technical training. They do not appear to have a broad view of the implications of their work. Florman insists that the humanities inspire and expand the technical imagination and, therefore, must be increased in engineering programs (Florman 1987). He notes that the engineer's creative thought processes would acquire flexibility and variety as a result of being exposed to art, music, poetry, stories, and myths.

Nevertheless, today, an engineering position is considered by many engineers and non-engineers to be basically a job. This is at variance with a professional attitude and image. In other regions of the world, however, engineers accept a great deal of responsibility for their work (Latvian Bridge Fails Test 1986). Consequently they also enjoy a relatively high standard of living and considerable respect from the society in which they reside.

Due to the broad scope of engineering work, and the practice of using non-registered engineers to perform the same technical tasks as registered engineers, it may be difficult to elevate the engineering position above the "just another job" level. To elevate the engineering position, the profession may have to follow paths involving formal education and stricter registration similar to that of other professions, such as medicine and law.

Unfortunately, there is also a recent trend in the United States to place non-engineers rather than engineers in positions of leadership (Guy 1987). In addition, there is a tendency for American engineers to abandon their technical "roots" when they become part of senior management. Senior managers often lose an active interest in and appreciation for the current knowledge and state-of-the-art of their field (Halpin 1987). This development may foster a lag in the growth of professional attitudes and may also inhibit the formation of a positive image of engineers.

Education

It has been written that modern education is generally limited to footnoting or to sharpening a student's critical and analytical ability. It, therefore, often does not provide an individual with the knowledge and synthesizing capacity essential for coping with a technological society (Bugliarello and Hockney 1985). This omission may be creating an image problem for the civil engineer. For example, a well known state legislator recently remarked that "engineers are only concerned with nuts and bolts." This comment was in regard to a possible conflict in schedule between a state water planning conference and an ASCE Section Meeting. The legislator indicated that engineers are generally not concerned with broad issues and are usually not the movers and shakers in a community. He strongly suggested that they were not the group that should dominate and/or participate in, if it were to be successful, a state water planning conference.

This perception of engineers is probably held, in varying degrees, by numerous legislators and other individuals throughout the country. It most likely affects, in a detrimental manner, directly or indirectly the image of the civil engineer and the overall engineering profession.

A noted historian remarked in a television course for college credit that "engineers are methodical in nature." This statement was made in a discussion concerning President Herbert Hoover. The historian also suggested that, Hoover, an engineer, viewed society, as most engineers do, in terms of a blueprint. Engineers were perceived, therefore, to have limited vision. These perceptions are at variance with the expectations of any profession. They are, however, probably held by numerous if not a majority of the individuals in the nation. Certainly it is difficult to refute these ideas when they are presented in a classroom setting on a television screen.

Recently, concern has developed involving the exposure of engineering students to the topics of ethics and professionalism. In response to this concern, the Student Activities Committee of the Texas Section/ASCE polled the contact members assigned to the 12 State ASCE student chapters to determine whether practicing engineers perceived the topics as important. It is noteworthy that 100% of the respondents indicated that there was a need to include the topics of ethics and professionalism in the engineering curriculum. There also was a unanimous response that the local ASCE branches could assist in these efforts.

The poll also attempted to ascertain whether there was a consensus as to how the information should be presented and the level of support for such a program. The findings related to method of presentation were mixed. Fifty percent of the respondents indicated that a separate course should be devoted to the topics, 25% suggested the topics be incorporated into existing courses while the remaining 25% indicated the topics should be presented outside the classroom through such medium as an ASCE Student Chapter meeting.

The data appears to indicate strong support for exposing students, at the university level, to ethics and professionalism. Surprisingly, only 17% of the respondents indicated that they would be willing to serve on an ethics and professionalism committee. This group's task was envisioned to consist of conducting a survey of available information on ethics and professionalism and assembling material which may be presented to civil engineering students. The apparent low interest of practicing engineers to become involved may be an indication that, due to lack of time, lack of support of an employer or other reasons, the engineer does not feel that, as a professional, he/she is responsible, at least in some part, for the education of future engineers.

Summary and Conclusions

Engineering and construction projects of all types and sizes have political, economic and sociological implications. In fact, large projects often have a great and/or complex influence on a city or region and may basically change the way of life in a community (Koehn). Engineers, especially civil engineers, are usually involved with these projects. Unfortunately, they often are perceived as not being concerned with and aware, as a professional, of the short and/or long term effects of the activities with which they are associated. This may suggest a severe professional image problem.

In general, engineers are not looked upon by the public as leaders of society. This may be attributable, in part, to a lack of communication and public relation skills, attitude of the engineering employee toward his profession, and the difficulty educational systems encounter preparing students for work as a professional.

There is much that can be done to correct the areas, as defined in this paper, which tend to suppress the public image of the engineer as a professional. However, before initiating a program to deal with the problems, the profession needs to define what level of professionalism should be attained. For example, while engineers in other parts of the world enjoy a relatively high standard of living and considerable respect from the society in which they reside, they also accept a great deal of responsibility for their work.

Financial benefits would most likely occur if the image of the American engineer is increased to the level of other learned professions. There would also probably be a greater number of top students enrolling in the field of engineering. With the benefits, however, responsibility must also be increased. Among the responsibilities to be incurred will be an increase in the formal education requirements, both at the university and continuing education level. Engineers will also need to become more active in a public leadership role and practicing engineers must become actively involved in the education of the engineering student.

Today, there is a great deal of interest in the engineering profession involving the public image issue. However, for any change to be successful it will take the active participation of not just a select group but all concerned professionals.

Appendix-References

Bugliarello, G., & Hockney, D. (1985). "A Liberal Arts Curriculum for Age of Tech." The Bent, 76(4), 50-52.

Conner, D. M. (1976). "Citizen Inputs to Public Works Projects." Issues in Engrg. - J. Prof. Activ., ASCE, 102(EI1), 29-39.

Florman, S. C. (1987). The Civilized Engineer. St. Martin's Press, New York, N.Y.

Godfrey, K. A., Jr. (1979). "Education in Civil Engineering: Boost Professional Orientation?" Civil Engineering, ASCE, 49(8), 64-69.

Guy, L. L, Jr. (1987). "Educating Engineers for the 21st Century." Engineering Times, NSPE, 9(1), 6-7.

Halpin, D. W., & Paulson, B. C., Jr., et al. (1987). "Constr. Education and Research in Central Europe." J. Constr. Engrg. and Mgmt., ASCE, 113(3), 468-479.

Hughes, T. P. (1985). "How did the Heroic Inventors do it?" American Heritage of Invention and Technology, American Heritage Press, Inc., 1(2), 18-26.

Koehn, E. & Winkleman, L. (1981). "Public Reaction to a Project." J. Constr. Div., ASCE, 107(CO2), 209-217.

Klotz, B. (1978). "Impact of Education & Communication on Productivity." Issues in Engrg.-J. Prof. Activ., ASCE, 104(EI1), 35-41.

"Latvian Bridge Fails Test." (1986). Engineering News-Record, 217(17), 13.

Salvadori, M. G. (1974). "Making Engineering Curricula More Relevant." Civil Engineering, ASCE, 44(6), 70-73.

Willeke, G. E. (1974). "Citizen Participation: Here to Stay." Civil Engineering, ASCE, 44(1), 78-82.

A Course in Engineering Professionalism and Ethics

Joseph R. Herkert[1]
and B. Vincent Viscomi[2], M. ASCE

Abstract

This paper describes the development of a new course, Engineering Professionalism and Ethics, required of all engineering majors at Lafayette College beginning with the Fall 1989 semester. The course is designed to provide an introduction to the role of the engineer in society, time value of money concepts and engineering economic decisions, engineering ethics, safety and risk assessment, the professional responsibilities of engineers and the legal aspects of engineering.

Introduction

A number of educators have recently called for increased emphasis in the engineering curriculum upon the relationship between technology and social, political, economic, ethical and legal concerns. Wenk (1988), for example, has suggested a radical restructuring of the curriculum in order to provide engineering students with a focus on the skills necessary to manage technology in contemporary society. Vanderburg (1987) has called for the development of a "complementary science," revolving around the interaction between technology and society, to augment traditional engineering approaches. Karbhari (1989), has promoted the establishment of an integrated humanities and social sciences core for engineering students, including courses that deal with the history and management of science and technology.

----

[1]Assistant Professor of Civil Engineering and Dana Faculty Fellow, Lafayette College, Easton, PA 18042.
[2]Professor of Civil Engineering, Lafayette College, Easton, PA 18042.

Despite their different points of emphasis, these and other observers (Andersen 1986; MacVicar 1987; Olds and Miller 1987) agree that the typical practice of selection of elective courses in the humanities and social sciences does not adequately facilitate the making of connections between engineering and other modes of inquiry. Even when non-engineering courses dealing explicitly with the connections between technology and society are offered on an elective basis, engineering students are not always willing or able to enroll in them (Herkert 1989). A solution that does not rely on humanities and social science courses for emphasizing the relationships between technology and society is the development of engineering courses which incorporate the study of these relationships (Andersen 1986; Olds and Miller 1987).

Course Background

An opportunity to develop such a course recently arose at Lafayette College, a private, undergraduate institution with programs of study in engineering, the natural and social sciences, the humanities, and the arts. Roughly twenty-two percent of the College's 2,000 students are enrolled in the Bachelor of Science degree programs in Chemical, Civil, Electrical, or Mechanical Engineering or in the novel Bachelor of Arts in Engineering degree program.

The development of the new course was initiated by the retirement of a faculty member who was responsible for a required Engineering Science course for freshmen which included units on engineering economics, modeling, statistics, and engineering design. It was the consensus of the engineering department heads that these topics, with the exception of engineering economics, could be better dealt with elsewhere in the curriculum, thus opening the door for the design of a new course, E.S. 225: Engineering Professionalism and Ethics, which considers the role of engineering with respect to social, economic, ethical and legal matters. Reorganization of the freshman-sophomore engineering sequence required that the course be shifted from the freshman to sophomore year, thus providing a full year in which to carefully plan the new course.

The course development also occured at a time when the College was disbanding the Engineering Science department and distributing the Engineering Science faculty lines and course responsibilities to the degree-granting departments. The new course in Professionalism and Ethics was assigned to the Department of Civil

Engineering, along with courses in Statics and Strength of Materials. The assignment of the new course to the C.E. department resulted from staffing considerations. The only engineering course at Lafayette which has previously dealt with such material is an upper level C.E. course taught by one of the authors (BVV), entitled Professional Aspects of Engineering, which includes substantial emphasis on engineering ethics and the legal aspects of engineering, as well as discussion of the role of the engineering profession in society. When the Engineering Science department was disbanded, the other author (JRH), who teaches Engineering Economics and an interdisciplinary course in Technology, Values and Society, was reassigned to the C.E. department. The department thus became the logical place for the development and staffing of the new course. During the summer of 1989, the authors received faculty grants from the College to support the final stages of course development.

Since the new course in Professionalism and Ethics includes some material previously covered in the existing Professional Aspects course, the C.E. department plans to reorient the upper level course towards systems analysis, thus enabling it to strengthen the course offerings for its own majors while contributing expertise in the area of professionalism and ethics to the core engineering curriculum.

Content

Beginning with the Fall 1989 semester, Engineering Professionalism and Ethics is required of all engineering majors (B.S. and A.B.) during their third semester. The course has four areas of emphasis: Engineering and Society, Engineering Economics, Engineering Ethics, and the Legal Aspects of Engineering.

The section on Engineering and Society, which grew out of material utilized in the Professional Aspects course and the Technology, Values and Society course, serves both as an overview of the role of the engineer in society and as an introduction to the other sections of the course. The focus of this section is on the historical development and current status of the profession with particular emphasis on what distinguishes engineering from other human endeavors. Specific topics covered include definitions of engineering, the engineering method, the engineer and business, and engineering and public policy.

The unit on Engineering Economics, the only holdover from the freshman-year predecessor course, is designed to provide students with an introduction to time value of money concepts and the fundamentals of economic decision making. Decision making is at the heart of most technological issues, and an appreciation of economic decision criteria is essential to a thorough understanding of such issues. Topics addressed are generally those covered in the engineering economics portion of the EIT exam, including use of interest factors, nominal and effective interest rates, present worth analysis, annual worth analysis and breakeven studies.

The section on Engineering Ethics is based in part upon material utilized in the Professional Aspects course. This section has also been developed in conjunction with a campus-wide "Ethics Project", funded by a private donor, the steering committee of which selected the topic of "ethics and the professions" as an initial area of emphasis. During the summer of 1988, one of the authors (JRH) and colleagues from the departments of anthopology and sociology, biology, economics and business, philosophy, and religion regularly met, with the support of the Ethics Project, to share insights and to brainstorm methods of exposing all of Lafayette's students to ethical issues. The Ethics Project also provides travel support for visits by faculty to major academic ethics centers. For example, both authors visited the Center for the Study of Ethics in the Professions at the Illinois Institute of Technology in May, 1988. The Ethics Project also sponsors an ongoing series of invited lectures on ethics in the professions, including recent talks on engineering ethics and computer ethics, and is planning a major symposium on professional misconduct for the spring of 1990.

The ethics portion of E.S. 225, the first actual curricular change resulting from the Ethics Project, includes such topics as safety and risk assessment, professionalism, codes of ethics, and the rights of engineers. This section of the course also features a guest lecture by a member of the philosophy department on the topic of "Moral Thinking and Moral Theories." We have found that the existence of a resource such as the Ethics Project greatly facilitates such cross-disciplinary exchange.

The Legal Aspects of Engineering section of the course grew out of the interest and experience of one of the authors (BVV) in how the work of engineers relates

Agency and Business Enterprises
Expert Evidence
Property
Student Panels
Summary and Course Evaluation

Final Exam

Appendix II--References

Andersen, D. A. (1986). "Liberalizing Aspects of a Civil Engineering Education," Journal of Professional Issues in Engineering, Vol. 112, No. 3, pp. 190-193.

Hassler, P. C. (1987). "Six Years Experience in Offering Courses in Ethics, Journal of Professional Issues in Engineering", Vol. 113, No. 1, pp. 10-15.

Herkert, J. R. (1989). "STS for Engineers in a Liberal Learning Environment," in Proceedings of the Annual Meeting of the St. Lawrence Section of the American Association for Engineering Education, Binghamton, New York, October 13-14.

Karbhari, V. M. (1989). "Quality Undergraduate Engineering Education--A Critical Perspective," Journal of Professional Issues in Engineering, Vol. 115, No. 3, pp. 241-251.

MacVicar, M. L. A. (1987). "General Education for Scientists and Engineers: Current Issues and Challenges," Bulletin of Science, Technology & Society, Vol. 7, Nos. 5 & 6, pp. 592-597.

Olds, B. M. and Miller, R. L. (1987)."Integrating the Two Literacies: Humanities in the Engineering Curriculum," Bulletin of Science, Technology & Society, Vol. 7, Nos. 5 & 6, pp. 875-882.

Vanderburg, W. H. (1987). "Engineering, Technology, and the University," IEEE Technology and Society, Vol. 6, No. 4, pp. 5-11.

Weil, V. (1985). "Ethics in Engineering Curricula," Research in Philosophy & Technology, Vol. 8, pp. 243-250.

Wenk, E., Jr. (1988). "Social, Economic, and Political Change: Portents for Reform in Engineering Curricula," Engineering Education, Vol. 79, No. 2, pp. 99-102.

Communications in the Civil Engineering Program

M. B. Lundstrom[1] and C. W. Lovell[2], Fellow ASCE

Abstract

Engineers are often perceived as poor communicators, more comfortable with numbers than with words. In the 1990s and beyond, however, heightened competition in the global market and an on-going knowledge explosion will place increasing demands on the communication skills of engineers. How to deal with the greater depth and breadth of information, how to convey ideas effectively to colleagues, and how to increase public awareness and support of engineering projects and of engineering itself as a career are questions which should be addressed by our nation's institutions of higher education. One answer is through strengthening the writing and speaking skills of future engineers. This paper will discuss the approach taken by the School of Civil Engineering at Purdue University.

Introduction

What is the most pervasive form of "illiteracy" among graduate civil engineers?...Mathematical? Scientific? Computer? Or is it just plain communications? The experiences of the authors reinforce the notion that many engineering students demonstrate serious deficiencies in oral and written communication. The following is a description of two courses that address the problem.

---

[1] Communications Specialist, Civil Engineering, Purdue University, West Lafayette, IN 47907.

[2] Professor of Civil Engineering, Purdue University, West Lafayette, IN 47907.

The Graduate Course: CE597B

An elective course has been initiated by the authors in their department of civil engineering aimed at graduate students who desire formal instruction in technical communication.

Course Structure

Students are a mix of foreign and domestic, which is a challenge to the authors in the selection of appropriate topics for practice sessions. The format is one of presentation of resource material followed by student practice. A typical course outline is included in Table 1. Practice takes the form of classroom exercises in the case of speaking skills and outside work in the case of writing. Ideal class size is perhaps ten students, which allows for necessary individual attention and interaction among teachers and students.

Course Content

Oral skills appear to require at least as much attention as written ones. The course begins with consideration of extemporaneous speaking skills, followed by the techniques of adversative dialogue. Next is the chalk talk of flip chart presentation, with emphasis on handling questions. The presentation portion of the course continues with the keys to successful chairing of a committee of peers and a presentation of primary elements of parliamentary procedure. The concluding topic is professional demeanor.

Writing skills are practiced in short form, including abstracts, extended abstracts, and short papers. The authors do not believe that it is necessary for the students to write lengthy terms paper in the course, although a complete listing of the rules of best practice is made available in lecture notes. Since practicing engineers often avoid peer review of technical papers, or perform it poorly, lecture-practice sessions are devoted to this subject, as well.

Course Evaluation

A summary and critique period has been very useful in revising course content to meet student interests and needs. Students are universally enthusiastic about this course and recommend it to their peers.

The Undergraduate Course: CE292

The communication skills of undergraduates are also

TABLE 1

CE597B
Speaking and Writing for Engineers

Credit: 1

Week	Topic
1	Course Overview; Motivation for Effective Communication
2	Informal or Extemporaneous Speaking
3	Practice
4	Adversative Speaking
5	Practice
6	Handling Questions
7	Practice
8	Chairing a Technical Meeting
9	Guidelines for Technical Writing
10	Abstracts; Extended Abstracts; Summaries; Short Reports
11	Professional Demeanor
12	Short Formal Talk; Extended Abstract
13	Short Formal Talk; Extended Abstract
14	Peer Review
15	Practice
16	Summary & Critique

addressed in the School of Civil Engineering in the form of another 1-credit course. This course is part of the School's core curriculum and is taught by the civil engineering technical communications consultant. It has three objectives:

1. To serve as a formal introduction to technical writing and speaking for 4th and 5th semester civil and construction management undergraduates.

2. To provide an opportunity for students to exercise their communication skills.

3. To determine if any students are in need of remedial assistance.

Course Structure

The 16-week semester is divided roughly into half, with the first 8 weeks devoted to technical writing and the remainder focusing on oral presentations. A brief report (5-7 pages) is required based on a technical topic chosen by the student. Because many civil engineering students have relevant work experience at this time, either through internships, co-op positions, or summer jobs, they are encouraged to use it for their reports for the semester.

This report then serves as the basis of a formal presentation by the student which is videotaped for review and feedback. Three students give talks each 50-minute class period, with a 5-minute question and answer session immediately following each talk. All students except the day's speakers serve as evaluators.

A technical writing handbook is required as reference for the course. A speaker from industry also presents one talk on communication in the corporate world. Class size is roughly 20 students per section to facilitate student participation and discussion.

Course Content

During the writing portion of the semester, students are introduced to technical writing style, report format, and audience analysis. One session is devoted to technical description and another to technical illustration. Students also exchange drafts and spend an hour in class critiquing each other's organization and structure. In addition, formal reports prepared for laboratory courses are used as an assignment in the communications course.

Attention then shifts to oral presentation. After one lecture which addresses technical speaking, students prepare a brief technical talk (10 minutes), complete with overhead transparencies or other visual aids.

Throughout the semester professional presentation is stressed. Use of the computer is encouraged, mainly for electronic mail and word processing. A laser printer is also available for use by the students.

Grading

To maintain objectivity and consistency, grading sheets are used which outline the distribution of points awarded to various components (tone, grammar, etc.) of a written report. These sheets help to keep the grader "on track" and also provide a quantitative measure of the student's performance. Students whose skills are deemed inadequate at this point are required to complete a remedial English course before graduation.

Evaluation sheets are also used to grade the oral presentations. The forms, containing ratings and comments of audience members, are given to the speaker for feedback, and points are awarded by the instructor for the final grade on the talk. Grades are recorded after students view their taped talks.

Results and Reactions

The course is evaluated near the end of each semester by the students. In general, they realize that communication skills are important and that they are in need of some improvement. Last semester, for instance, roughly 15% of the class required remedial assistance to strengthen their writing skills. Public speaking is particularly difficult for some; as one student said, "I hate speeches! But I did learn something".

Given Purdue's rigorous civil engineering curriculum, the workload for this 1-credit course also can pose a problem. "Too much work for 1 credit", one student put it concisely. Preparing a paper and a talk consumes valuable time, and because of its one-credit status, the course if often at the bottom of the list of a student's priorities.

However, the course does meet its objectives. It formally introduces students to technical writing and speaking, provides an opportunity for them to practice these skills, and is an effective tool for diagnosing skill levels. To quote yet another student: "I do feel that this course was beneficial and that other fields of engineering should have a similar communications course".

Closure

The authors believe that courses such as those described above can be part of the solution to the problem of inadequate writing and speaking skills of civil engineering students. If properly designed and effectively taught, they can pave the way for professional presentation by civil engineers of the future.

BUILDING AN ARTICULATE ENGINEER
A Laboratory Approach

Hillary Hart[1]

Abstract

Undergraduate technical writing and communication courses for engineers are proliferating all over this country as engineering educators respond to the need expressed by industry and the public sector for engineers who can write and present information skillfully. Most of these courses are still taught within academic departments traditionally associated with the teaching of writing: English, Rhetoric, Linguistics, and Communication. A few engineering colleges, however, are attempting to teach technical communication within their own curricula. This paper outlines the recent development of a technical communication course taught within the Civil Engineering department at the College of Engineering at The University of Texas at Austin. While it is certainly too early to make any substantial claims of success for this course, the evidence so far indicates that teaching Technical Communication from within an engineering college or department allows for a more focused and a more user-oriented approach, as well as a better use of resources, than can be achieved in an English or other non-engineering department.

Background

The course, CE333T, was developed in response to a simple but critical need: in order to graduate, every University of Texas engineering undergraduate has to pass a technical writing course, but by 1986-87 no such course was available to him/her on the UT campus. When,

---

[1]Lecturer, Dept. of Civil Engineering, College of Engineering, The University of Texas, Austin, TX 78712.

for various political, economic and philosophical reasons, the English department stopped teaching all writing-across-the-disciplines courses, the College of Engineering, among many other colleges on campus, was left high and dry. Within a year (1987-88), the Civil Engineering department, along with one other College department, hired its own instructor to teach writing and communication. This paper is a report on the first two years of life for CE 333T, Technical Communication: the logistics, the methodology, the opportunities and challenges of setting up an upper division writing course within an engineering college.

Objectives of Technical Communication Course

The primary objective of this course is to prepare students to write and present the kinds of documents required of civil engineers: reports of all kinds, proposals, documentation, instructions, correspondence. Since all of the students are upper division and have declared their major, they possess a rapidly expanding body of technical knowledge about which to write. They need not and do not write any non-engineering-related prose in this course: no essays or book reports. We want to approximate, as closely as is feasible, the communication situation in industry, where documents are not simply graded and returned to the writers. As Mathes and Stevenson point out, in contrast to the classroom, "in industry, a report is implemented. It sets people in motion, guides decisions, provides direction. That difference, perhaps more than any other, distinguishes the communication situation in the classroom from that in industry" (1976). Our goal is to narrow this gap between school and industry by using the resources at hand: our highly qualified civil engineering faculty and the technical companies here in Austin.

A secondary and very related goal is to prepare students to write and work in an environment similar to the one in which as professionals they will soon be working: as part of a collaborative group. Harold Goldstein points out that " . . . engineering is, by and large, a group activity. Engineers must be able to submerge themselves for the sake of the group effort. There have been few applications of this, or similar approaches, to traditional, lecture-oriented engineering courses" (1982). At the College of Engineering, we want to use the "group activity" approach within a fairly traditional academic context: without, in other words, completely flouting the lecture-orientation of students' other engineering courses.

Methods and Materials

Lectures and Labs

To achieve these two objectives, a combination of lectures and labs was set up: each student attends two hour-and-a-half lectures and one hour-and-a-half lab each week. All labs have half the number of students in the lectures, and they have two parts: the first hour or so is a workshop in which students discuss in small groups the week's writing assignment and read and comment upon each others' drafts of that assignment. After this peer review, students bring their drafts and peer feedback to the departmental writing lab, where they can begin revising their assignment. This lab contains ten Macintosh Plus computers and two printers networked together. Students can retrieve a draft of their work or a set of instructions from the server, a Macintosh II with a total of 250 megabytes on three hard disks. The server contains all the word processing and style-checking software students need to complete assignments: to outline, write, edit, revise, and save their work. When they work on group projects, they can create a private file for the group.

These lab sessions are a critical counterpoint to the more traditional lecture sessions. Aside from having the great advantage of introducing some students to word processing, the labs serve a purpose crucial to the goals of this course: they provide a forum for students to practice the twin skills of writing effectively and reading critically. These are skills that can only be mastered through practice. The lecture sessions are designed to instill certain theories about effective technical writing -- that you must analyze your audience before you start to write, for example -- but the students actually learn to write better only through the process of writing, revising, and revising again. There is a limited amount of factual information about style and grammar and organization to feed them. Beyond that limit lies the great, terrifying (for some) region of choices about language and style and format. Should I use headings in this memo? Will my reader understand these technical terms? Exactly how quickly should I try to get my main idea across? Students can best answer these questions by trying out various answers, and the workshop/lab gives them that opportunity <u>before</u> they have to hand a document in for grading.

Peer Review

One of the major complaints about the writing of engineers is that they cannot write for a mixed audience -- partly technical, partly semi-technical, partly non-technical -- and they cannot write for management. To teach students to write from the reader's rather than the writer's (their own) point of view, the lab sessions involve extensive peer review. Students, who have differing concentrations within civil engineering and some of whom come from other departments (especially aerospace), read each other's writing and tell each other how effective or ineffective that writing is. Most of them are not skilled enough to be technical editors, but they can tell each other what is clear, what is jargon, what is wordy or confusing, what is incomplete. Peer review can work as well for written documents and oral presentations as it can for project management and project design (see Miller and Hribar, 1989). Students often listen more readily to each other than to yet another authority figure; students learn faster and participate more fruitfully in small working groups than they do in large lecture halls. To ensure that each student gives his/her peer's draft a careful reading, we ask each reviewer to answer a set of questions about the document. The original author reads these comments, uses them to help him/her revise the document, and then hands in the comments along with the final draft.

Inter-Related Assignments for Mixed Audiences

To teach students to write for a variety of audiences, the writing assignments are geared toward increasingly management-level audiences. The first assignments are written for students' peers, the next for a "supervisor," the next for an engineering "client," the next for a non-engineering client. The final report is written for an upper-management-level audience.

Since the students in CE 333T are writing technical documents, they must have read or discovered the information they are transmitting. They must, in other words, have done or be able to do research on the subjects of their documents. This course is not, however, a research course. One way to keep the focus where it belongs -- on writing -- is to tie the assignments together such that the research for them is cumulative. At the beginning of the semester each student chooses a topic that he/she wishes to work on for a semester project. This topic is a problem or a need that students can research within a few weeks, evaluate solutions for, and then make recommendations about in a final report. Topics come from campus needs -- an evaluation of the drafting and computer-aided design facilities available to Architectural Engineering undergraduates, for example, -- or topics come from more technical work students are doing in other engineering courses. Many civil engineering faculty contribute topics that derive from current concerns in industry or the public sector. We are currently working on getting local industry involved in setting topics.

This semester-long project is then broken down into, usually, six writing assigments: a memo, a proposal, a progress report, a set of instructions, a draft of their final report, and the final report. To make the project as professionally realistic as possible, every student also presents his/her work and conclusions to the whole class (and invited guests) in an oral presentation. These presentations are critiqued by the student audience, as well as by the instructor. The presentations are videotaped so that each student has a chance to critique his/her own performance.

Group Project

To simulate the conditions of working in industry or in the public sector, one large assignment is set up as a group project. Students working on similar report topics are placed in small groups that plan, write and present a set of instructions related to their area of interest. One group might produce instructions on how to measure levels of sound in and around a house, another might work on how to use a particular software application. To prepare students to cope with the frustrations as well as the benefits of working within a group, we discuss strategies for encouraging full participation and productive meetings and for discouraging inequities in workload and time-wasting meetings. Each set of instructions receives a grade shared by all members of the group who produced it.

Results and Discussion

Currently, the only tool by which to measure the success of this course is the standard College of Engineering and University of Texas student evaluations. These evaluations do not ask students specific questions about the structure or methods of CE333T, but the comments on them have revealed the following: a good deal of enthusiasm for the lab sessions and the oral presentation, a sense of disappointment in the peer review process, and a generally favorable response to the inter-relatedness of the assignments and to the group project. The success of the writing lab with its networked computers has not yet been measured, since Fall 1989 is the first semester of its existence.

Of the almost 400 students who have taken this course thus far, a majority of those who commented on what they liked best cited the draft/revision process as the most valuable aspect. These students all said the course had taught them the value of writing drafts, and comments showed they were pleased to be able to receive feedback (even if inadequate) on their writing before they had to turn it in for grading. A majority of students also said they benefitted from the individual attention they received in the labs from their teaching assistants.

About half the students who offered comments, however, felt that the peer review process fell very short of their expectations. Their major complaint was that the person reviewing their draft simply didn't read it carefully and often didn't catch errors and problems. Currently we are implementing three possible remedies for this "sloppy reading" problem. One solution is to make sure that each student has a different reviewer each week. Another solution is to set up students' expectations very carefully at the beginning of the course. They must realize that peer review can never take the place of the author's own good old-fashioned editing and proofreading. And finally, the questions the reviewer must answer about the draft can be strengthened; ideally those questions could not be answered at all except after careful reading and thought. One further possibility is to penalize the sloppy reviewers by grading their comments.

Although students were not asked to evaluate the usefullness of their written assignments, about 25% of each class stated in their comments that they appreciated working on technical assignments that allowed them to use the information and faculty guidance offered in other engineering courses. Another 30% stated that they got greatest value from giving their oral presentation because no other course had prepared them for what they know will be a regular part of their professional lives.

The group project received very little comment on student evaluations, perhaps because it covers two weeks in the middle of the semester and evaluations come of course at the end. Verbally, however, several students each semester have expressed satisfaction to the instructor, the teaching assistants, and other engineering faculty. These students say they feel better prepared for the group work involved in their senior design course and/or in the workplace.

Conclusions

All of the perceived benefits of CE333T -- small lab sections and small-group projects, plenty of opportunity to critique and revise assignments, ongoing assistance of engineering faculty, and the opportunity to make a professional, videotaped presentation -- all of these benefits are made possible by the resources of the College of Engineering. It is difficult to imagine that an English department would be willing to dedicate the space and equipment necessary for a technical writing lab or for the two-week-long videotaping of technical presentations. Technical writing classes taught in non-engineering departments are often crowded, difficult to get into, and taught exclusively by teaching assistants or assistant instructors. Small lab sections and lots of individual attention would be generally an impossibility. And, finally, the most important resource of all -- a highly educated, dedicated, and proficient engineering faculty in close proximity -- is simply unavailable outside most colleges of engineering.

References

Goldstein, Harold (1982). "Learning through Cooperative Groups." Eng. Educ. 73(2), 171-174.

Mathes, J. C. and Stevenson, D. W. (1976). "Completing the Bridge: Report Writing in `Real Life' Engineering Courses." Eng. Educ. 67(8), 154-158.

Miller, Robert D. and Hribar, John P. (1989). "Quality through Project Peer Reviews." J. Prof. Issues Eng. 115(1), 1-7.

Engineering Ethics - Beyond Professionalism

R. J. Evans[1]

Abstract

It is argued that the branch of ethics relevant to civil engineering is environmental ethics. Some aspects of environmental ethics are described.

Introduction

Increasing attention is being paid to the importance of ethics in engineering education. The need for ethics education is not questioned here but is taken for granted. The question to be discussed is what the scope of engineering ethics should be. Given that there is only a limited time available for education in engineering ethics, it will be argued that matters under the general heading of "professionalism" are of relatively minor importance and that the broader topic of "environmental ethics" should be given priority treatment. This is particularly true for civil engineering education; since civil engineering works involve alteration of the natural environment for the "benefit" of humankind, it is important that ethics be considered from a point of view which considers the relationship between enhancement of human welfare and preservation of the natural environment.

Professional ethics, including engineering ethics, are categorized as "practical" in that they consider the question "how shall I act?" The more abstract ethical questions concerning the meaning of "good" and of "right" are beyond the scope of and irrelevant to professional ethics. Indeed, the First Fundamental Principle of the ASCE Code of Ethics has established the practical nature of engineering ethics in that it exhorts engineers to use their skill and knowledge to enhance human welfare.

In the following, ethical theory is briefly discussed in regard to person ethics, i.e., how one should act with regard to other people; these ideas are then extended to environmental ethics, how one should act with regard to the environment.

---

[1] Professor of Civil Engineering, University of Washington, Seattle, WA 98195, MASCE.

Ethical Theory

Ethical theory has two aspects, normative ethics and descriptive or meta-ethics. Normative ethics relates to rules of behavior while meta-ethics is concerned primarily with the analysis of moral concepts. Both have been discussed previously in regard to engineering ethics by Evans (1988) and only those terms which will be explicitly used in this paper are defined here.

The predominant normative theories of Western ethics are either deontological or teleological. In deontology one acts by duty whereas in teleology one acts to produce the best outcome. A clear discussion of normative ethics is given by Hoff (1986). It has been claimed by Nelson and Peterson (1982) that engineers are, in general, utilitarians (teleologists).

A person making ethical decisions is said to be the moral agent, those embraced by his actions are moral patients and together they make up the moral community. Moral communities are not fixed and, typically, their memberships expand with time. In the history of the USA, the moral community has widened (in theory) to include, in addition to white males, African Americans, Native Americans and women.

A moral agent ascribes value to those objects in which he has an interest. Values can be intrinsic (belong to the thing itself) or instrumental (leading to other values).

A term essential to the ideas discussed here is contextualism. This concerns the fact that ethical statements are about specific situations and that these statements can be justified only in the context of the situation.

Professionalism

Professionalism is defined here as that particular class of person ethics which relates to professional engineering issues, such topics as taking bribes, whistle blowing and even competence in design are examples of professionalism.

In a profession such as medicine, professional ethics is a unique and significant subject involving such topics as organ transplant allocation, euthanasia, when to remove life support systems and the very definition of death itself.

It is hard to argue, however, that the same is true for engineering professionalism and a strong case can be made that there is, in fact, no esoteric knowledge attributable to professional engineering ethics. On the contrary, only two kinds of questions appear to occur. The first kind is one of honesty. People's ethical values in this regard are established during upbringing and before they commence their careers. The second class involves decision making which affects the lives of others as in the management of projects with major environmental impact. There, however, calculus is used as a decision indicator and the engineer is relieved of value decisions. If this is so, then education in engineering ethics is a liberal art.

If we are truly serious about education of engineers in ethics, perhaps it would be appropriate to require a liberal arts minor which includes a series of courses in philosophy and ethics offered by a philosophy department. The engineering departments could contribute by offering a senior seminar with invited speakers.

Environmental Ethics

When one looks at the context in which modern engineering is practiced, one's attention is immediately drawn to the state of the environment, a state which has resulted from the impact of engineering works, works effected by the skill and knowledge of engineers. One sees degradation of the physical environment on a significant scale as evidenced by pollution of air, land and water, by depletion of ground water, soil and the ozone layer and by the accumulation of greenhouse gases. From this perspective, human welfare takes on an entirely different aspect, and questions of person ethics acquire diminished relevance when compared to the threats facing humankind.

Furthermore, because the present context has developed since person ethics was embedded in our culture, perhaps this indicates that person ethics is not able to address global environmental issues.

Ethicists favoring this point of view have extended person ethics to the field of environmental ethics.

The essence of environmental ethics is to extend the moral community beyond that of humans and beyond animals. Person ethics do in some extent extend the moral community. The hedonistic utilitarianism of Bentham includes all sentient creatures in the moral community since such creatures feel pain and pain, the opposite of pleasure, is to be diminished on aggregate. Kantian deontological ethics, on the other hand, includes only rational creatures within the moral community--abuse of animals is immoral only in that such practices encourage humans to abuse each other.

Environmental ethics provide a framework for both a philosophy and a political course of action. Thus Regan (1981) proposed that there are non-human beings which have moral standing and that some nonconscious being must be held to have moral standing. Callicot (1980) says that "environmental ethics may be conceived...as a theoretical formulation for the just distribution of the ultimate natural resource, solar energy among all earth's working partners in the biotic community..."

In any event, there is a broad spectrum of measures of the moral community, the logical extension of which, surely, is the deep ecology or ecosophy movement.

Deep ecology is, in a sense, similar to engineering ethics in that it is normative and has norms (i.e., principles) and principles (i.e., canons). The two norms are [Devall and Sessions (1985)] self realization, the identification of self which goes beyond humans to include the whole ecosphere and biotic egalitarianism, the equal right of all things in the biosphere to live and blossom. Golley (1987) has argued that the norms of deep ecology are consistent with scientific ecology. Deep ecology principles [Devall and Sessions (ibid.)] include the following:

"The flourishing of human life and cultures is compatible with a substantial decrease in human population. The flourishing of non-human life requires such a decrease"

and

"Policies must therefore be changed. These policies affect basic economic, technological and ideological structures. The resulting state of affairs will be deeply different from the present."

Deep ecology has been attacked from both a political and a social point of view. Richard Watson (1983) has claimed that ecosophy should be abandoned as a valid philosophy in favor of anthropocentrism (human chauvinism). To paraphrase Watson, deep ecology suffers from internal contradictions, the essential one being that ecosophy arguments are, in fact, anthropocentric. The very claim of ecosophers that men should curb their activities to preserve ecological equilibrium is anthropocentric; the only species to be thwarted in its natural behavior is the human species. Human nature is to disrupt the ecology of the planet and to thrive at the expense of other species. What humans do, in fact, is to alter things. A consistent deep ecologist would allow humans to live out their evolutionary potential even if this results in the destruction of the planet.

Only by thinking anthropocentrically, Watson, argues, and by accepting human nature for what it really is, can we hope to correct the damage that humans are inflicting on the environment.

More recently, deep ecology has been strongly criticized by both social ecologists [Bookchin (1986)] and by ecofeminists [Zimmerman (1987)].

Bookchin says that there is a relationship between the way people deal with each other and the way they deal with nature and that, in fact, our ecological problems have their ultimate roots in society and social problems. Deep ecologists, Bookchin argues, overlook the facts that human communities, in contrast with other communities in nature, are consciously formed and that human society constitutes a Second Nature, a cultural artifact, out of First Nature. The deep ecologists see this, not as a fact, but as a value, a negative value at that.

Social ecologists take the same position as Watson that humans must solve their problems within their own Second Nature, rather than revert to First Nature. Social ecologists also point out that deep ecologists comprise an elite of Caucasian intellectuals who seek to impose change by authoritarian measure, the cost of which will be borne by the poor of the Third World, the very group which is already paying the price of the exploitation of nature by the rich industrialized nations. They are particularly critical of arguments, based on ecosophy, against famine relief as being contrary to natural processes; they see this as "brutal and Fascist." Some ecosophers claim, for example, that in Ethiopia, it would be better to let nature take its course rather than aid the starving.

An even stronger argument against deep ecology has come from ecofeminists who regard ecosophers as being androcentric (male chauvinists) and who claim that environmental damage has been caused by men who treat nature as a resource to be exploited. God is a man, who gave men dominion over nature and over women. According to ecofeminists, until male domination is overcome, the human race will continue to exploit the planet.

Engineers and Environmental Ethics

Perhaps the most significant aspect of ethical decision making is the context. In a sense, the environment is the context; it is certainly true [Bookchin (ibid)] that ideas may utterly change when placed in a new context.

A dominant contextual reference frame likely in the near future is that brought on by global warming due to the greenhouse effect. [See Scientific American, September 1989, In Context, No. 22, 1989]. If global warming is, in fact, real or of sufficient likelihood that it be treated as real [Gilman, 1989] then its effect on the human infrastructure (Second Nature) will be significant. The predicted rise in sea level, for example, may be but one of many environmental changes which would occur.

Civil engineers are concerned with the infrastructure, its creation, its maintenance and its alteration. The First Fundamental Principle of the Code of Ethics states that engineers shall use their skill and knowledge to enhance human welfare. It seems that more consideration should be given to the proposition to limit, to the maximum extent possible, the continued discharge of greenhouse gases. Such limitations might require significant political change and such political change would be unlikely to occur peacefully without a fundamental change in human ideology.

As recently as 1917, William James described nature as a "harlot" and wrote that "...to such a harlot we have no allegiance; with her as a whole we can establish no moral communion; and we are free in our dealings with her several parts to obey or destroy; and to follow no law but that of prudence in coming to terms with such of her particular features as will help our private ends" (quoted in Callicot (ibid)).

Sixty years later such words would cause outrage. Now we suspect it is us who must change if our life, as we know it, is to flourish. The "separate parts of nature" are parts of a single system. The "second nature" of humans comprises another part. Our "private ends" require a vast consumption of fossil fuels. It seems unlikely that the current global rate of human caused carbon emission (7 billion tons per year) can continue to be consistent with any conceivable measure of human welfare.

At certain times in history, paradigm shifts occur [Kuhn (1970)]. Perhaps such a paradigm shift is occurring now as we reassess the human place in the planetary ecosystem. Environmental ethics may be descriptive of this paradigm. If so change has started and has not ended.

Engineering education is part theory, part practice. A sound theoretical understanding of one's subject is an important prerequisite to practice. If the future role of engineers is central to an attempt to develop a "sustainable" second nature then the theoretical basis for their practice may well be environmental ethics.

The civil engineering profession has not been inactive in this area. For example, the Environmental Impact Analysis Research Council of the ASCE Technical Council on Research recently (1984) proposed that an Eighth

Fundamental canon be added to the ASCE Code of Ethics that "Engineers shall perform services in such a manner as to husband the world's resources and the natural and cultured environment to the benefit of present and future generations. No action on this proposal has been taken by ASCE.

The time seems appropriate to devote more attention in the civil engineering curriculum to environmental ethics. I suspect that it might be the most important agenda item in the next five years, especially if the emphasis is in educating engineers who have the ability and courage to enhance human welfare.

Appendix - References

Bookchin, M. (1987) "Thinking Ecologically: A Dialectic Approach," Our Generation, 18, 2, 3-40.

Callicot, J. B. (1980) "The Search for an Environmental Ethic" in Matters of Life and Death, Beauchamp, T. L. (Ed.), Temple Press, Philadelphia, PA.

Devall, W. and Sessions, S. (1985) Deep Ecology, G. M. Smith, Salt Lake City, Utah.

Environmental Research Council (1984) "A Proposed Eighth Fundamental Canon for the ASCE Code of Ethics," J. Professional Issues in Engineering, ASCE 110, 3, 118-122.

Evans, R. J. (1988) "Commentary on the Code of Ethics," J. Professional Issues in Engineering, ASCE 114 (2), 148-156.

Gilman, R. (1989) "What's Wrong with the Climate," In Context, 22, 11-17.

Golley, F. (1987) "Deep Ecology from the Perspective of Ecological Science," Environmental Ethics, 9, 45-55.

Kuhn, T. (1975) The Structure of Scientific Revolutions, University of Chicago Press, Chicago.

Nelson, C. and Peterson, S. R. (1982) "If You're an Engineer, You're Probably a Utilitarian," J. Professional Issues in Engineering, ASCE 108 (1), 21-25.

Regan, T. (1981) "The Nature and Possibility of an Environmental Ethic," Environmental Ethics, 3, 19-34.

Sommers, C. H. (1986) Right and Wrong: Basic Readings in Ethics, Harcourt, Brace &Jovanovich, San Diego, CA.

Watson, R. A. (1983) "A Critique of Antic Anthropocentric Biocentrism," Environmental Ethics, 5, 245-256.

Zimmerman, M. E. (1987) "Feminism, Deep Ecology and Environmental Ethics," Environmental Ethics, 9, 21-44.

COMMUNICATION SKILLS OF NEW GRADUATES MUST BE IMPROVED

R. Noel Tolbert

Abstract

A case is made for engineering educators to examine their academic programs to determine if adequate attention is being directed toward providing the marketplace with graduates that can function in our current "information age."

Various improvement alternatives are examined by listing the typical pro and con arguments for the purpose of providing the framework for developing a communication improvement program.

Background

Kimel and Monsees (1979) surveyed a spectrum of companies who recruited engineers at the University of Missouri-Columbia (UMC) regarding capabilities of recent graduates. The 1979 study was structured to survey a broad sample of graduating engineers from a variety of degree-granting institutions, not just graduates from UMC. Areas of competence were listed and the respondents were asked to evaluate "the importance of the area" to their specific engineering practice and "the capability of recent engineering graduates." Writing and speaking were the capabilities selected as the most important to civil engineering practice. They were also designated as the capabilities in which recent graduates were most poorly prepared for the marketplace. Interestingly, social sciences and the humanities was the area of competence deemed least important, an indication that employers saw little if any connection between communication capability and general education requirements.

----

[1]Professor of Civil Engineering, Tennessee Technological University, Box 5015, Cookeville, TN 38505.

During this same period of time, Saul and Sigmund (1979) reported the results of an alumni survey of civil engineering graduates from the University of Wisconsin. Along with many other items, the survey included a section on the current content of the B.S. curriculum in civil engineering with respondent choices varying from "more emphasis needed" to "over emphasized." The most significant curricular items where the alumni indicated "more emphasis needed" were those related to communication. The percentages and curricular item headings were as follows: 58.4 percent for written engineering reports, 65.1 percent for oral reports, and 66.5 percent for writing skills. It was also interesting to note the response did not vary significantly as a function of years since graduation.

In 1985, Hardy, et. al. (1985) reported the results from surveys of civil engineering employers covering two different time periods (1977 and 1983) with similar results. Two of the three areas cited as most important overall were oral and written communication skills. Comparisons of other responses such as, "the relative importance of factors used in selecting employees," signified an increasing emphasis on communications between 1977 and 1983.

Another source for surveying the needs of civil engineering employers is by reviewing advertisements. A review of the Engineering Marketplace advertisements in Civil Engineering also indicates the growing employer needs for improved communication skills in their engineering staffs. In 1978, advertisements for employment had specific references to communication qualifications in eight percent of the ads. A decade later this reference to communication skills had more than tripled to 25 percent. A review of current issues produced a similar percentage. In addition to the frequency increase, the intensity of the communication requirement had likewise increased; recent ads contained such statements as: "excellent written and oral communication skills a must." This intensity is also reflected by recent trends in graduate recruitment such as seminar presentations by graduates during company interview trips. While seminar presentations have traditionally been used by universities to assist in evaluating prospective new faculties, their use in the recruitment of B.S. graduates is new, at least to our institution.

The Accreditation Board for Engineering and Technology, ABET, recognized the need for better communication skills in engineering graduates and adopted the following provision under Curricular Objective and Content, item 4h:

> Competency in written communication in the English language is essential for the engineering graduate. Although specific course work requirements serve as a foundation for such competency, the development and enhancement of writing skills must be demonstrated through student work in engineering courses as well as other studies. Oral communication skills in the English language must also be demonstrated within the curriculum by each engineering student.

As attested to by both employers and surveyed alumni, this ABET requirement has not had the desired effect of significantly improving communication skills of our graduates. A more rigid interpretation of the phrases "competency in written communication in the English language is essential," "the development and enhancement of writing skills <u>must</u> be demonstrated," and "oral communication <u>skills</u> in the English language must also be demonstrated" would expose serious deficits in many, if not most, civil engineering programs.

<u>Improvement Alternatives</u>

One method that is commonly used to improve communication skills, particularly those pertaining to writing engineering reports, is to require a formal course in technical writing. This has been the major change instituted by many programs in attempting to improve the writing skills of their graduates. The civil engineering program at Tennessee Technological University (TTU) elected this option in the early eighties but has since dropped the requirement due to curricular pressures. TTU's experience with the technical writing course had been excellent with current students and alumni that had taken the course commenting positively on its attributes. In addition, in a survey made of TTU civil engineering senior students asking them to select a current course in their curriculum that they expect will be of the least value to them, not one student selected the technical writing course.

This leaves one wondering why the civil engineering department at TTU would drop such a course when the demonstrated need had been thoroughly documented and past experience with the course had been positive.

A review of the period during which the requirement was dropped indicated several factors, part of which are common to most other civil engineering programs.

First, the institution switched to the semester system from the quarter system and a consolidation of courses was required. In making this transition, the faculty were most concerned with the transition's impact on their own speciality area. In addition, many of the engineering science requirements outside the department were three quarter hour courses that would have to be replaced with the new three semester hour courses if retained. While the previous curriculum generously exceeded minimum ABET requirements for engineering science and engineering design, the subsequent program changes increased the engineering science content 13 percent and the engineering design content 16 percent.

It was not unexpected for the civil engineering faculty to elect technical content over non-technical. A study by VanderMeer and Lyons (1980) of various professional field attitudes toward the liberal arts indicated engineering faculty, while generally rejecting the extreme postures on either end of the spectrum, had less favorable attitudes. In fact, out of nine fields surveyed, engineering was tied for last in one survey and next to last in the other. While liberal arts education is not the focus of this paper, this study does provide some insight of engineering faculty views toward course requirements with non-technical content.

Student performance on the Fundamentals Examination, F.E. (E.I.T.), or should one say, lack of performance, also creates pressures to increase technical content, particularly in those areas covered on the examination. Currently in Tennessee, a portion of institutional funding is based on such criteria as the percentage of civil engineering students that score higher than the national mean on the Fundamentals Examination. Furthermore, regardless of how good one can speak or write, you cannot become a registered professional engineer without passing the Fundamentals Examination. However, the objectives of preparing students with a strong foundation in engineering fundamentals and good communications skills should be compatible and one area need not be neglected to accommodate the other. No doubt, this does intensify curricular pressures.

ABET criteria itself, albeit indirectly, played

a part in the technical writing requirement being deleted in favor of technical content. While ABET has communication curricular requirements, the "integrated educational experience" is emphasized and specific courses in writing or speaking are not counted as part of the three years of designated course work. My observations have been that if "ABET doesn't count it," the course is vulnerable to deletion at any time. In spite of the course "not counting," many programs have opted to add the course as part of the "other" ABET category in an effort to improve their graduates writing skills and/or to deflect the ABET communications requirement.

Still another criticism of adding a technical writing course to the civil engineering curriculum was raised by a faculty member that said, "these students have had 12 years of English prior to attending college. One more course will not make good writers out of these students." While not defending the criticism of the poor English preparation of many of our entering students, this same comment could have been made about mathematics. Engineering programs require several additional courses in mathematics and provide substantial reinforcement of their mathematical skills through applications in their engineering course work. A similar approach may be needed in the communications area.

Another method of improving communications skills and that recommended by ABET is the "integrated educational experience" approach. This would require that student writing assignments would be a regular part of the student's engineering classes. Experience with efforts to integrate computers fully into engineering classes in the broad sense indicates that efforts to similarly integrate writing assignments would be difficult. A significant portion of existing faculty feel unprepared and uncomfortable in assigning and grading written assignments. However, additional integration is both desirable and possible.

Still another recommendation that would impact communication skills involves making a basic change in the requirements for the professional degree. While many have advocated that the master's degree should become the degree requirement for the professional degree, Friedland and Dorato (1987) advocated that to properly educate professionals, a three year professional program beyond the baccalaureate is required. Friedland and Dorato proposed that separate professional schools of engineering be established similar to those in medicine and law where admission would require a bachelor's degree in liberal arts with specified pre-engineering course work. Experience at TTU with students that had received a previous degree in a non-engineering field prior to beginning their civil engineering education indicates such graduates communications skills would be greatly enhanced. However, this almost doubling of the educational requirement is most unlikely in the near future but has merit to consider as a future goal. However it is possible and likely to expect increased educational requirements for the first professional degree. A three year professional school with a two year pre-admission requirement has merit as well as requiring the master's degree as the "professional degree."

Summary

The need for improvements in the communications skills of graduating civil engineering is well documented. We should not be debating the issue of whether we should improve these skills but instead start implementing changes for improvement.

Since current curricular pressures indicates that few if any courses can be added, additional integration of communications requirements into engineering course work is appropriate. By adding only one 3-5 page writing assignment in each civil engineering course, the student will have had at least ten additional progressive writing experiences. The instructor may find it to be of more value to ask for a half page or one paragraph summary concluding each weekly assignment. Students can show marked improvement with this type of practice in written communication.

Experience with requiring a technical writing course has been favorable and can be added to the curriculum but generally only at the expense of deleting technical content. Even when such a course is added, this still does not relegate the entire responsibility of communications improvement to another department. As is the case with mathematics, reinforcement of communications skills can and should take place in civil engineering courses.

Additional improvements may be possible through our elementary and secondary schools and we should be supportive of such efforts. More substantial improvements are probably possible through improving the instruction of English at the college level and careful attention to the writing requirements included in the social science and humanity electives.

Finally, substantial communications improvements would be possible by restructuring the requirements for the first professional degree. Broad curricular pressures exist, other than communications skills, which can only be accommodated by such a restructured program. However, even with such a restructuring, it is not expected that all communications learning would be pre-professional or relegated to another department.

While substantial barriers to adopting measures for improving communications skills exist, the need is too great to dismiss or disregard. We must take steps now to improve the communications skills of our graduates.

References

Friedland, Bernard and Dorato, Peter, "A Case for the Doctor of Engineering As a First Professional Degree," Engineering Education, April/May, 1987, pp. 707-713.

Hardy, Thomas B. et. al., "A Survey of Graduate Education in Environmental Engineering," Conference Proceedings, Challenges to Civil Engineering Educators and Practitioners/ASCE/Ohio State Univ., Columbus, OH, April 11-13, 1985, pp. 38-43.

Kimel, William R. and Monsees, Melford E., "Engineering Graduates: How Good Are They?," Engineering Education, Nov., 1979, pp. 210-212.

Saul, William E. and Sigmund, Thomas, "University of Wisconsin Civil Engineering Alumni Survey - 1979," Madison, Wisconsin.

VanderMeer, A. W. and Lyons, Mack, "What Engineering Faculty Believe About the Liberal Arts," Engineering Education, Nov., 1980, pp. 170-172, 181.

Ethics in Engineering

Raymond J. Krizek[1] and Bradley J. Hulbert[2]

Abstract

Although the exercise of ethical conduct constitutes a significant responsibility in the professional career of every engineer, the acceptable bounds imposed by ethics are rather nebulous and related to the setting in which a given action occurs. While law may be defined as a body of rules that is prescribed and enforced by a controlling authority to benefit society as a whole, ethics is a set of moral principles or values that provides a systematic approach for resolving questions of morality in the absence of coercion from an external source. In general, the standards of ethical behavior are beyond the minimum standards established by law and are thus unenforceable by the judicial system. Discussed here are the diverse relationships between ethical and legal behavior and a framework for resolving "conflicts of ethics" in professional interactions. The determination of an ethical course of action in an imperfect world imposes the burden of differentiating between right and wrong in the moral sense; this leads to a challenge which is deceptively simple in principle, but awesome and paradoxical in reality. Notwithstanding the apparent inadequacy of simplistic definitions of right and wrong for addressing the variety of ethical dilemmas faced by engineers, respect for our profession dictates that each individual must strive sincerely to develop a properly responsive conscience and then exercise the directives of this conscience in a consistent manner.

Introduction

Every engineer can be assured that ethics will play a dominant role in a multitude of career and life decisions. For example, a fledgling engineer might have to decide if a job making more effective nuclear weapons is ethically acceptable, or an older engineer might have to decide if it is ethical to remain

[1]M.ASCE, Chairman and Stanley F. Pepper Professor, Department of Civil Engineering, Northwestern University, Evanston, Illinois 60208.

[2]Partner, Allegretti & Witcoff, Ltd., Chicago, Illinois 60606.

silent on an impending job change so that certain benefits will be realized. Is it ethical to accept a commission from a company for recommending that company to a client? To what extent can excessive design and its associated higher costs be tolerated? At what point should an engineer express reservations about a technically feasible individual project that is felt to contribute to an unfavorable local environment? When serving as an expert witness, how far can one go to slant testimony in favor of a client, assuming that technical correctness is always maintained? It is clear that the future will present a barrage of problems and challenges that require a careful analysis of ethics--either consciously or subconsciously. Explored here is the nature of ethical considerations that an engineer will face, and a general framework is offered for making "correct" choices.

Definitions

Ethics may be defined as a set of moral principles or values, the moral quality of a course of action, or the standards that govern the conduct of members of a profession. An ethical framework provides a systematic approach to resolving questions of morality in the absence of any coercion from an external source. In contrast, law may be defined as a body of rules that is both prescribed and enforced by a controlling authority to benefit society as a whole. Laws are often created and maintained to reflect current or recent concepts of morality. While ethics and law frequently mandate the same behavior, the two are far from synonymous. Indeed, ethical behavior may be considered as that behavior which is imposed or desired by society or one's profession for the mutual benefit of all; it is often above and beyond the minimum standards of behavior established by the legislative process and therefore is unenforceable by the judicial system and not subject to its sanctions.

Ethical Versus Legal Behavior

The diverse relationship between ethical and legal behavior can be visualized by referring to Figure 1. If the area within the rectangle represents all courses of action that might be undertaken, it can readily be seen that most conduct is either ethical and legal or unethical and illegal. However, there are many cases where the distinction between legal, illegal, ethical, and unethical are not well defined. In particular, there are definitely situations in which an action may be ethical but illegal or unethical but legal, with the latter probably encompassing more actions than the former. One example of conduct considered by most individuals as ethical but illegal is breaking a law with the purpose of preventing some very serious societal misfortune (e.g., violating traffic laws to rush an injured person to the hospital). Alternatively, an example of unethical but legal conduct in the eyes of most persons involves accepting favors under conditions where personal judgments may be affected (e.g., an engineer accepting lavish entertainment from a contractor on whose work he must pass judgment). It is this latter body of conduct that comprises the prime concern of engineering ethics and constitutes the basis for the various codes of ethics that have been formulated.

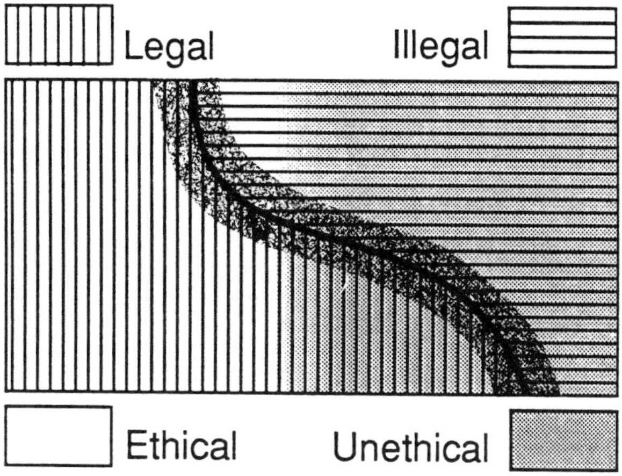

Figure 1. Schematic Relationship Among Various Modes of Behavior.

Of significance in Figure 1 is the nature of the boundaries between legal and illegal and between ethical and unethical. In assessing the distinction between legal and illegal, it is important to realize that our society places great value in the principle that the public should have the opportunity to become aware of a law before it is enforced, and a large quantity of enacted statutes, judicial decisions, administrative regulations, and advisory opinions exist for this purpose. Notwithstanding this plethora of background information, the entire range of possible circumstances and responsive human behavior cannot be covered specifically, and a cadre of lawyers is constantly employed to advise on matters of law and "proper" behavior. Accordingly, the boundary between legal and illegal is admittedly somewhat nebulous in many cases, and this is depicted by a shaded band; the darker line passing through the middle of the band represents court decisions that have been made in specific instances.

Despite the uncertainties associated with questions of legality, they are probably more amenable to definition than questions relating to ethics; this is depicted in Figure 1 by the broader and less well defined interface between ethical and unethical behavior. This situation prevails in large measure because there exists a large body of statutory and case law to offer guidance as to whether an act is legal or illegal, whereas the determination of whether an act is ethical or unethical rests largely on the interpretation of general principles. In the case of the engineering profession, such general principles are usually set forth in codes of conduct issued by state licensing boards and various professional societies.

Framework for Addressing Conflicts of Ethics

From the foregoing explanations it follows that ethical conduct may often not be intuitively obvious, and therefore it is necessary to develop a framework for resolving "conflicts of ethics" in our professional world. In so doing, however, we must fully understand that this framework will not consist of a set of rules postulated in some written code, but rather it must emanate from a set of principles that needs to be instilled in the young engineer. After satisfying the law and various standards set forth by professional societies, these principles normally have their bases in religious, cultural, family, and personal values. As such, ethical conduct can not normally be enforced in the same sense as legal conduct.

Perhaps the first notion to dispel is that an understanding of ethics is simple. This idea can begin to be appreciated if one accepts the premise that, if all persons were properly and completely developed in a perfect world, there would be total coincidence between their needs and interests. Accordingly, in the absence of conflict or even the possibility of conflict, the question of ethical conduct would be nonexistent for all practical purposes because there would be no need to differentiate between good and bad or between right and wrong. However, given that the world is not perfect and the needs and interests of all persons are not coincident, we are led to the task of distinguishing between good and bad and between right and wrong in the moral sense.

While deceptively simple in principle, the burden of differentiating between right and wrong is indeed awesome and challenging, as can be appreciated by the following brief arguments. If a given action is deemed to be right or wrong, it is logical to ask why it is right or why it is wrong; furthermore, if such an action is inherently right or wrong, then it should always be right or wrong. But one can think of many actions which might be termed wrong in most circumstances, but which may be right in special situations. For example, while it is generally accepted as wrong to lie, such an action might be right if it served to divert an impending theft.

This suggests that simplistic and firm definitions for right and wrong are inappropriate and that a more enlightened perspective is needed to better categorize actions. One such approach is to view actions as being morally right or morally wrong, but even this broadened outlook is inadequate for our purposes, as can be seen by the following argument. Since it is always wrong to perform an action which one truly believes to be wrong, even if such action is really right, one is led to the paradox of perhaps not taking the right course of action because it is believed to be wrong, or alternatively performing a wrong act which is believed to be right. A careful analysis of this reasoning suggests that the apparent contradiction is attributable to an ambiguity in the words "right" and "wrong", which must be interpreted differently in different parts of the argument. This paradox can be resolved by introducing the concept of moral justification instead of morally right or wrong. This leads to the modified rationale that one is not morally justified in performing an act that is believed to be wrong, or vice versa. To a large extent this places the burden of conforming to ethical conduct squarely on one's personal judgment with the attendant responsibility of striving to engender a properly responsive conscience.

Ethical Behavior Dependent on Context

It is important to recognize that ethical conduct relates to interactions among people, and not to standards of behavior that govern one's personal existence, except insofar as such existence affects society. For example, if one were isolated on an island and devoid of all present and future interactions with any element of society, all actions taken to regulate personal well-being would not be deemed within the realm of ethics because the consequences of any such actions would not affect the needs or interests of anyone else. In another vein we must appreciate the fact that there are no absolutes in ethical behavior--no fixed reference system against which to judge or evaluate a given action. Hence, any particular action must be viewed within the context of its occurrence when assessing whether or not it is ethical. For example, times change and a mode of behavior deemed unethical a generation or two ago may well be acceptable today. Or certain types of conduct considered entirely ethical in Asia might be viewed unethical in the United States. Or an act taken against an individual might satisfy ethical standards whereas the same act against a group of persons would be deemed improper. Clearly, the judgment of ethical conduct is relative to the setting in which it takes place, thereby rendering even more incumbent on each of us the responsibility to develop a proper conscience to guide the ethics of our behavior.

Hypothetical Example

Let us conclude this presentation by examining a hypothetical situation in terms of several typical systems of ethics that might be applied. Assume that an engineer works for a firm that employs many people to manufacture landing gear for military aircraft. While the engineer is convinced that a particular landing gear is safe, he/she suspects that others in the company may have altered published test data so that it would appear to satisfy the demanding standards set by the purchaser. The engineer must decide whether to ignore the situation or "blow the whistle." If the company loses the contract, many people will be out of work.

Under a model termed ethical egoism, each individual attempts to maximize personal happiness. Proponents of this approach argue that members of our capitalist society are economically better off because we are allowed to pursue our own self interests. Thus, an ethical egoist might reason that the engineer should serve his own best interests by not making enemies within the company and thus remain silent. Another model, called utilitarianism, is derived from economics and, in an extreme application, empirically weighs the pros and cons of each proposed action. In the example under consideration, it is unlikely that anyone will be harmed by ignoring the situation, whereas many will be harmed by "blowing the whistle"; hence, the utilitarian might decide that silence is the only ethical action. In perfectionism, often associated with Aristotle, one believes that cultivating virtues in young people will engender in old people the ability to naturally make correct decisions. A follower of this approach, confident in the morality of the prevailing culture, might intuitively decide to remain silent out of compassion for fellow workers, or to "blow the whistle" if honesty were deemed a more important factor. Under a deontic model, however, fixed rules should be followed; if the engineer followed this system, believed in the Ten Commandments, and felt that keeping quiet would be a violation of a commandment, he would be ethically obliged to expose the perceived fraud.

Summary

In light of the foregoing philosophical arguments and hypothetical example, it is clear that simplistic definitions of right and wrong are an inadequate and inappropriate response to the ethical dilemmas faced by engineers. The continued well-being of our profession and the enhancement in respect from those we serve can best be achieved by carefully articulating an informed systematic analysis of ethical problems and the consequences of alternative courses of action and then exercising the dictates of a proper conscience in a consistent manner.

ETHICS - TODAYS ACHILLES' HEEL IN THE PROFESSIONS

Roger M. Boisjoly, P.E.[1]
IEEE, Senior Member

Abstract

Unethical business practices emanating from the short term profit mindset and treating employees as a renewable resource has created a 30 year decline of ethics, productivity and the once competitive edge enjoyed by U.S. industry. The past and present professional career environment is explored with correctional recommendations to restore ethical conduct with the focus on the long term.

THE WAY IT WAS - YESTERDAY

As I embarked upon my career as a Mechanical Engineer in 1960, I observed a condition of basic cooperation between colleagues and departments within my company. I soon was helped along the professional path by informal mentors and quickly became acquainted with a steep learning curve but found willing help available for the asking at each plateau and as experience was assimilated, my confidence flourished and my capabilities expanded.

Those were the career opportunity times when a company understood and promoted the theme of steady and deliberate training of new technical employees and fostered the ideals that a new entry level engineer may not earn his keep for at least several years. Most new hires were initially assigned to review and check the work product of more senior people and by so doing learned the methods and procedures to use.

[1]Consulting Mechanical Engineer, 3047 E. Menlo Street, Mesa, Arizona 85213

We were also allowed and actually expected to investigate areas associated with but actually outside our specific work domain. For example, as a design engineer I often inquired and solicited help from other departments such as manufacturing, model shop, drafting, test lab and others, as I attempted to discover the best way to integrate all the specification requirements of the different but necessary disciplines into my designs. By receiving the support from my direct supervisor to engage in this experience process, I quickly became involved and proficient in the design layout review process which was the forum for interdepartmental discussions as to whether design, performance, manufacturability and quality specifications could be attained. In this way the company was indeed the sponsor and promoter of the long term stability of the work force and the ability to maintain high productivity of quality products delivered on time at a competitive price.

I would guesstimate, by my observations, that up to 90 percent of the technical workforce was actually engaged in productive output while perhaps 3 to 5 percent of the remainder were a drain on resources and would undoubtedly be laid-off at the first opportunity. To avoid misunderstanding, I am not stating that the industrial work environment was almost perfect, but I am stating that the basic ethics of conducting business by treating employees as a precious resource was highly understood and promoted along with long term goals for the company and its employees.

The general workforce composition seemed cooperative, supportive and ready and willing to be molded into a team effort which produced quality products. Team members inherently understood that each assignment within the team was an important part of the total effort and as such required the best performance from all involved. In my career, this included assignments such as designer, checker, analysts, liason, etc; all of which made me feel important to the team effort. The projects were generally structured to allow sufficient time for employees to think and assess what task needed to be performed at any point during the design, development, production phases in producing a quality product. Another important aspect of the professional work environment was to check each others work to avoid costly mistakes early in product design and development.

Many and perhaps most college faculties were comprised of instructors and professors who had a B.S. Degree with extensive applications experience along with others who had M.S. and PhD Degrees with or without industrial experience. This faculty mixture promoted application knowledge plus a sound foundation in theory for students to learn from and draw upon when entering the workforce.

It was during the first 5 years of my career that I received some of the best advice I have ever received. A quality control manager had observed me receiving tremendous pressure from a project manager as a result of my refusal to sign-off on a discrepant part because he considered the discrepancy to be just a minor nuisance instead of a potential problem requiring further engineering evaluation. The quality manager approached me with support after the project manager left and told me to ask myself the following question when faced with a tough decision of whether a product was acceptable: "would you allow your wife or children to use this product without any reservations?" if I could not answer that question with an unqualified "yes", he said, I should not sign off on that product for others to use. I have always used this as a surefire method to defuse an ethical dilemma concerning a product and it is a very simple and direct application of the golden rule - do unto others as you would have them do unto you.

THE WAY IT BECAME - TODAY

Some will argue that the transition towards lower ethical standards started in the early 1950's, but it matters not when it started precisely but rather that the transition took place. I observed that ethical standards were on a downward trend in the early 1960's at about the time some so called "think tanks" started offering specialized and concentrated 3 to 5 day courses to corporate management on how to maximize profits, but management was either not told or astute enough to realize the price that this practice would extract from their organizations in the form of reduced productivity, poor employee morale, increased absenteeism, increased employee turnover and a host of other ills. The beginning of this era of management by oppression was the precursor to todays MBA programs, which have attempted to legitimize by formal education the principles started in the early 1960's.

My best estimate by observation at many different companies, which I have confirmed with many others in recent years, is that 25 percent of the technical workforce is engaged in 75 percent of the productivity at any given company location. These figures are somewhat true for medium (2000 to 5000 employees) and larger (over 5000 employees) company divisions and are most accurate at the largest operations (over 15,000 employees). What, you may ask is the remaining 75 percent of the workforce engaged in? They simply report to work and produce mediocre output with many errors that go unchecked by others within a management system that promotes this attitude and work ethic. Employees today are well aware that employers use them as a renewable resource and therefore are simply interested in completing each day at work to receive their paycheck and nothing more. Most managers today are only interested in program milestones being met rather than if a product will perform to specifications and this management style adds to the widening gap between employer and employee resulting in more reduced productivity. Todays technical meetings are mostly dominated by managers in attendance as they present their views on a particular problem early in a meeting and then sit back and listen to their point of view being replayed back to them by the technical subordinates in the meeting. This is in direct contrast to past methods used by managers as they contently sat and listened to a thorough airing of problems by the technical people and then expressed their own views and set the stage for objective problem solution. Employees today are afraid to engage in truly objective and constructive discussion in meetings for fear of bringing disapproval upon themselves from upline management.

Most of our universities are no longer teaching institutions. They have become research facilities in search of funding and have become like branches of industry or government with all the attendant ills of business management style and treatment of their employees. PhD top heavy faculties dominate so fully that students come to industry finely tuned in theory but without much application skill and therefore have great difficulty in performing productive work. Our priorities in college accreditation promote this very ill by requiring the PhD type faculty rather than the former mix of experience and theory.

Now that some of the causes of our industrial deterioration have been stated, we need to make some tough decisions that will restore us towards ethical behavior which will automatically produce long term employee dedication and pride in workmanship which will lead us into the 21st century and again into world leadership.

THE WAY IT CAN BE - TOMORROW

Some of the more enlightened corporate CEO's

already have recognized the past ills that their organizations were sold and have begun the change back towards ethical conduct and positive employee treatment as a precious resource by installing and supporting ombudsman programs which are for the benefit of both the company and its employees. A sad commentary on this point is that some company executives see the ombudsman program as a new oppressive tool to identify "troublemakers"(their definition) and target them for punishment or dismissal. This practice must be identified and exposed via our professional societies if we are to force curtailment of the practice. I have no doubt that the ethical long term orientated companies will enter the world economy of the 21st century in a strong competitive position while the "business as usual" ones will self destruct from within.

A powerful corporation has an easy time of ruining an employees career even if the employee has well documented proof of wrong doing by the company as evidenced by my well documented example of grief from Morton Thiokol for my role in exposing the truth in testimony to the Presidential Commission investigating the Space Shuttle Challenger Disaster (Boisjoly 1987).

Change is also taking place at many universities in the form of a few very dedicated professors and administrators that offer courses showing the importance of integrating science, technology and society (STS). Students, by my observation, are thirsting for these courses to help them form a responsible direction for their careers and students believe as I do that college is indeed an acceptable place in the timetable of education to teach ethical and moral values relating to their future careers. The only disagreement I've heard comes from some purely academic professors who have never practiced in industry and seem to be oblivious to the realities of industrial business practices.

As a result of feedback that I have received from students and professors after my talks (65+), I can state that students have responded to the message I present and support a need for upgrading ethics in business and they will be our ticket for long term change for the better. The current system of management, by my estimate, will take between 10 to 15 years to fix, if we all start now to force a change.

We must require and support, as a first step, the passage of the E.I.T. exam as a condition to receive a B.S. engineering degree for the class of 1994. Also, all those employed and classified as engineers, after 5 years experience must take and pass the P.E. exam as a requirement for continued employment as an engineer. I suggest a 5 year maximum grace period for this requirement to be accomplished. All university professors who teach engineering courses must become registered by 1994 or they cannot teach engineering students.

Also, the boards responsible for engineering curcicula accreditation must take a leadership role and insist on balanced faculities with provisions for practitioners as well as those with PhD's. Provisions must also be defined for those already in industry but not currently licensed. My recommendation for discussion is that those who will attain the age of 50 by 1994 be grandfathered with all others given until the year 2000 to comply.

Many from industry, academia and government will classify these requirements as outrageous without giving much objective thought to the proposal as one which will definitely raise engineering to a true profession. Some will claim that engineering enrollment at universities will be decimated and technological advancement will be set back decades. To these I say baloney, the so-called engineering shortages that have been claimed for years are the manufactured product of self serving factions of industry and academia to provide cheap labor and larger supply of students than warranted.

We must put a stop to the paradox that currently exists, whereby engineers are licensed to build our roads, bridges and buildings for public safety while others who design and build our automobiles, aircraft, missiles and spacecraft are not required to be licensed. This doesn't even make good sound basic sense but our profession has been stalled because of it while industry laughs all the way to the bank at our expense.

I do not make these recommendations lightly or without the personal experience to back them up. I have just accomplished the very thing I am recommending, only I did it the hard way by waiting 29 years after graduation from college to take both the E.I.T. and P.E. exams to become licensed. It is my belief that a great benefit will result from these recommendations. Namely, the transfer of the professional code of ethics to corporate managers for ignoring technical recommendations and making pure business decisions based mostly on profit and schedule.

It is my belief that the courts would eventually uphold the transfer of the code of ethics to managers and thus provide the mechanism to extract accountability from managers and engineers alike for their actions.

This would change the shameful lack of accountability that currently exist in this country for unethical decision making practices.

The time is now for all engineering societies and professional societies to stop blinking because of the fear of loss of industrial support for their organization and start to serve the individual membership. Engineering societies must become professional societies for the benefit of the membership and seek out and recommed those companies that indeed conduct business ethically and have demonstrated their concern for the long term and their employees. ASCE has taken a lead position in this regard by producing a manual of professional practice titled, "Quality in the constructed project" which is currently being trial used by practitioners to obtain constructive review and comment under actual use conditions. This manual is a bold step and sets the example for other societies about the requirements needed to raise the level of the profession with the focus on quality and ethics. This should lessen the practice of focusing on lawsuit avoidance procedures and instead focus on good sound engineering practice which in itself will mitigate failures and law suits.

Let us all close ranks and remove our selfserving hats and replace them with hats of candor, objectivity, ethics, and determination to elevate engineering to a true profession.

APPENDIX

References:

Boisjoly, Roger, "Ethical Decisions - Morton Thiokol and the Space Shuttle Challenger Disaster", ASME paper 87-WA/TS-4, December 1987

Verbal Skill of Civil Engineering Graduate Students

Richard J. Heggen[1]

Abstract

A review of Graduate Record Examination data indicates that applicants to Civil Engineering graduate programs are among the least verbally equipped students in all academic disciplines. Potential reasons for this distinction are explored. Implications are discussed.

Introduction

The heritage of Civil Engineering is one of innovative social contribution. More than other engineers, Civil Engineers professionally interact with the spectrum of society. ASCE's "people serving" motto is apt.

As a participant in decisions of broad social scale, the Civil Engineer communicates in various manners. Computations justify designs. Charts summarize findings. Drawings convey layout. Bottom-line figures indicate conclusions. But it is verbally where ideas are communicated. It would thus seem that verbally-inclined students might pursue the Civil specialty more often than more publicly-secluded engineering branches.

Advanced-degree Civil Engineers, a group of enhanced professional stature, need superior verbal ability. It would thus seem that verbally-apt Civil Engineers would pursue graduate training.

The facts speak otherwise. Civil Engineers are among the least verbally-proficient graduates in the entire university. This article documents the problem, responds to possible justifications, and suggests that resolution may require change in Civil Engineering education.

----

[1]Assoc. Prof. and Grad. Advisor, Dept. of Civil Engrg., Univ. of New Mexico, Albuquerque, NM 87131.

The Graduate Record Examination

The Graduate Record Examination (GRE) is a standard instrument for assessing applicants to graduate school. The GRE consists of a general and an optional subject exam. The general exam contains three portions. The verbal portion deals with antonyms, analogies, sentence completion and reading comprehension. Subject matter is balanced between the arts, humanities, physical and biological sciences, social studies, everyday life and human relationships. The quantitative portion consists of discrete quantitative, data interpretative and quantitative comparison questions. Formal mathematics beyond first-year high school algebra is not required. The analytical portion pursues analytical and logical reasoning, testing structural understanding of arbitrary relationships and the handling of arguments. A maximum of 800 points is possible for each portion. Table 1. summarizes results. For the period 1981-84, mean scores were a few points lower, but the relationships were the same.

Table 1. GRE Score Summary, 1984-87 (GRE Board, 1989)

	All Examinees	All Engineers	Civil Engineers	s.d
Verbal	475	477	439	126
Quantitative	549	677	651	140
Analytical	519	582	539	128
n	844960	35945	4223	

The American Society of Engineering Education (1989) survey of 200 engineering graduate schools indicates that 46 percent require the GRE for all applicants. Another 28 percent employ test scores in more limited extent. Barber and Morgan (1988) report the GRE is used by 60 percent of engineering programs.

According to Barber and Morgan's (1987) survey of 441 engineering chairpersons, GRE scores are emphasized 2.80, 2.12 and 2.53 for U.S., foreign and financial-aid applicants respectively, 1 being strongly relied upon, 5 being not relied upon at all. For many engineering schools, a combined 1000 for the verbal and quantitative portions is sought. Dowell (1987) reports the average verbal and quantitative scores for students entering 17 leading engineering schools range from 1070 to 1410. As validity of the analytical portion is not yet fully established, most programs do not use this result.

The subject exam in engineering deals with mechanics, transfer mechanisms, circuits, chemistry, light and sound, computational methods, engineering economy, engineering judgement, knowledge of mathematical facts and manipulations and intuitive selection of mathematical methods. This exam is required by less than ten percent of engineering schools (ASEE, 1989).

<u>The Problem</u>

As Table 1 indicates, Civil Engineers rank below engineers in general in all scores. The difference is statistically significant at the 0.00005 level. Against other engineering specializations, Civil Engineers are lowest in all but one comparison, as shown in Fig. 1.

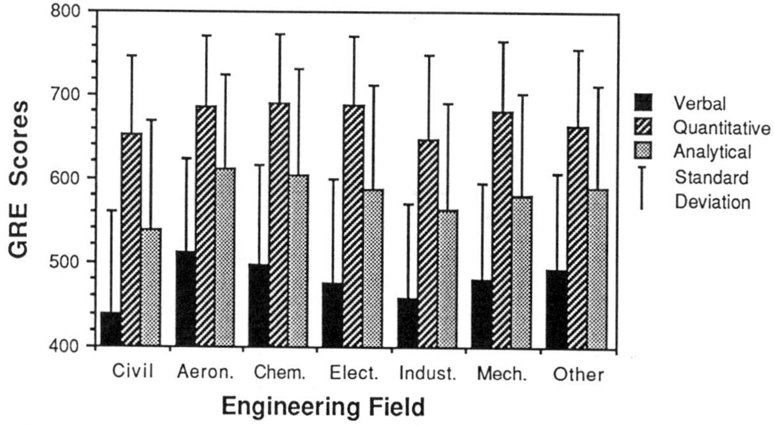

Figure 1. GRE Scores by Engineering Field, 1983-86 (GRE, 1989)

Of course, being at the bottom of an elite group may not be so bad. How do Civil Engineers rank against the 89 non-engineering disciplines categorized by the GRE? Civil Engineering is exceeded in quantitative scores only by biophysics, astronomy, oceanography, physics and the four subfields of mathematics. In the analytic portion, Civil Engineers are mediocre, outscoring students in education, health and other social sciences, but outscored by students in the arts, languages and other humanities, behavioral sciences, biosciences, mathematical and physical sciences.

Notoriety comes in the verbal results. Only graduate students in physical education, home economics and speech-language pathology rank below Civil Engineers. Before 1984, Civil Engineers outperformed educational administrators, but no longer is that the case.

<u>Is the Problem Real?</u>

Plausible reasons can be put forth for Civil Engineers' low verbal standing. The GRE instrument might be flawed. Civil Engineers might more-frequently pursue graduate degrees, diluting the exclusiveness of advanced schooling. Civil engineering graduate programs might draw from a larger part-time student body, students further removed from English instruction. Civil Engineering might have a disproportionately high foreign enrollment. Such justifications merit examination. Table 2 summarizes data for the six engineering disciplines categorized by the GRE.

Table 2. Disciplinary Data, Graduate Students

Engineering Discipline	GRE Verbal	Percent Grad/Undergrad	Percent Part-Time	Percent Foreign
Aeronautical	511	33.5	24.2	31.5
Chemical	498	47.2	19.1	30.3
Mechanical	479	28.7	35.0	33.8
Electrical	476	33.9	48.1	33.8
Industrial	457	49.4	52.4	39.5
Civil	439	45.9	31.9	39.6
Reference	GRE Board, 1989	Ellis, 1988	NSF 1985	Ellis, 1989

The GRE is a mature instrument annually subject to scores of internal and external critical analyses. There have been no reports indicating bias against Civil Engineers. If anything, a profession traditionally drawn from the white educated middle class, might find some advantage in the exam. There are no alternative verbal tests where Civil Engineers do relatively well. The problem cannot be dumped back at the feet of the GRE.

The lowest correlation between GRE verbal score and subsequent grades, 0.22, is achieved by engineers (GRE Board, 1989). For Civil Engineers, Olson (1980) found both with and without foreign students a 0.18 correlation between grade point and combined verbal and quantitative score. Thus, the verbal GRE may have little bearing on engineering academic success. The onus is complicated, however, by the possibility that graduate Civil Engineering programs simply ignore verbal skill.

A high graduate/undergraduate ratio might indicate a dilution of graduate student quality, lowering GRE performance. Simple rank-order comparison of the data does not support this hypothesis.

Practitioners pursuing an MS on a part-time basis may lower the results. To the degree that programs modify entrance standards, substituting experience for scores, this may be so. But if that is the problem, it should not appear as one of verbal difficulties. Clark (1984) found that while older students tended to have lower quantitative scores, verbal scores remained the same. Thus if part-time students were pruned from the data, the quantitative scores might rise, but the verbal scores would not change. From Table 2, there is no significant relation between part-time percentage and verbal score.

Wilson (1984) shows that GRE scores of non-native English speakers should not be compared to those from English-speaking nations. Is the data biased by foreign students? Within Engineering, Civil Engineering has the highest proportion of foreign students entering full-time studies. The differences of Figure 1 may in fact be attributed to nationality. With one slight switch, there is exact inverse rank-order correlation between percent foreign and verbal scores. Were foreign students eliminated from the comparison (scores by ethnicity are unavailable from the GRE Board), the Civil Engineering verbal scores would rise, but the quantitative scores would likely fall (Morris, 1987). While Civil Engineering would look verbally better within the University if the foreign student scores were removed, there is little basis to believe that a xenophobic graduate policy would make Civil Engineering appear strong within engineering. Most professors will agree that verbal difficulties are not restricted to noncitizens. Foreign student bias provides a plausible excuse, but not one sufficient to dismiss the problem.

<u>Resolution</u>

There are three levels within the realm of engineering training where verbal ability might be augmented: in undergraduate, in graduate, and in continuing education. The latter level may be in fact where many positive things occur, but delaying the endeavor to that stage is abrogating academic responsibility.

While language may be inadequately emphasized through high school, future engineers are not slighted. Friday

(1986) reports that verbal skills as measured by the Scholastic Aptitude Test are the same for beginning freshmen in engineering and non engineering fields. By graduation, engineers fall behind, 51.4 vs. 65.8 percent passing the Writing Proficiency Examination. The cause is clear; undergraduate engineers tend to take no more language arts that that required by ABET. Little change can be expected here.

Returning to Table 1, the undergraduate engineering verbal deficiency is erased in the data for engineering graduate students as a whole. The more verbally-able engineers do pursue advanced degrees. Civil Engineering has its unique problem, and that is where Civil Engineering innovation should be applied.

Civil Engineering graduate students scoring below a specified level on the verbal GRE should return to the English classroom. Coursework could draw upon their technical knowledge and professional aspirations. Unlike undergraduate programs where curricular requirements are largely set by ABET, graduate programs have relative autonomy in their standards. If the professional community wants verbal improvement, graduate programs can adjust.

The University of New Mexico has incorporated such verbal emphasis in its new program in water resources management. The professional community was surveyed to determine program objectives. In ranking necessary employee skills, respondents placed communications skills in first place, technical skills in second. UNM students take three one-hour classes in written, graphical and oral communication. It is yet too early to assess the benefits of this effort. One cost of upgrading verbal skills is student balking. Students view technical skill, not verbal ability, as the key to their professional future. It may be that students are not the best judge in this matter. A second cost is that of course design. There are few models from which to work.

The statistics in this paper serve as a point of focus, but the solution will not come by calculation. As with most significant problems in Civil Engineering, the key to resolution is through verbal pursuit. Is there a need? Should it be satisfied? How? Civil Engineers know the problem, and though they might not verbalize it well, will support change.

References

American Society of Engineering Education (1989). "Engineering college research and graduate study." *Engineering Education,* ASEE 79(2), 73-352

Barber, E.G., and Morgan, R.P. (1987). *Summary of responses to chairpersons questionnaire on the impact of foreign graduate students in U.S. engineering education.* IIE, New York, NY

Barber, E.G., and Morgan, R.P. (1988). *Boon or bane, foreign graduate students in U.S. engineering programs.* Research Report Series No. 15, IIE, New York, NY.

Clark, M.J. (1984). *Older and younger graduate students: a comparison of goals, grades, and GRE scores.* Report 81-17R, GRE Board, Princeton, NJ.

Dowell, E.H. (1987). "A survey of leading schools of engineering." *Engineering Education,* ASEE, 77(7&8), 756-758.

Ellis, R.A. (1988). "Engineering and engineering technology enrollments, 1987." *Engineering Education,* ASEE 78(9), 51-61.

Ellis, R.A. (1989). "Engineering and Engineering Technology Degrees, 1988." *Engineering Education,* ASEE, 79(4), 511-521.

Friday, C. (1986). "An evaluation of graduating engineers' writing proficiency." *Engineering Education,* ASEE, 77(2), 114-116.

Graduate Record Examinations Board (1989). *1989-90 GRE .guide to the use of the Graduate Record Examinations program.* Educational Testing Service, Princeton, NJ.

Morris, J.E. (1987). "GRE policies in electrical engineering." *Engineering Education,* ASEE 78(2), 125-129.

National Science Foundation (1986). *Foreign citizens in U.S. science and engineering: history, status, and outlook.* Special Report 86-305 Revised, NSF, Washington, DC.

Olson, R.E. (1980). "Prediction of success of graduate students in civil engineering." *Civil Engineering Education,* ASEE II(2), 7-12.

Engineering Ethics without Religion

by Otto J. Helweg,[1] Fellow, ASCE

ABSTRACT: It seems that writers on Engineering Ethics attempt to dissociate ethics from religion. There are philosophical reasons why this is not only bad strategy but fundamentally flawed logically. If each individual does not have an existential reason for being ethical, all the codes in the world cannot produce ethical behavior. This paper argues that a theistic presupposition is a sufficient, if not necessary, condition to supply the existential motivation. Moreover, professional societies should encourage rather than discourage engineers to integrate their theological inclinations with the appropriate ethical codes.

INTRODUCTION:

The field of ethics can be divided into two parts; first, what ought one to do?; and second, why should one do what one ought? Ethical literature concentrates, almost exclusively, on the first question as illustrated by the "Engineering Ethics" column in **The Professional Engineer** as well as the various codes (ASCE, 1962; Baier, 1965; Baum, 1980; Canfield & Bowman, 1954; Flores, 1980; Mantell; and Mock, 1969). However, most of the ethical code violations indicate that the problem is "not knowing what to do, but failing to do what one knows." (Alger, <u>et.al.</u>, 1965) Consequently, this study attempts to redress this imbalance by dealing with the individual's motivation to be ethical.

---

[1]Chair/Prof. Dept. of Civil Engrg., Memphis State University, Memphis, TN 38152

BACKGROUND

Laymen have been frustrated because philosophers do not agree on an ethical system. That is, philosophers do not have a consistent answer the two questions raised previously. Some engineers have interpreted this diversity of opinion as meaning there is no definition (Vesilind, 1988) or no definition is possible. Such an attitude is understandable for engineers who are used to statistical if not mathematical rigor. Nevertheless, two points may be made; first, disagreement concerning which ethical system is correct does not mean a correct system does not exist and second, in spite of seeming confusion, each person must choose the ethical system he or she will follow. In a sense to not choose is to choose by default.

The first question one encounters when deciding what one ought to do is whether ethics is normative (prescriptive) or empirical (descriptive). The former would accept an absolute standard of "good" while the latter would declare that "good" is relative, such as situational ethics. The following list of ethical systems will generally fall into one of these two camps.

ETHICAL SYSTEMS

First in considering what one ought to do, that is, which system of ethics is right (if any), it is helpful to list the alternatives. Though any taxonomy is somewhat arbitrary, the one suggested by Holmes (1984) will be followed.

Emotivist Ethics

Ethical emotivism holds that talk about ethics is based on emotion or feelings and as such has no objective content. For example, the logical positivists would allow ethical terms to be defined but would deny that emotive words have meaning because moral judgments are not definitions nor are they empirically verifiable. The emotive ethicist would not say, "stealing is wrong," but "people in Memphis feel stealing is not acceptable." Therefore, ethics is subjective (empirical, descriptive).

In support of this position, most would agree that the subject of ethics is highly emotional; notwithstanding, how does the emotive ethicists know ethical terms do not refer to something external to ourselves? To what would they attribute mankind's universal "judicial instinct" (Carnell, 1957).

Ethical Egoism

This ethical school is similar to utilitarianism (which follows) but ethical egoism bases "oughtness" on what is best for one's self while utilitarianism is based on what is best for society. Ethical egoism ranges from crass (destructive) selfishness to a "benign" selfishness (Scriven, 1966). The former would base "oughtness" on what ever satisfies one's wants, no matter how materialistic. Benign selfishness (really unselfishness) argues that seeking the good of society really maximizes our selfish desires because (among other things) it promotes a safe environment.

Of course strict selfishness would produce anarchy of the worst sort but unselfish behavior may, indeed, produce a desirable society. The problem would be convincing people that unselfishness would, in fact, benefit them more than crass selfishness. Such a theory would be severely tested when it came time to sacrifice one's own life for the sake of society. This seems to be the reverse of Adam Smith's "invisible hand." (Ferguson and Gould, 1975)

Utilitarianism

Utilitarianism, as mentioned above, defines "oughtness" as that which produces the best for society. This was recently popularized in the movie, "Startreck II" in which Dr. Spock said, "the good of the many out weigh the good of the few." The problem is how to define "the good of the many." Would it be the dictator in Huxley's, **Brave New World**, or some similar totalitarian system?

Again, we have our "oughtness" built on an empirical base. Without the ability to define the "good of the many" in a normative sense, our "oughtness" may not only be incorrect, but might not protect individual rights or minority interests. Moreover, we still have the problem of convincing people to be unselfish. This has proven to be the Achilles heel of Marxism. That is, why should I sacrifice now for a future that arrives (if it does) after I am dead?

Christian Ethics

Rather than define "religious ethics," Holmes chooses to deal with a subset, Christian ethics. Both religious ethics and Christian ethics are normative systems that defines "oughtness" in terms of God's laws. Something is good or right because God says it is.

There are two problems with religious ethics. First, as an argument, it may be begging the question. One must presuppose a God who has created an ethical order as well as physical universe. (The failure of the Eastern Religions to accept this does not have a practical effect at this point.) Second, the various religions seem to have the same problem philosophers do by disagreeing on what God's ethical laws are. This second problem is not as severe as it may seem at first because all religions have remarkable agreement on the basic principles of ethics. For example, some form of the "golden rule" can be found in them all. (Hume, 1959)

In spite of these common elements in the ethics of all the major religions, they (with the exception of Christianity) are built on a legal system which can easily degenerate into casuistry. The distinctive of Christianity is that it deals with principles (ie. the "law" of love) rather than individual commands. Though the operation of love is further defined in Christian writings (namely the Bible), even here there can be areas of disagreement in the actual outworkings. For example, some feel serving in the armed forces is wrong while others believe it is an obligation.

ETHICAL MOTIVATION

The main thrust of this paper resulted because the author has found no reference dealing with engineering ethics that examines the second question of ethics, "Why should one be ethical?" All the authors address what we ought to do and few, if any, even acknowledge the existence of the motivation to do what we ought. Many assume education produces motivation. The weakness of this assumption is obvious.

The study of economics give insight into motivation. The economic term "goods" was derived from what is good ethically. The economist assumes that the more goods, the better; both to society and the individual which (some economists acknowledge) means that man is basically selfish.

The results of this selfishness can be good or bad behavior (usually the latter is ignored). The issue is whether individual selfishness is good or bad for society. Some, calling upon Adam Smith's "invisible hand" say maximizing individual selfishness also maximizes social utility. However, others cite the "law of the commons" which says the opposite.

The law of the commons can be illustrated by ten shepherds each having 10 sheep on a meadow (commons) that can optimally support 100 sheep. One of the shepherds reasons that adding only one sheep will increase his income by 10% while hardly effecting the total output of the meadow. Of course the other shepherds quickly recognizes that unless they do likewise, they will lose out. The result is a ruined meadow.

The invisible hand might work where there are unlimited resources, but in a "zero sum game" the law of the commons will prevail. Moreover, Adam Smith seemed to presume that individuals maximizing their utility (selfishness) would do so fairly. Even Adam Smith recognized that common goods (water, air, etc.) and public goods (military service, education, etc.) were exceptions.

It may be further argued that individual selfishness will rarely, if ever, produce societal well being. In fact, one could argue that it is theoretically and practically impossible that integrating individual selfishness can somehow produce social good. Therefore, the problem is to motivate people to be unselfish.

This can be done by external motivation or internal motivation. Our law enforcement system is designed to supply external motivation and is primarily negative in that it punishes but does not reward. Society or culture may offer positive rewards, however.

The weakness with external motivation is the ability of some selfish individuals to not be caught so those who remain unselfish are like the shepherds on the meadow who do not react to the one increasing his herd. Also, an unacceptable burden is placed on the judicial system if a majority of individuals are not voluntarily "unselfish."

The theist's principal motivation is not to escape "being caught" but obeying God because in the final judgement, nothing is hid. There is, however, the problem of people fooling themselves into thinking they are obeying the divine command but in reality are engaging in casuistry. That is, they seek "legal ways" to be "illegal."

Christian ethics are based on internal motivation also, but with the difference that one is to obey the divine command "from the heart" as a result of having been forgiven, not for any negative or positive reward.

It is patently clear that even Christians, in spite of their internal motivation, do not always obey the law. This problem stems from the battle between the superego and the id (Freud,1952). Put in other terms, "we do not do what we want" (Romans 7:18-24). At least, however, the internal motivation is there which is a perquisite to conscious ethical behavior.

CONCLUSIONS

There are three conclusions to draw from the above arguments. First, there are persuasive arguments that say a theistic presupposition is a prerequisite to rational internal ethical motivation. Anscombe (1958) supports this by saying, "...it is not possible to have such a concept unless you believe in God as lawgiver..." The existence of a final judgement (completely fair with no hidden evidence) provides a minimum ethical motivation. Further motivation would be available to those who desire to obey God from gratitude rather than for reward or punishment. This is not to imply that atheists or non theists are not ethical, but only that they do not have a rational basis for their motivation.

Second, constructing a normative ethical system (which, by definition, codes of ethics are) from a relative foundation is a non sequitur. It would seem that only a normative (prescriptive) ethical system can supply the principles and standards on which we base our professional codes.

Finally, it is not necessary to agree that a theistic world view is the only foundation for ethical action or ethical motivation. But, if one concedes that the individual's theological presuppositions **can** produce the existential motivation to "do what one ought," then these presuppositions should at least be tolerated if not encouraged.

Some professional societies, in particular, ASCE, have taken a neutral if not negative stance toward religious groups. For example, the Christian Engineers have not been allowed to announce their luncheon meetings in the ASCE conference schedule. Instead, ASCE (and all professional organizations) should encourage members to integrate their faith as engineers with the practice of their profession. It may well be that this cadre of religious professionals is essential to support both the codes ("what ought one to do?") and the practice ("doing what one ought").

APPENDIX I, REFERENCES

P. L. Alger, N. A. Christensen, and S. P. Olmsted, **Ethical Problems in Engineering**, John Wiley: NY; 1965

G. E. M. Anscombe, "Modern Moral Philosophy," **Philosophy**, Vol. 33, 1958

ASCE, "Code of Practice of the American Society of Civil Engineers," Adopted, January 1927, reprinted 1962, ASCE, NY

K. Baier, **The Moral Point of View, Random House: NY; 1965**

R. J. Baum, ed., **Ethical Problems in Engineering, Vol. II: Cases**, Renselear Polytechnic Institute, Troy, NY, 1980

D. J. Bronstein, Y. H. Krikorian, and P. P. Wiener, **Basic Problems of Philosophy**, 2nd ed., Prentice-Hall: NY; 1955

D. T. Canfield and J. H. Bowman, **Business, Legal, and Ethical Phases of Engineering**, McGraw-Hill: NY; 1954

E. J. Carnell, **Christian Commitment; an Apologetic**, Macmillan: NY; 1957

C. E. Ferguson and J. P. Gould, **Microeconomic Theory**, 4th ed. Richard D. Irwin: Homewood; 1975

A. Flores, ed., **Ethical Problems in Engineering, Vol II: Readings**, Renselear Polytechnic Institute, Troy, NY, 1980

S. Freud, "The Ego and the Id," **Great Books of the Western World**, Encyclopaidia Britannica, Inc.: Chicago; 1952

A. F. Holmes, **Ethics: Approaching Moral Decisions**, InterVarsity Press: Downers Grove; 1984
R. E. Hume, **The World's Living Religions**, Charles Scribners Sons: NY 1959

R. F. Ladenson, J. Choromokos, E. d'Anjou, M. Pimsler, and H. Rosen, **A Selected Annotated Bibliography of Professional Ethics and Social Responsibility in Engineering**, Center for the Study of Ethics in the Professions, Illinois Inst. of Tech., Chicago, 1980

M. I. Mantell, **Ethics and Professionalism in Engineering**, Macmillan: NY; 1964

M. W. Martin and R. Schinzinger, **Ethics in Engineering**, McGraw-Hill: NY; 1983

J. Mock, ed., **The Engineer's Responsibility to Society**, ASME: NY; 1969

M. Scriven, **Primary Philosophy**, McGraw-Hill: NY; 1966

P. A. Vesilind, "Rules, Ethics, and Morals in Engineering Education," **Engineering Education**, Feb. 1988

THE ERA OF DISARMAMENT & ENGINEERING PROFESSIONS

Rusk Masih, M.ASCE[1]

Abstract

The world is moving with a strong momentum towards an era of disarmament and economic cooperation, away from the era of military build up and confrontation. Such an era will have a great influence on engineering professions in general and civil engineering profession in particular. Huge intercontinental engineering projects, which have no prospect in the era of confrontation, will become feasible. The horizon will be more open for the engineering professions. The demand for civil engineers is expected to increase. The new frontiers will dictate several changes in the education and profession of civil engineers. The technical societies will have a new task, mainly to develop international standards, to close the gaps between the different national codes in the industrial world to make technical cooperation easier, and to point out areas of intercontinental projects. Hence, international cooperation in engineering education will become a necessity.

Since USA is the most technically advanced country in the world, this author believes that ASCE should take the lead in forging international engineering cooperation, be it engineering education or technical and professional advancement. A first step towards that goal would be to hold a forum wholly dedicated to serve the era of disarmament from an engineering point of view.

Introduction

In order to have realistic planning for the future of the profession, one must look at the general economic, social and political picture of the country, then conclude what impact that could have on the profession. It is generally agreed that sooner or later, the superpowers will reach a point, either willingly or under pressure, at which the only solution to their local problems will be a massive cut in the military budgets and the reallocation of resources from the military build-up to the restructuring of their economies. This era will be signified by a change of priorities because of socio-economic changes. On our side, many problems require a wholesale overhaul of planning: the drug problem plaguing our society; the deterioration of the infrastructure because of long negligence; nuclear waste dumping; cleaning the nuclear weapons producing plants; pollution and other environmental problems; the saving and loan associations' crisis. On the other side, the Soviets face a low standard of living, ethnic problems, the democratization of their society, which can only be achieved through the higher production of consumer goods.

1-Assoc. Prof., The University of Conn., Avery Point, Groton, Ct. 06340

It is clear that the logical response to these problems on both sides is the massive allocation of billions of dollars, and the shift from an economy depending upon military spending to an economy depending upon peaceful prosperity.

Engineering education and engineering profession is faced with new tasks and challenges dictated by the era of disarmament. One thing is clear: that such an ara will be recognized by more economic and technical cooperation between the superpowers. Such cooperation will increase the volume of civil engineering projects considerably, as it will be discussed and explained later on; it will require more civil engineers to be graduated and more specialization in the CE profession. This change must be studied carefully to come up with certain technical goals, then to work towards that end. Planning for such era must start now to keep the USA, US engineering education and US engineering societies in a leading position world wide.

Which type of engineering will be the beneficiary of this era?

The top beneficiary of such a shift will be the civil engineering profession. It will benefit from the massive infusion of funds to repair the roads and the bridges; the use of funds for the construction of new needed ones; the programs to control pollution-whether it be air or water pollution; the efforts to solve the drought problem in our most fertile land in the midwest by building huge irrigation projects and dams; the need for new innovative housing projects to satisfy a big sector of the population, deprived from realizing the American Dream of owning a home, and the need to solve the problem of the homeless people.

The engineering professions which stand to lose the most from the era of disarmament are the electrical-electronic and aerospace engineering, since the defense projects are heavily geared towards these professions. Although such a setback will be temporary, these professions stand to reap a great benefit in the far distant futures. America could shock the world again by introducing new generations of powerful computers, robotics and other electronic equipment to increase the productivity of labor, which will push our export to a high mark on the scale. The peaceful economy will also try to fulfill man's ambitions in space by building space stations and space project to explore the universe; perhaps some of these projects will be joint ventures with the Soviet Union and other nations. Embarking on such ambitious plans will greatly enhance the importance of aerospace engineering and offset its loss suffered by the disarmament policy.

How can the CE profession benefit from such era?

We can see the signs of such a shift taking place, and much more to come: the students' enrollment in CE programs is gradually increasing, the graduate CE students are finding no difficulty in getting suitable jobs, while, in the meantime, the enrollment in electrical engineering is decreasing as are the demand and prospects for a job in that area.

The civil engineering profession has been in a slump for a long time. Even though the CE students go through the same difficulties experienced by other engineering students, however, their pay when they graduate is considerably less than that of their colleagues in other engineering disciplines.

Although it is believed that the market forces of demand and supply will work their way to put the CE engineering profession at par with other engineering disciplines, there is a lot to be done by the ASCE to upgrade the profession and to help the market forces to achieve the goals in a short period of time. Such actions by the ASCE include the following: stricter codes-specifically when it comes to safety, health and environmental requirements; stricter procedures in granting accreditation to engineering schools; guidelines for the degree of sophistication in engineering design and analysis, depending on the magnitude of the problem, and its effect on safety, health and environment; requirement of checking of the design to be done by firms separate from the ones who do the design as is the case in most of Europe; classification of the engineers as to what type of engineering design they can do, depending on their degree of education; demand that the competition for the design of large government projects be opened to medium size and even small size engineering firms-as long as they are qualified technically; and finally action to put teeth into ASCE guidelines to have the violaters punished by the court and the authorities.

Needless to say, professional development always depends on the number and varieties of research done in the field, so it is imperative that the ASCE should demand from the NSF an allocation of funds to CE research proportionate to the dollar volume of the CE projects done in the country.

<u>The era will have the following influence:</u>

<u>1-Will create CE Mega Projects</u>

If the superpowers reach an understanding and an agreement to disarm themselves simultaneously, then the next step will be to integrate two different economies in one worldwide economy, or at least to intermesh the different economies together, one way or another. One obvious result of this integration will be the planning of mega projects spanning through countries and continents. The already built projects, such as the gas pipe line which transports the Siberian gas to Western Europe, the tunnels accross the Alps, the tunnels between Britain and France, Bullwinkle offshore platform which is taller than Sears Tower (6 Robinson, 1989), all these projects may look like dwarfs compared to the expected projects which will be planned through the era of disarmament. Such projects are expected to start as soon as confidence in the durability of the disarmament takes hold and as soon as all sides believe that the era is not a temporary phenomena, but it is there to stay. Projects envisioned on the intercontinental scale could be the following:

(A) Projects to increase food production and eliminate hunger worldwide, projects such as the irrigation of the Saharas, by rerouting the rivers or transporting glacials (5 Morse 1989), which will have a secondary effect of modifying the deserts' climate and making them habitable. The projects will also solve the problem of high density population areas, because the migration rules between contries will be eased and people can move to the newly habitable deserts. What cannot be doubted is the great magnitude of CE projects that can be created from the rehabilitation of the deserts, ranging from town planning, roads, railroads and airports to irrigation, dredging, pipelines and other construciton projects.

(B) Money will be available for needed projects in our cities, such as "Resource Recovery Systems" with advanced combustion and emission control to relieve the conjested land fill areas from the ever increasing piles of garbage, and reduce the amount of pollution created by the present incinerator system.

(C) Projects to solve the housing crisis, which is becoming a world wide phenomena, where large sectors of the population cannot afford the ever escalating housing prices. A solution will be sought in one of the following three directions or by the combination of either two or the three of them together:

(a)-Search to discover cheaper construction material;
(b)-Search to automate the industry and reduce the labor cost;
(c)-Building new housing projects in the reclaimed remote areas to reduce the land sky rocketing cost. Land high prices have become a major cause of high cost of housing.

<u>2-Will create engineering work in space and under seas</u>

A lot of studies related to space and underseas have been achieved by both superpowers; unfortunately most of those studies were done for military purposes. However, when disarmament becomes a reality, then those valuable studies will be declassified and directed towards peaceful civilized purposes and away from starwars programs. Most of the long term budget of billions of dollars allocated to star-war purposes by both superpowers could be used to space explorations and undersea mining. It may not be too far distant in the future where space stations will be built or even the colonization of the moon and other planets can become a reality (2 Johnson et al, 1981). Technology has no limited boundaries if given sufficient investment. New technology will emerge, which will turn those dreams to reality.

<u>3-Will effect the CE education and practice</u>

Throughout history, technology has always influenced engineering education, and engineering education has always influenced technology (3 Masih 1989). Computer technology dictated that engineering education be revamped (4 Masih 1985), not only by teaching computer languages in every engineering discipline, but also by revamping all engineering courses to make them fit for computer usage. The same thing can happen again when new technology is used to perform civil engineering work in space or under sea. More diversification in CE curriculum may be required; new courses need to be added because of the new technology. It will ultimately lead to the specialization split in the CE programs: one program dealing with CE problems in space (where higher or lower gravity can have significant effect); another program dealing with the undersea engineering (where buoyancy and pressure increase with depth could influence the solutions); and the traditional on the earth surface civil engineering. In fact, some universities have already laid the foundation to deal with this problem (1 Clapp 1989).

Another influence that could be exercised is bringing together the different systems of engineering education in different countries. A global perspective in engineering education may become a reality (7 Simpson 1989). The diversified programs of different countries could be improved by the exchange of students and faculties.

The duties of the engineering societies

Engineering societies, in general, and ASCE, in particular, will be faced with new tasks compatible with the emerging new condition, which, as pointed out before, will dictate mega size intercontinental projects. Engineering societies, whose task is to organize the profession, will have to look beyond the national boundaries of their own countries to forge a better technical cooperation. Perhaps international technical specifications will have to be developed as a joint effort of several technical societies in the advanced countries. Such technical specifications will make it easier for American engineering firms working on Soviet projects or vice versa. It will be a wise step if our ASCE took a leading position in this respect; such a step can be accomplished by the following:

(1) Holding several joint meetings with the technical societies in other countries;

(2) Holding more international conferences dedicated to closing the gap between their national codes for design and construciton;

(3) Issuing a joint recommendation for feasible intercontinental projects.

One major benefit from such a policy will be the adoption in the national codes of different countries of the good points in the best of them. It is a known fact that the West European codes are more stringent in safety and environmental rules, while the Soviet codes are the most negligent in this respect.

Another good outcome of such a joint effort towards intercontinental projects study and recommendation is to draw the attention of the politicians and economic planners to the great economic opportunities which exist in the world. Only the technical societies and technical people can point to that direction. Such a challenge can be tackled successfully by the technical societies, and it will greatly contribute to lasting world peace. No one can doubt that the European economic cooperation has converted the feuding and warring nations of Europe into one large integrated market; the last thing to ever cross their minds is to go to war again.

References

1- Clapp, William G., Space technology education through the combined effort of industry, education and government, 1989 ASEE annual conference proceedings, pp 188-189.

2- Johnson, S.W. et al, Developing concepts for Lunar astronomical observatories: Interdisciplinary team experiences at the University of New Mexico, 1989 ASEE annual conference proceedings, pp 12-14.

3- Masih, R.Y., The pillars of excellence in engineering education. 1989 ASEE annual conference proceedings, pp 422-424.

4- Masih, R.Y., Automating the civil engineering profession and education. 1985 ASCE conference proceedings on civil engineering education, pp 410-414.

5- Morse, Dan, Tapping a glacier by bringing a glacier to a water shortage, civil engineering magazine, July 1989, pp 38-41.

6- Robinson, Rita, Bullwinkle, the outstanding civil engineering achievement of 1989, civil engineering magazine, July 1989, pp 34-37.

7- Simpson, I.R., Global perspective in engineering education: network among international institutions, 1989 ASEE annual conference proceedings, pp 139-141.

ARE TRANSPORTATION ENGINEERS PREPARED FOR WINTER ?

Task Committee on Transportation of the Education Committee of the Technical Council on Cold Regions Engineering

Abstract

Resource and other development in northern regions will mean increasing opportunities for transportation engineers. Many newly assigned engineers to northern projects will be from more temperate areas. A number of effects of cold and other aspects of northern regions upon several areas of civil engineering are discussed. The results of a survey of cold regions engineering in the civil enginering curricula are summarized. Several recommendations are made regarding greater emphasis on cold regions problems in the curricula.

Introduction

The past twenty years has been marked by a significant increase in development projects in northern areas, both in North America and Eurasia. Not only are the natural resources of the north being tapped at an increasing rate, but the settlements which are home to the indigenous people of the north are being provided with basic utility systems, community facilities and the associated transportation infrastructure. The increasing activity has resulted in a need for greater numbers of civil engineers, and many of the new engineers are migrants whose education and early training took place in more temperate areas. This forum provides a timely opportunity to examine the needs of the northern engineering community and to try to identify areas where some deficiencies in the educational process may exist.

Future of northern transportation

It might be said that the sub-arctic begins where the presence of a mature, integrated transportation system ends. Because of this the cold regions will offer many oportunities to the next few generations of transportation engineers. New airstrips will need to be located and constructed, and existing airports expanded. The highway system is fragmented, and new roads to development sites will be needed. Village streets need improvement, and existing gravelled roads will need to be paved. New airports will need roads to the town or site they are intended to serve. Extension of the few northern railroads is often discussed, and eventually some of this talk may turn into action. Water transportation is still important in the north, and dock and port improvements for ocean and major river shipping will be required. The Alyeska pipeline was the largest privately funded engineering project in history, and northern development may likely require similarly sized projects in the future.

Cold affects all aspects of life in the north. It also affects most of the disciplines which are the subject of the civil engineering curriculum. We shall examine several subject areas and briefly discuss some of the factors which are of concern to the northern practitioner.

Geotechnical and frost effects

It is unlikely that any graduate is unaware that the ground will freeze if the weather is sufficiently cold. For frozen ground, however, there may be a set of parameters which differ from those of conventional analysis involving measures of strength, consolidation, etc. Frost-susceptable soils may heave when freezing, and almost always lose strength when thawing. Frost action may result in significant upward movement of loaded piles. Fine grained soils often can not be used for embankment construction, and must be wasted and replaced by granular materials. Large areas may be underlain by organics soils (peat and muskeg), but still must be crossed by a highway.

Permafrost is perhaps the best known but not necessarily the best understood phenomenon in parts of the north. The term is actually a description of the thermal state of a soil, and analysis is dependent, among other things, upon what is frozen, its moisture (ice) content and possible changes in the thermal regime.

Heat transfer

Most engineers learned in high school physics of the three modes of heat transfer, and that conductive heat flow is from the warmer to the colder. Many civil engineers, however, are not prepared to apply these principles as a part of a design analysis. Any time the effects of freezing or thawing of a material must be taken into account it is important that the engineer be familiar with heat transfer including change of state. This is basic to an understanding of depth of frost penetration predictions, estimates of the extent of permafrost degradation, and the effects of adding insulation layers to reduce the rate of heat flow. Pipelines having contents which are either heated or chilled can present a special case. Not only must the temperature change of the contents be predicted, but often the temperature change of the surrounding soils must also be estimated.

Environment

It is axiomatic that whatever we build must be designed to withstand those natural forces to which it might be subjected. The effects of extreme and prolonged cold can result in natural forces and phenomena which differ from those in more temperate areas. Ice, snow and packed snow will cover most horizontal or sloping surfaces for several months each year unless it is removed. Facilities for both vehicles and pedestrians must be designed with a recognition of their being used under conditions of reduced surface traction. It may be possible to design highways, railroads and airports so that blowing snow will not form drifts. In some cases, however, it may be necessary to provide space specifically for the seasonal storage of snow. For most transportation facilities snow and ice control will be the major determinant in establishing personnel and equipment maintenance budgets.

Springs and seepages are common in parts of the north. Some freeze and stop flowing during the winter, but others do not. The resulting ice accumulations can completly engulf a highway or other facility. Similarly, many culverts become completely filled with ice and are rendered useless as drainage conduits. If these problems can not be avoided during the site or route selection process, then mitigating measures must provided.

Most northern rivers and streams freeze during the winter. This can radically alter the hydraulic regime of the river, and the conveyance of a particular reach may, when ice covered, differ significantly from the unfrozen condition. In extreme cases, the river may "freeze to the bottom", and the resulting frozen overflows cover the flood plain. On some rivers the transition periods, freeze-up and break-up, may be especially troublesome. Frazil ice may collect at obstructions and initiate ice jams. Ice floes during break-up may also jam at obstructions. An ice jam may also occur when a river thaws upstream and ice floes meet a still frozen reach downstream.

For marine structures, ice forces may be the major lateral load. Where there are water surface fluctuations, such as tides, ice may progressivly coat structural components during each fluctuation. The increasing ice thickness can add a significant vertical load to the structure, especially if there has been a liberal use of cross bracing.

Cold can also affect other properties of construction materials. Flexible pavements often develop a series of transverse cracks resulting from thermal contraction of the pavement. Steel generally becomes more brittle at low temperatures, and designers must be familiar with low temperature impact requirements for selected alloys under various loading conditions.

Construction and Operations

Most projects with which civil engineers are involved are built outdoors with minimal protection from the elements. This is especially the case with most transportation projects. It is no surprise that most construction in northern areas is during the summer. An added advantage is that the long hours of daylight often allow a two shift operation for at least a part of the season.

There is no guarantee that a proposed project will be accessable from the existing transportation network, or even from a remote barge landing. In these cases a project manager may want to consider a winter mobilization or even winter construction. Frozen ground and frozen river and lake ice provide a good foundation for vehicles and construction equipment, and winter mobilization has a successful history in the north. The same is true of winter embankment construction over areas which could not support equipment during the summer. There are negative aspects to such work. During periods of extreme cold production rates may become very low, and the situation is worsened by high winds. Reduced visibilities from blowing snow may bring work to a halt. The lack of winter daylight will usually mean artificial illumination of the work site. Controlled compaction of even dry granular materials is generally not possible. While winter work is expensive, designers and construction managers should be aware of the fact that it can be done. If winter work is to be considered, the designer's product must be capable of being built under winter conditions.

Because of the low population density, many northern areas can be described as remote, and local support services for the construction industry may not exist. The nearest spare parts vendor may be two thousand miles away. Accordingly, project planning should include an on-site inventory of commonly used parts. Estimates for projects which include installed mechanical equipment should also include a parts inventory.

Frost action can result in problems for both maintenance and operation of transportation facilities. Surface deformations from heaving of frost-susceptable soils or thaw settlements in permafrost can cause a reduction in safe operating speeds with an attendant increase in transport costs. Maintenance forces will eventually be required to re-grade and re-pave highways and shim or raise and level railroad tracks. Load restrictions occasioned by weakened subgrades during the spring thaw can effectivly halt all heavy trucking for several weeks, and this must be recognized in planning a mobilization or supply operation.

Present State of Cold Regions Education

During 1985 the Education Committee of the Technical Council on Cold Regions Engineering (TCCRE) canvassed 225 universities in the United States and 28 in Canada. A questionnaire requested information regarding, among other things, the extent of cold regions engineering in both the undergraduate and graduate curricula. The results were reported in Inventory of Education and Research Opportunities in the field of Cold Regions Engineering by the Education Committee, TCCRE.

Of the U.S. Universities, 161 or 72% responded. Of these, 39 or 17% reported significant educational activity in this area. As might be expected, the positive response was greater from the northern states. 19 of the 39 responses describing cold regions programs were from the 14 states which abut the border with Canada.

18 of the schools with cold regions programs reported course offerings at the undergraduate level. 31 offered graduate level courses, while 13 reported offering courses with significant cold regions content at both levels.

17 of the Canadian universities, or 61%, responded. In Canada, 15 or 54% of those queried reported significant cold regions educational opportunities. That this is a substantially higher percentage than in the U.S. should be no surprise. Six offered courses at the undergraduate level, ten at the graduate level and two at both levels. These are essentially the same ratios as in the U.S.

The reporting format combined with the relatively small data base makes it impractical to attempt to report results specifically for transportation engineering related courses. As, however, most of the undergraduate curriculum has some transportation related aspect, a reasonable assumption is that the same percentages would apply.

There is a strong possibility that some degree of under-reporting exists. Unsolicited questionnaires often do not receive the amount of attention the pollster would desire. In addition, the respondents at larger institutions might not be thoroughly familiar with the content of the entire catalogue of courses. Nevertheless, it is evident that the effects of cold climates on the work of the civil engineer are generally being ignored at many universities. This is particularily unfortunate in the transportation engineering field where forces related to the natural environment can be critically important. Engineers who becomes involved in any type of northern development may find that they are in need of some reeducation.

Current textbook content generally does not encourage

coverage of cold regions problems. An examination of the texts in use at one university discloses that many had no mention of cold regions problems. One, on transportation engineering, did devote 7.5 % of its contents to cold related problems, but the average for all textbooks was barely over 0.5 %.

Recommendations

We cannot with a clear conscience suggest that a course in cold regions engineering be made a required undergraduate course. We are very much aware of the pressures to add courses in the humanities, arts, computer sciences, etc. As it is possible that in some parts of the U.S. (e.g., the "sun belt" states) there may be little student interest in such a course, we are hesitant to recommend that an elective course in cold regions engineering be universally adopted. It is notable, however, that some southern schools which are close to the petroleum industry have recognized a need, and cold regions problems are included in some of their courses.

This committee does recommend that greater awareness of the occurrence and importance of cold regions problems in civil engineering be incorporated into the undergraduate curriculum. Cold regions material should be an integral part of such courses as geotechnical engineering, hydraulics, hydrology, transportation, construction management, and structures, recognizing that all of these impact on transportation to some degree.

Because of the increased degree of specialization at the post-graduate level, no blanket recommendation can be made. There are some universities which do not offer advance degrees in transportation engineering. For those that do, this committee offers the following recommendations:

1) One graduate level course specifically dealing with cold regions geotechnical engineering, presented with a strong slant toward transportation engineering.
2) One course in cold regions hydrology (basics of snow and ice) with specific applications to transportation systems.
3) A strong emphasis on cold regions engineering and remote area operations in courses in construction and construction management.

There are, of course, advanced degree candidates whose area of geographic interest may be one that would not require an understanding of cold regions engineering, e.g. the developing tropical nations. We would concur with a suggestion that these students not be required to take such courses. At the same time we would question the accuracy of a student's predictions of really knowing where he will spend his entire professional career.

The recognition of cold regions problems in engineering course work must be reflected in the textbooks used for the courses. The amount of a text devoted to cold regions problems will, of course, depend upon the subject. A minimum of two percent of the total pages on cold related phenomena is suggested as a starting point for debate. As transportation engineering is one of the areas most strongly affected by cold related problems, we recommend an early emphasis on texts in this field.

Certain minimum facilities are necessary to support course work, especially at the graduate level. It is the recommendation of this committee that universities having graduate programs in transportation engineering have at least a rudimentary cold room. It need not be large or elaborate, but should be equipped with the proper controls to maintain preselected temperatures and rates of temperature change.

The next several decades will see large scale development in many parts of the north. This will result in industry and large consulting firms mobilizing engineers and builders, often on an expedited schedule. These engineers will be drawn from from all parts of the country if not the world. If the lessons from similar activity during the recent past are not to be lost, and of necessity relearned at substantial cost, it is the duty of the education community to adequately prepare fledgling transportation engineers. The newly arrived engineer in the north must be equipped with more than warm clothing.

ASCE TECHNICAL COUNCIL ON COLD REGIONS MONOGRAPH SERIES

Rupert G. Tart, Jr.,[1] Member ASCE
Randy D. Crissman[2], Member ASCE

Abstract

The Technical Council on Cold Regions Engineering's monograph series on cold regions engineering was developed to disseminate state-of-the-practice information on engineering techniques and methods as applied to civil engineering in cold regions. Seven monographs have been published to date and several more are being developed and planned. This paper discusses how the monographs are being used by practicing engineers and in civil engineering education.

Introduction

Formal education rarely can extend beyond theory and basics. Learning how things work in the real world is frequently a product of experience. This certainly was the case, until a few years ago, in the field of cold regions engineering. This paper discusses an educational vehicle that was developed by the Technical Council on Cold Regions Engineering to elevate awareness of the impacts cold regions can have on engineered projects.

With the completion of the Trans Alaska Pipeline in 1977, the largest privately funded construction project in history, a new awareness of civil engineering problems that are unique to cold regions was born. In May 1977, a group of American Society of Civil Engineers (ASCE) members who had worked in Alaska and some of the northern

---

[1]Associate, Golder Associates Inc., 1813 East 1st Avenue, Anchorage, Alaska 99501
[2]Hydraulic Engineer, New York Power Authority, Niagara Power Project, P. O. Box 277, Niagara Falls, New York 14302

tier states convened in Anchorage for the first ASCE Cold Regions Engineering Specialty Conference. Over 300 people attended (176 more than had been anticipated). The participation in this conference demonstrated that civil engineers had a strong interest in learning more about the unique cold regions problems that were confronted during the construction of the Trans Alaska Pipeline.

At this conference, preliminary plans were formulated to establish the Technical Council on Cold Regions Engineering (TCCRE). On July 20, 1979, the ASCE Board of Directors authorized the formation of the TCCRE. The purpose of the Council is to "identify, assess, and report effects of cold region environment upon engineering design, construction, and operations and to make recommendations for the advancement of scientific knowledge and practice in engineering solution of cold regions problems".

One of the first actions of the Council was to form technical and administrative committees to address the concerns of civil engineers designing and building in cold regions. The Design, Construction, and Operations Committee was formed "to identify, assess, and report influences and effects on engineering concepts, methods, and facilities in environments where below freezing temperatures may occur for a significant period of time." This committee was later renamed the Design and Construction Committee (DCC).

Purpose of the Monograph Series

The organizational meeting of the Design and Construction Committee took place in Seattle, Washington in October 1981. At that meeting, the committee spent considerable time discussing how best to accomplish the goals inherent in the committee's purpose. The committee felt that hurried development of manuals or standards would not be appropriate, because of the lack of up-to-date documentation of the state-of-practice in the field of cold regions engineering. At that time, many of the techniques and methods that had been used during the construction of the Trans Alaska Pipeline had not yet been documented in the literature.

It was the consensus of the DCC that most ASCE publications stressed state-of-the-art techniques and methods, were heavily theoretical, and did not address the state-of-the-practice and how the practice was developing. This position was supported, and still is, by many practicing civil engineers. Concerns about dissemination of practical information in all disciplines of civil engineering have been presented to the ASCE Publications Committee on several occasions, primarily due to the strong theoretical and academic leaning of the ASCE Journals.

In order to provide a forum for disseminating state-of-the-practice information, the committee proposed the development of a monograph series to define the state-of-the-practice in cold regions engineering. The monographs were intended to summarize the state-of-the-practice so practicing engineers could review, critique, and update the information for eventual inclusion in standards and manuals of practice for cold regions engineering. The goal was to invite practicing engineers to prepare manuscripts for monograph topics that were limited to particular areas of interest in cold regions engineering.

Monograph Preparation

The first monograph was entitled "Design for Ice Forces" and addressed the current state-of-the-practice in design of structures to withstand forces exerted by floating and moving ice. The effort required to produce this monograph was significant, because it broke new ground in ASCE's publication procedures and policies. The project succeeded partly because the editors were not familiar with ASCE's publication policies. The proposal submitted to the ASCE encountered some resistance, but the editors finally prevailed and were able to publish a document that fairly represented the current practice.

Currently, monographs in the series are developed under the direction of a Task Committee, whose Chairman and Vice Chairman are also the editors. The Task Committee is responsible for developing the scope of the monograph, inviting authors to prepare manuscripts, tracking the progress of manuscript preparation, and submitting the final document to ASCE for publication. In preparing a monograph, the Task Committee and the editors follow guidelines developed by the editors that prepared the first few monographs (ASCE, 1989a). The invited authors are also provided with guidelines for preparing their manuscripts (ASCE, 1989b).

The monograph papers are normally presented at ASCE annual conventions or specialty conferences. However, presentation of the papers is not required. When the papers are presented at a convention or conference, there is no reference to the ASCE convention or conference printed on the cover or the title page of the monograph.

The Task Committee goes through the following steps in producing a monograph:

1. Outline the proposed content of the monograph.
2. Obtain approval for the monograph from the TCCRE Executive Committee.
3. Identify potential authors for each chapter in the monograph.
4. Solicit abstracts and outlines for each chapter from invited authors.
5. Review abstracts and outlines and provide comments to the authors.
6. Refine monograph table of contents and request draft manuscripts from the authors.
7. Receive draft monograph chapters from the authors.
8. Obtain at least two peer reviews of the manuscripts.
9. Transmit peer reviews to the authors for inclusion in the final, camera-ready version of the manuscript.
10. Submit camera-ready monograph manuscripts to ASCE publications staff for publication.

This process can be lengthy. Some monographs have been produced within 18 months. Most take two or more years. Table 1 is a list of the Cold Regions Engineering Monographs published to date.

Success of the Monographs

Number of sales is a measure that ASCE uses for success of a publication. By this standard, the monographs have done very well. Table 1 presents the number of monographs printed and the sales to date of each. On average, more than 500 copies of each of the monographs have been sold.

Another measure of success for the monographs is how they are used. Several of the monographs have been used as texts or supplementary texts for cold regions related courses taught at the University of Alaska. Many consulting firms use the monographs as primary references in their respective technical areas.

The popularity of the monograph concept for disseminating state-of-the-practice information to the civil engineering profession is expanding. The guidelines for editors and authors developed by the TCCRE have been requested by several other ASCE divisions and technical councils. It appears they are interested in adopting the approach initiated by the TCCRE.

Table 1

Cold Regions Engineering Monographs Published by ASCE

Title	Date Printed	No. of Papers	Number Printed	Number[3] Sold
Design for Ice Forces	1983	12	1,032	958
Cold Regions Construction	1983	8	1,047	644
Frost Action and Its Control	1984	4	1.046	734
Freezing and Thawing of Soil-Water Systems	1985	10	1,061	377
Thermal Design Considerations in Frozen Ground Engineering	1985	9	993	458
Embankment Design and Construction in Cold Regions	1988	7	846	295
Arctic Coastal Processes and Slope Protection Design	1988	11	807	222

Future Plans for Monographs

TCCRE continues to place a high level of importance on the preparation of monographs. The program has been very successful and is justifiable. There are two goals the TCCRE has relative to the monographs. The first is to continue producing new monographs. The second is to begin to revise and update older monographs. Because of the time required to cover all of the areas in civil engineering impacted by cold regions, it is likely that TCCRE will continue to develop new and revise old monographs for the next few years. Monographs that are planned or are in the process of being produced are listed in Table 2. Once all the areas of cold regions engineering are documented by the monographs, the TCCRE plans to begin the preparation of cold regions engineering standards and manuals.

---

[3]as of September 30, 1989.

Table 2

Cold Regions Engineering Monographs
In Production or Planned

Title	Planned Print Date
Cold Regions Hydrology and Hydraulics	Aug. 1990
Roadways and Airfields in Cold Regions	Feb. 1991
Pipelines in Cold Regions	May 1991
Arctic Foundations	Feb. 1991
Materials in Cold Regions	not set
Hazardous Waste Management in Cold Regions	not set
Revision of "Design for Ice Forces"	not set

Conclusion

The TCCRE found that increasing the awareness of civil engineers of the impacts of cold regions on engineered projects could be implemented through a series of Cold Regions Engineering Monographs that documented the state-of-the-practice. They found that many practitioners were willing to contribute and found that the documents produced were well received and have been highly utilized by civil engineers practicing in cold regions. The TCCRE is continuing to work hard to prepare new monographs to disseminate more information on cold regions engineering to the profession.

References

American Society of Civil Engineers, "Editor's Guide for Preparation of Monographs," Technical Council on Cold Regions Engineering, Cold Regions Engineering Monograph Series, 1989a.

American Society of Civil Engineers, "Author's Guide for Preparation of Manuscripts," Technical Council on Cold Regions Engineering, Cold Regions Engineering Monograph Series, 1989b.

Information Systems for Cold Regions Engineering

Bernard D. Alkire[1]

Abstract

Information systems available to the cold regions engineer are reviewed and sources are listed. Included are discussions of published and on-line abstract services as well as other written, video tape and short courses available.

Introduction

Information is a critical need for the engineer. When the information is related to a specializing area such as cold regions engineering, the information becomes even more critical because the pool of "knowledgeable" persons is limited.

As a seeker of information related to a specialized subject area there are several sources available. A typical search for information would begin with a look at your own resources. Textbooks from your college days, professional journals, manufacture's literature and other items may be at your finger tips. However, once this easy search for information has been explored, then it is necessary to expand your efforts into a more systematic and extensive search into what might be called the information systems.

In this paper, the information systems for cold regions engineering will be identified as abstracts and indexes, written, video tapes and short courses. The emphasis will be on cold regions engineering with the examples drawn from this area of specialty.

---

[1]M.ASCE, Professor of Civil and Environmental Engineering, Michigan Technological University, Houghton, MI 49931.

Indexes and Abstracts

Published

The best way of identifying written material is by using an engineering abstracting service. Coverage of engineering literature is extensive and an article by John Herling (Herling, 1961) lists over 40 services. Typical examples related to engineering are the Engineering Index, Applied Science and Technology and Science Abstract.

The Engineering Index is typical of this type of service. It is an annual volume arranged by subjects with an author and affiliation index. Abstracts are developed selectively on the basis of engineering significance. The list of journals, conference proceedings, reports, etc. number nearly 1500 titles from 80 countries. Abstracts are tied to the key words listed in the publication, SHE: Subject Headings for Engineering by Engineering Index Inc., 345 East 47th St., New York, NY 10017. Entries are made under one subject with cross reference from other subjects.

The ASCE Annual Combined Index contains subject and author indexes for all journal papers, technical notes, Civil Engineering - ASCE feature articles and books published during the previous year. For example, the 1988 index listed the article "Thermal Performance Degradation of Wet Insulation in Cold Regions" by Terry McFadden under the subject heading COLD REGIONS. ASCE's Publication Information contains the abstracts for the materials listed in the combined index.

Abstracting services more closely related to cold regions engineering are the Bibliography of Cold Regions Science and Technology and the Antarctic Bibliography. The former is prepared under the sponsorship of the U.S. Army Corps of Engineers Cold Regions Research and Engineering Laboratory (CRREL) at Hanover, NH. The bibliography is directed toward works in the area of snow, ice, frozen ground, climatology, military operations, navigation on ice, ice engineering, civil, chemical, construction and environmental engineering in cold regions. Items are indexed by subject and author and hard bound copies are available from the National Technical Information Service, Springfield, VA, 22161. The Antarctic Bibliography is sponsored by the National Science Foundation and is more general and includes material on expeditions, equipment and supplies, political geography, as well as biological, geological, chemical science and engineering. Material cited include maps, patents, dissertations, government reports, journal articles, conference proceedings, and monographs. The cited literature in both bibliographies is international in scope and includes Russian, Japanese, Canadian and others as well as U.S. sources.

Most abstracting services have shortcomings including duplication of periodicals indexed, insufficient coverage, long time lag between publication of journal articles and indexing, poor abstract quality, and poor or inadequate subject headings. However, abstracts are a vital link in the information system and any engineer seeking current information on subjects related to cold regions engineering would use these systems as a matter of routine.

On-Line Data Bases

An on-line data base means a collection of records called a data base held in a rapid access computer memory. The data base is searched to locate records of interest. The main advantages of an on-line search are: 1) the number of records that can be searched, 2) the user can interact with the data base expanding or contracting the subject list searched, 3) there is access to more than one data base, and 4) a hard copy of desired records can be easily obtained. The main disadvantage is cost and lack of access. To illustrate the depth of coverage over 100 million references were listed as on-line in the late 1980's (Hall and Brown, 1981).

A data base that is of particular interest to the cold regions engineer is known as COLD available from the ORBIT Search Service, 8000 West Park Dr., McLean, VA 22102. This data base provides coverage of the same sources cited in the Bibliography on Cold Regions Science and Technology and Antarctic Bibliography already discussed. Citations in the data base may be located by fields describing the subjects, authors and titles or any combination of these fields. For example, a search with the single key words Creep OR Ice would produce a longer list than the combination of Creep AND Ice. In fact, it is possible to AND/OR combinations of subjects, authors, titles, and dates to obtain the broadest possible or narrowest selections of sources (Cummings, 1983).

Once the correct combination of key words has been searched, it is necessary to select the type of output desired. Because cost of the service is determined by connect time ($80/hr) and number of citations (0.20

ea), it is to the users advantage to have the citations printed off-line and mailed to the user. However, on-line printing may be appropriate if the number of citations is limited or time is an important factor.

The ORBIT system also contains other complementary files that might be searched such as COMPENDEX (Engineering Index), EIMET (Engineering Meetings), NTIS (National Technical Information Service), and INSPEC (Physics Abstracts).

Written Sources

Books

Books related to cold regions engineering are not numerous. The 1988 Books in Print subject index lists the following books under the indicated heading:

Engineering-Cold Weather

Andersland, O. B. and Anderson, D. eds., Geotech-technical Engineering for Cold Regions," McGraw-Hill, 1978.
Phukan, A., Frozen Ground Engineering, Prentice-Hall, 1985.
Yakovlev, G. N., Studies in Ice Physics and Ice Engineering, Coronet Books, 1972.

Frozen Ground

Harris, S. A., The Permafrost Environment, B and N Imports, 1986.
Jumikis, A. R., Thermal Geotechnics, Rutgers University Press, 1977.
Finell, K. and Tedrow, C. F., Soil and Permafrost Surveys in the Arctic, Oxford University Press, 1981.
Pewe, T. L., Permafrost and Its Effect on Life in the North, Oregon State University Press, 1970.

These books can be purchased from the publisher or ordered through a bookstore.

ASCE Cold Regions Engineering Monograph Series*

This monograph series was developed by the Technical Council on Cold Regions Engineering (TCCRE) to provide state-of-the-practice engineering techniques and methods as applied to civil engineering in cold regions. Each volume addresses a major aspect of cold regions engineering through manuscripts prepared by invited authors. To date there are seven volumes in this series:

Caldwell, S. R. and Crissman, R. D. (Eds.), Design for Ice Forces, 1983.
Jordan, D. F. and McDonald, G. N. (Eds.), Cold Regions Construction, 1983.
Berg, R. L. and Wright, E. A. (Eds.), Frost Action and Its Control, 1984.
Anderson, D. M. and Williams, P. J. (Eds.), Freezing and Thawing of Soil-Water Systems, 1985.
Krzewinski, T. G. and Tart, R. G. (Eds.), Thermal Design Considerations in Frozen Ground Engineering, 1985.
Johnson, E. G., Phukan, A., and Haas, W. H. (Eds.), Embankment Design and Construction in Cold Regions, 1988.
Chen, A. T. and Leidersdorf, C. B. (Eds.), Arctic Coastal Processes and Slope Protection Design, 1988.

Proceedings Cold Regions Engineering Specialty Conferences

The Technical Council on Cold Regions Engineering sponsors a specialty conference on cold regions engineering on a semi-annual basis. The first specialty conference was held at Anchorage and the most recent (5th) was held in St. Paul, Minnesota.*

*Available from ASCE, 345 East 47th St., New York, NY 10017.

Other Major Conference Proceedings and Reports

International Permafrost Conferences: 1st in 1963 to 5th in 1988.
International Symposium on Ground Freezing: 1st in 1978 to 5th in 1988.
Polar Research Board, National Science Foundation
International Conference on Offshore Mechanics and Arctic Engineering (1st to 7th)
International Conference on Port and Ocean Engineering Under Arctic Conditions (1st to 9th)
Canadian Permafrost Conferences (1st to 4th)

Technical Journals

Many journals publish articles related to cold regions engineering as part of their coverage of civil engineering. However, there are only a few journals with a primary interest in cold regions. They are:

Journal of Cold Regions Engineering, ASCE, New York, NY.
Cold Regions Science and Technology, Elsevier Scientific Publishing Company, Amsterdam.
Arctic Research of the United States, National Science Foundation, Washington, D.C.

Video Tapes

A rapidly expanding source of information available to the engineer is video tapes. They are relatively easy to use and are particularly effective when used to train or inform a group. At present, the number of videos available to the cold regions engineering community is somewhat limited but is rapidly expanding. To the first time user with an interest in a specific subject, the best method of location of a possible source is to contact your state library and obtain a copy of their video catalog. Other good sources are:

The Video Source Book: 8th Edition, National Video Clearing House, Inc., 100 Lafayette Drive, Syosset, NY 11791.
PBS Video Catalog, PBS Video, 1320 Braddock Plane, Alexandria, VA 22314.
International Film and Video Resource Guide, National Film Board of Canada, 1251 Avenue of the Americas, New York, NY 10020.

Each of these sources has a subject index. For example, The Video Source Book lists the subject: Engineering, and under that heading the title "Arctic Engineering". Looking up this title provides the following information about the video tape:

Title: Arctic Engineering
Length, Color, Format: 60 min., color, VHS
No. of Programs in Series: 36
Program Description: A course in 35 hours involving engineering in polar regions.
Ancillary Material: Available on Request
Producer: University of Alaska-Fairbanks
Acquisition: Rent or Purchase

Other sources for video materials include governmental agencies (USA CRREL); product manufactures (National Salt Institute); and professional societies (ASCE).

Short Course

Many universities sponsor short courses that provide information to the cold regions engineer. An example is the course, "Ice Engineering for Rivers, Lakes and Oceans," sponsored by the University of Wisconsin Extension Service, Madison, WI. As an aid in identifying available university courses, TCCRE's Education Committee conducted an inventory of cold regions engineering courses taught by universities in the United States and Canada. The report from this effort, "Inventory of Education and Research Opportunities in the Field of Cold Regions Engineering," lists courses available from 39 institutions. Copies of this report are available through the secretary of the Education Committee.

Summary

There are many sources of information available to the cold regions engineer. The main types identified in this paper are abstracts and indexes, written, video tapes and short courses. Written information can be identified through library indexes or by using on-line computer assisted searches. The on-line searches are fast and comprehensive but may be expensive.

Video information is rapidly increasing and will be a primary source of engineering information in the future. Present sources of information can be located through video catalogs or source books.

Courses taught at universities are available and several efforts are being made to improve the number and availability. Overall, there are many excellent sources of information available to the cold regions engineer.

Appendix: References

1. Cummings, N. H., "Cold Regions Science and Technology Bibliography," Cold Regions Science and Technology, Vol. 4, 1983, pp. 73-75.

2. Hall, J. L. and Brown, M. J., "On-Line Bibliographic Databases," An International Directory, 2nd Edition, Aslib, 3 Belgrave Square, London, 1981.

3. Herling, J. P., "Engineering Abstracting Services," Special Libraries, Vol. 52, No. 10, Dec. 1961, pp. 560-565.

COLD REGIONS ENGINEERING, A RESEARCH AGENDA

Robert F. Carlson[1]

Abstract

The Arctic and cold regions of the United States present many unique and difficult engineering problems that demand a coordinated fundamental research program. A recent workshop discussed the type of fundamental research programs that would aid in the solution of long term cold regions engineering problems. Over 40 participants in a 2-1/2 day period suggested 14 research programs within four broad groupings--offshore technology; watersheds, rivers and coastal zones; facilities infrastructure technology; and transportation infrastructure technology. These programs, if implemented in an academic environment with strong graduate programs, will enhance the civil engineering profession's ability to solve cold region problems.

Cold Regions Technology in the United States

Cold regions have been important to the United States since its inception in 1776 and dramatically increased in significance with Seward's purchase of the Alaska Territory in 1867. In addition to Alaska, a significant portion of the population of the United States lives and works in areas impacted by cold and severe winter conditions. Low temperatures, snow, ice, freeze/thaw cycles and the accompanying special climatic and geographic features couple with both population centers and remote areas to present unique

1 Professor, Department of Civil Engineering, University of Alaska Fairbanks, Fairbanks, AK 99775-0660

and challenging engineering issues. Following statehood in 1958, the discovery of oil on the North Slope dramatically added to the strategic and economic importance of the northernmost state.

Research should address significant knowledge gaps, and new research programs should address those areas where a major return on investment can be realized. The cold regions engineering research outlined here fits these criteria well. It is extremely important to the nation, both from strategic defense and natural resource perspectives, to significantly improve the theoretical underpinning of our engineering capabilities in cold regions. The ability to cope with potential major climate variations, the continued environmentally compatible development of natural resources, the effective design and construction of future facilities and transportation systems in the face of rising energy costs, and the ability to contain, cleanup and reduce hazardous and toxic wastes are major national issues that will benefit from the research conducted. The proposed research program is a priority national need that is seldom recognized by a population that relies on the fruits of the high latitudes but seldom experiences the rigors and challenges.

Cold Regions Engineering Research

In response to a need to better define the nation's arctic research effort, the National Science Board issued a report ("Colwell Report") (National Science Board, 1987) which states "the arctic is a new and extremely difficult frontier environment with major engineering demands to provide facilities, to find and develop resources, to provide transportation and communications and to develop infrastructure. Conventional disciplines in engineering are faced with new challenges in cold regions." As a follow up, the National Science Foundation sponsored a workshop study (Carlson, Zarling, and Link, 1989) that develops four broad groups of research, under which 14 research programs are discussed.

Other discussions of arctic and cold regions engineering research include Carlson and Morgenstern (1978), Haneman and Carlson (1985), and the Interagency Arctic Research Policy Committee (1987).

An important activity for the research program would be to provide educational and training opportunities that would increase the number of engineering students graduating with expertise in cold regions engineering. Provision of educational opportunities could be made through graduate research support, fellowships, workshops, and short courses. Information transfer would be encouraged and would increase proportionately with the increase in research activities.

A Cold Regions Engineering Research Program

Offshore Technology. Offshore technology in cold regions is of great engineering importance, as offshore sites are often the site of large petroleum and other resource development structures which must be operated in an extremely harsh environment. Designers must consider ice forces, unique geotechnical problems, difficult material problems, and challenging pollution control and cleanup procedures. In spite of the strategic importance of offshore technology, very little long-term basic engineering research commitment has been made.

The workshop identified research projects in four areas:

1. Ice technology
2. Offshore geotechnology
3. Material science
4. Pollution control technology

Watersheds, Rivers and Coastal Zones. Watersheds, rivers and coastal areas in cold regions constitute complex and especially sensitive environments that pose numerous special engineering problems and provide opportunities for engineering and industrial activities. Frigid air, fresh and salt water, together with the formation, presence, movement and decay of ice in its various forms, and permafrost are, of course, the prominent features of these environments. The problems associated with these features do not simply add to problems to be overcome in "warm regions." These features radically alter environments by affecting or interacting with natural processes such as precipitation, water storage and flow, erosion, terrain morphology, and heat and salt transport in soils.

The group developed three research areas:

1. Watersheds and rivers
2. Nearshore and coastal processes
3. Nearshore engineered structures

Facilities Infrastructure Technology. The cold regions of the nation are subjected to natural phenomena that exert an extraordinary influence on the infrastructure needed for habitation, resource development, and the proper growth of society. Such phenomena include snow and ice, seasonal and permanently frozen soils, frost heave, low temperatures, and environmental corrosion.

The harsh environment of cold regions subjects highways, buildings, and utility systems to greater stress than they encounter elsewhere. It introduces new complications with foundations, buried utilities, building envelopes, and energy consumption. The large environmental value of many cold region areas places a great premium on engineering construction that proceeds with minimum impact. The group discussed three main areas:

1. Coping with the natural environment
2. Improving the built environment
3. Environmental protection

Transportation Infrastructure Technology. This group addressed the concerns of cold regions engineering in the area of transportation facilities including roadways, airfields, intermodal terminals and pipelines. These projects are often characterized by a large capital expenditure, large linear distance scales, and design and construction in a very uncertain information environment. Design failures are often catastrophic or incur a high maintenance cost.

In addition to transportation problems, this group also addressed general concerns associated with cold regions, such as changes in global climate, quality of environment, material science problems, and geophysical methods for site investigation in cold regions. Although aspects of these concerns touch the other research areas, they don't fit cleanly into any one of them and are sufficiently broad in scope to warrant mentioning in their own right. Anticipated global climate changes, for example, impact many, if not all, aspects of engineering and entail large distance scales. As a result of this discussion, the four priority areas that were developed are interdisciplinary but identify major research needs for transportation infrastructures:

1. Impact of global climatic change
2. Protection and restoration of the environment
3. Material science problems
4. Geophysical methods for site investigation

Conclusions

The civil engineering establishment gives little attention to cold regions phenomena, even though they cause numerous problems for design and operation of the nation's infrastructure. Both research and education need to be expanded. Adequate education programs can be most effectively addressed through graduate level research and education at selected universities and research centers. The National Science Foundation engineering directorate should instigate a coordinated program similar to that in the area of earthquake engineering. Such a program would focus on fundamental research and education, and supplement the several fine mission-oriented research programs such as that at the U.S. Army Cold Regions Research and Engineering Laboratory in Hanover, NH.

Acknowledgements

This paper is a brief summary of the results of a workshop at Hanover, New Hampshire on November 30, December 1-2, 1988. Group chairmen were S. Shyam Sunder, MIT, Offshore Technology; Robert Ettema, University of Iowa, Watersheds, Rivers and Coastal Zones; Virgil Lunardini, U.S. Army CRREL, Facilities Infrastructure Technology; and Ted Vinson, Oregon State University, Transportation Infrastructure Technology. John Zarling, University of Alaska Fairbanks and Lewis

E. Link co-edited the original workshop report. The staff at the U.S. Army Cold Regions Research and Engineering Laboratory oversaw much of the workshop logistics, and the publications staff at the Institute of Northern Engineering, University of Alaska Fairbanks prepared the final report. The National Science Foundation, Division of Polar Programs and the Engineering Directorate, provided funding through Grant No. ENG8814352.

Appendix I. References

Carlson, R., and N.R. Morgenstern. 1978. "Report of the joint U.S.-Canadian Northern Civil Engineering Research Workshop." Boreal Institute for Northern Studies, University of Alberta, Edmonton.

Carlson, R., J. Zarling, and L. Link. 1989. "Cold regions engineering research--a strategic plan." *J. Cold Regions Engrg.*, ASCE, (Dec.), and Report No. INE201, Institute of Northern Engineering, University of Alaska Fairbanks.

Haneman, V. and R. Carlson. 1985. "Arctic Research Needs in Civil Engineering." Report No. EES-1. School of Engineering, University of Alaska Fairbanks.

Interagency Arctic Research Policy Committee. 1987. "United States Arctic Research Plan." National Science Foundation, Washington, DC.

National Science Board. 1987. "The Role of the National Science Foundation in Polar Regions." Report No. NSB-87-128, Committee on the NSF role in Polar Region, National Science Foundation.

IN SEARCH OF LIBRARY EXCELLENCE IN COLD REGIONS RESEARCH

Nancy C. Liston[1]

In the age of the information explosion engineers and librarians must work together to achieve a value-added research product. The importance of the librarian's role in the research process was never more important than it is today in a world of diminishing resources and expanding demands. This paper presents a special engineering librarian's view of how the information specialist will enhance the research process in the 1990s.

In recent years we have experienced an information explosion which, compounded by the technological explosion, has left the end user of library and information services spinning wheels to keep on top of the current literature of his/her field. Those interested in science and engineering are also faced with the fastest growing field of technical literature as well as how to process this information into research products with as little rekeying as possible. Unfortunately, for as many engineers/scientists who know how to use a library or information center there are an equal number who do not. Graduate and undergraduate curricula still do not provide enough emphasis on the research process and the value of the information professional in that process. As a technical librarian in a specialized world class library I continue to find it amazing when professionals with inch-thick theses do not know how to operate in today's information centered environment. By better understanding and utilization of the capabilities of the library, the end user will be able to spend more time fulfilling his/her primary function, analyzing and evaluating the research results. The library personnel will be doing more of the hard-core retrieval of the information. The end-user will have much greater potential for higher productivity and, I dare say, higher quality results. No longer will the researcher be working in a vacuum of outdated or incomplete information.

Specialized collections such as the *Bibliography on Cold Regions Science and Technology*, which is contained in 44 volumes of print, one CD-ROM (compact disc-read only memory), or one online database, are excellent examples of where the

[1]Librarian, U.S. Army Cold Regions Research and Engineering Laboratory, Hanover, New Hampshire 03755.

future is headed in technology transfer. This bibliography, initiated in 1951, has been supported by the Army Corps of Engineers and prepared by a bibliographic staff at the Library of Congress. There are currently close to 140,000 citations covering the world's cold regions literature including over 55,000 Russian citations. The topical areas are civil engineering including winter construction, pipelines, utilities, road construction and maintenance, heat transfer, mobility, and water supply as a function of cold regions. The physical, chemical, and mechanical properties of snow, ice, and permafrost are also covered comprehensively. In 1978 this database was among the first technical databases to be converted to an online database, "COLD," for interactive literature searching by librarians or trained end-users. The search techniques are very user friendly but, for the most part, the end-users have chosen not to be trained for searching directly and rely on information professionals to accomplish this task for them. This requires careful dialogue between the librarian and the researcher to establish a search strategy including correct keywords, languages, document types of interest (patents, journal articles, theses, etc.), scope of topic, and depth of interest.

There have been times that I thought I had failed in my searching process, because I could not find any citations on a certain subject, only to learn that this delighted the researcher because this left open an excellent opportunity for him to receive funding for a new area of research. This literature search took five minutes on the computer and would have taken two months in the pre-1970s when printed indexes and manual card catalogs were the only tools available for conducting a literature review. The entire thrust of the library of the future will be different. Now the patron arrives in the library with problem in hand. In the future, the patron will FAX the search to the library or more likely hook up electronically from a personal computer and access the library's entire collection.

The U.S. Army Cold Regions Research and Engineering Laboratory library, Hanover, N.H., is on the forefront of technology with a newly automated online catalog, which contains all the non-cold regions material such as reference books and general science and engineering reports and monographs. This online catalog will soon be available via the local area network (LAN) to every desktop equipped with a PC at CRREL. Long-term plans will include dial-up access from PCs located anywhere. The research staff will simply hook up with the library via of the network and search by author, title or subject without ever leaving their desks. In the same process they can find out who has a certain book checked out, and they can in turn leave a message via electronic mail that they would like to borrow the book from the individual who has it.

With the arrival of satellite communications including short wavelength, high-frequency (18–23 GHz) transmission for voice and data communications, video transmission, and long wavelength low-frequency communications for most documents, information will be simply a switch of a dial away in the future. Telefacsimile is a technology whose time has finally arrived. It has been predicted for the last 20 years, but it will cause the revolution of information exchange for the immediate future. Satellite networks will allow for library networks in the future that will outstretch anything imagined currently. Cable television systems, digital transmission services, the Integrated Services Digital Network (ISDN) and optical recognition including page readers, document readers, and the combination of these will allow for not only immediate access but immediate transmission of technical information between library centers, individual colleagues, and mini- and mainframe computers.

The first step that the CRREL library has taken in preparing for the next 20 years is to produce a CD-ROM version of the *Bibliography on Cold Regions Science & Technology* entitled *Arctic and Antarctic*. This CD-ROM contains the previously described Bibliography of 140,000 items and the *Antarctic Bibliography*, produced by the National Science Foundation, containing 30,000 items, on one compact disc. This disc contains the search software retrieval program as well as all the citations and still has room for another 500,000 citations. The CD-ROM is searchable by all the same access points as the online version, but you avoid the telecommunication and vendor charges associated with the online version. An annual subscription fee is paid for the CD-ROM and you are sent two updates per year, updates to the documentation and software as released. The current drawbacks to "read-only discs" seem to be they are not as current as the online database which is updated four times a year and the search speed is not as fast as the online system. However, the psychology of not trying to beat the meter on the clock as it stacks up charges, and the additional time that people will take to practice without charges being incurred, may in the long term make the CD-ROM the version that the end user will feel more comfortable using. CD-ROM technology has already arrived that will allow for network access to any CD-ROM from a remote access point. In the aggressive and competitive environment of the future, engineers and scientists will be multigenerational, highly technical, highly trained and hands-on types of people. These people will be knowledge and information intensive and dependent. The skills needed will include computer literacy, communication abilities, analysis skills, and planning and negotiating talents. It will be the librarians who will be the gatekeepers of all the information systems, and the end users will need to recognize early on that the librarians can add value to their final research product if they bring them into the research cycle early in the process.

Document retrieval is currently handled by mail, microfiche, telephone, and personal visits to information centers. The CRREL library, for example, has every item cited in the *Bibliography on Cold Regions Science & Technology* in microfiche format. This technology was never fully satisfactory to the end user because the quality of filming and reproduction was never the quality of the original. Full text retrieval of journal articles from databases and CD-ROMs is now finally reaching that standard of quality. Full text and entire bibliographies can be downloaded directly into word processing systems and document retrieval software, which gives direct ownership and transmission of the information without rekeying. Optical scanning will allow direct entry of information into a database without rekeying. This is an area again where the satellite and video transmission will play a big role in the instant information transfer of the 1990s and beyond.

Some excellent examples of information needed immediately from the *Bibliography on Cold Regions Science & Technology* have occurred during three separate national emergencies. The search for the black box in the Air Florida Flight 90 crash in the Potomac River, Washington, D.C., in 1983, was aided by information searched online to support the use of radar above the ice. The Marine plane crash site in Gander, Newfoundland, Canada, required information immediately on how to secure an area prior to an impending snowstorm so that the victims' remains and the fuselage would not be lost. The COLD database provided this information. The latest use of our online bibliographic information came during the Alaska oil spill of 1989. The Exxon Company needed on-the-spot information on how to laser ignite an oil slick in very cold temperatures. On all three occasions the bibliographic utility at CRREL was able to provide real-time information for truly emergency situations. Possibly a bit dramatic, the point I am trying to make is how important knowing how and where to access information can be.

The new technology in libraries must be looked upon as a challenge and not a burden. Engineering schools must stand ready to incorporate the new technologies of libraries into the academic program of their students, including seminars, training sessions, assignments, and classes which teach hands-on library interaction. The challenge for librarians and educators in the next decade will be to make the technology humane while not decreasing the library's productivity. Librarians will have to have someone on the job all the time to keep that friendly atmosphere which people can count on in a library, even if the friendly atmosphere is at the other end of a computer hookup. New technologies will require experimentation and returning to the drawing board frequently to perfect the latest technology. This will keep the librarian and the library user in a state of dependence and flux. The innovative librarian's role is to give the engineer a more precise reference, in less time, with more alternatives on how to move that information into the final engineering product. Librarians won't be replaced but their responsibilities will shift. They must stand ready to instruct the user on how to use the technologies and, when called upon, become part of the research process. But without acceptance from the technical community this meshing of roles will not work and both sides will lose.

INFRASTRUCTURE AND CIVIL ENGINEERING EDUCATION

Neil S. Grigg and Darrell G. Fontane[1]

If civil engineering is the lead profession for infrastructure management why has there been a decline in CE enrollments? This indicator suggests that either infrastructure needs are not being met, or that they are being met with less civil engineering input than the profession would like. In this paper we suggest that although considerable effort is going into infrastructure management, civil engineering is not in the lead role it should be, and that the programs of the nation's CE departments could be improved to put the profession into more of a lead role.

The background facts for an analysis of this question are well established:

- US infrastructure facilities, depending on what is included, involve a capital investment approaching $10 trillion and an annual expenditure on the order of $500 to $1000 billion.

- Infrastructure, the built environment, provides the structures, transportation systems, energy, water and waste management that make the economy go.

- There is a steady call for more attention to public works education and training to enable the nation to deal with its infrastructure problems (National Council on Public Works Improvement, 1988).

Infrastructure management problems. One of the reasons that civil engineering is not in more of a lead role is the fact that the problem is not mainly technical; it is a management challenge, and managers do not have to be civil engineers. Infrastructure systems are complex technically and politically, and success in managing them requires special talents exercised in an

[1]Department of Civil Engineering, Colorado State University, Fort Collins, CO 80523

interdisciplinary environment by engineers, managers and policy makers. The need for an interdisciplinary approach was put well by Donald Stone, one of the founders of the profession of public works management:

"Engineering capability alone is insufficient for these multidimensional purposes. Engineering and other specialized skills must be complemented by public affairs and managerial competencies. These include capacity to deal with the gamut of social, economic, environmental and political factories inherent in planning, policy resolution and program implementation. Practitioners are needed who can integrate public works systems and subsystems into urban and national development programs." (Stone, 1974).

The tasks of engineers working on infrastructure are planning, designing, constructing and operating. But there is more to infrastructure management than just getting the facilities planned, built and operated. There is a high level of dependence on politics, finance and project management. ASCE recognizes this; a conference like this one recently concluded that "The principal conclusion ...from this...examination of our profession is that carefully crafted planning coupled with leadership and action at all levels is essential" (ASCE, 1988).

To train engineers for management and political tasks is a real challenge to civil engineering education. We do not suggest that technical subjects be downgraded; on the contrary, they are essential for some infrastructure management tasks. However, new and bold initiatives in technical management are required.

Current civil engineering programs. Existing educational programs that deal with infrastructure are concentrated in civil engineering; it might even be said that all of civil engineering education deals with infrastructure. For the most part, other academic areas, with the exception of public administration at the graduate level and political science and urban economics at all levels, do not really deal with infrastructure. However, civil engineering education has not really embraced the real challenges of infrastructure.

The University of Hartford made a survey about changes made in civil engineering curricula as a result of the infrastructure media attention. The results showed that faculty members were introducing the subject into existing courses, but there seemed to be only a minor impact on existing educational procedures, and no impact on enrollment (Kalika, 1985). This seems to be the case in most schools. It is not a problem of deliberate neglect; it is a problem that is similar in nature to others in engineering education that deals with complex, interdisciplinary problems; most faculty get their rewards from making advances in technical areas and regrettably, many do not have much experience in the real world of infrastructure management.

As it is, civil engineering curricula are often quite good, containing technical courses related to infrastructure, such as transportation engineering and structural design, and most programs contain a senior design course which allows the study of infrastructure problems. At Colorado State University, for example, we have the senior design sequence, a core undergraduate course on project management, and an elective on applied transportation engineering which is project-oriented. We believe that our curriculum is excellent, and that improvements can only come at the margin. Faculty hold strong beliefs about this and have worked hard to come up with the best possible combination of courses.

Graduate programs have received national attention from the American Public Works Association (APWA) which has stimulated master's programs in public works management, usually as a blend of programs of civil engineering and public administration. Enrollments in these programs have been steady, but not high. The earliest of these was at the University of Pittsburgh. Graduates of these programs have been well-qualified to become city engineers and directors of public works. Similar programs are now appearing within schools of public administration.

In terms of courses, at Colorado State University we introduced a master's level course entitled "Infrastructure Engineering and Management" (Grigg, 1988). It covers: elements of the infrastructure problem; planning and management skills; economics and finance of infrastructure; operational analysis; and case studies. Case studies are very applied. In a recent year they included: flood protection in a coastal city, rebuilding a war torn city (Beirut), the Chicago sewer program, reuse of drinking water, rehabilitation of highway systems, water supply problems in Portugal, transportation problems in Capetown, land subsidence in Taiwan, river lock rehabilitation, transportation problem analysis in Houston (and in Denver), wind problems in cities, urbanization problems in Brazil, military infrastructure problems, rehabilitation of the Statue of Liberty, landfills and groundwater pollution and replacing bridge decking. These topics could have been covered for the most part in existing civil engineering courses. The differences were that they were mostly presented by experienced and older graduate students, many from developing countries, and they emphasized management aspects rather than design or analysis.

Another plank in our CSU program has been seminars on public works which we conducted for three years. The seminars were co-sponsored by the Colorado Chapter of APWA, the Department of Civil Engineering and the CSU Student Chapter of ASCE. The purpose of the seminars was to foster interest of civil engineering students to pursue careers in public works areas. The seminars were presented by practicing engineers from the local area, and covered a variety of public works topics. A typical schedule of topics included the following: pavement management, railroad location through the city, interchange and bridge design, winter highway maintenance, water supply management, water utility system management, storm drainage utility operations, consultant roles in storm drainage planning, and computers in public works (water and wastewater, building management, transportation, and in consultant offices).

Infrastructure as a challenge to civil engineering education. While technical aspects of civil engineering in the US are excellent, we believe that infrastructure problems call for different skills. What is needed is to redirect parts of the curriculum into making civil engineers better leaders of interdisciplinary teams that deal with infrastructure. It is the same challenge referred to by Donald Stone. The key word is to "integrate", or in Stone's words: "Practitioners are needed who can integrate public works systems..." This is a real challenge and does not call for more technical courses but for what we call in engineering education "synthesis". In this case what is needed is not to synthesize mechanical systems but socio-technical systems.

The question is: can civil engineering education prepare graduates for this needed synthesis and integration? We believe that more emphasis on the "systems approach" can help. This would be an emphasis that combined computing with real world approaches to problem identification and solution with interdisciplinary teams. For some problems this may mean to deal with the development of new technologies, especially better and more appropriate technology for meeting infrastructure demands at reasonable costs. At the larger scale it may mean dealing with entire urban and regional systems, usually with a high level of political content.

To lead in infrastructure management, students need to learn about interdisciplinary efforts in areas of policy, finance and organizational arrangements. There is not time in a four-year undergraduate curriculum for special courses on these, but they can be introduced in interdisciplinary project-oriented courses.

Why have enrollments declined? Whether more civil engineers are needed for technical jobs would seem to depend on construction spending, and be cyclical. Rather than more civil engineers we may need better engineering leaders. Can engineering leadership be taught? This is the same question that is debated in business schools, and the consensus seems to be: let's keep trying.

References

American Society of Civil Engineers, Civil Engineering in the 21st Century, New York, 1988.

Grigg, Neil S., Infrastructure Engineering and Management, John Wiley & Sons, New York, 1988.

Kalika, Peter W., Impact of Infrastructure Crisis on Civil Engineering Education and Research, University of Harford, January 1985

National Council on Public Works Improvement, Fragile Foundations: A Report on America's Public Works, Washington, February 1988.

Stone, Donald C., Professional Education in Public Works/Environmental Engineering and Administration, American Public Works Association, Chicago, 1974.

PUTTING AESTHETICS INTO THE CIVIL ENGINEERING CURRICULUM

Stephen R. J. Sheppard, Affil. ASCE[1]
Edward S. Neumann, M. ASCE[2]

Abstract

Due to the importance of aesthetic considerations to public acceptance of civil engineering projects, courses on aesthetics should be designed specifically for civil engineers. The paper presents a curriculum syllabus that addresses three major components: the need and background for aesthetics in engineering; principles and concepts of aesthetics; and practical techniques which can be implemented.

Introduction

Civil engineering is, by definition, engineering for communities. These days, communities hold a strong interest in quality of life issues. One of the more common problems that civil engineering projects encounter in communities is public concern over aesthetic impacts; yet, most engineering curricula do not even address the subject. This usually leaves the typical civil engineer with little warning of the problems he or she may expect in dealing with aesthetic issues, no training or resources on which to rely when confronting such issues, and no systematic procedures to incorporate aesthetic principles into the project development process so as to improve public acceptance of engineering solutions.

The Argument for Aesthetics in Engineering

History has provided us with numerous examples of spectacular engineering projects which have become supreme aesthetic symbols of mankind's progress and creativity. Structures such as the Eiffel tower, the Golden Gate Bridge, and the arch at St. Louis have become world famous landmarks and tourist attractions. However, the products of every day engineering which are all around us exert no less of an influence, in that they are seen and used by people every day. There are many philosphical and psychological arguments for improving the quality of life through the quality of design. Here, however, we wish to concentrate on

---

[1] Director of Visual Services, Dames & Moore, 221 Main Street, Suite 600 San Franciso, CA. 94105
[2] Director, H.O. Staggers National Transportation Center, West Virginia University, Morgantown, WV 26506-6101

more practical reasons for training engineers to adopt aesthetic practices in their work.

For every spectacular engineering success, any number of examples which fail to meet basic aesthetic standards can be found. Uncompleted water front freeways, stopped in their tracks by public opposition in many of our larger cities, are among the better known examples. Everywhere, from small road-widening projects in the suburbs to dams on mountain streams, the engineer responsible for building new projects is facing community pressure: to select environmentally preferred sites, to modify designs to reflect local values, to build alternatives which may cost more to construct but which gain vital community support. We ignore community input and aesthetics at our peril. There have been countless lawsuits and multiple environmental studies in recent years, stemming from aesthetic issues. Major projects have been stalled for years and sometimes shelved as a result. The true construction costs need to reflect the price of inadequate treatment of issues which the community feels strongly about and which could delay the project or lead to wasteful re-design. For these reasons, we are beginning to see engineering practice gearing up for a different kind of process. Precedents abound. The agencies responsible for implementing new transportation projects, such as Santa Clara County Traffic Authority in California and TriMet in Oregon, are producing visual Design Guidelines as an integral part of their design process, as a means of keeping the local communities in support of much-needed projects (Roberts 1989). Engineers siting and designing new landfills are working closely with landscape architects to satisfy local residents and provide for well-planned post-closure use which benefits the community in the long term. Civil engineers laying out grading schemes for new residential developments are being forced to conform to stricter hillside development ordinances aimed at preserving views of ridgelines and open space. Traditional engineering agencies, such as the Corps of Engineers and California Department of Transportation, now train and maintain specialized staff with aesthetic expertise who can be called in to consult on visual design issues. Increasingly, engineers find themselves in the "hot-seat" communicating the design process to the public and local decision-makers.

A Proposal for Aesthetics in the Curriculum

A basic education in aesthetic principles would strengthen the practice of all engineers who have to deal with the public and local agencies, or who design facilities in urban or other sensitive areas.

It should not be assumed that aesthetics is an entirely new component of an engineering education. There are some precedents available. Engineering courses at Stanford, California Polytechnic at San Luis Obispo, and Southeastern Massachusetts

University among others, routinely integrate some aspects of aesthetics. Some professional and extension classes are available, for example at University of California, Berkeley Extension. ASCE itself has organized sessions and lectures on aesthetics at conferences, through its Aesthetics and Visual Design Committee. Several books on engineering provide some treatment of the subject, most notably those on bridge aesthetics by Billington (1977) and others (Kruckemeyer and Fleet, 1986). The vast majority of civil engineering students, nevertheless, are not systematically exposed to these sources of learning.

A survey conducted by the Aesthetics and Visual Design Committee of the Urban Planning and Development Division in 1973-74 revealed that civil engineering educators held inconsistent attitudes regarding the relationship between the importance of aesthetics and the need to include formal classroom experience at the undergraduate level (Neumann, 1978). Educators who felt that "The Civil Engineer should be trained to respond to design proposals from both technical and aesthetic points of view,"did not tend to feel that "Civil Engineers should be encouraged to take courses in architecture, landscape architecture, architectural history, etc." (In retrospect this response partially may have reflected a reluctance by civil engineering educators to require students to take a course outside their Department over which they had little control of content, particularly in view of the undesirability of adding more credit hours to already heavy curricula). A second interesting finding was that while many educators felt that "optimal design will inlcude aesthetic considerations", they also felt that "extra costs should be anticipated in order to satify esthetic considerations". Thus, it appeared as though esthetics were perceived as an "add on" cost, and not a basic goal of the design process itself. We contend that such thinking is misguided, particulary in view of the frequency with which proposed civil engineering projects are challenged by the public for reasons of aesthetics. What we propose is that rather than send civil engineering students to general appreciation or skills development courses outside their major, courses be tailored to address issues of highest relevance to the civil engineer.

It is possible to draw on the above sources to sketch a basic course content suitable for the engineering profession. There are also many other sources from outside the profession, drawn from the fields of planning, landscape architecture, architecture, and specifically visual resources analysis. Much in the way of content and materials can be re-used from texts and training procedures already well established in these disciplines (eg. Smardon et al, 1986; FHWA, undated; CORPS of Engineers, 1988; US Forest Service, 1974; Bureau of Land management, 1980; Sheppard, 1989). It would, however, be a mistake to assume that these resources need not be adapted to the special requirements of an engineering education.

The following outline of a proposed course curriculum is intended as the basis for either an undergraduate class in civil engineering or a professional extension class. Level of detail and actual teaching methods could vary widely; at this point, it would be inappropriate to attempt to specify these, since the core content may need to be integrated into other engineering subjects in ways suited to each institution or forum.

Proposed Curriculum Outline-Aesthetics In Engineering

Part I - Background

I Introduction
Course Objectives
Provide exposure to theories, methods, and examples of aesthetic analysis and visual design; Help recognize/avoid/overcome aesthetic related problems in project planning; Improve thr quality and consistency of visual design; Provide a reference source for future use.

2 Aesthetic Issues In Engineering
Typical Problems With Treatment of Visual Issues
Inadequate awareness of aesthetics and public perceptions; Poorly defined visual design criteria; Subjectivity and vagueness developing solutions.
Consequences For Project Planning (Illustrated with Project examples)
Uncertainty and controversy; construction delays and project denial; Adverse community reaction to built project; Expensive remedial mitigations; Lost opportunities for positive engineering statements which contribute to both the community and the profession; Poor image of the engineering profession within the community.

3 Trends In Aesthetics
Historical Background
History of aesthetic design in engineering and related professions; Visual resource management and analysis-a new discipline; The environmental movement and aesthetics in the planning/project approval process.
Regulation And Requirements For Aesthetics
Legislation which specifically calls for attention to aesthetic values: e.g. National Environmental Policy Act (NEPA) 1969; US Army Corps of Engineers Permit Program Regulations(1984); US Nuclear Regulatory Commission NUREG-0555; Federal Highway Administration Regulations 23 CFR (1984); Historic Preservation Act (1966); State and Local regulations and policies.
Recent Trends In Aesthetic Design

Part II - Principles and Concepts

1 Visual Elements of Perception (Reference: U.S. Forest Service, 1973; Smardon et al., 1986, Chapters 3-5)
  Elements:
    Scale,Position, Form, Edge/line, Color/reflectivity, Texture, Detail.
  Variables:
    Distance, Lighting, Atmosphere, Season, Motion, Time
  Patterns:
    Contrast, Balance, Sequence/rhythm, Focus/emphasis
2 Aesthetic Theories
    (Ref. Zube et al., 1975)

Part III - Techniques

1 Scoping Visual Issues On Projects
    Identifying Red Flags (real or perceived potential Problems)
    Problem Avoidance Strategies
    Aesthetic Opportunities- Potentially beneficial attributes of a plan or project which, if recognized early and built in the process, could offset environmental problems or meet visual objectives which would otherwise not be reached.
2 Inventory Of Site Conditions
  Existing Conditions and Constraints
      Landscape Types and Scenic Qualities; Viewer Exposure and Sensitivity; Visual Constraints
  Analysis Methods
    Field/Office Techniques-
      Photography; Field Notation; Video Footage; Land-use Mapping/ Aerial Interpretation; GIS Modeling; Line-of-sight Sections; Viewshed Mapping (manual/computerized).
      Forest Services (FS); Soil Conservation Service (SCS); Corps of Engineers (CORPS); Federal Highway Administration (FHWA); Housing and Urban Development (HUD)
    Public Input
3 Visual Design Analysis For Projects
    Visual Effects of Project
      Visibility-will people see the project?)
      Contrast/Dominance (will people notice the project)
      Compatibility (will people think the project is attractive, well-designed, and appropriate for its setting?)
    Project Analysis Methods-
      Line-of-sight Sections; Viewshed Mapping; Site Survey/Scale Markers; Expert Panels; Public Workshops/Surveys.
4 Visual Simulation - Visualizing The Appearance of Project Design Before Construction
  (Reference: BLM 1980b; Sheppared, 1989)

Simulation Types And Media (perspective rendering, Scale models, CADD, photomantage, video-imaging, etc.)

5 Visual Design Treatments And Mitigations

Siting/Location Considerations; Modified Grading Practices;

Structure Design Factors; Integrating with Landscaping; Costs of Aesthetic Treatment

6 Monitoring

Post-Construction Evaluation; Reclamation/Restoration; Image Enhancement for Projects and Areas;

Implementation of Aesthetic Courses

A course with content as described above should be combined with field trips and site visits, studio sessions to learn and practice techniques, and case studies of actual projects.

In an undergraduate or graduate curriculum, it may not always be necessary or even desirable to establish aesthetics as a separate course; providing training as a key part of a required course on environmental factors or design could be very beneficial. There are opportunities also to combine aspects of this course structure with other classes, such as computer-aided design, public relations, presentation techniques, and environmental regulations. Joint studios with landscape architecture or architecture students would also add to the learning opportunities of engineering students.

Conclusions

An infusion of aesthetics into the core surriculum for students and professionals promises, among other things, to add a stimulating and attractive component to the overall civil engineering education. The skills learned in this area may make life easier for the student once out in the real world, and prepare engineers to solve visual problems themselves before making needless mistakes or having to call on other disciplines to solve their problems for them. The intent of the aesthetics training would not be to make every engineer an instant expert, nor should it be just a specialty for a few in the profession. The objective should be to help engineers to identify red flags in good time, implement routine procedures to resolve aesthetic issues before they become problems, and know where to go for additional help if they need it. If this could be achieved through the educational process, it may become easier to maintain control of projects in sensitive environments and contribute more visibly to both the quality of the environment and the image of the engineering profession.

References

American Society of Landscape Architects, undated. Visual Impact Assessment for Highway Projects. ASLA. Washington, DC.
Billington, D.P., 1977. History & Aesthetics in Concrete Arch Bridges, Journal of the Structural Division, ASCE, Vol. 102, No. ST-11, Nov. 1977, pp. 2129-2143.
Bureau of Land Management, 1980a. Visual Resource Management Program. Washington, DC: U.S. Dept. of Interio (govt. Printing Office).
Bureau of Land Management, 1980b. Visual Simulation Techniques. Washington, DC: U.S. Dept. of Interior (Govt. Printing Office).
Corps of Engineers, 1988. Visual Resources Assessment Procedure. Prepared by R.C. Smardon, J..F. Palmer, A.Knopf, and K. Grinde. Waterways Exp. Sta., Vicksburg, MS.
Duerksen, C.J., 1986. Aesthetics and Land Use Controls. APA Planning and Advisory Service, Report Number 399.399.
Elsner, G.H. and R.C. Smardon (Technical Coordinators), 1979. Our National Landscape: A Conference on Applied Techniques for Analysis and Management of the Visual Resource. U.S. Forest Service General Technical Report PSW-35. Berkeley, CA: PSW Forest and Range Experiment Station.
Kruckemeyer, K.E., and A. Fleet (Eds.), 1986. Bridge Design-Aesthetics and Developing Technologies. Mass. Dept. of Public Works/Mass. Council on the Arts and Humanities.
Neumann, E.S., Lyles, R.W., 1978. Why Include Aesthetics in Civil Engineering Education? Journal of Professional Activities. ASCE January 1978; 1-8
Roberts, J.E., undated. Guidelines for Architectural and Visual Design. California Dept. of Transportation, Sacramento, CA.
Sheppard, S.R.J., 1989. Visual Simulation: A User's Guide for Architect, Engineers, and Planners. Van Nostrand Reinhold, New York.
Smardon, R.C.,J. Palmer, and J. Felleman, 1986. Foundations for Visual Project Analysis. J. Wiley and Sons: New York.
US Forest Service, 1973. National Forest Landscape Management, Volume 1, Washington, DC: US Dept. of Agriculture.
US Forest Service, 1974. National Forest Management, Volume 2, Chapter 1. The Visual Management System. Washington, DC: US Dept of Agriculture.
Zube, E.H., J.L. Sell, and J.G. Taylor, 1982. Landscape Perception: Research, Application, and Theory. Landscape Planning 9:1-23.

LARGE SPACE STRUCTURES AND SYSTEMS TECHNOLOGY EDUCATIONAL STATUS AND NEEDS

Milind Pimprikar[1]

ABSTRACT

Although the potential and scope for development of the rapidly evolving large space structures (LSS) and related systems technologies are vast, so far the civil engineering professionals, particularly educators, have not kept pace to develop appropriate educational programs and curricula to meet these needs for next decades and into the 21st century. Therefore, in order to meet the challenges of this multi-disciplinary engineering field it is essential to undertake specific measures at this stage.

This is a learning phase in which engineers and scientists from around the world are in the process of developing the necessary means and methods for constructing various large structures in low orbits. Civil, particularly structural engineers have always expressed special interest to participate in these challenging projects. In the decades to come, structural engineers will require *skills that are not currently provided by universities and which are not adequately represented in professional societies.*

This *Discussion Paper* intends to set a stage to examine various issues and concerns of adequacy & quality of current civil engineering educational programs to meet the evolving needs of space structures and related technologies.

In the first part, an assessment of current status of large space structures and related educational programs - at various levels - is proposed. This assessment is to be based on the technical needs for the current and projected (immediate and Long-range) large space structures as well as the likely characteristics of the future aerospace engineering work. This assessment will also help identify the deficiencies and the areas of concern in our educational system in order to meet the challenges ahead.

Finally, recommendations need to be presented to ABET and professional groups for enhancing the effectiveness of space structures and systems related educational programs. The assessment should also identify civil engineering educational programs that will support these initiatives.

---

[1]Chairman, National Task Group on Large Space Structures and Systems Technology
1455 de Maisonneuve Blvd. West, Montréal, Québec, Canada, H3G 1M8
Tel.: (514) 933-1494, Fax.: (514) 848-3494, Telex: 055-61256

INTRODUCTION

Recently, civil engineers, through ASCE's newly established Aerospace Division and the Journal of Aerospace Engineering, have undertaken a very important initiative - the first step - in bringing together many of the civil engineering related disciplines involved, interested and concerned with the space technology projects.

Dean Colby Ardis of the Southern Illinois University, in his recent article in the Civil Engineering-ASCE Sept. 89 issue, has also raised similar concerns and appealed that *civil engineering educators must be brought as close as possible to "current" as soon as possible if our students are going to be ready to tackle the problems of space exploration...... We must incorporate the concepts of civil engineering in space into our curriculums as soon as possible.*

Further, it is also encouraging to note the recent statement by the ASCE Board of Direction (Casey Dinges's article in the Oct. 89 issue of the Civil Engineering-ASCE) *to support the US national space policy as set forth in the Presidential Directive on National Space Policy on Feb. 11, 1988, as well as to support the funding of the appropriate projects required for the implementation of this policy.*

The Potential and Scope for Development of LSS

The potential and scope for development of diversified large space structures (LSS) and related systems are extensive. A variety of such structures - in various categories - with dimensions ranging from one hundred to several thousand meters have been identified and are being planned for missions during year 1990-2025. There will be major developments in this area from now and into the 21st century.

Over the years, substantial efforts and resources have been devoted in the U.S., Europe, Japan and the U.S.S.R., for developing many critical near - and - long - term LSS and related Systems technology. In the U.S., some of such recent initiatives by the NASA and other agencies include: establishment of number of Research Centres, Institutes, Specialized Testing Facilities, and related educational programs; and the $25.1 million increase in 1989 fiscal budget for LSS development under a *Pathfinder Technologies Program.*.

A review of recent developments in this field shows remarkable progress in the areas of innovative structural concepts and configurations; utilization of advanced materials and fabrication processes; determination of structural integrity by analysis & tests; improved modelling and scaling techniques; weightreliability-lift cycle consideration; environment & load definition; interaction between structural responses and control systems; design criteria's for long-life, thermally resistant structures; and efficient joining & erection techniques.

For example, countries from around the world will construct a variety of large space structures of various categories:

(i) Large space structures *to provide power* for space manufacturing, communications, ion or electric propulsion,

(ii) Large antennas *to receive and transmit energy* for a variety of applications, including: deep-space communications, multibeam communications satellites, earth & space observations, power transmission in various bands of spectrum,

(iii) Large space *platforms to provide areas of general utilization* for docking, assembling, checking out, resuppling smaller spacesraft, or space processing.

Over the decade, various structural concepts and systems for the design and assembly of efficient and flexible large structures, for the space station and number of multi-purpose platforms are being developed and tested by the NASA and others for their optimum and economic performance. Further, the U.S. National Commission on Space has identified Mars and Lunar Base projects for future growth.

It is anticipated that scientists, technical experts, engineers, space station workers and specialists will live, learn and work in space continuously, performing a wide variety of functions: conducting research and development programs, overseeing the repair and refurbishment of the space station, satellites and other structures, constructing and assembling large space structures, including more and large habitats, laboratories, automated factories and manufacturing facilities. These projections, however, are viewed with considerable caution particularly since experience in these activities is still quite limited.

Any country that wishes to integrate such activities into its economic, social and cultural life will have to integrate space technology into its educational system. However, because space technology have developed so recently and so rapidly, the importance of these fields is not reflected in educational curricula, both because educational system tend to be conversative and slow to change, and because there are not enough trained professionals in these disciplines.

The U.N. Committee on Science and Technology recently passed an resolution "Because of the relevance as well as the growing importance of space activities as indicated above, space technology should be included in educational programmes at all levels, from primary school to highly specialized graduate degree to adult education. Clearly, each level has its own appropriate subject matter and methods." (UN-A/AC.105/390-I.C).

NEED FOR DISCUSSION

The Multi Disciplinary Nature of the Subject:

The blossoming of the *space program* in the early 1950's produced a new aerospace discipline in engineering which combined the skills of the traditional civil, mechanical, and aeronautical engineering fields. Significant and far reaching advances have been made in the past three decades, which are having considerable impact on our lives and offering new dimensions to others technologies.

Large space structures is a field which include number of engineering disciplines: structural, mechanical, aeronautical, and alike - those concerned, interested, and involved in one way or another with the various aspects of the subject matter. However, so far, there has not been a corrcsponding emphasis to introduce appropriate educational programs. *Very few formal instructional courses, related to the analysis, design and construction of structures in space are currently offered, and this also at certain levels.*

This field is particularly of interest to those professionals currently or planning to be involved in the design and analysis of large structures in space, including structural designers, and stress/materials/dynamic analysts interested in structural configuration development, from initial requirements to configuration selection/definition. The type of projects considered are the typical specialization of structural engineers in the past and should be the inspiration for the future.

It is time to look to the future and plan long-range educational programs. With the building of space stations, and other projects underway, space research is no longer remains the domain of a small, elite group of engineers. Low earth orbit is becoming increasingly available to creative individuals, comprising many disciplines.

This is not to suggest of course that all civil engineering students should be given extensive education related to space technology. Rather, everyone should have a general knowledge of space mechanics and its importance in our lives, specialists in certain disciplines should have a working knowledge of the applications of space technology in their discipline, and a small number of people should have extensive and advanced knowledge of the various domains in the strict sense.

AREAS OF CONCERN

Large space structures is a new and challenging field that has not yet fully developed and matured. Yet, it has existed long enough, and now its practical applications are coming into reality. Enough research has been performed to provide a solid foundation upon which future progress will be built.

Design of these variety of large-scale orbital platforms and low-gravity facility structures involves a unique application of nonstandard ideas. The new "live loads" of gravity gradients, extreme thermal variations, micrometeorite impacts, and solar wind are not traditional structural engineering concerns, but may be analyzed with current available structural engineering tools.

There will always be the basic civil engineering subjects, which are being taught year after year, with some modifications, but if they occupy effort exclusively then our future progress may halt. Nobody will be asking why we should be involved in the area of large space structures in about 20 year's 'time'. *The challenge is whether current civil engineering educational content is adequate to meet future needs related to space technology programs.*

In order for our young and practicing civil engineers from universities and industries to participate and contribute in these developments, there is a need to introduce basic and advanced courses at the undergraduate and graduate level as well as through continuing engineering education resources.

This rapidly evolving technology necessitates to upgrade our educational and professional skills to meet these challenges. Clearly, learning how to design and erect large structures in space routinely and cheaply presents a major challenge to the engineering community over the next few decades. David Carrier and James Van Hoften two civil engineers and astronauts involved in the early stages of space program envisions a near-term "hey-day" for structural engineers, however, researchers and engineers need to develop competitive expertise to achieve these goals.

At present only few universities in the country offer selected space-related programs at undergraduate and graduate levels. To achieve our potential -- indeed, to even become aware of our potential -- our engineers, business leaders, and policy-makers must coordinate efforts and resources.

There is a need to **develop new text material**, such as the one I am preparing on - "Analysis and Design of Large Structures in Space" - for most of the new courses and educational programs. Eventually, we may package the program for use at universities and professional organisations around the country. This could take the form of textbooks and audio/visual aids or even real-time satelliet links.

Need for Conferences and Corporate Seminars: The space structural engineering educational program cannot be locked away in an "ivory tower", out of touch with the real-world community and thereby irrelevant. Much of the experience of the faculty lies in direct participation in that community. To make sure its continuity, up-to-date conference projects such as Space 90, bringing together ideas, proposals, and experimental results, are vital.

Role of Educators, Practitioners and Professional Groups: A professional organisation such as the ASCE has the responsibility to bring together representatives from universities, industries, concerned research centers and the accreditation commission in anticipating and seeking solutions to the emerging technological challenges.

Development of Currriculum and Courses (Accreditation Issues): In the new and emerging areas such as this, the student is entitled to know that a program or a course, only minimally described in brochures and pamphlets, will meet an established requirement and standard. In a similar vein, the individual or corporation seeking to employ a graduate for the purpose of professional practice or train an employee for particular needs, is entitled to know that there is some specific body of knowledge which the individual has been required to master. Professional accreditation fills both of these requirements. Another argument for the accreditation of this proposed program at the advanced elevl is that it establishes a requirement for periodic review and visitation of the program and courses.

In the past, a professional society has assumed responsibility for developing criteria to be used in evaluating programs submitted for accreditation in that discipline. In order to be eligible for accreditation by ABET as an undergraduate or graduate level program, the curriculum must meet certain general criteria regardless of discipline ("Criteria for Accrediting Programs in Engineering", ABET Inc., New York, 1987).

"Of course, in the zero gravity of space it would be completely self supporting."

The option has been provided by ABET that any institution may accredit a program in a given discipline at the basic level or at the advanced level. Additional program criteria for the advanced level may be specified by the cognizant professional society; however, this option has not been exercised most of the time.

As far as the nature and level of subject matter is concerned, courses in space structural engineering could be offered at both the undergraduate and graduate as well as continuing educational level. Perhaps, at the initial stages, a graduate level program directed toward these particular areas should be introduced. In terms of accreditation, the AIAA (American Institute of Aeronautics and Astronautics) holds that no programs should be ABET-accredited at the master's level. Consequently, it has not aproved the use of advanced level criteria for any of the aerospace programs.

There remains, however, an alternative provision made by ABET for the accreditation of programs in non-traditional areas. One might interpret the position of the cognizant society for aerospace matters to be that graduate level programs in space-related areas are non-traditional. As such, they automatically become eligible for accreditation, by different criteria.

Considering the importance and urgency to undertake appropriate educational programs in this field, at all levels, there is a need for fruitful discussion at this forum. I believe that this proposed in-depth examination and evaluation of space technology related civil engineering educational programs, to be undertaken in consultation and discussion with educators, practitioners, experts from the concerned industries, and representatives of various professional organizations and societies concerned, involved and interested in the development and advancement of this field, will help us to offer concrete recommendations for educational programs for the next decades and into the twenty-first century.

OBJECTIVES

The subject - to those not well acquainted with the area of large space structures - may appear to be rather exotic and mysterious, however, any engineer who understands the basic concepts of structural mechanics and analysis, as might be acquired in an undergraduate engineering program, should find little conceptual difficulty with it.

In the decades to come, civil engineers will require skills that are not currently provided by universities and which are not adequately represented in our professional societies. Structural engineers are and will be playing a significant role in design requirements, engineering, testing, assembly, and operation of various structures in space in. Space structures and related systems technology education requires an inter-disciplinary approach to the subject matters involved. The growth of our aerospace industry, particularly various long-term projects demands attracting more young people into these fields.

It is up to our universities and professional organizations such as ASCE to insure that future engineers are prepared to meet these challenge. This paper topic, proposed as part of "SETTING THE AGENDA" is an endeavor attempting to address the challenges of space structures related educational needs for the next decade and into the 21st century. This initiative, I believe, will also inspire many of our young and practicing civil-structural-engineers to take an active part in this next generation exploration.

The proposed forum on this topic is a first attempt to assess the educational status and technical contents of space structures related programs. The **proposed plan of action** needs to be treated in four parts:

(a) assess the current status of civil engineering educational programs related to space technology being offered or being planned at various universities, continuing educational institutions, and professional organisations.
(b) based on the assessment conducted in the first part - identify the needs in order to meet the challenges ahead.
(c) present recommendations for increasing the effectiveness of aerospace, particularly, space structures related educational programs at various levels.
(d) prepare outline of the goals and key elements of the educational programs, i.e. development of educational contents, role of educational institutions and concerned organizations, cross-disciplinary instructional teams, and new instructional technology.

Initially, this proposed study needs to evaluate the current status of existing civil engineering educational programs at various levels considering the space technology related developments and identify the need & potentials for undertaking concrete initiatives and programs. The assessment should be based on the technical needs for the new and projected (immediate and Long-range) space structures related projects being developed around the world, as well as the likely characteristics of future aerospace engineering work.

This study, to be prepared in cooperation and consultation with the educators, business leaders, and the experts in this field, will also help substantially in revitalizing student interest in civil, particularly structural engineering education.

The initiative presented herein will have similar appeal to other nations of the world involved in the civil engineering educational reorientation to introduce programs related to space technology.

An in-depth discussion among educators, practicing engineers from the concerned industries, and representatives of various professional organizations involved and interested in the development and advancement of this field, will help us to offer concrete recommendations for the multi-disciplinary structural engineering education for the next decades.

Introducing Civil Engineering in Space Into Our Curriculum?

Colby V. Ardis, P.E., Ph.D., F. ASCE,[1]

Introduction

How can we bring ourselves as civil engineering educators closer to the creative civil engineering activity currently going on in the exploration and development of space? I feel we must understand what impact this activity will have on our civil engineering curriculum, our advising, and our teaching of civil engineering students -- they are the future of our profession.

Civil engineers must become more involved in space exploration. The exploration and development of space cannot take place without civil engineers. The civilization we know today is supported by an infrastructure designed and built by civil engineers. The exploration and development of outer space will provide opportunities for civil engineers of the future not unlike that encountered since the exploration and settlement of the West. The only problem is that the rate of exploration and development will be greatly accelerated. At the same time, civil engineers are rebuilding America and aiding in the development of other countries. Therefore, our curriculum must be enriched with examples of what is happening today and what is expected tomorrow if our students are going to be tomorrow's leaders on Earth and in space. These well educated leaders will be in high demand due to the extreme high cost associated with the engineering, construction, and operations needed for putting and maintaining humans in space.

---

[1]Dean of Engineering, Southern Illinois University at Edwardsville, Edwardsville, IL 62026-1804, FAX 618/692-3374, PH 618/692-2541

Current State of Practice

As an indication of the rate at which civil engineering is advancing in space engineering, I will note several recent activities that have taken place that I am aware of. I am sure there are others that I am not aware of.

1. A paper entitled "Civil Engineers in Space" was written by the Aerospace Division (four contributing authors) for the American Society of Civil Engineers' invitational workshop on "Civil Engineering in the 21st Century" at Williamsburg, VA, November 11-14, 1987. I was an invited participant. The paper pointed out, "by citing a few examples, how the science, technology and practice of civil engineering will be needed by mankind as we pioneer the space frontier." For example, what measurement techniques will be required to align 10 kilometer structures floating in space? What geotechnical designs will be required for the foundations of observatories, habitats, mining equipment and process facilities? What is the impact on civil engineering design due to gravitational pulls much less than those on Earth and in no atmosphere. How will we handle, process, treat, and re-cycle water and waste on a two-year trip to Mars? How will we protect the Martian environment, for example, while trying to establish a self-sufficient base of operations? Can we design zero-discharge and complete re-cycling waste systems? What materials will be used for the extremely large structures required in space? How will we account for the active control of these structures, gravity gradients, extreme thermal gradients, and the effects of body forces? How will we build these structures? New robotic techniques? New joining methods? Since most of the material will be brought from Earth, or moved from another planet, weight will be more important to civil engineers than ever before. What will be the problems? Optimization concepts and systems analysis in engineering design and construction will be even more important than now. How must our current thinking change?

2. The **Journal of Aerospace Engineering** commenced publication by the American Society of Civil Engineers in January 1988. This journal was the first to define the role of the civil engineer in space. "Emphasis is given to practical applications of civil engineering in space and on earth. Topics include lunar soil mechanics, environmental factors for the design of

inhabited space facilities, aerodynamics of low earth orbit structures, aerospace structures and materials, advanced energy systems, extra-terrestrial construction and resources development, remote sensing applications, computational fluid dynamics, and real-time data collection systems. The issue of adapting robotics to the needs of civil engineers in space, and on earth in support of space projects, is also examined." Do all of our civil engineering educators understand the concepts covered in this journal?

3. I attended the annual meeting of the American Society of Civil Engineers in St. Louis October 23-27, 1988. A significant portion of the program, including the plenary session and others, was devoted to civil engineering in space. Speakers frequently cited analogies to TVA, the Panama Canal, Apollo, Columbus' discovery of the Americas, Lewis and Clark, and other pioneering examples. These concepts I could understand. When speakers talked about the economic opportunities of space, I felt that I could clearly understand the logic, but I felt I didn't understand enough about the civil engineering necessary to make it happen. That was not a comfortable feeling. For example, our development of space could be an instrument of foreign policy and an instrument of world peace. An international "island in the sky" could be used for worldwide communication, surveillance, educational TV, energy distribution, and research. Energy as foreign aid? Global rural electrification? Drought relief (and therefore famine relief)? Educational TV as foreign aid? Should we be thinking about economic deterrence and not nuclear deterrence? Can civil engineering educators discuss these concepts with students? Can we educate them about how civil engineers will be needed to make the exploration and development of space happen? Can we advise them of the opportunities?

4. "Engineering, Construction, and Operations in Space" edited by Stewart W. Johnson and John P. Wetzel, appeared as the Proceedings of **Space 88** held in Albuquerque, New Mexico. "This book includes 125 papers providing in-depth discussions of space policy, extra-terrestrial basing, space stations and orbiting structures, and areas of special interest." The book was published by the American Society of Civil Engineers. Do all of our civil engineering educators understand the concepts covered in this book? How many have read it? Next week (April 23 - 26, 1990), **Space 90** will be held,

again in Albuquerque. How many civil engineering educators present at this ASCE Education Forum will be present?

5. I recently completed a desktop published textbook on **Engineering Mechanics: Statics.** My purpose in writing the text was to improve student learning and do so at half the cost. My goals to date have been achieved. However, there are many opportunities for improvement. The text already has several unique characteristics that other authors do not have that I will not go into here. But, the text only includes a few examples and problems that relate to statics of structures in outer space. I do not feel confident enough about the subject at this time to do any more on the subject. I had planned to personally get "up-to-speed" on a sabbatical this year, but a change in jobs is delaying that development.

Professional Development of Faculty

Most civil engineering faculty understand the questions raised in the previous section, but can they answer them? Can our students? As civil engineering educators, we must become more knowledgeable about the answers to these questions. I believe we need to take time to learn how to understand and solve the problems associated with exploring space. Then we can become more effective in helping our students focus on the same problems. Maybe then they will take an interest in learning how to solve the problems associated with space exploration and development.

We need to find ways to enhance the development of faculty in order to bring them closer to the edge of the state-of-the-art in understanding the concepts of civil engineering in space. How can this be done? Independent study, funded research, sabbatical leaves, attendence at professional society meetings, and specialty conferences are all possibilities. All of these ways of getting up-to-speed are effective, but they are also slow. In my opinion, we cannot continue to wait. Civil engineering professors must be brought as close as possible to "current" as soon as possible if our future students are going to be ready to tackle the problems of space exploration.

I feel that the only way we can get the civil engineering faculty of our country current in a hurry is via funded specialty conferences dedicated to that purpose. We must bring the current experts together from government and industry to teach us the current state-of-the-art. I am not sure yet just how much time that will take, but I envision at least a two-day overview for all civil engineering faculty and a three-day breakout session for each of the major areas of civil engineering like structural, environmental, transportation, and geotechnical. A planning group must be selected to work out the details. I would like to help that planning group.

Methodology

First, I envision the planning group would assign a selected list of references that all faculty would be expected to read before the "University Professors Workshop on Civil Engineering in Space."

Second, I envision that the planning group would involve a few civil engineering faculty that are current and key researchers from NASA and industry like those at Bechtel National, Inc.

Third, since the NASA research centers are where the government's creative space engineering research is underway, I envision that the workshops should take place at each of the NASA research centers around the country. The headquarters offices of Bechtel National in San Francisco could also be used as a workshop site since this is where much of the "leading-edge" work relative to the private practice of civil engineering in space is underway. To minimize travel expenses and optimize faculty involvement, I envision a schedule that would require essentially the same teaching team rotating through the NASA research centers around the country. This is an ambitious, but necessary schedule.

Since there are 214 civil engineering departments with roughly 2000 faculty members, it seems logical to develop a plan that would be a partnership between civil engineering faculty members at each university. This plan might lead to an arrangement where 1000 faculty members would be brought up-to-speed as described above and then they would go back to their departments and teach the others. Depending on the size of each workshop and the number that could be offered each year, the planning group could design a schedule that could be completed in as little as one year, but hopefully no longer that three years.

During faculty visits to NASA, Bechtel, et al., relationships might develop that could lead to funded research and further faculty development opportunities back at their universities.

Conclusion

Five centuries after Columbus found our "new world" we are ready to begin again. Just like the promise of new land and the opportunity to live in freedom inspired our forefathers, the challenges of space are intriguing to most civil engineering educators. With our current knowledge of space engineering, I do not feel that we can make the contributions we are capable of making and should be making on behalf of the civil engineering profession. Civil engineers will be required to help make space exploration and development happen. As advisors and professors of civil engineering students, I feel we have an obligation to help students receive the best possible civil engineering education. We will not be able to do that if we do not get up-to-date and stay up-to-date. Civil engineering educators must incorporate the concepts of civil engineering in space as soon as possible into our curriculums. The ideas expressed in this paper will give us that opportunity.

SUBJECT INDEX

Page number refers to first page of paper.

AUTHOR INDEX

Page number refers to first page of paper.